AF801530

Springer Textbooks in Earth Sciences, Geography and Environment

The Springer Textbooks series publishes a broad portfolio of textbooks on Earth Sciences, Geography and Environmental Science. Springer textbooks provide comprehensive introductions as well as in-depth knowledge for advanced studies. A clear, reader-friendly layout and features such as end-of-chapter summaries, work examples, exercises, and glossaries help the reader to access the subject. Springer textbooks are essential for students, researchers and applied scientists.

More information about this series at http://www.springer.com/series/15201

Chung Fang

An Introduction to Fluid Mechanics

Springer

Chung Fang
Department of Civil Engineering
National Cheng Kung University
Tainan, Taiwan

ISSN 2510-1307 ISSN 2510-1315 (electronic)
Springer Textbooks in Earth Sciences, Geography and Environment
ISBN 978-3-319-91820-4 ISBN 978-3-319-91821-1 (eBook)
https://doi.org/10.1007/978-3-319-91821-1

Library of Congress Control Number: 2018948619

© Springer International Publishing AG 2019
This work is subject to copyright. All rights are reserved by the Publisher, whether the whole or part of the material is concerned, specifically the rights of translation, reprinting, reuse of illustrations, recitation, broadcasting, reproduction on microfilms or in any other physical way, and transmission or information storage and retrieval, electronic adaptation, computer software, or by similar or dissimilar methodology now known or hereafter developed.
The use of general descriptive names, registered names, trademarks, service marks, etc. in this publication does not imply, even in the absence of a specific statement, that such names are exempt from the relevant protective laws and regulations and therefore free for general use.
The publisher, the authors and the editors are safe to assume that the advice and information in this book are believed to be true and accurate at the date of publication. Neither the publisher nor the authors or the editors give a warranty, express or implied, with respect to the material contained herein or for any errors or omissions that may have been made. The publisher remains neutral with regard to jurisdictional claims in published maps and institutional affiliations.

Cover illustration: Picture credit for Olga Nikonova (Shutterstock)

This Springer imprint is published by the registered company Springer Nature Switzerland AG
The registered company address is: Gewerbestrasse 11, 6330 Cham, Switzerland

To Yen-I, Kolli and Meimei

Preface

In the past decade, I have been teaching fluid mechanics from the fundamental to advanced levels at Department of Civil Engineering at National Cheng Kung University in Taiwan, and at School of Aeronautics and Astronautics at Zhejiang University in China, and have an impression that a textbook encompassing the topics from the fundamental disciplines to more advanced treatments of fluid mechanics with a balanced discussion between the mathematics and underlying physics of fluid motion is not available. This became the motivation of present work.

The book comprises 12 chapters. Chapter 1 deals with the mathematical prerequisites including tensor analysis, integral theorems, and theory of complex variables. A clear understanding of the mathematical knowledge provides not only a better access to understand the underlying physics of fluid motion, but also is essential to other branches of science and technology. The fundamental concepts of fluid motion are introduced in Chap. 2. Specifically, the distinction between solids, liquids and gases, method of analysis, continuum hypothesis, and Newton's law of viscosity are the main topics. The disciplines devoting to the fluid behavior in static circumstance are discussed in Chap. 3, with the focus on the hydrostatic pressure distribution, hydrostatic forces on submerged surfaces, phenomena of surface tension and buoyancy, and liquids in rigid-body motion. Flow kinematics without referring to the dynamics foundation such as flow lines and the concepts of circulation, vorticity, stream tube and stream filament, and vortex tube and vortex filament are introduced in Chap. 4, which are used intensively for flow visualization. The fundamental physical laws of classical physics, specifically the balances of mass, linear momentum, angular momentum, energy and entropy, are formulated in the base of a general balance statement of an extensive variable in either the integral or differential form in Chap. 5, for which the basic concepts of continuum mechanics and a simplified introduction to the theory of material equations are given. These balance equations are important, because they are valid not only for fluids, but also for other deformable bodies within the continuum hypothesis, provided that the material equations can appropriately be formulated. The study of fluid motion uses model test intensively, and the complete similarity between a model and a prototype needs a priori to be established. For this purpose, the theory of dimensional analysis and model similitude is discussed in Chap. 6. The flows of ideal, incompressible viscous, and compressible inviscid fluids are discussed separately in the forthcoming three chapters. For ideal-fluid flows, the discussions on the Euler and Bernoulli equations, Kelvin's theorem, two- and three-dimensional potential flows, and surface liquid waves are given in Chap. 7. For incompressible viscous flows, the vorticity equation, exact and low-Reynolds-number solutions to the Navier-Stokes equation, boundary-layer and buoyancy-driven flows, and a brief discussion on turbulent flows with applications to pipe-flow problems are presented in Chap. 8. For compressible inviscid flows, the Crocco equation, propagations of sound and shock waves, and some selected topics in one- and multi-dimensional circumstances are discussed in Chap. 9. Chapter 10 deals with open-channel flows, which is provided particularly for students in civil and hydraulic engineering. The essential knowledge of classical thermodynamics is summarized in Chap. 11, which provides an energy perspective in parallel to the mechanics perspective in understanding the physics of fluid motion. The last chapter concerns with some features of granular flows, which is used to illustrate the applications of the mature disciplines of fluid mechanics and thermodynamics to complex problems.

The chapter arrangement follows the sequence of statics, kinematics, and dynamics of deformable materials, which is the common lecture sequence used in many university-level education facilities. This was done in order to let students to understand the disciplines of fluid mechanics in a coherent manner. Although not explicitly accomplished, the book can be divided into three parts. Part I contains the first six chapters for the fundamental disciplines of fluid mechanics. Part II comprises the next four chapters devoting to an advanced treatment of fluid mechanics. Part III consists of the last two chapters, which may be used to show the applications of fluid mechanics in various problems of interest. At the end of each chapter, some problems are given for exercises or testing materials of the introduced disciplines. The detailed solutions to selected problems are provided in Appendix B, while the orthogonal curvilinear coordinates introduced in the first chapter are represented in a more concise manner in Appendix A for reference. Associated with each chapter, a list for further reading is provided for those readers who want to know more about the related topics. The book can be used for one- or two-semester lectures to deliver a broad and deep discussion on fluid mechanics with balanced mathematical treatments and physical understanding.

I would like to express my sincere gratitude to Prof. Kolumban Hutter for the constant encouragement of writing the book. Miss Annett Buettner, Miss Helen Rachner, Miss Raghavy Krishnan and Mr. Karthik Raj Selvaraj from Springer Verlag are greatly acknowledged for their great care of managing all administrative and publishing issues of the book. John Wiley & Sons Inc., is greatly acknowledged for the kind permission to use the figures quoted from its published books. A large part of this book was carried out during a sabbatical semester at School of Aeronautics and Astronautics, Zhejiang University, China, and I should like to thank Prof. Weiqiu Chen and Prof. Zhaosheng Yu for their hospitality throughout this stay. There will be errors remaining in the book, and for these I alone am responsible.

Before I finish this Preface, I would like to say that writing a book can never be finished, and a finished book has to be abandoned. This is I am now going to do, well knowing that a book bears intrinsically its weaknesses, that I would know now how to do it better. While writing this book through all its stages needed isolation and separation from the beloved family members, who all deserve my deepest gratitude.

Tainan, Taiwan
March 2018

Chung Fang

Contents

1 Mathematical Prerequisites

Fluid mechanics is the mechanics of fluids embracing liquids and gases and is the discipline within a broad field of applied mechanics concerned with the behavior of liquids and gases at rest and in motion. Knowledge of ordinary and partial differential equations, linear algebra, vector calculus, and integral transforms is a fundamental prerequisite. However, to better access the underlying physical interpretations and mechanisms of fluid motions, additional mathematical knowledge is required, which is introduced in this chapter. First, the index notation with free and dummy indices are discussed, followed by the elementary theory of the Cartesian tensor, including tensor algebra and tensor calculus. Based on these, field quantities and mathematical operations which are essential to fluid mechanics in orthogonal curvilinear coordinate systems can be expressed in a coherent manner. Useful integral theorems in establishing the theory of fluid mechanics, such as Gauss's divergence theorem, Green's and Stokes' theorems are summarized as an outline. A review of complex analysis which is used intensively in discussing two-dimensional potential-flow theory of fluid mechanics is provided at the end. Detailed derivations and proofs of most equations and theorems are absent. They provide additional exercises for readers to become familiar with the topics introduced in this chapter.

1.1 Index Notation

1.1.1 Summation Convention, Dummy and Free Indices

The summation

$$\mathcal{S} = a_1 x_1 + a_2 x_2 + \cdots + a_n x_n, \tag{1.1.1}$$

can be expressed alternatively by using the summation symbol, viz.,

$$\mathcal{S} = \sum_{i=1}^{n} a_i x_i = \sum_{j=1}^{n} a_j x_j = \sum_{k=1}^{n} a_k x_k, \tag{1.1.2}$$

© Springer International Publishing AG 2019

C. Fang, *An Introduction to Fluid Mechanics*, Springer Textbooks in Earth Sciences, Geography and Environment, https://doi.org/10.1007/978-3-319-91821-1_1

in which i, j, and k are the repeated indices that the summation is independent of the letter used. Similarly, the equations

$$\begin{aligned} a_{11}x_1 + a_{12}x_2 + \cdots + a_{1n}x_n &= c_1, \\ a_{21}x_1 + a_{22}x_2 + \cdots + a_{2n}x_n &= c_2, \\ &\vdots \\ a_{m1}x_1 + a_{m2}x_2 + \cdots + a_{mn}x_n &= c_m, \end{aligned} \tag{1.1.3}$$

can be recast in the form of

$$\sum_{i=1}^{m}\sum_{j=1}^{n} a_{ij}x_j = c_i. \tag{1.1.4}$$

The summations in Eqs. (1.1.2) and (1.1.4) are further simplified if *Einstein's summation convention* is applied[1]; i.e., whenever an index is repeated once, it is a *dummy index* indicating a summation with the index running through the integers $1, 2, \cdots$ in its possible variation range, while an index is called a *free index* if it appears only once in each term of an equation, in which its value takes on the integral number $1, 2, \cdots$ one at a time. Thus, Eqs. (1.1.2) and (1.1.4), by using the index notation, are recast respectively as

$$S = a_i x_i, \qquad a_{ij}x_j = c_i, \tag{1.1.5}$$

in which i in the first equation is a dummy index, while i and j in the second equation are respectively a free and a dummy indices. Expressions such as $a_i b_i x_i$ or $a_{ij}x_j = c_k$ are meaningless, for a dummy index should never be repeated more than once, and a free index appearing in every term of an equation must be the same. The letter used to represent a dummy index is irrelevant and that for a free index should follow the summation convention.

Conventionally, possible variation range of an index is $\{1, 2, 3\}$ in three-dimensional circumstance, unless stated otherwise, and each integer may represent a linear-independent direction in general. Thus,

$$a_i, \qquad a_{ij}, \qquad a_{ijk}, \qquad a_{ijkl}, \tag{1.1.6}$$

have respectively 3, 9, 27 and 81 components, for all i, j, k, and l are free indices.

1.1.2 The Kronecker Delta and Permutation Symbol

The *Kronecker delta*, δ_{ij}, is defined as[2]

$$\delta_{ij} \equiv \begin{Bmatrix} 0, \ i \neq j \\ 1, \ i = j \end{Bmatrix}, \tag{1.1.7}$$

[1] Albert Einstein, 1879–1955, a German-born theoretical physicist. The summation convention was first introduced in 1916 in his "*The Foundation of the General Theory of Relativity*".

[2] Leopold Kronecker, 1823–1891, a German mathematician. His viewpoint of mathematics is reflected by the famous motto, which reads: "Die ganzen Zahlen hat der liebe Gott gemacht, alles andere ist Menschenwerk" (God made the integers, all else is the work of man).

whose matrix representation corresponds to the identity matrix, $\boldsymbol{I}$, i.e.,

$$[\delta_{ij}] = \begin{bmatrix} \delta_{11} & \delta_{12} & \delta_{13} \\ \delta_{21} & \delta_{22} & \delta_{23} \\ \delta_{31} & \delta_{32} & \delta_{33} \end{bmatrix} = \begin{bmatrix} 1 & 0 & 0 \\ 0 & 1 & 0 \\ 0 & 0 & 1 \end{bmatrix} = \boldsymbol{I}, \qquad (1.1.8)$$

in which $[\alpha]$ denotes the matrix representation of α. Let $\{\boldsymbol{e}_1, \boldsymbol{e}_2, \boldsymbol{e}_3\}$ form an orthonormal base; then $\boldsymbol{e}_i \cdot \boldsymbol{e}_j = \delta_{ij}$. The Kronecker delta possesses the following properties:

$$\begin{aligned} &\delta_{ii} = 3, && \delta_{ij}\delta_{ij} = 3, \ \delta_{ij}\delta_{jk} = \delta_{ik}, \\ &\delta_{im}\delta_{mn}\delta_{nj} = \delta_{ij}, && \delta_{ij}a_j = a_i, \ \delta_{ij}t_{jk} = t_{ik}. \end{aligned} \qquad (1.1.9)$$

The *permutation symbol*, or the *Levi-Cività ε-tensor*, denoted by ε_{ijk}, is defined by[3]

$$\varepsilon_{ijk} \equiv \begin{Bmatrix} 1, & \text{even } p \\ -1, & \text{odd } p \\ 0, & \text{otherwise} \end{Bmatrix}, \qquad (1.1.10)$$

where p is a permutation of the set $\{i, j, k\}$. Specifically, p is even if $\{i, j, k\} = \{1, 2, 3\}$, $\{2, 3, 1\}$, $\{3, 1, 2\}$ and odd if $\{i, j, k\} = \{1, 3, 2\}$, $\{3, 2, 1\}$, $\{2, 1, 3\}$, and

$$\varepsilon_{ijk} = \varepsilon_{jki} = \varepsilon_{kij} = -\varepsilon_{jik} = -\varepsilon_{ikj} = -\varepsilon_{kji}. \qquad (1.1.11)$$

Obviously, $\varepsilon_{ijk} = 0$ when any two of the set $\{i, j, k\}$ are identical. Let $\{\boldsymbol{e}_1, \boldsymbol{e}_2, \boldsymbol{e}_3\}$ be an orthonormal base in a right-handed triad; then $\boldsymbol{e}_i \times \boldsymbol{e}_j = \varepsilon_{ijk}\boldsymbol{e}_k$. The relations

$$\begin{aligned} \varepsilon_{ijk}\varepsilon_{lmn} &= \det \begin{bmatrix} \delta_{il} & \delta_{im} & \delta_{in} \\ \delta_{jl} & \delta_{jm} & \delta_{jn} \\ \delta_{kl} & \delta_{km} & \delta_{kn} \end{bmatrix}, \\ \varepsilon_{ijk}\varepsilon_{imn} &= \delta_{jm}\delta_{kn} - \delta_{jn}\delta_{km}, \end{aligned} \qquad (1.1.12)$$

known as the *δ-ε identities* with "det" standing for determinant can be applied to derive the following identities:

$$\delta_{ij}\varepsilon_{ijk} = 0, \quad \varepsilon_{ijk}\varepsilon_{mjk} = 2\delta_{im}, \quad \varepsilon_{ijk}\varepsilon_{ijk} = 6. \qquad (1.1.13)$$

There exist some manipulation rules associated with the index notation. If $a_i = U_{ij}b_j$ and $b_i = V_{ij}c_j$, it follows that

$$a_i = U_{ij}V_{jk}c_k, \qquad (1.1.14)$$

called the *substitution rule*. If $x = a_ib_i$ and $y = c_id_i$, then

$$xy = a_ib_ic_jd_j, \qquad (1.1.15)$$

called the *multiplication rule*. If $t_{ij}n_j - \lambda n_i = 0$, then

$$(t_{ij} - \lambda\delta_{ij})n_j = 0, \qquad (1.1.16)$$

[3]Tullio Levi-Civitá, 1873–1941, an Italian mathematician. The permutation symbol is intensively used in linear algebra, tensor analysis, and differential geometry.

called the *factoring rule*. Last, it follows that $\delta_{ij}t_{ij} = t_{ii}$. For example, if $t_{ij} = \lambda\theta\delta_{ij} + 2\mu E_{ij}$, then

$$t_{ii} = 3\lambda\theta + 2\mu E_{ii}, \tag{1.1.17}$$

called the *contraction rule*.

The index notation can be used to conduct various vector operations. For example, if $\boldsymbol{a} = a_i\boldsymbol{e}_i$, $\boldsymbol{b} = b_i\boldsymbol{e}_i$, and $\boldsymbol{c} = c_i\boldsymbol{e}_i$, then

$$\boldsymbol{a}\cdot\boldsymbol{b} = a_ib_i, \qquad \boldsymbol{a}\times\boldsymbol{b} = \varepsilon_{ijk}a_ib_j\boldsymbol{e}_k, \qquad \boldsymbol{a}\cdot(\boldsymbol{b}\times\boldsymbol{c}) = \varepsilon_{ijk}a_ib_jc_k. \tag{1.1.18}$$

1.2 Tensor Analysis

1.2.1 Definition and Components of a Tensor

A *tensor*, or alternatively a *second-order tensor*, $\boldsymbol{T}$, is defined to be a linear transformation, which transforms any vector into another vector satisfying the linear property given by

$$\boldsymbol{T}(\alpha\boldsymbol{a} + \beta\boldsymbol{b}) = \alpha\boldsymbol{T}\boldsymbol{a} + \beta\boldsymbol{T}\boldsymbol{b}, \tag{1.2.1}$$

where $\{\alpha, \beta\}$ and $\{\boldsymbol{a}, \boldsymbol{b}\}$ are arbitrary scalars and vectors, respectively. For example, if $\boldsymbol{T}$ is a linear transformation which transforms every vector into a fixed vector, it is not a tensor. On the contrary, if $\boldsymbol{T}$ transforms every vector into its mirror image with respect to a fixed plane, it is a tensor. If two tensors, $\boldsymbol{T}$ and $\boldsymbol{U}$, transform any arbitrary vector $\boldsymbol{a}$ in an identical manner, they are the same; i.e., if $\boldsymbol{T}\boldsymbol{a} = \boldsymbol{U}\boldsymbol{a}$, then $\boldsymbol{T} = \boldsymbol{U}$.

Let $\{\boldsymbol{e}_1, \boldsymbol{e}_2, \boldsymbol{e}_3\}$ be the orthonormal base in the directions of the $\{x_1, x_2, x_3\}$-axes and $\boldsymbol{T}$ be a tensor; it follows that

$$\begin{aligned}
\boldsymbol{T}\boldsymbol{e}_1 &= T_{11}\boldsymbol{e}_1 + T_{21}\boldsymbol{e}_2 + T_{31}\boldsymbol{e}_3,\\
\boldsymbol{T}\boldsymbol{e}_2 &= T_{12}\boldsymbol{e}_1 + T_{22}\boldsymbol{e}_2 + T_{32}\boldsymbol{e}_3,\\
\boldsymbol{T}\boldsymbol{e}_3 &= T_{13}\boldsymbol{e}_1 + T_{23}\boldsymbol{e}_2 + T_{33}\boldsymbol{e}_3,
\end{aligned} \tag{1.2.2}$$

which are expressed alternatively by using the index notation, viz.,

$$\boldsymbol{T}\boldsymbol{e}_i = T_{ji}\boldsymbol{e}_j, \tag{1.2.3}$$

for $\boldsymbol{T}$ transforms every unit vector into a vector which can be expressed by using the orthonormal base. The components of $\boldsymbol{T}$, based on Eq. (1.2.3), are then identified to be

$$T_{ij} = \boldsymbol{e}_i\cdot\boldsymbol{T}\boldsymbol{e}_j, \tag{1.2.4}$$

with the corresponding matrix representation given by

$$[\boldsymbol{T}] = \begin{bmatrix} T_{11} & T_{12} & T_{13}\\ T_{21} & T_{22} & T_{23}\\ T_{31} & T_{32} & T_{33} \end{bmatrix}, \tag{1.2.5}$$

known as the matrix of tensor $\boldsymbol{T}$. The first, second, and third columns of Eq. (1.2.5) correspond to the components of $\boldsymbol{T}\boldsymbol{e}_1$, $\boldsymbol{T}\boldsymbol{e}_2$, and $\boldsymbol{T}\boldsymbol{e}_3$ shown in Eq. (1.2.2). For example, let $\boldsymbol{T}$ be a counterclockwise rotation of a rigid body about the x_3-axis by an angle of θ; its matrix is identified to be

$$[\boldsymbol{T}] = \begin{bmatrix} \cos\theta & -\sin\theta & 0 \\ \sin\theta & \cos\theta & 0 \\ 0 & 0 & 1 \end{bmatrix}. \tag{1.2.6}$$

1.2.2 Tensor Algebra

Let $\{\alpha, \beta\}$ be arbitrary scalars, $\{\boldsymbol{a}, \boldsymbol{b}, \boldsymbol{c}, \boldsymbol{d}\}$ be arbitrary vectors, and $\{\boldsymbol{T}, \boldsymbol{U}, \boldsymbol{V}, \boldsymbol{W}\}$ be arbitrary second-order tensors unless stated otherwise. If

$$\boldsymbol{b} = \boldsymbol{T}\boldsymbol{a}, \tag{1.2.7}$$

it follows that

$$b_i = T_{ij}a_j, \tag{1.2.8}$$

or

$$[\boldsymbol{b}] = [\boldsymbol{T}][\boldsymbol{a}], \qquad \begin{bmatrix} b_1 \\ b_2 \\ b_3 \end{bmatrix} = \begin{bmatrix} T_{11} & T_{12} & T_{13} \\ T_{21} & T_{22} & T_{23} \\ T_{31} & T_{32} & T_{33} \end{bmatrix} \begin{bmatrix} a_1 \\ a_2 \\ a_3 \end{bmatrix}. \tag{1.2.9}$$

Thus, the components of a transformed vector can be computed directly by using the matrix multiplication.

The *sum* of two tensors $\boldsymbol{T}$ and $\boldsymbol{U}$, denoted by $\boldsymbol{V} = \boldsymbol{T} + \boldsymbol{U}$, is given by

$$\boldsymbol{V}\boldsymbol{a} = (\boldsymbol{T} + \boldsymbol{U})\,\boldsymbol{a} = \boldsymbol{T}\boldsymbol{a} + \boldsymbol{U}\boldsymbol{a}, \tag{1.2.10}$$

from which $\boldsymbol{V}$ is also a tensor with its components given by

$$V_{ij} = T_{ij} + U_{ij}, \qquad [\boldsymbol{V}] = [\boldsymbol{T}] + [\boldsymbol{U}]. \tag{1.2.11}$$

Thus, the sum of tensors follows exactly the sum of matrices.

The *products* of two tensors, $\boldsymbol{TU}$ and $\boldsymbol{UT}$, are defined by

$$(\boldsymbol{TU})\,\boldsymbol{a} \equiv \boldsymbol{T}\,(\boldsymbol{U}\boldsymbol{a}), \qquad (\boldsymbol{UT})\,\boldsymbol{a} \equiv \boldsymbol{U}\,(\boldsymbol{T}\boldsymbol{a}), \tag{1.2.12}$$

which are equally a tensor, with the components given by

$$\begin{aligned} (\boldsymbol{TU})_{ij} &= T_{im}U_{mj}, \qquad & [\boldsymbol{TU}] &= [\boldsymbol{T}][\boldsymbol{U}], \\ (\boldsymbol{UT})_{ij} &= U_{im}T_{mj}, \qquad & [\boldsymbol{UT}] &= [\boldsymbol{U}][\boldsymbol{T}]. \end{aligned} \tag{1.2.13}$$

Thus, the product of two tensors follows exactly the matrix multiplication, and a tensor product is not commutative, i.e., $\boldsymbol{TU} \neq \boldsymbol{UT}$. Making a product of more than two tensors can be conducted by using Eq. (1.2.12), e.g.

$$\begin{aligned} (\boldsymbol{TUV})\boldsymbol{a} &= \boldsymbol{T}((\boldsymbol{UV})\boldsymbol{a}) = \boldsymbol{T}(\boldsymbol{U}(\boldsymbol{V}\boldsymbol{a})), \\ (\boldsymbol{TU})(\boldsymbol{V}\boldsymbol{a}) &= \boldsymbol{T}(\boldsymbol{U}(\boldsymbol{V}\boldsymbol{a})), \end{aligned} \tag{1.2.14}$$

giving rise to

$$\boldsymbol{T}(\boldsymbol{UV}) = (\boldsymbol{TU})\boldsymbol{V}, \tag{1.2.15}$$

indicating that a tensor product is associative. The associative rule is applied to establish the integral positive powers of a tensor by simple products, e.g. $\boldsymbol{T}^2 = \boldsymbol{T}\boldsymbol{T}$, $\boldsymbol{T}^3 = \boldsymbol{T}\boldsymbol{T}\boldsymbol{T}$, etc.

The *transpose* of a tensor is denoted by using the superscript T, which is defined by

$$\boldsymbol{a} \cdot \boldsymbol{T}\boldsymbol{b} \equiv \boldsymbol{b} \cdot \boldsymbol{T}^{\mathrm{T}}\boldsymbol{a}. \tag{1.2.16}$$

It follows from Eq. (1.2.1) that the transpose of a tensor is also a tensor, whose components are given by

$$T_{ij} = T_{ji}^{\mathrm{T}}, \qquad [\boldsymbol{T}]^{\mathrm{T}} = [\boldsymbol{T}^{\mathrm{T}}], \tag{1.2.17}$$

indicating that the matrix of $\boldsymbol{T}^{\mathrm{T}}$ is the transpose matrix of $\boldsymbol{T}$. Eqs. (1.2.16) and (1.2.17) are extended to obtain the following identities:

$$\boldsymbol{T} = (\boldsymbol{T}^{\mathrm{T}})^{\mathrm{T}},\ (\boldsymbol{T}\boldsymbol{U})^{\mathrm{T}} = \boldsymbol{U}^{\mathrm{T}}\boldsymbol{T}^{\mathrm{T}},\ (\boldsymbol{T}\boldsymbol{U}\cdots\boldsymbol{W})^{\mathrm{T}} = \boldsymbol{W}^{\mathrm{T}}\cdots\boldsymbol{U}^{\mathrm{T}}\boldsymbol{T}^{\mathrm{T}}. \tag{1.2.18}$$

The *dyadic product* of two vectors $\boldsymbol{a}$ and $\boldsymbol{b}$, denoted by $\boldsymbol{a}\boldsymbol{b}$ or $\boldsymbol{a} \otimes \boldsymbol{b}$, is defined by

$$(\boldsymbol{a}\boldsymbol{b})\boldsymbol{c} \equiv \boldsymbol{a}(\boldsymbol{b} \cdot \boldsymbol{c}), \tag{1.2.19}$$

where $\boldsymbol{c}$ is a third vector, by which the relation

$$(\boldsymbol{a}\boldsymbol{b})(\alpha\boldsymbol{c} + \beta\boldsymbol{d}) = \alpha(\boldsymbol{a}\boldsymbol{b})\boldsymbol{c} + \beta(\boldsymbol{a}\boldsymbol{b})\boldsymbol{d}, \tag{1.2.20}$$

is satisfied. Thus, the dyadic product $\boldsymbol{a}\boldsymbol{b}$ plays exactly the role as a second-order tensor, with the components given by

$$(\boldsymbol{a}\boldsymbol{b})_{ij} = a_i b_j, \qquad [\boldsymbol{a}\boldsymbol{b}] = [\boldsymbol{a}][\boldsymbol{b}]^{\mathrm{T}}. \tag{1.2.21}$$

The dyadic product can be used to establish the "base" of a second-order tensor, e.g.

$$[\boldsymbol{e}_1\boldsymbol{e}_1] = \begin{bmatrix} 1\,0\,0 \\ 0\,0\,0 \\ 0\,0\,0 \end{bmatrix}, \qquad [\boldsymbol{e}_1\boldsymbol{e}_2] = \begin{bmatrix} 0\,1\,0 \\ 0\,0\,0 \\ 0\,0\,0 \end{bmatrix}, \quad \cdots, \tag{1.2.22}$$

with which a second-order tensor $\boldsymbol{T}$ can be expressed as

$$\boldsymbol{T} = T_{ij}(\boldsymbol{e}_i\boldsymbol{e}_j) = T_{ij}(\boldsymbol{e}_i \otimes \boldsymbol{e}_j). \tag{1.2.23}$$

The *trace* of a dyadic product $\boldsymbol{a}\boldsymbol{b}$ is defined by

$$\mathrm{tr}\,(\boldsymbol{a}\boldsymbol{b}) \equiv \boldsymbol{a} \cdot \boldsymbol{b}, \tag{1.2.24}$$

which fulfills the relation

$$\mathrm{tr}\,(\alpha\,\boldsymbol{a}\boldsymbol{b} + \beta\,\boldsymbol{c}\boldsymbol{d}) = \alpha\,\mathrm{tr}\,(\boldsymbol{a}\boldsymbol{b}) + \beta\,\mathrm{tr}\,(\boldsymbol{c}\boldsymbol{d}). \tag{1.2.25}$$

It follows that

$$\mathrm{tr}\,\boldsymbol{T} = T_{ii},\ \mathrm{tr}\,(\boldsymbol{T}^{\mathrm{T}}) = \mathrm{tr}\,\boldsymbol{T},\ \mathrm{tr}\,(\boldsymbol{T}\boldsymbol{U}) = \mathrm{tr}\,(\boldsymbol{U}\boldsymbol{T}), \tag{1.2.26}$$

and

$$\begin{aligned} \boldsymbol{T}\boldsymbol{a} &= T_{ij}a_j\boldsymbol{e}_i, & \boldsymbol{a}\boldsymbol{T} &= a_i T_{ij}\boldsymbol{e}_j, \\ \boldsymbol{T}\boldsymbol{U} &= T_{im}U_{mj}(\boldsymbol{e}_i\boldsymbol{e}_j), & \boldsymbol{T} \cdot \boldsymbol{U} &= T_{ij}U_{ji}, \end{aligned} \tag{1.2.27}$$

in which Eqs. (1.2.19) and (1.2.24) have been used, where $\boldsymbol{TU}$ corresponds to a matrix product of $\boldsymbol{T}$ and $\boldsymbol{U}$, while $\boldsymbol{T} \cdot \boldsymbol{U}$ indicates a scalar product, which is denoted alternatively by $\boldsymbol{T} : \boldsymbol{U}$.

An *identity tensor* is defined to be a linear transformation which transforms every vector into itself, conventionally denoted by $\boldsymbol{I}$, viz.,

$$\boldsymbol{I}\boldsymbol{a} \equiv \boldsymbol{a}, \tag{1.2.28}$$

with the components given by

$$I_{ij} = \delta_{ij}, \qquad [\boldsymbol{I}] = \begin{bmatrix} 1 & 0 & 0 \\ 0 & 1 & 0 \\ 0 & 0 & 1 \end{bmatrix}. \tag{1.2.29}$$

If $\boldsymbol{T}\boldsymbol{a} = \boldsymbol{a}$ for any vector $\boldsymbol{a}$, then $\boldsymbol{T} = \boldsymbol{I}$.

A tensor $\boldsymbol{U}$ is called the *inverse* of a tensor $\boldsymbol{T}$ if

$$\boldsymbol{U}\boldsymbol{T} = \boldsymbol{I}, \tag{1.2.30}$$

is satisfied, for which $\boldsymbol{U}$ is denoted by $\boldsymbol{U} = \boldsymbol{T}^{-1}$. The components of the inverse of a tensor $\boldsymbol{T}$ are determined by using the inverse matrix of $[\boldsymbol{T}]$, provided that it is non-singular, i.e., $\det \boldsymbol{T} \neq 0$. The inverse of a tensor $\boldsymbol{T}$ satisfies the reciprocal relation, namely

$$\boldsymbol{T}^{-1}\boldsymbol{T} = \boldsymbol{T}\boldsymbol{T}^{-1} = \boldsymbol{I}, \tag{1.2.31}$$

with which the following relations can be obtained:

$$\begin{aligned} &(\boldsymbol{T}^{-1})^{-1} = \boldsymbol{T}, \qquad (\boldsymbol{T}^{\mathrm{T}})^{-1} = (\boldsymbol{T}^{-1})^{\mathrm{T}}, \\ &(\boldsymbol{T}\boldsymbol{U})^{-1} = \boldsymbol{U}^{-1}\boldsymbol{T}^{-1}, \ (\boldsymbol{T}\boldsymbol{U}\cdots\boldsymbol{W})^{-1} = \boldsymbol{W}^{-1}\cdots\boldsymbol{U}^{-1}\boldsymbol{T}^{-1}, \end{aligned} \tag{1.2.32}$$

corresponding to the matrix operations. If $\boldsymbol{T}\boldsymbol{a} = \boldsymbol{b}$, then $\boldsymbol{a} = \boldsymbol{T}^{-1}\boldsymbol{b}$, provided that $\boldsymbol{T}$ is invertible, for a one-to-one mapping between $\boldsymbol{a}$ and $\boldsymbol{b}$ is established. On the contrary, it is not the case if $\boldsymbol{T}$ does not have an inverse.

The *symmetry* and *antisymmetry* (or *skew symmetry*) of a second-order tensor $\boldsymbol{T}$ are defined by

$$\left\{ \begin{aligned} &\boldsymbol{T} = \boldsymbol{T}^{\mathrm{T}}, \quad \boldsymbol{T} \text{ is symmetric,} \\ &\boldsymbol{T} = -\boldsymbol{T}^{\mathrm{T}}, \ \boldsymbol{T} \text{ is anti-symmetric.} \end{aligned} \right\} \tag{1.2.33}$$

It follows that $T_{ij} = T_{ji}$ and $T_{ij} = -T_{ji}$ for symmetric and antisymmetric tensors, respectively. Thus, the off-diagonal components of a symmetric tensor are symmetric with respect to the diagonal line, giving rise to six independent components. For an antisymmetric tensor, the three components on the diagonal line vanish, and only three off-diagonal components are independent. It is always possible to decompose any second-order tensor $\boldsymbol{T}$ into a sum of a symmetric tensor, $\boldsymbol{T}^{\mathrm{s}}$, and an antisymmetric tensor, $\boldsymbol{T}^{\mathrm{a}}$, viz.,

$$\boldsymbol{T} = \boldsymbol{T}^{\mathrm{s}} + \boldsymbol{T}^{\mathrm{a}}, \qquad \boldsymbol{T}^{\mathrm{s}} = \frac{1}{2}\left(\boldsymbol{T} + \boldsymbol{T}^{\mathrm{T}}\right), \qquad \boldsymbol{T}^{\mathrm{a}} = \frac{1}{2}\left(\boldsymbol{T} - \boldsymbol{T}^{\mathrm{T}}\right). \tag{1.2.34}$$

It can be shown that the trace of a product of a symmetric and an antisymmetric tensor vanishes.

An antisymmetric tensor $\boldsymbol{W}$ behaves like a vector and can be expressed by using its *dual vector*, $\boldsymbol{a}^{\mathrm{W}}$, which is defined by

$$\boldsymbol{W}\boldsymbol{a} \equiv \boldsymbol{a}^{\mathrm{W}} \times \boldsymbol{a}, \tag{1.2.35}$$

where $\boldsymbol{a}$ is any arbitrary vector. Equation (1.2.35) indicates that the linear transformation of $\boldsymbol{a}$ through $\boldsymbol{W}$ is identified by the cross product of $\boldsymbol{a}^{\mathrm{W}}$ and $\boldsymbol{a}$. In terms of the index notation, $\boldsymbol{a}^{\mathrm{W}}$ is expressed as

$$2\boldsymbol{a}^{\mathrm{W}} = -\varepsilon_{ijk} W_{jk} \boldsymbol{e}_i. \tag{1.2.36}$$

1.2.3 Orthogonal Tensor and Transformation Laws

An *orthogonal tensor*, denoted by $\boldsymbol{Q}$, is defined to be a linear transformation, by which the transformed vectors preserve their lengths and angles, i.e.,

$$\boldsymbol{Q}\boldsymbol{a} \cdot \boldsymbol{Q}\boldsymbol{b} = \boldsymbol{a} \cdot \boldsymbol{b}, \tag{1.2.37}$$

for any vectors $\boldsymbol{a}$ and $\boldsymbol{b}$, with $\| \boldsymbol{Q}\boldsymbol{a}\| = \|\boldsymbol{a}\|$ and $\| \boldsymbol{Q}\boldsymbol{b}\| = \|\boldsymbol{b}\|$, where $\|\alpha\|$ indicates the norm of α. It follows from Eqs. (1.2.12) and (1.2.16) that

$$\boldsymbol{Q}^{\mathrm{T}} \boldsymbol{Q} = \boldsymbol{Q} \boldsymbol{Q}^{\mathrm{T}} = \boldsymbol{I}, \tag{1.2.38}$$

or alternatively in the component and matrix representations,

$$Q_{ki} Q_{kj} = Q_{ik} Q_{jk} = \delta_{ij}, \qquad [\boldsymbol{Q}]^{\mathrm{T}}[\boldsymbol{Q}] = [\boldsymbol{Q}][\boldsymbol{Q}]^{\mathrm{T}} = [\boldsymbol{I}]. \tag{1.2.39}$$

Equation $(1.2.39)_2$ indicates that the determinant of $\boldsymbol{Q}$ satisfies $\det \boldsymbol{Q} = \pm 1$, where $+1$ and -1 correspond respectively to rotation and reflection. For example, the determinant of the counterclockwise rotation of a rigid body given in Eq. (1.2.6) is $+1$.

Let $\boldsymbol{e}_i$ and $\boldsymbol{e}'_i$ be the orthonormal bases of two different Cartesian coordinates. The relations between $\boldsymbol{e}_i$ and $\boldsymbol{e}'_i$ are established by using the orthogonal tensor, viz.,

$$\boldsymbol{e}'_i = \boldsymbol{Q}\boldsymbol{e}_i = Q_{ji} \boldsymbol{e}_j, \tag{1.2.40}$$

where Q_{ij} is the direction cosine between $\boldsymbol{e}_i$ and $\boldsymbol{e}'_j$ given by

$$Q_{ij} = \cos(\boldsymbol{e}_i, \boldsymbol{e}'_j), \tag{1.2.41}$$

with the matrix representation

$$[\boldsymbol{Q}] = \begin{bmatrix} Q_{11} & Q_{12} & Q_{13} \\ Q_{21} & Q_{22} & Q_{23} \\ Q_{31} & Q_{32} & Q_{33} \end{bmatrix}. \tag{1.2.42}$$

Thus, the transformation between two rectangular Cartesian coordinates can be conducted by using the orthogonal tensor. For example, let $\boldsymbol{e}'_i$ be obtained by rotating counterclockwise $\boldsymbol{e}_i$ about the x_3-axis through an angle of θ, for which $\boldsymbol{Q}$ is determined to be

$$[\boldsymbol{Q}] = \begin{bmatrix} \cos\theta & -\sin\theta & 0 \\ \sin\theta & \cos\theta & 0 \\ 0 & 0 & 1 \end{bmatrix}. \tag{1.2.43}$$

Within coordinate transformations, the components of vectors and tensors can be related to the orthogonal tensor. Let $\boldsymbol{a}$ and $\boldsymbol{T}$ be an arbitrary vector and tensor, respectively. Their components are given by

$$a_i = \boldsymbol{a} \cdot \boldsymbol{e}_i, \; T_{ij} = \boldsymbol{e}_i \cdot \boldsymbol{T}\boldsymbol{e}_j, \tag{1.2.44}$$

under the orthonormal base $\boldsymbol{e}_i$, or alternatively

$$a'_i = \boldsymbol{a} \cdot \boldsymbol{e}'_i, \; T'_{ij} = \boldsymbol{e}'_i \cdot \boldsymbol{T}\boldsymbol{e}'_j, \tag{1.2.45}$$

under the orthonormal base $\boldsymbol{e}'_i$. Since $\boldsymbol{e}'_i = Q_{ji}\boldsymbol{e}_j$, it follows that

$$a'_i = Q_{ji}a_j, \; T'_{ij} = Q_{mi}Q_{nj}T_{mn}, \tag{1.2.46}$$

with the corresponding matrix representations given by

$$[\boldsymbol{a}'] = [\boldsymbol{Q}]^{\mathrm{T}}[\boldsymbol{a}], \; [\boldsymbol{T}'] = [\boldsymbol{Q}]^{\mathrm{T}}[\boldsymbol{T}][\boldsymbol{Q}], \tag{1.2.47}$$

or

$$\begin{bmatrix} a'_1 \\ a'_2 \\ a'_3 \end{bmatrix} = \begin{bmatrix} Q_{11} & Q_{21} & Q_{31} \\ Q_{12} & Q_{22} & Q_{32} \\ Q_{13} & Q_{23} & Q_{33} \end{bmatrix} \begin{bmatrix} a_1 \\ a_2 \\ a_3 \end{bmatrix},$$
$$\begin{bmatrix} T'_{11} & T'_{12} & T'_{13} \\ T'_{21} & T'_{22} & T'_{23} \\ T'_{31} & T'_{32} & T'_{33} \end{bmatrix} = \begin{bmatrix} Q_{11} & Q_{21} & Q_{31} \\ Q_{12} & Q_{22} & Q_{32} \\ Q_{13} & Q_{23} & Q_{33} \end{bmatrix} \begin{bmatrix} T_{11} & T_{12} & T_{13} \\ T_{21} & T_{22} & T_{23} \\ T_{31} & T_{32} & T_{33} \end{bmatrix} \begin{bmatrix} Q_{11} & Q_{12} & Q_{13} \\ Q_{21} & Q_{22} & Q_{23} \\ Q_{31} & Q_{32} & Q_{33} \end{bmatrix}. \tag{1.2.48}$$

On the other hand, one can reverse the derivations to obtain

$$\begin{aligned} a_i &= Q_{ij}a'_j, & [\boldsymbol{a}] &= [\boldsymbol{Q}][\boldsymbol{a}'], \\ T_{ij} &= Q_{im}Q_{jn}T'_{mn}, & [\boldsymbol{T}] &= [\boldsymbol{Q}][\boldsymbol{T}'][\boldsymbol{Q}]^{\mathrm{T}}. \end{aligned} \tag{1.2.49}$$

Equations (1.2.44)–(1.2.49) form a unique one-to-one mapping between the components of a vector and a tensor from one orthonormal base to another. A scalar, a vector, and a tensor can then be defined by using the transformations laws relating the components with respect to different bases. The Cartesian components of tensors of different orders, within the transformation laws, are then defined viz.,

$$\begin{aligned} & a' = a, && 0^{\text{th}}-\text{order tensor (scalar)} \\ & a'_i = Q_{ji}a_j, && 1^{\text{st}}-\text{order tensor (vector)} \\ & a'_{ij} = Q_{mi}Q_{nj}a_{mn}, && 2^{\text{nd}}-\text{order tensor (tensor)} \\ & a'_{ijk} = Q_{mi}Q_{nj}Q_{rk}a_{mnr}, && 3^{\text{rd}}-\text{order tensor} \\ & \vdots \end{aligned} \tag{1.2.50}$$

where the primed quantities are referred to the $\boldsymbol{e}'_i$ base and unprimed quantities to the $\boldsymbol{e}_i$ base, and $\boldsymbol{Q}$ represents the orthogonal transformation with $\boldsymbol{e}'_i = \boldsymbol{Q}\boldsymbol{e}_i$. The definition (1.2.50) is based on the number of free index. That is, a scalar is one without any free index; a vector is one with a single free index; a second-order tensor is one with two free indices; and higher-order tensors are those with more than two free indices.

Three manipulation rules associated with the transformation laws are given in the following:

- *The addition rule*. The components of a third tensor are determined by adding the corresponding components of any other two tensors of the same order. For example, if $V_{ijk} = T_{ijk} + U_{ijk}$, it follows that

$$T'_{ijk} = Q_{mi} Q_{nj} Q_{rk} T_{mnr}, \qquad U'_{ijk} = Q_{mi} Q_{nj} Q_{rk} U_{mnr}, \tag{1.2.51}$$

giving rise to

$$T'_{ijk} + U'_{ijk} = Q_{mi} Q_{nj} Q_{rk}(T_{mnr} + U_{mnr}) = Q_{mi} Q_{nj} Q_{rk} V_{mnr} = V'_{ijk}, \tag{1.2.52}$$

indicating that V_{ijk} are components of a third-order tensor.
- *The multiplication rule*. Many kinds of products can be conducted from the components of any vectors and tensors. Depending on the number of free index in the products, they are classified as scalars, vectors, tensors, or higher-order tensors. For example, the product $a_i a_i$ forms a scalar, while $a_i a_j a_k$ is a third-order tensor. That is,

$$a'_i = Q_{ji} a_j, \; a'_i a'_i = Q_{ji} Q_{ji} a_j a_j = a_j a_j, \; a'_i a'_j a'_k = Q_{mi} Q_{nj} Q_{rk} a_m a_n a_r. \tag{1.2.53}$$

- *The quotient rule*. If a_i and T_{ij} are components of any two vector and tensor, respectively, and $a_i = T_{ij} b_j$, then b_i represents the components of a vector. Similarly, if T_{ij} and E_{ij} are the components of any two tensors, and $T_{ij} = C_{ijkl} E_{kl}$, then C_{ijkl} are the components of a fourth-order tensor. The proof of the first statement is given here, while that of the second statement is left as an exercise. Since $a_i = T_{ij} b_j$, it follows that

$$a_i = Q_{im} a'_m, \qquad T_{ij} = Q_{im} Q_{jn} T'_{mn}, \qquad Q_{im} a'_m = Q_{im} Q_{jn} T'_{mn} b'_j. \tag{1.2.54}$$

Since $a_i = T_{ij} b_j$ holds for all coordinates, it is concluded that

$$a'_m = T'_{mn} b'_n, \qquad \longrightarrow \qquad Q_{im} T'_{mn} b'_n = Q_{im} Q_{jn} T'_{mn} b'_j. \tag{1.2.55}$$

With $Q_{ik} Q_{im} = \delta_{km}$, multiplying Eq. (1.2.55)$_2$ with Q_{ik} leads to

$$b'_n = Q_{jn} b_j, \tag{1.2.56}$$

showing that b_i are the components of a vector.

1.2.4 Eigenvalues and Eigenvectors of a Tensor

Let $\boldsymbol{T}$ be a second-order tensor. If for any vector $\boldsymbol{a}$, $\boldsymbol{T}$ satisfies

$$\boldsymbol{T}\boldsymbol{a} = \lambda \boldsymbol{a}, \qquad \lambda \in \mathcal{R}, \tag{1.2.57}$$

then $\boldsymbol{a}$ is an *eigenvector* (*eigen direction*) of $\boldsymbol{T}$ with the corresponding *eigenvalue* λ. Eq. (1.2.57) indicates that $\boldsymbol{T}$ transforms every vector into a vector which is parallel to the original one. Obviously, $\beta \boldsymbol{a}$ is also an eigenvector corresponding to the same eigenvalue, where β is a scalar. Thus, all eigenvectors ought to be expressed per unit length. A special case is the identity tensor $\boldsymbol{I}$, for which all vectors are its eigenvectors, corresponding to the same eigenvalue $\lambda = 1$.

To find the eigenvectors and eigenvalues, the non-trivial solutions to the *characteristic equation* of $\boldsymbol{T}$ given by

$$\det(\boldsymbol{T} - \lambda \boldsymbol{I}) = 0, \tag{1.2.58}$$

need to be found. The roots of Eq. (1.2.58) are the eigenvalues. The corresponding eigenvectors are determined by substituting the solutions to Eq. (1.2.58) into Eq. (1.2.57) for the non-trivial solutions of $\boldsymbol{a}$.

For Newtonian fluids, the stress and stretching tensors are real and symmetric.[4] It has been demonstrated from linear algebra that the eigenvalues of real symmetric tensors are all real, and there exist at least three real eigenvectors. The eigenvectors are mutually orthogonal if the corresponding eigenvalues are distinct. In this case, the eigenvectors are called the *principal directions* with the corresponding eigenvalues termed the *principal values*. Since three principal directions are mutually orthogonal, they are used to construct a coordinate system, termed the *principal coordinate system*.

Let $\boldsymbol{T}$ be a real and symmetric tensor, with the corresponding principal directions denoted by $\{\boldsymbol{n}_1, \boldsymbol{n}_2, \boldsymbol{n}_3\}$, corresponding respectively to the principal values $\{\lambda_1, \lambda_2, \lambda_3\}$. The components of $\boldsymbol{T}$ in the principal coordinate system satisfy

$$(T_{ij} - \lambda \delta_{ij}) n_j = 0, \tag{1.2.59}$$

which gives the matrix representation of $\boldsymbol{T}$ as

$$[\boldsymbol{T}]|_{\boldsymbol{n}_i} = \begin{bmatrix} \lambda_1 & 0 & 0 \\ 0 & \lambda_2 & 0 \\ 0 & 0 & \lambda_3 \end{bmatrix}. \tag{1.2.60}$$

It can be demonstrated that the maximum/minimum of the principal values of $\boldsymbol{T}$ are the maximum/minimum of the diagonal elements of all $[\boldsymbol{T}]|_{\boldsymbol{n}_i}$.

1.2.5 Tensor Invariants and the Cayley-Hamilton Theorem

For every second-order tensor $\boldsymbol{T}$, there exist three scalar invariants $I_{\boldsymbol{T}}^1$, $I_{\boldsymbol{T}}^2$, and $I_{\boldsymbol{T}}^3$, called the *tensor invariants*, which are defined by

$$I_{\boldsymbol{T}}^1 \equiv \operatorname{tr} \boldsymbol{T}, \qquad I_{\boldsymbol{T}}^2 \equiv \frac{1}{2}\left\{ (\operatorname{tr} \boldsymbol{T})^2 - \operatorname{tr} (\boldsymbol{T})^2 \right\}, \qquad I_{\boldsymbol{T}}^3 \equiv \det \boldsymbol{T}. \tag{1.2.61}$$

Let $\boldsymbol{e}_i$ be an orthonormal base and $\boldsymbol{n}_i$ be the principal direction of $\boldsymbol{T}$. The matrix representations of $\boldsymbol{T}$ are given by

$$[\boldsymbol{T}]|_{\boldsymbol{e}_i} = \begin{bmatrix} T_{11} & T_{12} & T_{13} \\ T_{21} & T_{22} & T_{23} \\ T_{31} & T_{32} & T_{33} \end{bmatrix}, \qquad [\boldsymbol{T}]|_{\boldsymbol{n}_i} = \begin{bmatrix} \lambda_1 & 0 & 0 \\ 0 & \lambda_2 & 0 \\ 0 & 0 & \lambda_3 \end{bmatrix}, \tag{1.2.62}$$

[4] Stretching tensor is the symmetric part of velocity gradient, which will be discussed in Sect. 5.1.5.

respectively in the $\boldsymbol{e}_i$ and $\boldsymbol{n}_i$ bases, with which the three invariants are expressed explicitly as

$$\begin{aligned} I_T^1 &= T_{11} + T_{22} + T_{33}, \\ I_T^2 &= \begin{vmatrix} T_{11} & T_{12} \\ T_{21} & T_{22} \end{vmatrix} + \begin{vmatrix} T_{22} & T_{23} \\ T_{32} & T_{33} \end{vmatrix} + \begin{vmatrix} T_{11} & T_{13} \\ T_{31} & T_{33} \end{vmatrix}, \\ I_T^3 &= \begin{vmatrix} T_{11} & T_{12} & T_{13} \\ T_{21} & T_{22} & T_{33} \\ T_{31} & T_{32} & T_{33} \end{vmatrix}, \end{aligned} \tag{1.2.63}$$

under the $\boldsymbol{e}_i$ base, and

$$\begin{aligned} I_T^1 &= \lambda_1 + \lambda_2 + \lambda_3, \\ I_T^2 &= \lambda_1\lambda_2 + \lambda_2\lambda_3 + \lambda_1\lambda_3, \\ I_T^3 &= \lambda_1\lambda_2\lambda_3, \end{aligned} \tag{1.2.64}$$

under the $\boldsymbol{n}_i$ base.

The characteristic equation of a tensor $\boldsymbol{T}$ is in connection with the *Cayley-Hamilton theorem*,[5] which states that the characteristic equation is not only fulfilled by the eigenvalues, but also by the tensor itself, i.e.,

$$\boldsymbol{T}^3 - I_T^1\boldsymbol{T}^2 + I_T^2\boldsymbol{T} - I_T^3\boldsymbol{I} = \boldsymbol{0}. \tag{1.2.65}$$

The Cayley-Hamilton theorem is useful; for example, the inverse of $\boldsymbol{T}$ is obtained by multiplying the two sides of Eq. (1.2.65) by $\boldsymbol{T}^{-1}$, followed by some simple mathematical operations, provided that the three invariants are determined.

1.2.6 Isotropic Tensor

A tensor is termed *isotropic* if its components assume the same values in all coordinates. Thus, a scalar is an isotropic tensor of zeroth order, while the Kronecker delta and permutation symbol are respectively the isotropic tensors of second and third orders. The proofs are given in the following.

It follows from the transformation laws that

$$\delta'_{ij} = Q_{mi}Q_{nj}\delta_{mn} = Q_{mi}Q_{mj} = \delta_{ij}, \tag{1.2.66}$$

leading to the definition of second-order isotropic tensor. To prove that ε_{ijk} is a third-order isotropic tensor, the transformation laws that would have to hold if ε_{ijk} were a tensor are first written down, and the interpretation of this equation will be given. Thus,

$$\varepsilon'_{ijk} = Q_{mi}Q_{nj}Q_{rk}\varepsilon_{mnr} = (\det \boldsymbol{Q})\varepsilon_{ijk}, \tag{1.2.67}$$

[5] Arthur Cayley, 1821–1895, a British mathematician. Sir William Rowan Hamilton, 1805–1865, an Irish physicist, astronomer, and mathematician. Cayley helped found the modern British school of pure mathematics. The main contributions of Hamilton are in the fields of classical mechanics, optics, and algebra.

with det $\boldsymbol{Q}$ given by

$$\det \boldsymbol{Q} = \begin{vmatrix} Q_{11} & Q_{12} & Q_{13} \\ Q_{21} & Q_{22} & Q_{33} \\ Q_{31} & Q_{32} & Q_{33} \end{vmatrix}. \tag{1.2.68}$$

Since interchanging rows and columns of Eq. (1.2.68) does not affect the value of det $\boldsymbol{Q}$, it follows that

$$(\det \boldsymbol{Q})^2 = \begin{vmatrix} Q_{11} & Q_{12} & Q_{13} \\ Q_{21} & Q_{22} & Q_{33} \\ Q_{31} & Q_{32} & Q_{33} \end{vmatrix} \begin{vmatrix} Q_{11} & Q_{21} & Q_{31} \\ Q_{12} & Q_{22} & Q_{32} \\ Q_{13} & Q_{23} & Q_{33} \end{vmatrix} = 1, \tag{1.2.69}$$

in which the multiplication rule of determinant has been used. Letting $\boldsymbol{Q}$ be an identity transformation, i.e., $\boldsymbol{e}'_i$ coincides to $\boldsymbol{e}_i$, gives rise to det $\boldsymbol{Q} = +1$. Substituting it into Eq. (1.2.67) results in $\varepsilon'_{ijk} = \varepsilon_{ijk}$, indicating that ε_{ijk} is a third-order isotropic tensor.

It can be shown that any second-order isotropic tensor must be of the form of a constant times δ_{ij}, and any third-order isotropic tensor must be of the form of a constant times ε_{ijk}. The most general formulation of a fourth-order isotropic tensor is given by

$$a_{ijkl} = \alpha\delta_{ij}\delta_{kl} + \beta\delta_{ik}\delta_{jl} + \gamma\delta_{il}\delta_{jk}, \tag{1.2.70}$$

where α, β, and γ are constants. Generally, any even-order isotropic tensor possesses a form analogous to Eq. (1.2.70), in which all possible combinations of δ_{ij} involve.

1.3 Tensor Calculus

1.3.1 Time Rate of Change of a Tensor

Let $\boldsymbol{T}$ be a tensor depending on a scalar t (e.g. time), viz., $\boldsymbol{T} = \boldsymbol{T}(t)$. The derivative of $\boldsymbol{T}$ with respect to t is defined by

$$\frac{\mathrm{d}\boldsymbol{T}}{\mathrm{d}t} \equiv \lim_{\Delta t \to 0} \frac{\boldsymbol{T}(t+\Delta t) - \boldsymbol{T}(t)}{\Delta t}, \tag{1.3.1}$$

yielding a second-order tensor, with which the following identities are obtained:

$$\begin{aligned} &\frac{\mathrm{d}}{\mathrm{d}t}(\boldsymbol{T}+\boldsymbol{U}) = \frac{\mathrm{d}\boldsymbol{T}}{\mathrm{d}t} + \frac{\mathrm{d}\boldsymbol{U}}{\mathrm{d}t}, \quad \frac{\mathrm{d}}{\mathrm{d}t}(\alpha\boldsymbol{T}) = \frac{\mathrm{d}\alpha}{\mathrm{d}t}\boldsymbol{T} + \alpha\frac{\mathrm{d}\boldsymbol{T}}{\mathrm{d}t}, \\ &\frac{\mathrm{d}}{\mathrm{d}t}(\boldsymbol{T}\boldsymbol{U}) = \frac{\mathrm{d}\boldsymbol{T}}{\mathrm{d}t}\boldsymbol{U} + \boldsymbol{T}\frac{\mathrm{d}\boldsymbol{U}}{\mathrm{d}t}, \quad \frac{\mathrm{d}}{\mathrm{d}t}(\boldsymbol{T}\boldsymbol{a}) = \frac{\mathrm{d}\boldsymbol{T}}{\mathrm{d}t}\boldsymbol{a} + \boldsymbol{T}\frac{\mathrm{d}\boldsymbol{a}}{\mathrm{d}t}, \\ &\frac{\mathrm{d}\boldsymbol{T}^{\mathrm{T}}}{\mathrm{d}t} = \left(\frac{\mathrm{d}\boldsymbol{T}}{\mathrm{d}t}\right)^{\mathrm{T}}, \end{aligned} \tag{1.3.2}$$

in which α, $\boldsymbol{a}$, and $\boldsymbol{U}$ are respectively arbitrary scalar, vector, and tensor all depending on t.

1.3.2 Gradient, Divergence, and Curl

Let ϕ be a scalar function depending on a vector argument $\boldsymbol{a}$, i.e., $\phi = \phi(\boldsymbol{a})$. The *gradient* of ϕ at the point $\boldsymbol{a}$, denoted by grad ϕ, is defined to be a vector such that its dot product with $\mathrm{d}\boldsymbol{a}$ yields the difference in the values of ϕ at $\boldsymbol{a} + \mathrm{d}\boldsymbol{a}$ and $\boldsymbol{a}$, viz.,

$$\mathrm{d}\phi = \phi(\boldsymbol{a} + \mathrm{d}\boldsymbol{a}) - \phi(\boldsymbol{a}) \equiv \operatorname{grad}\phi \cdot \mathrm{d}\boldsymbol{a}, \tag{1.3.3}$$

which, by choosing the unit vector $\boldsymbol{e}$ to be in the direction of $\boldsymbol{a}$, is recast alternatively as

$$\frac{\mathrm{d}\phi}{\|\mathrm{d}\boldsymbol{a}\|} = \operatorname{grad}\phi \cdot \boldsymbol{e}. \tag{1.3.4}$$

Hence, the explicit expression of grad ϕ, by choosing $\boldsymbol{a}$ to be the position vector, is obtained as

$$\operatorname{grad}\phi = \frac{\partial \phi}{\partial x_i}\boldsymbol{e}_i. \tag{1.3.5}$$

Let $\boldsymbol{v}$ be a vector function depending on a vector argument $\boldsymbol{a}$, i.e., $\boldsymbol{v} = \boldsymbol{v}(\boldsymbol{a})$. The gradient of $\boldsymbol{v}$ at the point $\boldsymbol{a}$, denoted by grad $\boldsymbol{v}$, is defined to be a second-order tensor which gives the difference in the values of $\boldsymbol{v}$ at $\boldsymbol{a} + \mathrm{d}\boldsymbol{a}$ and $\boldsymbol{a}$ when operating on $\mathrm{d}\boldsymbol{a}$, viz.,

$$\mathrm{d}\boldsymbol{v} = \boldsymbol{v}(\boldsymbol{a} + \mathrm{d}\boldsymbol{a}) - \boldsymbol{v}(\boldsymbol{a}) \equiv (\operatorname{grad}\boldsymbol{v})\mathrm{d}\boldsymbol{a}. \tag{1.3.6}$$

By using a similar procedure described previously, the explicit expression of grad $\boldsymbol{v}$ is given by

$$\operatorname{grad}\boldsymbol{v} = \frac{\partial v_i}{\partial x_j}(\boldsymbol{e}_i\boldsymbol{e}_j), \tag{1.3.7}$$

with the matrix representation

$$[\operatorname{grad}\boldsymbol{v}] = \begin{bmatrix} \dfrac{\partial v_1}{\partial x_1} & \dfrac{\partial v_1}{\partial x_2} & \dfrac{\partial v_1}{\partial x_3} \\ \dfrac{\partial v_2}{\partial x_1} & \dfrac{\partial v_2}{\partial x_2} & \dfrac{\partial v_2}{\partial x_3} \\ \dfrac{\partial v_3}{\partial x_1} & \dfrac{\partial v_3}{\partial x_2} & \dfrac{\partial v_3}{\partial x_3} \end{bmatrix}. \tag{1.3.8}$$

The gradients of second- or higher-order tensors can be obtained in a similar manner. It is recognized that the gradient operation increases the number of free index of the operated quantity. Applying gradient to a scalar yields a vector, and applying gradient to a vector gives rise to a second-order tensor, etc.

The gradient operation has physical interpretations. For example, if $\boldsymbol{a}$ denotes the position vector of a mass particle m, whose temperature is denoted by ϕ, then grad ϕ indicates the temperature variation in space at $\boldsymbol{a}$, whose direction is perpendicular to the surface described by $\phi =$ constant. The maximum temperature variation occurs if $\mathrm{d}\boldsymbol{a}$ is in the same direction of grad ϕ. In this case, $\mathrm{d}\phi/\|\mathrm{d}\boldsymbol{a}\| = \|\operatorname{grad}\phi\|$. If $\boldsymbol{v}$ is a fluid velocity, grad $\boldsymbol{v}$ represents the "deformation rate" and rigid body rotation of a fluid element.

Let $\boldsymbol{v}$ be a vector function. The *divergence* of $\boldsymbol{v}$, denoted by div $\boldsymbol{v}$, is defined to be a scalar satisfying the relation

$$\operatorname{div} \boldsymbol{v} \equiv \operatorname{tr}(\operatorname{grad} \boldsymbol{v}), \tag{1.3.9}$$

which is expressed alternatively by using the index notation, viz.,

$$\operatorname{div} \boldsymbol{v} = \frac{\partial v_i}{\partial x_i}. \tag{1.3.10}$$

Similarly, let $\boldsymbol{T}$ be a second-order tensor, whose divergence is denoted by div $\boldsymbol{T}$, which is defined to be a vector, so that

$$(\operatorname{div} \boldsymbol{T}) \cdot \boldsymbol{a} \equiv \operatorname{div} \left(\boldsymbol{T}^{\mathrm{T}} \boldsymbol{a}\right) - \operatorname{tr} \left(\boldsymbol{T}^{\mathrm{T}} (\operatorname{grad} \boldsymbol{a})\right), \tag{1.3.11}$$

for any vector $\boldsymbol{a}$, with its explicit expression given by

$$\operatorname{div} \boldsymbol{T} = \frac{\partial T_{ij}}{\partial x_j} \boldsymbol{e}_i. \tag{1.3.12}$$

The divergence operations for higher-order tensors can be obtained in a similar manner. It decreases the number of free index of the operated quantity. Applying divergence to a vector yields a scalar, and applying divergence to a second-order tensor gives a vector, etc. Divergence of a scalar is, however, not defined. A physical interpretation of divergence, for example, is that if $\boldsymbol{v}$ is the velocity of a fluid, then div $\boldsymbol{v}$ yields the volume flow rate across a specific surface in space.

If $\boldsymbol{v}$ is a vector function, its *curl*, denoted by curl $\boldsymbol{v}$, is defined to be a vector satisfying

$$\operatorname{curl} \boldsymbol{v} \equiv 2\boldsymbol{a}^{\mathrm{w}}, \tag{1.3.13}$$

where $\boldsymbol{a}^{\mathrm{w}}$ is the dual vector of the antisymmetric part of grad $\boldsymbol{v}$, with its explicit expression given by

$$\operatorname{curl} \boldsymbol{v} = \left(\frac{\partial v_3}{\partial x_2} - \frac{\partial v_2}{\partial x_3}\right) \boldsymbol{e}_1 + \left(\frac{\partial v_1}{\partial x_3} - \frac{\partial v_3}{\partial x_1}\right) \boldsymbol{e}_2 + \left(\frac{\partial v_2}{\partial x_1} - \frac{\partial v_1}{\partial x_2}\right) \boldsymbol{e}_3, \tag{1.3.14}$$

under the $\boldsymbol{e}_i$ base. The curl operation does not change the number of free index of the operated quantity, and its physical interpretation, for example, is twice the angular velocity of a fluid if $\boldsymbol{v}$ is the fluid velocity.

1.3.3 Nabla and the Laplacian Operators

With the orthonormal base $\boldsymbol{e}_i$, the *Nabla* and *Laplacian operators*,[6] ∇ and ∇^2, are given respectively by

$$\nabla = \frac{\partial}{\partial x_i} \boldsymbol{e}_i, \qquad \nabla^2 = \nabla \cdot \nabla = \frac{\partial^2}{\partial x_i^2} \equiv \text{lap}. \tag{1.3.15}$$

---

[6] The name "Nabla" comes from the Hellenistic Greek word for a Phoenician harp based on the symbol's shape. Pierre-Simon Laplace, 1749–1827, a French scholar whose main contributions are the development of mathematics, statistics, physics, and astronomy.

The Nabla operator is frequently used to express the gradient, divergence, and curl operations. Let ϕ and $\boldsymbol{v}$ be arbitrary scalar and vector, respectively, for which

$$\begin{aligned}\operatorname{grad}\phi &= \nabla\phi = \left(\frac{\partial}{\partial x_i}\boldsymbol{e}_i\right)\phi = \frac{\partial\phi}{\partial x_i}\boldsymbol{e}_i,\\ \operatorname{div}\boldsymbol{v} &= \nabla\cdot\boldsymbol{v} = \left(\frac{\partial}{\partial x_j}\boldsymbol{e}_j\right)\cdot(v_i\boldsymbol{e}_i) = \frac{\partial v_i}{\partial x_j}\delta_{ji} = \frac{\partial v_i}{\partial x_i},\\ \operatorname{curl}\boldsymbol{v} &= \left(\frac{\partial}{\partial x_i}\boldsymbol{e}_i\right)\times\left(v_j\boldsymbol{e}_j\right) = \frac{\partial v_j}{\partial x_i}\left(\boldsymbol{e}_i\times\boldsymbol{e}_j\right) = \frac{\partial v_j}{\partial x_i}\varepsilon_{ijk}\boldsymbol{e}_k,\end{aligned} \tag{1.3.16}$$

corresponding exactly to Eqs. (1.3.5), (1.3.10) and (1.3.14), respectively. The Nabla operator can equally be used to conduct various vector operations. For example, let ϕ be any scalar; it follows that

$$\nabla\times(\nabla\phi) = \left(\frac{\partial}{\partial x_i}\boldsymbol{e}_i\right)\times\left(\frac{\partial\phi}{\partial x_j}\boldsymbol{e}_j\right) = \frac{\partial^2\phi}{\partial x_i\partial x_j}(\boldsymbol{e}_i\times\boldsymbol{e}_j) = \frac{\partial^2\phi}{\partial x_i\partial x_j}\varepsilon_{ijk}\boldsymbol{e}_k = \boldsymbol{0}, \tag{1.3.17}$$

provided that ϕ is continuous subject to its second derivatives. The operations conducted in Eqs. (1.3.16) and (1.3.17) in terms of the Nabla operator are referred to as the *symbolic representation*.

However, caution must be made when applying the Nabla operator to conduct the gradient of a vector, which is given by

$$\operatorname{grad}\boldsymbol{v} = (\nabla\otimes\boldsymbol{v})^{\mathrm{T}} = \frac{\partial v_i}{\partial x_j}\boldsymbol{e}_i\otimes\boldsymbol{e}_j, \qquad (\operatorname{grad}\boldsymbol{v})^{\mathrm{T}} = \nabla\otimes\boldsymbol{v} = \frac{\partial v_i}{\partial x_j}\boldsymbol{e}_j\otimes\boldsymbol{e}_i, \tag{1.3.18}$$

for any vector $\boldsymbol{v}$. Similarly, for a second-order tensor $\boldsymbol{T}$, it follows that

$$\operatorname{div}\boldsymbol{T} = \nabla\cdot\boldsymbol{T}^{\mathrm{T}} = \frac{\partial T_{ij}}{\partial x_j}\boldsymbol{e}_i, \qquad \operatorname{div}\boldsymbol{T}^{\mathrm{T}} = \nabla\cdot\boldsymbol{T} = \frac{\partial T_{ij}}{\partial x_i}\boldsymbol{e}_j. \tag{1.3.19}$$

In calculating gradients and divergences of higher-order tensors by using the symbolic representation, care has to be taken with respect to which indices these should be differentiated.

1.4 Orthogonal Curvilinear Coordinates

Let ϕ, $\boldsymbol{v}$, and $\boldsymbol{T}$ be any scalar, vector, and tensor, respectively, and $\{x_i\}$ be a set of right-handed orthogonal curvilinear coordinates with $\{\boldsymbol{e}_i\}$ the corresponding orthonormal base. Define the position vector $\boldsymbol{r}$ in the form

$$\boldsymbol{r} = x\boldsymbol{e}_x + y\boldsymbol{e}_y + z\boldsymbol{e}_z, \tag{1.4.1}$$

where $\boldsymbol{e}_x, \boldsymbol{e}_y, \boldsymbol{e}_z$ are fixed in space. Define the orthonormal base vectors $\boldsymbol{e}_i$, metric scale factors h_i, and line element $\mathrm{d}\boldsymbol{r}\cdot\mathrm{d}\boldsymbol{r}$ as

$$\boldsymbol{e}_i = \frac{\partial\boldsymbol{r}}{\partial x_i}\Big/\left\|\frac{\partial\boldsymbol{r}}{\partial x_i}\right\|, \qquad h_i = \left\|\frac{\partial\boldsymbol{r}}{\partial x_i}\right\|, \qquad \mathrm{d}\boldsymbol{r}\cdot\mathrm{d}\boldsymbol{r} = h_i^2(\mathrm{d}x_i)^2. \tag{1.4.2}$$

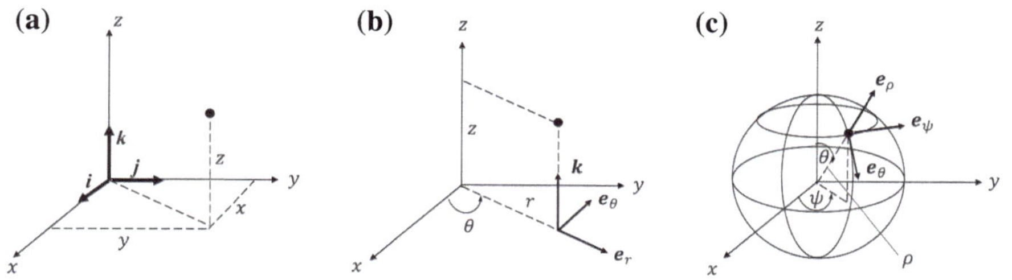

Fig. 1.1 Orthogonal curvilinear coordinate systems. **a** The rectangular coordinates. **b** The cylindrical coordinates. **c** The spherical coordinates

1.4.1 Rectangular Coordinates

Consider the rectangular coordinates shown in Fig. 1.1a, for which

$$\begin{aligned} \{x_1, x_2, x_3\} &= \{x, y, z\}, \ \boldsymbol{r} = x\boldsymbol{i} + y\boldsymbol{j} + z\boldsymbol{k}, \\ \{\boldsymbol{e}_1, \boldsymbol{e}_2, \boldsymbol{e}_3\} &= \{\boldsymbol{i}, \boldsymbol{j}, \boldsymbol{k}\}, \ \mathrm{d}\boldsymbol{r} = \mathrm{d}x\boldsymbol{i} + \mathrm{d}y\boldsymbol{j} + \mathrm{d}z\boldsymbol{k}, \end{aligned} \tag{1.4.3}$$

and

$$\boldsymbol{v} = [v_x, v_y, v_z], \ [\boldsymbol{T}] = \begin{bmatrix} T_{xx} & T_{xy} & T_{xz} \\ T_{yx} & T_{yy} & T_{yz} \\ T_{zx} & T_{zy} & T_{zz} \end{bmatrix}. \tag{1.4.4}$$

With these, the gradient, divergence, curl, Laplacian operations, and *Lagrangian derivative* are given by[7]

- Gradient of ϕ:

$$\operatorname{grad} \phi = \frac{\partial \phi}{\partial x}\boldsymbol{i} + \frac{\partial \phi}{\partial y}\boldsymbol{j} + \frac{\partial \phi}{\partial z}\boldsymbol{k}. \tag{1.4.5}$$

- Gradient of $\boldsymbol{v}$:

$$[\operatorname{grad} \boldsymbol{v}] = \begin{bmatrix} \dfrac{\partial v_x}{\partial x} & \dfrac{\partial v_x}{\partial y} & \dfrac{\partial v_x}{\partial z} \\ \dfrac{\partial v_y}{\partial x} & \dfrac{\partial v_y}{\partial y} & \dfrac{\partial v_y}{\partial z} \\ \dfrac{\partial v_z}{\partial x} & \dfrac{\partial v_z}{\partial y} & \dfrac{\partial v_z}{\partial z} \end{bmatrix}. \tag{1.4.6}$$

- Divergence of $\boldsymbol{v}$:

$$\operatorname{div} \boldsymbol{v} = \frac{\partial v_x}{\partial x} + \frac{\partial v_y}{\partial y} + \frac{\partial v_z}{\partial z}. \tag{1.4.7}$$

[7] Joseph-Louis Lagrange, 1736–1813, an Italian Enlightenment Era mathematician and astronomer, who contributed to the fields of analysis, number theory, and both classical and celestial mechanics. The Lagrangian derivative is the convection part of *material derivative*, which will be discussed in Sect. 5.1.3.

- Divergence of $\boldsymbol{T}$:

$$
\begin{aligned}
(\operatorname{div}\boldsymbol{T})|_x &= \frac{\partial T_{xx}}{\partial x} + \frac{\partial T_{yx}}{\partial y} + \frac{\partial T_{zx}}{\partial z},\\
(\operatorname{div}\boldsymbol{T})|_y &= \frac{\partial T_{xy}}{\partial x} + \frac{\partial T_{yy}}{\partial y} + \frac{\partial T_{zy}}{\partial z},\\
(\operatorname{div}\boldsymbol{T})|_z &= \frac{\partial T_{xz}}{\partial x} + \frac{\partial T_{yz}}{\partial y} + \frac{\partial T_{zz}}{\partial z}.
\end{aligned}
\tag{1.4.8}
$$

- Curl of $\boldsymbol{v}$:

$$
\operatorname{curl}\boldsymbol{v} = \left(\frac{\partial v_z}{\partial y} - \frac{\partial v_y}{\partial z}\right)\boldsymbol{i} + \left(\frac{\partial v_x}{\partial z} - \frac{\partial v_z}{\partial x}\right)\boldsymbol{j} + \left(\frac{\partial v_y}{\partial x} - \frac{\partial v_x}{\partial y}\right)\boldsymbol{k}. \tag{1.4.9}
$$

- Laplacian of ϕ:

$$
\operatorname{lap}\phi = \frac{\partial^2\phi}{\partial x^2} + \frac{\partial^2\phi}{\partial y^2} + \frac{\partial^2\phi}{\partial z^2}. \tag{1.4.10}
$$

- Laplacian of $\boldsymbol{v}$:

$$
\begin{aligned}
(\operatorname{lap}\boldsymbol{v})|_x &= \frac{\partial^2 v_x}{\partial^2 x} + \frac{\partial^2 v_x}{\partial y^2} + \frac{\partial v_x^2}{\partial z^2},\\
(\operatorname{lap}\boldsymbol{v})|_y &= \frac{\partial^2 v_y}{\partial x^2} + \frac{\partial^2 v_y}{\partial y^2} + \frac{\partial^2 v_y}{\partial z^2},\\
(\operatorname{lap}\boldsymbol{v})|_z &= \frac{\partial^2 v_z}{\partial x^2} + \frac{\partial^2 v_z}{\partial y^2} + \frac{\partial^2 v_z}{\partial z^2}.
\end{aligned}
\tag{1.4.11}
$$

- Lagrangian derivative of $\boldsymbol{v}$:

$$
\begin{aligned}
(\boldsymbol{v}\cdot\nabla)\boldsymbol{v}|_x &= v_x\frac{\partial v_x}{\partial x} + v_y\frac{\partial v_x}{\partial y} + v_z\frac{\partial v_x}{\partial z},\\
(\boldsymbol{v}\cdot\nabla)\boldsymbol{v}|_y &= v_x\frac{\partial v_y}{\partial x} + v_y\frac{\partial v_y}{\partial y} + v_z\frac{\partial v_y}{\partial z},\\
(\boldsymbol{v}\cdot\nabla)\boldsymbol{v}|_z &= v_x\frac{\partial v_z}{\partial x} + v_y\frac{\partial v_z}{\partial y} + v_z\frac{\partial v_z}{\partial z}.
\end{aligned}
\tag{1.4.12}
$$

1.4.2 Cylindrical Coordinates

Consider the cylindrical coordinates shown in Fig. 1.1b, for which

$$
\begin{aligned}
&\{x_1, x_2, x_3\} = \{r, \theta, z\}, \quad && \boldsymbol{r} = \|\boldsymbol{r}\|\cos\theta\boldsymbol{i} + \|\boldsymbol{r}\|\sin\theta\boldsymbol{j} + z\boldsymbol{k},\\
&\{\boldsymbol{e}_1, \boldsymbol{e}_2, \boldsymbol{e}_3\} = \{\boldsymbol{e}_r, \boldsymbol{e}_\theta, \boldsymbol{k}\}, && \mathrm{d}\boldsymbol{r} = \mathrm{d}r\boldsymbol{e}_r + (r\mathrm{d}\theta)\boldsymbol{e}_\theta + \mathrm{d}z\boldsymbol{k},\\
&\boldsymbol{e}_r = \cos\theta\boldsymbol{i} + \sin\theta\boldsymbol{j}, && \boldsymbol{e}_\theta = -\sin\theta\boldsymbol{i} + \cos\theta\boldsymbol{j},
\end{aligned}
\tag{1.4.13}
$$

and

$$
\boldsymbol{v} = [v_r, v_\theta, v_z],\ [\boldsymbol{T}] = \begin{bmatrix} T_{rr} & T_{r\theta} & T_{rz}\\ T_{\theta r} & T_{\theta\theta} & T_{\theta z}\\ T_{zr} & T_{z\theta} & T_{zz} \end{bmatrix}. \tag{1.4.14}
$$

The corresponding expressions of Eqs. (1.4.5)–(1.4.12) are given by

- Gradient of ϕ:

$$\text{grad}\,\phi = \frac{\partial \phi}{\partial r}\boldsymbol{e}_r + \frac{1}{r}\frac{\partial \phi}{\partial \theta}\boldsymbol{e}_\theta + \frac{\partial \phi}{\partial z}\boldsymbol{k}. \tag{1.4.15}$$

- Gradient of $\boldsymbol{v}$:

$$[\text{grad}\,\boldsymbol{v}] = \begin{bmatrix} \dfrac{\partial v_r}{\partial r} & \dfrac{1}{r}\left(\dfrac{\partial v_r}{\partial \theta} - v_\theta\right) & \dfrac{\partial v_r}{\partial z} \\ \dfrac{\partial v_\theta}{\partial r} & \dfrac{1}{r}\left(\dfrac{\partial v_\theta}{\partial \theta} + v_r\right) & \dfrac{\partial v_\theta}{\partial z} \\ \dfrac{\partial v_z}{\partial r} & \dfrac{1}{r}\dfrac{\partial v_z}{\partial \theta} & \dfrac{\partial v_z}{\partial z} \end{bmatrix}. \tag{1.4.16}$$

- Divergence of $\boldsymbol{v}$:

$$\text{div}\,\boldsymbol{v} = \frac{1}{r}\frac{\partial}{\partial r}(r v_r) + \frac{1}{r}\frac{\partial v_\theta}{\partial \theta} + \frac{\partial v_z}{\partial z}. \tag{1.4.17}$$

- Divergence of $\boldsymbol{T}$:

$$\begin{aligned} (\text{div}\,\boldsymbol{T})|_r &= \frac{\partial T_{rr}}{\partial r} + \frac{1}{r}\frac{\partial T_{r\theta}}{\partial \theta} + \frac{T_{rr} - T_{\theta\theta}}{r} + \frac{\partial T_{rz}}{\partial z}, \\ (\text{div}\,\boldsymbol{T})|_\theta &= \frac{\partial T_{\theta r}}{\partial r} + \frac{1}{r}\frac{\partial T_{\theta\theta}}{\partial \theta} + \frac{T_{r\theta} + T_{\theta r}}{r} + \frac{\partial T_{\theta z}}{\partial z}, \\ (\text{div}\,\boldsymbol{T})|_z &= \frac{\partial T_{zr}}{\partial r} + \frac{1}{r}\frac{\partial T_{z\theta}}{\partial \theta} + \frac{\partial T_{zz}}{\partial z} + \frac{T_{zr}}{r}. \end{aligned} \tag{1.4.18}$$

- Curl of $\boldsymbol{v}$:

$$\text{curl}\,\boldsymbol{v} = \left(\frac{1}{r}\frac{\partial v_z}{\partial \theta} - \frac{\partial v_\theta}{\partial z}\right)\boldsymbol{e}_r + \left(\frac{\partial v_r}{\partial z} - \frac{\partial v_z}{\partial r}\right)\boldsymbol{e}_\theta + \left(\frac{\partial v_\theta}{\partial r} + \frac{v_\theta}{r} - \frac{1}{r}\frac{\partial v_r}{\partial \theta}\right)\boldsymbol{k}. \tag{1.4.19}$$

- Laplacian of ϕ:

$$\text{lap}\,\phi = \frac{1}{r}\frac{\partial}{\partial r}\left(r\frac{\partial \phi}{\partial r}\right) + \frac{1}{r^2}\frac{\partial^2 \phi}{\partial \theta^2} + \frac{\partial^2 \phi}{\partial z^2}. \tag{1.4.20}$$

- Laplacian of $\boldsymbol{v}$:

$$\begin{aligned} (\text{lap}\,\boldsymbol{v})|_r &= \frac{1}{r}\frac{\partial}{\partial r}\left(r\frac{\partial v_r}{\partial r}\right) - \frac{v_r}{r^2} + \frac{1}{r^2}\frac{\partial^2 v_r}{\partial \theta^2} - \frac{2}{r^2}\frac{\partial v_\theta}{\partial \theta} + \frac{\partial v_r^2}{\partial z^2}, \\ (\text{lap}\,\boldsymbol{v})|_\theta &= \frac{1}{r}\frac{\partial}{\partial r}\left(r\frac{\partial v_\theta}{\partial r}\right) - \frac{v_\theta}{r^2} + \frac{1}{r^2}\frac{\partial^2 v_\theta}{\partial \theta^2} + \frac{2}{r^2}\frac{\partial v_r}{\partial \theta} + \frac{\partial^2 v_\theta}{\partial z^2}, \\ (\text{lap}\,\boldsymbol{v})|_z &= \frac{1}{r}\frac{\partial}{\partial r}\left(r\frac{\partial v_z}{\partial r}\right) + \frac{1}{r^2}\frac{\partial^2 v_z}{\partial \theta^2} + \frac{\partial^2 v_z}{\partial z^2}. \end{aligned} \tag{1.4.21}$$

- Lagrangian derivative of $\boldsymbol{v}$:

$$\begin{aligned} (\boldsymbol{v}\cdot\nabla)\boldsymbol{v}|_r &= v_r\frac{\partial v_r}{\partial r} + \frac{v_\theta}{r}\frac{\partial v_r}{\partial \theta} - \frac{v_\theta^2}{r} + v_z\frac{\partial v_r}{\partial z}, \\ (\boldsymbol{v}\cdot\nabla)\boldsymbol{v}|_\theta &= v_r\frac{\partial v_\theta}{\partial r} + \frac{v_\theta}{r}\frac{\partial v_\theta}{\partial \theta} + \frac{v_r v_\theta}{r} + v_z\frac{\partial v_\theta}{\partial z}, \\ (\boldsymbol{v}\cdot\nabla)\boldsymbol{v}|_z &= v_r\frac{\partial v_z}{\partial r} + \frac{v_\theta}{r}\frac{\partial v_z}{\partial \theta} + v_z\frac{\partial v_z}{\partial z}. \end{aligned} \tag{1.4.22}$$

1.4.3 Spherical Coordinates

Consider the spherical coordinates shown in Fig. 1.1c, for which

$$\begin{aligned}
&\{x_1, x_2, x_3\} = \{\rho, \theta, \psi\}, && \{\boldsymbol{e}_1, \boldsymbol{e}_2, \boldsymbol{e}_3\} = \{\boldsymbol{e}_\rho, \boldsymbol{e}_\theta, \boldsymbol{e}_\psi\},\\
&\boldsymbol{r} = \|\boldsymbol{r}\| \sin\theta\cos\psi\boldsymbol{i} + \|\boldsymbol{r}\|\sin\theta\sin\psi\boldsymbol{j} && \mathrm{d}\boldsymbol{r} = \mathrm{d}\rho\,\boldsymbol{e}_\rho + (\rho\mathrm{d}\theta)\boldsymbol{e}_\theta + (\rho\sin\theta\mathrm{d}\psi)\boldsymbol{e}_\psi,\\
&\quad +\|\boldsymbol{r}\|\cos\theta\boldsymbol{k}, \\
&\boldsymbol{e}_\rho = \sin\theta\cos\psi\boldsymbol{i} + \sin\theta\sin\psi\boldsymbol{j} + \cos\theta\boldsymbol{k},\\
&\boldsymbol{e}_\theta = \cos\theta\cos\psi\boldsymbol{i} + \cos\theta\sin\psi\boldsymbol{j} - \sin\theta\boldsymbol{k},\\
&\boldsymbol{e}_\psi = -\sin\psi\boldsymbol{i} + \cos\psi\boldsymbol{j},
\end{aligned} \tag{1.4.23}$$

and

$$\boldsymbol{v} = [v_\rho, v_\theta, v_\psi],\ [\boldsymbol{T}] = \begin{bmatrix} T_{\rho\rho} & T_{\rho\theta} & T_{\rho\psi}\\ T_{\theta\rho} & T_{\theta\theta} & T_{\theta\psi}\\ T_{\psi\rho} & T_{\psi\theta} & T_{\psi\psi}\end{bmatrix}. \tag{1.4.24}$$

With the procedures conducted previously, the corresponding expressions of Eqs. (1.4.5)–(1.4.12) are given by

- Gradient of ϕ:

$$\operatorname{grad}\phi = \frac{\partial\phi}{\partial\rho}\boldsymbol{e}_\rho + \frac{1}{\rho}\frac{\partial\phi}{\partial\theta}\boldsymbol{e}_\theta + \frac{1}{\rho\sin\theta}\frac{\partial\phi}{\partial\psi}\boldsymbol{e}_\psi. \tag{1.4.25}$$

- Gradient of $\boldsymbol{v}$:

$$[\operatorname{grad}\boldsymbol{v}] = \begin{bmatrix}
\dfrac{\partial v_\rho}{\partial\rho} & \dfrac{1}{\rho}\left(\dfrac{\partial v_\rho}{\partial\theta} - v_\theta\right) & \dfrac{1}{\rho\sin\theta}\left(\dfrac{\partial v_\rho}{\partial\psi} - v_\psi\sin\theta\right)\\
\dfrac{\partial v_\theta}{\partial\rho} & \dfrac{1}{\rho}\left(\dfrac{\partial v_\theta}{\partial\theta} + v_\rho\right) & \dfrac{1}{\rho\sin\theta}\left(\dfrac{\partial v_\theta}{\partial\psi} - v_\psi\cos\theta\right)\\
\dfrac{\partial v_\psi}{\partial\rho} & \dfrac{1}{\rho}\dfrac{\partial v_\psi}{\partial\theta} & \dfrac{1}{\rho\sin\theta}\dfrac{\partial v_\psi}{\partial\psi} + \dfrac{v_\rho}{\rho} + \dfrac{v_\theta\cot\theta}{\rho}
\end{bmatrix}. \tag{1.4.26}$$

- Divergence of $\boldsymbol{v}$:

$$\operatorname{div}\boldsymbol{v} = \frac{1}{\rho^2}\frac{\partial(\rho^2 v_\rho)}{\partial\rho} + \frac{1}{\rho\sin\theta}\frac{\partial(v_\theta\sin\theta)}{\partial\theta} + \frac{1}{\rho\sin\theta}\frac{\partial v_\psi}{\partial\psi}. \tag{1.4.27}$$

- Divergence of $\boldsymbol{T}$:

$$\begin{aligned}
(\operatorname{div}\boldsymbol{T})|_\rho &= \frac{1}{\rho^2}\frac{\partial(\rho^2 T_{\rho\rho})}{\partial\rho} + \frac{1}{\rho\sin\theta}\frac{\partial(T_{\rho\theta}\sin\theta)}{\partial\theta} - \frac{T_{\theta\theta} + T_{\psi\psi}}{\rho} + \frac{1}{\rho\sin\theta}\frac{\partial T_{\rho\psi}}{\partial\psi},\\
(\operatorname{div}\boldsymbol{T})|_\theta &= \frac{1}{\rho^3}\frac{\partial(\rho^3 T_{\theta\rho})}{\partial\rho} + \frac{1}{\rho\sin\theta}\frac{\partial(T_{\theta\theta}\sin\theta)}{\partial\theta} + \frac{T_{\rho\theta} - T_{\theta\rho} - T_{\psi\psi}\cot\theta}{\rho}\\
&\quad + \frac{1}{\rho\sin\theta}\frac{\partial T_{\theta\psi}}{\partial\psi},\\
(\operatorname{div}\boldsymbol{T})|_\psi &= \frac{1}{\rho^3}\frac{\partial(\rho^3 T_{\psi\rho})}{\partial\rho} + \frac{1}{\rho\sin\theta}\frac{\partial(T_{\psi\theta}\sin\theta)}{\partial\theta} + \frac{T_{\rho\psi} - T_{\psi\rho} + T_{\theta\psi}\cot\theta}{\rho}\\
&\quad + \frac{1}{\rho\sin\theta}\frac{\partial T_{\psi\psi}}{\partial\psi}.
\end{aligned} \tag{1.4.28}$$

- Curl of $\boldsymbol{v}$:

$$\operatorname{curl}\boldsymbol{v} = \left(\frac{1}{\rho\sin\theta}\frac{\partial(v_\psi\sin\theta)}{\partial\theta} - \frac{1}{\rho\sin\theta}\frac{\partial v_\theta}{\partial\psi}\right)\boldsymbol{e}_\rho + \left(\frac{1}{\rho\sin\theta}\frac{\partial v_\rho}{\partial\psi} - \frac{1}{\rho}\frac{\partial(\rho v_\psi)}{\partial\rho}\right)\boldsymbol{e}_\theta + \left(\frac{1}{\rho}\frac{\partial(\rho v_\theta)}{\partial\rho} - \frac{1}{\rho}\frac{\partial v_\rho}{\partial\theta}\right)\boldsymbol{e}_\psi. \tag{1.4.29}$$

- Laplacian of ϕ:

$$\operatorname{lap}\phi = \frac{1}{\rho^2}\frac{\partial}{\partial\rho}\left(\rho^2\frac{\partial\phi}{\partial r}\right) + \frac{1}{\rho^2\sin\theta}\frac{\partial}{\partial\theta}\left(\sin\theta\frac{\partial\phi}{\partial\theta}\right) + \frac{1}{\rho^2\sin^2\theta}\frac{\partial^2\phi}{\partial\psi^2}. \tag{1.4.30}$$

- Laplacian of $\boldsymbol{v}$:

$$\begin{aligned}
(\operatorname{lap}\boldsymbol{v})|_\rho &= \nabla^2 v_\rho - \frac{2}{\rho^2}\left(v_\rho + \frac{\partial v_\theta}{\partial\theta} + v_\theta\cot\theta + \frac{1}{\sin\theta}\frac{\partial v_\psi}{\partial\psi}\right),\\
(\operatorname{lap}\boldsymbol{v})|_\theta &= \nabla^2 v_\theta + \frac{2}{\rho^2}\frac{\partial v_\rho}{\partial\theta} - \frac{1}{\rho^2\sin^2\theta}\left(u_\theta + 2\cos\theta\frac{\partial v_\psi}{\partial\psi}\right),\\
(\operatorname{lap}\boldsymbol{v})|_\psi &= \nabla^2 v_\psi + \frac{1}{\rho^2\sin^2\theta}\left(2\sin\theta\frac{\partial v_\rho}{\partial\psi} + 2\cos\theta\frac{\partial v_\theta}{\partial\psi} - v_\psi\right),
\end{aligned} \tag{1.4.31}$$

with

$$\nabla^2 = \frac{1}{\rho^2}\frac{\partial}{\partial\rho}\left(\rho^2\frac{\partial}{\partial\rho}\right) + \frac{1}{\rho^2\sin\theta}\frac{\partial}{\partial\theta}\left(\sin\theta\frac{\partial}{\partial\theta}\right) + \frac{1}{\rho^2\sin^2\theta}\frac{\partial^2}{\partial\psi^2}. \tag{1.4.32}$$

- Lagrangian derivative of $\boldsymbol{v}$:

$$\begin{aligned}
(\boldsymbol{v}\cdot\nabla)\boldsymbol{v}|_\rho &= v_\rho\frac{\partial v_\rho}{\partial\rho} + \frac{v_\theta}{\rho}\frac{\partial v_\rho}{\partial\theta} + \frac{v_\psi}{\rho\sin\theta}\frac{\partial v_\rho}{\partial\psi} - \frac{v_\theta^2 + v_\psi^2}{\rho},\\
(\boldsymbol{v}\cdot\nabla)\boldsymbol{v}|_\theta &= v_\rho\frac{\partial v_\theta}{\partial\rho} + \frac{v_\theta}{\rho}\frac{\partial v_\theta}{\partial\theta} + \frac{v_\psi}{\rho\sin\theta}\frac{\partial v_\theta}{\partial\psi} + \frac{v_\rho v_\theta}{\rho} - \frac{v_\psi^2}{\rho}\cot\theta,\\
(\boldsymbol{v}\cdot\nabla)\boldsymbol{v}|_\psi &= v_\rho\frac{\partial v_\psi}{\partial\rho} + \frac{v_\theta}{\rho}\frac{\partial v_\theta}{\partial\theta} + \frac{v_\psi}{\rho\sin\theta}\frac{\partial v_\psi}{\partial\psi} + \frac{v_\rho v_\psi}{\rho} + \frac{v_\psi v_\theta}{\rho}\cot\theta.
\end{aligned} \tag{1.4.33}$$

It is noted that although three orthogonal coordinate systems and the corresponding operations are given explicitly, they can be formulated in a concise manner, in which the metric scale factor plays a role as a "generator" for different coordinate systems. The concise formulation is provided in Appendix A.

1.5 Integral Theorems

Let V be any volume in space which is enclosed by the surface A, $\boldsymbol{n}$ be the unit outward normal on A, and ϕ and $\boldsymbol{v}$ be respectively any scalar and vector defined in V and A. Under sufficiently continuous conditions of V, A, ϕ, and $\boldsymbol{v}$, there exist two theorems relating a surface integral to a volume integral given by

$$\int_A \boldsymbol{v}\cdot\boldsymbol{n}\,\mathrm{d}a = \int_V \operatorname{div}\boldsymbol{v}\,\mathrm{d}v, \qquad \int_A v_i n_i \mathrm{d}a = \int_V \frac{\partial v_i}{\partial x_i}\mathrm{d}v, \tag{1.5.1}$$

known as *Gauss's divergence theorem*,[8] or simply as the *divergence theorem*, and

$$\int_A \phi \frac{\partial \phi}{\partial n} da = \int_V \{\text{grad}\, \phi \cdot \text{grad}\, \phi + \phi \text{lap}\, \phi\}\, dv,$$
$$\int_A \phi \frac{\partial \phi}{\partial n} da = \int_V \left\{ \left(\frac{\partial \phi}{\partial x_i} \right)^2 + \phi \frac{\partial^2 \phi}{\partial x_i^2} \right\} dv, \tag{1.5.2}$$

known as *Green's theorem*.[9] There exists a theorem, known as *Stokes' theorem*,[10] which relates a line integral to an equivalent surface integral given by

$$\oint_\ell \boldsymbol{v} \cdot \mathrm{d}\boldsymbol{\ell} = \int_A (\text{curl}\, \boldsymbol{v}) \cdot \boldsymbol{n} \mathrm{d}a, \qquad \oint_\ell v_i \mathrm{d}\ell_i = \int_A 2a_i^{\mathrm{w}} n_i \mathrm{d}a, \tag{1.5.3}$$

where A is an arbitrary surface which must terminate on the line $\boldsymbol{\ell}$, and a_i^{w} are the components of the dual vector of the antisymmetric part of grad $\boldsymbol{v}$.

The Stokes theorem is used in the potential theory of ideal-fluid flows, while the Gauss and Green theorems are used to relate the production and surface flux of a physical quantity in a space enclosed by a surface. For example, it follows from Eq. $(1.5.1)_2$ that

$$\int_A \phi_{,i} n_i \, da = \int_V \phi_{,ii} \, dv, \tag{1.5.4}$$

which is expressed alternatively as

$$\int_A \left(\frac{\partial \phi}{\partial x_1} n_1 + \frac{\partial \phi}{\partial x_2} n_2 + \frac{\partial \phi}{\partial x_3} n_3 \right) da = \int_A \frac{\mathrm{d}\phi}{\mathrm{d}n} \, da = \int_V \nabla^2 \phi \, \mathrm{d}v, \tag{1.5.5}$$

where $\mathrm{d}\phi/\mathrm{d}n$ is the normal flux of ϕ, with

$$n_1 = \frac{\mathrm{d}x_1}{\mathrm{d}n}, \qquad n_2 = \frac{\mathrm{d}x_2}{\mathrm{d}n}, \qquad n_3 = \frac{\mathrm{d}x_3}{\mathrm{d}n}. \tag{1.5.6}$$

Equilibrium problem involving $\nabla^2 \phi = 0$ can then be satisfied only if

$$\int_A \frac{\mathrm{d}\phi}{\mathrm{d}n} \, da = 0, \tag{1.5.7}$$

showing that the resultant normal flux must vanish to ensure the validity of equilibrium condition.

---

[8] Johann Carl Friedrich Gauss, 1777–1855, a German mathematician, who contributed to many fields such as statistics, analysis, differential geometry, geophysics, mechanics, electrostatics, astronomy, matrix theory, and optics.

[9] George Green, 1793–1841, a British mathematical physicist, whose main contributions were summarized in "An Essay on the Application of Mathematical Analysis to the Theories of Electricity and Magnetism" in 1828.

[10] Sir George Gabriel Stokes, 1819–1903, an Irish physicist and mathematician. He made not only contributions to fluid dynamics and physical optics, but also to the *theory of asymptotic expansions*.

1.6 Complex Analysis

1.6.1 Complex Numbers, Complex and Analytic Functions

The *imaginary number* i is defined as the square root of -1, denoted by $i \equiv \sqrt{-1}$. In a two-dimensional rectangular coordinate system $\{x, y\}$, a *complex number* z is defined viz., $z \equiv x + iy$, where the x-axis is referred to as the *real axis*, while the y-axis is referred to as the *imaginary axis*. The two-dimensional plane spanned by the real and imaginary axes are termed the *complex plane*. The real part of a complex number is usually denoted as $Re(z) = x$, while its imaginary part is represented by $Im(z) = y$. The *complex conjugate* of a complex number z is denoted by $\bar{z}$, which is given by $\bar{z} = x - iy$. With this, the magnitude of z is simply identified as $|z| = x^2 + y^2 = z\bar{z}$. It is conventionally to express a complex number by using the polar coordinates defined on the complex plane. The polar representations of a complex number z and its complex conjugate $\bar{z}$ are given by

$$z = x + iy = re^{i\theta}, \qquad \bar{z} = x - iy = re^{-i\theta}, \tag{1.6.1}$$

where

$$r = \sqrt{x^2 + y^2}, \qquad \theta = \tan^{-1}\left(\frac{y}{x}\right). \tag{1.6.2}$$

A *complex function* is a function that acts on complex numbers and produces complex numbers. A complex function $F(z)$ is said to be *analytic* if the derivative $\mathrm{d}F/\mathrm{d}z$ exists at a point z_0 and in some neighborhood of z_0 and if the value of $\mathrm{d}F/\mathrm{d}z$ is independent of the direction by which it is evaluated. Specifically, if $F(z)$ is analytic, its derivative with respect to z exists and may be determined in any direction, so that

$$\frac{\mathrm{d}F}{\mathrm{d}z} = \frac{\partial F}{\partial x} = -i\frac{\partial F}{\partial y}. \tag{1.6.3}$$

On the contrary, a complex function $F(z)$ may not be analytic at a specific point z_0, and this point is referred to as a *singular point*. If $F(z)$ is analytic in some neighborhood of z_0, but not at z_0 itself, then this point is termed an *isolated singular point*.

1.6.2 The Cauchy-Riemann Equations and Multi-valued Functions

If $F(z)$ is analytic and has the form of $F(z) = \phi(x, y) + i\psi(x, y)$, then its real and imaginary parts must satisfy the *Cauchy-Riemann equations* given by[11]

$$\frac{\partial \phi}{\partial x} = \frac{\partial \psi}{\partial y}, \qquad \frac{\partial \phi}{\partial y} = -\frac{\partial \psi}{\partial x}. \tag{1.6.4}$$

[11] Augustin-Louis Cauchy, 1789–1857, a French mathematician and physicist. Georg Friedrich Bernhard Riemann, 1826–1866, a German mathematician. Cauchy almost singlehandedly founded complex analysis. Riemann contributed to complex analysis by the introduction of the *Riemann surfaces*.

It is noted that the Cauchy-Riemann equations are a necessary, but not a sufficient condition for an analytic function. Eliminating first ϕ and then ψ from the Cauchy-Riemann equations shows that both ϕ and ψ are *harmonic functions*; namely, they must satisfy *Laplace's equation*.

Many functions are analytic but assume more than one value at any point $z = re^{i\theta}$ on the complex plane as θ increases by multiples of 2π, which are called *multi-valued functions*. The difficulty is overcome by replacing the single complex plane by a series of the *Riemann sheets* which are connected to each other along a *branch cut* which runs between two *branch points*, usually along the negative real axis from $z = 0$ to $z \to \infty$, which are singular points of the function.

1.6.3 The Cauchy-Goursat Theorem and Cauchy Integral Formula

If $F(z)$ is analytic at all points inside and on a closed contour C, it must satisfy

$$\int_C F(z)\mathrm{d}z = 0, \tag{1.6.5}$$

which is referred to as the *Cauchy-Goursat theorem*, or simply *Cauchy's integral theorem*.[12] Furthermore, if z_0 is any point inside C, then

$$\frac{\mathrm{d}^n F}{\mathrm{d}z^n}(z_0) = \frac{n!}{2\pi i}\int_C \frac{F(z)}{(z - z_0)^{n+1}}\mathrm{d}z, \tag{1.6.6}$$

for $n \geq 1$, where $F(z_0)$ is given by

$$F(z_0) = \frac{1}{2\pi i}\int_C \frac{F(z)}{(z - z_0)}\mathrm{d}z. \tag{1.6.7}$$

These two equations are termed the *Cauchy integral formula*.

1.6.4 The Taylor, Maclaurin, and Laurent Series

If $F(z)$ is analytic at all points within a circle $r < r_0$ whose center locates at z_0, then $F(z)$ may be represented by the series given by

$$\begin{aligned} F(z) =& F(z_0) + (z - z_0)\frac{\mathrm{d}F}{\mathrm{d}z}(z_0) + \frac{(z - z_0)^2}{2!}\frac{\mathrm{d}^2 F}{\mathrm{d}z^2}(z_0) + \frac{(z - z_0)^3}{3!}\frac{\mathrm{d}^3 F}{\mathrm{d}z^3}(z_0) \\ &+ \cdots , \end{aligned} \tag{1.6.8}$$

where the radius of convergence r_0 is the distance from z_0 to the nearest singularity. This equation is known as the *Taylor series* of $F(z)$ at $z = z_0$, and the special case with $z_0 = 0$ is known as the *Maclaurin series*.[13]

---

[12] Édouard Jean-Baptiste Goursat, 1858–1936, a French mathematician. He sets a standard for the high-level teaching of mathematical analysis, especially of complex analysis.

[13] Brook Taylor, 1685–1731, a British mathematician with his most known works as the *Taylor theorem* and Taylor series. Colin Maclaurin, 1698–1746, a Scottish mathematician, with contributions to geometry and algebra.

If $F(z)$ is analytic at all points within the annual region $r_0 < r < r_1$ whose center is at z_0, then the *Laurent series* of $F(z)$ at z_0 is given by[14]

$$F(z) = \cdots + \frac{b_2}{(z-z_0)^2} + \frac{b_1}{z-z_0} + a_0 + a_1(z-z_0) + a_2(z-z_0)^2 + \cdots, \tag{1.6.9}$$

with

$$a_n = \frac{1}{2\pi i}\int_{C_0} \frac{F(\varepsilon)}{(\varepsilon - z_0)^{n+1}} \mathrm{d}\varepsilon, \qquad b_n = \int_{C_1} \frac{F(\varepsilon)}{(\varepsilon - z_0)^{-n+1}} \mathrm{d}\varepsilon, \tag{1.6.10}$$

where $n = 0, 1, 2, \cdots$. The contours C_0 and C_1 correspond respectively to $r = r_0$ and $r = r_1$. The series is convergent from the smallest radius r_0 and the largest radius r_1, and there exist no singular points in the annular region. The part of series containing the b_n coefficients is known as the *principal part*. For the special case in which $r_0 = 0$, Eq. (1.6.9) reduces to the Taylor series.

1.6.5 Residues and Residue Theorem

The *residue* of a function $F(z)$ at point z_0 is defined as the coefficient b_1 in its Laurent series about the same point, i.e., the coefficient of the term $1/z$ in the Laurent series of the function written about the point z_0. If $F(z)$ is analytic within and on a closed contour C except for a finite number of singular points $z_1, z_2, \cdots$, then

$$\int_C F(z)\mathrm{d}z = 2\pi i(R_1 + R_2 + \cdots), \tag{1.6.11}$$

where $R_1, R_2, \cdots$ are the residues of $F(z)$ at $z_1, z_2, \cdots$, respectively. This equation is known as the *residue theorem*, or alternatively as *Cauchy's residue theorem*.[15]

To evaluate the residues of a function at some point, it is useful to identify the type of singularity which exists at that point. Singularities are conventionally classified in the following categories:

- *Branch points*. The singular points exist at the end of each branch cut of a multi-valued function. In this circumstance, the residue theorem cannot be applied.
- *Essential singular points*. If the principal part of the Laurent series of a function about some point contains an infinite number of terms, that point is an essential singular point.
- *Pole of order m*. If the principal part of the Laurent series of a function about some point contains only terms up to $(z - z_0)^m$, that point is a pole of order m. In this case, the expression $(z - z_0)^m F(z)$ becomes analytic at that point.

---

[14]Pierre Alphonse Laurent, 1813–1854, a French mathematician and military officer. He is best known as the discoverer of the Laurent series.

[15]From the mathematical view point, the residue theorem is a generalization of the Cauchy integral theorem and Cauchy integral formula.

- *Simple pole*. If the principal part of the Laurent series of a function about some point contains only a term proportional to $(z - z_0)$, that point is a simple pole, and it follows that $(z - z_0)F(z)$ becomes analytic at that point.

The methods of determining the residue R of a function $F(z)$ at a singular point z_0 is summarized in the following:

- Expand $F(z)$ in a series about z_0, and obtain the coefficient of the term $1/(z - z_0)$. This is the fundamental method by the definition of residue and is valid for all types of singularities.
- If the point z_0 is a pole of order m, R is given by

$$R = \lim_{z \to z_0} \frac{1}{(m-1)!} \frac{\mathrm{d}^{m-1}}{\mathrm{d}z^{m-1}} \left[(z - z_0)^m F(z) \right]. \tag{1.6.12}$$

- If the point z_0 is a simple pole, R is obtained as

$$R = \lim_{z \to z_0} (z - z_0) F(z). \tag{1.6.13}$$

- If $F(z)$ may be recast in the form $F(z) = p(z)/q(z)$, where $q(z_0) = 0$, but $\mathrm{d}q/\mathrm{d}z(z_0) \neq 0$ and $p(z_0) \neq 0$, R may be determined by

$$R = \lim_{z \to z_0} \frac{p(z)}{\mathrm{d}q(z)/\mathrm{d}z}. \tag{1.6.14}$$

1.6.6 Conformal Transformation

A *conformal transformation* is a one-to-one mapping from the z-plane (one complex plane) to the ζ-plane (another complex plane) via

$$\zeta = f(z), \qquad z = f^{-1}(\zeta), \tag{1.6.15}$$

where f is an analytic function of z, and the z- and ζ-planes are spanned respectively by $z = x + iy$ and $\zeta = \xi + i\eta$, as shown in Fig. 1.2. With this, any geometric shape body in the z-plane can be transformed into other shape body in the ζ-plane, and vice versa. Conformal transformations preserve angles between small arcs except at points where $\mathrm{d}f^{-1}/\mathrm{d}\zeta = 0$, which are termed the *critical points* of the transformation, through which smooth curves in the ζ-plane may give angular corners in the z-plane.

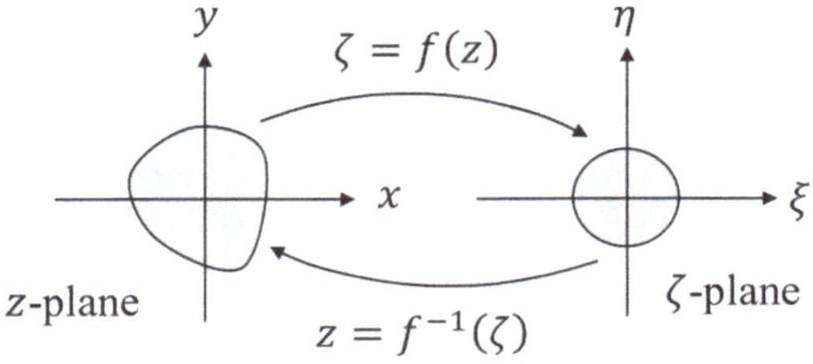

Fig. 1.2 A conformal transformation between two complex planes

The mapping is proposed to determine the solutions to two-dimensional potential flows of complex bodies in a complex plane if the corresponding solutions in another complex plane are known. Typical applications of conformal transformation will be given in Sect. 7.5.

Let ϕ be an analytic function in the z-plane satisfying the Laplace equation. Its Laplace equation in the ζ-plane, by using the chain rule of differentiation and Eq. (1.6.15), is obtained as

$$\left(a_1^2 + a_2^2\right) \frac{\partial^2 \phi}{\partial \xi^2} + \left(a_3^2 + a_4^2\right) \frac{\partial^2 \phi}{\partial \eta^2} + 2\left(a_1 a_3 + a_2 a_4\right) \frac{\partial^2 \phi}{\partial \xi \partial \eta} + \left(\frac{\partial a_1}{\partial x} + \frac{\partial a_2}{\partial y}\right) \frac{\partial \phi}{\partial \xi} + \left(\frac{\partial a_3}{\partial x} + \frac{\partial a_4}{\partial y}\right) \frac{\partial \phi}{\partial \eta} = 0, \tag{1.6.16}$$

for any transformation $\zeta = f(z)$, where

$$a_1 = \frac{\partial \xi}{\partial x}, \qquad a_2 = \frac{\partial \xi}{\partial y}, \qquad a_3 = \frac{\partial \eta}{\partial x}, \qquad a_4 = \frac{\partial \eta}{\partial y}. \tag{1.6.17}$$

If $\zeta = f(z)$ is a conformal transformation, then the mapping f is analytic, so that the real and imaginary parts of ζ should be harmonic, i.e.,

$$\frac{\partial a_1}{\partial x} + \frac{\partial a_2}{\partial y} = \frac{\partial^2 \xi}{\partial x^2} + \frac{\partial^2 \xi}{\partial y^2} = 0, \qquad \frac{\partial a_3}{\partial x} + \frac{\partial a_4}{\partial y} = \frac{\partial^2 \eta}{\partial x^2} + \frac{\partial^2 \eta}{\partial y^2} = 0, \tag{1.6.18}$$

which is supplemented by that $\xi(x, y)$ and $\eta(x, y)$ should fulfill the Cauchy-Riemann equations given by

$$a_1 = \frac{\partial \xi}{\partial x} = a_4 = \frac{\partial \eta}{\partial y}, \qquad a_2 = \frac{\partial \xi}{\partial y} = -a_3 = -\frac{\partial \eta}{\partial x}. \tag{1.6.19}$$

Substituting these results into Eq. (1.6.16) yields

$$\left(a_1^2 + a_2^2\right) \frac{\partial^2 \phi}{\partial \xi^2} + \left(a_3^2 + a_4^2\right) \frac{\partial^2 \phi}{\partial \eta^2} = 0, \tag{1.6.20}$$

which, by using the Cauchy-Riemann equations, is recast alternatively as

$$\left(a_3^2 + a_4^2\right) \left(\frac{\partial^2 \phi}{\partial \xi^2} + \frac{\partial^2 \phi}{\partial \eta^2}\right) = 0, \qquad \left(a_1^2 + a_2^2\right) \left(\frac{\partial^2 \phi}{\partial \xi^2} + \frac{\partial^2 \phi}{\partial \eta^2}\right) = 0. \tag{1.6.21}$$

Since these two equations must be satisfied for all analytic mapping functions f, it follows that

$$\frac{\partial^2 \phi}{\partial \xi^2} + \frac{\partial^2 \phi}{\partial \eta^2} = 0. \tag{1.6.22}$$

Thus, the Laplace equation of any complex function is *indifferent* through any conformal transformation between two complex planes.

Let $F(z)$ be an analytic function in the z-plane. Its derivative with respect to z through a conformal transformation is obtained as

$$W(z) = \frac{\mathrm{d}F(z)}{\mathrm{d}z} = \frac{\mathrm{d}F(\zeta)}{\mathrm{d}\zeta} \frac{\mathrm{d}\zeta}{\mathrm{d}z} = \frac{\mathrm{d}\zeta}{\mathrm{d}z} W(\zeta), \qquad \frac{\mathrm{d}F(\zeta)}{\mathrm{d}\zeta} \equiv W(\zeta), \tag{1.6.23}$$

which indicates that the derivatives of an analytic function do not in general satisfy a one-to-one mapping. However, they are proportional to each other, and the proportional factor depends on the mapping function $\zeta = f(z)$. Consider any closed contour C in the z-plane, on which two scalar functions m and Γ are defined by

$$m = \int_C \boldsymbol{u} \cdot \boldsymbol{n} \mathrm{d}\ell = \int_C (u\mathrm{d}y - v\mathrm{d}x), \qquad \Gamma = \int_C \boldsymbol{u} \cdot \mathrm{d}\boldsymbol{\ell} = \int_C (u\mathrm{d}x + v\mathrm{d}y), \tag{1.6.24}$$

where $\mathrm{d}\ell$ is a line element on C with positive slope for simplicity, and $\boldsymbol{u} = u\boldsymbol{i} + v\boldsymbol{j}$, which is any vector defined in the z-plane. These two expressions can be combined into a single integral of $W(z)$, viz.,

$$\int_C W(z)\mathrm{d}z = \int_C (u - iv)(\mathrm{d}x + i\mathrm{d}y) = \Gamma + im. \tag{1.6.25}$$

Applying the conformal transformation to the above equation gives

$$\Gamma_\zeta + im_\zeta = \int_{C_\zeta} W(\zeta)\mathrm{d}\zeta = \int_C W(z)\frac{\mathrm{d}z}{\mathrm{d}\zeta}\mathrm{d}\zeta = \int_C W(z)\mathrm{d}z = \Gamma + im, \tag{1.6.26}$$

where the subscript "ζ" is used to denote that the indexed quantities are evaluated on the ζ-plane. Thus, the values of these two scalars remain unchanged under the conformal transformation.

The quantities ϕ, $F(z)$, $W(z)$, m, and Γ have physical interpretations in two-dimensional potential flows. For a given flow field, ϕ is the velocity potential function, $F(z)$ represents the complex potential, $W(z)$ is the complex velocity, while m and Γ are respectively the source and circulation strengths. Detailed discussions on these quantities will be provided in Sect. 7.5.

1.7 Exercises

In the following, let $\{\boldsymbol{a}, \boldsymbol{b}, \boldsymbol{c}, \boldsymbol{d}\} \in \mathcal{R}^3$, and $\{\boldsymbol{T}, \boldsymbol{U}, \boldsymbol{V}\} \in \mathcal{R}^{3\times 3}$, unless stated otherwise.

1.1 Use the index notation to prove the following identities:

(a) $\boldsymbol{a} \cdot (\boldsymbol{b} \times \boldsymbol{c}) = \boldsymbol{b} \cdot (\boldsymbol{c} \times \boldsymbol{a}) = \boldsymbol{c} \cdot (\boldsymbol{a} \times \boldsymbol{b})$,
(b) $\boldsymbol{a} \times (\boldsymbol{b} \times \boldsymbol{c}) = (\boldsymbol{a} \cdot \boldsymbol{c})\boldsymbol{b} - (\boldsymbol{a} \cdot \boldsymbol{b})\boldsymbol{c}$,
(c) $(\boldsymbol{a} \times \boldsymbol{b}) \cdot ((\boldsymbol{b} \times \boldsymbol{c}) \times (\boldsymbol{c} \times \boldsymbol{a})) = (\boldsymbol{a} \cdot (\boldsymbol{b} \times \boldsymbol{c}))^2$,
(d) $(\boldsymbol{a} \times \boldsymbol{b}) \cdot (\boldsymbol{c} \times \boldsymbol{d}) + (\boldsymbol{b} \times \boldsymbol{c}) \cdot (\boldsymbol{a} \times \boldsymbol{d}) + (\boldsymbol{c} \times \boldsymbol{a}) \cdot (\boldsymbol{b} \times \boldsymbol{d}) = 0$.

1.2 Prove Eqs. (1.1.9) and (1.1.13) for the properties of the Kronecker delta and permutation symbol. Furthermore, if

$$\mathcal{P} = \varepsilon_{ijk}\varepsilon_{mij}\sigma_{km}, \qquad \mathcal{Q} = \sigma_{kk},$$

show that $\mathcal{P} = 2\mathcal{Q}$.

1.3 Let $\boldsymbol{T}$ and $\boldsymbol{U}$ be a symmetric and an antisymmetric second-order tensor, respectively. Show that $\text{tr}\,(\boldsymbol{TU}) = 0$ in terms of the index notation.

1.4 If $\boldsymbol{r}$ denotes the position vector of a material point, with $\boldsymbol{r} = x_i \boldsymbol{e}_i$ and $r^2 = x_i x_i$, show that

$$\text{lap}\,(r^n) = n(n-1)r^{n-2},$$

where n is an integer.

1.5 Show that the symmetry of a second-order tensor is unaffected by the transformation laws; i.e., if $T_{ij} = T_{ji}$ under the $\boldsymbol{e}_i$ base, then $T'_{ij} = T'_{ji}$ under the $\boldsymbol{e}'_i$ base.

1.6 Show that if T_{ij} and E_{ij} are second-order tensors and $T_{ij} = C_{ijkl}E_{kl}$, then C_{ijkl} represents a fourth-order tensor.

1.7 Verify that Eq. (1.2.70), which is the most general form of fourth-order isotropic tensor, remains unchanged by the transformation laws.

1.8 Let $\boldsymbol{T}$ be a rotation tensor given by $\boldsymbol{T} = (1 - \cos\theta)(\boldsymbol{a}^{\text{w}} \otimes \boldsymbol{a}^{\text{w}}) + \cos\theta\,\boldsymbol{I} + \sin\theta\,\boldsymbol{U}$, where $\boldsymbol{a}^{\text{w}}$ is the dual vector of the antisymmetric tensor $\boldsymbol{U}$, and θ represents the rotation angle.

(a) Find the antisymmetric part of $\boldsymbol{T}$, which is denoted by $\boldsymbol{T}^{\text{a}}$.
(b) Show that the dual vector of $\boldsymbol{T}^{\text{a}}$ is given by $(\sin\theta)\boldsymbol{a}^{\text{w}}$.

1.9 Let $\boldsymbol{T} = \boldsymbol{UV}$, and both $\boldsymbol{U}$ and $\boldsymbol{V}$ have the same eigenvector $\boldsymbol{n}$ corresponding to the eigenvalues U_1 and V_1, respectively. Find an eigenvalue and the corresponding eigenvector of $\boldsymbol{T}$.

1.10 The *inertia tensor* $\boldsymbol{J}$ of a rigid body with respect to a point $\mathcal{O}$ is given by

$$\boldsymbol{J} = \int_V \left(r^2\boldsymbol{I} - \boldsymbol{r} \otimes \boldsymbol{r}\right) \rho\, \mathrm{d}v,$$

where $\boldsymbol{r}$ denotes the position vector with $\|\boldsymbol{r}\| = r$ and ρ is the *mass density* of the body. The *moment of inertia* with respect to an axis passing through $\mathcal{O}$ is given by $\boldsymbol{J}_{nn} = \boldsymbol{n} \cdot \boldsymbol{J}\boldsymbol{n}$ (no summation over n), where $\boldsymbol{n}$ is a unit vector in the direction of the axis of interest.

(a) Show that $\boldsymbol{J}$ is symmetric.
(b) Let $\boldsymbol{r} = x_i\boldsymbol{e}_i$, find all components of $\boldsymbol{J}$.
(c) The diagonal and off-diagonal components of $\boldsymbol{J}$ are the *moments of inertia* and *products of inertia*, respectively. For what axes will the products of inertia vanish? For which axes will the moments of inertia be greatest or least?

1.11 Use the symbolic representation to prove the following identities:

(a) $\nabla(\boldsymbol{a} \cdot \boldsymbol{b}) = (\boldsymbol{a} \cdot \nabla)\boldsymbol{b} + (\boldsymbol{b} \cdot \nabla)\boldsymbol{a} + \boldsymbol{a} \times (\nabla \times \boldsymbol{b}) + \boldsymbol{b} \times (\nabla \times \boldsymbol{a})$,
(b) $\nabla \cdot (\boldsymbol{a} \times \boldsymbol{b}) = -\boldsymbol{a} \cdot (\nabla \times \boldsymbol{b}) + \boldsymbol{b} \cdot (\nabla \times \boldsymbol{a})$,
(c) $\nabla \times (\boldsymbol{a} \times \boldsymbol{b}) = \boldsymbol{a}(\nabla \cdot \boldsymbol{b}) - \boldsymbol{b}(\nabla \cdot \boldsymbol{a}) + (\boldsymbol{b} \cdot \nabla)\boldsymbol{a} - (\boldsymbol{a} \cdot \nabla)\boldsymbol{b}$,
(d) $\nabla \times (\nabla \times \boldsymbol{a}) = \nabla(\nabla \cdot \boldsymbol{a}) - \nabla^2\boldsymbol{a}$.

1.12 Use the index notation to derive that

$$\det \boldsymbol{T} = \frac{1}{6}\varepsilon_{ijk}\varepsilon_{rst}T_{ir}T_{js}T_{kt}, \quad \left[\boldsymbol{T}^{-1}\right]_{ij} = \frac{1}{2\det \boldsymbol{T}}\varepsilon_{ikl}\varepsilon_{jmn}T_{km}T_{ln}.$$

1.13 Use the Cayley-Hamilton theorem to deduce that

$$I_T^3 = \frac{1}{3}\left(I_{T^3}^1 - I_T^1 I_{T^2}^1 + I_T^2 I_T^1\right) = \frac{1}{3}\left[I_{T^3}^1 - \left(I_T^1\right)^3\right] + I_T^2 I_T^1,$$

and

$$\frac{\partial I_T^1}{\partial \boldsymbol{T}} = \boldsymbol{I}, \qquad \frac{\partial I_T^2}{\partial \boldsymbol{T}} = I_T^1 \boldsymbol{I} - \boldsymbol{T}^{\mathrm{T}}, \qquad \frac{\partial I_T^3}{\partial \boldsymbol{T}} = \boldsymbol{T}^{-\mathrm{T}} I_T^3.$$

1.14 Consider the relations given by

$$t_{ij} = s_{ij} + \frac{1}{3}t_{kk}\delta_{ij}, \qquad J_2 = \frac{1}{2}s_{ij}s_{ji}, \qquad J_3 = \frac{1}{3}s_{ij}s_{jk}s_{ki},$$

where both t_{ij} and s_{ij} are symmetric second-order tensors. Show that

$$s_{ij} = 0, \qquad \frac{\partial J_2}{\partial t_{ij}} = s_{ij}, \qquad \frac{\partial J_3}{\partial t_{ij}} = s_{ik}s_{kj} - \frac{2}{3}J_2\delta_{ij}.$$

1.15 Verify all equations given in Sect. 1.4, namely the expressions of gradient, divergence, curl, Laplacian and Lagrangian derivatives in the rectangular, cylindrical, and spherical coordinate systems.

1.16 Evaluate the line integral of the following complex function:

$$f(z) = \frac{2iz - \cos z}{z^3 + z},$$

along any closed contour which does not pass through a singularity of $f(z)$.

Further Reading

R. Aris, *Vectors, Tensors, and the Basic Equations of Fluid Mechanics* (Dover, New York, 1962)

R.V. Churchill, J.W. Brown, *Complex Variables and Applications*, 5th edn. (McGraw-Hill, Singapore, 1990)

I.G. Currie, *Fundamental Mechanics of Fluids*, 2nd edn. (McGraw-Hill, Singapore, 1993)

F.B. Hildebrand, *Methods of Applied Mathematics*, 2nd edn. (Prentice-Hill, New Jersey, 1965)

M. Itskov, *Tensor Algebra and Tensor Calculus for Engineers: With Applications to Continuum Mechanics* (Springer, Berlin, 2015)

J.P. Keener, *Principles of Applied Mathematics* (Addison-Wesley, New York, 1988)

W.M. Lai, D. Rubin, E. Krempl, *Introduction to Continuum Mechanics*, 3rd edn. (Pergamon Press, New York, 1993)

J.E. Marsden, *Basic Complex Analysis* (W.H. Freeman and Company, San Francisco, 1973)

D.E. Neuenschwander, *Tensor Calculus for Physics* (Johns Hopkins University Press, New York, 2014)

K.F. Riley, M.P. Hobson, S.J. Bence, *Mathematical Methods for Physics and Engineering* (Cambridge University Press, Cambridge, 1998)

I.S. Sokolnikoff, *Tensor Analysis: Theory and Applications to Geometry and Mechanics of Continua*, 2nd edn. (Wiley, New York, 1969)

D.V. Widder, *Advanced Calculus*, 2nd edn. (Prentice-Hill, New Jersey, 1961)

2 Fundamental Concepts

Fluids at rest or in motion exhibit distinct characteristics from those of solids. Fundamental concepts which are essential to the understanding of fluid motions are explored in this chapter. First, distinctions between common fluids and solids with their underlying physical features are discussed. The Deborah number is introduced in order to take into account the rheological characteristics of matter under different external excitations. Equations in applied mechanics and fluid mechanics are classified into two categories to demonstrate their intrinsic features, followed by the method of analysis used in describing physical process. The assumption of fluid as a continuum plays a crucial role in defining fluid properties, with which theory of fluid motions may be established. Among the properties of a fluid are the viscosity and pressure relatively important. While the former is explored by using Newton's law of viscosity, the latter is discussed by using Pascal's law. Characteristics of fluid flows such as ideal flows versus viscous flows, incompressible flows versus compressible flows, and laminar flows versus turbulent flows are introduced, with their detailed discussions provided in the forthcoming chapters. A structural classification is given at the end to show the main topics of the book, which will be discussed separately in different chapters.

2.1 Fluids, Solids, and Fluid Mechanics

2.1.1 Classifications of Matter

In ancient time, people roughly differentiated and classified different matters by using their external and observable appearances, e.g. the shape, surface color, even the hardness if it was not dangerous to have a touch with the matter. Although this classification is not precise and lacks scientific foundation, fair definitions of fluids and solids may as a first step be given by

© Springer International Publishing AG 2019

C. Fang, *An Introduction to Fluid Mechanics*, Springer Textbooks in Earth Sciences, Geography and Environment, https://doi.org/10.1007/978-3-319-91821-1_2

- *Solids*: a piece of solid material which has a definite shape, and that shape changes only when there is a change in the external conditions.
- *Fluids*: a portion of fluid does not have a preferred shape, and different elements of a homogeneous fluid may be rearranged freely without affecting the macroscopic properties of the portion of fluid.

Based on experiences, ancient people further realized that there existed two categories of fluids. Liquids were those fluids which could hardly be compressed, while gases were those which could easily be compressed.

Nearly by the middle of seventh century BC, the Greeks abandoned the religious interpretations of physical world and started to develop a rational thinking of knowledge, marking the beginning of science. Among various branches of science became the theory of mechanics earliest mature. Without loss of generality, mechanics is the disciplines relating to the external excitations of a system to the reactions of that system. Thus, it becomes possible to differentiate and classify matters by using the mechanics perspective. Consider a *Gedankenexperiment* shown in Fig. 2.1a,[1] in which the test matter is placed on a fixed rigid plate. A shear force F_s (or equivalently a shear stress) is applied on the upper surface of material, which triggers an angular deformation θ. In a very short period of time, an equilibrium state is reached, and θ assumes a stable and invariant value. The material is called a solid if θ remains unchanged unless F_s varies.[2] If, on the contrary, θ depends not only on F_s but also on the time duration, namely $\theta = \theta(t)$, then the material is classified as a kind of fluid, as shown in Fig. 2.1b. For a solid, the angular deformation remains fixed when the applied shear force is unchanged, despite the time duration. For a fluid, θ still varies under a constant F_s and assumes different values at different times. The definitions of fluids and solids in the context of mechanics are thus summarized as

- *Solids*: a material which will not continuously change its shape when subject to a given stress.
- *Fluids*: a material that will continuously change its shape, i.e., will flow, when subject to a given stress, irrespective how small the stress may be.

Consequently, in an equilibrium state, the deformations in solids do not depend on time, while those in fluids do depend on time. This marks the first significant

[1]Gedankenexperiment is a German word, which means a thought experiment. This idea was first introduced by Galileo Galilei, 1564–1642, an Italian polymath, in his "*Discourses and Mathematical Demonstrations*" in 1638. Pioneered thought experiments are *Schrödinger's cat* in quantum mechanics and *Maxwell's demon* in second law of thermodynamics. Erwin Rudolf Josef Alexander Schrödinger, 1887–1961, a Nobel Prize-winning Austrian physicist. James Clerk Maxwell, 1831–1879, a Scottish scientist in mathematical physics, who formulated the classical theory of electromagnetic radiation.

[2]The applied shear force is the external excitation, while the angular deformation is the response of system in the context of mechanics. Material is classified according to the relations between external excitations and system responses. This experiment is called a *simple plane shear*, which is a standard method of mechanics to test material performance.

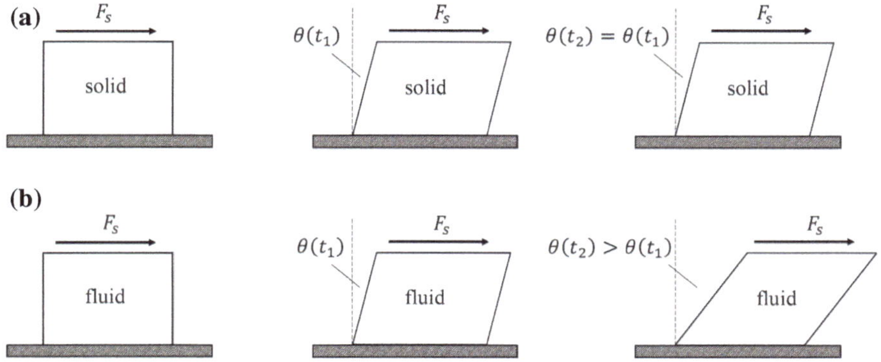

Fig. 2.1 Gedankenexperiments (simple plane shears) of solids and fluids. **a** Solids, in which $\theta(t_1) = \theta(t_2)$. **b** Fluids, in which $\theta(t_2) > \theta(t_1)$

difference between solids and fluids: For solids, the displacements, or alternatively deflections, are important, while for fluids the time rates of change of displacements, or alternatively velocities in the most general sense, play a central role in the mechanics of fluids. The importance of velocity for fluids corresponds exactly to that of deflection for solids.

Nearly during 1930–1940, quantum mechanics was established, and people realized that all matters are composed of atoms and molecules.[3] The most precise and scientific classification of matter ought to be developed by atomic and molecular structures and natures of interactions in-between. A matter is called a kind of solid, when its consisting atoms or molecules, e.g. the atoms of metals, are arranged regularly in a long-term ordered structure, with relatively strong intermolecular interactions. On the contrary, it is called a *gas* if its consisting molecules, e.g. the molecules of air, disperse randomly in space with significant velocities and negligible intermolecular interactions. In-between is the matter called a *liquid*, possessing "chain-like" molecular structures. The molecules on each chain behave like those in the solid structures with strong intermolecular interactions, while weak intermolecular interactions take place among different chains.[4] Table 2.1 summarizes the matter classifications in the context of atomic and molecular structures and the corresponding statistics that are needed to describe the material properties, in which d_0 denotes the equilibrium position between two molecules at which the intermolecular interaction changes the sign, while d_t is the amplitude of random thermal movement of molecules.

---

[3]The latest research outcome of particle physics indicates that all matters are composed of the *Higgs bosons*. However, to simplify the discussions, it is assumed that all matters are composed of atoms and molecules.

[4]A direct scientific evidence of solid structure is, for example, the scanning electron microscope image of copper. The indirect evidence of gas molecular structure is provided by using the technique entitled "Development of methods to cool and trap atoms with laser light", proposed by Steven Chu,

Table 2.1 Classifications of matter in terms of atomic and molecular structures with the corresponding statistics

	Intermolecular force	d_t/d_0	Molecular arrangement	Type of statistics needed
Solids	Strong	$\ll 1$	Ordered	Quantum
Liquids	Medium	Of order unity	Partially ordered	Quantum + classical
Gases	Weak	$\gg 1$	Disordered	Classical

2.1.2 The Deborah Number

For engineering applications, the classifications of matter in the context of mechanics prevail. However, a supplementary information needs to be provided. There exist two interesting examples challenging the mechanics classifications of matter. First, consider a metal bar in pure tension. If the metal bar is perfectly isotropic and linearly elastic, and the applied tensile normal stress is under the elastic limit, the normal tensile strain is immediately determined by using *Hooke's law*.[5] The length of metal bar restores to its initial value, giving rise to a vanishing tensile normal strain when the applied tensile normal stress is removed. However, if the metal bar is kept in the same circumstance for a sufficiently long period of time, it does not restore to its initial length even after the removal of the tensile normal stress, yielding a non-vanishing tensile normal strain. The longer the time duration is, the larger the tensile normal strain will be. In this case, the metal bar, although considered conventionally a solid, behaves like a fluid, for its deformation depends on time. Second, consider a stone impacting a water surface. When the stone with appropriate shape is thrown carefully to the water surface, it is bounced back, called a skipping stone. Although a repel hardly takes place between a solid and a fluid, the water in this case, considered conventionally a kind of liquid, does behave like a solid.

Thus, material response depends additionally on how the material is excited, and the definitions of solids and fluids in the context of mechanics are supplemented by the *Deborah number* given by[6]

$$\mathcal{D}_e \equiv \frac{\tau}{T}, \tag{2.1.1}$$

where τ is the *stress relaxation time* of matter, while T refers to the *timescale of observation*. With this, the classifications of matter in the context of mechanics are revised as

---

Claude Cohen-Tannoudji, and William D. Phillips, who were the winners of The Nobel Prize in Physics 1997.

[5]Robert Hooke, 1653–1703, a British polymath. He came near to an experimental proof that gravity follows an inverse square law and hypothesized that such a relation governs the motions of planets.

[6]This number was originally proposed by Markus Reiner, 1886–1976, an Israeli scientist and a major figure in rheology. The name was inspired by a verse in the Bible, which reads: "The mountains flowed before the Lord" in a song by the prophet Deborah.

- For smaller values of $\mathcal{D}_e$: A material behaves more like a fluid.
- For larger values of $\mathcal{D}_e$: A material behaves more like a solid.
- For medium values of $\mathcal{D}_e$: A material exhibits fluid- and solid-like characteristics simultaneously.

Instead of classifying matters directly into a kind of solids or fluids, it is more appropriate to state that under what circumstances the considered material behaves like a solid or a fluid. A material, despite its physical nature, may exhibit fluid-like characteristics in some circumstances, while exhibiting solid-like characteristics in other circumstances.[7] However, in most common operation circumstances and timescales, it is not necessary to consider the Deborah number, and a fluid and a solid materials behave like what they appear to behave. For example, for a Hookean elastic solid, the relaxation time τ will be infinite, while it will vanish for a Newtonian viscous fluid. For liquid water, τ is typically 10^{-12} s; for lubricating oils passing through gear teeth at high pressure, it is of an order of 10^{-6} s; for polymers undergoing plastic processing, τ will be of an order of a few seconds. The last two liquids, departing from purely viscous behavior, may exhibit solid-like features under specific external excitations.

2.1.3 Fluid Mechanics as a Fundamental Discipline

The theory of mechanics is the earliest branch of physics developed as an exact science. It is the study of motions of material bodies and is divided conventionally into three subdisciplines: (I) statics, (II) kinematics, and (III) dynamics of rigid and deformable bodies. Fluid mechanics is understood as the mechanics of fluids and is that discipline within the broad field of applied mechanics concerned with the behavior of liquids and gases at rest and in motion. Although it may be thousands of years old, fluid mechanics is still highly relevant in various branches of traditional and novel sciences and technologies.

2.2 Equations in Mechanics

Mechanics is an exact science, in which various quantities can be defined, and most of the time be observed and measured. Relations among quantities are described by using mathematical equations, which may be classified into the following two categories:

[7] In the limiting case of $T \to \infty$, all materials behave like fluids. This idea was first introduced by Heraclitus of Ephesus, c. 535–475 BC. a pre-Socratic Greek philosopher. A related proverb reads: "Everything flows if you wait long enough", so that "It is impossible to step twice into the same river", which is stated in another motto.

- *General balance equations*: the balances of mass, linear momentum, angular momentum, energy and entropy in classical physics,[8]
- *Specific constitutive/closure equations*: equations relating external excitations to material reactions, for example, elasticity, plasticity, viscosity, viscoelasticity, hyperelasticity, hypoplasticity.

Since general balance equations are nothing else but physical laws, they are valid for all materials. On the contrary, specific constitutive equations are oriented to specific materials and are not universal. For example, Hooke's law can be used to determine the stress and strain for perfectly linear elastic materials, while it is inappropriate to use it to relate the stress and strain of a collection of sands. Combinations of general balance equations and specific constitutive equations for a specific material give rise to the *governing equations* of that material under the considered circumstances.

The mathematical equations in fluid mechanics are equally classified in the same manner. The formulations of general balance and specific constitutive equations for fluids will be given in Sects. 5.2 and 5.6.

2.3 Methods of Analysis

2.3.1 System, Surrounding, Closed and Open Systems

A *system*, in the most general sense, comprises a device or a combination of devices containing a quantity of matter that is studied; i.e., a system is defined and chosen by interest. Everything outside a system is called the *surrounding* of that system. A system and its surrounding are separated by an imaginary or a real interface which is movable or fixed, called the *boundary*, as shown in Fig. 2.2a. For example, if a falling ball with constant speed is of interest, it is chosen as a system, and everything outside the ball is its surrounding. The ball surface represents the boundary and is a real physical interface. If the deformation of a portion of a concrete column is of interest, then that portion is chosen as a system. The imaginary surface used to isolate the portion of concrete column from the other parts is a boundary. Practically, a surrounding is chosen as that which has significant interactions with the system.

A system is called a *closed system*, or alternatively a *control-mass system (CM)*, or simply a *system*, if the interface permits energy transfer between the system and its surrounding while mass transfer is prohibited. From this perspective, the system is closed to its surrounding with respect to mass transfer, giving rise to the term of closed system. Similarly, the matter quantity contained inside the system remains fixed and identifiable, yielding the term of control-mass system. The interface is called specifically the *system boundary*. For example, air inside a stable ascending

[8]The energy and entropy balances are officially called *first and second laws of thermodynamics*, respectively, which will be discussed in a detailed manner in Sects. 11.4 and 11.5.

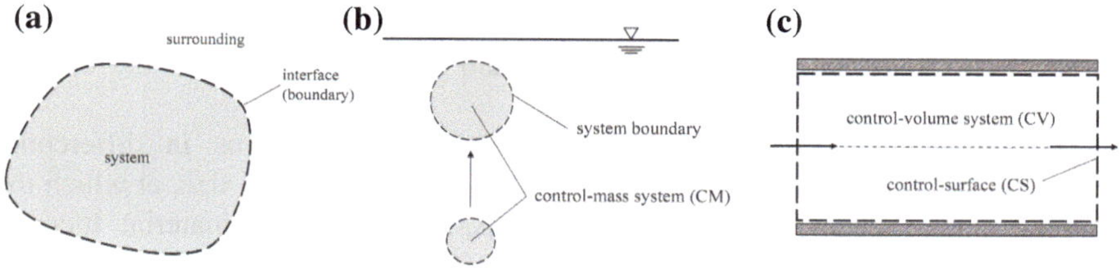

Fig. 2.2 System, surrounding, and interface. **a** General definitions. **b** Control-mass system and system boundary. **c** Control-volume system and control-surface

air bubble in a still water, shown in Fig. 2.2b, is a closed system, and the surface of air bubble is the system boundary. The concept of closed system is used frequently in Newtonian mechanics of particles, in which a mass particle is considered a control-mass system. In fluid mechanics, the concept of closed system is used essentially to establish the fundamental equations, which are transformed subsequently by using the concept of control-volume system.

A system is called an *open system*, or alternatively a *control-volume system (CV)*, if the interface permits not only mass but also energy transfers between the system and its surrounding. The interface is called the *control-surface (CS)*. The term of open system derives from the fact that the system has no restriction on its surrounding with respect to mass and energy transfers. In practice, an open system is accomplished by locating an arbitrary volume with prescribed shape and size in space through which matter flows. The definite prescriptions of volume size, shape, and location deliver the term of control-volume system. For example, a circular pipe section is shown in Fig. 2.2c, in which an open system in constructed near the inner surface of pipe section. The property changes of an air flowing through the pipe can be estimated by using the constructed control-volume system. The concept of open system is intensively used in describing fluid motions.

2.3.2 Differential and Integral Approaches

An *integral approach* is that the mathematical equations of fluid mechanics are formulated in terms of finite control-mass or control-volume system, in which the whole fluid quantity is taken into consideration, giving rise to the integral forms of equations. This approach delivers the gross behavior of a flow without a detailed knowledge. For example, the overall lift of an airfoil can be estimated by using the integral forms of equations without a detailed information of pressure and shear stress distributions over the airfoil surface. On the other hand, the mathematical equations can be formulated in terms of infinitesimal control-mass or control-volume system, called the *differential approach*, yielding the equations in differential forms. When compared with the integral approach, the differential approach delivers a detailed knowledge of a flow to describe its motion in a precise manner. The integral approach is used frequently in the theories of statics and dynamics of rigid body, while the differential approach is used intensively in e.g. the mechanics of materials or elasticity. Both approaches are used in fluid mechanics to derive the balance equations in integral and differential forms.

The infinitesimal control-mass and control-volume systems in differential approach ought to be the minimum undividable material sample size, at which the material sample assumes the same properties as the original bulk material. It corresponds exactly to the concept of *material point* in the *continuum hypothesis*, which will be discussed in Sect. 2.4.1.

2.3.3 The Lagrangian and Eulerian Descriptions

Applications of control-mass or control-volume system depend on the problems under consideration. If it is easy to keep track of an identifiable material point for the descriptions of its motion, such a concept is referred to as the *Lagrangian description*. If it is not the case, an infinitesimal control-volume system is used, which gives rise to the *Eulerian description*.[9] Without loss of generality, it may be stated that the Lagrangian description is a combination of differential approach and control-mass system, while the Eulerian description is that of differential approach and control-volume system.

Let ϕ be any quantity of matter. Its functional dependency is given by

$$\phi = \phi(\boldsymbol{X}, t), \qquad \phi = \phi(\boldsymbol{x}, t), \tag{2.3.1}$$

in the Lagrangian and Eulerian descriptions, respectively, where t is a time measure. The quantity $\boldsymbol{X}$ in the first equation represents the position vector of an identified material point with fixed mass, followed which ϕ is described. The quantity $\boldsymbol{x}$ in the second equation is the position vector of a differential control-volume at which the variation in ϕ is studied. Thus, in the Lagrangian description all quantities are only functions of time, while those in the Eulerian description depend on the spatial and time coordinates simultaneously. More detailed discussions on two descriptions are provided in Sect. 5.1.2.

2.4 Continuum Hypothesis

2.4.1 Continuum, Material Point, and Field Quantity

All matters are composed of atoms and molecules in regular or irregular pattern. The gross behavior as well as physical quantities of a matter can be considered the average behavior of consisting atoms and molecules, resulting in the *microscopic point of view*. Although logically possible, this point of view is hardly accomplished in practice. For example, consider a cubic box with side length of 25 mm, which

[9] Leonhard Euler, 1707–1783, a Swiss mathematician, physicist, and engineer, who made influential discoveries in many branches of mathematics and is also known for the work in mechanics, fluid dynamics, optics, and music theory.

is filled with a monatomic gas at 1 atmospheric pressure and room temperature. It follows that there exists an amount of 10^{20} gas molecules, and the gross behavior of gas is the average behavior of these 10^{20} molecules. For simplification, let a gas molecule be a sphere. To determine the motion of a sphere in space, at least 6 variables are necessary: 3 for the position components and 3 for the velocity components. It results in an amount of 6×10^{20} variables, for which 6×10^{20} equations need to be formulated to make the problem mathematically well-posed. Although the formulations of equations can follow Newton's second law of motion, provided that the interactions among different spheres are appropriately established, such a huge calculation is hardly accomplished even by the modern computer technology.[10] An alternative would be to ignore the atomic and molecular structures to describe the behavior in terms of macroscopically sensible and perceivable quantities, called the *macroscopic point of view*.

Consider a matter in space with m the mass and V the volume, as shown in Fig. 2.3a. A Gedankenexperiment is conducted as follows: At point C, a portion of material is taken to estimate its mass δm and volume δV, with which the value of $\delta m/\delta V$ is calculated. The procedure is repeated, and each time with more or less material content. The relation of all the calculated values of $\delta m/\delta V$ with respect to the sample material size δV is illustrated graphically in Fig. 2.3b, in which two regions are identified.[11]

In region I, the fraction $\delta m/\delta V$ experiences significant fluctuations with respect to the sample material size, while it approaches a finite and stable value in region II if the sample material size is larger than the minimum sample size, $\delta V'$. The minimum sample size $\delta V'$ marks a criterion of stable and smooth variations of $\delta m/\delta V$. Since in region II the value of $\delta m/\delta V$ becomes stable and definable, it is plausible to define that

$$\rho \equiv \lim_{\delta V \to \delta V'} \frac{\delta m}{\delta V}, \tag{2.4.1}$$

Fig. 2.3 Continuum hypothesis and concept of material point. **a** Illustration of the Gedankenexperiment. **b** Influence of the atomic and molecular agitations on the values of quantities

[10]However, if the number of spheres approaches infinite, statistical methods can be applied to conduct the calculations, giving rise to the theory of *statistical mechanics* or *statistical thermodynamics*.
[11]Region I in Fig. 2.3b is very close to the vertical axis in real scale. It is enlarged here to simplify the discussions.

known as the *density* or *mass density* of material. The fluctuation on the values of $\delta m/\delta V$ in region I results from the influence of atomic and molecular agitations if the sample material size is smaller than $\delta V'$. The minimum material sample size $\delta V'$ is referred to as a *material point*, or alternatively a *material particle* or a *material element*. A material point should be sufficiently larger enough than $\delta V'$ for negligible influence of atomic and molecular agitations on the values of quantities and sufficiently smaller enough to become a representable element of matter. It corresponds exactly to the infinitesimal volume in the differential approach. Other physical properties and quantities of matter can be defined in a similar manner, e.g.

$$\sigma \equiv \lim_{\delta l \to \delta l'} \frac{\delta F}{\delta l}, \qquad t_{11} \equiv \lim_{\delta A_1 \to \delta A'} \frac{\delta F_1}{\delta A_1}, \tag{2.4.2}$$

for the surface tension σ and normal stress t_{11}, respectively, where $\delta l'$ and $\delta A'$ are the corresponding line and surface of $\delta V'$.

A material is said to become a *continuum* if it is described in terms of material point, with which all properties and quantities become definable and are continuous functions in space and time. Let ϕ be any quantity of the material, it follows that

$$\begin{Bmatrix} \phi = \phi(\boldsymbol{X}, t), \\ \boldsymbol{X} = x_I \boldsymbol{e}_I, \end{Bmatrix}, \qquad \begin{Bmatrix} \phi = \phi(\boldsymbol{x}, t), \\ \boldsymbol{x} = x_i \boldsymbol{e}_i, \end{Bmatrix}, \tag{2.4.3}$$

in the Lagrangian and Eulerian descriptions, respectively, where $\boldsymbol{e}_I$ is the orthonormal base used for the coordinates of $\boldsymbol{X}$, while $\boldsymbol{e}_i$ is that for the coordinates of $\boldsymbol{x}$. Thus, with the continuum hypothesis, properties and quantities become *fields*. For example, density becomes density field; velocity becomes velocity field, etc. It is noted that the continuum hypothesis is only a mathematical assumption, with which the discrete material properties, resulted from the physically discrete atomic and molecular structures of matter, are transformed to mathematically continuously distributed functions in space and time. It is not used to denote a specific material class. There exists the continuum hypothesis, instead of a continuum material. From now on, the continuum hypothesis is used throughout the book in discussing the motions of fluids.

2.4.2 The Knudsen Number

The size of a material point is different in different materials. The applications of continuum hypothesis to specific materials is indicated by the *Knudsen number* k_n,[12] defined by

$$k_n \equiv \frac{\lambda}{L}, \tag{2.4.4}$$

[12]Martin Hans Christian Knudsen, 1871–1949, a Danish physicist, who is known for his study of molecular gas flow and the development of the *Knudsen cell*, which is a primary component of molecular beam epitaxy systems.

where λ is the molecular mean free path of material and L represents the representative physical length of the problem that matters the material behavior. The validity of continuum hypothesis is summarized in the following:

- $k_n \ll \mathcal{O}(10^{-1})$: Continuum hypothesis validates.
- $\mathcal{O}(10^{-1}) < k_n < \mathcal{O}(1)$: Transition region.
- $k_n \gg \mathcal{O}(1)$: Continuum hypothesis fails.

As similar to the Deborah number, instead of stating that the continuum hypothesis validates for a specific material, it is more appropriate to state that the continuum hypothesis validates for a specific material in a specific circumstance. For example, a geosynchronous satellite locates nearly at 36000 km above the earth's surface, at which the air is highly rarefied, possessing a molecular mean free path nearly of an order of 10^1 m, while the representative length of a satellite is equally of the same order. It follows that $k_n \sim \mathcal{O}(1)$, indicating that the continuum hypothesis fails for the air around the satellite. On the other hand, air at standard conditions possesses a molecular mean free path nearly of an order of 10^{-9} m, indicating that the continuum hypothesis can be applied for air in most engineering applications with conventional physical length scales.[13]

2.5 Velocity and Stress Fields

2.5.1 Velocity Field

The role played by the velocity of a fluid is similar to the role played by the deflection of a solid. Conventionally, the velocity field of a fluid is denoted by $\boldsymbol{v}$ or $\boldsymbol{u}$, with the corresponding index notations given by

$$\boldsymbol{v} = v_i \boldsymbol{e}_i, \qquad \boldsymbol{u} = u_i \boldsymbol{e}_i, \tag{2.5.1}$$

under the orthonormal base $\boldsymbol{e}_i$. For a rectangular Cartesian coordinate system, $\boldsymbol{u}$ is conventionally decomposed as $\boldsymbol{u} = u\boldsymbol{i} + v\boldsymbol{j} + w\boldsymbol{k}$, with $\{u, v, w\}$ reserved for the velocity components in the x-, y- and z-axes, respectively.

All flows are essential three-dimensional, with field quantities depending on three spatial coordinates and time in the Eulerian description. In occasions, simplifications to flow fields can be made with sufficient accuracy. A flow is called *one-, two- or three-dimensional* if its field quantities depend respectively on one, two or three spatial coordinates.[14]

---

[13] An exception emerges for nano-structures, in which the physical lengths are of an order of 10^{-9} m, yielding $k_n \sim \mathcal{O}(1)$.

[14] Based on the molecular structures of fluids, the simplification of two-dimensional flow is exact and physically justified. On the other hand, two-dimensional formulations of solids, e.g. plane stress and plane strain theories, are only approximations, for non-vanishing out-of-plane strain and stress exist due to the conservation of mass.

If the field quantities at every point in a flow field do not depend on time, the flow is termed *steady*, defined mathematically by

$$\frac{\partial \eta}{\partial t} \equiv 0, \qquad \longrightarrow \qquad \eta = \eta(x_1, x_2, x_3), \tag{2.5.2}$$

in the Eulerian description, where η represents any fluid quantity. Although in a steady flow the field quantities do not change with time, they are still functions of spatial coordinates.[15] A flow is termed *uniform* if its field quantities at every point on a specific surface assume the same values, but may change with time. The term *uniform flow* ought to be referred to a specific surface.[16]

2.5.2 Stress Field

There exists an internal force inside a material to maintain an equilibrium state if the material is initially in equilibrium under external excitation. The internal force is expressed as a force per unit area at a specific point on the internal surface, known as the *traction vector*, or the *Cauchy stress vector*, $\boldsymbol{t}^{\mathrm{n}}$, which is defined by

$$\boldsymbol{t}^{\mathrm{n}} \equiv \lim_{\delta A \to 0} \frac{\delta \boldsymbol{F}}{\delta A}, \tag{2.5.3}$$

where the superscript "n" is used to denote that $\boldsymbol{t}^{\mathrm{n}}$ is evaluated in the direction of $\boldsymbol{n}$ which is the unit outward normal of surface element δA, upon which the force $\delta \boldsymbol{F}$ acts. Essentially, the functional dependency of $\boldsymbol{t}^{\mathrm{n}}$ may be given by

$$\boldsymbol{t}^{\mathrm{n}} = \boldsymbol{t}^{\mathrm{n}} \left(\boldsymbol{x}, \boldsymbol{n}, t, \xi \right), \tag{2.5.4}$$

where $\boldsymbol{x}$ denotes the position of evaluation point, t is the time measure, and ξ represents the differential geometric properties, e.g. the *mean* or *Gaussian curvature*, of δA at that point. It follows from the *Cauchy stress principle* and the *Cauchy lemma* that $\boldsymbol{t}^{\mathrm{n}}$ can be expressed as a product of a second-order tensor with the vector $\boldsymbol{n}$ given by[17]

$$\boldsymbol{t}^{\mathrm{n}} = \boldsymbol{t}\boldsymbol{n}, \tag{2.5.5}$$

where $\boldsymbol{t}$ is called the *Cauchy stress*. Physically, at a specific surface there exist three forces per unit area; one being normal to the surface and the other two being parallel to the surface, which are called respectively the normal and shear stresses. The magnitudes of three stresses depend on the orientation and magnitude of the area vector, and direction and magnitude of the force vector. Thus, two free indices are required to index a stress, indicating that the stress inside a material is a second-order tensor, as implied by the linear transformation in Eq. (2.5.5).

---

[15] On the contrary, a quantity is a constant if its time rate of change vanishes in the Lagrangian description.

[16] An inconsistency is the term *uniform flow field*, which is used to denote a flow whose velocity is constant throughout the entire space.

[17] The Cauchy stress principle and the Cauchy lemma will be discussed in a detailed manner in Sect. 5.2.2.

The stress state at a specific point inside a fluid is then given by

$$\boldsymbol{t} = t_{ij}(\boldsymbol{e}_i\boldsymbol{e}_j), \tag{2.5.6}$$

where the index i represents the orientation of surface element, while the index j denotes the direction of force. The matrix representation of Eq. (2.5.6) is given by

$$[\boldsymbol{t}] = \begin{bmatrix} t_{11} & t_{12} & t_{13} \\ t_{21} & t_{22} & t_{23} \\ t_{31} & t_{32} & t_{33} \end{bmatrix}, \tag{2.5.7}$$

where the first column denotes the three stress components pointing to the x_1-axis while lying on different surface elements, and the first row represents the three stress components acting on the same surface element with different directions, etc.

Consider a *Newtonian fluid* in static equilibrium.[18] Since in a static equilibrium no relative motion takes place between any two material points, there exist no shear stresses, with which Eq. (2.5.7) reduces to a diagonal matrix. The "compressive feeling" of a human hand immersed into a still water suggests that the normal stresses are compressive, giving rise to

$$[\boldsymbol{t}] = \begin{bmatrix} -p_1 & 0 & 0 \\ 0 & -p_2 & 0 \\ 0 & 0 & -p_3 \end{bmatrix}, \tag{2.5.8}$$

in which $p_1 = -t_{11}$, $p_2 = -t_{22}$, and $p_3 = -t_{33}$, known as the *pressures* acting along the x_1-, x_2-, and x_3-axes, respectively. Thus, the pressures of a Newtonian fluid in static equilibrium are compressive normal stresses. Consider an infinitesimal cubic box with vanishing volume size but finite surface area as the differential control-volume system at a specific point. Applying Newton's second law of motion to the control-volume yields that $p_1 = p_2 = p_3$, indicating that the pressures are invariant with respect to the directions. This conclusion holds equally for a Newtonian fluid in motion and is summarized as *Pascal's law*[19].

2.1 (Pascal's law) *The pressure at a point in a Newtonian fluid at rest, or in motion, is independent of the direction as long as there are no shearing stresses present.*

By using Pascal's law, the Cauchy stress of a Newtonian fluid is conventionally decomposed into

$$\boldsymbol{t} = -p\boldsymbol{I} + \boldsymbol{T}, \qquad t_{ij} = -p\delta_{ij} + T_{ij}, \tag{2.5.9}$$

where $\boldsymbol{T}$ is the *extra stress tensor*, and p is specifically termed the *thermodynamic pressure*.[20] If the fluid is *incompressible*, i.e., $\rho =$ constant, p can be defined as

$$p \equiv -\frac{1}{3}\operatorname{tr}\boldsymbol{t}, \qquad \operatorname{tr}\boldsymbol{T} = 0. \tag{2.5.10}$$

---

[18] A Newtonian fluid is that satisfies *Newton's law of viscosity*, to be discussed in Sect. 2.6.1.

[19] Blaise Pascal, 1623–1662, a French mathematician, physicist, and Catholic theologian, who contributed to the study of fluids and clarified the concepts of pressure and vacuum by generalizing the work of Evangelista Torricelli.

[20] There exists another pressure, called the *mechanical pressure*. The difference between thermodynamic and mechanical pressures of the Newtonian fluids will be discussed in Sect. 5.6.3.

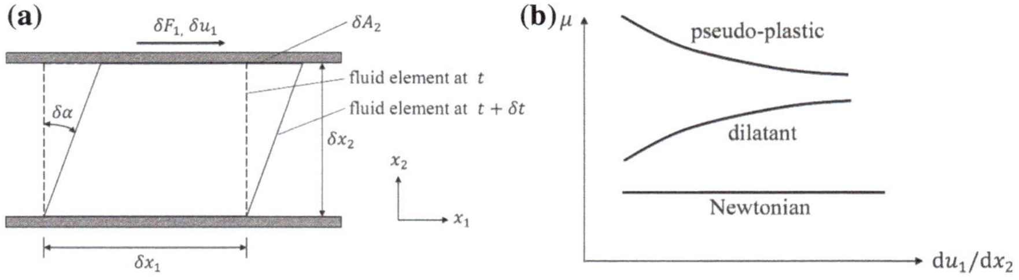

Fig. 2.4 Illustrations of the Newtonian and non-Newtonian fluids. **a** Simple plane shear experiment. **b** Dynamic viscosities of the Newtonian and non-Newtonian fluids with respect to shear rate

2.6 Viscosity and Other Fluid Properties

2.6.1 Newton's Law of Viscosity

Consider a simple plane shear Gedankenexperiment shown in Fig. 2.4a, in which a fluid element is placed between two rigid plates, with the lower plate fixed and upper plate movable. Applying a shear force δF_1 on the upper plate drives the plate to move with a constant speed δu_1, triggering an angular deformation $\delta\alpha$ of the fluid underneath in the time span δt. The shear stress t_{21} acting on the upper surface of fluid element and the corresponding angular deformation rate $\dot{\gamma}$ (shear strain rate) are obtained respectively as

$$t_{21} = \frac{\delta F_1}{\delta A_2}, \qquad \dot{\gamma} = \lim_{\delta t \to 0} \frac{\delta\alpha}{\delta t} \sim \lim_{\delta t \to 0} \frac{\tan(\delta\alpha)}{\delta t} = \frac{\delta u_1}{\delta x_2} = \frac{\mathrm{d}u_1}{\mathrm{d}x_2}, \tag{2.6.1}$$

where δA_2 is the upper surface area of fluid element. The experiment can be repeated by changing the parameters such as the size of fluid element or the magnitudes of δF_1 and δA_2 to obtain sequences of t_{21} and $\dot{\gamma}$.

A fluid is called a *Newtonian fluid* if all the t_{21}s are proportional to the corresponding $\dot{\gamma}$s, for which an explicit expression is given by

$$t_{21} = \mu\dot{\gamma}, \qquad \mu \neq \mu(\dot{\gamma}), \tag{2.6.2}$$

known as *Newton's law of viscosity*,[21] with the proportional constant μ termed the *absolute* or *dynamic viscosity*. Another viscosity frequently used is the *kinematic viscosity* ν, defined by $\nu \equiv \mu/\rho$. The dynamic viscosity is a real physical property and has a physical interpretation in relation to the molecular structures of liquids and gases, whereas the kinematic viscosity is defined only for convenience.

[21] Sir Isaac Newton, 1642–1726, a British mathematician, astronomer, and physicist. His book entitled "*Mathematical Principles of Natural Philosophy*", first published in 1687, laid the foundations of classical mechanics. In the same book, the property of viscosity was also defined, and the original statement reads: "The resistance which arises from the lack of slipperiness of the parts of the liquid, other things being equal, is proportional to the velocity with which the parts of the liquid are separated from one another". Instruments for viscosity measurements are called *viscometers*.

The dynamic viscosity of a gas results from the momentum exchanges of gas molecules in collisions. Increasing the gas temperature increases the momentums of gas molecules and enhances the momentum exchanges in collisions in due course, giving rise to larger values of μ. On the other hand, the dynamic viscosity of a liquid lies in the molecular attractions between different chain structures. Increasing the temperature of a liquid enlarges the distance between two chain structures, resulting in smaller molecular attractions with lower values of μ. For example, an engine oil such as SAE 10W becomes easier to flow if its temperature is higher. The air surrounding a space shuttle in the returning flight from space is heated by the friction between the space shuttle and itself, resulting in an increasing μ, which in due course enhances the friction and forms a "Teufelskreis" (devil's circle). Macroscopically, the dynamic viscosity depends on other properties of a fluid, e.g. the pressure, temperature, density, etc. The three-dimensional generalization of Eq. (2.6.2) will be provided in Sect. 5.6.3.

Defining a Newtonian fluid in experiments conducted at constant temperature and pressure needs to satisfy the following requirements:

- In a simple shear flow, only the shear stress takes place, with vanishing two normal stress differences.
- The dynamic viscosity is not a function of shear rate.
- The dynamic viscosity is constant with respect to the time of shearing. Stress in the liquid vanishes immediately the shearing is absent. Any subsequent shearing, despite the period of resting between measurements, yields the viscosity measured previously.
- The dynamic viscosity measured by different deformations is always in simple proportional to one another.

A fluid is called *non-Newtonian* if its measured sequences of t_{21}s are not proportional to the corresponding sequences of $\dot{\gamma}$s, for which Eq. (2.6.2) can still be used to describe the relation between the shear stress and shear strain rate, viz.,

$$t_{21} = \mu\dot{\gamma}, \qquad \mu = \mu(\cdot, \dot{\gamma}), \tag{2.6.3}$$

indicating that the major difference between a non-Newtonian and a Newtonian fluids lies in the dependency of the dynamic viscosity on the shear strain rate and other factors such as time. The behavior of a non-Newtonian fluid depends equally on how it is excited. For example, a mixture of water and corn starch, when placed on a flat surface, flows as a thick, viscous fluid. However, if the mixture is rapidly disturbed, it appears to fracture and behave more like a solid. The theory of the non-Newtonian fluids is referred to as *Mechanics of non-Newtonian Fluids*, or alternatively as *Rheology*.

There exist essentially two categories of non-Newtonian fluids, as shown in Fig. 2.4b. The *pseudo-plastic fluids*, or alternatively *shear-thinning fluids*, are those in which the dynamic viscosity decreases as the deformation rate increases. The

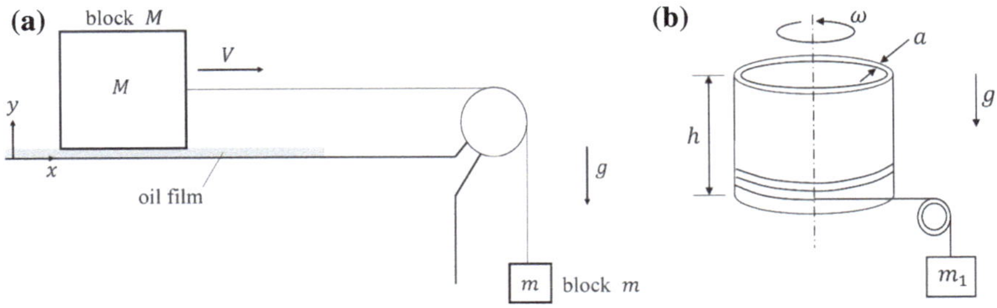

Fig. 2.5 Applications of Newton's law of viscosity. **a** A mass block sliding above an oil film. **b** A concentric-cylinder viscometer

dilatant fluids, or alternatively *shear-thickening fluids*, exhibit a reverse tendency.[22] For example, tomato sauce is a kind of shear-thinning fluid, while maltose in liquid form is a kind of shear-thickening fluid. Air and water, the most encountered fluids in conventional engineering applications, behave very closely to the Newtonian fluids.

An important implication of viscosity is that when a fluid is in contact with a solid, the fluid velocity on the solid boundary corresponds to the velocity of that solid boundary, which is called the *no-slip boundary condition*, i.e.,

$$\boldsymbol{u} = \boldsymbol{u}_{\mathrm{w}}, \tag{2.6.4}$$

at every point on the solid boundary, where $\boldsymbol{u}$ is the fluid velocity and $\boldsymbol{u}_{\mathrm{w}}$ is the velocity of solid boundary. In the case of an infinite expanse of fluid, one common form of Eq. (2.6.4) is that $\boldsymbol{u} \to \boldsymbol{0}$ as $\boldsymbol{x} \to \infty$. Specifically, let $\boldsymbol{n}^{\mathrm{t}}$ and $\boldsymbol{n}^{\mathrm{n}}$ denote respectively the unit vectors tangential and normal to the solid surface. It follows from Eq. (2.6.4) that

$$\boldsymbol{u} \cdot \boldsymbol{n}^{\mathrm{t}} = \boldsymbol{u}_{\mathrm{w}} \cdot \boldsymbol{n}^{\mathrm{t}}, \qquad \boldsymbol{u} \cdot \boldsymbol{n}^{\mathrm{n}} = \boldsymbol{u}_{\mathrm{w}} \cdot \boldsymbol{n}^{\mathrm{n}}, \tag{2.6.5}$$

for a moving solid surface, and

$$\boldsymbol{u} \cdot \boldsymbol{n}^{\mathrm{t}} = \boldsymbol{u}_{\mathrm{w}} \cdot \boldsymbol{n}^{\mathrm{t}}, \qquad \boldsymbol{u} \cdot \boldsymbol{n}^{\mathrm{n}} \neq \boldsymbol{u}_{\mathrm{w}} \cdot \boldsymbol{n}^{\mathrm{n}}, \tag{2.6.6}$$

for a moving porous solid surface, through which fluid flows.

As an illustration of Newton's law of viscosity, consider a block of mass M sliding on a thin film of oil with ρ the density and μ the dynamic viscosity, as shown in Fig. 2.5a. The oil film thickness is h, which is a constant, and the area of block in contact with the oil film is A. The block M is connected to another block with mass m via a rope through a pulley. When released, block m exerts a tension on the rope, accelerating in turn block M. It is required to derive an expression of the viscous force of the oil that acts on block M when it moves at a constant speed V, for which

[22]The classification is based on the relation between μ and $\dot{\gamma}$. The non-Newtonian fluids can also be classified as *thixotropic* and *rheopectic* (*antithixotropic*) *fluids*. In thixotropic fluids, the dynamic viscosity decreases with time under a constant applied shear stress, while rheopectic fluids exhibit a reverse tendency.

the friction in the pulley and air resistance are neglected. It is also required to derive a differential equation for V as a function of time and find its terminal value.

For the steady circumstance, in which block M moves with a constant speed V, applying Newton's law of viscosity to the oil film yields

$$t_{yx} = \mu \frac{\mathrm{d}u}{\mathrm{d}y} \sim \mu \frac{V}{h}, \tag{2.6.7}$$

by which the viscous force F_v acting on block M is given by

$$F_v = t_{yx} A = \frac{\mu A}{h} V, \tag{2.6.8}$$

which points to the negative x-axis. For the unsteady case in which V is a function of time, applying Newton's second law of motion to block M in the x-direction gives

$$mg - \frac{\mu A}{h} V = (M+m)a = (M+m)\frac{\mathrm{d}V}{\mathrm{d}t}, \qquad a = \frac{\mathrm{d}V}{\mathrm{d}t}, \tag{2.6.9}$$

where a is the acceleration of block M in the x-direction, and it is noted that both blocks move coherently with the same acceleration. The solution to the above equation is obtained as

$$V = \frac{mgh}{\mu A}\left\{1 - \exp\left[-\frac{\mu A}{(M+m)h} t\right]\right\}. \tag{2.6.10}$$

As $t \to \infty$, the speed of block M approaches the terminal value $V_{max} = mgh/(\mu A)$.

Consider a concentric-cylinder viscometer shown in Fig. 2.5b as another example of Newton's law of viscosity, in which the outer cylinder is very thin in thickness with mass m_2 and radius R. It is connected via a rope to a mass block m_1 through a pulley. The clearance between two cylinders is a, which is filled by a liquid, whose viscosity is to be measured. It is required to obtain an algebraic expression for the torque due to the viscous shear that acts on the outer cylinder rotating at a constant angular speed ω, if the bearing friction, pulley and air resistances, and the influence of liquid mass are assumed to be negligible. It is also required to derive and solve a differential equation for the angular speed of outer cylinder as a function of time in an unsteady circumstance.

For the steady circumstance, it follows from Newton's law of viscosity that the shear stress acting on the surface of outer cylinder is given by

$$\tau = \mu \frac{\mathrm{d}u}{\mathrm{d}y} \sim \mu \frac{U}{a} = \mu \frac{R\omega}{a}, \tag{2.6.11}$$

in which a linearization has been used due to the fact that $a/R \ll 1$. The torque due to the shear stress is obtained as

$$T = [\tau (2\pi R h)] R = \frac{2\pi R^3 \mu h}{a} \omega. \tag{2.6.12}$$

For the unsteady circumstance, the tension of rope is denoted by F_c, and Newton's second law of motion reads

$$F_c R - T = m_2 R^2 \frac{\mathrm{d}\omega}{\mathrm{d}t}, \qquad m_1 g - F_c = m_1 R \frac{\mathrm{d}\omega}{\mathrm{d}t}, \tag{2.6.13}$$

for the outer cylinder and block m_1, respectively. Substituting Eqs. (2.6.12) and $(2.6.13)_2$ into Eq. $(2.6.13)_1$ yields

$$m_1 g R - \frac{2\pi R^3 \mu h}{a}\omega = (m_1 + m_2)\, R^2 \frac{\mathrm{d}\omega}{\mathrm{d}t}, \tag{2.6.14}$$

whose solution is given by

$$\omega = \frac{m_1 g a}{2\pi R^2 \mu h}\left\{1 - \exp\left[-\frac{2\pi R \mu h}{(m_1 + m_2) a} t\right]\right\}. \tag{2.6.15}$$

As $t \to \infty$, the maximum value of ω is obtained as

$$\omega_{max} = \frac{m_1 g a}{2\pi R^2 \mu h}. \tag{2.6.16}$$

2.6.2 Other Fluid Properties

For fluids within the continuum hypothesis, the density is defined as the mass per unit volume, which depends on the pressure and temperature for *simple compressible substances* given by $\rho = \rho(T, p)$.[23] The *specific volume* v is defined to be the inverse of density given by $v \equiv 1/\rho$, i.e., the volume per unit mass, which is used frequently in e.g. gas- and aerodynamics, and in power plants for estimating the characteristics of high-pressure and high-temperature water steams. For incompressible fluids, ρ is a constant.[24] The *specific weight* γ is defined as the product of density and gravitational acceleration given by $\gamma \equiv \rho g$, i.e., the fluid weight per unit volume. It is used frequently in calculating the force exerted by a liquid on a surface. The *specific gravity* s is defined as the ratio of fluid density divided by that of water at 1 atmospheric pressure and 4 °C, given by $s \equiv \rho/\rho_{\mathrm{H_2O},\,4\,^\circ\mathrm{C}}$. It is used, for example, to estimate the buoyant force of a body immersed in a still fluid and plays a significant role in boiling process.

The *bulk compressibility modulus* E_v, or *modulus of elasticity*, is defined by

$$E_v \equiv \frac{\mathrm{d}p}{(\mathrm{d}\rho/\rho)}, \tag{2.6.17}$$

corresponding to *Young's modulus* of solid.[25] The *speed of sound* c in a fluid is given by

$$c = \sqrt{\frac{E_v}{\rho}} = \sqrt{\gamma_s R T}, \tag{2.6.18}$$

[23] Simple compressible substances are a subset of *simple materials*, whose states are determined by prescribing the values of two independent intensive properties. A detailed discussion will be provided in Sect. 11.1.4.

[24] Rigorous mathematical conditions of incompressibility of fluids will be provided in Sect. 5.3.1.

[25] Thomas Young, 1773–1829, a British polymath and physician, who made contributions to the fields of vision, light, solid mechanics, energy, etc., and has been described as "The Last Man Who Knew Everything".

where γ_s is the *specific-heat ratio*, R denotes the *gas constant*, and T represents the *Kelvin temperature scale*.[26] The first equality of Eq. (2.6.18) is a general expression for both gases and liquids, while the second equality only holds for ideal gases. The *Mach number* $\mathcal{M}_a$ of an object is defined as the ratio of the object speed V divided by the sound speed c in the fluid surrounding the object, viz.,[27]

$$\mathcal{M}_a \equiv \frac{V}{c}. \tag{2.6.19}$$

Equations (2.6.17)–(2.6.19) are used in compressible fluid flows such as gas- and aerodynamics to estimate the influence of fluid compressibility.

Let a liquid be placed in a closed container with a completely vacuum space above the liquid surface. Some liquid molecules at the liquid surface may have sufficient momentum to overcome the intermolecular attractions and escape into the empty space, called the *evaporation*. The momentum exchange between the container surface and evaporated liquid molecules per unit time and per unit area results in the macroscopic property, called the *vapor pressure* p_v of liquid. The *saturated vapor pressure*, $p_{v,\,\mathrm{sat}}$ develops when an equilibrium condition is reached so that the number of liquid molecules leaving the surface is equal to the number entering. The values of p_v and $p_{v,\,\mathrm{sat}}$ depend significantly on the pressure and temperature of liquid and are used to determine, e.g. in the atmospheric science, the locations at which a cloud may form, or the cavitation locations of liquids in pipe flows.

At the interface between a liquid and a gas, or between two immiscible liquids, the molecules experience unbalanced molecular attractions from different sides, giving rise macroscopically to forces acting at the interface. Expressing the force per unit length yields the *surface tension* σ. The force can equivalently be expressed in terms of unit area or unit volume, known as the *surface energy*. Surface tension plays a significant role e.g. in water droplet formation from a leakage, in the manufacturing process of liquid-crystal display devices,[28] or in the foam formation. A quantitative discussion on surface tension and the corresponding *capillary effect* will be provided in Sect. 3.3.

---

[26] Sir William Thomson, or Lord Kelvin, 1824–1907, a Scots-Irish mathematical physicist and engineer, who contributed not only to the mathematical analysis of electricity and formulation of first and second laws of thermodynamics, but also did much to unify the emerging discipline of physics in its modern form.

[27] The exact definition of the Mach number is the square root of the ratio of inertia force divided by compressibility force of a fluid, as will be discussed in Sect. 6.5.2. Ernst Waldfried Josef Wenzel Mach, 1838–1916, an Austrian physicist and philosopher, who contributed to the study of *shock waves*. Through his criticism of Newton's theories of space and time, he foreshadowed *Einstein's theory of relativity*.

[28] Influence of surface tension on flow behavior is described by the dimensionless *Weber number*, which will be discussed in Sect. 6.5.2. Moritz Gustav Weber, 1871–1951, a German engineer and university professor, who is known for his work on the systematic study of model similarity.

Table 2.2 Common properties of air and pure water at 1 atmospheric pressure and 20 °C in SI unit

	ρ (kg/m^3)	γ (kN/m^3)	μ (Ns/m^2)	ν (m^2/s)	σ (N/m)	p_v (Pa)	c (m/s)
Air	1.204	$11.81{\cdot}10^{-3}$	$1.82{\cdot}10^{-5}$	$1.51{\cdot}10^{-5}$	–	–	343.3
Water	998.2	9.789	$1.002{\cdot}10^{-3}$	$1.004{\cdot}10^{-6}$	$7.28{\cdot}10^{-2}$	$2.338{\cdot}10^{3}$	1481
Water (4 °C)	1000	9.807	$1.519{\cdot}10^{-3}$	$1.519{\cdot}10^{-6}$	$7.49{\cdot}10^{-2}$	$8.722{\cdot}10^{2}$	1427

Table 2.2 summarizes the values of common properties of air and pure water at standard conditions in SI unit.[29]

2.7 State Equation of Ideal Gas

Newtonian fluid is a kind of *simple material*, whose state is determined by prescribing the values of two independent intensive properties. For pure liquids considered a kind of *pure substance*, the state is determined conventionally by using the *phase diagram*. For example, the state of a pure water is determined by using the p–v–T diagram or *steam table*, which is established essentially by experiments. In fact, every material has its own *state equation*; the problem is that if one has discovered it or not.

For gases, or mixtures of non-reactive gases at low density and high temperature, their states can be described approximately by using the *ideal gas state equation* for considerable accuracy. The ideal gas state equation is given by

$$pV = n\bar{R}T, \qquad p\bar{v} = \bar{R}T, \tag{2.7.1}$$

where p is the pressure, V represents the volume, $\bar{v}$ becomes the gas volume per unit mole, n denotes the number of moles of the gas, $\bar{R}$ is the universal gas constant with $\bar{R} = 8.3143$ kJ/kmol-K, and T is the absolute (Kelvin) temperature scale. The above equation is expressed in the *mole-base*, and can be converted to the *mass-base*, viz.,

$$pV = mRT, \qquad pv = RT, \tag{2.7.2}$$

where m is the mass of gas, v becomes the specific volume (volume per unit mass), and R denotes the gas constant given by $R = \bar{R}/M$, with M the molecular weight of gas. For air, the value of R is given conventionally by $R = 0.287$ kJ/kg-K.

The state equation of ideal gas is a macroscopic description and was established on the foundations of *Charlies' law*, *Gay-Lussac's law*, and *Boyle's law*.[30] Although

[29] Data quoted from Blevins, R.D., *Applied Fluid Dynamics Handbook*, Van Norstrand Reinhold Co. Inc., New York, 1984; and *Handbook of Chemistry and Physics*, 69th ed., CRC Press, New York, 1988.

[30] Jacques Alexandre César Charles, 1746–1823, a French scientist, who formulated the original law in his unpublished work from the 1780s. Joseph Louis Gay-Lussac, 1778–1850, a French chemist and physicist. This law can refer to several discoveries made by Gay-Lussac and other scientists

purely phenomenological, the ideal gas state equation can also be derived by using the kinetic theory of gas, which will be discussed in Sect. 11.1.6. The ideal gas state equation is employed frequently e.g. for the determinations of states of dry air, unsaturated moist air and saturated air before the condensation of water vapor in atmospheric science.

For gases at high density and low temperature, the theoretical foundation of ideal gas state equation is no longer valid, and the results predicted by using the ideal gas state equation are not accurate. To solve the problem, different state equations are proposed, e.g. the *van der Waals equation*[31] or the *Benedict-Webb-Rubin equation*, which are semi-empirical state equations. Alternatively, based on the concept of ideal gas state equation, the behavior of real gases can be accounted for by using the *compressibility factor* Z defined by

$$Z \equiv \frac{pv}{RT}. \tag{2.7.3}$$

The compressibility factor Z assumes a value of unity for ideal gases. Larger the deviation of Z from unity is, larger the deviation of the gas response from that of ideal gas will be. Different real gases have their own compressibility factors. However, when evaluated by the *reduced pressure* and *reduced temperature*, all real gases behave similarly, yielding the *generalized chart of compressibility factor*, which will be discussed in Sect. 11.1.5.

2.8 Flow Characteristics

2.8.1 Ideal and Viscous Flows

All real fluids have viscosities, and the shear forces (or shear stresses) play a significant role in the behavior of fluid motions. However, in some circumstance it is possible to simplify the flow field by assuming that the fluid is *incompressible* and *frictionless*, yielding a flow of an *ideal fluid*. The difference between ideal and real fluids becomes obvious when they are in contact with solid boundaries. For example, consider a fluid passing through a two-dimensional circular cylinder. If the fluid is an ideal one, the flow field around the cylinder is shown in Fig. 2.6a, while that of a real fluid is shown in Fig. 2.6b. It follows from the *potential-flow theory* in Sect. 7.5 that the flow field of an ideal fluid is symmetric with respect to both x- and y-axes, yielding no *drag* and *lift forces* acting on the cylinder. On the contrary, the flow field of a real fluid is not symmetric essentially with respect to the y-axis due

in the late eighteenth and early nineteenth centuries. Robert William Boyle, 1627–1691, an Anglo-Irish natural philosopher. This law was first noted by Richard Towneley and Henry Power in the seventeenth century and was confirmed by Boyle through experiments.

[31] Johannes Diderik van der Waals, 1837–1923, a Dutch theoretical physicist and thermodynamicist, who is known for his work on an equation of state for gases and liquids.

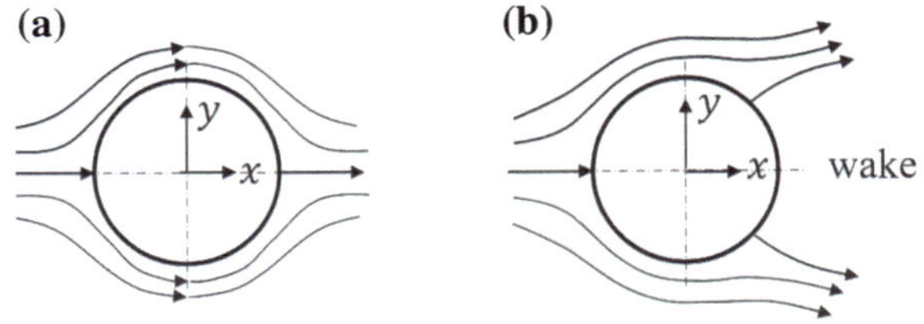

Fig. 2.6 Illustrations of flow fields of ideal and viscous fluids. **a** The flow field of an ideal fluid around a two-dimensional circular cylinder. **b** The flow field of a viscous fluid in the same situation

to the presence of *flow separation*, resulting in a *wake region* immediately behind the cylinder, causing a non-vanishing drag force acting on the cylinder along the x-axis. The inconsistency between the prediction on the vanishing drag force from ideal fluid and the non-vanishing drag force of real fluid in known as *d'Alembert's paradox*.[32] It took about 150 years for an answer after the paradox first appeared, which was obtained by Ludwig Prandtl,[33] by using the concept of *boundary layer*, to be discussed in Sect. 8.4.1. The boundary layer is a very thin layer on the surface of cylinder, in which the fluid friction is significant and across the layer width the fluid velocity increases rapidly from zero, as implied by the no-slip condition, to the value that is predicted by the theory of ideal fluid.

Although the theory of ideal fluid fails to predict the drag force acting on a solid body, it delivers rather accurate predictions on the lift forces and also acts as an ideal limit that needs to be matched by viscous flow theory outside the boundary layer.

2.8.2 Compressible and Incompressible Flows

All real fluids are more or less *compressible*, i.e., the density of a fluid is not a constant and may vary from time to time or at different points in space. For liquids, although they are physically compressible to some extent, they can be approximated to be incompressible with satisfied accuracy of the solutions to the flow field. On the contrary, gases are compressible, and the influence of compressibility needs to be taken into account in describing the flow behavior of a gas. The most significant phenomenon of compressible flows is the existences of *normal shock waves* and *oblique shock waves*, which will be discussed in Sects. 9.2.4 and 9.2.5. A shock wave is a very thin layer in space immediately in the vicinity before a body in a supersonic flow, across which the fluid pressure and density increase dramatically to very large values at the expense of the kinetic energy of fluid. Most aircraft design depends significantly on the formation of shock waves around the aircraft. However,

[32]Jean-Baptiste le Rond d'Alembert, 1717–1783, a French mathematician, mechanician, and physicist, who also contributed to *d'Alembert's equation* for obtaining solutions to the wave equations.
[33]Ludwig Prandtl, 1875–1953, a German aerodynamicist. He was a pioneer in the development of rigorous systematic mathematical analyses which he used for underlying the science of aerodynamics and is recognized as "Father of Modern Fluid Mechanics".

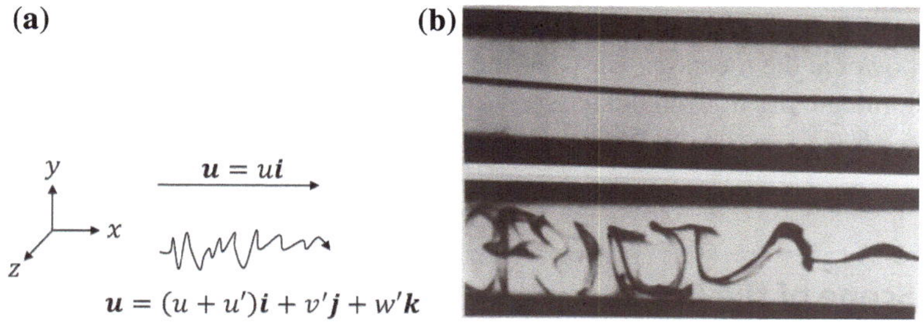

Fig. 2.7 Illustrations of laminar and turbulent flows. **a** Velocity measurements for laminar and turbulent flows. **b** Experimental outcomes of laminar and turbulent flows in a circular pipe, quoted from Hutter, K., *Fluid- und Thermodynamik: Eine Einführung*, Springer Verlag, Berlin Heidelberg, New York, 2003. Reprint permission by Springer Verlag. Original reference: Frauenfelder, P., Huber, P., *Einführung in die Physik*, Ernst Reinhardt Verlag AG, Basel, 1968

it will be shown in Sect. 9.3.4 that the maximum density variation is less than 5% if the Mach number of a flow is smaller than 0.3. Practically, gas flows with $\mathcal{M}_a < 0.3$ can be approximated as incompressible.

2.8.3 Laminar and Turbulent Flows

Let the flow velocity at a specific point in space be measured by a *Pitot tube*,[34] to be discussed in Sect. 7.3.2, in a time duration Δt. The experimental outcomes are illustrated graphically in Fig. 2.7a for a one-dimensional flow along the x-axis. The flow is termed *laminar* if the velocity component u is a constant during the time duration Δt. In laminar flows, the fluid particles move in smooth layers with the interactions between different layers induced mainly via diffusion. On the other hand, the flow is termed *turbulent* if the velocity component experiences significant fluctuations u' during the time duration Δt. Although the flow is one-dimensional, non-vanishing fluctuations v' and w' exist equally in other two directions. The coupled influence among the fluctuations of velocity components causes the flow field to be highly in chaos, which is characterized by the *turbulent eddies* at different time and length scales. An experimental outcome of laminar and turbulent flows inside a circular pipe is shown in Fig. 2.7b.

The most important implication of a turbulent flow is that its shear stress cannot be determined solely by Newton's law of viscosity. Additional stress contributions, known as *Reynolds' stresses*,[35] need to be accounted for by introducing e.g. the concept of *turbulent viscosity*. Equally, the kinetic energy carried by the turbulent

---

[34] Henri Pitot, 1695–1771, a French hydraulic engineer, who invented the original pitot tube in the early eighteenth century.

[35] Osborne Reynolds, 1842–1912, a British prominent innovator in the understanding of fluid dynamics. He most famously studied the conditions in which the fluid state in pipes transitioned from laminar to turbulent flows.

eddies at different time and length scales and the subsequent energy cascade from the stress power at the mean scale toward the turbulent dissipation at the smallest length scale dominate the flow characteristics dramatically. These topics will be discussed in Sect. 8.6.

2.9 Scope of the Book

The structure of the book follows the conventional sequence of statics, kinematics, and dynamics of classical mechanics and is divided into the following chapters to cover the main topics which are essential to an introduction to fluid mechanics:

- Chapter 1: Mathematical Prerequisites,
- Chapter 2: Fundamental Concepts,
- Chapter 3: Hydrostatics,
- Chapter 4: Flow Kinematics,
- Chapter 5: Balance Equations,
- Chapter 6: Dimensional Analysis and Model Similitude,
- Chapter 7: Ideal-Fluid Flows,
- Chapter 8: Incompressible Viscous Flows,
- Chapter 9: Compressible Inviscid Flows,
- Chapter 10: Open-Channel Flows,
- Chapter 11: Essentials of Thermodynamics, and
- Chapter 12: Granular Flows.

The continuum hypothesis is used a priori all discussions, with the focus mainly on the Newtonian fluids. The topics are selected and oriented in accordance with two most important properties of fluids, namely the viscosity and compressibility. After discussing the theories of hydrostatics, flow kinematics, balance equations, and dimensional analysis and model similitude, the focus is shifted to the dynamics of fluid flows. Ideal-fluid flows, incompressible viscous flows, compressible inviscid flows, and open-channel flows consist the main discussions of the book. It is intended to enable readers to have a complete and clear understanding of the fundamentals and applications of fluid mechanics with balanced mathematical foundations and physical features. Essential topics of classical thermodynamics are provided to deepen the understanding of the characteristics of fluid motions from the energy perspective. The last chapter is devoted to the fundamentals of granular flows to demonstrate the applications of the mature disciplines of fluid mechanics to a new branch of scientific study in environmental fields. Two appendices are provided. The formulation of generalized curvilinear coordinate system is given in the first appendix, while the second appendix contains detailed solutions to selected exercises in each chapter.

Although not accomplished explicitly, the book can be divided into three parts. The first part contains the first six chapters, which forms the essential disciplines of fluid mechanics. These disciplines are used in the second part, embracing the forthcoming four chapters, to describe the motions of the Newtonian fluids in different circumstances. The third part consists of the last two chapters, serving as a supplementary knowledge to deepen the understanding of fluid motions.

2.10 Exercises

2.1 How to differentiate fluids and solids? What is the Deborah number? For what purpose do we need to evaluate the value of the Deborah number for a specific material?

2.2 What is the continuum hypothesis? What does it mean by referring to a material point? How to verify the validity of continuum hypothesis for a specific material?

2.3 A flow is called incompressible if its velocity satisfies $\nabla \cdot \boldsymbol{u} = 0$. Consider a one-dimensional radial flow in the $(r\theta)$-plane, which is given by $u_r = f(r)$ and $u_\theta = 0$, where f is any differentiable function. Determine the restrictions of $f(r)$, so that the condition of incompressibility is satisfied.

2.4 A flow is called *irrotational* if its velocity satisfies $\nabla \times \boldsymbol{u} = \boldsymbol{0}$. Consider again a one-dimensional radial flow in the $(r\theta)$-plane, which is described by $u_r = 0$ and $u_\theta = f(r)$. Determine the restrictions of $f(r)$, so that the condition of irrotationality is satisfied.

2.5 Let $\boldsymbol{t}$ be the Cauchy stress of a Newtonian fluid, and $\boldsymbol{D}$ be the symmetric part of velocity gradient given by $\boldsymbol{D} = \text{sym}(\text{grad}\, \boldsymbol{v})$. Derive an expression of $\boldsymbol{t} \cdot \boldsymbol{D}$ in a rectangular Cartesian coordinate system. This scalar product is called the *stress power*, i.e., the power done by the stresses, which will be used to formulate a general energy balance of fluids.

2.6 Two Newtonian fluids with ρ_1, μ_1 and ρ_2, μ_2 are placed between two horizontal rigid plates in parallel, where $\rho_2 > \rho_1$. The thicknesses of two fluid layers are denoted respectively by h_1 and h_2. Let the lower rigid plate be fixed, while the upper rigid plate be movable in parallel to the lower plate. Determine the force required to move the upper plate with a constant speed V, and the fluid velocity at the interface between two fluid layers.

2.7 A block with mass m slides down a smooth inclined surface as shown in the figure. Determine the terminal velocity V of the block if the gap thickness between the block and inclined surface is h, in which an incompressible liquid film with density ρ and dynamic viscosity μ presents. The velocity distribution of the liquid in the gap is assumed to be linear, and the area of block in contact with the liquid is A.

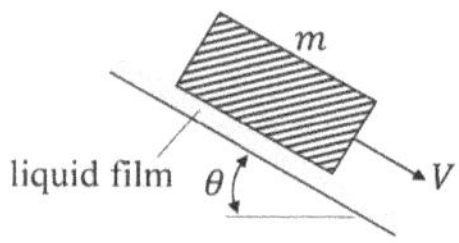

2.8 Consider the concentric-cylinder viscometer shown in Fig. 2.5b. Initially, the outer cylinder rotates with a constant angular speed ω_0. Unfortunately, the rope snaps during the experiment. How long will it take that the angular speed of outer cylinder becomes only one percent of ω_0? For simplicity, the initial condition can be approximated by $\omega(t = 0) = \omega_0$.

2.9 Consider a pair of two parallel disks shown in the figure, in which a gap with constant thickness h is maintained. The gap is filled by a liquid with density ρ and dynamic viscosity μ, and the upper disk is driven to rotate at a constant angular speed ω. Derive an expression for the torque needed to turn the upper disk, if the lower disk is stationary.

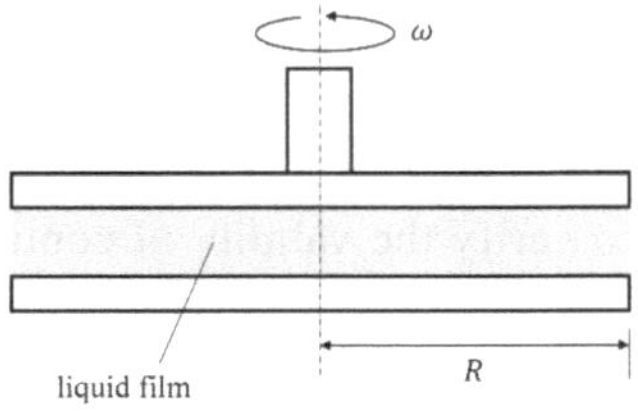

2.10 Consider a cone-and-plate viscometer shown in the figure, with a fixed plate and a rotating cone with a very obtuse angle. The apex of cone just touches the surface of plate and forms a narrow gap which is filled by a liquid with density ρ and dynamic viscosity μ. Derive an expression for the shear rate in the liquid, and determine the torque on the driven cone.

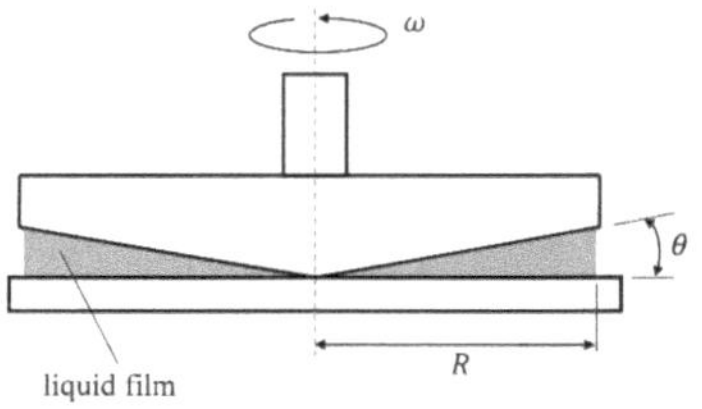

2.11 The figure shows a spherical bearing, in which the gap between the spherical component and housing assumes a constant thickness h, which is filled by an oil with density ρ and dynamic viscosity μ. Derive an expression for the dimensionless torque on the spherical component as a function of angle α.

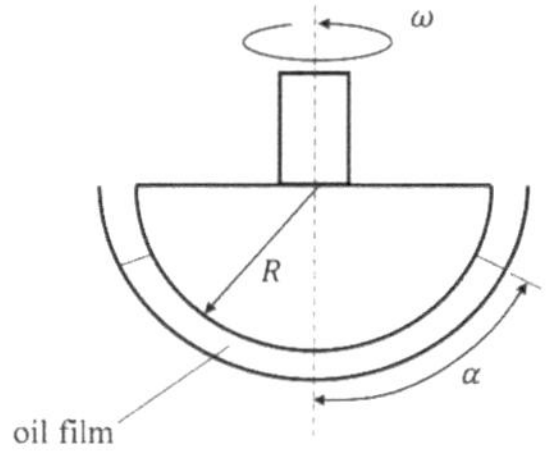

Further Reading

H.A. Barnes, J.F. Hutton, K. Walters, *An Introduction to Rheology* (Elsevier, Amsterdam, 1989)
G.K. Batchelor, *An Introduction to Fluid Dynamics* (Cambridge University Press, Cambridge, 1992)
C. Cercignani, *Rarefied Gas Dynamics: From Basic Concepts to Actual Calculations* (Cambridge University Press, Cambridge, 2000)
D.F. Elger, B.C. Williams, C.T. Crowe, J.A. Roberson, *Engineering Fluid Mechanics*, 10th edn. (Wiley, New York, 2014)
R.W. Fox, P.J. Pritchard, A.T. McDonald, *Introduction to Fluid Mechanics*, 7th edn. (Wiley, New York, 2009)
P.M. Gerhart, R.J. Gross, *Fundamentals of Fluid Mechanics* (Addison-Wesley, New York, 1985)
L.D. Landau, E.M. Lifshitz, *Fluid Mechanics*, 2nd edn. (Elsevier, Amsterdam, 2005)
E.A. Moelwyn-Hughnes, *States of Matter* (Oliver and Boyd, New York, 1961)
B.R. Munson, D.F. Young, T.H. Okiishi, *Fundamentals of Fluid Mechanics*, 3rd edn. (Wiley, New York, 1990)
P. Oswald, *Rheophysics: The Deformation and Flow of Matter* (Cambridge University Press, Cambridge, 2009)
R.H.F. Pao, *Fluid Mechanics* (Wiley, New York, 1961)
W.R. Schowalter, *Mechanics of Non-Newtonian Fluids* (Pergamon Press, Oxford, 1978)
A.J. Smith, *A Physical Introduction to Fluid Mechanics* (Wiley, New York, 2000)
D. Tabor, *Gases, Liquids and Solids, and Other States of Matter*, 3rd edn. (Cambridge University Press, Cambridge, 1993)
R.I. Tanner, *Engineering Rheology*, revised edn. (Oxford University Press, Oxford, 1992)

Hydrostatics 3

The knowledge about the characteristics of fluids at rest is referred to as fluid statics, or alternatively as hydrostatics, which is derived from the fact that the disciplines are used frequently for water at rest. The pressure of a still fluid in the gravitational field experiences a spatial variation, and the pressure distribution over the surface of a solid body with finite volume results in a net force acting on the body. This net force is termed differently as the hydrostatic or buoyant force in different circumstances, which are discussed in separate sections of this chapter. Specifically, the pressure distribution in a still fluid is discussed, followed by the estimations on the hydrostatic forces on a plane and a curved surface. Formation of the free surface of a still fluid with relations to the surface tension and capillary effect is presented. Buoyancy and stability analysis of a floating and submerged bodies in a still fluid are introduced by using the relative positions between the centers of gravity and buoyancy. Last, due to the same Cauchy stress state as that of a still fluid, the pressure variation of a fluid in rigid body motion is discussed.

3.1 Thermodynamic Pressure

3.1.1 Equations of Pressure Distribution

Based on Pascal's law, the pressure at a specific point in a still incompressible Newtonian fluid is defined to be the average of three normal stress components of the Cauchy stress and is a compressive force per unit area, with its value unchanged with respect to different directions. It is termed officially the *thermodynamic pressure*, or simply the pressure, which experiences a spatial variation if the gravitational field or other acceleration fields present.

Consider an infinitesimal cubic box with volume $\mathrm{d}v$ as the differential control-volume system (the material point), and the pressure p at the center of $\mathrm{d}v$ is known. The six rectangular planes of $\mathrm{d}v$, denoted by $\mathrm{d}\boldsymbol{a} = \mathrm{d}x_i \mathrm{d}x_j \varepsilon_{ijk} \boldsymbol{e}_k$, are acted by the corresponding pressures there, with $\mathrm{d}v = \varepsilon_{ijk} \mathrm{d}x_i \mathrm{d}x_j \mathrm{d}x_k$. The infinitesimal

© Springer International Publishing AG 2019
C. Fang, *An Introduction to Fluid Mechanics*, Springer Textbooks in Earth Sciences, Geography and Environment, https://doi.org/10.1007/978-3-319-91821-1_3

body and surface forces acting on the material point, denoted respectively by $\mathrm{d}\boldsymbol{F}^{\mathrm{B}}$ and $\mathrm{d}\boldsymbol{F}^{\mathrm{S}}$, by using the Taylor series expansion, are obtained as

$$\mathrm{d}\boldsymbol{F}^{\mathrm{B}} = \rho \boldsymbol{g}\,\mathrm{d}v, \qquad \mathrm{d}\boldsymbol{F}^{\mathrm{S}} = -\nabla p\,\mathrm{d}v, \tag{3.1.1}$$

with $\boldsymbol{g}$ the gravitational acceleration. Applying Newton's second law of motion to the material point yields

$$\rho \boldsymbol{g} - \nabla p = \rho \boldsymbol{a}, \tag{3.1.2}$$

per unit volume, where $\boldsymbol{a}$ is the resulting acceleration experienced by the fluid. Equation (3.1.2) will be used to discuss the pressure distribution in a fluid subject to a rigid body motion in Sect. 3.5. If a fluid is at rest, this equation reduces to

$$\rho \boldsymbol{g} - \nabla p = \boldsymbol{0}, \tag{3.1.3}$$

corresponding to $\sum \boldsymbol{F} = \boldsymbol{0}$ per unit volume.

Equation (3.1.3) indicates that a change in pressure can only take place in the direction parallel to that of the gravitational acceleration. An increase in pressure is obtained if the direction from one point to another is the same as that of $\boldsymbol{g}$, and vice versa. If the direction is orthogonal to the direction of $\boldsymbol{g}$, no pressure change takes place. Thus, the free surface of water on the earth's surface is always perpendicular to the gravitational direction, for the pressure on the water free surface assumes a constant value, i.e., the pressure of the air above. This equation also delivers a physical interpretation for the pressure variation in a still fluid in the gravitational field. Consider an infinitesimal cylinder with unit cross-sectional area and finite height, and the long axis being parallel to the gravitational acceleration as the control-volume system. The pressure on the lower cross-section is larger than that on the upper cross-section in order to maintain a static equilibrium of the cylinder. Alternatively, it is just the weight of the fluid contained in the cylinder that a pressure difference between the upper and lower cross-sections is caused. The interpretation can be extended equally to Eq. (3.1.2), if other acceleration fields present.

Consider an application of Eq. (3.1.3) on the earth's surface under the rectangular Cartesian coordinates $\{x, y, z\}$. Let x and y be on the horizontal plane, z point vertically upwards and the gravitational acceleration point vertically downwards. With these, Eq. (3.1.3) reduces to

$$\frac{\mathrm{d}p}{\mathrm{d}z} = -\rho g = -\gamma, \tag{3.1.4}$$

indicating that moving upwards yields a decrease in pressure, and vice versa. Integrating this equation for incompressible fluids, e.g. water, yields[1]

$$p - p_{\mathrm{ref}} = -\rho g(z - z_{\mathrm{ref}}) = \rho g h = \gamma h, \tag{3.1.5}$$

where p_{ref} and z_{ref} are the pressure and elevation of a reference point, usually taken at the fluid free surface, and h is the distance between any point and the reference point. Equation (3.1.5) indicates that the pressure difference between two points is nothing

[1]The gravitational acceleration g is considered a constant.

else than the weight of fluid contained in a cylinder with unit cross-sectional area and height h in-between. A positive pressure increase is obtained if the evaluated point is below the reference point, and vice versa. For compressible fluids, the dependency of ρ on z needs to be determined a priori the integration of Eq. (3.1.4). Consider the air in the troposphere, in which the temperature decreases linearly as z increases, which is described by $T = T_0 - mz$, with T_0 the earth surface temperature and m the temperature decreasing rate.[2] Let the state of air be described by the ideal gas state equation viz., $p = \rho RT$, with R the gas constant of air and T the Kelvin temperature scale.[3] Combining these with Eq. (3.1.4) and integrating the resulting equation give

$$p = p_0\left(1 - \frac{mz}{T_0}\right)^{g/mR} = p_0\left(\frac{T}{T_0}\right)^{g/mR}, \tag{3.1.6}$$

with p_0 the air pressure on the earth's surface, and $\{p, T\}$ represent the pressure and temperature at the altitude z. It follows from Eqs. (3.1.5) and (3.1.6) that the pressure changes linearly with respect to elevation in incompressible fluids, while it varies nearly exponentially in compressible fluids. Let one stand on the sea surface surrounded by air at 1 atmospheric pressure, p_{atm}, and be free to move vertically upwards or downwards. The pressure that the experiences at the location 1 km below the sea surface are $p/p_{\text{atm}} \sim 98$, while the pressure at the same height above the sea surface is $p/p_{\text{atm}} \sim 0.89$.[4]

Other practical applications of Eq. (3.1.3) can be found e.g. in canal construction connecting rivers at different water levels, water level identification in geotechnical engineering, transmission of fluid pressure and power in pneumatic and hydraulic engineering, and hydraulic braking system in automobile industry.

As an illustration of Eq. (3.1.4), consider four containers in a serial connection, as shown in Fig. 3.1a. Water is supplied to tank A and flows through the connecting conducts subsequently to containers B, C, and D. Since Eq. (3.1.4) implies that the pressures at the same elevation must be the same, containers C and D should be the first two which are filled completely by water, followed by containers D and B. Container A is the last one which is filled completely by water. Thus, the sequence of complete filling of water is $C \cup D \to B \to A$.

Another illustration is the inclined-tube *manometer* shown in Fig. 3.1b, which is used to measure a small pressure difference between points A and B. It follows from Eq. (3.1.4) that

$$p_A - p_B = -\gamma_1 h_1 + \gamma_2 \ell \sin\theta + \gamma_3 h_3. \tag{3.1.7}$$

[2]The value of m is 9.8 °C/km for completely dry air. It takes the value of 6.5 °C/km if the water vapor in the air does not condense to liquid water during ascending, while the value of $m = 5.5$ °C/km is used when condensation takes place.

[3]Although air is a gas mixture consisting of nearly 78% nitrogen, nearly 21% oxygen and less than 1% minor gases and water vapor, it is a simple compressible substance in the macroscopic point of view.

[4]The calculation is conducted by that the density of seawater is 1000 kg/m^3, $g = 9.8$ m/s^2, $m =$ 9.8 °C/km, $T_0 = 25$ °C, and $R = 0.287$ kJ/kg-K.

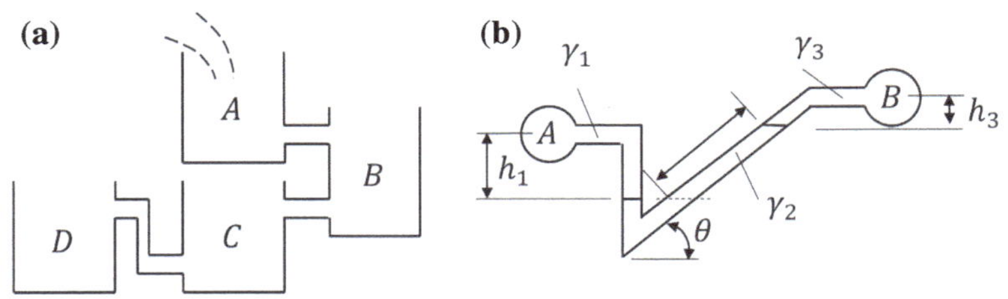

Fig. 3.1 Illustrations of pressure variation. **a** Four containers in a serial connection. **b** An inclined-tube manometer

If the fluids at points A and B are gases, whose specific weights γ_1 and γ_3 are so small when compared to the specific weight γ_2 of the liquid inside the inclined-tube, the above equation may be reduced to

$$p_A - p_B \sim \gamma_2 \ell \sin\theta, \quad \longrightarrow \quad \ell = \frac{p_A - p_B}{\gamma_2 \sin\theta}, \quad \Longleftrightarrow \quad \ell_U = \frac{p_A - p_B}{\gamma_2}, \tag{3.1.8}$$

indicating that the differential reading ℓ of the inclined-tube manometer for a given pressure difference $p_A - p_B$ can be increased over ℓ_U obtained with a conventional U-tube manometer by a factor of $1/\sin\theta$. Larger values of ℓ for better reading of small pressure differences are accomplished by letting $\theta \to 0$.

3.1.2 Reference Level of Pressure

It follows from the molecular theory of gas and statistical mechanics that the pressure exerted by a gas on a solid boundary results from the momentum exchange of gas molecules per unit time per unit solid boundary surface area. A pressure is called an *absolute pressure*, denoted by p_{abs}, if it is measured on the reference of absolute empty and vacuum level. A pressure is called a *gage pressure*, denoted by p_{gage}, if it is measured based on the pressure in the ambient environment, which is called the *surrounding pressure*, p_{surr}. Most of the time, the pressure in the ambient environment is the *atmospheric pressure*, p_{atm}. Thus, a gage pressure is essentially a pressure difference. If a gage pressure assumes a negative value, it is denoted frequently by using its absolute magnitude with the term "suction" or "vacuum" behind called respectively a *suction pressure* or a *vacuum pressure*. The relations between p_{abs}, p_{gage}, p_{surr} and suction and vacuum pressures are summarized in the following:

$$p_{\text{gage}} = p_{\text{abs}} - p_{\text{surr}}, \qquad \begin{cases} p_{\text{gage}} > 0, \; p = p_{\text{gage}}, \\ \\ p_{\text{gage}} < 0, \; p = p_{\text{gage}} = |p_{\text{gage}}| \text{ suction}, \\ \qquad\qquad\qquad\quad = |p_{\text{gage}}| \text{ vacuum}. \end{cases} \tag{3.1.9}$$

For example, if the air inside a closed bottle has an absolute pressure of $1.8\ p_{\text{atm}}$, its pressure is denoted by $p_{\text{abs}} = 1.8\ p_{\text{atm}}$, or alternatively by $p_{\text{gage}} = 0.8\ p_{\text{atm}}$. If, on the other hand, the absolute pressure is $0.6\ p_{\text{atm}}$, it is denoted by $p_{\text{abs}} = 0.6\ p_{\text{atm}}$, or $p_{\text{gage}} = -0.4\ p_{\text{atm}} = 0.4\ p_{\text{atm}}$ suction $= 0.4\ p_{\text{atm}}$ vacuum.

Table 3.1 Standard atmospheric properties at sea level in SI unit

Temperature, T	Pressure, p_{atm}	Density, ρ	Specific weight, γ	Dynamic viscosity, μ
288.15 K (15 °C)	101.325 kPa (abs)	1.225 kg/m^3	12.014 N/m^3	$1.789 \cdot 10^{-5}$ N · s/m^2

3.1.3 Standard Atmospheric Properties

Most applications of engineering disciplines are nearly on the earth's surface, of which the standard atmospheric properties at sea level are summarized in Table 3.1.[5] Specifically, the values of p_{atm} in different SI units are given in the following for convenience of conversion:

$$p_{atm} = 101.325\,\text{kPa} = 1013.25\,\text{mb} = 10.34\,\text{m}\,\text{H}_2\text{O} = 760\,\text{mm}\,\text{Hg}, \qquad (3.1.10)$$

where "mb" stands for the millibar, which is used frequently in atmospheric science and meteorology. The atmospheric pressure on the sea level is nothing else than the total weight of the above air till the edge of atmosphere per unit area. Devices used to measure pressure are called *pressure transducers*; e.g. the *Bourdon gage* is the most encountered pressure transducer in everyday life. Devices used to measure the atmospheric pressure are called specifically the *barometers*.

3.2 Hydrostatic Forces on Submerged Surfaces

3.2.1 Force on Plane

Consider an arbitrarily bounded and oriented plane with an inclined angle θ with respect to the free surface of a liquid, which is fully wetted by the liquid, as shown in Fig. 3.2a. The origin of coordinates x_i is placed at the centroid of plane, with x_3 being normal to the plane, x_1 being parallel to the plane, and x_2 being on the plane and parallel to the free surface, forming a right-handed tripod, with the corresponding orthonormal base $\{\boldsymbol{e}_1, \boldsymbol{e}_2, \boldsymbol{e}_3\}$. The gravitational acceleration $\boldsymbol{g}$ is given by

$$\boldsymbol{g} = -g \sin\theta \boldsymbol{e}_1 - g \cos\theta \boldsymbol{e}_3, \qquad \|\boldsymbol{g}\| = g. \qquad (3.2.1)$$

It follows from Eqs. (3.1.3) and (3.2.1) that the pressure at a specific point on the plane with the position denoted by $\{x_1, x_2\}$ from the centroid is given by

$$p = p_c + \nabla p \cdot \mathrm{d}\boldsymbol{r} = p_c - \rho g x_1 \sin\theta, \qquad (3.2.2)$$

[5] Data quoted from: *The U.S. Standard Atmosphere (1976)*, Washington, D.C., U.S. Government Printing Office, 1976.

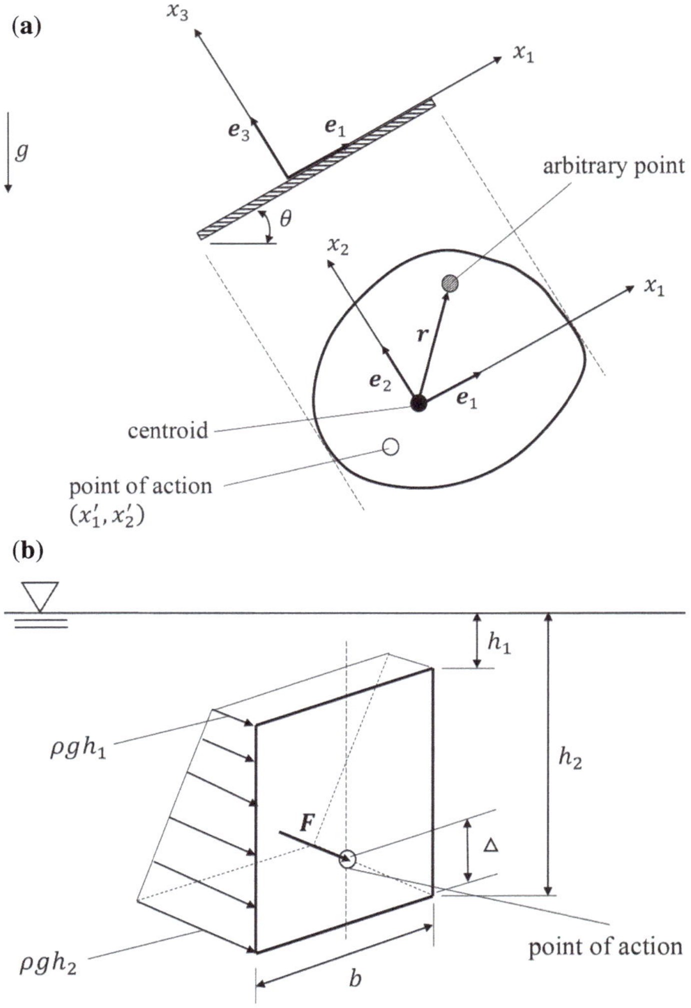

Fig. 3.2 Hydrostatic forces on submerged planes. **a** Coordinates of the general formulation. **b** Illustration of the rapid formulation for rectangular planes

where $\mathrm{d}\boldsymbol{r} = x_1\boldsymbol{e}_1 + x_2\boldsymbol{e}_2$ and p_c is the pressure at the centroid. The *hydrostatic force* $\boldsymbol{F}$ exerted by the liquid on the plane is the summation of local force at each point over the entire plane given viz.,

$$\boldsymbol{F} = \int_A -p\,\mathrm{d}\boldsymbol{a} = \int_A (-p\,\mathrm{d}a)\boldsymbol{e}_3, \tag{3.2.3}$$

indicating that $\boldsymbol{F}$ is perpendicular to the plane and always points to the plane. Substituting Eq. (3.2.2) into Eq. (3.2.3) yields

$$\boldsymbol{F}=-\int_A\Big[(p_c-\rho g x_1\sin\theta)\mathrm{d}a\Big]\boldsymbol{e}_3=-p_cA\boldsymbol{e}_3+\rho g\sin\theta\int_A(x_1\mathrm{d}a)\,\boldsymbol{e}_3=-p_cA\boldsymbol{e}_3, \tag{3.2.4}$$

with

$$\int_A(x_1\mathrm{d}a)\,\boldsymbol{e}_3=x_1^cA\boldsymbol{e}_3=\boldsymbol{0}, \tag{3.2.5}$$

where x_1^c is the x_1-coordinate of the centroid, which vanishes in the context of the used coordinate system. Equation (3.2.4) indicates that the magnitude of hydrostatic force is the product of the pressure at the centroid of plane and plane area.

The moment with respect to the centroid, $\boldsymbol{M}$, generated by the pressure distribution on the plane, is obtained as

$$\begin{aligned}\boldsymbol{M}&=-\int_A p(\boldsymbol{r}\times\mathrm{d}\boldsymbol{a})=-\int_A\Big[(x_1\boldsymbol{e}_1+x_2\boldsymbol{e}_2)\times\boldsymbol{e}_3\Big](p_c-\rho g x_1\sin\theta)\mathrm{d}a\\&=-\int_A\Big[(p_cx_1-\rho g x_1x_1\sin\theta)\varepsilon_{132}\mathrm{d}a\Big]\boldsymbol{e}_2-\int_A\Big[(p_cx_2-\rho g x_1x_2\sin\theta)\varepsilon_{231}\mathrm{d}a\Big]\boldsymbol{e}_1\\&=\int_A\Big[(p_cx_1-\rho g x_1x_1\sin\theta)\mathrm{d}a\Big]\boldsymbol{e}_2-\int_A\Big[(p_cx_2-\rho g x_1x_2\sin\theta)\mathrm{d}a\Big]\boldsymbol{e}_1.\end{aligned} \tag{3.2.6}$$

The point of action of $\boldsymbol{F}$, denoted by $\boldsymbol{r}'=x_1'\boldsymbol{e}_1+x_2'\boldsymbol{e}_2$, is determined if the moment of $\boldsymbol{F}$ with respect to the centroid is the same as $\boldsymbol{M}$, i.e.,

$$\begin{aligned}\boldsymbol{r}'\times\boldsymbol{F}&=(x_1'\boldsymbol{e}_1+x_2'\boldsymbol{e}_2)\times(-p_cA)\boldsymbol{e}_3\\&=-\varepsilon_{132}(p_cAx_1')\boldsymbol{e}_2-\varepsilon_{231}(p_cAx_2')\boldsymbol{e}_1=(p_cAx_1')\boldsymbol{e}_2-(p_cAx_2')\boldsymbol{e}_1\\&=\int_A\Big[(p_cx_1-\rho g x_1x_1\sin\theta)\mathrm{d}a\Big]\boldsymbol{e}_2-\int_A\Big[(p_cx_2-\rho g x_1x_2\sin\theta)\mathrm{d}a\Big]\boldsymbol{e}_1,\end{aligned} \tag{3.2.7}$$

giving rise to

$$\begin{aligned}p_cAx_1'&=-\rho g\sin\theta I^c_{x_1x_1},\longrightarrow x_1'=-\frac{\rho g\sin\theta I^c_{x_1x_1}}{p_cA},\\p_cAx_2'&=-\rho g\sin\theta I^c_{x_1x_2},\longrightarrow x_2'=-\frac{\rho g\sin\theta I^c_{x_1x_2}}{p_cA},\end{aligned} \tag{3.2.8}$$

in which the moment of inertia relative to the x_1-axis, $I^c_{x_1x_1}$, and the mixed moment of inertia, $I^c_{x_1x_2}$, have been used, with their definitions given by

$$I^c_{x_1x_1}=\int_A x_1x_1\,\mathrm{d}a,\qquad I^c_{x_1x_2}=\int_A x_1x_2\,\mathrm{d}a. \tag{3.2.9}$$

Since $I^c_{x_1x_1}$ is always positive, it follows that x_1' always locates in the negative x_1-axis, indicating the difference between the centroid of plane and the point of action of hydrostatic force. However, such a conclusion does not hold for x_2', for the value of $I^c_{x_1x_2}$ may vary depending on the plane shape.

Equations (3.2.4) and (3.2.8)–(3.2.9) are called the *general formulation* of hydrostatic force on a submerged plane, whose conclusions are summarized in the following for convenience:

- Physical mechanism: The hydrostatic force results from the non-uniform pressure distribution on a plane.
- Direction: Based on the compressive nature of pressure, the hydrostatic force always points perpendicularly to a plane.
- Magnitude: The magnitude of hydrostatic force is the product of pressure at the centroid of a plane and the area of that plane. To take away the influence of pressure in the surrounding (e.g. the atmospheric pressure), the pressure at the centroid of plane ought to be expressed as a gage one.
- Point of action: The point of action of hydrostatic force is determined by using Eqs. (3.2.8) and (3.2.9) within the coordinates used in Fig. 3.1a.

If the plane is rectangular, as shown in Fig. 3.2b, the hydrostatic force can be determined in an easier manner, which is summarized as the *rapid formulation* given in the following:

- Along the edge of a rectangular plane, plot a diagram of the gage pressure distribution, known as the *pressure distribution diagram.*
- The magnitude of hydrostatic force is given by the product of the area of pressure distribution diagram and the width of rectangular plane.
- The point of action of hydrostatic force locates at the centroid of pressure distribution diagram.

For example, the pressure distribution diagram of the rectangular plane in Fig. 3.2b is a trapezoid. The magnitude of hydrostatic force is then given by

$$\|\boldsymbol{F}\| = \frac{\rho g(h_1 + h_2)}{2}(h_2 - h_1)b. \tag{3.2.10}$$

The point of action lies on the vertical center line of plane, with the vertical position determined by the centroid of trapezoid, viz.,

$$\Delta = \frac{2\rho g h_1 + \rho g h_2}{3(\rho g h_1 + \rho g h_2)}(h_2 - h_1) = \frac{2h_1 + h_2}{3(h_1 + h_2)}(h_2 - h_1). \tag{3.2.11}$$

Nevertheless, for a rectangular plane, the hydrostatic force can be determined by using either the general or rapid formulation. The validity of rapid formulation for rectangular plane lies in the fact that the pressure distribution remains unchanged at different edges of the plane.

3.2.2 Force on Curved Surface

Consider a curved surface submerged in a still fluid, as shown in Fig. 3.3a. The hydrostatic force acting on the curved surface originates from the same physical mechanism as before, i.e., due to the non-uniform pressure distribution over the curved surface, which is given by

$$\boldsymbol{F} = \int_A -p\, \mathrm{d}\boldsymbol{a} = F_i \boldsymbol{e}_i, \tag{3.2.12}$$

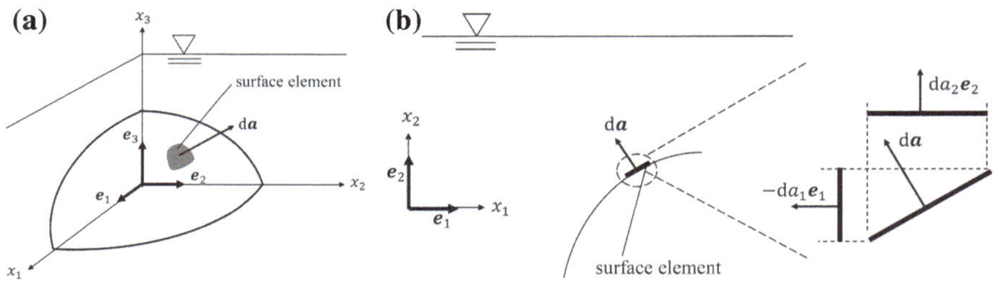

Fig. 3.3 Hydrostatic forces on curved surfaces. **a** Coordinates of the formulation. **b** A two-dimensional curved surface with an enlargement of a surface element

where $\mathrm{d}\boldsymbol{a}$ is a surface element. Taking inner product of this equation with the orthonormal base $\boldsymbol{e}_i$ yields

$$F_i = -\int_A p\,\mathrm{d}\boldsymbol{a}\cdot\boldsymbol{e}_i = -\int_A p\,(\mathrm{d}a_j\boldsymbol{e}_j)\cdot\boldsymbol{e}_i = -\int_{A_i} p\,\mathrm{d}a_i, \tag{3.2.13}$$

indicating that the components of hydrostatic force are in the reverse directions of the projection planes. These force components can be calculated by using the disciplines described in Sect. 3.2.1, with which the direction, magnitude, and point of action of $\boldsymbol{F}$ are consequently determined.

An application of Eq. (3.2.13) is taken for a simple two-dimensional curved surface shown in Fig. 3.3b. It follows that

$$\begin{aligned} F_1 &= -\int_A p\,\mathrm{d}\boldsymbol{a}\cdot\boldsymbol{e}_1 = -\int_A p\,(-\mathrm{d}a_1\boldsymbol{e}_1 + \mathrm{d}a_2\boldsymbol{e}_2)\cdot\boldsymbol{e}_1 = \int_{A_1} p\,\mathrm{d}a_1,\\ F_2 &= -\int_A p\,\mathrm{d}\boldsymbol{a}\cdot\boldsymbol{e}_2 = -\int_A p\,(-\mathrm{d}a_1\boldsymbol{e}_1 + \mathrm{d}a_2\boldsymbol{e}_2)\cdot\boldsymbol{e}_2\\ &= -\int_{A_2} p\,\mathrm{d}a_2 = -\int_{A_2} \rho g h\,\mathrm{d}a_2 = -\int_V \rho g\,\mathrm{d}v. \end{aligned} \tag{3.2.14}$$

While Eq. $(3.2.14)_1$ indicates that F_1 is nothing else than the hydrostatic force acting on the projection area A_1 in the x_1-direction with a sign change indicating that F_1 acts in the reverse direction of $\mathrm{d}a_1$ (i.e., along the positive x_1-axis), Eq. $(3.2.14)_2$ delivers that F_2 is simply the weight of fluid above the curved surface till the free surface, and the minus sign indicates that F_2 points to the reverse direction of $\mathrm{d}a_2$ (i.e., along the negative x_2-axis). Since F_2 is the weight of fluid in the region above the curved surface till the free surface, it acts at the center of gravity of that region. It reduces to the center of mass if the gravitational acceleration is a constant, and subsequently to the centroid if the fluid is homogeneous with constant density.

To illustrate the disciplines, consider first a two-dimensional inclined gate shown in Fig. 3.4a, which is hinged along edge A and with width b perpendicular to the page. It follows from the general formulation that the magnitude of hydrostatic force $\boldsymbol{F}$ acting on the gate by water is given by

$$\|\boldsymbol{F}\| = p_c A = \gamma\left(D + \frac{L}{2}\sin\theta\right)Lb, \tag{3.2.15}$$

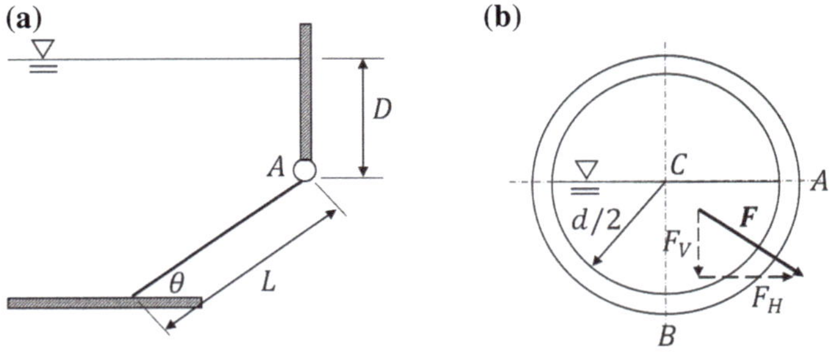

Fig. 3.4 Illustrations of the hydrostatic forces on planes and curved surfaces. **a** A rectangular gate in contact with a still water. **b** A drainage conduit which is half full of water at rest

which points perpendicularly to the gate. The point of action, by using Eq. (3.2.8), is determined to be

$$x_1' = -\frac{\rho g \sin\theta I^c_{x_1x_1}}{p_c A} = -\frac{1}{6}\frac{L^2\sin\theta}{2D + L\sin\theta}, \qquad x_2' = -\frac{\rho g \sin\theta I^c_{x_1x_2}}{p_c A} = 0, \tag{3.2.16}$$

under the coordinate system used in Fig. 3.2a. Since the gate is a rectangular plane, $\|\boldsymbol{F}\|$ can also be determined by using the rapid formulation, viz.,

$$\|\boldsymbol{F}\| = \frac{1}{2}\left[\gamma D + \gamma(D + L\sin\theta)\right] Lb = \gamma\left(D + \frac{L}{2}\sin\theta\right) Lb, \tag{3.2.17}$$

which coincides to Eq. (3.2.15). The rapid formulation shows that the point of action lies in the centerline of gate (i.e., $x_2' = 0$), and the centroid of pressure distribution diagram is identified to be

$$\Delta = \frac{2h_1 + h_2}{3(h_1 + h_2)}(h_2 - h_1) = \frac{3D + L\sin\theta}{3(2D + L\sin\theta)}L, \tag{3.2.18}$$

which is measured from the gate bottom. The corresponding x_1-coordinate of Eq. (3.2.18) is then obtained as

$$x_1' = -\frac{L}{2} + \Delta = -\frac{1}{6}\frac{L^2\sin\theta}{2D + L\sin\theta}, \tag{3.2.19}$$

which coincides to Eq. $(3.2.16)_1$.

Consider a drainage conduit with diameter d shown in Fig. 3.4b, which is half-full of water at rest. The section length of conduit perpendicular to the page is denoted by b. The hydrostatic force $\boldsymbol{F}$ acting on curved surface AB of the conduit is decomposed into the horizontal component F_H and vertical component F_V. It follows from the rapid formulation that

$$|F_H| = \frac{\gamma b d^2}{8}, \qquad \Delta = \frac{d}{6}, \tag{3.2.20}$$

for the horizontal force component, where Δ is measured from the bottom of conduit, and F_H points to the right. The vertical force component is nothing else than the water weight above curved surface AB till the free water surface, which is given by

$$|F_V| = \frac{\gamma b \pi d^2}{16}, \tag{3.2.21}$$

pointing vertically downwards through the centroid of volume ABC. Since the directions, magnitudes and points of action of F_H and F_V are known, the direction, magnitude, and point of action of $\boldsymbol{F}$ can immediately be determined.

3.3 Free Surface of a Liquid

3.3.1 Surface Tension and Capillary Effect

Due to the unbalanced molecular attractions described in Sect. 2.6.2, the number of liquid molecules on the interface surface between two dissimilar liquids assumes a minimum value necessary for the formation of surface. Macroscopically, this manifests itself as if a tension were acting at the interface. Consider a line element δl of a surface boundary shown in Fig. 3.5a, upon which a force $\delta \boldsymbol{F}$ acts, resulted from the unbalanced molecular attractions. The *surface tension* $\boldsymbol{\sigma}$ is defined as the intensity of unbalanced molecular attractions per unit length along any line lying at the interface surface, viz.,

$$\boldsymbol{\sigma} = \lim_{\delta l \to 0} \frac{\delta \boldsymbol{F}}{\delta l} = \frac{\mathrm{d}\boldsymbol{F}}{\mathrm{d}l}, \tag{3.3.1}$$

which is called alternatively the *stress vector* of surface tension. Essentially, $\boldsymbol{\sigma}$ has the components in the normal and tangential directions of a line element. If the liquid particles which form the free surface are at rest, the tangential component vanishes, with which Eq. (3.3.1) reduces to

$$\boldsymbol{\sigma} = C\boldsymbol{m}, \tag{3.3.2}$$

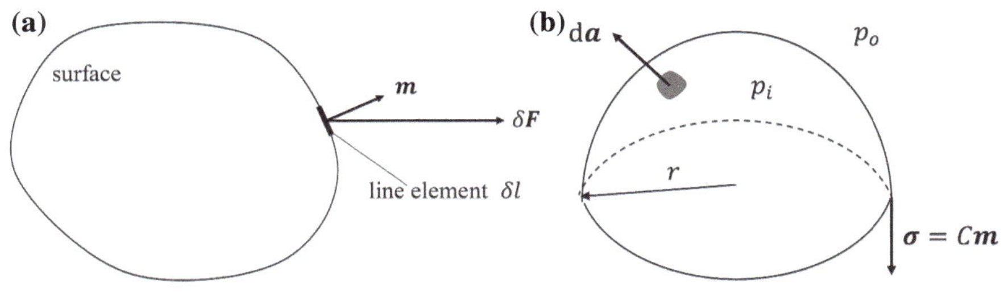

Fig. 3.5 Illustrations of surface tension. **a** Surface tension on a line element of a surface boundary. **b** Force balance on the free surface of a liquid drop

where $\boldsymbol{m}$ is the unit normal vector to δl, and C is termed the *capillary constant*, which is the magnitude of $\boldsymbol{\sigma}$ and is independent of $\boldsymbol{m}$ but depends on the fluid properties forming the interface surfaces, e.g. liquid-gas, or liquid-liquid interface.

An application of Eq. (3.3.2) is the surface tension of a spherical shape of a small drop of liquid, as shown in Fig. 3.5b, in which there exist a pressure p_i inside the drop and a pressure p_o outside the drop. The static equilibrium of the drop requires that

$$2\pi r C \boldsymbol{m} - \int_A (p_i - p_o)\boldsymbol{n}\,\mathrm{d}a = \boldsymbol{0}, \tag{3.3.3}$$

which reduces to

$$\Delta p = p_i - p_o = \frac{2C}{r}. \tag{3.3.4}$$

This equation can be extended for a tiny small drop liquid with a general interface surface. It can be shown that the pressure drop over the surface is given by

$$\Delta p = C\Phi, \qquad \Phi = \frac{1}{r_1} + \frac{1}{r_2}, \tag{3.3.5}$$

where r_1 and r_2 are the principal radii of curvature and Φ is termed the *mean curvature* of surface. For a plane surface, both r_1 and r_2 approach infinite, giving rise to a vanishing Δp. This result indicates that surface tension can never take place on planes.

Non-vanishing curvatures of interface surface often take place on boundaries when three fluids meet, or two fluids and a solid wall meet. Consider two fluids and a solid wall which are in contact in a two-dimensional circumstance, as shown in Fig. 3.6a, in which fluid 1 is below fluid 2, and the solid wall is identified by the number 3. Both fluids are immiscible, and α is called the *angle of contact*, or alternatively the *wetting angle*, which is defined as the angle that the tangent to the surface of fluid 1 makes with the solid surface at the point of contact. The interface surface is explicitly described by $z = z(x_1)$. It follows from Eq. (3.1.5) that the pressure drop across the interface surface is given by

$$\Delta p = p_2 - p_1 = (\rho_1 - \rho_2)gz, \tag{3.3.6}$$

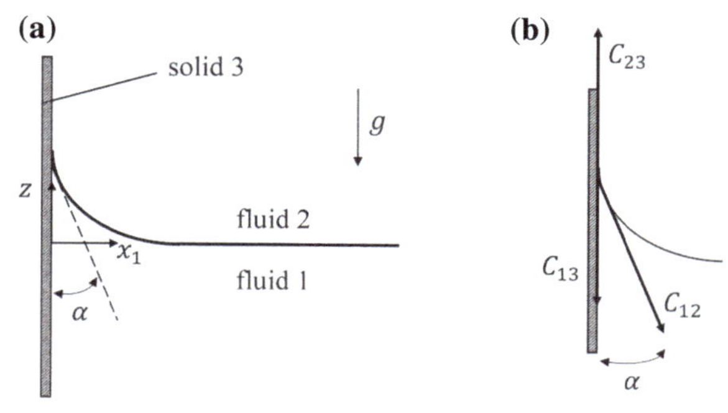

Fig. 3.6 Capillary effect at the contact point between two fluids and a solid. **a** Curvature of the interface surface with the angle of contact. **b** Balance between the capillary stresses at the contact point

which, with Eq. (3.3.5), is recast alternatively as

$$\mathcal{C}_{12}\left(\frac{1}{r_1}+\frac{1}{r_2}\right)=(\rho_1-\rho_2)gz, \tag{3.3.7}$$

where $\mathcal{C}_{12}$ is the capillary constant between fluids 1 and 2. In the considered two-dimensional circumstance, $r_1 \to \infty$. Substituting this and the assumption that fluid 2 is a gas with $\rho_2 \ll \rho_1$ into Eq. (3.3.7) yields

$$\frac{\mathcal{C}_{12}}{r_2}=\rho_1 gz, \tag{3.3.8}$$

from which *Laplace's length*, a, is defined as

$$a \equiv \sqrt{\frac{\mathcal{C}_{12}}{\rho_1 g}}. \tag{3.3.9}$$

The Laplace length provides an estimation on the importance of capillary effect in a still liquid. It needs to be taken into account if the characteristic length of problem assumes a similar order of magnitude of a. For example, for liquid water in contact with air, a assumes a value of nearly 3 mm. Water can thus flow easily in a pipe if the pipe diameter is much larger than a, whilst flow can hardly take place in a capillary tube, whose diameter is nearly of the same order of magnitude of a. Substituting the known expression

$$\frac{1}{r_2}=\frac{z''}{\left(z'^2+1\right)^{3/2}}, \qquad z'=\frac{\mathrm{d}z}{\mathrm{d}x_1}, \tag{3.3.10}$$

for the curvature r_2 of interface surface $z(x_1)$ and Eq. (3.3.9) into Eq. (3.3.8) gives

$$\frac{z''}{(z'^2+1)^{3/2}}-\frac{z}{a^2}=0. \tag{3.3.11}$$

Two boundary conditions are required to integrate Eq. (3.3.11). First, the condition that $z(x_1 \to \infty)=0$ is used, as motivated by the physical observation. Second, consider an equilibrium state of the capillary stresses at the contact point shown in Fig. 3.6b, where $\mathcal{C}_{13}$ and $\mathcal{C}_{23}$ are respectively the capillary constants between fluid 1 and solid wall 3, and fluid 2 and solid wall 3. Requiring a balance of the capillary stresses in the direction parallel to the wall yields[6]

$$\cos\alpha=\frac{\mathcal{C}_{23}-\mathcal{C}_{13}}{\mathcal{C}_{12}}, \quad\longrightarrow\quad \frac{\mathrm{d}z}{\mathrm{d}x_1}(x_1=0)=-\cot\alpha. \tag{3.3.12}$$

With these, an implicit solution to Eq. (3.3.11) is obtained as

$$\begin{aligned}\frac{x_1}{a}&=\cosh^{-1}\left(\frac{2a}{z}\right)-\cosh^{-1}\left(\frac{2a}{h}\right)+\sqrt{4-\left(\frac{h}{a}\right)^2}-\sqrt{4-\left(\frac{z}{a}\right)^2},\\ h^2&=z|_{x_1=0}=2a^2(1-\sin\alpha),\end{aligned} \tag{3.3.13}$$

[6]However, an equilibrium state cannot be maintained if $(\mathcal{C}_{23}-\mathcal{C}_{13}) \gg \mathcal{C}_{12}$, in which fluid 1 coats the whole wall, e.g. petrol in metal containers.

where h represents the maximum climbing height of fluid 1 on the solid wall.

For example, if fluid 1 is a pure water, fluid 2 is air and solid wall 3 is a clean soda-lime glass of a container with characteristic length of nearly 1 m. It follows that at 1 atmospheric pressure and 20 °C, $\mathcal{C}_{12} \sim 0.073$ N/m with nearly vanishing values of $\mathcal{C}_{23}$ and $\mathcal{C}_{13}$ when compared with $\mathcal{C}_{12}$, giving rise to $\alpha \sim 0°$. Alternatively, if fluid 1 is replaced by a mercury, the values of $\mathcal{C}_{12}$ and $\mathcal{C}_{13}$ are given respectively by 0.44 N/m and 0.283 N/m with nearly vanishing value of $\mathcal{C}_{23}$, giving rise to $\alpha \sim 130°$. In these circumstances, water is said to *wet the solid surface* for its $\alpha < 90°$, while mercury is said *not to wet the solid surface* for its $\alpha > 90\,°$. Moreover, it follows from Eqs. (3.3.9) and (3.3.13)$_2$ that $h \sim 3.86$ mm for pure water, a *climbing effect*, and $h \sim -1.24$ mm for mercury, a *sliding effect*. The small climbing and sliding heights result from the fact that the ratios of the Laplace lengths over the characteristic length of container in the considered two cases nearly vanish, indicating an insignificant capillary effect. Capillary effect becomes more significant if the characteristic size of container becomes small, e.g. a capillary tube with its diameter corresponding to $a \sim 3$ mm for pure water, or to $a \sim 2$ mm for mercury. In such a case, the magnitude of surface tension can immediately be determined by using a simple force balance between the surface tension force and weight of the fluid that is displaced.

3.3.2 Free Surface of a Still Liquid

When a liquid is in contact with a gas with its density much larger than the gas density, its surface is called a *free surface* with the pressure corresponding to that of the gas above. For example, when a water is in contact with air, the pressure on the water free surface in a distance far away from solid boundaries, i.e., in a distance much larger than the Laplace length of water, corresponds exactly to the pressure of the air above, most of the time the atmospheric pressure. This statement can be proven by the following arguments.[7] Consider an infinitesimal cubic box on the water surface as the differential control-volume, with the upper and lower planes of equal area in contact with air and water, respectively. The height of box approaches null, while the upper and lower planes remain non-vanishing. If the water pressure is larger than the air pressure, a force balance in the vertical direction cannot be reached, causing water to mover upwards, a phenomenon similar to water spring. On the contrary, if the water pressure is smaller than the air pressure, an unbalanced vertical force causes water to move downwards, leading to a phenomenon similar to water sink. Since these two situations are not observed, it follows that the water pressure on the free surface corresponds to the pressure of the air above. With these, it follows from Eqs. (3.3.6) and (3.3.7) that $z = 0$, $r_1 \to \infty$ and $r_2 \to \infty$, indicating that the water free surface is exactly perpendicular to the gravitational acceleration, as already verified by Eq. (3.1.3).

[7] The analysis is termed *jump conditions* in continuum mechanics.

The free surface is denoted graphically by using a straight horizontal solid line, with an inverse triangle immediately above and two shorter line segments underneath, as shown in the left-up corner in Fig. 3.2b.

3.4 Buoyancy and Stability

3.4.1 Buoyant Force

The *buoyancy* of a submerged body with finite volume in a still fluid results from the influence of non-uniform pressure distribution over the surface of body. Let the volume and surface of body be denoted by V and A, respectively, and ρ be the density of surrounding fluid. The *buoyant force* $\boldsymbol{F}$ is given by

$$\boldsymbol{F} = -\int_A p \, \mathrm{d}\boldsymbol{a} = -\int_V \nabla p \, \mathrm{d}v = -\int_V \rho \boldsymbol{g} \, \mathrm{d}v = -\rho \boldsymbol{g} V, \tag{3.4.1}$$

in which the Gauss theorem has been used. Equation (3.4.1) indicates not only the direction of $\boldsymbol{F}$, which is reverse to the gravitational acceleration, but also its magnitude as the weight of displaced fluids, which is known as *Archimedes' principle*.[8] Let $\boldsymbol{r}$ be the position vector of a point on the body surface. The point of action of buoyant force, called the *center of buoyancy*, denoted by $\boldsymbol{r}'$, must satisfy

$$\boldsymbol{r}' \times \boldsymbol{F} = -\int_A \boldsymbol{r} \times (p \, \mathrm{d}\boldsymbol{a}) = -\int_V (\boldsymbol{r} \times \nabla p) \mathrm{d}v = \int_V \boldsymbol{r} \times (\rho \boldsymbol{g}) \mathrm{d}v, \tag{3.4.2}$$

indicating that the center of buoyancy coincides to the center of gravity of displaced volume. It reduces subsequently to the center of mass under a constant gravitational acceleration, and to the centroid if the surrounding fluid is homogeneous with a constant density. Obviously, the buoyant force of a floating body is exactly the same as its own weight.

The finite volume of a body is crucial to buoyancy. Instead of the buoyant force, Pascal's law will be reproduced if the volume of body is infinitesimal as that of a material point in the differential approach.

As an illustration, consider a ball with diameter d and density ρ_b shown in Fig. 3.7. The ball is completely immersed into a still water with density ρ_w and is connected to the ground via a rope. The ball and rope are in a static equilibrium state. It is required to determine the tension of rope, which is denoted by T. The ball is considered a control-mass system, with its free body diagram also shown in the figure. Applying Newton's second law of motion to the ball along the vertical direction yields

$$T + W = B, \tag{3.4.3}$$

[8] Archimedes of Syracuse, c. 287–212 BC., a Greek polymath, who is regarded as one of the leading scientists in classical antiquity. The original statement of Archimedes' principle reads: "A body in a fluid experiences an apparent reduction in weight equal to the weight of the displaced fluid."

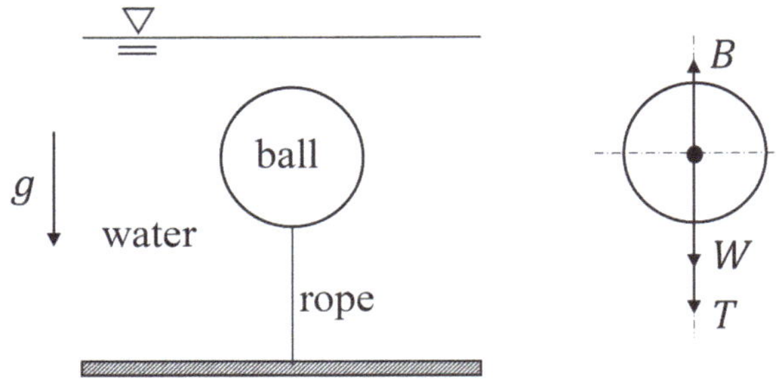

Fig. 3.7 Illustration of the buoyant force acting on a ball which is immersed into a still water

where W and B represent the weight of ball and buoyant force acting on the ball, respectively, which are given by

$$W = \rho_b g \frac{\pi d^3}{6}, \qquad B = \rho_w g \frac{\pi d^3}{6}. \tag{3.4.4}$$

Substituting these into Eq. (3.4.3) gives

$$T = \frac{\pi g d^3}{6} (\rho_w - \rho_b). \tag{3.4.5}$$

Thus, the tension of rope is positive if $\rho_w > \rho_b$. However, the tension of rope becomes negative if $\rho_w < \rho_b$. Such a circumstance is not justified if a rope is used to connect the ball and ground, for which the rope should be replaced by a solid rod.

3.4.2 Stabilities of Submerged and Floating Bodies

A system is recognized to be in a *stable equilibrium state* if it restores to its initial equilibrium state when disturbed. Conversely, it is in an *unstable equilibrium state* if it moves to a new equilibrium state when disturbed even slightly. It is also possible that a system is in a *neutral equilibrium state*, if it is always in an equilibrium state when disturbed.

The stability of a submerged body depends essentially on the relative positions between its center of gravity and center of buoyancy. For example, consider a body which is completely immersed into a still fluid, with its weight and buoyant force acting at the centroid of body if the densities of body and surrounding fluid are constant. In a static equilibrium, two forces are the same in magnitude but reverse in direction. The body is obvious in neutral equilibrium when disturbed. If, however, the body is heavier in its lower part, the center of gravity is lower than the center of buoyancy, causing itself in a stable equilibrium state, for the couple generated by the weight and buoyant force, termed the *righting moment*, restores itself to its initial equilibrium position when disturbed. On the contrary, the body is in an unstable equilibrium state if the body is heavier in its upper part due to the non-restoring righting moment when disturbed.

The stability of a floating body depends equally on the relative positions between the center of gravity and center of buoyancy; however, the location of the center of

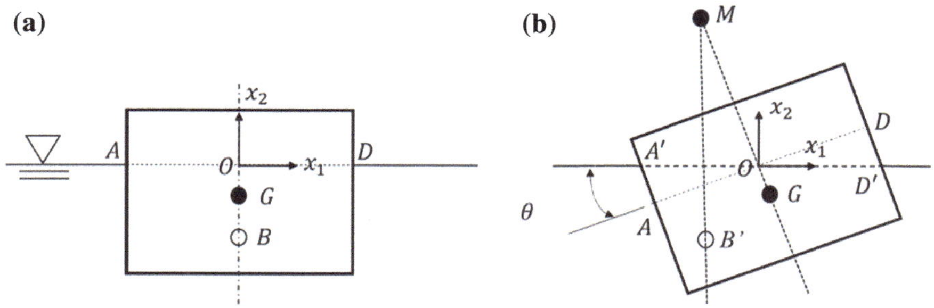

Fig. 3.8 Stability of a floating body. **a** Initial configuration. **b** Configuration of the body tilted by an angle θ

buoyancy may vary when disturbed. Consider a two-dimensional floating body with the coordinates shown in Fig. 3.8a, in which B denotes the center of buoyancy and G is the center of gravity. When disturbed, e.g. the body is titled by an angle θ, point B shifts to a new position B' due to the volume increase of wedge AOA' in the left and volume decrease of wedge ODD' in the right, as shown in Fig. 3.8b. The vertical line through point B' intersects the straight line connecting points B and G at point M, which is marked as the *metacenter*. The righting moment C is then identified to be

$$C = \|\boldsymbol{F}\| L_{MG} \sin\theta \sim \rho g V L_{MG}\,\theta, \tag{3.4.6}$$

for a small value of θ, where L_{MG} represents the length between points M and G, and $\|\boldsymbol{F}\|$ is the magnitude of buoyant force which remains unchanged when tilted, for the displaced volume V is the same during rolling. Alternatively, the righting moment can also be obtained by the moment generated by two buoyant forces of wedges AOA' and DOD', viz.,

$$C = 2\rho g\theta \int_A x_1^2 \mathrm{d}a = \rho g\theta\mathcal{I} = \|\boldsymbol{F}\| L_{MB}\,\theta, \qquad \mathcal{I} = 2\int_A x_1^2 \mathrm{d}a, \tag{3.4.7}$$

where $\mathrm{d}a = \ell \mathrm{d}x_1$, which is a differential area element in the plane of water line area with ℓ the extension in the x_3-direction, L_{MB} represents the length between points B and M, and $\mathcal{I}$ is the moment of inertia of water line area about its longitudinal axis (i.e., the x_3-axis). It follows immediately from Eq. $(3.4.7)_1$ that

$$L_{MB} = \frac{\mathcal{I}}{V}. \tag{3.4.8}$$

Consequently, the righting moment is then given by

$$C = \rho g V L_{GM}\theta, \qquad L_{GM} = L_{BM} - L_{BG}, \tag{3.4.9}$$

where L_{GM} is termed the *transverse meta-centric height*. Since a rolling of a floating body may cause point M to locate above or below point G, the stability of a floating body may be identified by the value of L_{GM}, viz.,

- $L_{GM} > 0$: a stable equilibrium with restoring righting moment;
- $L_{GM} < 0$: an unstable equilibrium with non-restoring righting moment;
- $L_{GM} = 0$: a neutral equilibrium with vanishing righting moment.

3.5 Liquids in Rigid Body Motion

A liquid which is subject to a prescribed acceleration $\boldsymbol{a}$ in addition to the gravitational acceleration $\boldsymbol{g}$ in rigid body motion is similar to a liquid in a static circumstance, for all shear stresses vanish due to the fact that there exist no relative motions between any two points inside the liquid. It follows immediately from Eq. (3.1.2) that

$$\nabla p = \rho\,(\boldsymbol{g} - \boldsymbol{a})\,, \tag{3.5.1}$$

which governs the pressure distribution in the liquid. This equation indicates that a change in pressure can only be accomplished in the direction parallel to $(\boldsymbol{g} - \boldsymbol{a})$, and with a pressure increase or a pressure decrease if the direction is the same or reverse to $(\boldsymbol{g} - \boldsymbol{a})$, respectively. If the liquid carries a free surface in contact with the atmospheric air, the free surface must be perpendicular to the direction of $(\boldsymbol{g} - \boldsymbol{a})$.

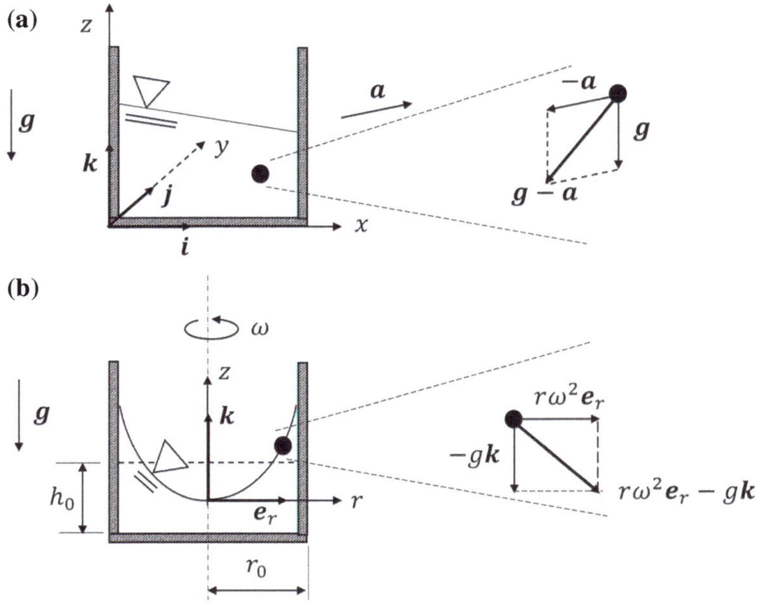

Fig. 3.9 Liquids in rigid body motion. **a** A liquid in a rectangular container. **b** A liquid in a rotating cylindrical container

For example, consider a rectangular box filled with a liquid, which is subject to the gravitational acceleration $\boldsymbol{g}$ and a prescribed acceleration $\boldsymbol{a}$, as shown in Fig. 3.9a. The origin of rectangular coordinate system $\{x, y, z\}$ is located at the corner of box. With this, the total pressure change $\mathrm{d}p$ between any two points, which are separated by a distance $\mathrm{d}\boldsymbol{r}$, is obtained as

$$\mathrm{d}p = \nabla p \cdot \mathrm{d}\boldsymbol{r} = \rho(g_x - a_x)\mathrm{d}x + \rho(g_y - a_y)\mathrm{d}y + \rho(g_z - a_z)\mathrm{d}z, \tag{3.5.2}$$

which reduces to

$$\rho(g_x - a_x)\mathrm{d}x + \rho(g_y - a_y)\mathrm{d}y + \rho(g_z - a_z)\mathrm{d}z = 0, \tag{3.5.3}$$

if the considered two points are lying on the free surface. Equation (3.5.3) yields the slopes of free surface on the three coordinate planes, viz.,

$$\left.\frac{\mathrm{d}y}{\mathrm{d}x}\right|_{z=0} = -\frac{g_x - a_x}{g_y - a_y}, \qquad \left.\frac{\mathrm{d}z}{\mathrm{d}x}\right|_{y=0} = -\frac{g_x - a_x}{g_z - a_z}, \qquad \left.\frac{\mathrm{d}z}{\mathrm{d}y}\right|_{x=0} = -\frac{g_y - a_y}{g_z - a_z}. \tag{3.5.4}$$

Specifically, if $\boldsymbol{a} = a_x\boldsymbol{i}$ and $\boldsymbol{g} = -g\boldsymbol{j}$ for a two-dimensional rectangular box in the (xy)-plane, the slope of liquid surface is identified to be

$$\frac{\mathrm{d}y}{\mathrm{d}x} = -\frac{a_x}{g}. \tag{3.5.5}$$

Additionally, consider a cylindrical container filled initially with a liquid to the height h_0, as shown in Fig. 3.9b. The container and liquid rotate coherently with a constant angular speed ω along the longitudinal axis (i.e., the z-axis). The origin of cylindrical coordinate system $\{r, \theta, z\}$ is located at the lowest point of curved free surface, corresponding to $r = 0$. The considered problem is essentially axis-symmetric, and the acceleration that a liquid particle experiences on the curved free surface at (r, z) is identified to be

$$(\boldsymbol{g} - \boldsymbol{a}) = r\omega^2\boldsymbol{e}_r - g\boldsymbol{k}, \tag{3.5.6}$$

to which the "infinitesimal straight free surface" must be perpendicular. Since this infinitesimal straight free surface represents a local tangential line of the curved liquid surface and is proportional to r, the curved free surface must be a function of r^2, which is a parabola. This verifies that the lowest point of curved free surface locates at $r = 0$, which is used previously. The quantitative determination of free surface is given in the following.

It follows from Eq. (3.5.1) that

$$\mathrm{d}p = \nabla p \cdot \mathrm{d}\boldsymbol{r} = \frac{\partial p}{\partial r}\mathrm{d}r + \frac{\partial p}{\partial z}\mathrm{d}z = \rho\left(r\omega^2\mathrm{d}r - g\mathrm{d}z\right), \qquad \frac{\partial p}{\partial \theta} = 0, \tag{3.5.7}$$

where the second equation verifies the previous statement that the problem is axis-symmetric. Integrating the first equation gives rise to

$$p - p_{\mathrm{ref}} = p - p_0 = \frac{\rho(r\omega)^2}{2} - \rho g z, \tag{3.5.8}$$

where p_{ref} is the reference pressure, which is chosen to be the pressure at $r = 0$ and $z = 0$, corresponding to the atmospheric pressure p_0. Applying this and Eq. (3.5.8) to the curved free surface yields

$$z = \frac{(r\omega)^2}{2g}. \tag{3.5.9}$$

The location at which Eq. (3.5.9) assumes an extreme value is identified viz.,

$$\frac{\mathrm{d}z}{\mathrm{d}r} = \frac{r\omega^2}{g} = 0, \qquad \longrightarrow \qquad r = 0, \tag{3.5.10}$$

showing that the minimum value of z occurs at $r = 0$. It verifies again that the lowest point of curved free surface locates at the origin of cylindrical coordinate system.

As an illustration of the analysis, consider the cylindrical container in Fig. 3.9b again. It is required to determine (a) the angular speed ω_1, at which the lowest point of liquid free surface just touches the bottom of cylinder, if the liquid content remains unchanged during the rotation, and (b) the angular speed ω_2, at which there exists a circular plane with radius r_2 on the bottom of container which is not wetted by the liquid, if the liquid content is only 70% of its initial volume before rotation. The liquid elevations at the edge of cylinder in both cases should be determined equally.

For the first case, locate the origin of cylindrical coordinate system at the lowest point of the free surface of liquid. The liquid content before rotation is given by

$$V_0 = \pi r_0^2 h_0, \tag{3.5.11}$$

while the liquid content during rotation is obtained as

$$V_1 = \int_0^{r_0} 2\pi r \mathrm{d}r \frac{(r\omega_1)^2}{2g} = \int_0^{r_0} \frac{\pi\omega_1^2}{g} r^3 \mathrm{d}r = \frac{1}{4}\frac{\pi\omega_1^2}{g} r_0^4. \tag{3.5.12}$$

Since the liquid content remains unchanged, it follows that

$$V_0 = V_1, \qquad \longrightarrow \qquad \omega_1 = \frac{2}{r_0}\sqrt{h_0 g}, \tag{3.5.13}$$

and the elevation of liquid free surface on the container sidewall is determined as

$$z_1 = \frac{(\omega_1 r_0)^2}{2g} = 2h_0. \tag{3.5.14}$$

Thus, the difference in the elevations of liquid free surface on the container sidewalls before and after the rotation is $\Delta z = h_0$.

For the second case, locate the origin of cylindrical coordinate system still at the lowest point of liquid free surface, which is outside the container. The liquid content remaining in the container is assumed to be unchanged first, which is given by

$$V_2 = \int_0^{r_0} \frac{\pi\omega_2^2}{g} r^3 \mathrm{d}r - \int_0^{r_2} \frac{\pi\omega_2^2}{g} r^3 \mathrm{d}r - \pi(r_0^2 - r_2^2)\frac{(r_2\omega_2)^2}{2g} = \frac{1}{4}\frac{\pi\omega_2^2}{g}\left(r_0^2 - r_2^2\right)^2. \tag{3.5.15}$$

Since the liquid content remaining in the container is only 70% of the original content, it follows immediately that

$$\frac{7}{10}\pi r_0^2 h_0 = \frac{1}{4}\frac{\pi\omega_2^2}{g}\left(r_0^2 - r_2^2\right)^2, \qquad \longrightarrow \qquad \omega_2 = \frac{r_0}{r_0^2 - r_2^2}\sqrt{\frac{14}{5}h_0 g}. \tag{3.5.16}$$

The elevation of liquid free surface on the container sidewall is then obtained as

$$z_2 = \frac{(r_0\omega_2)^2}{2g} = \frac{7}{5}\frac{r_0^4}{(r_0^2 - r_2^2)^2}h_0, \tag{3.5.17}$$

with which the difference in the elevation of liquid free surface on the container sidewall is given by

$$\Delta z = \left(\frac{7}{5}\frac{r_0^4}{(r_0^2 - r_2^2)^2} - 1\right)h_0, \tag{3.5.18}$$

which must be greater than zero, for $(r_0^2 - r_2^2) < r_0^2$.

3.6 Exercises

3.1 An inclined-tube manometer is connected to a reservoir, as shown in the figure, in which the initial liquid free surface is displayed by the dashed line. Let a pressure difference Δp be applied on the reservoir. Derive a general expression of the liquid deflection ℓ, and an expression of the manometer sensitivity which depends on D, d, θ, and s, the specific gravity of liquid.

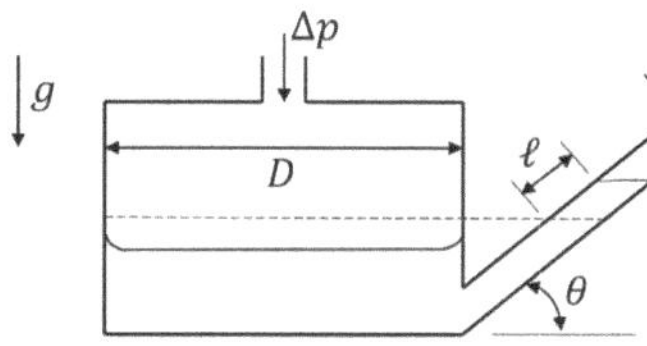

3.2 It is supposed that you have a barometer and a thermometer. How to use these instruments to estimate the height of the Khalifa Tower (The Burj Khalifa) in Dubai? State the possible methods to estimate the height by using these instruments as many as you can.

3.3 A rectangular gate with width b is shown in the figure, which is connected to a mass M via a rope and is in contact with a liquid with density ρ. Determine the liquid depth d so that the gate is in a static equilibrium state.

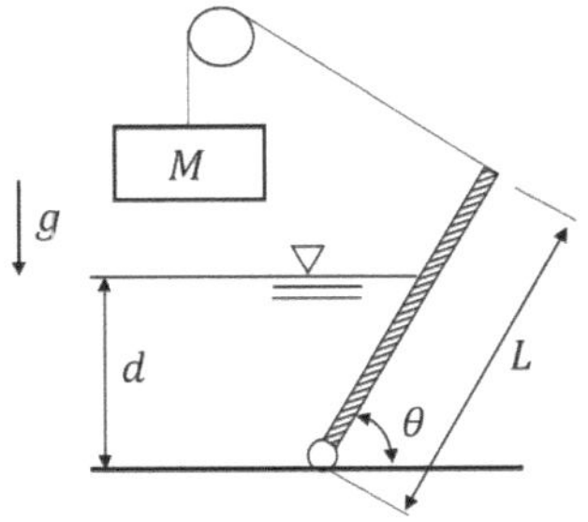

3.4 The gate shown in the figure has a width b and is pivoted at O. Determine (a) the magnitudes of the horizontal and vertical components of hydrostatic force, and their moments with respect to O, and (b) the magnitude of the horizontal force pointing to the right, which needs to be applied at point A in order to hold the gate in position.

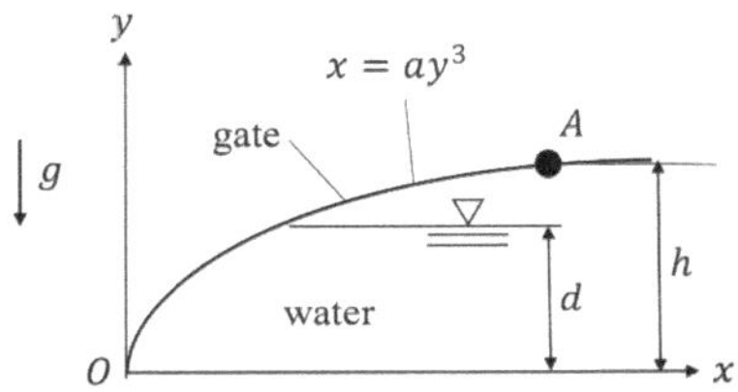

3.5 A gate in the shape of a quarter cylinder, shown in the figure, is hinged at A and sealed at B. The gate is rectangular and has a width b. Determine the force at point B if the gate is made of a material with specific gravity s.

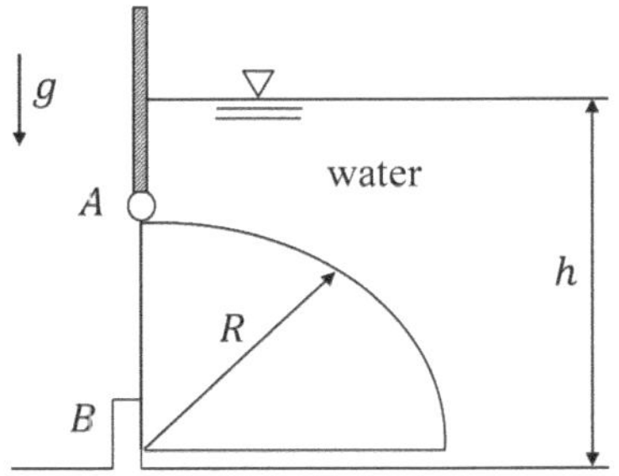

3.6 What diameter of a clean glass tube should be, if the rise of a pure water at 20 °C due to the capillary effect in this tube is less than 1 mm?

3.7 Consider a submerged body with thickness b in a still liquid, as shown in the figure. Show that the center of buoyancy coincides to the centroid of displace volume in the upper figure, and to the center of mass of displaced volume in the lower figure.

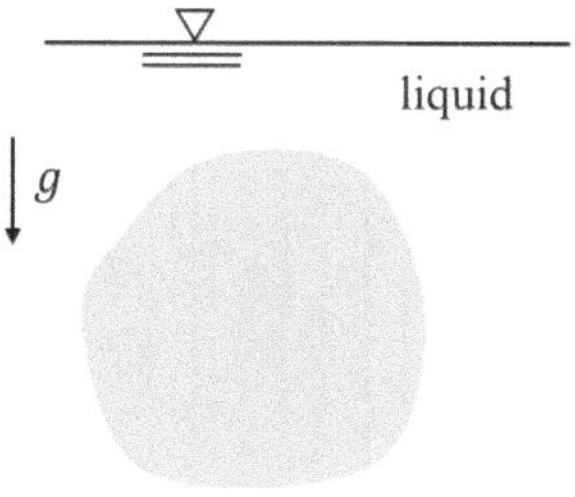

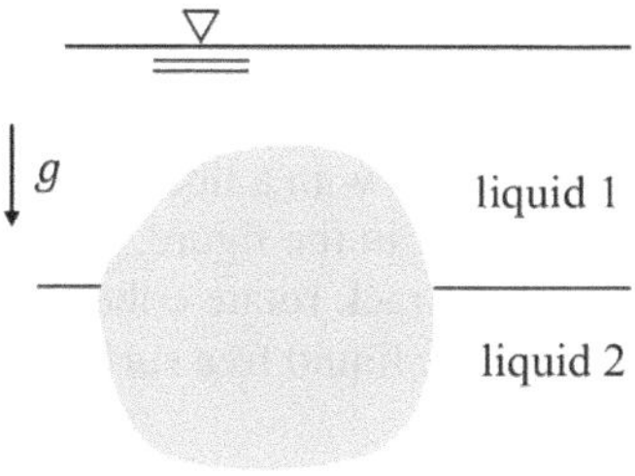

3.8 A cubic box with mass M and volume V is allowed to sink in water, as shown in the figure. A circular rod with length L and diameter d is attached to the cubic box and the wall. Determine the equilibrium angle θ if the mass of rod is m.

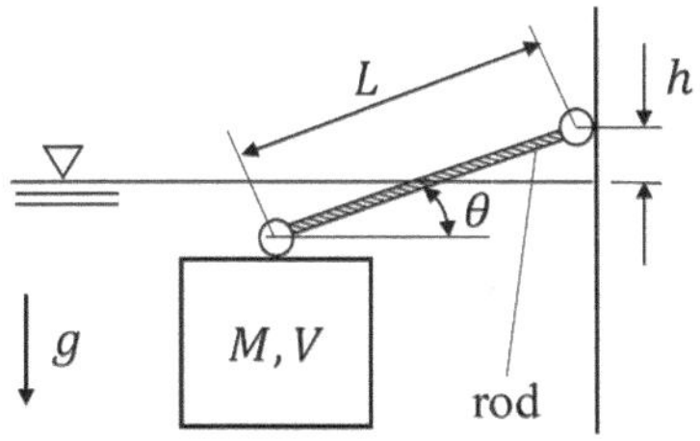

3.9 A rectangular container filled with water undergoes a constant acceleration down an inclined plane, as shown in the figure. Determine the slope of water free surface in terms of the given rectangular coordinates $\{x, y\}$.

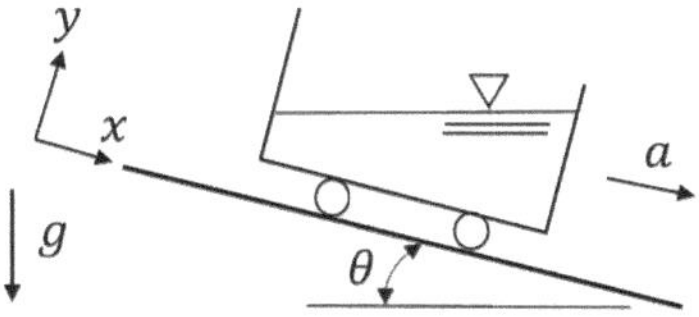

3.10 The U-tube shown in the figure is filled with a liquid. It is sealed at point A and open to the atmosphere at point D. The tube is rotating with respect to the vertical axis AB. Determine the maximum angular speed if the minimum liquid pressure reaches to its vapor pressure p_v.

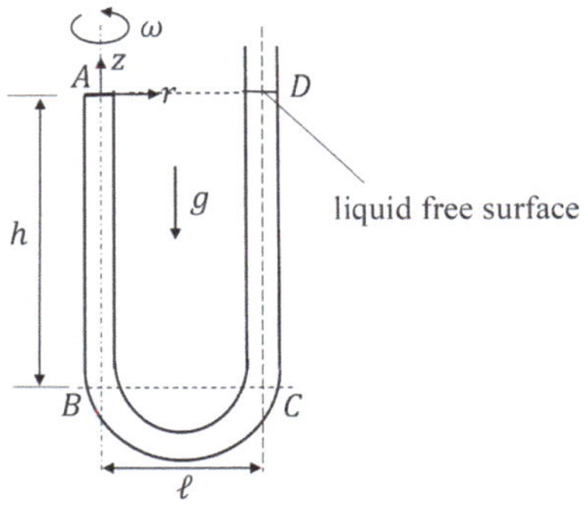

3.11 A rectangular container filled with a liquid slides freely down an inclined plane by a sliding track, as shown in the figure. During sliding, the container, liquid, inclined plane and sliding track rotate coherently with respect to the axis AB. Obtain an expression for the liquid free surface in terms of the fixed coordinates $\{x, y\}$.

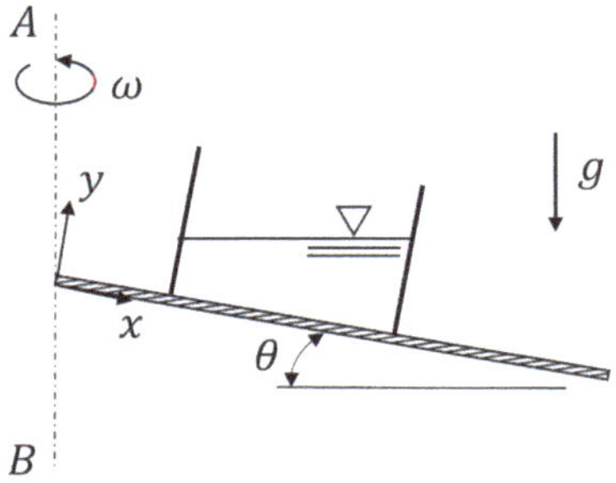

Further Reading

Y.A. Cengel, J.M. Cimbala, *Fluid Mechanics: Fundamentals and Applications*, 3rd edn. (McGraw-Hill, New York, 2014)

S. Chandrasekhar, *Hydrodynamic and Hydromagnetic Stability* (Dover, New York, 1961)

D.F. Elger, B.C. Williams, C.T. Crowe, J.A. Roberson, *Engineering Fluid Mechanics*, 10th edn. (Wiley, Singapore, 2014)

R.W. Fox, P.J. Pritchard, A.T. McDonald, *Introduction to Fluid Mechanics*, 7th edn. (Wiley, New York, 2009)

B.R. Munson, D.F. Young, T.H. Okiishi, *Fundamentals of Fluid Mechanics*, 3rd edn. (Wiley, New York, 1990)

P. Oswald, *Rheophysics: The Deformation and Flow of Matter* (Cambridge University Press, Cambridge, 2009)

R.H.F. Pao, *Fluid Mechanics* (Wiley, New York, 1961)

L. Prandtl, O.G. Tietjens, *Fundamentals of Hydro- and Aeromechanics* (Dover, New York, 1934)

L. Prandtl, O.G. Tietjens, *Applied Hydro- and Aeromechanics* (Dover, New York, 1934)

A.J. Smith, *A Physical Introduction to Fluid Mechanics* (Wiley, New York, 2000)

J. Spurk, *Fluid Mechanics* (Springer, Berlin, 1997)

C.S. Yih, *Fluid Mechanics: A Concise Introduction to The Theory* (McGraw-Hill, New York, 1969)

Flow Kinematics

4

The topics which may be deduced about the nature of a flowing fluid without referring to the dynamics of continuum are explored in this chapter. First, the flow lines embracing streamlines, pathlines, and streaklines are discussed. These flow lines are not only useful for flow visualization, but also supply means by which solutions to governing equations of flow problems may be interpreted physically. Second, the concepts of circulation and vorticity are introduced, with their full usefulness becoming apparent in discussing the balance equations of fluid motion. Streamline and vorticity lead to the concepts of stream tube and stream filament, and vortex tube and vortex filament, respectively. Discussions on the kinematics of stream and vortex filaments are provided at the end, which consists part of the Helmholtz equations. The other part, i.e., the dynamics of vorticity, will be discussed in Sect. 8.1.

4.1 Flow Lines

4.1.1 Streamline

Streamlines are those curves whose tangents are everywhere parallel to the fluid velocities at all points. Let x_i be the coordinate with the corresponding orthonormal base $\boldsymbol{e}_i$, with which a streamline is defined to be a curve satisfying

$$\frac{\mathrm{d}x_1}{u_1} = \frac{\mathrm{d}x_2}{u_2} = \frac{\mathrm{d}x_3}{u_3} = \mathrm{d}s, \qquad \frac{\mathrm{d}x_i}{\mathrm{d}s} = u_i(\boldsymbol{x}, t), \tag{4.1.1}$$

where ds is an infinitesimal arc length from a reference point. Since a flow depends in general on time, a streamline becomes meaningful when it is referred to a specific time instant. That is, integration of Eq. (4.1.1) ought to be accomplished at a fixed value of t, and the expression of a streamline may be obtained as

© Springer International Publishing AG 2019
C. Fang, *An Introduction to Fluid Mechanics*, Springer Textbooks in Earth Sciences, Geography and Environment, https://doi.org/10.1007/978-3-319-91821-1_4

$$x_i = x_i(\boldsymbol{x}^0, t = t_0, s), \tag{4.1.2}$$

in which $\boldsymbol{x}^0$ marks the point through which the streamline passes at $t = t_0$. Conventionally, $s = 0$ is chosen at $\boldsymbol{x} = \boldsymbol{x}^0$. As s takes all real values, the required streamline is traced out. It follows equally from Eq. (4.1.1) that a streamline behaves like a "wall" in a flow field, and fluid particles are unable to penetrate a streamline.

4.1.2 Pathline

A *pathline* is a curve which is traced out in time by a prescribed identifiable fluid particle with fixed mass as it moves, and corresponds exactly to the moving trajectory of a mass particle in Newtonian mechanics. It is described mathematically by

$$\frac{\mathrm{d}x_i}{\mathrm{d}t} = u_i(\boldsymbol{x}, t). \tag{4.1.3}$$

Integrating this equation yields an expression of a pathline given viz.,

$$x_i = x_i(\boldsymbol{x}^0, t), \tag{4.1.4}$$

in which $\boldsymbol{x}^0$ marks the position which is occupied by the prescribed identifiable fluid particle at $t = 0$.

4.1.3 Streakline

A *streakline* is a curve which is traced out by a neutrally buoyant marker fluid which is continuously injected into a flow field at a fixed point in space. In other words, a streakline at a specific time t is a curve connecting all fluid particles which have passed a fixed point in space in an earlier time τ. The mathematical description of a streakline is the same as that of a pathline of a single fluid particle, i.e.,

$$\frac{\mathrm{d}x_i}{\mathrm{d}t} = u_i(\boldsymbol{x}, t). \tag{4.1.5}$$

Integrating Eq. (4.1.5) yields

$$x_i = x_i(\boldsymbol{x}^0, t, \tau), \tag{4.1.6}$$

in which $\tau \leq t$ and $\boldsymbol{x}^0$ marks the point that has been passed through by a fluid particle in an earlier time τ. Taking all possible values in the range of $-\infty \leq \tau \leq t$ gives the positions of all fluid particles on the streakline, yielding the streakline through the point $\boldsymbol{x} = \boldsymbol{x}^0$ at time t.

Essentially, three flow lines are different from one another if a given flow field is unsteady. Equation (4.1.1) can be expressed alternatively as

$$\frac{\mathrm{d}x_i}{\mathrm{d}t} = \frac{\mathrm{d}s}{\mathrm{d}t} u_i. \tag{4.1.7}$$

Comparing this equation with Eqs. (4.1.3) and (4.1.5) indicates that a streamline, a pathline, and a streakline are different from one another in an unsteady flow, even though they may pass through the same point in space at the same initial time, for

$\mathrm{d}s/\mathrm{d}t$ does not vanish in general and depends on time. However, if the flow is steady, the velocity components u_i are constant with respect to time. With this, integrating Eq. (4.1.7) yields

$$x_i - x_i^0 = \int_0^t C\frac{\mathrm{d}s}{\mathrm{d}t}\mathrm{d}t = \int_0^s C\mathrm{d}s, \tag{4.1.8}$$

for a streamline, where C is a constant. Integrating Eqs. (4.1.3) and (4.1.5) also yields respectively

$$x_i - x_i^0 = \int_0^t C\mathrm{d}t, \qquad x_i - x_i^0 = \int_\tau^t C\mathrm{d}t, \tag{4.1.9}$$

for a pathline and a streakline. Equations (4.1.8) and (4.1.9) are mathematically identical if it is required that the streamline and pathline pass through x_i^0 at $t = 0$ (hence $s = 0$) and the streakline passes through x_i^0 at $t = \tau$. This indicates that the streamline, pathline, and streakline passing through the same point in space are the same if the flow is steady.

As an illustration of the concepts of streamlines, streaklines, and pathlines, consider a two-dimensional flow field described by

$$u = x(1 + 2t), \qquad v = y, \tag{4.1.10}$$

in the (xy)-plane. It is required to determine (a) the streamline which passes the point $(x, y) = (1, 1)$ at $t = 0$, (b) the pathline of the particle locating at the same point at $t = 0$, and (c) the streakline of the fluid particles passing the same point at $t = \tau$. For the streamline, it follows that

$$\frac{\mathrm{d}x}{\mathrm{d}s} = x(1 + 2t), \qquad \frac{\mathrm{d}y}{\mathrm{d}s} = y. \tag{4.1.11}$$

Integrating these two equations yields

$$x = C_1 \exp[(1 + 2t)s], \qquad y = C_2 \exp(s), \tag{4.1.12}$$

where the integration constants C_1 and C_2 are determined by using the initial condition given by $(x, y) = (1, 1)$ at $s = 0$, leading to $C_1 = C_2 = 1$. Thus, the parametric equations of streamline become

$$x = \exp[(1 + 2t)s], \qquad y = \exp(s), \tag{4.1.13}$$

which describe the streamline passing through point $(x, y) = (1, 1)$. Since in an unsteady flow it is meaningful to discuss a streamline at a specific instant, applying $t = 0$ to the above equations gives

$$x = \exp(s), \qquad y = \exp(s), \qquad \longrightarrow \qquad x = y, \tag{4.1.14}$$

which is the required streamline passing through point $(x, y) = (1, 1)$ at $t = 0$.

For the pathline, it follows that

$$\frac{\mathrm{d}x}{\mathrm{d}t} = x(1 + 2t), \qquad \frac{\mathrm{d}y}{\mathrm{d}t} = y, \tag{4.1.15}$$

which is integrated to obtain

$$x = C_1 \exp[t(1 + t)], \qquad y = C_2 \exp(t), \tag{4.1.16}$$

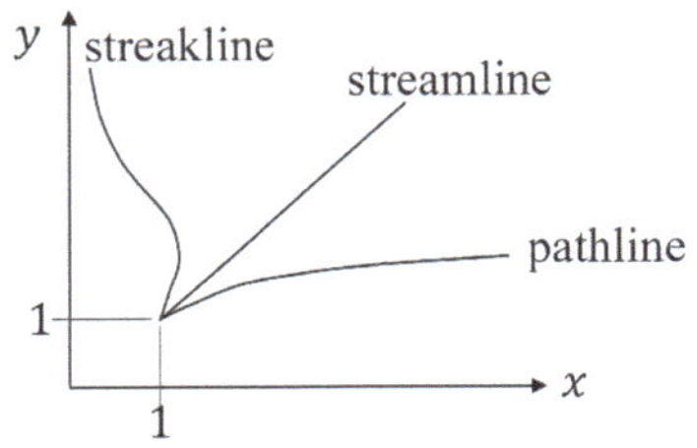

Fig. 4.1 Comparison of the streamline, pathline, and streakline passing through point $(x, y) = (1, 1)$ at $t = 0$ for a two-dimensional flow field described by $u = x(1 + 2t)$ and $v = y$

where the integration constants C_1 and C_2 are determined by using the initial condition $(x, y) = (1, 1)$ at $t = 0$, yielding $C_1 = C_2 = 1$. With these, the parametric equations of pathline become

$$x = \exp[t(1+t)], \quad y = \exp(t), \quad \longrightarrow \quad x = y^{1+\ln y}, \tag{4.1.17}$$

which describe the required pathline passing through point $(x, y) = (1, 1)$ at $t = 0$. For the streakline, the governing equation is the same as that of the pathline. Thus, Eq. (4.1.16) is valid for the streakline, except that it passes point $(x, y) = (1, 1)$ at an earlier time τ. Using this as the initial condition to determine the integration constants, the equations of streakline become

$$x = \exp[t(1+t) - \tau(1+\tau)], \quad y = \exp(t - \tau). \tag{4.1.18}$$

These parametric equations are valid for the streakline passing through point $(x, y) = (1, 1)$ for all times t. Applying $t = 0$ to the above equations results in

$$x = \exp[-\tau(1+\tau)], \quad y = \exp(-\tau), \quad \longrightarrow \quad x = y^{1-\ln y}, \tag{4.1.19}$$

which describe the required streakline passing through point $(x, y) = (1, 1)$ at $t = 0$. The obtained results are displayed graphically in Fig. 4.1.

4.2 Circulation and Vorticity

The *circulation* Γ contained within a closed contour C in a fluid is defined by the line integral around the contour of the velocity components, i.e.,[1]

$$\Gamma \equiv \oint_C \boldsymbol{u} \cdot \mathrm{d}\boldsymbol{\ell} = \oint_C u_i \mathrm{d}\ell_i, \tag{4.2.1}$$

with $\mathrm{d}\boldsymbol{\ell}$ representing an element of C. The integration ought to be conducted counterclockwise around C, giving rise to a positive value of Γ if the integral is positive. The *vorticity* $\boldsymbol{\zeta}$ of a fluid element is defined as the curl of velocity given by

$$\boldsymbol{\zeta} \equiv \nabla \times \boldsymbol{u}. \tag{4.2.2}$$

[1] The velocity vector is locally tangent to the contour.

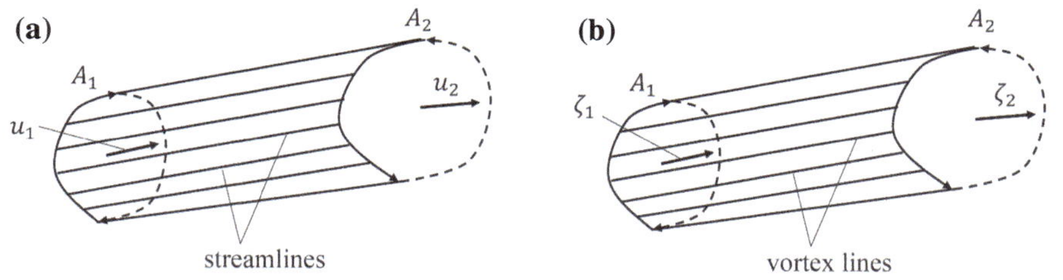

Fig. 4.2 Stream and vortex tubes with varying cross-sections in a fluid. **a** Stream tube. **b** Vortex tube

It should be noted that a fluid element may travel on a circular streamline with vanishing vorticity, for the vorticity is proportional to the curl of velocity of a fluid element about its principal axis, not that of the center of gravity of the element about some reference point. The free vortex described in Exercise 4.5 is an example.

It follows from the Stokes theorem that Eq. (4.2.1) can be recast alternatively as

$$\Gamma = \oint_C \boldsymbol{u} \cdot \mathrm{d}\boldsymbol{\ell} = \int_A (\nabla \times \boldsymbol{u}) \cdot \boldsymbol{n}\, \mathrm{d}a = \int_A \boldsymbol{\zeta} \cdot \boldsymbol{n}\, \mathrm{d}a, \tag{4.2.3}$$

where A is the surface defined by the closed contour C, around which the circulation is conducted, and $\boldsymbol{n}$ is the unit normal to A. Equation (4.2.3) shows that for an arbitrarily chosen contour C with the corresponding enclosing surface A, $\Gamma = 0$ if $\boldsymbol{\zeta} = \boldsymbol{0}$ and vice versa. A flow is termed *irrotational* if $\boldsymbol{\zeta} = \boldsymbol{0}$ and termed *rotational* if $\boldsymbol{\zeta} \neq \boldsymbol{0}$. Vorticity and circulation are useful concepts in calculating the lift of an airfoil, and classifications of rotational and irrotational flows provide an important simplification to the shear stresses of a fluid. Both topics will be discussed in Sect. 7.1.

4.3 Stream and Vortex Tubes

A *stream tube* is defined as a region in a fluid whose sidewalls are made up of streamlines. For example, let C be a closed contour in a flow field. At each point on C, a streamline passes through. By considering all points on C, series of streamlines are obtained, which form a surface. This surface and the two end cross-sections form a stream tube, as shown in Fig. 4.2a, in which the cross-sectional area A_1 is in general different from the cross-sectional area A_2 in a finite-length stream tube. A stream tube is called a *stream filament* if its cross-sectional area is infinitesimal.

A similar concept of the streamline is the *vortex line*, which is defined as a curve whose tangents are everywhere parallel to the fluid vorticities at all points. Thus, for any closed contour C in a flow field, each point on C has a vortex line passing through it. A *vortex tube* for the contour is defined as the region enclosed by the vortex lines with two end cross-sections, as shown in Fig. 4.2b. As similar to a stream tube, the cross-sectional areas of a vortex tube (e.g. A_1 and A_2) are different at different locations. A *vortex filament* is obtained if the cross-sectional area of a vortex tube is infinitesimal.

4.4 Kinematics of Stream and Vortex Tubes

Consider the stream tube shown in Fig. 4.2a as the integral control-volume with finite length and $A_1 \neq A_2$ in general. The *flow rate* Q at a specific cross-section A of the stream tube is defined as the fluid volume crossing A per unit time, viz.,

$$Q \equiv \int_A \boldsymbol{u} \cdot \mathrm{d}\boldsymbol{a}, \tag{4.4.1}$$

with a negative and a positive sign representing an intake and a discharged volume flow rates, respectively. Since the volume of stream tube remains fixed, if the fluid passing through the stream tube is assumed to be incompressible, applying Eq. (4.4.1) to cross-sections A_1 and A_2 yields

$$\int_{A_1} \boldsymbol{u} \cdot \mathrm{d}\boldsymbol{a} + \int_{A_2} \boldsymbol{u} \cdot \mathrm{d}\boldsymbol{a} = Q_1 + Q_2 = 0, \tag{4.4.2}$$

which is a special form of the integral conservation of mass, termed the *continuity equation*. This equation indicates that for an incompressible fluid, the fluid volumes entering into a control-volume per unit time must be the same as those leaving the control-volume. This result holds equally if the control-volume is associated with multi-intake and discharged surfaces. The same analysis can be extended to a stream filament, giving rise to

$$\int_A \boldsymbol{u} \cdot \mathrm{d}\boldsymbol{a} = \int_V (\nabla \cdot \boldsymbol{u})\, \mathrm{d}v = 0, \qquad \longrightarrow \qquad \nabla \cdot \boldsymbol{u} = 0, \tag{4.4.3}$$

indicating that the velocity of an incompressible flow is *divergent-free*. Equation (4.4.3)$_2$ is a special form of the differential conservation of mass. A detailed discussion on the integral and differential conservations of mass will be provided in Sect. 5.3.1.

A similar analysis can be made to the vortex tube shown in Fig. 4.2b.[2] It follows from Eq. (4.2.2) that

$$\nabla \cdot \boldsymbol{\zeta} = 0, \tag{4.4.4}$$

indicating that the vorticity is equally divergent-free, which implies that there can be no sources and sinks of vorticity in the fluid itself. Vortex lines must either form closed loops or terminate on the boundaries of the fluid, which may be either solid or free surfaces. It follows from Eq. (4.4.4) that

$$\int_V (\nabla \cdot \boldsymbol{\zeta})\, \mathrm{d}v = 0, \qquad \longrightarrow \qquad \int_A \boldsymbol{\zeta} \cdot \mathrm{d}\boldsymbol{a} = 0, \tag{4.4.5}$$

[2] The present analysis forms part of the *Helmholtz theorems of vorticity*. Hermann Ludwig Ferdinand von Helmholtz, 1821–1894, a German physicist, with contributions to several scientific fields. The largest German association of research institutions, the Helmholtz Association, is named after him.

where V and A are the volume and entire surface of a vortex tube, respectively. Applying Eq. (4.4.5)$_2$ to a vortex tube with two end cross-sectional areas A_1 and A_2 yields

$$\int_{A_1} \zeta \cdot \mathrm{d}\boldsymbol{a} + \int_{A_2} \zeta \cdot \mathrm{d}\boldsymbol{a} = 0, \tag{4.4.6}$$

which, by using Eq. (4.2.3), is expressed alternatively as

$$\Gamma_1 + \Gamma_2 = 0, \tag{4.4.7}$$

in which Γ_1 assumes a negative value and Γ_2 assumes a positive value. Equation (4.4.7) shows that the circulation around the limiting contour on A_1 is equal to that around A_2. Alternatively, the circulation at each cross-section of a vortex tube in the same. It means that if the cross-section of a vortex tube varies, the average value of vorticity across that cross-section must vary correspondingly, which is similar to the velocity variation implied by the continuity equation. Since vorticity is divergent-free, it follows that vortex tubes must terminate on themselves, at a solid boundary or at a free surface. For example, smoke rings terminate on themselves, while a vortex tube in a free surface flow may have one end at the solid boundary forming the bottom and the other end at the free surfaces.

4.5 Exercises

4.1 For a water flowing from a two-dimensional oscillating slit, its flow field is described by

$$\boldsymbol{u} = u_0 \sin\left[\omega\left(t - \frac{y}{v_0}\right)\right]\boldsymbol{i} + v_0\boldsymbol{j},$$

where u_0 and v_0 are constants. Determine (a) the streamlines passing through the origin at $t = 0$ and $t = \pi/2\omega$, (b) the pathlines of the fluid particles which locate at the origin at $t = 0$ and $t = \pi/2\omega$, and (c) the shape of the streakline that passes through the origin.

4.2 For most unsteady flows, the streamlines and streaklines are not the same. However, there are unsteady flows in which streamlines and streaklines are the same. Describe a flow field for which this statement holds.

4.3 Show that the streamlines and pathlines are the same in the flow field described by

$$u_i = \frac{x_i}{1+t}.$$

4.4 Consider a two-dimensional flow field with its velocity given by

$$\boldsymbol{u} = -\frac{x}{x^2+y^2}\boldsymbol{i} + \frac{y}{x^2+y^2}\boldsymbol{j}.$$

Calculate the circulation around the square contour, whose four vertices are given by $x = \pm 1$ and $y = \pm 1$. Furthermore, determine (a) the circulation and vorticity for the whole flow field and (b) the divergence of vorticity.

4.5 Determine the vorticity for the following two flow fields: (a) $u_r = 0, u_\theta = a/r$, and (b) $u_r = 0$, $u_\theta = ar$, where a is a constant and r represents the radius. Determine also the circulations on the circular contour with radius $r = 1$ for the given two flow fields. The flow field in (a) is called a *free vortex*, while that in (b) is called a *forced vortex*.

Further Reading

R.S. Brodkey, *The Phenomena of Fluid Motions* (Dover, New York, 1967)
Y.A. Cengel, J.M. Cimbala, *Fluid Mechanics: Fundamentals and Applications*, 3rd edn. (McGraw-Hill, New York, 2014)
S. Chandrasekhar, *Hydrodynamic and Hydromagnetic Stability* (Dover, New York, 1961)
I.G. Currie, *Fundamental Mechanics of Fluids*, 2nd edn. (McGraw-Hill, Singapore, 1993)
D.F. Elger, B.C. Williams, C.T. Crowe, J.A. Roberson, *Engineering Fluid Mechanics*, 10th edn. (Wiley, Singapore, 2014)
R.W. Fox, P.J. Pritchard, A.T. McDonald, *Introduction to Fluid Mechanics*, 7th edn. (Wiley, New York, 2009)
K. Hutter, Y. Wang, *Fluid and Thermodynamics. Volume 1: Basic Fluid Mechanics* (Springer, Berlin, 2016)
H. Lamb, *Hydrodynamics*, 6th edn. (Dover, New York, 1945)
B.R. Munson, D.F. Young, T.H. Okiishi, *Fundamentals of Fluid Mechanics*, 3rd edn. (Wiley, New York, 1990)
R.H.F. Pao, *Fluid Mechanics* (Wiley, New York, 1961)
A.J. Smith, *A Physical Introduction to Fluid Mechanics* (Wiley, New York, 2000)
C.S. Yih, *Fluid Mechanics: A Concise Introduction to the Theory* (McGraw-Hill, New York, 1969)

5 Balance Equations

The motions of a fluid can be described by using the time rates of change of physical variables defined on the fluid. To reach this end, within the continuum hypothesis, fluid as a continuum should a priori be assumed and the fundamentals of continuum mechanics need to be introduced, including the concepts of material body, reference and present configurations, and motion of a fluid element. Based on these, the material derivative of physical variable and deformation of a material may be defined to obtain the expressions of velocity and acceleration of a fluid element.

The time rate of change of a physical variable is accomplished by formulating a general balance statement in relation with possible external excitations, which may cause a variation in the physical variable in a process. The formulations are conducted separately for a fluid as a whole and a fluid element, giving rise respectively to the global and local balance equations of physical variable. The global balance equations are the balance statements corresponding to the integral approach, while the local balance equations correspond to the differential approach. Specifically, twofold balance statements are used for the physical variables of mass, linear momentum, angular momentum, energy and entropy of a fluid to obtain the global and local balance equations of these physical variables. These balance equations are universal because they are nothing else than the physical laws and are valid for all materials. Selected problems are explored to illustrate the applications of the global and local balance equations of physical laws.

Although the global and local balance equations of physical laws are universal for all materials, different materials behave differently even under the same circumstance. The difference in the material responses is accounted for by using the concept of material or constitutive equations. Material equations can be formulated either experimentally or theoretically; however, in most cases individual approach is insufficient. The specific rules which need to be followed in the theoretical formulation are outlined. With the aid of theoretical formulation supplemented by experimental outcomes, the material equations of the Newtonian fluids are obtained and the local

© Springer International Publishing AG 2019

C. Fang, *An Introduction to Fluid Mechanics*, Springer Textbooks in Earth Sciences, Geography and Environment, https://doi.org/10.1007/978-3-319-91821-1_5

balance equations of physical laws for the Newtonian fluids are derived. Selected problems of motions of the Newtonian fluids are discussed to illustrate the applications of local balances of mass and linear momentum.

With the knowledge of classical thermodynamics, the discussions on the balance statements of energy and entropy will be completed in Sect. 11.6.

5.1 Motion of a Fluid Continuum

5.1.1 Material Body, Reference and Present Configurations

Within the continuum hypothesis, a fluid is considered a *material body* $\mathcal{B}$, which consists of an infinite number of *fluid elements*, $\mathcal{X}$, viz.,

$$\mathcal{B} = \{\mathcal{X}\}, \tag{5.1.1}$$

where the symbol "$\{\cdot\}$" represents a set, and the notation $\{\mathcal{X}\}$ means a set of $\mathcal{X}$. A material body is an abstract concept, which does not necessary correspond to a real physical material. In order to describe the motion of a fluid as a material body, all fluid elements must be allocated a position. It is accomplished by using the concept of *position vector*. For every fluid element $\mathcal{X} \in \mathcal{B}$, there exists a vector space $\mathcal{V}_R^3$, so that there is assigned a position vector $\boldsymbol{X}$ to every fluid element, viz.,

$$\boldsymbol{X} : \mathcal{B} \longrightarrow \mathcal{V}_R^3, \qquad \mathcal{X} \mapsto \boldsymbol{X} = \boldsymbol{X}(\mathcal{X}), \tag{5.1.2}$$

with which each individual fluid element is identified by using its own position vector $\boldsymbol{X}$.

The *reference configuration* of $\mathcal{B}$, denoted by $\mathcal{B}^R$, is defined by the set of all position vectors $\boldsymbol{X}$s defined in itself, i.e.,

$$\mathcal{B}^R \equiv \{\boldsymbol{X}(\mathcal{X}) \mid \mathcal{X} \in \mathcal{B}\}. \tag{5.1.3}$$

Without loss of generality, the reference configuration of a fluid body is its extent in the physical space at a fixed or initial time. The position vector $\boldsymbol{X}$ in $\mathcal{B}^R$ is represented by using the *material coordinates*, viz.,

$$\boldsymbol{X} = X_I \boldsymbol{e}_I, \tag{5.1.4}$$

where $\boldsymbol{e}_I$ represents the orthonormal base of material coordinate system.

When a fluid body $\mathcal{B}$ moves or deforms with time, its fluid element $\mathcal{X}$ takes a new position at time $t \in \mathcal{R}^+$. To every fluid element at a given time t, there exists a vector space $\mathcal{V}_t^3$, such that there is assigned a position vector $\boldsymbol{x}$ to every fluid element, i.e.,

$$\boldsymbol{x} : \mathcal{B} \longrightarrow \mathcal{V}_t^3, \qquad \mathcal{X} \mapsto \boldsymbol{x} = \boldsymbol{x}(\mathcal{X}, t), \tag{5.1.5}$$

followed which $\mathcal{X}$ can be identified by using $\boldsymbol{x}$. The set of all position vectors $\boldsymbol{x}$s defines the *present configuration* or *actual configuration* of $\mathcal{B}$ at time t, denoted by $\mathcal{B}^P$, which is given by

$$\mathcal{B}^P \equiv \left\{\boldsymbol{x}(\mathcal{X}, t) \mid \mathcal{X} \in \mathcal{B},\ t \in \mathcal{R}^+\right\}. \tag{5.1.6}$$

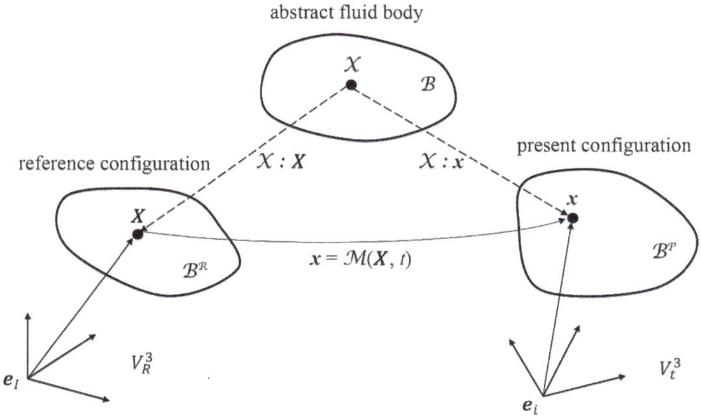

Fig. 5.1 Relation between an abstract fluid body $\mathcal{B}$, its reference configuration $\mathcal{B}^R$ and present configuration $\mathcal{B}^P$

Similarly, the present configuration of $\mathcal{B}$ is its extent in the physical space at time t, with the position vector $\boldsymbol{x}$ of each fluid element $\mathcal{X}$ expressed by using the *spatial coordinates* given by

$$\boldsymbol{x} = x_i \boldsymbol{e}_i, \tag{5.1.7}$$

where $\boldsymbol{e}_i$ represents the orthonormal base of spatial coordinate system. In short, the reference configuration of a fluid body is its initial occupied space before the motion, while the present configuration is its occupied space at time t after the motion has taken place, and two orthonormal bases $\boldsymbol{e}_I$ and $\boldsymbol{e}_i$ are different in general. Figure 5.1 illustrates the concepts of material body and its reference and present configurations.

5.1.2 Motion and Physical Variable

The *motion* of a fluid body is defined as the succession of positions that a fluid element $\mathcal{X}$ transverses with time. Since $\mathcal{X}$ assumes the position $\boldsymbol{X}$ in the reference configuration and $\boldsymbol{x}$ in the present configuration at a specific time $t \in \mathcal{R}^+$, the motion $\boldsymbol{\mathcal{M}}$ of a fluid element is described by the mapping

$$\boldsymbol{\mathcal{M}} : \mathcal{B}^R \times \mathcal{R}^+ \longrightarrow \mathcal{B}^P, \qquad (\boldsymbol{X}, t) \mapsto \boldsymbol{x} = \boldsymbol{\mathcal{M}}(\boldsymbol{X}, t), \tag{5.1.8}$$

in which the position of a fluid element in the present configuration is expressed as a motion function which depends on its position in the reference configuration and time. The motion $\boldsymbol{\mathcal{M}}$ in Eq. (5.1.8) is assumed to be continuously differentiable in the entire fluid body, so that the mapping is invertible, with its inverse given by

$$\boldsymbol{X} = \boldsymbol{\mathcal{M}}^{-1}(\boldsymbol{x}, t), \tag{5.1.9}$$

indicating that all positions $\boldsymbol{X}$s in $\mathcal{B}^R$ can be determined, provided that all positions $\boldsymbol{x}$s in $\mathcal{B}^P$ and the motion $\boldsymbol{\mathcal{M}}(\boldsymbol{X}, t)$ are known, and vice versa, as shown in Fig. 5.1.

Let $\aleph$ be any physical variable defined on an identifiable fluid element $\mathcal{X}$ at a certain time t given by[1]

$$\aleph \equiv \aleph(\mathcal{X}, t), \tag{5.1.10}$$

which can be expressed by using either the material coordinates in $\mathcal{B}^R$ or spatial coordinates in $\mathcal{B}^P$. It follows from Eqs. (5.1.2) and (5.1.5) that

$$\aleph = \aleph(\mathcal{X}, t) = \aleph(\boldsymbol{X}^{-1}(\boldsymbol{X}), t) = \aleph^R(\boldsymbol{X}, t) = \aleph(\boldsymbol{x}^{-1}(\boldsymbol{x}, t), t) = \aleph^P(\boldsymbol{x}, t),$$

with $\aleph^R$ and $\aleph^P$ the expressions of $\aleph$ in terms of the material and spatial coordinates, respectively. Each expression can be transformed into the other, and both $\aleph^R$ and $\aleph^P$ assume the same value, although they may have different mathematical forms. This statement can be verified by using the motion $\mathcal{M}$. It follows from Eqs. (5.1.8) and (5.1.9) that

$$\begin{aligned} \aleph^R(\boldsymbol{X}, t) &= \aleph^P\left(\mathcal{M}^{-1}(\boldsymbol{x}, t), t\right) = \aleph^P(\boldsymbol{x}, t), \\ \aleph^P(\boldsymbol{x}, t) &= \aleph^R\left(\mathcal{M}(\boldsymbol{X}, t), t\right) = \aleph^R(\boldsymbol{X}, t), \end{aligned} \tag{5.1.11}$$

provided that $\mathcal{M}$ is invertible. Representing $\aleph$ as a function of the material coordinates and time, i.e., $\aleph = \aleph^R(\boldsymbol{X}, t)$, is called the *Lagrangian description*, while $\aleph = \aleph^P(\boldsymbol{x}, t)$, in which $\aleph$ is expressed in terms of the spatial coordinates and time, is called the *Eulerian description*. Nevertheless,

$$\frac{\partial \aleph^R}{\partial X_I} \neq \frac{\partial \aleph^P}{\partial x_i}, \tag{5.1.12}$$

holds essentially, for both $\aleph^R$ and $\aleph^P$ are *coordinate dependent* and may have different mathematical forms with respect to different coordinates. In the following, the superscript R is used to denote that the indexed quantity is expressed by using the Lagrangian description (the material coordinates defined in $\mathcal{B}^R$), and the superscript P is used to denote that the indexed quantity is expressed by using the Eulerian description (the spatial coordinates defined in $\mathcal{B}^R$ at time t). These denotations are used throughout the chapter, unless stated otherwise.

Among the variables of a fluid body is its mass most important, based on which other physical variables could be defined. Within the continuum hypothesis, the mass per unit volume assigned to every fluid element $\mathcal{X}$ is termed the *density* or *mass density*, which is a positive quantity and is denoted by ρ^R and ρ^P in the Lagrangian and Eulerian descriptions, respectively. With these, the mass m of a fluid body $\mathcal{B}$ is then given by

$$m = \int_{V^R} \rho^R \, \mathrm{d}v^R = \int_{V^P} \rho^P \, \mathrm{d}v^P, \tag{5.1.13}$$

where V^R and V^P are the volumes occupied by $\mathcal{B}$ in $\mathcal{B}^R$ and $\mathcal{B}^P$, respectively, with the corresponding infinitesimal volume elements denoted by $\mathrm{d}v^R$ and $\mathrm{d}v^P$. Equation (5.1.13) can be extended to define other physical variables and implies that all extensive variables associated with $\mathcal{B}$ are additive, which is called the *additive assumption*.

[1]Specifically, $\aleph$ is a physical variable per unit mass of the fluid element, called the *specific variable*.

5.1.3 Material Derivative

The *material derivative* of a physical variable $\aleph$ is nothing else than its time rate of change. Since $\aleph$ is defined on a fixed identifiable fluid element $\mathcal{X}$ and is expressed differently in the Lagrangian and Eulerian descriptions, it follows that

$$\dot{\aleph} \equiv \frac{\mathrm{d}\aleph(\mathcal{X},t)}{\mathrm{d}t} = \frac{\mathrm{d}\aleph^R(\boldsymbol{X}^{-1}(\boldsymbol{X}),t)}{\mathrm{d}t} = \frac{\partial \aleph^R(\boldsymbol{X},t)}{\partial t}, \tag{5.1.14}$$

in the Lagrangian description, and

$$\dot{\aleph} \equiv \frac{\mathrm{d}\aleph(\mathcal{X},t)}{\mathrm{d}t} = \frac{\mathrm{d}\aleph^P(\boldsymbol{x}^{-1}(\boldsymbol{x},t),t)}{\mathrm{d}t} = \frac{\partial \aleph^P(\boldsymbol{x},t)}{\partial t} + \frac{\partial \aleph^P(\boldsymbol{x},t)}{\partial x_i}\dot{x}_i, \tag{5.1.15}$$

in the Eulerian description, where $\dot{x}_i$ is the velocity component, which will be discussed later. The material derivative of $\aleph$ in the Eulerian description is frequently expressed as

$$\dot{\aleph} = \frac{\mathrm{D}\aleph}{\mathrm{D}t} = \frac{\partial \aleph^P(\boldsymbol{x},t)}{\partial t} + \frac{\partial \aleph^P(\boldsymbol{x},t)}{\partial x_i}\dot{x}_i, \tag{5.1.16}$$

where $\mathrm{D}\alpha/\mathrm{D}t$ simply represents the material derivative of any quantity α in the Eulerian description to distinguish the symbol used for the time rate of change of α, i.e., $\mathrm{d}\alpha/\mathrm{d}t$, in the Lagrangian description.

5.1.4 Deformation Gradient

Consider the fluid body $\mathcal{B}$ in Fig. 5.1 again. Let a line element in $\mathcal{B}^R$ be denoted by $\mathrm{d}\boldsymbol{X}$. This line element moves or deforms via the motion $\boldsymbol{\mathcal{M}}$ and is represented by $\mathrm{d}\boldsymbol{x}$ in $\mathcal{B}^P$ at time t. The *deformation gradient* $\boldsymbol{F}$ is defined to satisfy

$$\mathrm{d}\boldsymbol{x} \equiv \boldsymbol{F}\mathrm{d}\boldsymbol{X}, \qquad \mathrm{d}x_i = F_{iI}\mathrm{d}X_I, \tag{5.1.17}$$

or alternatively,

$$\boldsymbol{F} = \frac{\partial \boldsymbol{x}}{\partial \boldsymbol{X}} = \frac{\partial \boldsymbol{\mathcal{M}}(\boldsymbol{X},t)}{\partial \boldsymbol{X}} = \mathrm{Grad}\,\boldsymbol{\mathcal{M}}(\boldsymbol{X},t), \qquad F_{iI} = \frac{\partial \mathcal{M}_i(\boldsymbol{X},t)}{\partial X_I}, \tag{5.1.18}$$

where "Grad" stands for the gradient operation with respect to X_I. It follows from Eq. (5.1.18) that $\boldsymbol{F}$ is a linear transformation which maps vectors in $\mathcal{B}^R$ onto vectors in $\mathcal{B}^P$, and is known as a *two-point tensor*. By using the index notation, $\boldsymbol{F}$ is expressed as[2]

$$\boldsymbol{F} = F_{iI}\left(\boldsymbol{e}_i\boldsymbol{e}_I\right), \tag{5.1.19}$$

where $\boldsymbol{e}_I$ and $\boldsymbol{e}_i$ are the orthonormal bases in the material and spatial coordinates, respectively. The term two-point tensor derives from the above equation, i.e., one of the two free indices of $\boldsymbol{F}$ comes from the material coordinates and the other comes

[2]The deformation gradient $\boldsymbol{F}$ can further be decomposed into a product of two tensors by using the *polar decomposition*, from which various strain measures of deformable materials can be defined.

from the spatial coordinates. While $\mathrm{d}\boldsymbol{x}$ in $\mathcal{B}^P$ is determined by using Eq. (5.1.17), $\mathrm{d}\boldsymbol{X}$ in $\mathcal{B}^R$ is simply determined by

$$\mathrm{d}\boldsymbol{X} = \boldsymbol{F}^{-1}\mathrm{d}\boldsymbol{x}, \qquad \mathrm{d}X_I = F_{Ii}^{-1}\mathrm{d}x_i, \tag{5.1.20}$$

with $\boldsymbol{F}^{-1} = F_{Ii}^{-1}(\boldsymbol{e}_I\boldsymbol{e}_i)$, if the motion $\boldsymbol{\mathcal{M}}$ is invertible, yielding a non-singular $\boldsymbol{F}$. Thus, the determinant of $\boldsymbol{F}$, which is denoted by J, is always non-vanishing, i.e.,

$$J \equiv \det \boldsymbol{F} \neq 0. \tag{5.1.21}$$

Let $\mathrm{d}\boldsymbol{a}$ and $\mathrm{d}v$ be an infinitesimal surface and volume elements in $\mathcal{B}$, respectively, which are expressed as $\mathrm{d}\boldsymbol{a}^R$ and $\mathrm{d}v^R$ in $\mathcal{B}^R$, and $\mathrm{d}\boldsymbol{a}^P$ and $\mathrm{d}v^P$ in $\mathcal{B}^P$. It follows that

$$\begin{aligned} \mathrm{d}\boldsymbol{a}^R &= \mathrm{d}\boldsymbol{X}_1 \times \mathrm{d}\boldsymbol{X}_2, \; \mathrm{d}v^R = \mathrm{d}\boldsymbol{X}_1 \cdot (\mathrm{d}\boldsymbol{X}_2 \times \mathrm{d}\boldsymbol{X}_3), \\ \mathrm{d}\boldsymbol{a}^P &= \mathrm{d}\boldsymbol{x}_1 \times \mathrm{d}\boldsymbol{x}_2, \; \mathrm{d}v^P = \mathrm{d}\boldsymbol{x}_1 \cdot (\mathrm{d}\boldsymbol{x}_2 \times \mathrm{d}\boldsymbol{x}_3), \end{aligned} \tag{5.1.22}$$

which, by using Eq. (5.1.17), may be reduced to

$$\mathrm{d}\boldsymbol{a}^P = J\boldsymbol{F}^{-\mathrm{T}}\mathrm{d}\boldsymbol{a}^R, \qquad \mathrm{d}v^P = J\mathrm{d}v^R, \tag{5.1.23}$$

which are the transformation rules of surface and volume elements between the Lagrangian and Eulerian descriptions. The derivations are left as an exercise.

5.1.5 Velocity, Acceleration, and Velocity Gradient

The *velocity* of a fixed identifiable fluid element $\mathcal{X}$ is defined as the time rate of change of its position given by

$$\boldsymbol{u} \equiv \dot{\boldsymbol{x}} = \frac{\mathrm{d}\boldsymbol{x}(\mathcal{X}, t)}{\mathrm{d}t}, \tag{5.1.24}$$

which reduces to

$$\boldsymbol{u} = \boldsymbol{u}^R(\boldsymbol{X}, t) = \frac{\partial \boldsymbol{\mathcal{M}}(\boldsymbol{X}, t)}{\partial t}, \tag{5.1.25}$$

in the Lagrangian description, and

$$\boldsymbol{u} = \boldsymbol{u}^R(\boldsymbol{X}, t) = \boldsymbol{u}^P(\boldsymbol{\mathcal{M}}^{-1}(\boldsymbol{x}, t), t) = \boldsymbol{u}^P(\boldsymbol{x}, t), \tag{5.1.26}$$

in the Eulerian description, where the fluid element occupying the position $\boldsymbol{x}$ at time t is held fixed.[3] The *acceleration* of a fluid element $\mathcal{X}$ is defined in a similar manner. It is the time rate of change of velocity, viz.,

$$\boldsymbol{a} \equiv \dot{\boldsymbol{u}} = \frac{\mathrm{d}\boldsymbol{u}(\mathcal{X}, t)}{\mathrm{d}t} = \frac{\mathrm{d}\boldsymbol{u}^R(\boldsymbol{X}^{-1}(\boldsymbol{X}), t)}{\mathrm{d}t} = \frac{\partial \boldsymbol{u}^R(\boldsymbol{X}, t)}{\partial t}, \tag{5.1.27}$$

[3]It is possible to obtain the velocity in the Eulerian description by using the material derivative, viz.,

$$\boldsymbol{u} = \dot{\boldsymbol{x}} = \frac{\partial \boldsymbol{x}(\boldsymbol{x}, t)}{\partial t} + \frac{\partial \boldsymbol{x}(\boldsymbol{x}, t)}{\partial \boldsymbol{x}}\dot{x}_i = \boldsymbol{0} + \boldsymbol{I}\boldsymbol{u} = \boldsymbol{u},$$

which holds identically.

in the Lagrangian description, and

$$\boldsymbol{a} \equiv \dot{\boldsymbol{u}} = \frac{\mathrm{d}\boldsymbol{u}(\mathcal{X}, t)}{\mathrm{d}t} = \frac{\mathrm{d}\boldsymbol{u}^P(\boldsymbol{x}^{-1}(\boldsymbol{x}, t), t)}{\mathrm{d}t} = \frac{\partial \boldsymbol{u}^P}{\partial t} + \boldsymbol{L}\boldsymbol{u}^P, \tag{5.1.28}$$

in the Eulerian description, where $\boldsymbol{L}$ is the spatial *velocity gradient* given by

$$\boldsymbol{L} = \operatorname{grad} \boldsymbol{u}^P = L_{ij}\left(\boldsymbol{e}_i \boldsymbol{e}_j\right), \tag{5.1.29}$$

which is a second-order tensor, where "grad" stands for the gradient operation with respect to $\boldsymbol{e}_i$.[4]

The velocity gradient $\boldsymbol{L}$ is decomposed into a sum of a symmetric tensor $\boldsymbol{D}$ and an antisymmetric tensor $\boldsymbol{W}$, viz.,

$$\boldsymbol{L} = \frac{1}{2}\left(\boldsymbol{L} + \boldsymbol{L}^{\mathrm{T}}\right) + \frac{1}{2}\left(\boldsymbol{L} - \boldsymbol{L}^{\mathrm{T}}\right) \equiv \boldsymbol{D} + \boldsymbol{W}, \tag{5.1.30}$$

where $\boldsymbol{D}$ is termed the *stretching tensor*,[5] and $\boldsymbol{W}$ is called the *vorticity* or *spin tensor*, or *tensor of rotational velocity*. The dual vector of $\boldsymbol{W}$ corresponds exactly to the vorticity $\boldsymbol{\zeta}$ defined in Sect. 4.2. It follows from Eqs. (5.1.18) and (5.1.25) that

$$\operatorname{Grad} \boldsymbol{u}^R(\boldsymbol{X}, t) = \operatorname{Grad}\left(\frac{\partial \boldsymbol{\mathcal{M}}(\boldsymbol{X}, t)}{\partial t}\right) = \frac{\partial}{\partial t}\left[\operatorname{Grad} \boldsymbol{\mathcal{M}}(\boldsymbol{X}, t)\right] = \dot{\boldsymbol{F}}, \tag{5.1.31}$$

which is recast alternatively as

$$\begin{aligned} \dot{\boldsymbol{F}} &= \operatorname{Grad} \boldsymbol{u}^R(\boldsymbol{X}, t) = \operatorname{Grad} \boldsymbol{u}^P(\boldsymbol{\mathcal{M}}^{-1}(\boldsymbol{x}, t), t) = \operatorname{Grad} \boldsymbol{u}^P(\boldsymbol{x}, t) \\ &= \operatorname{grad} \boldsymbol{u}^P(\boldsymbol{x}, t) \operatorname{Grad} \boldsymbol{\mathcal{M}}(\boldsymbol{X}, t) = \boldsymbol{L}(\boldsymbol{x}, t)\boldsymbol{F}(\boldsymbol{X}, t), \end{aligned} \tag{5.1.32}$$

or

$$\boldsymbol{L} = \dot{\boldsymbol{F}}\boldsymbol{F}^{-1} = \operatorname{grad} \boldsymbol{u}^P, \qquad L_{ij} = \frac{\partial u_i^P}{\partial x_j}. \tag{5.1.33}$$

It follows also from Eqs. (5.1.18) and (5.1.21) that

$$\dot{J} = J(\operatorname{div} \boldsymbol{u}^P), \qquad (\mathrm{d}v^P)^{\cdot} = (\operatorname{div} \boldsymbol{u}^P)\mathrm{d}v^P. \tag{5.1.34}$$

Equation $(5.1.34)_2$ delivers a relation between the time rate of change of an infinitesimal volume element in $\mathcal{B}^P$ and the divergence of velocity field, which will be used later in the discussions of balance equations. The derivation of Eq. (5.1.34) is left as an exercise.

In practice, it is hardly possible to describe a fluid motion by tracing a fixed identifiable fluid element during the flow (i.e., in terms of the Lagrangian description), for the initially identifiable fluid element becomes un-identifiable when the flow starts, for which the Eulerian description is more appropriate. Thus, from now on, all discussions are based on the Eulerian description, unless stated otherwise. That

---

[4]Although $\boldsymbol{u}^P$ and $\boldsymbol{u}^R$ are in general different in their mathematical forms, the velocity is differentiated with respect to $\boldsymbol{x}$ for almost all circumstances, so that the velocity gradient always means the spatial gradient.

[5]The stretching tensor does not correspond to the strain rate tensor, for the integration of the latter does not correspond exactly to the former in general.

is, the focus is on the present configuration $\mathcal{B}^P$ at time t of a fluid body $\mathcal{B}$. The superscripts used previously to distinguish the Lagrangian and Eulerian descriptions are abandoned, and the fluid volume occupied by $\mathcal{B}$ in $\mathcal{B}^P$ at time t is denoted by V with its surface denoted by A. Similarly, $\mathrm{d}v$ is used to denote an infinitesimal volume element of V, with an infinitesimal surface element denoted by $\mathrm{d}\boldsymbol{a}$. All quantities are functions of spatial coordinates and time in the Eulerian description.

5.2 Balance Equations in Global and Local Forms

5.2.1 General Formulation

Consider the present configuration $\mathcal{B}^P$ of $\mathcal{B}$ in Fig. 5.1 again. Let ϕ be any extensive physical variable of the whole fluid body $\mathcal{B}$ at time t. Its *specific property*, i.e., the value of ϕ per unit mass, is denoted by $\aleph_\phi$. The total amount of ϕ, by using the additive assumption, is given by

$$\phi(t) = \int_V \aleph_\phi\, \rho\, \mathrm{d}v. \tag{5.2.1}$$

The variable ϕ may change with time due to the influence of external excitations and internal processes inside the fluid body. External excitations are classified into two categories: (a) the excitations taking place in the entire fluid body, e.g. the gravitational or magnetic force, which is termed *body force* or *body excitation*, and (b) the excitations taking place over the surface enclosing the body, e.g. frictional force or heat flux, which is termed *surface force* or *surface excitation*. In addition, ϕ may also experience a time variation due to internal process, e.g. heat source or heat sink inside the material. The possible contributions to the time rate of change of ϕ are summarized in the following:

- Production $\mathcal{P}$: The quantity is produced within V, with its specific property denoted by $\pi_\aleph$, i.e., $\pi_\aleph$ is the value of $\mathcal{P}$ per unit mass. For example, the production of heat in a fluid body due to a radioactive decay.
- Supply $\mathcal{S}$: The quantity is supplied to the fluid body from its surrounding via body excitation, with its specific property denoted by $\sigma_\aleph$ in V. For example, the gravitational field or radiation heat from a furnace.
- Flux $\mathcal{F}$: The quantity takes place over the surface of body and is supplied from the surrounding to the fluid body as a surface phenomenon, with its *surface density* denoted by $\Psi_\aleph$, i.e., $\Psi_\aleph$ is the value of $\mathcal{F}$ per unit area. Specifically, $\Psi_\aleph = \Psi_\aleph(\boldsymbol{x}, t, \boldsymbol{n}, \xi)$, where $\boldsymbol{n}$ represents the unit outward normal of the surface of body, and ξ represents the differential geometric properties of A at $\boldsymbol{x}$, e.g. the *mean* or *Gaussian curvature*. For example, the stress on the surface of a fluid body, the heat flux or electrical current through the surface of a fluid body.

These contributions and their mass or surface densities are summarized in Table 5.1.

Table 5.1 Contributions to the time rate of change of ϕ with their densities

	Volume V, surface A
Any quantity ϕ	$\aleph_\phi(\boldsymbol{x}, t)$ (mass density)
Production $\mathcal{P}$	$\pi_\aleph(\boldsymbol{x}, t)$ (mass density)
Supply $\mathcal{S}$	$\sigma_\aleph(\boldsymbol{x}, t)$ (mass density)
Flux $\mathcal{F}$	$\Psi_\aleph(\boldsymbol{x}, t, \boldsymbol{n}, \xi)$ (surface density)

With these, the general statement of time rate of change of ϕ is established as

$$\frac{\mathrm{d}\phi}{\mathrm{d}t} = \mathcal{P} + \mathcal{S} + \mathcal{F}, \tag{5.2.2}$$

with

$$\mathcal{P} = \int_V \pi_\aleph\, \rho\, \mathrm{d}v, \qquad \mathcal{S} = \int_V \sigma_\aleph\, \rho\, \mathrm{d}v, \qquad \mathcal{F} = \int_A \Psi_\aleph\, \mathrm{d}a. \tag{5.2.3}$$

Substituting these expressions into Eq. (5.2.2) yields

$$\frac{\mathrm{d}}{\mathrm{d}t}\int_V \aleph_\phi\, \rho\, \mathrm{d}v = \int_V (\pi_\aleph + \sigma_\aleph)\, \rho\, \mathrm{d}v + \int_A \Psi_\aleph\, \mathrm{d}a, \tag{5.2.4}$$

which is a statement of the general balance equation of any extensive variable ϕ.

5.2.2 Cauchy's Stress Principle and Lemma

It is assumed that the traction vector resulted from a surface density at any given point, and time has a common value on all parts of material having a common tangent plane at that point and lying on the same side of it. This statement is termed the *Cauchy assumption* or the *Cauchy stress principle*, with which the surface density $\Psi_\aleph(\boldsymbol{x}, t, \boldsymbol{n}, \xi)$ becomes independent on the differential geometric properties of A, i.e.,

$$\Psi_\aleph = \Psi_\aleph(\boldsymbol{x}, t, \boldsymbol{n}). \tag{5.2.5}$$

The above expression is further simplified by using the *Cauchy lemma* given in the following:

5.1 (The Cauchy lemma) *If the surface density $\Psi_\aleph(\boldsymbol{x}, t, \boldsymbol{n})$ depends on the normal $\boldsymbol{n}$ at the surface, this dependency is a linear contraction given viz.,*

$$\Psi_\aleph = -\boldsymbol{\psi}_\aleph(\boldsymbol{x}, t)\boldsymbol{n}, \tag{5.2.6}$$

where $\boldsymbol{\psi}_\aleph$ is termed the surface flux.

It is not difficult to prove the Cauchy lemma by applying a balance statement to an infinitesimal tetrahedron, which is left as an exercise.

5.2.3 Global Balance Equation

With the Cauchy assumption and lemma, Eq. (5.2.4) becomes

$$\frac{\mathrm{d}}{\mathrm{d}t}\int_V \aleph_\phi\, \rho\, \mathrm{d}v = \int_V (\pi_\aleph + \sigma_\aleph)\, \rho\, \mathrm{d}v - \int_A \boldsymbol{\psi}_\aleph \boldsymbol{n}\, \mathrm{d}a. \tag{5.2.7}$$

The left-hand-side of this equation is further explored as

$$\frac{\mathrm{d}}{\mathrm{d}t}\int_V \aleph_\phi\, \rho\, \mathrm{d}v = \int_V \left[(\aleph_\phi\, \rho)^{\cdot}\, \mathrm{d}v + \aleph_\phi\, \rho (\mathrm{d}v)^{\cdot}\right] = \int_V \left[(\aleph_\phi\, \rho)^{\cdot} + \aleph_\phi\, \rho (\operatorname{div} \boldsymbol{u})\right] \mathrm{d}v, \tag{5.2.8}$$

in which Eq. (5.1.34)$_2$ has been used, where $\dot{\alpha}$ represents the time rate of change of α. With the material derivative and Gauss theorem, Eq. (5.2.8) is recast alternatively as

$$\frac{\mathrm{d}}{\mathrm{d}t}\int_V \aleph_\phi\, \rho\, \mathrm{d}v = \frac{\partial}{\partial t}\int_V \aleph_\phi\, \rho\, \mathrm{d}v + \int_A \aleph_\phi\, \rho (\boldsymbol{u}\cdot\boldsymbol{n})\mathrm{d}a. \tag{5.2.9}$$

Equations (5.2.8) and (5.2.9) indicate that the time rate of change of ϕ in the Eulerian description is a sum of its temporal change within V and its change induced by the change of integration domain, resulted from the influence of flux contributions. Equation (5.2.8) or (5.2.9) is termed *Reynolds' transport theorem*,[6] which is a mathematical rule to express the time rate of change of any physical variable from the Lagrangian to Eulerian descriptions.

With Reynolds' transport theorem, Eq. (5.2.7) becomes

$$\int_V \left[(\aleph_\phi\, \rho)^{\cdot} + \aleph_\phi\, \rho (\operatorname{div} \boldsymbol{u})\right] \mathrm{d}v = \int_V (\pi_\aleph + \sigma_\aleph)\, \rho\, \mathrm{d}v - \int_A \boldsymbol{\psi}_\aleph \boldsymbol{n}\, \mathrm{d}a, \tag{5.2.10}$$

or

$$\frac{\partial}{\partial t}\int_V \aleph_\phi\, \rho\, \mathrm{d}v + \int_A \aleph_\phi\, \rho (\boldsymbol{u}\cdot\boldsymbol{n})\, \mathrm{d}a = \int_V (\pi_\aleph + \sigma_\aleph)\, \rho\, \mathrm{d}v - \int_A \boldsymbol{\psi}_\aleph \boldsymbol{n}\, \mathrm{d}a. \tag{5.2.11}$$

These two equations are the statements of global balance equation of any extensive variable ϕ.

5.2.4 Local Balance Equation

With the Gauss theorem, Eq. (5.2.10) is expressed alternatively as

$$\int_V \left[\frac{\mathrm{d}(\aleph_\phi\, \rho)}{\mathrm{d}t} + \aleph_\phi\, \rho\, (\operatorname{div} \boldsymbol{u}) - \rho\, \pi_\aleph - \rho\, \sigma_\aleph + \operatorname{div} \boldsymbol{\psi}_\aleph\right] \mathrm{d}v = 0. \tag{5.2.12}$$

[6]The one-dimensional analogue of Reynolds' transport theorem is the *Leibniz integration rule*. The relation between Reynolds' transport theorem and the material derivative will be discussed in Sect. 5.3.6. Gottfried Wilhelm von Leibniz, 1646–1716, a German polymath, who developed differential and integral calculus independent of Newton.

Table 5.2 Global and local balance equations of any extensive variable ϕ in the Eulerian description

$$\frac{\partial}{\partial t}\int_V \aleph_\phi\,\rho\,\mathrm{d}v + \int_A \aleph_\phi\,\rho\,(\boldsymbol{u}\cdot\boldsymbol{n})\mathrm{d}a = \int_V (\pi_\aleph + \sigma_\aleph)\,\rho\,\mathrm{d}v - \int_A \boldsymbol{\psi}_\aleph \boldsymbol{n}\,\mathrm{d}a$$

$$\frac{\partial(\aleph_\phi\,\rho)}{\partial t} + \mathrm{div}\left(\aleph_\phi\,\rho\,\boldsymbol{u}\right) = -\mathrm{div}\,\boldsymbol{\psi}_\aleph + \rho\,\pi_\aleph + \rho\,\sigma_\aleph$$

V	Volume of fluid body	A	Surface of V, with normal $\boldsymbol{n}$
$\aleph_\phi(\boldsymbol{x},t)$	Mass density of ϕ	$\pi_\aleph(\boldsymbol{x},t)$	Mass density of $\mathcal{P}$
$\sigma_\aleph(\boldsymbol{x},t)$	Mass density of $\mathcal{S}$	$\boldsymbol{\psi}_\aleph(\boldsymbol{x},t)$	Surface flux of $\mathcal{F}$

Since within the continuum hypothesis, $\mathrm{d}v$ does not vanish in general, this equation cannot be satisfied unless

$$\frac{\mathrm{d}(\aleph_\phi\,\rho)}{\mathrm{d}t} + \aleph_\phi\,\rho\,(\mathrm{div}\,\boldsymbol{u}) - \rho\,\pi_\aleph - \rho\,\sigma_\aleph + \mathrm{div}\,\boldsymbol{\psi}_\aleph = 0, \tag{5.2.13}$$

which is expressed as

$$\frac{\mathrm{d}(\aleph_\phi\,\rho)}{\mathrm{d}t} + \aleph_\phi\,\rho\,(\mathrm{div}\,\boldsymbol{u}) = -\mathrm{div}\,\boldsymbol{\psi}_\aleph + \rho\,\pi_\aleph + \rho\,\sigma_\aleph. \tag{5.2.14}$$

This equation, by using the material derivative, is rewritten as

$$\frac{\partial(\aleph_\phi\,\rho)}{\partial t} + \mathrm{div}\left(\aleph_\phi\,\rho\,\boldsymbol{u}\right) = -\mathrm{div}\,\boldsymbol{\psi}_\aleph + \rho\,\pi_\aleph + \rho\,\sigma_\aleph. \tag{5.2.15}$$

Equations (5.2.14) and (5.2.15) are the statements of local balance equation of any extensive variable ϕ. The global and local balance equations are conventionally termed the balance equations in *integral* and *differential forms*, respectively, which are summarized in Table 5.2.

5.3 Balance Equations of Physical Laws

The fundamental laws in classical physics are the balances of mass, linear momentum, angular momentum, and first and second laws of thermodynamics, which need to be satisfied by all materials simultaneously. The balance statements of these physical laws in integral and differential forms may be established by prescribing different densities in the global and local balance equations derived previously. Table 5.3 summarizes the densities used to derive the balance statements of physical laws in this section.

5.3.1 Balance of Mass

To every fluid element $\mathcal{X}$ of a fluid body a (mass) density is allocated, which is denoted by ρ. It is assumed that mass is a physical quantity which can neither flow

Table 5.3 Prescribed densities in the balance statements of physical laws

	$\aleph_\phi$	$\pi_\aleph$	$\sigma_\aleph$	$\psi_\aleph$
Mass	1	0	0	$\mathbf{0}$
Linear momentum	$\boldsymbol{u}$	$\mathbf{0}$	$\boldsymbol{b}$	$-\boldsymbol{t}$
Angular momentum	$\boldsymbol{x} \times \boldsymbol{u}$	$\mathbf{0}$	$\boldsymbol{x} \times \boldsymbol{b}$	$-\boldsymbol{x} \times \boldsymbol{t}$
Energy	$\frac{1}{2}\boldsymbol{u} \cdot \boldsymbol{u} + \epsilon$	0	$\boldsymbol{b} \cdot \boldsymbol{u} + \zeta$	$-\boldsymbol{u}\boldsymbol{t} + \boldsymbol{q}$
Entropy	η	$\pi_\eta \geq 0$	s_η	$\boldsymbol{\phi}_\eta$

through a surface, nor be produced or supplied from the surrounding to the fluid body if chemical or nuclear reaction is not taken into account. Thus, the densities are prescribed by

$$\aleph_\phi = 1, \qquad \pi_\aleph = 0, \qquad \sigma_\aleph = 0, \qquad \boldsymbol{\psi}_\aleph = \mathbf{0}. \tag{5.3.1}$$

Substituting these expressions into the equations in Table 5.2 gives the global mass balance as

$$\frac{\partial}{\partial t} \int_V \rho \, \mathrm{d}v + \int_A \rho \, (\boldsymbol{u} \cdot \boldsymbol{n}) \, \mathrm{d}a = 0, \tag{5.3.2}$$

while the local balance statement is given by

$$\dot{\rho} + \rho \, \mathrm{div} \, \boldsymbol{u} = 0, \qquad \frac{\partial \rho}{\partial t} + \mathrm{div} \, (\rho \boldsymbol{u}) = 0. \tag{5.3.3}$$

Equation (5.3.2) shows that for a finite control-volume, the time change of fluid mass contained within the CV is balanced by the net fluid mass across the CS of CV per unit time. For example, consider the water in a sealed bottle. If the water content is described by using the control-mass system (i.e., the Lagrangian description), it is naturally a constant. On the contrary, the bottle cap is opened to allow water exchange to the surrounding. In this circumstance, the bottle is considered the CV with the bottle surface as the CS. The change in the water content per unit time inside the bottle is nothing else than the net amount of water entering/leaving the bottle per unit time, corresponding exactly to Eq. (5.3.2). The interpretation of Eq. (5.3.3) is the same for a differential CV.

A fluid is called *density preserving* or *incompressible* if the density of a fluid element does not change with time, i.e.,

$$\frac{\mathrm{d}\rho(\mathcal{X}, t)}{\mathrm{d}t} = 0. \tag{5.3.4}$$

Since $\rho(\mathcal{X}, t)$ is the mass divided by the volume of a fluid element, it follows that $\mathrm{d}v^P = \mathrm{d}v^R$ for a fixed identifiable fluid element, implying that $J = \det \boldsymbol{F} = 1$. With this, the density preservation delivers

$$\dot{J} = J \, (\mathrm{div} \, \boldsymbol{u}) = 0, \longleftrightarrow \mathrm{div} \, \boldsymbol{u} = 0, \longleftrightarrow \dot{\rho} = 0, \tag{5.3.5}$$

indicating that the density in the Eulerian description is a constant. A fluid is also termed *volume preserving* if the above equation is satisfied. However, it should be noted that Eq. (5.3.5) can also be fulfilled if $J = \det \boldsymbol{F} = \text{constant}$. Flows with $J = \text{constant}$ are volume preserving and are called *isochoric flows*. Those with $J = 1$ are termed *unimodular flows*. For volume-preserving flows, the velocity $\boldsymbol{u}$ in the Eulerian description is a *solenoidal field* and vice versa. With the assumption of density preservation, the global and local mass balances reduce respectively to

$$\int_A \boldsymbol{u} \cdot \boldsymbol{n} \, \mathrm{d}a = 0, \qquad \operatorname{div} \boldsymbol{u} = 0. \tag{5.3.6}$$

The first equation is used to define the *(volume) flow rate* Q across a specific surface, and shows that the net flow rate vanishes for an incompressible flow, i.e., the fluid volumes entering a CV should be the same as those leaving the CV per unit time, as already described in Sect. 4.4. Equation $(5.3.6)_2$ has the same physical interpretation and can be derived directly from Eq. $(5.3.6)_1$ by using the Gauss theorem.

Similarly, for *steady flows*, the global and local mass balances read the forms

$$\int_A \rho \boldsymbol{u} \cdot \boldsymbol{n} \, \mathrm{d}a = 0, \qquad \operatorname{div} (\rho \boldsymbol{u}) = 0, \tag{5.3.7}$$

with which the *mass flow rate* $\dot{m}$ across a specific surface is defined by

$$\dot{m} \equiv \int_A \rho \boldsymbol{u} \cdot \boldsymbol{n} \, \mathrm{d}a. \tag{5.3.8}$$

Equation $(5.3.7)_1$ indicates that the net mass flow rate crossing the CS of a finite CV vanishes, while Eq. $(5.3.7)_2$ has the same interpretation for a differential CV and can equally be derived from Eq. $(5.3.7)_1$ by using the Gauss theorem.

5.3.2 Balance of Linear Momentum in Inertia Frame

To every fluid element $\mathcal{X}$ of a fluid body, a liner momentum is allocated, with its mass density denoted by $\boldsymbol{u}$. The linear momentum of a material is a conservative quantity, which can be neither created nor destroyed. However, it can be changed via external volume and surface excitations as supply and flux, respectively, as motivated by Newton's second law of motion. Since the expressions of linear momentum depend on the coordinate systems, the balance of linear momentum is discussed here for an inertial coordinate system. The balance of linear momentum in non-inertia coordinate systems will be discussed in Sect. 5.4. Thus, the densities in the balance statement are prescribed as

$$\aleph_\phi = \boldsymbol{u}, \qquad \pi_\aleph = \boldsymbol{0}, \qquad \sigma_\aleph = \boldsymbol{b}, \qquad \psi_\aleph = -\boldsymbol{t}, \tag{5.3.9}$$

where $\boldsymbol{b}$ represents the body force per unit mass, which equals the gravitational acceleration $\boldsymbol{g}$ if the fluid body experiences only the gravitational field. It can be generalized to take into account other possible body forces.[7] The linear momentum flux on the surfaces is the negative Cauchy stress tensor $\boldsymbol{t}$.

With these, the global balance of linear momentum is obtained as

$$\frac{\partial}{\partial t}\int_V \rho\boldsymbol{u}\,\mathrm{d}v + \int_A \boldsymbol{u}\,(\rho\boldsymbol{u}\cdot\boldsymbol{n})\,\mathrm{d}a = \int_V \rho\boldsymbol{b}\,\mathrm{d}v + \int_A \boldsymbol{t}\boldsymbol{n}\,\mathrm{d}a. \tag{5.3.10}$$

The right-hand-side of this equation is the sum of all external body forces acting on the finite CV and surface forces acting on the CS, which can be generalized as[8]

$$\int_V \rho\boldsymbol{b}\,\mathrm{d}v + \int_A \boldsymbol{t}\boldsymbol{n}\,\mathrm{d}a = \sum\mathcal{F}_{CV} + \sum\mathcal{F}_{CS}, \tag{5.3.11}$$

with which Eq. (5.3.10) becomes

$$\frac{\partial}{\partial t}\int_V \rho\boldsymbol{u}\,\mathrm{d}v + \int_A \boldsymbol{u}\,(\rho\boldsymbol{u}\cdot\boldsymbol{n})\,\mathrm{d}a = \sum\mathcal{F}_{CV} + \sum\mathcal{F}_{CS}, \tag{5.3.12}$$

which is the global balance equation of linear momentum. This equation shows that the time change of linear momentum of the fluids contained within a finite CV plus the linear momentum change induced by the fluids entering and leaving the CS per unit time is balanced by the total external body forces acting on the CV and surface forces acting on the CS. For example, consider a bottle filled with a high-pressure air. The bottle is initially sealed and placed on a horizontal table which is perpendicular to the gravitational field. When the bottle cap is removed, there exists an air jet from the bottle, causing the bottle to move in the reverse direction of air jet. The time change of linear momentum of the air remaining inside the bottle is balanced by the linear momentum carried by the air jet per unit time, resulting in a vanishing resultant force acting on the bottle in the horizontal direction.

The local balance of linear momentum is given by[9]

$$(\rho\boldsymbol{u})^{\cdot} + (\rho\boldsymbol{u})\operatorname{div}\boldsymbol{u} = \operatorname{div}\boldsymbol{t} + \rho\boldsymbol{b}, \tag{5.3.13}$$

which, by using the local mass balance, reduces to

$$\rho\dot{\boldsymbol{u}} = \operatorname{div}\boldsymbol{t} + \rho\boldsymbol{b}, \qquad \rho\dot{u}_i = t_{ij,j} + \rho b_i. \tag{5.3.14}$$

[7]Caution must be made for the formulations of $\boldsymbol{b}$ if other body forces present, or the material under consideration is not homogeneous, in which $\boldsymbol{b}$ may be different for different material elements, even though $\boldsymbol{b}$ is the constant gravitational acceleration.

[8]It is noted that

$$\sum\mathcal{F}_{CV} = \int_V \rho\boldsymbol{b}\,\mathrm{d}v, \qquad \sum\mathcal{F}_{CS} \neq \int_A \boldsymbol{t}\boldsymbol{n}\,\mathrm{d}a,$$

for $\sum\mathcal{F}_{CS}$ becomes now the sum of all surface forces external to the fluid body, and the stress traction on CS consists only a part of $\sum\mathcal{F}_{CS}$.

[9]Equation (5.3.13) and its general form are called the *Cauchy equations of motion*, which have been derived first by Cauchy, and are applied to study the motions of elastic solid bodies.

The material derivative can be used to express the above equation in an alternatively form given by

$$\frac{\partial}{\partial t}(\rho \boldsymbol{u}) + \operatorname{div}(\rho \boldsymbol{u}\boldsymbol{u}) = \operatorname{div} \boldsymbol{t} + \rho \boldsymbol{b}, \qquad (\rho u_i)_{,t} + (\rho u_i u_j)_{,j} = t_{ij,j} + \rho b_i, \tag{5.3.15}$$

with

$$\dot{\boldsymbol{u}} = \frac{\partial \boldsymbol{u}}{\partial t} + (\operatorname{grad} \boldsymbol{u})\, \boldsymbol{u} = \frac{\partial \boldsymbol{u}}{\partial t} + \operatorname{grad}\left(\frac{\|\boldsymbol{u}\|^2}{2}\right) - \boldsymbol{u} \times \operatorname{curl} \boldsymbol{u}. \tag{5.3.16}$$

Equation (5.3.14) or (5.3.15) is the local balance equation of linear momentum. Equation (5.3.16) can be derived by using the index notation and is left as an exercise.

Unlike its counterpart in integral form, Eq. (5.3.14) or (5.3.15) cannot be used at this moment, although it is a physical law, i.e., Newton's second law of linear motion, which should be satisfied for all materials, for a definite prescription of $\boldsymbol{t}$ needs to be conducted a priori, which is a kind of the *material* or *constitutive equations*. The topic of material equation will be discussed in Sect. 5.6. Furthermore, it is not possible to derive Eq. (5.3.13) directly from Eq. (5.3.12) as what has been done previously for the mass balance. It is so, because the global balance of liner momentum is related to all excitations external to the finite control-volume, while the local balance of linear momentum deals with all its surface forces resulted from the surrounding fluid elements as stresses which are internal to a finite control-volume.

5.3.3 Balance of Angular Momentum in Inertia Frame

In general, the angular momentum of a body with respect to an arbitrarily fixed point in space consists of the *moment of momentum* and *spin*. The latter is the body's angular momentum relative to its center of mass, while the former is the moment of momentum of body's center of mass with respect to the arbitrarily fixed point in space. To simplify the analysis, it is assumed that the fluid body has no spin and there exist no *volume moments* such as the magnetic polarization and no *surface couple stresses* on the surface of fluid body. With these, the balance of angular momentum reads "the time rate of change of angular momentum of a body with respect to a fixed point in space equals the resultant moments acting on the body with respect to the same point."[10] It is simply Newton's second law in rotational motion. As similar to linear momentum, angular momentum is a conservative quantity which can only be changed by external excitations in forms of moments. Thus, the densities of angular momentum balance are prescribed by

$$\aleph_\phi = \boldsymbol{x} \times \boldsymbol{u}, \qquad \pi_\aleph = \boldsymbol{0}, \qquad \sigma_\aleph = \boldsymbol{x} \times \boldsymbol{b}, \qquad \psi_\aleph = -\boldsymbol{x} \times \boldsymbol{t}, \tag{5.3.17}$$

[10]The balance of angular momentum is one of the basic axioms of the *Galilean physics* and has been formulated first by Euler for rigid bodies, which is termed the *Euler equation of dynamics*.

where $\boldsymbol{x}$ is the position vector of a fluid element, and only the moments generated by volume and surface forces are taken into account.

Substituting Eq. (5.3.17) into the global balance statement yields the global balance of angular momentum given by

$$\frac{\partial}{\partial t}\int_V (\boldsymbol{x}\times\boldsymbol{u})\rho\,\mathrm{d}v + \int_A (\boldsymbol{x}\times\boldsymbol{u})(\rho\boldsymbol{u}\cdot\boldsymbol{n})\,\mathrm{d}a = \int_V (\boldsymbol{x}\times\boldsymbol{b})\rho\,\mathrm{d}v + \int_A (\boldsymbol{x}\times\boldsymbol{t}\boldsymbol{n})\,\mathrm{d}a. \tag{5.3.18}$$

As similar to the linear momentum balance, the right-hand-side of this equation can be generalized to include all possible external moments acting on the CV and CS, viz.,

$$\int_V (\boldsymbol{x}\times\boldsymbol{b})\rho\,\mathrm{d}v + \int_A (\boldsymbol{x}\times\boldsymbol{t}\boldsymbol{n})\,\mathrm{d}a = \sum \boldsymbol{x}\times\boldsymbol{\mathcal{F}}_{CV} + \sum \boldsymbol{x}\times\boldsymbol{\mathcal{F}}_{CS} + \sum \boldsymbol{M}_{shaft}, \tag{5.3.19}$$

where the first term on the right-hand-side represents all moments generated by the external body forces, and the second term are those by all external surface forces, while $\sum \boldsymbol{M}_{shaft}$ denotes other possible external moments which are provided mainly via shafts into the CV. For example, an external moment is provided to a cup of water if the water is swirled by using a spoon. With these, Eq. (5.3.18) becomes

$$\frac{\partial}{\partial t}\int_V (\boldsymbol{x}\times\boldsymbol{u})\rho\,\mathrm{d}v + \int_A (\boldsymbol{x}\times\boldsymbol{u})(\rho\boldsymbol{u}\cdot\boldsymbol{n})\,\mathrm{d}a = \sum \boldsymbol{x}\times\boldsymbol{\mathcal{F}}_{CV} + \sum \boldsymbol{x}\times\boldsymbol{\mathcal{F}}_{CS} + \sum \boldsymbol{M}_{shaft}, \tag{5.3.20}$$

which is the global balance of angular momentum. The equation shows that the applied external moments to a finite CV is the same as the time change of angular momentum of the fluids contained within the CV plus the change in angular momentum of the fluids entering and leaving the CS per unit time. For example, consider a sprinkler used in garden which is initially at rest. When water is supplied to the sprinkler through its center, it rotates in the direction which is reverse to the direction of moments generated by the linear momentums of leaving water jets to ensure a vanishing angular momentum of the sprinkler during rotation.

The local balance of angular momentum is given by

$$(\boldsymbol{x}\times\rho\boldsymbol{u})^{\cdot} + (\boldsymbol{x}\times\rho\boldsymbol{u})\,\mathrm{div}\,\boldsymbol{u} = \mathrm{div}\,(\boldsymbol{x}\times\boldsymbol{t}) + \boldsymbol{x}\times\rho\boldsymbol{b}, \tag{5.3.21}$$

which, by using the local balances of mass and linear momentum, reduces to

$$\boldsymbol{t}^* = \boldsymbol{0}, \qquad t_i^* = \varepsilon_{ijk}\frac{\partial x_j}{\partial x_l}t_{kl} = \varepsilon_{ijk}t_{kj} = 0, \tag{5.3.22}$$

showing that

$$t_{kj} = t_{jk}, \qquad \boldsymbol{t} = \boldsymbol{t}^{\mathrm{T}}. \tag{5.3.23}$$

Thus, the local balance of angular momentum delivers that the Cauchy stress tensor is symmetric. The derivation of Eq. (5.3.23) is left as an exercise.[11] As similar to the linear momentum balance, it is not possible to derive Eq. (5.3.23) directly from Eq. (5.3.20) for the same reason.

[11]The balance of linear momentum can also be formulated in the Lagrangian description, in which the stress is the *first Piola-Kirchhoff stress tensor* $\boldsymbol{T}$. Formulating the balance of angular momentum

5.3.4 Balance of Energy

The balance of energy is termed officially *first law of thermodynamics*, which states that the mechanical and thermal energies (and all other possible energies) of a material body are conserved altogether.[12] Conventionally, a material body has three forms of energy: the kinetic energy KE relating to the body velocity, the potential energy PE induced by the conservative force fields, e.g. the gravitational potential energy, and the *internal energy* U which is a collection of other energies that cannot be classified as kinetic or potential energies, which depends essentially on the body temperature. The KE and PE are termed the *mechanical energies*, while U is referred to as the *thermal energy*. In a more general sense, the PE is considered the work done by the external conservative body forces, e.g. the gravitational force and is regarded as a kind of energy supplied from the surrounding. Pure energy supplies possibly exist, e.g. heat radiation source. Moreover, there exists work done by the surface forces, which is considered a kind of mechanical energy flux (surface energy flux). Equally, pure energy fluxes exist on the surface of body, e.g. heat flux. On the contrary, it is a physical postulate that there exist no energy productions within the body or in the surrounding. Thus, the densities of energy balance are prescribed by

$$\aleph_\phi = \frac{1}{2}\boldsymbol{u}\cdot\boldsymbol{u} + \epsilon, \qquad \pi_\aleph = 0, \qquad \sigma = \boldsymbol{b}\cdot\boldsymbol{u} + \zeta, \qquad \boldsymbol{\psi}_\aleph = -\boldsymbol{ut} + \boldsymbol{q}, \tag{5.3.24}$$

where ϵ is the specific internal energy, ζ represents the specific energy supply, $-\boldsymbol{ut}$ is the power done by the stress as a surface energy flux, and $\boldsymbol{q}$ denotes other energy fluxes, including heat flux.

The first law of thermodynamics states that although the total energy of a system is a conserved quantity, it can be transformed between different energy forms, and the time rate of change of total energy of a system equals all the powers done by the surrounding. With this, the global energy balance reads

$$\begin{aligned}&\frac{\partial}{\partial t}\int_V\left(\frac{1}{2}\boldsymbol{u}\cdot\boldsymbol{u}+\epsilon\right)\rho\,\mathrm{d}v + \int_A\left(\frac{1}{2}\boldsymbol{u}\cdot\boldsymbol{u}+\epsilon\right)(\rho\boldsymbol{u}\cdot\boldsymbol{n})\,\mathrm{d}a\\ &= \int_V(\boldsymbol{b}\cdot\boldsymbol{u}+\zeta)\rho\,\mathrm{d}v + \int_A(\boldsymbol{ut}-\boldsymbol{q})\cdot\boldsymbol{n}\,\mathrm{d}a.\end{aligned} \tag{5.3.25}$$

As similar to the linear and angular momentum balances, the right-hand-side of this equation can be generalized to include all possible external powers done on the finite CV. First, let $\boldsymbol{b}$ consist of $\boldsymbol{g}$ and $\boldsymbol{b}^*$, where $\boldsymbol{g}$ is the gravitational acceleration and $\boldsymbol{b}^*$ represents other body forces per unit mass. The work done by the gravitational force per unit time can be incorporated into the left-hand-side as a potential energy. Second, the stress power $\boldsymbol{ut}$ is decomposed into the power done by the pressures,

---

in the Lagrangian description shows that $\boldsymbol{T}$ is not symmetric, but $\boldsymbol{T}\boldsymbol{F}^{\mathrm{T}}$ is symmetric. Gabrio Piola, 1794–1850, an Italian mathematician and physicist. Gustav Robert Kirchhoff, 1824–1887, a German physicist, who also contributed to the fundamental understanding of electrical circuits and the emission of *blackbody radiation* by heated objects.

[12] The first law of thermodynamics will be explored in a detailed manner in Sect. 11.4.

$-p\boldsymbol{u}\cdot\boldsymbol{n}\mathrm{d}a$, and the power done by the shear stresses $\boldsymbol{uT}$, where $\boldsymbol{T}$ represents the extra stress tensor. With these, Eq. (5.3.25) becomes

$$\frac{\partial}{\partial t}\int_V\left(\frac{1}{2}\boldsymbol{u}\cdot\boldsymbol{u}+\epsilon+gz\right)\rho\,\mathrm{d}v+\int_A\left(\frac{1}{2}\boldsymbol{u}\cdot\boldsymbol{u}+\epsilon+gz+pv\right)(\rho\boldsymbol{u}\cdot\boldsymbol{n})\,\mathrm{d}a$$
$$=\int_V(\boldsymbol{b}^*\cdot\boldsymbol{u}+\zeta)\rho\,\mathrm{d}v+\int_A(\boldsymbol{uT}-\boldsymbol{q})\cdot\boldsymbol{n}\mathrm{d}a, \tag{5.3.26}$$

where $v=1/\rho$, which is the specific volume, and z is the elevation. The term pv is called the *specific flow work*, which is the work done by a fluid element per unit mass in order to push the neighboring fluid elements to accomplish a flow motion. The right-hand-side of the above equation is generalized to be[13]

$$\int_V(\boldsymbol{b}^*\cdot\boldsymbol{u}+\zeta)\rho\,\mathrm{d}v+\int_A(\boldsymbol{uT}-\boldsymbol{q})\cdot\boldsymbol{n}\,\mathrm{d}a$$
$$=\sum\dot{Q}+\sum\dot{W}_{shear}+\sum\dot{W}_{shaft}+\sum\dot{E}_s, \tag{5.3.27}$$

where $\sum\dot{Q}$ denotes the powers supplied to the CV due to the external energy fluxes, $\sum\dot{W}_{shear}$ represents the powers done by the shear stresses, $\sum\dot{W}_{shaft}$ embraces all other possible external powers supplied to the CV, mostly via the shaft works per unit time, and $\sum\dot{E}_s$ is the powers supplied by the external energy sources. Substituting these into Eq. (5.3.26) yields

$$\frac{\partial}{\partial t}\int_V\left(\frac{1}{2}\boldsymbol{u}\cdot\boldsymbol{u}+\epsilon+gz\right)\rho\,\mathrm{d}v+\int_A\left(\frac{1}{2}\boldsymbol{u}\cdot\boldsymbol{u}+\epsilon+gz+pv\right)(\rho\boldsymbol{u}\cdot\boldsymbol{n})\,\mathrm{d}a$$
$$=\sum\dot{Q}+\sum\dot{W}_{shear}+\sum\dot{W}_{shaft}+\sum\dot{E}_s, \tag{5.3.28}$$

which is the global balance equation of energy.

Define the *specific total energy* e and *specific enthalpy* h as

$$e\equiv\frac{1}{2}\boldsymbol{u}\cdot\boldsymbol{u}+\epsilon+gz,\qquad h\equiv\epsilon+pv, \tag{5.3.29}$$

with which Eq. (5.3.28) is recast as

$$\frac{\partial}{\partial t}\int_V\rho\,e\,\mathrm{d}v+\int_A\left(h+\frac{1}{2}\boldsymbol{u}\cdot\boldsymbol{u}+gz\right)(\rho\boldsymbol{u}\cdot\boldsymbol{n})\,\mathrm{d}a$$
$$=\sum\dot{Q}+\sum\dot{W}_{shear}+\sum\dot{W}_{shaft}+\sum\dot{E}_s, \tag{5.3.30}$$

which is an alternative form of the global balance of energy. This equation indicates that for a finite control-volume, the time change of total energy of the fluids within the CV plus the total energies and flow works of the fluids entering and leaving the CS per unit time should be balanced by all the external powers done on the CV. For example, consider again a bottle filled with water which is located inside a microwave. The time rate of change of internal energy of water is nothing else

[13]The specific form of Eq. (5.3.27) depends on the definitions of the positivenesses of $\dot{Q}$, $\dot{W}$, and $\dot{E}_s$ in thermodynamics. Here they are defined to be positive if they are provided to the system by the surrounding.

than the power delivered to the water by the microwave radiation, which is a kind of energy supply per unit time.

The local balance of energy is obtained as

$$\left(\frac{1}{2}\rho\,\boldsymbol{u}\cdot\boldsymbol{u}+\rho\,\epsilon\right)^{\cdot}+\rho\left(\frac{1}{2}\boldsymbol{u}\cdot\boldsymbol{u}+\epsilon\right)\operatorname{div}\boldsymbol{u}=-\operatorname{div}\boldsymbol{q}+\operatorname{div}(\boldsymbol{u}\boldsymbol{t})+\rho(\boldsymbol{u}\cdot\boldsymbol{b}+\zeta), \tag{5.3.31}$$

which, by using the local balances of mass and linear momentum, reduces to

$$\rho\,\dot{\epsilon}=-\operatorname{div}\boldsymbol{q}+\operatorname{tr}(\boldsymbol{D}\boldsymbol{t})+\rho\,\zeta, \tag{5.3.32}$$

which is termed alternatively as the *balance of internal energy*, where $\operatorname{tr}(\boldsymbol{D}\boldsymbol{t})$ is the power per unit area that the Cauchy stress acts on the velocity gradient, termed the *stress power*. It follows that the frictional stress power provides a positive influence to increase the internal energy as reflected by a temperature increase and is considered a production to the internal energy. As similar to the linear momentum balance, it is at the moment not possible to use Eq. (5.3.32), for the internal energy should be prescribed a priori by using a material equation. It is equally not possible to derive Eq. (5.3.32) directly from Eq. (5.3.25) due to the shrinking of system boundary from the integral to differential domains of interest.

The energetic perspective of a fluid can also be established by formulating the *balance of kinetic energy*. Taking inner product of the local balance of linear momentum with the velocity yields

$$\boldsymbol{u}\cdot(\rho\,\dot{\boldsymbol{u}}-\operatorname{div}\boldsymbol{t}-\rho\,\boldsymbol{b})=0, \tag{5.3.33}$$

which reduces to

$$\rho\left(\frac{\boldsymbol{u}\cdot\boldsymbol{u}}{2}\right)^{\cdot}=\operatorname{div}(\boldsymbol{u}\boldsymbol{t})-\operatorname{tr}(\boldsymbol{D}\boldsymbol{t})+\boldsymbol{u}\cdot\rho\,\boldsymbol{b}, \tag{5.3.34}$$

in which the stress power has a negative sign when compared to Eq. (5.3.32). This means that the stress power acts as an annihilation of the kinetic energy. This is quite natural for the annihilated energy generates heat and provides a production to the internal energy.

5.3.5 Balance of Entropy

The balance of entropy corresponds to *second law of thermodynamics*. As similar to the internal energy that is implied by the first law of thermodynamics, the second law of thermodynamics implies the existence of a physical property, called the *entropy* S, which acts as a measure of the *irreversibility* of a physical process.[14] At the present stage, the entropy and temperature of a material body are defined as the

[14]The entropy of a material is microscopically interpreted as a measure of the disorder of atomic and molecular structures of that material, first proposed by Boltzmann. This topic will be explored in a detailed manner in Sect. 11.5.5. Without loss of generality, a *reversible process* is that in which the system and surrounding restore to their initial states if the process is reversed without any net change to the surrounding. If it is not the case, the process is referred to as an *irreversible process*. Ludwig Eduard Boltzmann, 1844–1906, an Austrian physicist, whose contribution was in the development of statistical mechanics and statistical thermodynamics.

primitive variables to simplify the analysis and the second law of thermodynamics reads: "during a *physically admissible process* the production of entropy should be nonnegative."[15] Thus, the densities of global balance statement are prescribed by

$$\aleph_\phi = \eta, \qquad \pi_\aleph = \pi_\eta \geq 0, \qquad \sigma_\aleph = s_\eta, \qquad \boldsymbol{\psi}_\aleph = \boldsymbol{\phi}_\eta, \tag{5.3.35}$$

where η is the *specific entropy*, π_η represents the mass density of entropy production, s_η stands for the mass density of entropy supply, and $\boldsymbol{\phi}_\eta$ denotes the entropy flux.

With these, the global balance of entropy is obtained as

$$\frac{\partial}{\partial t}\int_V \eta\,\rho\,\mathrm{d}v + \int_A \eta\,(\rho\boldsymbol{u}\cdot\boldsymbol{n})\,\mathrm{d}a = \int_V (\pi_\eta + s_\eta)\,\rho\,\mathrm{d}v - \int_A \boldsymbol{\phi}_\eta\cdot\boldsymbol{n}\,\mathrm{d}a, \tag{5.3.36}$$

or alternatively,

$$\frac{\partial}{\partial t}\int_V \eta\,\rho\,\mathrm{d}v + \int_A \eta\,(\rho\boldsymbol{u}\cdot\boldsymbol{n})\,\mathrm{d}a - \int_V s_\eta\,\rho\,\mathrm{d}v + \int_A \boldsymbol{\phi}_\eta\cdot\boldsymbol{n}\,\mathrm{d}a = \int_V \pi_\eta\,\rho\,\mathrm{d}v \geq 0, \tag{5.3.37}$$

where the conditions of ">0" and "$=0$" are assigned for reversible and irreversible processes, respectively. These two equations indicate that the time change of entropy of the fluids within a finite CV plus the change in entropy of the fluids entering and leaving the CS per unit time should be balanced by all possible external entropy supplies, entropy fluxes and the most important contribution, the entropy productions inside the CV. Conversely, the entropy production of a fluid body during a physically admissible process should always be nonnegative. Applications of Eq. (5.3.36) or (5.3.37) are not possible at the moment, for the entropy of a material needs to be described by a material equation. However, a simple illustration can be given.

Consider a bottle filled with water, which is sealed and placed on a horizontal table, and a heat flux $\boldsymbol{q}$ is supplied to the bottle from its surrounding without other entropy/energy supplies and fluxes. In the considered circumstance, the water in the bottle is exactly a control-mass system. Let the entropy of water after the heating process be denoted by S_2, and that before the heating process be denoted by S_1, and the relation between heat and entropy fluxes be given by

$$\boldsymbol{\phi}_\eta = \frac{\boldsymbol{q}}{\theta}, \tag{5.3.38}$$

which is known as the *Duhem-Truesdell relation*, where θ is an absolute temperature scale.[16] With this, Eq. (5.3.37) reduces to

$$(S_2 - S_1) + \int_A \frac{\boldsymbol{q}\cdot\boldsymbol{n}}{\theta}\mathrm{d}a = \int_V \rho\pi_\eta\mathrm{d}v \geq 0, \tag{5.3.39}$$

[15] A physically admissible process is one in which all balances of mass, linear, and angular momentums, energy and entropy are satisfied simultaneously.

[16] Another Duhem-Truesdell relation is the relation between entropy supply s_η and energy supply ζ given by

$$s_\eta = \frac{\zeta}{\theta}.$$

More general formulations on the entropy flux and entropy supply can be accomplished by using the Müller-Liu entropy principle, which will be discussed in Sect. 11.6.1. Pierre Maurice Marie Duhem, 1861–1916, a French physicist and mathematician, who is best known for his works on chemical thermodynamics, hydrodynamics, and the theory of elasticity. Clifford Ambrose Truesdell, 1919–2000, an American mathematician, natural philosopher, and historian of science, who, together with Noll, contributed to foundational rational mechanics.

where A denotes the surface, across which the heat transfer takes place. If the heating process is accomplished at a constant temperature, i.e., a reversible heating process, Eq. (5.3.39) becomes

$$S_2 - S_1 + \frac{1}{\theta}\int_A \boldsymbol{q}\cdot\boldsymbol{n}\,\mathrm{d}a = \int_V \rho\,\pi_\eta\,\mathrm{d}v, \tag{5.3.40}$$

which reduces to

$$\int_V \rho\,\pi_\eta\,\mathrm{d}v = 0, \tag{5.3.41}$$

for

$$S_2 - S_1 = -\frac{1}{\theta}\int_A \boldsymbol{q}\cdot\boldsymbol{n}\,\mathrm{d}a, \tag{5.3.42}$$

which is the classical definition of *entropy change* for a control-mass system in classical thermodynamics. However, if the heating process takes place at a finite temperature difference between the system and surrounding, the entropy change between any two states of the system cannot be determined by the above equation. The irreversibility generated by a finite temperature difference will result in a positive entropy production of the system and its surrounding.

The local balance of entropy is given by

$$(\rho\,\eta)^{\cdot} + \rho\,\eta\,\mathrm{div}\,\boldsymbol{u} = -\mathrm{div}\,\boldsymbol{\phi}_\eta + \rho\,\pi_\eta + \rho\,s_\eta, \tag{5.3.43}$$

which, by using the material derivative and local mass balance, is expressed alternatively as

$$\rho\pi_\eta = \rho\dot{\eta} + \mathrm{div}\left(\frac{\boldsymbol{q}}{\theta}\right) - \rho\frac{\zeta}{\theta} \geq 0, \tag{5.3.44}$$

in which the Duhem-Truesdell relations have been used to express the entropy flux and entropy supply. Equation (5.3.44) is termed the *Clausius-Duhem inequality*[17] and is used frequently to derive the material equations mathematically in the context of *continuum thermodynamics*.

5.3.6 Reynolds' Transport Theorem and Material Derivative

Let ϕ be any extensive quantity of a fluid body $\mathcal{B}$, and its mass density be denoted by $\aleph_\phi$. The time rate of change of ϕ in $\mathcal{B}$ under the Eulerian description, by using Reynolds' transport theorem, is given viz.,

$$\phi = \int_V \aleph_\phi\,\rho\,\mathrm{d}v, \qquad \frac{\mathrm{d}\phi}{\mathrm{d}t} = \frac{\partial}{\partial t}\int_V \aleph_\phi\,\rho\,\mathrm{d}v + \int_A \aleph_\phi\,(\rho\,\boldsymbol{u}\cdot\boldsymbol{n})\,\mathrm{d}a. \tag{5.3.45}$$

Equation $(5.3.45)_2$, by using the Gauss theorem, is recast alternatively as

$$\frac{\mathrm{d}\phi}{\mathrm{d}t} = \frac{\partial}{\partial t}\int_V \aleph_\phi\,\rho\,\mathrm{d}v + \int_V \mathrm{div}\,(\aleph_\phi\,\rho\,\boldsymbol{u})\,\mathrm{d}v = \int_V \left[\frac{\partial(\aleph_\phi\,\rho)}{\partial t} + \mathrm{div}\,(\aleph_\phi\,\rho\,\boldsymbol{u})\right]\mathrm{d}v. \tag{5.3.46}$$

[17]Rudolf Julius Emanuel Clausius, 1822–1888, a German physicist and mathematician, who is considered one of the central founders of the science of thermodynamics.

Table 5.4 Time rate of change of any extensive variable ϕ and the balance statements of physical laws in inertial frame in integral and differential forms

	Integral form	Differential form
$\dot{\phi}$	$\dot{\phi}=\frac{\partial}{\partial t}\int_V \aleph_\phi \rho\,\mathrm{d}v+\int_A \aleph_\phi(\rho\boldsymbol{u}\cdot\boldsymbol{n})\,\mathrm{d}a$	$\dot{\aleph}_\phi=\frac{\mathrm{d}\aleph_\phi}{\mathrm{d}t}=\frac{\partial\aleph_\phi}{\partial t}+\operatorname{grad}\aleph_\phi\cdot\boldsymbol{u}$ $\aleph_\phi=\frac{\phi}{m}$
Mass	$\frac{\partial}{\partial t}\int_V \rho\,\mathrm{d}v+\int_A \rho(\boldsymbol{u}\cdot\boldsymbol{n})\,\mathrm{d}a=0$	$\dot{\rho}+\rho\operatorname{div}\boldsymbol{u}=0$
Linear momentum	$\frac{\partial}{\partial t}\int_V \rho\boldsymbol{u}\,\mathrm{d}v+\int_A \boldsymbol{u}(\rho\boldsymbol{u}\cdot\boldsymbol{n})\,\mathrm{d}a$ $=\sum\mathcal{F}_{CV}+\sum\mathcal{F}_{CS}$	$\rho\dot{\boldsymbol{u}}=\operatorname{div}\boldsymbol{t}+\rho\boldsymbol{b}$
Angular momentum	$\frac{\partial}{\partial t}\int_V (\boldsymbol{x}\times\boldsymbol{u})\rho\,\mathrm{d}v+\int_A (\boldsymbol{x}\times\boldsymbol{u})(\rho\boldsymbol{u}\cdot\boldsymbol{n})\,\mathrm{d}a$ $=\sum\boldsymbol{x}\times\mathcal{F}_{CV}+\sum\boldsymbol{x}\times\mathcal{F}_{CS}+\sum\boldsymbol{M}_{shaft}$	$\boldsymbol{t}=\boldsymbol{t}^{\mathrm{T}}$
Energy	$\frac{\partial}{\partial t}\int_V \rho e\,\mathrm{d}v+\int_A\left(h+\frac{1}{2}\boldsymbol{u}\cdot\boldsymbol{u}+gz\right)(\rho\boldsymbol{u}\cdot\boldsymbol{n})\,\mathrm{d}a$ $=\sum\dot{Q}+\sum\dot{W}_{shear}+\sum\dot{W}_{shaft}+\sum\dot{E}_s$	$\rho\dot{\epsilon}=-\operatorname{div}\boldsymbol{q}+\operatorname{tr}(\boldsymbol{Dt})+\rho\zeta$
Entropy	$\frac{\partial}{\partial t}\int_V \eta\rho\,\mathrm{d}v+\int_A \eta(\rho\boldsymbol{u}\cdot\boldsymbol{n})\,\mathrm{d}a-\int_V s_\eta\rho\,\mathrm{d}v$ $+\int_A \phi_\eta\cdot\boldsymbol{n}\,\mathrm{d}a=\int_V \pi_\eta\rho\,\mathrm{d}v\geq 0$	$\rho\pi_\eta=\rho\dot{\eta}+\operatorname{div}\left(\frac{\boldsymbol{q}}{\theta}\right)-\rho\frac{\zeta}{\theta}\geq 0$

The time rate of change of $\aleph_\phi$, in terms of the material derivative, is given by

$$\frac{\mathrm{d}\aleph_\phi}{\mathrm{d}t}=\frac{\partial\aleph_\phi}{\partial t}+\operatorname{grad}\aleph_\phi\cdot\boldsymbol{u}=\frac{\partial(\aleph_\phi\,\rho)}{\partial t}+\operatorname{div}(\aleph_\phi\,\rho\,\boldsymbol{u}), \tag{5.3.47}$$

in which the local mass balance has been used. The last two equations show that Reynolds' transport theorem and the material derivative are essentially the same. They both represent the time rate of change of a quantity from the Lagrangian description to the Eulerian description. While Reynolds' transport theorem is a global expression, the material derivative is a local expression. Alternatively, the global expression of material derivative is Reynolds' transport theorem and vice versa.

Table 5.4 summarizes the results of the time rate of change of any extensive variable ϕ and the balance equations of five physical laws in inertia frame in the integral and differential forms, in which the Duhem-Truesdell relations are used in the differential balance of entropy.

5.4 Moving Reference Frame

5.4.1 Transformations of Position Vector, Velocity and Acceleration

Consider the present configuration $\mathcal{B}^P$ at time t of a fluid body $\mathcal{B}$, as shown in Fig. 5.2. Every fluid element inside $\mathcal{B}^P$ is identified by using its position vector, which can be represented by $\boldsymbol{x}$ relative to a fixed reference frame $\mathcal{O}$ with the orthonormal base $\boldsymbol{e}_i$, or by $\boldsymbol{y}$ relative to a moving reference frame $\mathcal{O}'$ with the orthonormal base $\boldsymbol{e}'_i$, which has an arbitrary motion relative to $\mathcal{O}$, e.g. a translation and/or a rotation. The transformation from $\boldsymbol{e}_i$ to $\boldsymbol{e}'_i$ is described by the orthogonal tensor $\boldsymbol{Q}$ with det $\boldsymbol{Q} = 1$, i.e., the right-handed oriented base is followed. The relation between $\boldsymbol{x}$ and $\boldsymbol{y}$ is given by

$$\boldsymbol{x} = \boldsymbol{y} + \boldsymbol{c}, \tag{5.4.1}$$

where $\boldsymbol{c}$ is the position vector of $\mathcal{O}'$ relative to $\mathcal{O}$ and is described by using the orthonormal base $\boldsymbol{e}_i$. Since $\boldsymbol{y}$ can be expressed in terms of either $\boldsymbol{e}_i$ or $\boldsymbol{e}'_i$, it follows that

$$\boldsymbol{y}' = \boldsymbol{Q}^{\mathrm{T}}\boldsymbol{y}, \qquad \boldsymbol{y} = \boldsymbol{Q}\boldsymbol{y}', \tag{5.4.2}$$

where $\boldsymbol{y}'$ and $\boldsymbol{y}$ are expressed in terms of the reference frames $\mathcal{O}'$ and $\mathcal{O}$, respectively.[18]

It follows from the above two equations that

$$\boldsymbol{x} = \boldsymbol{Q}\boldsymbol{y}' + \boldsymbol{c}, \qquad \boldsymbol{y}' = \boldsymbol{Q}^{\mathrm{T}}\boldsymbol{x} - \boldsymbol{Q}^{\mathrm{T}}\boldsymbol{c}, \tag{5.4.3}$$

which is called the *Euclidean transformation*,[19] delivering the transformation of position vector between different orthonormal bases.

The time rate of change of $\boldsymbol{y}$, by using Eq. $(5.4.2)_2$, is given viz.,

$$\dot{\boldsymbol{y}} = \left(\boldsymbol{Q}\boldsymbol{y}'\right)^{\cdot} = \dot{\boldsymbol{Q}}\boldsymbol{y}' + \boldsymbol{Q}\left(\boldsymbol{y}'\right)^{\cdot}, \tag{5.4.4}$$

---

[18]For example, consider a two-dimensional Cartesian coordinate system which is spanned by the fixed orthonormal bases $\boldsymbol{e}_1$ and $\boldsymbol{e}_2$, and a new coordinate system $\{\boldsymbol{e}'_1, \boldsymbol{e}'_2\}$ is obtained by rotating the $\{\boldsymbol{e}_1, \boldsymbol{e}_2\}$ counterclockwise by an angle 30°. In this case, $\boldsymbol{Q}$ is given by

$$[\boldsymbol{Q}] = \begin{bmatrix} \cos 30^\circ & -\sin 30^\circ \\ \sin 30^\circ & \cos 30^\circ \end{bmatrix} = \begin{bmatrix} \sqrt{3}/2 & -1/2 \\ 1/2 & \sqrt{3}/2 \end{bmatrix}.$$

A point is described by $\boldsymbol{y} = [2, 2]^{\mathrm{T}}$ in the $\{\boldsymbol{e}_1, \boldsymbol{e}_2\}$ system, the vector $\boldsymbol{y}'$ of the same point is then obtained as

$$[\boldsymbol{y}'] = [\boldsymbol{Q}^{\mathrm{T}}][\boldsymbol{y}] = \begin{bmatrix} \sqrt{3}/2 & 1/2 \\ -1/2 & \sqrt{3}/2 \end{bmatrix}\begin{bmatrix} 2 \\ 2 \end{bmatrix} = \begin{bmatrix} \sqrt{3}+1 \\ \sqrt{3}-1 \end{bmatrix}.$$

[19]Euclid of Alexandria, c. Mid-fourth century to Mid-third century BC., a Greek mathematician, whom is often referred to as "Father of Geometry."

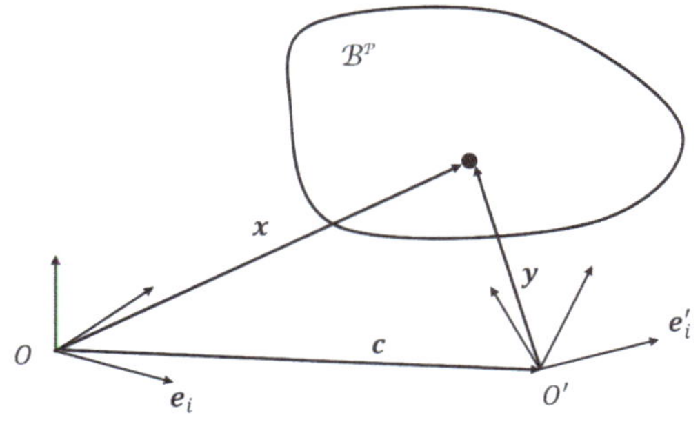

Fig. 5.2 A fixed reference frame $\mathcal{O}$ with the orthonormal base e_i, and a moving reference frame $\mathcal{O}'$ with the orthonormal base e'_i, with the corresponding position vectors x and y of a material point in the present configuration $\mathcal{B}^P$

where the first term on the right-hand-side represents the (absolute) time rate of change of y in the moving reference frame, while the second term denotes the change in y in the moving reference frame. Combining the above equation with Eq. (5.4.2)$_1$ yields

$$\dot{y} = (\dot{Q}\,Q^{\mathrm{T}})\,y + Q\,(y')^{\cdot}\,, \tag{5.4.5}$$

with which the time rate of change of x in Eq. (5.4.1) is obtained as

$$\dot{x} = \dot{c} + (\dot{Q}\,Q^{\mathrm{T}})\,y + Q\,(y')^{\cdot}\,. \tag{5.4.6}$$

This equation describes the velocities of a material point in different reference frames. Specifically, it is rewritten as

$$\dot{x} = u_f + u_{rel}, \qquad u_f = \dot{c} + (\dot{Q}\,Q^{\mathrm{T}})\,y, \qquad u_{rel} = Q\,(y')^{\cdot}\,, \tag{5.4.7}$$

where $\dot{x}$ is the velocity of material point expressed in terms of the fixed reference frame, while u_f and u_{rel} are called the *frozen velocity* and *relative velocity*, respectively. The relative velocity is that of the material point one measures if one moves coherently with the moving reference frame. On the contrary, the frozen velocity is that of the material point that is measured in terms of the fixed reference frame if the material point is *momentarily frozen* in the moving reference frame. In other words, it is the velocity of $\mathcal{O}'$ relative to $\mathcal{O}$, which consists of two contributions: the *translation velocity* $\dot{c}$ and *rotation velocity* $(\dot{Q}\,Q^{\mathrm{T}})$ of $\mathcal{O}'$. It follows from the property of orthogonal tensor that

$$(Q\,Q^{\mathrm{T}})^{\cdot} = \dot{Q}\,Q^{\mathrm{T}} + Q\,\dot{Q}^{\mathrm{T}} = \dot{I} = 0, \tag{5.4.8}$$

giving rise to

$$\dot{Q}\,Q^{\mathrm{T}} = -Q\,\dot{Q}^{\mathrm{T}} = -(\dot{Q}\,Q^{\mathrm{T}})^{\mathrm{T}}. \tag{5.4.9}$$

Thus, the rotation velocity is a skew-symmetric tensor and can be expressed by using its dual vector ω, viz.,

$$(\dot{Q}\,Q^{\mathrm{T}})\,a = \omega \times a, \qquad \omega_i = -\frac{1}{2}\varepsilon_{ijk}\,(\dot{Q}\,Q^{\mathrm{T}})_{jk}\,, \tag{5.4.10}$$

for any vector a.

Conducting again the time rate of change of Eq. (5.4.6) results in

$$\ddot{\boldsymbol{x}} = \ddot{\boldsymbol{c}} + \left(\dot{\boldsymbol{Q}}\,\boldsymbol{Q}^{\mathrm{T}}\right)^{\cdot}\,\boldsymbol{y} + \left(\dot{\boldsymbol{Q}}\,\boldsymbol{Q}^{\mathrm{T}}\right)\left(\dot{\boldsymbol{Q}}\,\boldsymbol{Q}^{\mathrm{T}}\right)\boldsymbol{y} + 2\left(\dot{\boldsymbol{Q}}\,\boldsymbol{Q}^{\mathrm{T}}\right)\boldsymbol{Q}\left(\boldsymbol{y}'\right)^{\cdot} + \boldsymbol{Q}\left(\boldsymbol{y}'\right)^{\cdot\cdot}, \tag{5.4.11}$$

in which Eq. (5.4.5) has been used. This equation is specifically rewritten as

$$\ddot{\boldsymbol{x}} = \boldsymbol{a}_f + \boldsymbol{a}_c + \boldsymbol{a}_{rel}, \tag{5.4.12}$$

with

$$\begin{aligned} \boldsymbol{a}_f &= \ddot{\boldsymbol{c}} + \left(\dot{\boldsymbol{Q}}\,\boldsymbol{Q}^{\mathrm{T}}\right)^{\cdot}\,\boldsymbol{y} + \left(\dot{\boldsymbol{Q}}\,\boldsymbol{Q}^{\mathrm{T}}\right)\left(\dot{\boldsymbol{Q}}\,\boldsymbol{Q}^{\mathrm{T}}\right)\boldsymbol{y}, \\ \boldsymbol{a}_c &= 2\left(\dot{\boldsymbol{Q}}\,\boldsymbol{Q}^{\mathrm{T}}\right)\boldsymbol{Q}\left(\boldsymbol{y}'\right)^{\cdot}, \quad \boldsymbol{a}_{rel} = \boldsymbol{Q}\left(\boldsymbol{y}'\right)^{\cdot\cdot}. \end{aligned} \tag{5.4.13}$$

The term $\boldsymbol{a}_f$ is called the *acceleration frozen to the moving reference frame*, $\boldsymbol{a}_c$ is termed the *Coriolis acceleration*,[20] and $\boldsymbol{a}_{rel}$ represents the *relative acceleration*. The acceleration $\ddot{\boldsymbol{x}}$ of a material point in terms of fixed reference frame consists of three contributions: $\boldsymbol{a}_f$, the acceleration that is measured if the material point is momentarily frozen to the moving reference frame, in other words, the acceleration of $\mathcal{O}'$ relative to $\mathcal{O}$; $\boldsymbol{a}_c$, the acceleration that is induced by the rotation of velocity in the moving reference frame, i.e., rotation of $\boldsymbol{u}_{rel}$ relative to the fixed reference frame; and $\boldsymbol{a}_{rel}$, the acceleration that one measures in the moving reference frame.

By using the dual vector $\boldsymbol{\omega}$, Eqs. (5.4.6) and (5.4.11) are expressed alternatively as

$$\begin{aligned} \dot{\boldsymbol{x}} &= \dot{\boldsymbol{c}} + \boldsymbol{\omega} \times \boldsymbol{y} + \boldsymbol{u}_{rel}, \\ \ddot{\boldsymbol{x}} &= \ddot{\boldsymbol{c}} + \dot{\boldsymbol{\omega}} \times \boldsymbol{y} + \boldsymbol{\omega} \times (\boldsymbol{\omega} \times \boldsymbol{y}) + 2\boldsymbol{\omega} \times \boldsymbol{u}_{rel} + \boldsymbol{a}_{rel}, \end{aligned} \tag{5.4.14}$$

where $\boldsymbol{x}$ and $\boldsymbol{y}$ are the position vectors of a material point in $\mathcal{O}$ and $\mathcal{O}'$, respectively, with the corresponding velocities $\dot{\boldsymbol{x}}$ and $\boldsymbol{u}_{rel}$, and accelerations $\ddot{\boldsymbol{x}}$ and $\boldsymbol{a}_{rel}$. The term $\boldsymbol{\omega}$ becomes the rotational velocity of $\mathcal{O}'$ relative to $\mathcal{O}$; $\dot{\boldsymbol{\omega}} \times \boldsymbol{y}$ and $\boldsymbol{\omega} \times (\boldsymbol{\omega} \times \boldsymbol{y})$ are termed specifically the *Euler* and *centrifugal accelerations*, respectively.

5.4.2 Invariance and Indifference of Variables and Equations

Consider a fluid body subject to two reference frames, one is fixed and the other is moving. All physical variables of fluid body and the mathematical equations describing the relations among the physical variables can in principle be expressed in terms of either the fixed reference frame or the moving reference frame. Essentially, the mathematical forms (expressions) are different, which are called *frame dependent*.

Physical variables which do not explicitly involve frame dependency are termed *objective*, or termed *non-objective* if it is not the case. Equally, an equation is termed *invariant*, if it does not change its form under a transformation of reference frame.

[20]Gaspard-Gustave de Coriolis, 1792–1843, a French mathematician, who is best known for his work on the supplementary forces that are detected in a rotating reference frame, leading to the *Coriolis effect*.

However, when applied, this concept complies that various terms appearing in the equation are interpreted differently. For example, consider Newton's second law of motion, $\sum \boldsymbol{F} = m\boldsymbol{a}$, as applied in a moving reference frame, in which $\sum \boldsymbol{F}$ comprises not only all external forces, but also those induced by the influence of moving reference frame, termed the *virtual forces* $\boldsymbol{F}_{virt}$, since they act on the material body without doing work. In this circumstance, Newton's second law of motion is recast as $\sum \boldsymbol{F} + \boldsymbol{F}_{virt} = m\boldsymbol{a}_{rel}$, where $\boldsymbol{a}_{rel}$ is the acceleration that one measures in the moving reference frame, and $\boldsymbol{F}_{virt} = -m(\ddot{\boldsymbol{c}} + \dot{\boldsymbol{\omega}} \times \boldsymbol{y} + \boldsymbol{\omega} \times (\boldsymbol{\omega} \times \boldsymbol{y}) + 2\boldsymbol{\omega} \times \boldsymbol{u}_{rel})$ in the most general case. Thus, although Newton's second law of motion is invariant, it is still frame dependent. An equation is termed *indifferent* if no frame dependency among various terms of the equation appears under a transformation of reference frame. In other words, an indifferent equation has the same mathematical form in different reference frames.

An equation may be indifferent for a group of frame transformation, but not for other groups. For example, Newton's second law of motion is indifferent when subject to the *Galilean transformation*, but it is not indifferent with respect to the Euclidean transformation; however, it is invariant in both transformations. Since the Galilean transformation only deals with relative motions with constant relative velocity, the discussions on the invariance of balance equations of the physical laws in the next subsection will be based on the Euclidean transformation. Let a, $\boldsymbol{a}$, and $\boldsymbol{t}$ be any scalar, vector, and second-order tensor, respectively. They are called *objective scalar*, *objective vector*, and *objective tensor*, if the following relations are satisfied:

$$a' = a, \qquad \boldsymbol{a}' = \boldsymbol{Q}^{\mathrm{T}}\boldsymbol{a}, \qquad \boldsymbol{t}' = \boldsymbol{Q}^{\mathrm{T}}\boldsymbol{t}\,\boldsymbol{Q}, \tag{5.4.15}$$

under the Euclidean transformation of reference frame.

5.4.3 Balance Equations of Physical Laws in Moving Reference Frame

The scalar quantities in the five physical laws are physical properties assuming the same values in different reference frames, satisfying Eq. $(5.4.15)_1$, and are all objective. The heat flux $\boldsymbol{q}$ and Cauchy stress tensor $\boldsymbol{t}$ belong to the material equations and can be made to be objective by choosing appropriate material descriptions.[21] However, the velocity $\boldsymbol{u}$ and acceleration $\boldsymbol{a}$, in view of Eq. (5.4.14), are not objective vectors. They have different expressions in different reference frames.[22] Specifically, the velocity, by using the Euclidean transformation, is rewritten for convenience as

$$\boldsymbol{u}' = \boldsymbol{Q}^{\mathrm{T}}\boldsymbol{u} + \dot{\boldsymbol{Q}}^{\mathrm{T}}\boldsymbol{x} - \left(\boldsymbol{Q}^{\mathrm{T}}\boldsymbol{c}\right)^{\cdot}, \qquad \boldsymbol{u} = \boldsymbol{Q}\boldsymbol{u}' + \dot{\boldsymbol{Q}}\boldsymbol{x}' + \dot{\boldsymbol{c}}. \tag{5.4.16}$$

Various terms in the balance statements of physical laws involve time and spatial derivatives, which need to be explored and are discussed in the following. For convenience, the prime is used to denote the quantities and mathematical operations in

[21] A detailed discussion on the topic will be provided in Sect. 5.6.

[22] But they are objective under the Galilean transformation.

a moving reference frame, while unprimed quantities and mathematical operations are referred to a fixed reference frame.

First, since $\boldsymbol{q}$ is an objective vector, it follows that

$$\begin{aligned} \operatorname{div}' \boldsymbol{q}' &= \frac{\partial q_i'}{\partial x_i'} = \frac{\partial}{\partial x_i'}\left(Q_{ji}q_j\right) = Q_{ji}\frac{\partial q_j}{\partial x_k}\frac{\partial x_k}{\partial x_i'} \\ &= Q_{ji}Q_{ki}\frac{\partial q_j}{\partial x_k} = \delta_{jk}\frac{\partial q_j}{\partial x_k} = \frac{\partial q_j}{\partial x_j} = \operatorname{div}\boldsymbol{q}, \end{aligned} \tag{5.4.17}$$

indicating that the divergence of heat flux is indifferent, resulting in an objective scalar. Second, the divergence of the Cauchy stress tensor reads

$$\begin{aligned} [\operatorname{div}' \boldsymbol{t}']_i &= \frac{\partial t_{ij}'}{\partial x_j'} = \frac{\partial}{\partial x_j'}\left(Q_{ji}Q_{ij}t_{ij}\right) = Q_{ji}Q_{ij}\frac{\partial t_{ij}}{\partial x_k}\frac{\partial x_k}{\partial x_j'} \\ &= Q_{ji}Q_{ij}Q_{kj}\frac{\partial t_{ij}}{\partial x_k} = Q_{ji}\delta_{ik}\frac{\partial t_{ij}}{\partial x_k} = Q_{ij}\frac{\partial t_{ij}}{\partial x_j} = \left[\boldsymbol{Q}^{\mathrm{T}}(\operatorname{div}\boldsymbol{t})\right]_i, \end{aligned} \tag{5.4.18}$$

showing that

$$\operatorname{div}' \boldsymbol{t}' = \boldsymbol{Q}^{\mathrm{T}}(\operatorname{div}\boldsymbol{t}). \tag{5.4.19}$$

That is, the divergence of an objective symmetric tensor is an objective vector. Furthermore, the divergence of velocity reads

$$\begin{aligned} \operatorname{div}' \boldsymbol{u}' &= \frac{\partial u_i'}{\partial x_i'} = \frac{\partial}{\partial x_i'}\left(Q_{ji}u_j + \dot{Q}_{ji}x_j - (Q_{ji}c_j)^{\cdot}\right) = Q_{ji}\frac{\partial u_j}{\partial x_k}\frac{\partial x_k}{\partial x_i'} + \dot{Q}_{ji}\frac{\partial x_j}{\partial x_i'} \\ &= Q_{ji}Q_{ki}\frac{\partial u_j}{\partial x_k} + \dot{Q}_{ji}Q_{ij} = \delta_{jk}\frac{\partial u_j}{\partial x_k} = \frac{\partial u_j}{\partial x_j} = \operatorname{div}\boldsymbol{u}, \end{aligned} \tag{5.4.20}$$

for $(Q_{ji}c_j)^{\cdot} = 0$, since it is not frame dependent, and $\dot{Q}_{ji}Q_{ij} = 0$, because $\dot{Q}_{ji}Q_{ij} = \operatorname{tr}(\dot{\boldsymbol{Q}}^{\mathrm{T}}\boldsymbol{Q}) = 0$. Equation (5.4.20) shows that the divergence of velocity is indifferent. Let ϕ be any scalar quantity in the physical laws, e.g. the density, specific internal energy, temperature, etc. Since it is an objective scalar and the time measure remains unchanged in different reference frames, it follows that

$$\frac{\mathrm{d}\phi'}{\mathrm{d}t} = \frac{\mathrm{d}\phi}{\mathrm{d}t}. \tag{5.4.21}$$

Last, the transformation of the time rate of change of velocity (i.e., the acceleration) reads

$$\dot{\boldsymbol{u}} = \boldsymbol{Q}\dot{\boldsymbol{u}}' + 2\dot{\boldsymbol{Q}}\boldsymbol{u}' + \ddot{\boldsymbol{Q}}\boldsymbol{x}' + \ddot{\boldsymbol{c}}, \tag{5.4.22}$$

and it is not difficult to show that the transformation of velocity gradient $\boldsymbol{L}$ is obtained as

$$\boldsymbol{L}' = \boldsymbol{Q}^{\mathrm{T}}\boldsymbol{L}\boldsymbol{Q} + \dot{\boldsymbol{Q}}^{\mathrm{T}}\boldsymbol{Q}, \tag{5.4.23}$$

from which the stretching tensor $\boldsymbol{D}$ is shown to be an objective tensor. The derivation of the above equation is left as an exercise. With these, the transformation of stress power reads

$$\operatorname{tr}(\boldsymbol{D}'\boldsymbol{t}') = \operatorname{tr}\left(\boldsymbol{Q}^{\mathrm{T}}\boldsymbol{D}\boldsymbol{Q}\boldsymbol{Q}^{\mathrm{T}}\boldsymbol{t}\boldsymbol{Q}\right) = \operatorname{tr}\left(\boldsymbol{Q}^{\mathrm{T}}\boldsymbol{D}\boldsymbol{t}\boldsymbol{Q}\right) = \operatorname{tr}(\boldsymbol{D}\boldsymbol{t}), \tag{5.4.24}$$

showing that it is an objective scalar.

With the results derived previously, the global and local balance equations of physical laws in a moving reference frame are summarized in the following:

- Mass balance:

$$\frac{\partial}{\partial t}\int_{V'}\rho' \mathrm{d}v' + \int_{A'}\rho'(\boldsymbol{u}'\cdot\boldsymbol{n}')\mathrm{d}a' = 0, \tag{5.4.25}$$
$$\dot{\rho}' + \rho' \mathrm{div}'\,\boldsymbol{u}' = 0,$$

  showing that the balance of mass is indifferent.
- Linear momentum balance:

$$\begin{aligned}\sum\mathcal{F}'_{CV} + \sum\mathcal{F}'_{CS} - \int_{V'}\left(\ddot{\boldsymbol{c}} + \dot{\boldsymbol{\omega}}\times\boldsymbol{y} + \boldsymbol{\omega}\times(\boldsymbol{\omega}\times\boldsymbol{y}) + 2\boldsymbol{\omega}\times\boldsymbol{u}'\right)\rho'\mathrm{d}v' \\ = \frac{\partial}{\partial t}\int_{V'}\rho'\boldsymbol{u}'\mathrm{d}v' + \int_{A'}\boldsymbol{u}'(\rho'\boldsymbol{u}'\cdot\boldsymbol{n}')\mathrm{d}a', \\ \rho'\dot{\boldsymbol{u}}' = \mathrm{div}'\,\boldsymbol{t}' + \rho'\boldsymbol{b}' - \rho'\left(\ddot{\boldsymbol{c}} + \dot{\boldsymbol{\omega}}\times\boldsymbol{y} + \boldsymbol{\omega}\times(\boldsymbol{\omega}\times\boldsymbol{y}) + 2\boldsymbol{\omega}\times\boldsymbol{u}'\right),\end{aligned} \tag{5.4.26}$$

  where $\boldsymbol{b}$ is assumed to be an objective vector, and $\boldsymbol{y}$ represents the position vector in the moving reference frame. Equation (5.4.26) shows that the global and local balances of linear momentum are not indifferent, in which the influence of moving reference frame is taken into account by including the virtual forces.
- Angular momentum balance:

$$\begin{aligned}\sum\boldsymbol{y}\times\mathcal{F}'_{CV} + \sum\boldsymbol{y}\times\mathcal{F}'_{CS} + \sum\mathcal{M}'_{shaft} - \int_{V'}(\boldsymbol{u}'\times\dot{\boldsymbol{c}} + \dot{\boldsymbol{c}}\times\boldsymbol{u}')\rho'\mathrm{d}v' \\ - \int_{V'}(\boldsymbol{c}\times\dot{\boldsymbol{u}}')\rho'\mathrm{d}v' - \int_{V'}(\boldsymbol{c}+\boldsymbol{y})\left(\ddot{\boldsymbol{c}} + \dot{\boldsymbol{\omega}}\times\boldsymbol{y} + \boldsymbol{\omega}\times(\boldsymbol{\omega}\times\boldsymbol{y}) + 2\boldsymbol{\omega}\times\boldsymbol{u}'\right)\rho'\mathrm{d}v' \\ = \frac{\partial}{\partial t}\int_{V'}(\boldsymbol{y}\times\rho'\boldsymbol{u}')\mathrm{d}v' + \int_{A'}(\boldsymbol{y}\times\boldsymbol{u}')(\rho'\boldsymbol{u}'\cdot\boldsymbol{n}')\mathrm{d}a', \\ (\boldsymbol{t}')^{\mathrm{T}} = \boldsymbol{t}',\end{aligned} \tag{5.4.27}$$

  showing that the global balance statement is frame-dependent, but the symmetry of the Cauchy stress tensor is indifferent.
- Energy balance:

$$\begin{aligned}\sum\dot{Q}' + \sum\dot{W}'_{shear} + \sum\dot{W}'_{shaft} + \sum\dot{E}'_s \\ = \frac{\partial}{\partial t}\int_{V'}e'\rho'\mathrm{d}v + \int_{A'}\left(h' + \frac{1}{2}\boldsymbol{u}'\cdot\boldsymbol{u}' + g'z'\right)(\rho'\boldsymbol{u}'\cdot\boldsymbol{n}')\mathrm{d}a', \\ \rho'\dot{\epsilon}' = -\mathrm{div}'\,\boldsymbol{q}' + \mathrm{tr}(\boldsymbol{D}'\boldsymbol{t}') + \rho'\zeta',\end{aligned} \tag{5.4.28}$$

  in which $g' = \|\boldsymbol{g}'\|$ and z' is the elevation. The above equations show that both global and local balances of energy are indifferent.

- Entropy balance:

$$\frac{\partial}{\partial t}\int_{V'}\rho'\eta'\mathrm{d}v'+\int_{A'}\rho'\eta'(\boldsymbol{u}'\cdot\boldsymbol{n}')\mathrm{d}a'-\int_{V'}\rho s'\mathrm{d}v'+\int_{A'}\boldsymbol{\phi}'_{\eta}\cdot\boldsymbol{n}'\mathrm{d}a'=\int_{V'}\rho'\pi'_{\eta}\mathrm{d}v'\geq 0, \tag{5.4.29}$$

$$\rho'\pi'_{\eta}=\rho'\dot{\eta}'+\mathrm{div}'\left(\frac{\boldsymbol{q}'}{\theta'}\right)-\rho'\frac{\zeta'}{\theta'}\geq 0,$$

showing that the global and local balances of entropy are indifferent.

The global and local balance equations in an inertia reference frame derived previously can now be obtained by simplifying Eqs. (5.4.25)–(5.4.29) directly.

5.5 Illustrations of Global Physical Laws

The global balance statements of physical laws are valid for all materials. In this section, they are applied to study selected problems to demonstrate their applications in describing fluid motions.

5.5.1 Mass Balance

Consider a two-dimensional air flow with constant velocity U passing a fixed horizontal plate, as shown in Fig. 5.3a. The shear stress on the plate prohibits the air flow near the plate, giving rise to a very thin region in which the air velocity is retarded. The thin layer is termed the *boundary layer*, with its thickness denoted by δ, which increases as one moves downstream along the plate. The velocity of air inside the boundary layer is given by

$$\frac{u}{U}=2\left(\frac{y}{\delta}\right)-\left(\frac{y}{\delta}\right)^{2}. \tag{5.5.1}$$

It is required to determine the air flow rate across the edge of boundary layer and its flow direction.

For simplicity, construct the finite control-volume $ABCD$ with the fixed coordinate system $\{x, y\}$ on the plate, as shown in the figure. It is assumed that air is incompressible, so that the m ass balance reduces to

$$\int_{A}\boldsymbol{u}\cdot\boldsymbol{n}\,\mathrm{d}a=0, \tag{5.5.2}$$

which is the continuity equation. Applying this equation to the CS of control-volume $ABCD$ yields

$$\int_{A_{AB}}\boldsymbol{u}\cdot\boldsymbol{n}\,\mathrm{d}a+\int_{A_{BC}}\boldsymbol{u}\cdot\boldsymbol{n}\,\mathrm{d}a+\int_{A_{CD}}\boldsymbol{u}\cdot\boldsymbol{n}\,\mathrm{d}a=0, \tag{5.5.3}$$

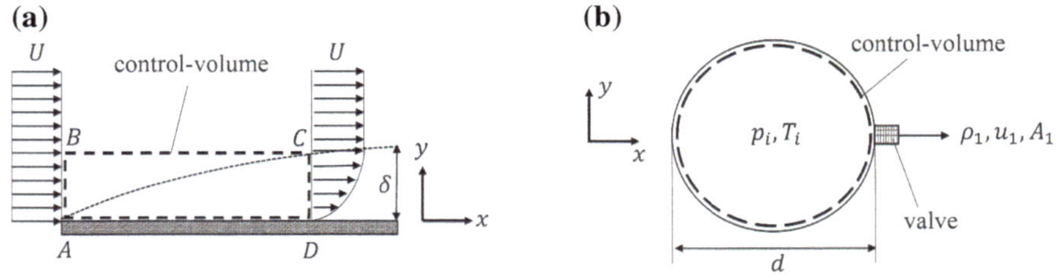

Fig. 5.3 Applications of the global balance of mass. **a** A boundary-layer flow of air passing a horizontal solid plate. **b** Air exhausts from a spherical rigid container

from which the flow rate Q_{BC} across the surface A_{BC} per unit thickness is obtained as

$$Q_{BC} = \int_{A_{BC}} \boldsymbol{u} \cdot \boldsymbol{n}\, \mathrm{d}a = -(-U)\int_0^\delta \mathrm{d}y - U\int_0^\delta \left\{2\left(\frac{y}{\delta}\right) - \left(\frac{y}{\delta}\right)^2\right\} \mathrm{d}y = \frac{1}{3}U\delta. \tag{5.5.4}$$

This result is justified, for the continuity equation implies that the volume of air entering the CV per unit time should be the same as those leaving the CV. Since the velocity on the surface A_{CD} is smaller than that on the surface A_{AB}, there should be an amount of air flowing out of the surface A_{BC}, and its magnitude is simply the difference between the flow rates on A_{CD} and A_{AB}.

Consider a spherical rigid tank filled with air at pressure $p = p_i$ and temperature $T = T_i$, which is connected to a valve, as shown in Fig. 5.3b. The tank is fixed to the ground and initially the valve is closed. At time $t = 0$ the valve is opened, triggering an air jet leaving the tank at speed $u = u_1$ with density ρ_1, which are momentarily constant. The constant cross-sectional area of valve is denoted by A_1. It is required to determine the instantaneous rate of change of density of the air inside the tank at $t = 0$.

Construct the finite control-volume system with the fixed coordinate system $\{x, y\}$ on the ground, as shown in the figure. Applying the mass balance to the CV gives

$$\frac{\partial}{\partial t}\int_V \rho\, \mathrm{d}v + \int_A \rho\boldsymbol{u} \cdot \boldsymbol{n}\, \mathrm{d}a = 0, \tag{5.5.5}$$

which reduces to

$$\frac{\partial}{\partial t}\int_V \rho\, \mathrm{d}v = -\int_A \rho\boldsymbol{u} \cdot \boldsymbol{n}\, \mathrm{d}a = -\int_{A_1} \rho\boldsymbol{u} \cdot \boldsymbol{n}\, \mathrm{d}a. \tag{5.5.6}$$

Since at $t = 0$ the air jet assumes constant velocity u_1 and constant density ρ_1 on the constant cross-sectional area A_1, the right-hand-side of Eq. (5.5.6) becomes

$$-\int_{A_1} \rho\boldsymbol{u} \cdot \boldsymbol{n}\, \mathrm{d}a = -\rho_1 u_1 A_1. \tag{5.5.7}$$

On the other hand, the air jet triggers a sequence of *pressure waves* traveling from the valve toward the end of tank with the speed of sound in air, which is denoted by c. The time needed for the pressure wave to reach the bottom of tank is estimated approximately as $\Delta t \sim d/c$. If $\mathcal{O}(\Delta t) \ll \mathcal{O}(10^{-3})$, the air density inside the tank changes approximately only with time with negligible spatial variations. Taking this as a first engineering approximation to the left-hand-side of Eq. (5.5.6) yields

$$\frac{\partial}{\partial t}\int_V \rho\, \mathrm{d}v \sim \frac{\partial}{\partial t}\left(\rho \int_V \mathrm{d}v\right) = \frac{\partial(\rho V)}{\partial t} = V\frac{\partial \rho}{\partial t}, \tag{5.5.8}$$

where V is the volume of CV (and hence the volume of spherical tank). Substituting the last two equations into Eq. (5.5.6) results in

$$\frac{\partial \rho}{\partial t} = -\frac{\rho_1 u_1 A_1}{V}, \tag{5.5.9}$$

which is justified, for the air jet decreases the amount of air in the tank, as reflected by a negative time rate of air density inside the tank. It is noted that the above result assumes its best accuracy immediately after the opening of valve. As time increases, the assumptions and approximations used in the analysis loss their validities gradually.

5.5.2 Linear Momentum Balance

Consider a water jet striking horizontally a stationary vane, as shown in Fig. 5.4a. The water jet leaving the nozzle assumes constant density ρ and constant velocity u_1 with constant cross-sectional area A_1 and is deflected through an angle θ by the vane. It is required to determine the force acting on the water jet by the vane.

Since the vane is stationary, construct the finite control-volume and locate the fixed coordinates $\{x, y\}$ on the ground, as shown in the figure. To simplify the analysis, the gravitational force and the friction between the water jet and the vane surface are neglected. Thus, the speed of entire water jet remains unchanged. After the water jet is deflected by the vane and leaves the control-volume, the flow reaches a steady state. With the incompressible assumption, the mass balance reduces to

$$\int_A \boldsymbol{u}\cdot\boldsymbol{n}\, \mathrm{d}a = \int_{A_1} \boldsymbol{u}\cdot\boldsymbol{n}\, \mathrm{d}a + \int_{A_2} \boldsymbol{u}\cdot\boldsymbol{n}\, \mathrm{d}a = 0, \tag{5.5.10}$$

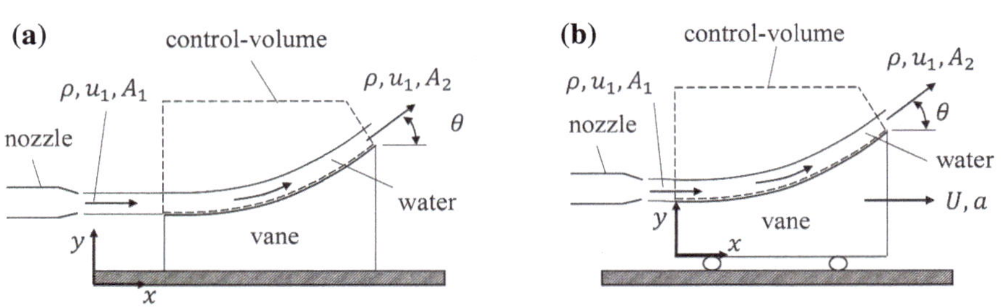

Fig. 5.4 Applications of the global balance of linear momentum. **a** A water jet striking a stationary vane. **b** A water jet striking a moving or an accelerating vane

where A_1 and A_2 are the cross-sectional areas of water jet entering and leaving the CV, respectively. This equation, by using the uniform flow assumption, is simplified to

$$-u_1 A_1 + u_1 A_2 = 0, \qquad \longrightarrow \qquad A_1 = A_2. \tag{5.5.11}$$

This result cannot be obtained if the friction between the water jet and vane surface is taken into account. Furthermore, let the force acting on the water jet by the vane be denoted by $\boldsymbol{f} = -f_x\boldsymbol{i} + f_y\boldsymbol{j}$. Applying the linear momentum balance to the CV yields

$$\begin{aligned} -f_x\boldsymbol{i} + f_y\boldsymbol{j} &= \int_{A_1} \boldsymbol{u}(\rho\boldsymbol{u}\cdot\boldsymbol{n})\,\mathrm{d}a + \int_{A_2} \boldsymbol{u}(\rho\boldsymbol{u}\cdot\boldsymbol{n})\,\mathrm{d}a \\ &= (u_1\boldsymbol{i})(-\rho u_1 A_1) + (u_1\cos\theta\boldsymbol{i} + u_1\sin\theta\boldsymbol{j})(\rho u_1 A_1), \end{aligned} \tag{5.5.12}$$

giving rise to

$$f_x = \rho u_1^2 A_1(1-\cos\theta), \qquad f_y = \rho u_1^2 A_1 \sin\theta. \tag{5.5.13}$$

The problem can equally be solved by using a simple conservation of linear momentum in classical physics. The water jet entering the CV has a horizontal linear momentum of $\rho u_1^2 A_1$ per unit time. When leaving the CV, the horizontal linear momentum reduces to $\rho u_1^2 A_1 \cos\theta$. Since the linear momentum is a conserved quantity and the flow is steady, in which no linear momentum changes inside the CV take place, the decrease in the horizontal linear momentum can only be accomplished by an external force acting on the water jet by the vane in the negative x-direction, and the force magnitude is simply the horizontal linear momentum difference per unit time, i.e., the impulse. Equally, the water jet at the intake surface has no vertical linear momentum. When leaving the CV, it has a vertical linear momentum per unit time. There is an increase in the vertical linear momentum, which can only be accomplished by an external force acting in the y-direction. The force magnitude is the vertical linear momentum difference per unit time.

Consider again the vane in the previous case with everything the same, except that the vane is now moving at a constant speed U in the x-direction, as shown in Fig. 5.4b. To fulfill the definition of control-volume, the coordinates $\{x, y\}$ are located on the vane which move coherently with it, yielding a moving reference frame. However, the balance of linear momentum remains indifferent in this moving reference frame, except that the velocity of water jet entering the CV reduces to $u_1 - U$. With this, the forces acting on the water jet by the moving vane are obtained directly from Eq. (5.5.13), i.e.,

$$f_x = \rho(u_1 - U)^2 A_1(1-\cos\theta), \qquad f_y = \rho(u_1 - U)^2 A_1 \sin\theta, \tag{5.5.14}$$

which are smaller than their counterparts in the stationary case.

Consider another case, in which the vane is initially at rest, and moves with constant acceleration a in the x-direction due to the impact of water jet. The coordinates $\{x, y\}$ located on the vane are no longer an inertia reference frame, whose influence needs to be taken into account in the linear momentum balance via the virtual forces. It follows from Eq. $(5.4.26)_1$ that

$$-f_x\boldsymbol{i} + f_y\boldsymbol{j} - \int_V (\rho a\boldsymbol{i})\mathrm{d}v = \int_{A_1} \boldsymbol{u}(\rho\boldsymbol{u}\cdot\boldsymbol{n})\,\mathrm{d}a + \int_{A_2} \boldsymbol{u}(\rho\boldsymbol{u}\cdot\boldsymbol{n})\,\mathrm{d}a, \tag{5.5.15}$$

where the third term on the left-hand-side represents the inertial force (virtual force), and $\boldsymbol{u}$ is the velocity of water jet measured in the moving reference frame. Let the mass of vane be denoted by M, the mass of water in the CV be denoted by m. It is assumed that the water jet moves coherently with the vane, and the instantaneous speed of vane is denoted by u^*. It follows from Newton's third law of motion that there exists a reaction acting on the accelerating vane by the water jet. Applying Newton's second law of motion to the vane in the x-direction gives

$$f_x = Ma, \qquad \frac{\mathrm{d}u^*}{\mathrm{d}t} = a. \tag{5.5.16}$$

Substituting this equation into Eq. (5.5.15) yields the linear momentum balances in the x- and y-directions given respectively by

$$-Ma - ma = \rho(u_1 - u^*)^2 A_1(\cos\theta - 1), \qquad f_y = \rho(u_1 - u^*)^2 A_1(\sin\theta). \tag{5.5.17}$$

It $M \gg m$, Eq. $(5.5.17)_1$ reduces to

$$Ma + ma \sim Ma = M\frac{\mathrm{d}u^*}{\mathrm{d}t} = \rho(u_1 - u^*)^2 A_1(1 - \cos\theta), \tag{5.5.18}$$

to which the solution of u^*, subject to the initial condition $u^*(t = 0) = 0$, is obtained as

$$\frac{u^*}{u_1} = \frac{u_1\alpha t}{1 + u_1\alpha t}, \qquad \alpha = \frac{(1 - \cos\theta)\rho A_1}{M}. \tag{5.5.19}$$

The forces f_x and f_y are determined by substituting the above expressions into Eq. (5.5.17). The calculations in this example are used e.g. to evaluate the forces acting on the blades of a water turbine in stationary, constantly rotational, and accelerating regions in a power plant.

5.5.3 Angular Momentum Balance

Consider a small lawn sprinkler shown in Fig. 5.5, to which a flow rate Q of water is provided through the sprinkler pivot in the center, and water flows out of the sprinkler

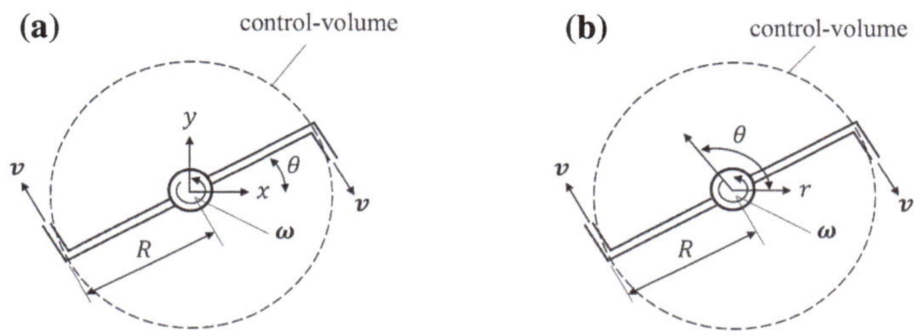

Fig. 5.5 Applications of the global balance of angular momentum. **a** A rotating sprinkler with a fixed reference frame at the center. **b** The same sprinkler with a rotating reference frame at the center

through two arm tubes having diameter d with right angles. The sprinkler rotates with respect to the axis passing the pivot counterclockwise at a constant rotational speed ω. It is required to evaluate the torque acting on the pivot.

The problem is solved first by locating a fixed reference frame $\{x, y, z\}$ at the center of sprinkler pivot, with the corresponding orthonormal base $\{\boldsymbol{i}, \boldsymbol{j}, \boldsymbol{k}\}$, as shown in Fig. 5.5a. Construct the finite control-volume with thickness perpendicular to the page as that of the diameter of arm tube, i.e., the established CV is a three-dimensional cylindrical volume with height d. Since in the CV water involves, the incompressible assumption is used. Applying the mass balance to the CV yields the magnitude of water jet velocity $\boldsymbol{v}$ given by

$$v = \|\boldsymbol{v}\| = \frac{2Q}{\pi d^2}. \tag{5.5.20}$$

This is so, because the water content inside the CV remains unchanged with time, although the arm tubes are rotating.

For simplicity, it is assumed that the gravitational acceleration is perpendicular to the page, which induces no moments acting on the CV. Next, the surface forces on the CS of CV result from the atmospheric pressure, with the resultant forces passing through the center of sprinkler pivot, yielding no moments. Moreover, it follows from the physical observations that the angular momentum of water contained in the CV is constant with respect to the fixed reference frame, for the sprinkler rotates at constant angular speed. The velocity $\boldsymbol{v}$ of water jet leaving the arm tubes needs to be expressed in terms of the fixed coordinates, which is given by

$$\begin{aligned} \boldsymbol{v} &= -\omega R \sin\theta \boldsymbol{i} + \omega R \cos\theta \boldsymbol{j} + v \sin\theta \boldsymbol{i} - v \cos\theta \boldsymbol{j} \\ &= (v - \omega R)\sin\theta \boldsymbol{i} - (v - \omega R)\cos\theta \boldsymbol{j}, \end{aligned} \tag{5.5.21}$$

and the position vector $\boldsymbol{R}$ of the point where the water jet leaves the arm tubes is identified to be

$$\boldsymbol{R} = R\cos\theta \boldsymbol{i} + R\sin\theta \boldsymbol{j}. \tag{5.5.22}$$

Substituting the last two equations into the angular momentum balance yields

$$-T\boldsymbol{k} = \int_A \rho(R\cos\theta \boldsymbol{i} + R\sin\theta \boldsymbol{j}) \times \big[(v - \omega R)\sin\theta \boldsymbol{i} - (v - \omega R)\sin\theta \boldsymbol{j}\big]\, da, \tag{5.5.23}$$

where T is the frictional torque acting at the sprinkler pivot. It follows immediately that

$$T = \rho Q R(v - \omega R). \tag{5.5.24}$$

The problem is now solved by using a rotating reference frame, as shown in Fig. 5.5b, in which the origin of coordinate system $\{r, \theta, z\}$ locates at the center of sprinkler pivot with the orthonormal base $\{\boldsymbol{e}_r, \boldsymbol{e}_\theta, \boldsymbol{k}\}$. The coordinate system rotates coherently with the sprinkler. With these, the water jet velocity is expressed in terms of the rotating reference frame given by

$$\boldsymbol{v} = -v\boldsymbol{e}_\theta, \tag{5.5.25}$$

with the position vector $\boldsymbol{y}$ of the point where the water jet leaves the arm tubes identified to be

$$\boldsymbol{y} = R\boldsymbol{e}_r. \tag{5.5.26}$$

Substituting the above two equations into the angular momentum balance in a rotational reference frame gives

$$-T\boldsymbol{k} - \int_V \rho R\boldsymbol{e}_r \times (2\omega\boldsymbol{k} \times v\boldsymbol{e}_r)\,\mathrm{d}v = \int_A [R\boldsymbol{e}_r \times (-v\boldsymbol{e}_\theta)]\,\rho Q, \tag{5.5.27}$$

resulting in

$$-T\boldsymbol{k} - \rho Q R^2\omega\boldsymbol{k} = -\rho Q R v\boldsymbol{k}, \quad \longrightarrow \quad T = \rho Q R(v - \omega R), \tag{5.5.28}$$

which is the same as that given in Eq. (5.5.24).

5.5.4 Energy Balance

Consider an air compressor which is fixed on the ground, as shown in Fig. 5.6. The state of air entering the compressor is characterized by $p = p_1$, $T = T_1$, $u = u_1$ and $\rho = \rho_1$, and with the properties of $p_2 > p_1$, $T_2 > T_1$ and $\rho_2 > \rho_1$ when leaving the compressor. The intake and discharge pipes of compressor are characterized by the diameters d_1 and d_2, respectively. To operate the compressor, a power is supplied via a shaft work per unit time, denoted by $\dot{W}$. It is required to determine the heat transfer rate $\dot{Q}$ of compressor.

Construct the finite control-volume and locate the fixed coordinate system on the ground. After operating the compressor in a sufficient period of time, the air flow becomes steady, with which the mass balance reduces to

$$\int_A \rho\boldsymbol{u}\cdot\boldsymbol{n}\,\mathrm{da} = 0, \quad \longrightarrow \quad \int_{A_1} \rho\boldsymbol{u}\cdot\boldsymbol{n}\,\mathrm{da} + \int_{A_2} \rho\boldsymbol{u}\cdot\boldsymbol{n}\,\mathrm{da} = 0, \tag{5.5.29}$$

where A_1 and A_2 represent the cross-sectional areas of intake and discharge pipes, respectively. With the uniform-flow assumption, this equation is simplified to

$$4\dot{m} = \rho_1 u_1 \pi d_1^2 = \rho_2 u_2 \pi d_2^2, \quad \longrightarrow \quad u_2 = \frac{\rho_1}{\rho_2}\left(\frac{d_1}{d_2}\right)^2 u_1, \tag{5.5.30}$$

where $\dot{m}$ denotes the mass flow rate of air. The force acting on the compressor by the air flow is determined by using the linear momentum balance, viz.,

$$f = \dot{m}(u_2 - u_1) = \dot{m}u_1\left[\frac{\rho_1}{\rho_2}\left(\frac{d_1}{d_2}\right)^2 - 1\right], \tag{5.5.31}$$

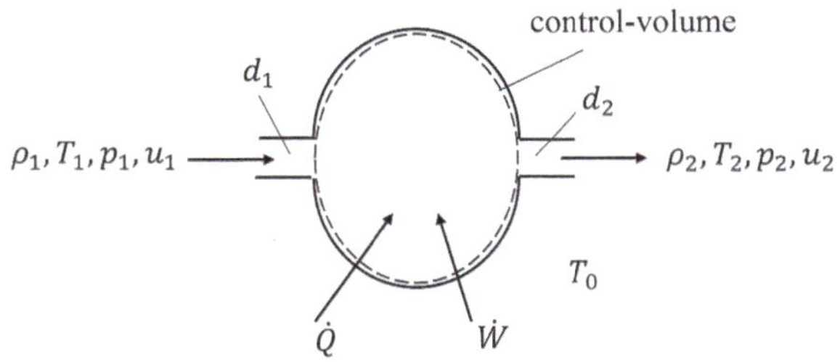

Fig. 5.6 Applications of the global balances of energy and entropy for an air compressor

which points to the reverse direction of air flow, since $u_2 > u_1$ in general. Thus, the compressor needs to be fixed to the ground to prevent sliding.

With the steady-flow assumption, the balance of energy of the CV reads

$$\int_A \left(h + \frac{1}{2}\boldsymbol{u}\cdot\boldsymbol{u} + gz\right)(\rho\boldsymbol{u}\cdot\boldsymbol{n})\,\mathrm{d}a = \dot{Q} + \dot{W}, \tag{5.5.32}$$

where the powers of shear stresses and external energy sources are not taken into consideration for simplicity. With the uniform-flow assumption, this equation reduces to

$$\dot{Q} = -\dot{W} + \dot{m}\left[(h_2 - h_1) + \frac{1}{2}u_1^2\left(\frac{\rho_1^2\, d_1^4}{\rho_2^2\, d_2^4}\right) + g(z_2 - z_1)\right]. \tag{5.5.33}$$

A heat transfer to the compressor is identified if a positive $\dot{Q}$ is obtained and vice versa. The above equation can further be simplified if the ideal gas state equation is used to express the specific enthalpy change of air, i.e.,

$$\dot{Q} = -\dot{W} + \dot{m}\left[c_p(T_2 - T_1) + \frac{1}{2}u_1^2\left(\frac{\rho_1^2\, d_1^4}{\rho_2^2\, d_2^4}\right) + g(z_2 - z_1)\right], \tag{5.5.34}$$

where c_p is the *specific heat at constant pressure* of air. Physically, $\dot{Q}$ must be negative, for physical observations indicate that the air temperature is increased during compression, which is larger than the ambient temperature T_0.

5.5.5 Entropy Balance

Consider again the air compressor in Fig. 5.6. It is required to determine the entropy production of air inside the compressor.

Applying the balance of entropy to the CV yields

$$\int_{A_1} \rho\eta\,(\boldsymbol{u}\cdot\boldsymbol{n})\,\mathrm{d}a + \int_{A_2} \rho\eta\,(\boldsymbol{u}\cdot\boldsymbol{n})\,\mathrm{d}a + \int_A \frac{\boldsymbol{q}}{\theta}\cdot\boldsymbol{n}\,\mathrm{d}a = \int_V \rho\,\pi_\eta\,\mathrm{d}v \geq 0, \tag{5.5.35}$$

in which no external entropy supply is assumed for simplicity, and the Duhem-Truesdell relation is used to express the entropy flux, i.e., $\boldsymbol{q}$ represents the heat flux on the CS of CV. In this equation, θ denotes the temperature and A is the portion of CS, at which the heat transfer takes place. Let the temperature of surrounding be denoted by T_0. It is assumed that the heat transfer takes place at the temperature difference $(T^* - T_0)$ as a *lump analysis*, where T^* is the average temperature given by $T^* = (T_1 + T_2)/2$, and $T^* > T_0$. In this regard, an amount of heat is delivered to the surrounding from the CV. With this and the uniform-flow assumption, Eq. (5.5.35) reduces to

$$\dot{m}\,(\eta_2 - \eta_1) + \frac{\dot{Q}}{T^* - T_0} = \int_V \rho\,\pi_\eta\,\mathrm{d}v \geq 0. \tag{5.5.36}$$

This result, by using the ideal gas state equation to express the specific entropy change of air within the CV, is recast alternatively as

$$\dot{m}\left[c_p \ln\left(\frac{T_2}{T_1}\right) - R\ln\left(\frac{p_2}{p_1}\right)\right] + \frac{\dot{Q}}{T^* - T_0} = \int_V \rho\,\pi_\eta\,\mathrm{d}v \geq 0, \tag{5.5.37}$$

where R is the gas constant of air, and $\dot{Q}$ takes a negative value due to its definition.

5.6 Material Equations

The local balances of mass, linear, and angular momentums in inertia frame, and energy and entropy derived previously are summarized in the following:

$$\begin{aligned}
0 &= \dot{\rho} + \rho \operatorname{div} \boldsymbol{u}, \\
\boldsymbol{0} &= \rho \dot{\boldsymbol{u}} - \operatorname{div} \boldsymbol{t} - \rho \boldsymbol{b}, \\
\boldsymbol{0} &= \boldsymbol{t} - \boldsymbol{t}^{\mathrm{T}}, \\
0 &= \rho \dot{\epsilon} + \operatorname{div} \boldsymbol{q} - \operatorname{tr}(\boldsymbol{D}\boldsymbol{t}) - \rho \zeta, \\
0 &= \rho \dot{\eta} + \operatorname{div} \boldsymbol{\phi}_\eta - \rho s_\eta - \rho \pi_\eta.
\end{aligned} \tag{5.6.1}$$

These equations need to be integrated simultaneously to obtain the field variables ρ, $\boldsymbol{u}$ and θ, totally five scalar unknowns. While the balance of angular momentum is an expression of the symmetry of the Cauchy stress tensor, the balance of entropy is used to indicate the admissibility of a physical process. Thus, the independent equations which can be used to obtain the unknown fields are the balances of mass, linear momentum and energy, totally five independent equations. It seems that the problem is mathematically well posed, for the number of independent equations corresponds to that of unknown fields. However, this is true only if it is possible to express the quantities $\boldsymbol{t}$, $\boldsymbol{b}$, ϵ, $\boldsymbol{q}$, ζ, η, $\boldsymbol{\phi}_\eta$, s_η and π_η as functions of unknown fields, which are called the *material equations*. These equations are sometimes called the *closure conditions* from the mathematical perspective.

From the physical perspective, on the other hand, the derived local statements of physical laws embrace all material behavior. However, different materials behave differently when subject to the same external excitations, although they indeed satisfy the physical laws at the same time. There must therefore also exist some laws which can describe different material responses that apparently separate the various materials from one another. These laws are the *material* or *constitutive equations*, which are different for different materials.

Once the materials equations of a specific material are prescribed, substituting the material equations into the local physical laws results in the *field equations* or *governing equations* of that material, by which it may be possible to determine the field variables by solving the resulting field equations.

5.6.1 General Formulation

In view of Eq. (5.6.1), the field variables

$$\rho = \rho(\mathcal{X}, t), \qquad \mathcal{M}(\mathcal{X}, t), \qquad \theta = \theta(\mathcal{X}, t), \tag{5.6.2}$$

are called the *basic fields*, on which the material equations should depend, and the motion $\mathcal{M}$ is used to replace the velocity for generality. The density, motion, and temperature are defined for a fluid element $\mathcal{X}$ of a material body $\mathcal{B}$. Associated with three basic fields are the balances of mass, linear momentum, and internal energy, in which the specific external body force $\boldsymbol{b}$ and energy supply ζ are considered the known quantities, which can be determined from the surrounding that the material body encounters. On the contrary, the stress tensor $\boldsymbol{t}$, specific internal energy ϵ, and heat flux $\boldsymbol{q}$ are the *material* or *constitutive quantities* which should be expressed as functions of basic fields.[23]

The validity of material model of a specific material is verified by experiments on the results it predicts. Conversely, experiments may suggest certain functional dependency of the material equations on the arguments to within a reasonable satisfaction for certain materials. Experiments alone, however, are rarely sufficient to determine the material equations of a material body. There are some universal requirements that a material model should obey lest its consequences be contradictory to some well-known experiences. Specifically, the universal requirements are summarized as follows:

- Principle of determinism;
- Principle of material objectivity;
- Material symmetry; and
- Thermodynamic considerations,

which are discussed separately in the following. For convenience, the material element $\mathcal{X}$ is expressed in terms of its position vector $\boldsymbol{X}$ in the Lagrangian description.

Principle of determinism. Essentially, the material response at a specific point $\boldsymbol{X}$ depends on the temporal successions that the basic fields experience, called the *memory effect*, and the states of material at all other points of the body, termed the *non-local effect*. This statement is summarized as the *principle of determinism*. Specifically, let $\mathcal{C}$ denote a constitutive quantity. Its functional dependency is then given by

$$\begin{aligned} &\mathcal{C}(\boldsymbol{X},t) = \sum_{s,\boldsymbol{Y}} \mathcal{F}\Big(\rho(\boldsymbol{Y},t-s), \boldsymbol{\mathcal{M}}(\boldsymbol{Y},t-s), \theta(\boldsymbol{Y},t-s), \boldsymbol{Y}, t\Big), \\ &s \in [0,\infty), \quad \boldsymbol{Y} \in \mathcal{B}, \end{aligned} \tag{5.6.3}$$

where $\mathcal{F}$ is called the *material* or *constitutive function* of $\mathcal{C}$, the expression $s \in [0,\infty)$ denotes the memory effect, while the expression $\boldsymbol{Y} \in \mathcal{B}$ represents the non-local effect, where $\boldsymbol{Y}$ are the position vectors of all other material points in $\mathcal{B}$. The dependency of $\mathcal{F}$ on $\boldsymbol{X}$ denotes the *effect of inhomogeneity*, and the summation symbol is used to denote the possible ranges of s and $\boldsymbol{Y}$.

In practice, it is hardly possible to take into account the influence of infinite memory and non-local effect to a large extent, and certain restrictions must be imposed on Eq. (5.6.3). A body is called a *simple material* if the material responses at $\boldsymbol{X}$

[23]The quantities in the entropy balance and the equation itself are used to accomplish the admissibility of a physical process and are not taken into account at the present stage. A detailed discussion will be provided in Sect. 11.6.2.

depend on the histories of basic fields in its immediate neighborhood. This can be accomplished by using the Taylor series expansions of ρ, θ, and $\boldsymbol{\mathcal{M}}$, viz.,

$$\mathcal{C}(\boldsymbol{X},t)=\sum_{s}\mathcal{F}\begin{pmatrix}\rho(\boldsymbol{X},t-s),\,\mathrm{Grad}\,\rho(\boldsymbol{X},t-s),\,\boldsymbol{\mathcal{M}}(\boldsymbol{X},t-s),\,\boldsymbol{F}(\boldsymbol{X},t-s),\\ \theta(\boldsymbol{X},t-s),\,\mathrm{Grad}\,\theta(\boldsymbol{X},t-s),\,\boldsymbol{X},t\end{pmatrix},\tag{5.6.4}$$

in which the second- and higher-order terms of the Taylor series expansions are neglected, and $\boldsymbol{F}$ is the deformation gradient. In this regard, only the local effect is considered, i.e., the material responses are considered to be local. It follows from the properties of $\boldsymbol{F}$ that

$$\mathrm{d}v^{P}=J\mathrm{d}v^{R},\quad\longrightarrow\quad J=\frac{\mathrm{d}v^{P}}{\mathrm{d}v^{R}}=\frac{\rho^{R}}{\rho^{P}},\quad\longrightarrow\quad\rho^{P}=\frac{\rho^{R}}{J},\tag{5.6.5}$$

with which Eq. (5.6.4) reduces to

$$\mathcal{C}(\boldsymbol{X},t)=\sum_{s}\mathcal{F}\Big(\boldsymbol{\mathcal{M}}(\boldsymbol{X},t-s),\,\boldsymbol{F}(\boldsymbol{X},t-s),\,\theta(\boldsymbol{X},t-s),\,\mathrm{Grad}\,\theta(\boldsymbol{X},t-s),\,\boldsymbol{X},t\Big),\tag{5.6.6}$$

for the density gradient involves the second derivative of $\boldsymbol{\mathcal{M}}$, which is neglected for simple materials, and the influence of density is incorporated into $\boldsymbol{\mathcal{M}}$.

The restrictions on memory effect are accomplished by using the Taylor series expansion with respect to time of the basic fields, in which the derivatives of basic fields with different orders appear. A material is said to be *rate dependent of degree* N, if the derivatives of basic fields with the orders smaller than N are considered in the functional arguments in Eq. (5.6.6), which is termed the *bounded memory*. Essentially, a material is of rate type with different degrees in each basic field. Specifically, a *viscous thermoelastic body* is the material depending additionally on the time rate of change of the deformation gradient, for which its constitutive function reduces to

$$\mathcal{C}(\boldsymbol{X},t)=\mathcal{F}\left(\boldsymbol{\mathcal{M}},\,\boldsymbol{F},\,\boldsymbol{L},\,\theta,\,\mathrm{Grad}\,\theta,\,\boldsymbol{X},t\right),\tag{5.6.7}$$

in which $\boldsymbol{L}$ is the velocity gradient, and $\boldsymbol{L}=\dot{\boldsymbol{F}}\boldsymbol{F}^{-1}$ has been used. The summation symbol is removed since the considered viscous thermoelastic body is a simple material with bounded memory.

Principle of material objectivity. Let $\mathcal{O}$ be a reference, and Eq. (5.6.7) may have different forms in different reference frames. Thus, Eq. (5.6.7) needs to be supplemented by the information of evaluated reference frame and is rewritten as

$$\mathcal{C}(\boldsymbol{X},t;\mathcal{O})=\mathcal{F}_{\mathcal{O}}\left(\boldsymbol{\mathcal{M}},\,\boldsymbol{F},\,\boldsymbol{L},\,\theta,\,\mathrm{Grad}\,\theta,\,\boldsymbol{X},t\right),\tag{5.6.8}$$

to denote that this expression is established in the reference frame $\mathcal{O}$. The physical postulate of *observer invariance* or *material objectivity* states that the material responses of a specific material should be independent of the choice of reference frame or observer. That is, changing a reference frame does not change the material response. Thus, constitutive functions are not only invariant, but also indifferent. This statement is summarized as the *principle of material objectivity*. Hence, the constitutive function $\mathcal{C}$ must be independent of the reference frame, namely

$$\mathcal{F}_{\mathcal{O}}(\cdot)=\mathcal{F}_{\mathcal{O}'}(\cdot),\tag{5.6.9}$$

for any two reference frames $\mathcal{O}$ and $\mathcal{O}'$. Applying this principle to Eq. (5.6.7) yields

$$\begin{aligned} \mathcal{C}'(\boldsymbol{X}, t; \mathcal{O}') &= \mathcal{C}(\boldsymbol{X}, t; \mathcal{O}), \\ \mathcal{C}'(\boldsymbol{X}, t; \mathcal{O}') &= \mathcal{F}_{\mathcal{O}'}\left(\boldsymbol{\mathcal{M}}', \boldsymbol{F}', \boldsymbol{L}', \theta', \operatorname{Grad}\theta', \boldsymbol{X}, t\right), \\ \mathcal{C}(\boldsymbol{X}, t; \mathcal{O}) &= \mathcal{F}_{\mathcal{O}}\left(\boldsymbol{\mathcal{M}}, \boldsymbol{F}, \boldsymbol{L}, \theta, \operatorname{Grad}\theta, \boldsymbol{X}, t\right), \end{aligned} \tag{5.6.10}$$

with the functional arguments under the Euclidean transformation given by[24]

$$\begin{aligned} &\boldsymbol{\mathcal{M}}' = \boldsymbol{Q}^{\mathrm{T}}\boldsymbol{\mathcal{M}} - \boldsymbol{Q}^{\mathrm{T}}\boldsymbol{c}, \quad \boldsymbol{F}' = \boldsymbol{Q}^{\mathrm{T}}\boldsymbol{F}, \qquad \boldsymbol{L}' = \boldsymbol{Q}^{\mathrm{T}}\boldsymbol{L}\boldsymbol{Q} + \dot{\boldsymbol{Q}}\boldsymbol{Q}^{\mathrm{T}}, \\ &\theta' = \theta, \qquad\qquad \operatorname{Grad}\theta' = \operatorname{Grad}\theta, \end{aligned} \tag{5.6.11}$$

where $\boldsymbol{Q}$ is the orthogonal tensor, and $\boldsymbol{c}$ is the translation of $\mathcal{O}'$ relative to $\mathcal{O}$.

Since Eq. (5.6.10) must be valid with respect to any arbitrary $\boldsymbol{Q}$, applying $\boldsymbol{Q} = \boldsymbol{I}$ and $\boldsymbol{c}' = \boldsymbol{\mathcal{M}}$ to it gives

$$\mathcal{F}_{\mathcal{O}'}\left(\boldsymbol{0}, \boldsymbol{F}, \boldsymbol{L}, \theta, \operatorname{Grad}\theta, \boldsymbol{X}, t\right) = \mathcal{F}_{\mathcal{O}}\left(\boldsymbol{\mathcal{M}}, \boldsymbol{F}, \boldsymbol{L}, \theta, \operatorname{Grad}\theta, \boldsymbol{X}, t\right), \tag{5.6.12}$$

which indicates that the constitutive functions are not allowed to explicitly depend on the motion $\boldsymbol{\mathcal{M}}$. In addition, condition (5.6.12) cannot be fulfilled in a general case, because $\boldsymbol{L}$ is not an objective tensor, although the other arguments are objective. To fulfill this condition, the stretching tensor $\boldsymbol{D}$ is used to replace $\boldsymbol{L}$, for $\boldsymbol{D}$ is objective. With these, Eq. (5.6.7) reduces to

$$\mathcal{C}(\boldsymbol{X}, t) = \mathcal{F}\left(\boldsymbol{F}, \boldsymbol{D}, \theta, \operatorname{Grad}\theta, \boldsymbol{X}, t\right), \tag{5.6.13}$$

for viscous thermoelastic bodies.

Material symmetry. Principle of material objectivity describes the indifference of material equations under a change of reference frame. Material equations should also reflect, depending on the structures of materials, that material responses remain invariant in different configurations. This means, if a material is described in terms of different configurations, its material responses should be the same in all configurations. This requirement is summarized as the *material symmetry*.[25]

Thus, Eq. (5.6.13) needs to be supplemented by which configuration it is referred to. Consider two reference configurations shown in Fig. 5.7, with their mutual relations and relations to the present configuration. The motion of a material point $\boldsymbol{x}$ can be described in terms of the material points in $\mathcal{B}^R$ and $\mathcal{B}^{\mathcal{R}*}$, i.e.,

$$\boldsymbol{x}: \begin{cases} \boldsymbol{\mathcal{M}}(\boldsymbol{X}, t) = \boldsymbol{\mathcal{M}}^*(\kappa^{-1}(\boldsymbol{X}^*, t), t) = \boldsymbol{\mathcal{M}}^*(\boldsymbol{X}^*, t), \\ \boldsymbol{\mathcal{M}}^*(\boldsymbol{X}^*, t) = \boldsymbol{\mathcal{M}}(\kappa(\boldsymbol{X}, t), t) = \boldsymbol{\mathcal{M}}(\boldsymbol{X}, t), \end{cases} \tag{5.6.14}$$

where κ is a one-to-one mapping between $\boldsymbol{X}$ and $\boldsymbol{X}^*$ given by $\boldsymbol{X}^* = \kappa(\boldsymbol{X})$, and $\boldsymbol{\mathcal{M}}^*$ is the motion between the $*$-reference configuration and present configuration. The deformation gradient in the $*$-reference configuration is defined similarly as before, namely

$$\boldsymbol{F}^* \equiv \frac{\partial \boldsymbol{\mathcal{M}}^*}{\partial \boldsymbol{X}^*}, \tag{5.6.15}$$

[24]Every column of the deformation gradient $\boldsymbol{F}$ transforms as an objective vector. Hence, $\boldsymbol{F}$ transforms as three objective vectors.

[25]This requirement is not so universal as the previous two, that it is not addressed as a principle.

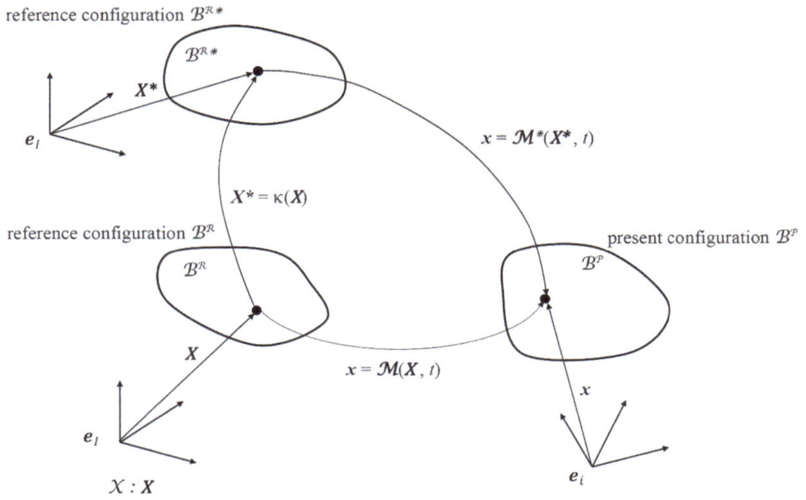

Fig. 5.7 Two reference configurations and the present configuration with their mutual motions

with which

$$\boldsymbol{F} = \frac{\partial \boldsymbol{\mathcal{M}}}{\partial \boldsymbol{X}} = \frac{\partial \boldsymbol{\mathcal{M}}^*}{\partial \boldsymbol{X}^*}\frac{\partial \boldsymbol{X}^*}{\partial \boldsymbol{X}} = \frac{\partial \boldsymbol{\mathcal{M}}^*}{\partial \boldsymbol{X}^*}\frac{\partial \kappa}{\partial \boldsymbol{X}} = \boldsymbol{F}^*\boldsymbol{P}, \qquad \boldsymbol{P} = \operatorname{Grad}\kappa. \tag{5.6.16}$$

With these, Eq. (5.6.13), by using twofold reference configurations, is expressed as

$$\begin{aligned} \mathcal{C}(\boldsymbol{X}, t) &= \mathcal{F}\left(\boldsymbol{F}, \boldsymbol{D}, \theta, \operatorname{Grad}\theta, \boldsymbol{X}, t\right), \\ \mathcal{C}^*(\boldsymbol{X}^*, t) &= \mathcal{F}^*\left(\boldsymbol{F}^*, \boldsymbol{D}, \theta^*, \operatorname{Grad}^*\theta^*, \boldsymbol{X}^*, t\right), \end{aligned} \tag{5.6.17}$$

in which $\boldsymbol{D}$ remains unchanged, for it is defined in the present configuration. As required by the material symmetry, it follows that

$$\theta^*(\boldsymbol{X}^*, t) = \theta(\boldsymbol{X}, t) = \theta, \qquad \operatorname{Grad}^*\theta^* = \operatorname{Grad}\theta\boldsymbol{P}^{-1}, \tag{5.6.18}$$

with which Eq. (5.6.17) reduces to

$$\begin{aligned} \mathcal{C}(\boldsymbol{X}, t) &= \mathcal{F}\left(\boldsymbol{F}, \boldsymbol{D}, \theta, \operatorname{Grad}\theta, \boldsymbol{X}, t\right), \\ \mathcal{C}^*(\boldsymbol{X}^*, t) &= \mathcal{F}^*\left(\boldsymbol{F}\boldsymbol{P}^{-1}, \boldsymbol{D}, \theta, \operatorname{Grad}\theta\, \boldsymbol{P}^{-1}, \boldsymbol{X}^*, t\right). \end{aligned} \tag{5.6.19}$$

To simplify the analysis, only those materials for which there exists a global reference configuration, in which the same material equations hold at all material points, are considered. These materials are termed to be *homogeneous*, and the corresponding global reference configuration is called *natural*. In this natural configuration, the dependency on $\boldsymbol{X}$ and $\boldsymbol{X}^*$ in Eq. (5.6.19) is not necessary, so is the dependency on t, for it is already included implicitly in other arguments. With these, the material symmetry requires that

$$\mathcal{F}\left(\boldsymbol{F}, \boldsymbol{D}, \theta, \operatorname{Grad}\theta\right) = \mathcal{F}\left(\boldsymbol{F}\boldsymbol{P}^{-1}, \boldsymbol{D}, \theta, \operatorname{Grad}\theta\, \boldsymbol{P}^{-1}\right), \tag{5.6.20}$$

in which $\mathcal{F}^*$ is replaced by $\mathcal{F}$ to meet the symmetry requirement.

Obviously, Eq. (5.6.20) cannot be fulfilled by arbitrary κ and $\boldsymbol{P}$. A transformation of reference configuration by which the body volume is preserved is called a *unimodular transformation* with $\det \boldsymbol{P} = 1$. Orthogonal transformations such as rotations or mirror reflections are typical examples. The set of all unimodular transformations is termed the *unimodular group*, which is a part of the *symmetry group*. Based on these, fluids are defined as those materials whose symmetry group is the unimodular one, which possess a very high degree of symmetry.[26] For example, consider a cup filled with water being initially at rest. The water is then strongly stirred so that a water molecule will nearly impossible occupy its initial position when the water is brought again to rest. Although the configuration of water has been changed, one cannot recognize a difference in the physical behavior of water and water is the same material as before. It follows that every configuration, including the present configuration, can be a reference configuration, and it becomes possible to express Eq. (5.6.20) in the present configuration by replacing all functional arguments by the corresponding notations in the present configuration.

Applying the principle of material objectivity to Eq. (5.6.20) results in

$$\left\{s,\, \boldsymbol{Q}^{\mathrm{T}}\boldsymbol{v},\, \boldsymbol{Q}^{\mathrm{T}}\boldsymbol{t}\,\boldsymbol{Q}\right\}(\boldsymbol{F},\, \boldsymbol{D},\, \theta,\, \mathrm{Grad}\,\theta) = \{s,\, \boldsymbol{v},\, \boldsymbol{t}\}\left(\boldsymbol{Q}^{\mathrm{T}}\boldsymbol{F}\,\boldsymbol{Q},\, \boldsymbol{Q}^{\mathrm{T}}\boldsymbol{D}\,\boldsymbol{Q},\, \theta,\, \boldsymbol{Q}^{\mathrm{T}}\mathrm{Grad}\,\theta\right), \tag{5.6.21}$$

for all orthogonal transformations $\boldsymbol{Q}$ representing the elements of the unimodular group of $\boldsymbol{P}$ which is temporally a constant. The quantities s, $\boldsymbol{v}$, and $\boldsymbol{t}$ belong to $\mathcal{C}$ and are called respectively *scalar*, *vectorial*, and *tensorial constitutive quantities*, which should be *isotropic functions* of their functional arguments. Equation (5.6.21) can be expressed in a general form in the present configuration given by

$$\mathcal{C} = \mathcal{C}(\rho,\, \boldsymbol{D},\, T,\, \mathrm{grad}\, T), \tag{5.6.22}$$

to denote the functional dependency of a constitutive variable for *viscous thermoelastic fluids*, where the temperature θ is replaced by the conventional symbol T, and $\boldsymbol{F}$ is represented by the density ρ. The most important implication of Eq. (5.6.22) is that $\mathcal{C}$ should be expressed as isotropic functions of the functional arguments to meet the universal requirement discussed previously.

Thermodynamic considerations. Equation (5.6.22) only shows the functional dependency of a material quantity, whose explicit expression can further be identified by using the *thermodynamic considerations*. That is, substituting this equation and the balances of mass, linear momentum, and energy into the balance of entropy to derive analytically, if possible, the explicit expressions of material equations. This can be conducted by using either the *Coleman-Noll* or the *Müller-Liu approach*. Since this procedure involves knowledge of thermodynamics, it is not explored at the moment. The topic will be discussed in Sect. 11.6.2.

5.6.2 Physical Interpretations of Stretching and Spin Tensors

In the previous derivations, uses have been made to the stretching tensor $\boldsymbol{D}$ and spin tensor $\boldsymbol{W}$, which need to be explored before proceeding to the material equations of

[26]The definition is given by Noll based on the rules of symmetry transformation. Walter Noll, 1925–2017, an American mathematician, who contributed to the mathematical tools of classical mechanics and thermodynamics.

the Newtonian fluids. It follows from the deformation gradient $\boldsymbol{F}$ that

$$(\mathrm{d}\boldsymbol{x})^{\cdot} = \dot{\boldsymbol{F}}\mathrm{d}\boldsymbol{X} + \boldsymbol{F}(\mathrm{d}\boldsymbol{X})^{\cdot} = \boldsymbol{L}\boldsymbol{F}\mathrm{d}\boldsymbol{X} = \boldsymbol{L}\mathrm{d}\boldsymbol{x}, \tag{5.6.23}$$

where $\mathrm{d}\boldsymbol{x}$ represents a line segment vector in the present configuration and $\boldsymbol{L}$ is the velocity gradient. Consider an infinitesimal surface element shown in Fig. 5.8a, whose horizontal side is denoted by $\mathrm{d}\boldsymbol{x}$ with $\mathrm{d}\boldsymbol{x} = \mathrm{d}x\boldsymbol{e}_1$. Taking inner product of the time rate of change of $\mathrm{d}\boldsymbol{x}$ with $\boldsymbol{e}_1$ yields

$$\boldsymbol{e}_1 \cdot (\mathrm{d}\boldsymbol{x})^{\cdot} = (\mathrm{d}x)^{\cdot} = L_{11}\mathrm{d}x, \qquad \longrightarrow \qquad L_{11} = \frac{\partial u_1}{\partial x_1} = \frac{(\mathrm{d}x)^{\cdot}}{\mathrm{d}x}, \tag{5.6.24}$$

showing that L_{11} represents the time rate of change of the length of line segment per unit length in the x_1-direction. Similar expressions and interpretations are also found for L_{22} and L_{33}. Next, consider two line segment vectors $\mathrm{d}\boldsymbol{x}_1 = \mathrm{d}x\boldsymbol{e}_1$ and $\mathrm{d}\boldsymbol{x}_2 = \mathrm{d}x\boldsymbol{e}_2$, which are initially perpendicular to each other as the horizontal and vertical sides of a surface element, as shown in Fig. 5.8b. It is found that

$$(\mathrm{d}\boldsymbol{x}_1)^{\cdot} = L_{11}(\mathrm{d}x)\boldsymbol{e}_1 + L_{12}(\mathrm{d}x)\boldsymbol{e}_2, \qquad (\mathrm{d}\boldsymbol{x}_2)^{\cdot} = L_{21}(\mathrm{d}x)\boldsymbol{e}_1 + L_{22}(\mathrm{d}x)\boldsymbol{e}_2. \tag{5.6.25}$$

It follows from Fig. 5.8b that for small values of the angles α and β,

$$\dot{\alpha} + \dot{\beta} \sim \frac{\partial u_2}{\partial x_1} + \frac{\partial u_1}{\partial x_2} = L_{21} + L_{12}, \tag{5.6.26}$$

showing that the time rate of change of the angular deformation of surface element on the (x_1x_2)-plane is the sum of L_{12} and L_{21}. Similar expressions and interpretations are equally found on the (x_1x_3)- and (x_2x_3)-planes. On the other hand, the time rate of change of rigid body rotation is given by

$$\frac{1}{2}(\dot{\alpha} - \dot{\beta}) = \frac{1}{2}\left(L_{21} - L_{12}\right), \tag{5.6.27}$$

on the (x_1x_2)-plane, with similar expressions on the (x_1x_3)- and (x_2x_3)-planes.

Thus, the stretching tensor $\boldsymbol{D}$, which is the symmetric part of $\boldsymbol{L}$, represents the time rate of change of the deformation of a fluid element, including linear and shear deformations, while the spin tensor $\boldsymbol{W}$, which is the antisymmetric part of $\boldsymbol{L}$, represents the rotational velocity of a fluid element.

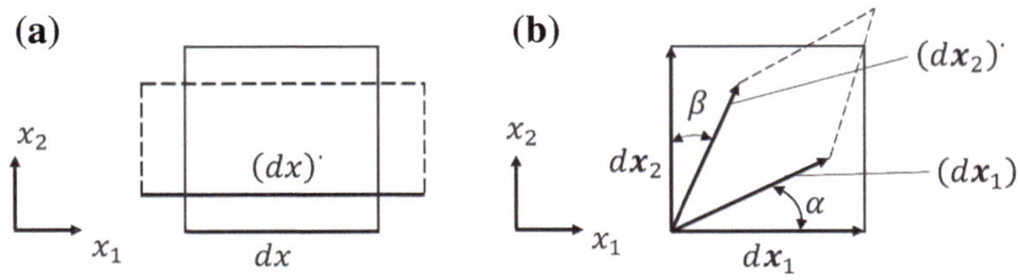

Fig. 5.8 Deformations of an infinitesimal surface element in terms of the stretching and spin tensors in the (x_1x_2)-plane. (**a**) A horizontal line segment. (**b**) Two mutually orthogonal line segments

5.6.3 Material Equations of the Newtonian Fluids

It can be shown by using the thermodynamic analysis that for viscous thermoelastic fluids, the functional dependencies of specific internal energy ϵ, heat flux $\boldsymbol{q}$, and the Cauchy stress tensor $\boldsymbol{t}$ reduce to[27]

$$\epsilon = \epsilon(\rho, T), \qquad \boldsymbol{q} = \boldsymbol{q}(\rho, T, \text{grad}\, T), \qquad \boldsymbol{t} = \boldsymbol{t}(\rho, \boldsymbol{D}), \tag{5.6.28}$$

in which $\boldsymbol{q}$ and $\boldsymbol{t}$ should be expressed respectively as an isotropic vectorial and an isotropic tensorial functions of the arguments. Different viscous thermoelastic fluids can be derived from the above equation, e.g. the *Reiner-Rivlin fluid* or the *Bingham fluid*.[28] Since the book is concerned with the fundamentals of fluid mechanics, only the Newtonian fluids are to be considered directly.

Without further mathematical or thermodynamic analysis of Eq. (5.6.28), the propositions of the material equations of the Newtonian fluids should be made based on certain experimental outcomes and observations. The operational definitions of the Newtonian fluids described in Sect. 2.6 are slightly revised to meet the purpose, which are given in the following:

- When the fluid is at rest, the stress is hydrostatic and the pressure exerted by the fluid is the thermodynamics pressure.
- The stress tensor $\boldsymbol{t}$ depends only linearly on the stretching tensor $\boldsymbol{D}$.
- No shear stresses take place when a fluid is in rigid body motion.

It follows form the first two statements that $\boldsymbol{t}$ may be given by

$$\boldsymbol{t} = -p\boldsymbol{I} + \boldsymbol{T}, \qquad \boldsymbol{T} = \boldsymbol{T}(\boldsymbol{D}), \qquad t_{ij} = -p\delta_{ij} + T_{ij}, \tag{5.6.29}$$

where p is the thermodynamic pressure, and $\boldsymbol{T}$ is the extra stress tensor, or equivalently the *shear stress tensor*, whose isotropic expression is given by

$$\boldsymbol{T} = \frac{1}{2}\boldsymbol{a}\boldsymbol{D}, \qquad T_{ij} = \frac{1}{2}a_{ijkl}\left(\frac{\partial u_k}{\partial x_l} + \frac{\partial u_l}{\partial x_k}\right), \tag{5.6.30}$$

where $\boldsymbol{a}$ is an isotropic tensor of fourth order. Substituting the most general form of $\boldsymbol{a}$, i.e., Eq. (1.2.70), into the above equation yields

$$T_{ij} = \frac{1}{2}\left(\alpha\delta_{ij}\delta_{kl} + \beta\delta_{ik}\delta_{jl} + \gamma\delta_{il}\delta_{jk}\right)\left(\frac{\partial u_k}{\partial x_l} + \frac{\partial u_l}{\partial x_k}\right), \tag{5.6.31}$$

which reduces to

$$T_{ij} = \lambda\delta_{ij}\frac{\partial u_k}{\partial x_k} + \mu\left(\frac{\partial u_i}{\partial x_j} + \frac{\partial u_j}{\partial x_i}\right), \qquad \lambda = \alpha, \;\; \mu = \frac{1}{2}(\beta + \gamma), \tag{5.6.32}$$

[27] It is also possible to obtain Eq. (5.6.28) by imposing certain *internal constraints* on Eq. (5.6.22).
[28] Ronald Samuel Rivlin, 1915–2005, a British-American physicist, mathematician and rheologist, who is known for his works on rubber. Eugene Cook Bingham, 1878–1945, an American chemist, whose contributions are mainly in rheology.

where λ and μ are scalars depending on the state of fluid. With Eq. (5.6.32), the Cauchy stress tensor $\boldsymbol{t}$ becomes

$$t_{ij} = -p\delta_{ij} + \lambda\delta_{ij}\frac{\partial u_k}{\partial x_k} + \mu\left(\frac{\partial u_i}{\partial x_j} + \frac{\partial u_j}{\partial x_i}\right), \qquad \boldsymbol{t} = -p\boldsymbol{I} + (\lambda \operatorname{div}\boldsymbol{u})\boldsymbol{I} + 2\mu\boldsymbol{D}. \tag{5.6.33}$$

Consider the simple shear flow shown in Fig. 2.4a. Applying the above equation to the simple shear experiment gives

$$t_{11} = t_{22} = t_{33} = -p, \qquad t_{13} = t_{31} = t_{23} = t_{32} = 0, \qquad t_{12} = t_{21} = \mu\frac{\mathrm{d}x_1}{\mathrm{d}x_2}, \tag{5.6.34}$$

indicating that μ is nothing else than the dynamic viscosity, and Eq. (5.6.33) is the three-dimensional generalization of Newton's law of viscosity. The scalar λ is referred to as the *second viscosity coefficient*.

Taking trace of Eq. (5.6.33) leads to

$$t_{11} + t_{22} + t_{33} = -3p + (3\lambda + 2\mu)\frac{\partial u_k}{\partial x_k}, \tag{5.6.35}$$

by which the *mechanical pressure* $\bar{p}$ is defined as the average of three normal stress components given by

$$\bar{p} \equiv -\frac{1}{3}\operatorname{tr}\boldsymbol{t}, \qquad -\bar{p} = -p + \left(\lambda + \frac{2}{3}\mu\right)\frac{\partial u_k}{\partial x_k}. \tag{5.6.36}$$

This equation indicates that the the thermodynamic and mechanical pressures are essentially different, for the mechanical pressure is either purely hydrostatic or hydrostatic plus a component induced by the stresses which result form the motion of fluid. The difference between the thermodynamic and mechanical pressures is proportional to the divergence of fluid velocity, and the proportional factor is usually referred to as the *bulk viscosity* κ, so that Eq. $(5.6.36)_2$ becomes

$$p - \bar{p} = \kappa(\operatorname{div}\boldsymbol{u}), \qquad \kappa \equiv \lambda + \frac{2}{3}\mu. \tag{5.6.37}$$

Up to this point, three viscosities of the Newtonian fluids exist: μ, λ and κ, and any two are independent. It is common to choose κ and μ as the two independent ones, which cannot be determined analytically, but should be determined experimentally.

While the physical interpretation of μ has already been discussed in Sect. 2.6, the physical interpretation of κ is given here from the kinetic theory of gas. The mechanical pressure is only a measure of the translational energy of molecules, while the thermodynamic pressure is that of the total energy of molecules, including the vibrational and rotational energy modes as well as the translational mode. For liquids, other energy modes exist equally, e.g. the intermolecular attraction. Different energy modes possess different relaxation times and permit themselves to be transformed into one another. The bulk viscosity κ is thus understood as a measure of the energy transfer from the translational mode to other modes. Its influence becomes significant for compressible flows, in which shock waves take place at the expense of translational energies, yielding non-vanishing κ. For monatomic gases, the translational mode is the only energy mode of molecules, giving rise to a vanishing κ.

Thus, for monatomic gases the thermodynamic and mechanical pressures are the same, and

$$\kappa = 0, \qquad \lambda = -\frac{2}{3}\mu, \tag{5.6.38}$$

hold, where the second equation is known as *Stokes' relation*, with which only one viscosity is independent, mostly chosen as μ. For polyatomic gases and liquids, the departure of κ from null is frequently small, so that it is possible to use Stokes' relation directly in the material equation of the Cauchy stress of the Newtonian fluids. For incompressible fluids, Stokes' relation is satisfied identically, for κ always vanishes, and no distinction between the thermodynamic and mechanical pressures is made.

The material equation of heat flux $\boldsymbol{q}$ is based on the *Fourier law of heat conduction*,[29] which states that the heat flux by conduction is proportional to the temperature gradient, viz.,

$$\boldsymbol{q} = -k\,(\operatorname{grad} T), \qquad q_i = -k\frac{\partial T}{\partial x_i}, \tag{5.6.39}$$

where k is the *thermal conductivity* of fluid, with $k = k(\rho, T)$ for simple materials. Last, for the specific internal energy ϵ, Eq. (5.6.28)$_1$ shows that it is a function of ρ and T. Further specifications involve atomic and molecular structures of materials, e.g. the internal energy of monatomic gases from the kinetic theory of gas. However, macroscopic prescriptions of ϵ are sometimes possible, e.g. the ideal gas state equation provides a starting point to derive the expression of specific internal energy change which is valid for almost all gases with relatively high temperature and low pressure. Since the Newtonian fluids are simple materials, Eq. (5.6.28)$_1$ verifies again the statement that the state of a simple material is determined by prescribing the values of any two independent specific properties.

5.6.4 Local Physical Laws of the Newtonian Fluids

With Eq. (5.6.33), the divergence of the Cauchy stress tensor is obtained as

$$\begin{aligned}
\frac{\partial t_{ij}}{\partial x_j} &= \frac{\partial}{\partial x_j}\left[-p\delta_{ij} + \lambda\delta_{ij}\frac{\partial u_k}{\partial x_k} + \mu\left(\frac{\partial u_i}{\partial x_j} + \frac{\partial u_j}{\partial x_i}\right)\right] \\
&= -\frac{\partial p}{\partial x_i} + \frac{\partial}{\partial x_i}\left(\lambda\frac{\partial u_k}{\partial x_k}\right) + \frac{\partial}{\partial x_j}\left[\mu\left(\frac{\partial u_i}{\partial x_j} + \frac{\partial u_j}{\partial x_i}\right)\right], \\
\operatorname{div}\boldsymbol{t} &= -\operatorname{grad}(p - \lambda\operatorname{div}\boldsymbol{u}) + \operatorname{div}\left[\mu\left(\operatorname{grad}\boldsymbol{u} + (\operatorname{grad}\boldsymbol{u})^{\mathrm{T}}\right)\right],
\end{aligned} \tag{5.6.40}$$

in which μ and λ are state functions of fluids. Substituting these equations into the local balance of linear momentum yields

$$\begin{aligned}
\rho\dot{u}_i &= -\frac{\partial p}{\partial x_i} + \frac{\partial}{\partial x_i}\left(\lambda\frac{\partial u_k}{\partial x_k}\right) + \frac{\partial}{\partial x_j}\left[\mu\left(\frac{\partial u_i}{\partial x_j} + \frac{\partial u_j}{\partial x_i}\right)\right] + \rho b_i, \\
\rho\dot{\boldsymbol{u}} &= -\operatorname{grad}(p - \lambda\operatorname{div}\boldsymbol{u}) + \operatorname{div}\left[\mu\left(\operatorname{grad}\boldsymbol{u} + (\operatorname{grad}\boldsymbol{u})^{\mathrm{T}}\right)\right] + \rho\boldsymbol{b},
\end{aligned} \tag{5.6.41}$$

[29] Jean-Baptiste Joseph Fourier, 1768–1830, a French mathematician and physicist, who contributed to the *Fourier series*, theory of heat transfer and discovered the *greenhouse effect*.

which is termed the *Navier-Stokes equation*,[30] i.e., the local balance of linear momentum of the Newtonian fluids. The Navier-Stokes equation is just Newton's second law of motion per unit volume in the Eulerian description, where the first two terms on the right-hand-side are the surface forces, while the last term is the body force. The surface forces are divided into two parts: The first term represents the normal forces, and the second term denotes the *shear (viscous) forces*. For the *incompressible Newtonian fluids with constant dynamic viscosity*, Eq. (5.6.41) is simplified to

$$\begin{aligned} \rho\frac{\partial u_i}{\partial t} + \rho u_j\frac{\partial u_i}{\partial x_j} &= -\frac{\partial p}{\partial x_i} + \mu\frac{\partial^2 u_i}{\partial x_j \partial x_j} + \rho b_i, \\ \rho\frac{\partial \boldsymbol{u}}{\partial t} + \rho(\operatorname{grad} \boldsymbol{u})\boldsymbol{u} &= -\operatorname{grad} p + \mu \operatorname{lap} \boldsymbol{u} + \rho \boldsymbol{b}, \end{aligned} \tag{5.6.42}$$

where the left-hand-side of Eq. (5.6.41) is expressed by using the material derivative. It follows from Eq. (5.6.42) that a fluid motion may be triggered by the pressure force, gravity force, or viscous force. Since the viscous force always prohibits the motion, only the pressure and gravity forces deliver the possible driven mechanism of fluid motion. The Navier-Stokes equation is a time dependent, nonlinear partial differential equation of second order, whose solutions in real physical circumstances can rarely be found analytically, and numerical calculations are needed for almost all engineering applications. It should be noted that either Eq. (5.6.41) or (5.6.42) suits for *laminar flows*. For *turbulent flows*, the Navier-Stokes equation needs to be supplemented in order to account for the influence induced by the statistically temporal and spatial variations of physical variables, e.g. *Reynolds' stresses* as additional stress contributions, which will be discussed in Sect. 8.6.3.

The stress power tr $(\boldsymbol{D}\boldsymbol{t})$ in the local balance of internal energy is determined viz.,

$$\begin{aligned} \operatorname{tr}(\boldsymbol{D}\boldsymbol{t}) = D_{ij}t_{ji} &= \frac{1}{2}\left(\frac{\partial u_i}{\partial x_j} + \frac{\partial u_j}{\partial x_i}\right)\left[-p\delta_{ji} + \lambda\delta_{ji}\frac{\partial u_k}{\partial x_k} + \mu\left(\frac{\partial u_j}{\partial x_i} + \frac{\partial u_i}{\partial x_j}\right)\right], \\ &= -p\frac{\partial u_k}{\partial x_k} + \lambda\left(\frac{\partial u_k}{\partial x_k}\right)^2 + \frac{1}{2}\mu\left(\frac{\partial u_i}{\partial x_j} + \frac{\partial u_j}{\partial x_i}\right)^2, \end{aligned} \tag{5.6.43}$$

which is recast alternatively as

$$\operatorname{tr}(\boldsymbol{D}\boldsymbol{t}) = -p(\operatorname{div}\boldsymbol{u}) + \Phi, \qquad \Phi = \lambda(\operatorname{div}\boldsymbol{u})^2 + 2\mu\operatorname{tr}(\boldsymbol{D}\boldsymbol{D}), \tag{5.6.44}$$

where Φ is the *dissipation function*. The stress power is the work done by the stresses per unit time, and the first term on the right-hand-side of Eq. $(5.6.44)_1$ represents a *reversible transfer of energy* due to the compressions of fluids, while Φ is a measure of the rate at which the mechanical energy is converted into the thermal energy. With positive values of λ and μ, Φ always works to increase *irreversibly* the internal energy, and hence acts as an energy source for the internal energy balance.

[30]Claude-Louis Navier, 1785–1836, a French engineer and physicist who specialized in mechanics. The Navier-Stokes equation was first derived by Navier in 1823 and later perfected by Stokes in 1845.

Substituting Eqs. (5.6.39) and (5.6.43) into the local balance of internal energy results in

$$\begin{aligned}\rho\dot{\epsilon} &= \frac{\partial}{\partial x_i}\left(k\frac{\partial T}{\partial x_i}\right) - p\frac{\partial u_k}{\partial x_k} + \lambda\left(\frac{\partial u_k}{\partial x_k}\right)^2 + \frac{1}{2}\mu\left(\frac{\partial u_i}{\partial x_j} + \frac{\partial u_j}{\partial x_i}\right)^2 + \rho\zeta, \\ \rho\dot{\epsilon} &= \operatorname{div}(k \operatorname{grad} T) - p(\operatorname{div}\boldsymbol{u}) + \lambda(\operatorname{div}\boldsymbol{u})^2 + 2\mu\operatorname{tr}(\boldsymbol{DD}) + \rho\zeta,\end{aligned} \tag{5.6.45}$$

for the Newtonian fluids, which is used to determined the temperature field. For the *incompressible Newtonian fluids with constant thermal conductivity*, Eq. (5.6.45) reduces to

$$\begin{aligned}\rho\frac{\partial\epsilon}{\partial t} + \rho u_i\frac{\partial\epsilon}{\partial x_i} &= k\frac{\partial^2 T}{\partial x_i^2} + \frac{1}{2}\mu\left(\frac{\partial u_i}{\partial x_j} + \frac{\partial u_j}{\partial x_i}\right)^2 + \rho\zeta, \\ \rho\frac{\partial\epsilon}{\partial t} + (\rho \operatorname{grad}\epsilon)\cdot\boldsymbol{u} &= k(\operatorname{lap} T) + 2\mu\operatorname{tr}(\boldsymbol{DD}) + \rho\zeta,\end{aligned} \tag{5.6.46}$$

where the left-hand-side of Eq. (5.6.45) has been expressed by using the material derivative. The local balance of angular momentum remains unchanged, for it delivers the result of the symmetry of the Cauchy stress tensor which is valid equally for the Newtonian fluids. The local balance of entropy involves the material equations of specific entropy η, entropy flux ϕ_η, and specific entropy supply s_η. These topics will be discussed later in Sect. 11.6.1. For convenience of application, the global balance equations of physical laws in inertia reference frame for the Newtonian fluids are summarized in Table 5.5, with the corresponding local balance statements

Table 5.5 Global balance equations of physical laws for the Newtonian fluids in inertia reference frame

Mass	$\frac{\partial}{\partial t}\int_V \rho\,\mathrm{d}v + \int_A (\rho\boldsymbol{u}\cdot\boldsymbol{n})\,\mathrm{d}a = 0$
	$\int_A (\boldsymbol{u}\cdot\boldsymbol{n})\,\mathrm{d}a = 0$ (continuity equation, $\rho = \text{constant}$)
	$\int_A (\rho\boldsymbol{u}\cdot\boldsymbol{n})\,\mathrm{d}a = 0$ (steady-flow assumption)
Linear momentum	$\frac{\partial}{\partial t}\int_V \rho\boldsymbol{u}\,\mathrm{d}v + \int_A \boldsymbol{u}\,(\rho\boldsymbol{u}\cdot\boldsymbol{n})\,\mathrm{d}a = \sum\boldsymbol{\mathcal{F}}_{CV} + \sum\boldsymbol{\mathcal{F}}_{CS}$
Angular momentum	$\frac{\partial}{\partial t}\int_V (\boldsymbol{x}\times\boldsymbol{u})\rho\,\mathrm{d}v + \int_A (\boldsymbol{x}\times\boldsymbol{u})(\rho\boldsymbol{u}\cdot\boldsymbol{n})\,\mathrm{d}a$
	$= \sum\boldsymbol{x}\times\boldsymbol{\mathcal{F}}_{CV} + \sum\boldsymbol{x}\times\boldsymbol{\mathcal{F}}_{CS} + \sum\boldsymbol{M}_{shaft}$
Energy	$\frac{\partial}{\partial t}\int_V \rho e\,\mathrm{d}v + \int_A \left(h + \frac{1}{2}\boldsymbol{u}\cdot\boldsymbol{u} + gz\right)(\rho\boldsymbol{u}\cdot\boldsymbol{n})\,\mathrm{d}a$
	$= \sum\dot{Q} + \sum\dot{W}_{shear} + \sum\dot{W}_{shaft} + \sum\dot{E}_s$
Entropy	$\frac{\partial}{\partial t}\int_V \eta\rho\,\mathrm{d}v + \int_A \eta\,(\rho\boldsymbol{u}\cdot\boldsymbol{n})\,\mathrm{d}a - \int_V s_\eta\rho\,\mathrm{d}v + \int_A \boldsymbol{\phi}_\eta\cdot\boldsymbol{n}\,\mathrm{d}a$
	$= \int_V \pi_\eta\rho\,\mathrm{d}v \geq 0$

Table 5.6 Local balance equations of physical laws for the Newtonian fluids in inertia reference frame

Mass	$\dfrac{\partial \rho}{\partial t} + \text{div}\ (\rho \boldsymbol{u}) = 0$
	$\text{div}\ \boldsymbol{u} = 0\ (\rho = \text{constant})$
	$\text{div}\ (\rho \boldsymbol{u}) = 0$ (steady-flow assumption)
Linear momentum	$\rho \dfrac{\partial \boldsymbol{u}}{\partial t} + \rho (\text{grad}\ \boldsymbol{u})\boldsymbol{u} = -\text{grad}\ p + \mu\ \text{lap}\ \boldsymbol{u} + \rho \boldsymbol{b}\ (\rho, \mu = \text{constant})$
Internal energy	$\rho \dfrac{\partial \epsilon}{\partial t} + \rho\ \text{grad}\ \epsilon \cdot \boldsymbol{u} = k\ (\text{lap}\ T) + 2\mu\ \text{tr}(\boldsymbol{DD}) + \rho \zeta\ (\rho, k = \text{constant})$
Entropy	$\rho \pi_\eta = \rho \dot{\eta} + \text{div}\ \left(\dfrac{\boldsymbol{q}}{\theta}\right) - \rho \dfrac{\zeta}{\theta} \geq 0$ (with the Duhem-Truesdell relations)

summarized in Table 5.6. For moving reference frames, the balance equations of linear and angular momentums in both tables need to be supplemented by introducing the inertia forces as the virtual forces.

For the incompressible Newtonian fluids at rest, the Navier-Stokes equation reduces to

$$\boldsymbol{0} = -\nabla p + \rho \boldsymbol{b} = -\nabla p + \rho \boldsymbol{g}, \tag{5.6.47}$$

corresponding exactly to Eq. (3.1.3), if the gravitational acceleration is the only body force. This result can be extended to the non-Newtonian fluids, although their pressures may be defined differently as that of the Newtonian fluids, for the material responses of different rheological fluids deviate from one another when there is a relative motion between any two points inside the fluids. For the Newtonian fluids in rigid-body motion, the Navier-Stokes equation equally reduces to Eq. (3.1.2), i.e.,

$$\rho \boldsymbol{a} = -\nabla p + \rho \boldsymbol{b} = -\nabla p + \rho \boldsymbol{g}. \tag{5.6.48}$$

Similarly, the local balance of internal energy of the incompressible Newtonian fluids at rest reduces to

$$\rho \frac{\partial \epsilon}{\partial t} = \rho \frac{\partial h}{\partial t} = \text{div}(k\ \text{grad}\ T), \tag{5.6.49}$$

in which no external energy sources present, and ϵ is replaced by h, for $h = \epsilon + p/\rho$. This equation can further be simplified to

$$\rho \frac{\partial h}{\partial t} = \rho \frac{\partial h}{\partial T} \frac{\partial T}{\partial t} = \rho c_p \frac{\partial T}{\partial t} = \text{div}(k\ \text{grad}\ T), \tag{5.6.50}$$

if the fluid is *thermally perfect*, i.e., $h = h(T)$ only. The above equation corresponds exactly to the *equation of heat conduction* and may be integrated to obtain the temperature field.

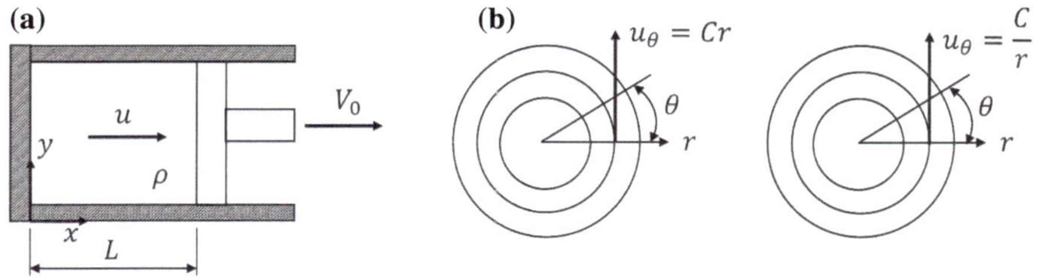

Fig. 5.9 Applications of the local balance of mass. **a** A gas in a piston-cylinder device. **b** Two-dimensional rotational flows

5.7 Illustrations of Local Physical Laws

In this section, selected problems are studied to show the applications of local mass balance and the Navier-Stokes equation. The applications of local balances of energy and entropy will be provided in Sects. 11.4.7 and 11.5.8 after the discussions on the fundamental knowledge of thermodynamics.

5.7.1 Mass Balance

Consider a piston-cylinder device filled with a high-pressure gas, as shown in Fig. 5.9a. Initially, the device is at rest, the gas assumes pressure p_1 and density ρ_1, and the distance from the end of cylinder to the piston is L_0. At $t = 0$, the piston starts to move at a constant velocity V_0, inducing a gas flow inside the cylinder which is approximated as a one-dimensional flow with velocity $u = V_0 x/L$. It is required to determine the time change of density of the gas inside the cylinder and obtain an expression of density as a function of time.

Construct the coordinate system shown in the figure. It follows from the local mass balance that

$$\frac{\partial \rho}{\partial t} + \operatorname{div}(\rho \boldsymbol{u}) = \frac{\partial \rho}{\partial t} + \frac{\partial (\rho u)}{\partial x} = 0, \tag{5.7.1}$$

which reduces to

$$\frac{\partial \rho}{\partial t} = -\rho \frac{\partial u}{\partial x} - u \frac{\partial \rho}{\partial x} = -\frac{\rho V_0}{L} - u \frac{\partial \rho}{\partial x}. \tag{5.7.2}$$

When the piston starts to move, sequences of pressure waves are triggered, which move from the piston toward the end of cylinder with the speed of sound c in the gas, and reach the cylinder end at the time duration $\Delta t = L/c$. If the order of magnitude of Δt satisfies $\mathcal{O}(\Delta t) < 10^{-3}$, it is plausible to assume that the densities at different points inside the cylinder change nearly correspondingly, so that $\rho \sim \rho(t)$. With these, Eq. (5.7.2) is simplified to

$$\frac{\partial \rho}{\partial t} \sim -\frac{\rho V_0}{L}. \tag{5.7.3}$$

Integrating this equation yields

$$\int_{\rho_1}^{\rho} \frac{\mathrm{d}\rho}{\rho} = -\int_0^t \frac{V_0}{L_0 + V_0\xi}\mathrm{d}\xi, \quad \longrightarrow \quad \rho = \rho_1 \left(\frac{1}{1 + V_0 t / L_0} \right). \tag{5.7.4}$$

This solution assumes its best accuracy immediately after the movement of piston. As time increases, the assumption of vanishing spatial variation in density losses the validity.

Figure 5.9b illustrates two two-dimensional rotational flows, one is characterized by the tangential velocity $u_\theta = Cr$, the other is by $u_\theta = C/r$, where C is a constant. It is required to determine the velocity components in the r-direction of these two-dimensional flows and the associated circulations.

For the left rotational flow, the local balance of mass in the cylindrical coordinate system reads

$$\frac{1}{r}\frac{\partial(ru_r)}{\partial r} + \frac{1}{r}\frac{\partial u_\theta}{\partial \theta} = 0, \tag{5.7.5}$$

which reduces to

$$\frac{\partial(ru_r)}{\partial r} = 0, \quad \longrightarrow \quad u_r = \frac{f(\theta, t)}{r}, \tag{5.7.6}$$

where $f(\theta, t)$ is an undetermined function, which needs to be explored by other physical laws. The circulation Γ for any closed contour $\mathcal{C}$ in the flow field is obtained as

$$\Gamma = \int_{\mathcal{C}} \boldsymbol{u} \cdot \mathrm{d}\boldsymbol{\ell} = \int_0^{2\pi} Cr^2 \mathrm{d}\theta = 2\pi C r^2 = 2CA, \tag{5.7.7}$$

where $\mathcal{C}$ is chosen as a circle with radius r centered at the origin, whose area is denoted by A. Similarly, the analysis of the right rotational flow leads to

$$u_r = \frac{f(\theta, t)}{r}, \qquad \Gamma = 2\pi C. \tag{5.7.8}$$

The left and right rotational flows are referred respectively to as the *forced vortex* and *free vortex*, if u_r vanishes. It is easy to show that a forced vortex is rotational, i.e., curl $\boldsymbol{u} \neq \boldsymbol{0}$, while a free vortex is irrotational (curl $\boldsymbol{u} = \boldsymbol{0}$).

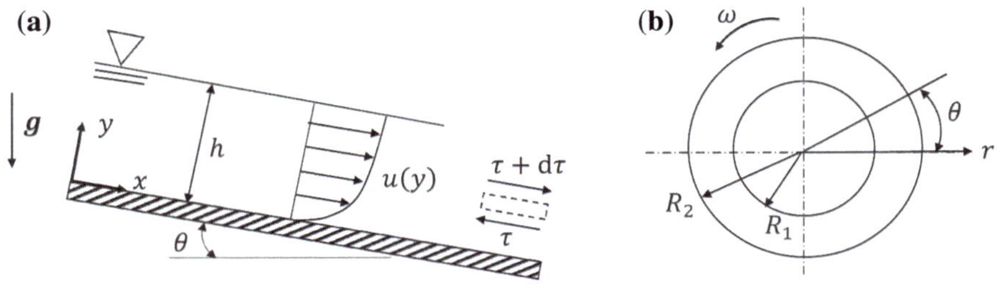

Fig. 5.10 Applications of the Navier-Stokes equation. **a** A steady, fully developed laminar flow of a Newtonian fluid down an incline. **b** A Newtonian fluid in a two-dimensional rotational cylindrical container

5.7.2 The Navier-Stokes Equation

Consider an incompressible Newtonian fluid down an inclined plane, as shown in Fig. 5.10a. It is assumed that the considered flow is two-dimensional and is in a region far away from the onset. The circumstance is then approximated to be a steady, *fully developed laminar flow*, i.e., the velocity component in the x-direction depends only on the y-coordinate. It is required to determine the velocity components in the x- and y-directions.

Since the fluid is incompressible and the flow is steady, the local balance of mass reads

$$\operatorname{div} \boldsymbol{u} = 0, \qquad \longrightarrow \qquad \frac{\partial u}{\partial x} + \frac{\partial v}{\partial y} = 0. \tag{5.7.9}$$

With the assumption of fully developed flow, the above equation reduces to

$$\frac{\partial v}{\partial y} = 0, \qquad \longrightarrow \qquad v = v(x). \tag{5.7.10}$$

Applying the no-slip boundary condition of velocity on the plane yields $v(x, y = 0) = 0$. Thus, a vanishing velocity component in the y-direction is obtained.

The Navier-Stokes equation is given by

$$\rho\left(\frac{\partial u}{\partial t} + u\frac{\partial u}{\partial x} + v\frac{\partial u}{\partial y}\right) = -\frac{\partial p}{\partial x} + \rho g_x + \mu\left(\frac{\partial^2 u}{\partial x^2} + \frac{\partial^2 u}{\partial y^2}\right), \tag{5.7.11}$$

in the x-direction, where $g_x = g\sin\theta$, and

$$\rho\left(\frac{\partial v}{\partial t} + u\frac{\partial v}{\partial x} + v\frac{\partial v}{\partial y}\right) = -\frac{\partial p}{\partial y} + \rho g_y + \mu\left(\frac{\partial^2 v}{\partial x^2} + \frac{\partial^2 v}{\partial y^2}\right), \tag{5.7.12}$$

in the y-direction, where $g_y = -g\cos\theta$. Substituting $v = 0$ into Eq. (5.7.12) yields

$$\frac{\partial p}{\partial y} = -\rho g\cos\theta, \tag{5.7.13}$$

which indicates that the pressure variation in the y-direction depends on the gravitational acceleration in the same direction, corresponding to the hydrostatic equation given in Sect. 3.1. Integrating this equation gives

$$\int_p^{p_0} \partial p = -\rho g\cos\theta \int_y^h \partial\xi, \qquad \longrightarrow \qquad p = p_0 + (\rho g\cos\theta)(h - y), \tag{5.7.14}$$

for the pressure on the free surface assumes the atmospheric pressure p_0. With these, Eq. (5.7.11) is simplified to

$$\frac{\mathrm{d}^2 u}{\mathrm{d}y^2} = -\frac{\rho g\sin\theta}{\mu}. \tag{5.7.15}$$

Integrating this equation needs two boundary conditions, which are given by

$$u|_{y=0} = 0, \qquad \left.\frac{\mathrm{d}u}{\mathrm{d}y}\right|_{y=h} = 0. \tag{5.7.16}$$

While the first condition results from the no-slip boundary condition, the second one is motivated by the assumption that the shear stress on the free surface is negligible. With these, the velocity component u is obtained as

$$u = \frac{\rho g \sin\theta}{\mu}\left(hy - \frac{y^2}{2}\right). \tag{5.7.17}$$

It is noted that the assumption of fully developed flow is crucial to the problem. Without it, it is impossible to deduce that $v = 0$, and the velocity component in the y-direction should retain its general form $v = v(x, y)$. In such a circumstance, the simplifications of mathematical analysis would never be made. This reflects the mathematical complexity of coupled local mass balance and the Navier-Stokes equation for real physical problems.

The problem can also be solved by using a simple physical concept. Consider an infinitesimal volume element $\mathrm{d}v = \mathrm{d}x\mathrm{d}y$ shown in the figure, with $\mathrm{d}x$ and $\mathrm{d}y$ the linear dimensions in the x- and y-directions, respectively. Applying Newton's second law of motion to $\mathrm{d}v$ in the x-direction yields

$$\sum F_x = (\rho g \sin\theta)\mathrm{d}x\mathrm{d}y + (\tau + \mathrm{d}\tau - \tau)\mathrm{d}x = 0, \tag{5.7.18}$$

for there exist no net pressure forces acting on $\mathrm{d}v$ because $p = p(y)$ only, and the acceleration component in the x-direction vanishes. The above equation shows that a force equilibrium in the x-direction should be maintained for $\mathrm{d}v$. It follows form Newton's law of viscosity that

$$\mathrm{d}\tau = \frac{\partial\tau}{\partial y}\mathrm{d}y = \frac{\mathrm{d}}{\mathrm{d}y}\left(\mu\frac{\mathrm{d}u}{\mathrm{d}y}\right)\mathrm{d}y = \mu\frac{\mathrm{d}^2u}{\mathrm{d}y^2}\mathrm{d}y, \tag{5.7.19}$$

where μ is assumed to be a constant. Substituting this equation into Eq. (5.7.18) results in

$$\frac{\mathrm{d}^2u}{\mathrm{d}y^2} = -\frac{\rho g \sin\theta}{\mu}, \tag{5.7.20}$$

which corresponds exactly to Eq. (5.7.15) and governs the distribution of u.

The dynamic viscosity of a Newtonian fluid is frequently estimated by using a concentric cylindrical viscometer, as shown in Fig. 5.10b, in which the outer cylinder is rotating at constant rotational speed ω, while the inner cylinder is held fixed. The fluid is placed in the annual gap between the outer and inner cylinders. It is required to determine the velocity profile of fluid, the shear stress at the surface of inner cylinder, and compare the latter with that obtained by using a planar approximation.

It is assumed that the Newtonian fluid is incompressible, and the gravitational acceleration points perpendicularly to the page for simplicity. The local mass balance in the cylindrical coordinate system then reads

$$\frac{1}{r}\frac{\partial(ru_r)}{\partial r} + \frac{1}{r}\frac{\partial(u_\theta)}{\partial\theta} = 0, \qquad \longrightarrow \qquad ru_r = c, \tag{5.7.21}$$

where c is a constant, for the axis-symmetry yields that $\partial\alpha/\partial\theta = 0$ for any physical variable α. It follows from the no-slip boundary condition on the surfaces of outer and inner cylinders that $c = 0$, implying that $u_r = 0$. With the steady-flow assumption, the Navier-Stokes equation is given by

$$\rho\left(u_r\frac{\partial u_r}{\partial r}+\frac{u_\theta}{r}\frac{\partial u_r}{\partial \theta}-\frac{u_\theta^2}{r}\right)=-\frac{\partial p}{\partial r}+\mu\left[\frac{\partial}{\partial r}\left(\frac{1}{r}\frac{\partial(ru_r)}{\partial r}\right)+\frac{1}{r^2}\frac{\partial^2 u_r}{\partial \theta^2}-\frac{2}{r^2}\frac{\partial u_\theta}{\partial \theta}\right],$$
$$\rho\left(u_r\frac{\partial u_\theta}{\partial r}+\frac{u_\theta}{r}\frac{\partial u_\theta}{\partial \theta}+\frac{u_r u_\theta}{r}\right)=-\frac{\partial p}{\partial \theta}+\mu\left[\frac{\partial}{\partial r}\left(\frac{1}{r}\frac{\partial(ru_\theta)}{\partial r}\right)+\frac{1}{r^2}\frac{\partial^2 u_\theta}{\partial \theta^2}+\frac{2}{r^2}\frac{\partial u_\theta}{\partial \theta}\right],$$
(5.7.22)

in the r- and θ-directions, respectively. With the previous assumptions and $u_r = 0$, Eq. (5.7.22)$_2$ reduces to

$$\frac{\mathrm{d}}{\mathrm{d}r}\left(\frac{1}{r}\frac{\mathrm{d}(ru_\theta)}{\mathrm{d}r}\right)=0, \quad \longrightarrow \quad u_\theta = c_1\frac{r^2}{2}+\frac{c_2}{r}, \tag{5.7.23}$$

where c_1 and c_2 are constants. Applying the no-slip boundary conditions

$$u_\theta(r=R_1)=0, \qquad u_\theta(r=R_2)=\omega R_2, \tag{5.7.24}$$

to Eq. (5.7.23)$_2$ yields an expression of u_θ given by

$$u_\theta=\frac{\omega R_1^2}{1-\lambda^2}\left(\frac{r}{R_1}-\frac{R_1}{r}\right), \qquad \lambda=\frac{R_1}{R_2}. \tag{5.7.25}$$

Substituting this expression into the Navier-Stokes equation in the r-direction gives

$$\frac{\partial p}{\partial r}=\frac{u_\theta^2}{r}=\left[\frac{\omega^2 R_1^4}{(1-\lambda^2)^2}\frac{1}{r}\left(\frac{r}{R_1}-\frac{R_1}{r}\right)^2\right]. \tag{5.7.26}$$

One can integrate this equation to obtain an expression of p. However, a simpler method can be used for the same purpose. Consider the force balance of a fluid element in the r-direction, which is given by

$$\frac{\partial p}{\partial r}\mathrm{d}r(r\mathrm{d}\theta)=\rho\mathrm{d}r(r\mathrm{d}\theta)r\omega^2, \tag{5.7.27}$$

where $r\omega^2$ is the centrifugal acceleration experienced by the fluid element at the distance r from the rotational axis. Integrating this equation immediately yields

$$p=p_i+\frac{\rho\omega^2}{2}\left(r^2-R_1^2\right), \quad \longrightarrow \quad p_o-p_i=\frac{\rho\omega^2}{2}\left(R_2^2-R_1^2\right), \tag{5.7.28}$$

where p_i and p_o are the pressures on the surfaces of inner and outer cylinders, respectively. This result shows that the pressure variation in the r-direction is induced via the centrifugal acceleration and assumes finite values if the thickness of annular gap is finite. On the contrary, $p_i \sim p_o$ if the thickness of annular gap approaches null.

By using Newton's law of viscosity, the shear stress is obtained as

$$\tau_{r\theta}=\mu r\frac{\mathrm{d}}{\mathrm{d}r}\left(\frac{u_\theta}{r}\right)=\frac{2\mu\omega R_1^2}{r^2(1-\lambda^2)}, \quad \longrightarrow \quad \tau_{r\theta}(r=R_1)=\frac{2\mu\omega}{1-\lambda^2}. \tag{5.7.29}$$

With a planar approximation, the shear stress on the surface of inner cylinder is approximated by using a linear velocity profile in Newton's law of viscosity, which is given by

$$\tau_{planar}=\mu\frac{\mathrm{d}u}{\mathrm{d}y}\sim\mu\frac{R_2\omega}{R_2-R_1}, \tag{5.7.30}$$

with which the shear stress ratio becomes

$$\frac{\tau_{r\theta}}{\tau_{planar}} = \frac{2}{1+\lambda}, \tag{5.7.31}$$

showing that τ_{planar} approaches $\tau_{r\theta}$ when $\lambda \to 1$, which is an estimation on the validity of planar approximation, i.e., it is only valid if the thickness of annual gap is extremely small.

The solutions to the problems shown in Fig. 5.10 were obtained by integrating the coupled local balance of mass and the Navier-Stokes equation simultaneously, which are referred to as the *exact solutions*. The exact solutions to more flow problems will be discussed in Sect. 8.2.

Remarks on the Integral and Differential Approaches:

The global balance equations correspond to the integral approach, while the local balance equations correspond to the differential approach. Although fluid behavior can be described by using both approaches, solutions with different accuracies are obtained. For example, consider air passing through a high building, and the net wind force exerting on the building by the air needs to be determined. The net wind force can easily be obtained by using the integral approach; however, the pressure and shear stress distributions on the surfaces of building cannot be delivered by the integral approach, although it is a physical fact that the net wind force results from the pressure and shear stress distributions on the surfaces. Such a detailed description of pressure and shear stress distributions needs to be provided by using the differential approach. Besides, even for the same problem, different finite control-volumes may be used by different investigators, resulting in small variations in the results of integral approach. This reflects the insufficient accuracy of integral approach. In the remaining part of the book, the discussions will be based on the differential approach, unless stated otherwise.

5.8 Exercises

5.1 Derive Eq. (5.1.23), namely the transformation rules for a surface and a volume elements between the reference and present configurations.

5.2 Derive Eq. (5.1.34), i.e., the time rate of change of determinant of the deformation gradient, and the time rate of change of a volume element in relation with the divergence of velocity in the present configuration.

5.3 Use a simple force balance to an infinitesimal tetrahedron to prove the Cauchy lemma.

5.4 Use the index notation to prove that for any vector $\boldsymbol{u}$, the following expression holds:

$$\dot{\boldsymbol{u}} = \frac{\partial \boldsymbol{u}}{\partial t} + (\text{grad}\, \boldsymbol{u})\boldsymbol{u} = \frac{\partial \boldsymbol{u}}{\partial t} + \text{grad}\left(\frac{\|\boldsymbol{u}\|^2}{2}\right) - \boldsymbol{u} \times \text{curl}\, \boldsymbol{u}.$$

5.5 Use the index notation and local balance of angular momentum to show that the Cauchy stress tensor is symmetric.

5.6 Complete the derivation from Eq. (5.3.31) to Eq. (5.3.32) for the local balance of internal energy.

5.7 Multiply the local balance of linear momentum dyadically with the velocity to show that

$$\frac{\partial}{\partial t}\left(\frac{\rho u_i u_j}{2}\right)+\frac{\partial}{\partial x_k}\left(\frac{\rho u_i u_j}{2}u_k\right)=\operatorname{sym}\left[\frac{\partial(u_i t_{jk})}{\partial x_k}\right]+\operatorname{sym}\left(\rho u_i b_j\right)$$
$$+\operatorname{sym}\left(\frac{\partial u_i}{\partial x_k}t_{jk}\right).$$

The above equation has the same structure as the general balance equation. Identify the terms of $\aleph_\phi$, $\pi_\aleph$, $\sigma_\aleph$, and $\psi_\aleph$.

5.8 Show that the stretching tensor $\boldsymbol{D}$ is an object tensor of second order under the Euclidean transformation.

5.9 Let s, $\boldsymbol{u}$, $\boldsymbol{t}$, and $\boldsymbol{T}$ be respectively arbitrarily isotropic scalar, vector, symmetric, and antisymmetric tensors of second order, whose functional dependencies are given by

$$\mathcal{C}=\mathcal{C}(\boldsymbol{v},\boldsymbol{A},\boldsymbol{B}),\qquad \mathcal{C}\in\{s,\boldsymbol{u},\boldsymbol{t},\boldsymbol{T}\},$$

where $\boldsymbol{v}$ is a vector, $\boldsymbol{A}$ and $\boldsymbol{B}$ are symmetric and skew-symmetric tensors of second order, respectively. Obtain the general isotropic expressions of s, $\boldsymbol{u}$, $\boldsymbol{t}$ and $\boldsymbol{T}$.

5.10 Consider an incompressible fluid down an inclined plane, as shown in the figure. The free surface of fluid is described by $y=h(x,t)$. Let the velocity components of fluid in the x- and y-directions be denoted respectively by u and v. Show that $h(x,t)$ can be described by using the kinematic equation given by

$$\frac{\partial h}{\partial t}+\frac{\partial h}{\partial x}u-v=0.$$

Further, show that h and the flow rate Q across a specific section satisfy the relation

$$\frac{\partial h}{\partial t}+\frac{\partial Q}{\partial x}=0,$$

which can be recast alternatively as

$$\frac{\partial h}{\partial t}+C(h)\frac{\partial h}{\partial x}=0,\qquad C(h)=\frac{\mathrm{d}Q}{\mathrm{d}h},$$

if Q is expressed as a function of h, i.e., $Q=Q(h)$. The above equation is a *one-dimensional wave equation* of $h(x,t)$, whose general solution is expressed as $h(x,t)=F\left(x-C(h)t\right)$, where F is any differentiable function.

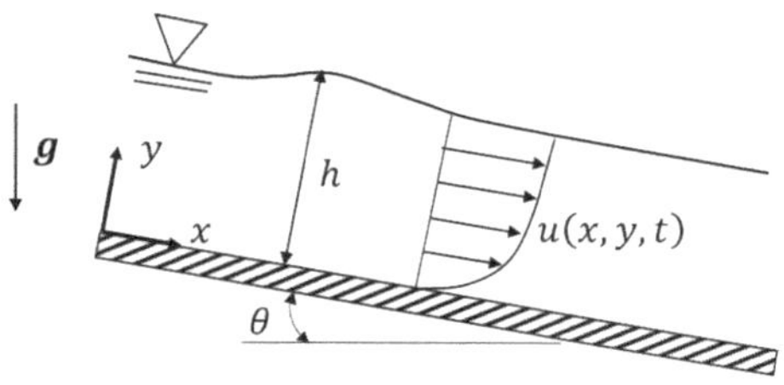

5.11 A jet of an incompressible fluid from a nozzle in an infinite two-dimensional space is shown in the figure, which is essentially an unsteady flow. It is assumed that the jet is symmetric with respect to the x-axis. By using the mass balance, show that

$$\frac{\partial h}{\partial t} + \frac{\partial Q}{\partial x} = q\sqrt{1 + \left(\frac{\partial h}{\partial x}\right)^2}, \qquad Q = \int_0^h u(x, y, t)\mathrm{d}y,$$

where q is the amount of fluid entering through the jet boundary (the dashed lines) per unit time and length. For simplicity, the gravitational acceleration is assumed to be in the z-axis.

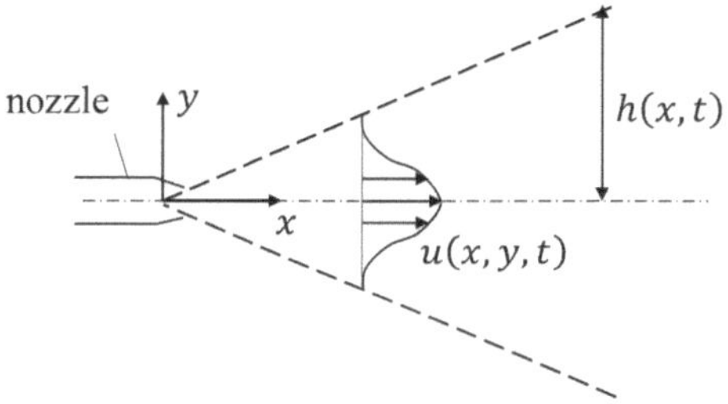

5.12 Water in an open channel is held by a two-dimensional sluice gate shown in the figure. Compare the horizontal forces exerted by the water on the gate for (a) the gate is closed, and (b) the gate is opened, after the water flow reaches a steady state.

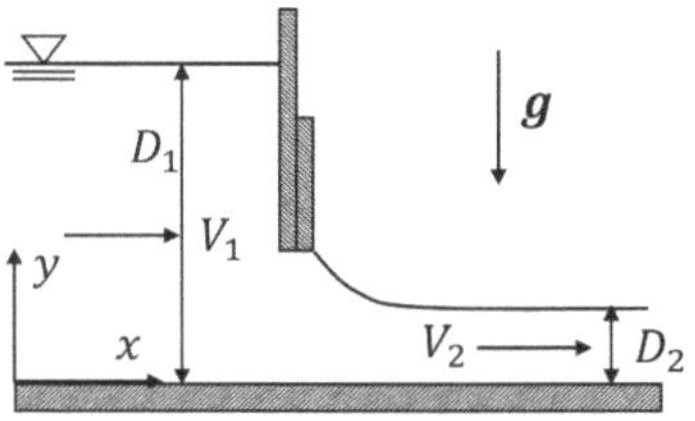

5.13 A free jet of water with constant cross-section A is deflected by a hinged plate with length L supported by a spring with spring constant k and un-compressed length y_0, as shown in the figure. Find the deflection angle θ as a function of jet speed V. For simplicity, the gravitational acceleration is assumed to be perpendicular to the page.

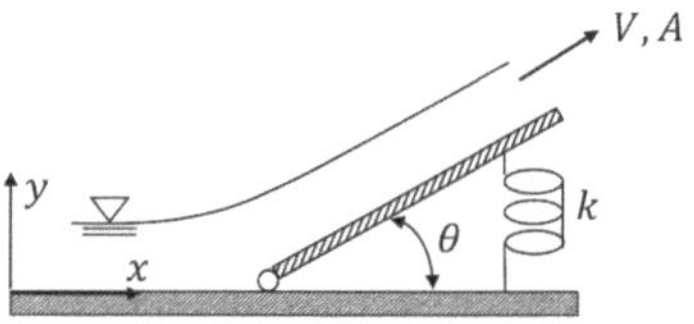

5.14 A small sphere is tested in a wind tunnel with diameter d, as shown in the figure. The absolute pressures are uniform in the cross-sections a and b, which are denoted by p_a and p_b, respectively. The air speed in the wind tunnel is denoted by V. The air velocity profile at cross-section a is uniform, while it is linear at cross-section b. Determine (a) the mass flow rate of air in the wind tunnel, (b) the maximum velocity V_{max}, and (c) the drag acting on the sphere. For simplicity, the viscous force on the wind tunnel wall is neglected.

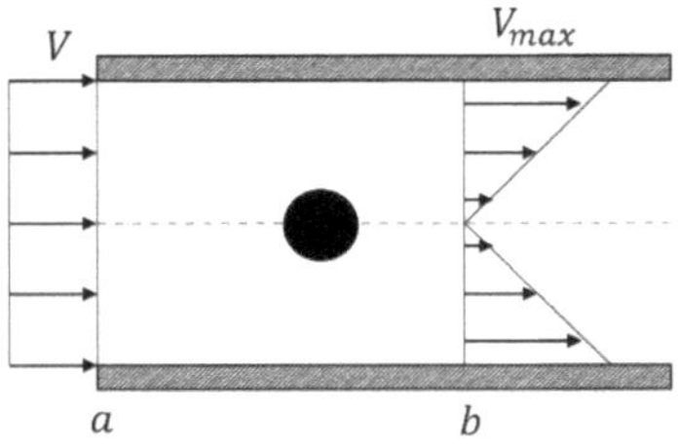

5.15 A block with mass M moves under the impact of a water jet, as shown in the figure. The dynamic frictional coefficient between the block and ground is μ_k. Determine the terminal speed of block.

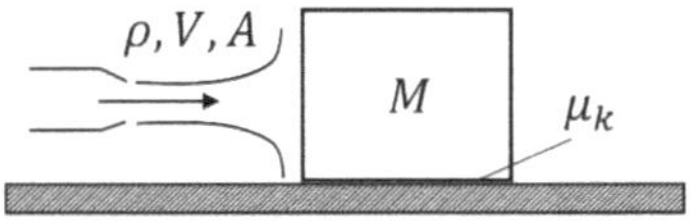

5.16 A cart is propelled by a water jet issuing from a tank shown in the figure. The tank is horizontal, and all the resistances to the motion of cart are neglected for simplicity. The tank is pressurized so that the water jet speed may be considered a constant, with the initial water mass in the tank M_0. Obtain an expression of the cart speed as it accelerates from rest.

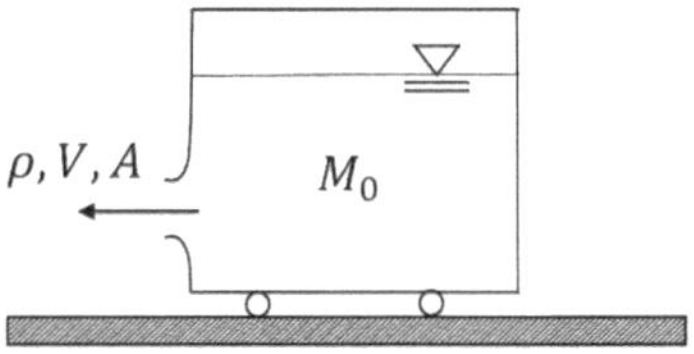

5.17 The tank shown in the figure rolls with negligible resistance along a horizontal track. The tank is to be accelerated from rest with initial mass M_0 by an external water jet with constant cross-section A and speed V that strikes the vane, which is deflected into the tank. Derive the expressions of tank speed and tank mass as functions of time.

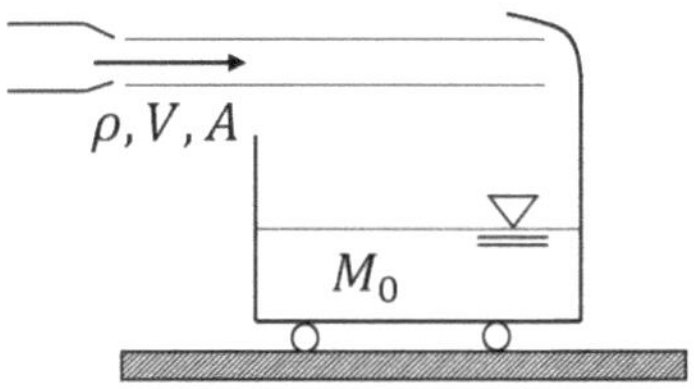

5.18 Water flows uniformly out of the slots with thickness h via two tubes of a rotating spray system shown in the figure. The diameter of rotating tube is d and the flow rate is Q. Find the torque required to hold the spray system stationary, and the steady-state rotational speed after it is released.

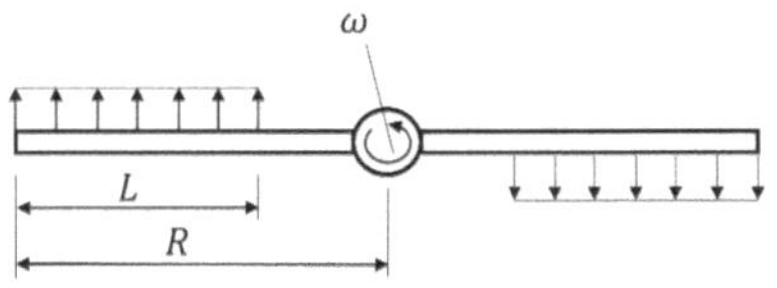

5.19 A pipe branches symmetrically into two legs of length L, and the whole system rotates with angular speed ω with respect to its symmetric axis. Each branch is inclined at angle θ. An incompressible liquid enters the pipe steadily, with vanishing angular momentum. The volume flow rate is denoted by Q. The pipe diameter D is much smaller than the length L. Obtain an expression for the external torque to turn the pipe. For simplicity, the cross-sectional areas of the pipe and two branches are assumed to be the same.

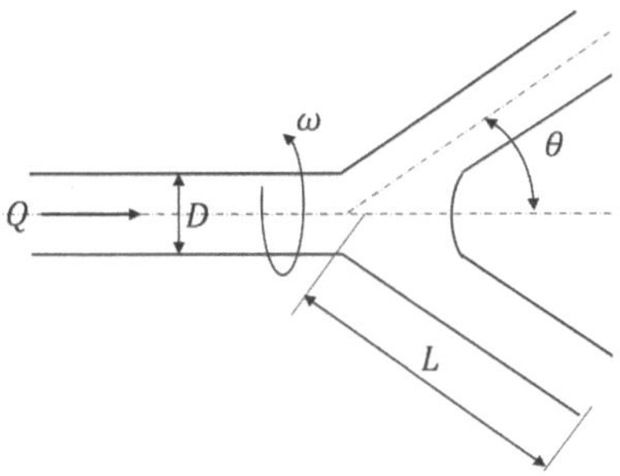

5.20 A tank with volume V is connected to a high-pressure air supply line with constant pressure p_0 and temperature T_0 shown in the figure. The tank is initially at a uniform temperature T_0. The initial absolute pressure in the tank is $p_i < p_0$. After the valve is opened, the tank temperature rises at the rate of α. Determine the instantaneous flow rate of air into the tank if heat transfer with the surrounding is neglected. Also obtain an expression for the entropy production of air inside the tank.

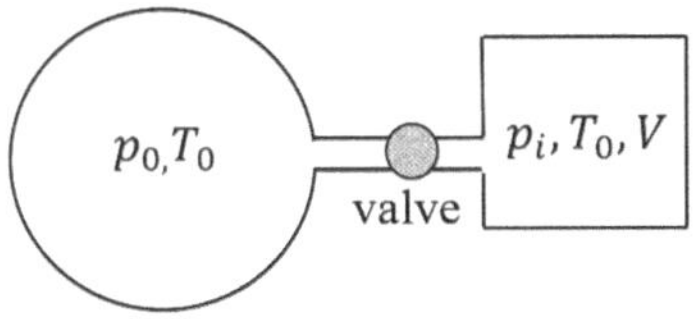

5.21 A moving belt passes through a container of an incompressible viscous liquid and moves vertically upwards with constant speed V_0, as shown in the figure. There exists a thin liquid film on the surface of belt. The liquid film tends to be driven downwards by gravity. It is assumed that the flow is laminar, steady and fully developed, determine the profiles of velocity components in the x- and y-directions.

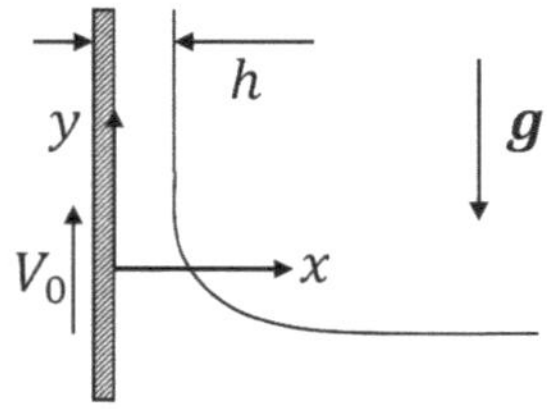

5.22 An incompressible Newtonian liquid is placed between two parallel solid plates, as shown in the figure. The upper plate is held fixed, while the lower plate moves at constant speed U in the x-direction. The motion of liquid is driven by the movement of lower plate and the non-vanishing pressure gradient $\partial p/\partial x$. Determine the relation between U and $\partial p/\partial x$, so that the shear stress acting on the lower plate vanishes.

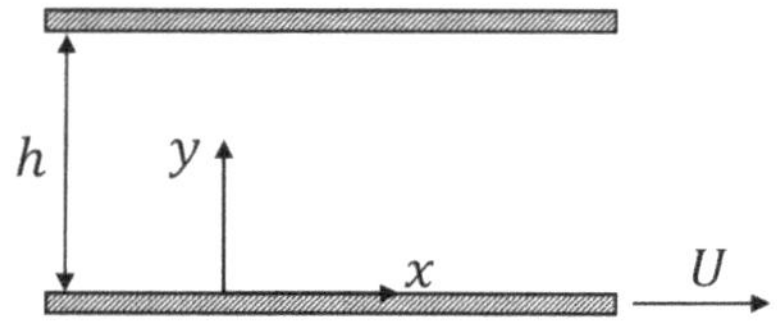

Further Reading

R. Aris, *Vectors, Tensors, and the Basic Equations of Fluid Mechanics* (Dover, New York, 1962)
G.K. Batchelor, *An Introduction to Fluid Dynamics* (Cambridge University Press, Cambridge, 1992)
P. Chadwick, *Continuum Mechanics* (Dover, New York, 1976)
A.J. Chorin, J.E. Marsden, *A Mathematical Introduction to Fluid Mechanics*, 2nd edn. (Springer, Berlin, 1990)
I.G. Currie, *Fundamental Mechanics of Fluids*, 2nd edn. (McGraw-Hill, Singapore, 1993)
K. Hutter, K. Jönk, *Continuum Methods of Physical Modeling* (Springer, Berlin, 2004)
I.S. Liu, *Continuum Mechanics* (Springer, Berlin, 2002)
J.E. Marsden, T.S. Ratiu, *Introduction to Mechanics and Symmetry*, 2nd edn. (Springer, Berlin, 1999)
I. Müller, W.H. Müller, *Fundamentals of Thermodynamics and Applications* (Springer, Berlin, 2009)
C. Truesdell, *A First Course in Rational Continuum Mechanics, Volume 1* (Academic Press, New York, 1977)
C. Truesdell, R.G. Muncaster, *Fundamentals of Maxwell's Kinetic Theory of a Simple Monatomic Gas* (Academic Press, New York, 1980)
C. Truesdell, W. Noll, *The Non-Linear Field Theories of Mechanics* (Springer, Berlin, 1992)

Dimensional Analysis and Model Similitude

6

Dimensional analysis is one of the most important mathematical tools in the study of fluid motion. It is a mathematical technique which makes use of dimensions of physical quantities as an aid to the solutions to many engineering problems. The main advantage of a dimensional analysis of a problem is that it reduces the number of variables by combining dimensional variables to form dimensionless products. Dimensional analysis has been found useful in both analytical and experimental work in the study of fluid mechanics and is closely related to the model similitude which is required for conducting experiments in laboratory. To explore the idea of dimensional analysis and model similitude, the discussion on dimensions and units of physical variables is introduced, followed by the Buckingham theorem and a suggested procedure in conducting dimensional analysis. The mathematical foundations of dimensional analysis and the theory of physical model, specifically the modeling law, are outlined, and the differential equations of fluid motion in dimensional forms are brought to dimensionless forms to illustrate the significant dimensionless products. The physical interpretations of obtained dimensionless products are given to show their influence in achieving a complete model similarity of a physical process.

6.1 Dimensions and Units of Physical Quantities

Physical quantities are characterized by their *dimensions*, which represent the intrinsic characteristics of quantities. For example, the height h and width w of a prismatic bar are two physical quantities and both represent a certain length. The length is thus the dimension of two quantities. To express the magnitudes of physical quantities, specific *units* must be allocated for rational values. Taking the prismatic bar again as an example, the height is e.g. 100 m. In this expression, the magnitude of height is evaluated in the base of meter. The height can equally be given as 0.1 km if the kilometer is used as the evaluation base. Quantities having no units are dimensionless quantities with unity dimension.

© Springer International Publishing AG 2019

C. Fang, *An Introduction to Fluid Mechanics*, Springer Textbooks in Earth Sciences, Geography and Environment, https://doi.org/10.1007/978-3-319-91821-1_6

The dimensions introduced as the *basic* or *fundamental dimensions* are used to express the dimensions of physical quantities. The units of fundamental dimensions are termed the *basic* or *fundamental units*. There exist four fundamental unit systems, which are introduced in the following:

- The *CGS-System*: The fundamental dimensions are length L, mass M and time t with the corresponding units as centimeter, gram and second. According to Newton's second law of motion, the dimension of fore is ML/t^2, having the unit of *dyne* as dyne $=$ g $\cdot$ cm/s^2. The energy dimension is ML^2/t^2 with the unit *erg* corresponding to erg $=$ dyne $\cdot$ cm.[1]
- The *MKS-System*: It is similar to the CGS-System, except that the units of fundamental dimensions L, M and t are given by meter, kilogram and second, respectively. Based on these, the force unit is *Newton*, denoted by N with N = kg $\cdot$ m/s^2. The energy unit is termed *Joule*,[2] denoted by J with J $=$ N $\cdot$ m. The power is the time rate of change of energy and is expressed by *Watt*,[3] denoted by W with W $=$ J/s $=$ N $\cdot$ m/s.
- The *MKS-Force-System*: Instead of choosing the mass as one of the fundamental dimensions, force is used together with length and time as the fundamental dimensions. Thus, the fundamental dimensions are length L, force F and time t, with the corresponding units given by meter, Newton and second. The dimension of mass is derived from Newton's second law of motion and is given by Ft^2/L.
- The *International System of Units*: The SI system is an extension of the MKS-System, with more fundamental dimensions with the corresponding units introduced. It is the modern form of the metric system and is the most widely used system of measurement, which compiles a coherent system of units of measurement built on seven basic units. The system also establishes a set of *prefixes* to the unit names and unit symbols that may be used when specifying multiples and fractions of the units.

Table 6.1 summarizes the fundamental dimensions and units in four unit systems described previously. When expressing very large or very small magnitudes of quantities, the standard prefix-notations from the SI system are suggested to be used, which are summarized in Table 6.2.

[1] The names "dyne" and "erg" were first proposed as the units of force and energy in 1861 by Joseph David Everett, 1831–1904, a British physicist.

[2] James Prescott Joule, 1818–1889, a British physicist and mathematician, who discovered the relation between *heat* and *mechanical work*, leading to the law of conservation of energy and the development of first law of thermodynamics.

[3] James Watt, 1736–1819, a Scottish inventor and mechanical engineer, who improved Thomas Newcomen's 1712 "*Newcomen steam engine*" by his "*Watt steam engine*" in 1781, which was a key part to *Industrial Revolution*.

Table 6.1 Fundamental dimensions and units in different unit systems

	Fundamental dimensions	Units
CGS-system	Length (L), Mass (M), Time (t)	L: centimeter, M: gram, t: second
MKS-system	Length (L), Mass (M), Time (t)	L: meter, M: kilogram, t: second
MKS-force-system	Length (L), Force (F), Time (t)	L: meter, F: Newton, t: second
SI system	Length (L), Mass (M), Time (t)	L: meter, M: kilogram, t: second
	Absolute temperature (T)	T: Kelvin
	Electric current (A), Substance (s)	A: Ampère, s: mole
	Light intensity (C)	C: Candela

Table 6.2 Prefix notations in the SI system

Prefix	Symbol	Power	Prefix	Symbol	Power
Exa	E	10^{18}	Deci	d	10^{-1}
Peta	P	10^{15}	Centi	c	10^{-2}
Trea	T	10^{12}	Milli	m	10^{-3}
Giga	G	10^{9}	Micro	μ	10^{-6}
Mega	M	10^{6}	Nano	n	10^{-9}
Kilo	K	10^{3}	Pico	p	10^{-12}
Hecto	h	10^{2}	Femto (Fermi)	f	10^{-15}
Deca	da	10^{1}	Atto	a	10^{-18}

6.2 Theory of Dimensional Analysis

6.2.1 Dimensional Homogeneity

An equation is called *homogeneous in its dimensions* or *dimensionally homogeneous* if its form does not depend upon the choice of basic units. For example, consider a mathematical pendulum, whose motion is described by

$$t = 2\pi \frac{L}{g}, \tag{6.2.1}$$

where t is the time period, L denotes the pendulum length, and g represents the gravitational acceleration. This equation is indifferent irrespective in which units of L and g are used, and the value of t is always correctly obtained in units of the dimensions that were chosen. On the other hand, the above equation can be expressed alternatively as

$$t = \frac{2\pi}{\sqrt{9.81}}\sqrt{L}, \tag{6.2.2}$$

if the value of g is substituted. This equation, however, is no longer indifferent in the choices of units, for L and t must be expressed in meter and second, respectively. Thus, Eq. (6.2.1) is dimensionally homogeneous, while Eq. (6.2.2) is not.

Dimensionally homogeneous functions are a special class of functions. The applications of dimensional analysis are based on the hypothesis that the solutions to a problem lead to dimensionally homogeneous functions only if the independent variables are correctly chosen. The hypothesis is justified by the fact that the fundamental physical laws are dimensionally homogeneous, thus deductions of these laws again give rise to dimensionally homogeneous equations. However, this holds true only if all the independent variables describing a physical process are completely considered. If it is not the case, there is no foundation to assume that the unknown equations are dimensionally homogeneous. For example, the drag force F acting on a sphere immersed completely in a fluid may be given by $F = f(V, d)$, where V is the fluid velocity and d is the diameter of sphere. This expression is not dimensionally homogeneous, for any combination of V and d would never lead to the dimension of force, although it is still meaningful.

6.2.2 Buckingham's Theorem and Dimensional Analysis

If an equation is constructed by the terms which are all dimensionally homogeneous, this equation is dimensionally homogeneous, for it does not depend upon the basic units that chosen. A sufficient condition for an equation to be dimensionally homogeneous is that this equation can be reduced to an equation of dimensionless products, which is known as the *Buckingham theorem* given by[4]

6.1 (The Buckingham Theorem) *If an equation is dimensionally homogeneous, it can be reduced to a relation of dimensionless products.*

It should be noted that the set of dimensionless products of given variables in the Buckingham theorem is complete, for each product is independent of any other products, and any product which does not belong to the set can be expressed as a product of powers of the dimensionless products of the set. It will be shown in Sect. 6.3.5 that the Buckingham theorem is not only a sufficient but also a necessary condition. The analysis of reducing a dimensionally homogeneous equation of given variables to a dimensionless equation of dimensionless products is called the *dimensional analysis*.

In conducting the dimensional analysis of an equation of given variables, or the dimensional analysis of a physical process in which certain physical quantities involve, a suggested procedure is outlined in the following:

- Step 1: List all dimensional quantities which are relevant, including all the dependent and independent variables. This step is crucial, for if the pertinent quantities are not all included, a dimensionless relation may still be obtained, but it does not give the complete story. For demonstration, let $Q_1, Q_2, \ldots Q_n$ be the

[4]Edgar Buckingham, 1867–1940, an American physicist, who contributed to the fields of soil physics, gas properties, acoustics, fluid mechanics, and blackbody radiation.

n-dimensional quantities which involve in a dimensional equation or a physical process.

- Step 2: Select a set of fundamental dimensions, e.g. the MKS-System or MKS-Force-System. The same dimensionless products can still be obtained even different fundamental dimensions are used, if the dimensional analysis is correctly conducted. For demonstration, the MKS-System is chosen as the fundamental dimensions for the quantities $Q_1, Q_2, \ldots Q_n$ selected in Step 1.
- Step 3: List the dimensions of all quantities in terms of the chosen fundamental dimensions. The matrix of dimensions of all quantities is called the *dimensional matrix* having rank r. For demonstration, the dimensions of dimensional quantities $Q_1, Q_2, \ldots Q_n$ in Step 1 in terms of the chosen MKS-System in Step 2 are given by

	Q_1	Q_2	Q_3	$\cdots$	Q_n
M	a_{11}	a_{12}	a_{13}	$\cdots$	a_{1n}
L	a_{21}	a_{22}	a_{23}	$\cdots$	a_{2n}
t	a_{31}	a_{32}	a_{33}	$\cdots$	a_{3n}

,

by which the dimensional matrix is identified to be

$$\begin{bmatrix} a_{11} & a_{12} & a_{13} & \cdots & a_{1n} \\ a_{21} & a_{22} & a_{23} & \cdots & a_{2n} \\ a_{31} & a_{32} & a_{33} & \cdots & a_{3n} \end{bmatrix}, \tag{6.2.3}$$

whose rank is r, representing the number of independent rows or columns.

- Step 4: Select a set of r-dimensional quantities that includes all the fundamental dimensions. The choice is not unique, but should satisfy that the selected r-dimensional quantities contain all the dimensions appearing in the dimensions of all physical variables. The selected dimensional quantities will be used later as the base in generating the dimensionless products. For demonstration, the quantities $Q_{n-r+1}, Q_{n-r}, \ldots Q_n$ are selected as the base of dimensional analysis, whose dimensions contain all dimensions of $Q_1, Q_2, \ldots Q_n$.
- Step 5: Setup dimensional equations by using the product of powers of the selected dimensional quantities in Step 4 with each of the remaining dimensional quantities in turn to form the dimensionless products. Solve the setup dimensional equations to obtain $n-r$ dimensionless products. For demonstration, the dimensionless product is constructed by using the powers of selected dimensional quantities such as

$$\begin{aligned} \Pi &= Q_1^{k_1} Q_2^{k_2} \cdots Q_n^{k_n}, \\ \longrightarrow \quad [\Pi] &= \left[M^{a_{11}} L^{a_{21}} t^{a_{31}}\right]^{k_1} \left[M^{a_{12}} L^{a_{22}} t^{a_{32}}\right]^{k_2} \cdots \left[M^{a_{1n}} L^{a_{2n}} t^{a_{3n}}\right]^{k_n}, \end{aligned} \tag{6.2.4}$$

where $[\beta]$ denotes the dimension of any quantity β, and Π is a dimensionless product. The above equation gives rise to a system of linear equations of the powers k_is, viz.,

$$\begin{aligned} k_1 a_{11} + k_2 a_{12} + \cdots + k_n a_{1n} &= 0, \\ k_1 a_{21} + k_2 a_{22} + \cdots + k_n a_{2n} &= 0, \\ k_1 a_{31} + k_2 a_{32} + \cdots + k_n a_{3n} &= 0, \end{aligned} \tag{6.2.5}$$

so that the solutions to k_is make Π be a dimensionless quantity. Since Q_{n-r+1}, $Q_{n-r}, \ldots Q_n$ are selected as the base of dimensional analysis, the above system of linear equations possesses $(n-r)$ linearly independent solutions. If one substitutes for k_1 to k_{n-r} the linearly independent arbitrarily choices of

$$\begin{aligned} &k_1 = 1, \quad k_2 = k_3 = \cdots k_{n-r} = 0, \\ &k_2 = 1, \quad k_1 = k_3 = \cdots k_{n-r} = 0, \\ &\vdots \\ &k_{n-r} = 1,\ k_1 = k_2 = \cdots k_{n-r-1} = 0, \end{aligned} \tag{6.2.6}$$

then the remaining k_js can be determined. The dimensionless products can then be represented by the array given by

	k_1	k_2	k_3	$\cdots$	k_{n-r}	k_{n-r+1}	$\cdots$	k_n
Π_1	1	0	0	$\cdots$	0	$\alpha_{1,n-r+1}$	$\cdots$	$\alpha_{1,n}$
Π_2	0	1	0	$\cdots$	0	$\alpha_{2,n-r+1}$	$\cdots$	$\alpha_{2,n}$
Π_3	0	0	1	$\cdots$	0	$\alpha_{3,n-r+1}$	$\cdots$	$\alpha_{3,n}$
$\vdots$	$\vdots$	$\vdots$	$\vdots$		$\vdots$	$\vdots$		$\vdots$
Π_{n-r}	0	0	0	$\cdots$	1	$\alpha_{n-r,n-r+1}$	$\cdots$	$\alpha_{n-r,n}$

,

in which totally $(n-r)$ dimensionless products are obtained, where all αs are the results of manipulations of all as in Eq. (6.2.5).

- Step 6: Check if each obtained dimensionless product is indeed dimensionless. For demonstration, the obtained dimensionless products $\Pi_1, \Pi_2, \ldots \Pi_{n-r}$ from the last step need to be verified.

The first step of dimensional analysis, i.e., the setup of a proper functional relationship among all the pertinent dimensional variables that enter a problem is decisive to a successful dimensional analysis. Unfortunately, there exist no definite rules which can be followed for a proper selection of variables which need to be included in any problem. Rather, the success of any investigation depends on the ability of operator to predict correctly the variables to be included in the problem. However, three simple rules may be summarized in the following for a suggested guidance of the inclusion of physical variables that need to be taken into account in the dimensional analysis of a problem of fluid motion:

1. Fluid properties, such as density, specific weight, dynamic viscosity, bulk modulus, compressibility, and surface tension strength.
2. Kinematic and dynamic characteristics of fluid motion, e.g. fluid velocity and pressure difference.
3. Boundary geometry in the flow field, frequently represented by some linear dimensions.

Dimensional analysis can also be done by using the *Rayleigh method*.[5] The Buckingham and Rayleigh methods are intrinsically the same. However, in using the Buckingham method for performing dimensional analysis one is free from the indiscriminate use of infinite series called for in the Rayleigh method. The construction of an infinite series is logically an indispensable step in the Rayleigh method, although in simple problems with relatively few variables an approximation is often made to equate arbitrarily the dependent variable in a physical process to a product of powers of the independent variables with numerical constant.

6.2.3 Illustrations of Dimensional Analysis

Consider a uniform rectilinear flow through a two-dimensional square shown in Fig. 6.1a. Due to the viscous effect of fluid, there exists a sequence of vortices behind the square, known as the *von Kármán vortex street*,[6] and the vortices take place alternatively with a definite frequency ω. For this physical process, the fluid density ρ, velocity V and dynamic viscosity μ, and the width of square b are relevant. The five physical quantities are described by a dimensionally homogeneous function given by

$$f(\rho, V, \mu, b, \omega) = 0. \tag{6.2.7}$$

Choose the MKS-System as the fundamental dimension system, with which the dimensions of five physical quantities are given by

	ρ	V	μ	b	ω
M	1	0	1	0	0
L	−3	1	−1	1	0
t	0	−1	−1	0	−1

with the dimensional matrix as

$$\begin{bmatrix} 1 & 0 & 1 & 0 & 0 \\ -3 & 1 & -1 & 1 & 0 \\ 0 & -1 & -1 & 0 & -1 \end{bmatrix}. \tag{6.2.8}$$

It is readily verified that the rank of this dimensional matrix is 3. Thus, three quantities are chosen as the base in generating the dimensionless products. Specifically, ρ, V, b are selected, for their dimensions include all dimensions appearing in all the

[5]John William Strutt, or Lord Rayleigh, 1842–1919, a British physicist, who, together with William Ramsay, earned the Nobel Prize for Physics in 1904 for his contribution to the discovery of argon. Sir William Ramsay, 1852–1916, a British chemist, who discovered the noble gases and received the Nobel Prize in Chemistry in 1904.

[6]Theodore von Kármán, 1881–1963, a Hungarian-American mathematician and physicist, who contributed to many key advances in aerodynamics and is recognized as "Father of Aerodynamics."

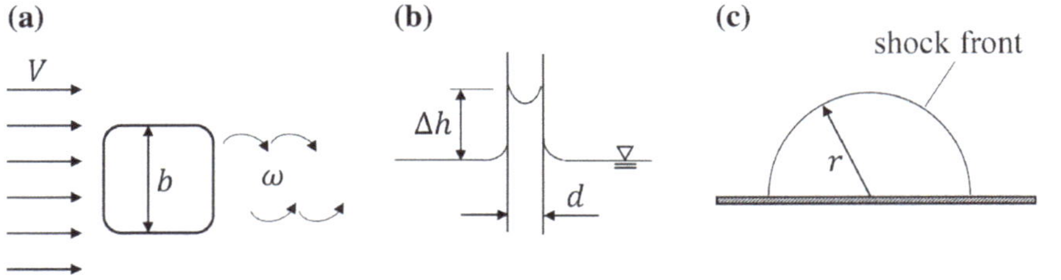

Fig. 6.1 Illustrations of dimensional analysis. **a** A uniform rectilinear flow through a two-dimensional square. **b** Free surface rise of a liquid in a capillary tube. **c** Semi-spherical shock wave traveling in a stationary air

five variables. It follows that there exist two dimensionless products, which are determined as

$$\Pi_1 = \mu\rho^{k_1}V^{k_2}b^{k_3}, \rightarrow \Pi_1 = \frac{\rho V b}{\mu}; \quad \Pi_2 = \omega\rho^{k_1}V^{k_2}b^{k_3} \rightarrow \Pi_2 = \frac{\omega b}{V}, \quad (6.2.9)$$

with which the original dimensional Eq. (6.2.7) may be brought to a relation of dimensionless products given by

$$f(\Pi_1, \Pi_2) = 0, \quad \longrightarrow \quad f\left(\frac{\rho V b}{\mu}, \frac{\omega b}{V}\right) = 0. \quad (6.2.10)$$

Consider the rise of free surface of a liquid in a capillary tube due the effect of surface tension shown in Fig. 6.1b. The rise of free surface, denoted by Δh, is related to the diameter d of capillary tube, the specific weight γ, and capillary constant σ of the liquid, for which a dimensionally homogeneous functional

$$f(\Delta h, d, \gamma, \sigma) = 0, \quad (6.2.11)$$

is constructed. Choose the MKS-Force-System as the fundamental dimension system to express the dimensions of these four quantities, so that the dimensional matrix is obtained as

$$\begin{array}{c|cccc} & \Delta h & d & \gamma & \sigma \\ \hline F & 0 & 0 & 1 & 1 \\ L & 1 & 1 & -3 & 1 \\ t & 0 & 0 & 0 & 0 \end{array}, \quad \longrightarrow \quad \begin{bmatrix} 0 & 0 & 1 & 1 \\ 1 & 1 & -3 & -1 \end{bmatrix}, \quad (6.2.12)$$

whose rank is 2. It follows that there exist two dimensionless products obtained as

$$\Pi_1 = \Delta h \gamma^{k_1} d^{k_2} = \frac{\Delta h}{d}; \quad \Pi_2 = \sigma\gamma^{k_1} d^{k_2} = \frac{\sigma}{d^2\gamma}, \quad (6.2.13)$$

if $\{d, \gamma\}$ are chosen as the base in generating the dimensionless products. Thus, dimensional Eq. (6.2.11) is brought to a dimensionless equation given by

$$f\left(\frac{\Delta h}{d}, \frac{\sigma}{d^2\gamma}\right) = 0. \quad (6.2.14)$$

Consider the motion of a shock front after an atomic explosion which was close to the ground shown in Fig. 6.1c. Let at a specific point on the ground a large quantity of energy E be released at time $t = 0$. As an idealization, it is assumed that the energy is released completely at $t = 0$ within an infinitesimal volume whose size is negligible. As a consequence of the explosion, a half spherical shock wave takes place, which travels outwards into the surrounding still air. The radius of shock wave, r, is related to the density of still air, ρ_0, time duration t and released energy E described by a dimensional equation given by

$$f(r, \rho_0, t, E) = 0. \tag{6.2.15}$$

Choose the MKS-System to express the dimensions of these quantities, with which the dimensional matrix is obtained as

$$\begin{array}{c|cccc} & r & \rho_0 & t & E \\ \hline M & 0 & 1 & 0 & 1 \\ L & 1 & -3 & 0 & 2 \\ t & 0 & 0 & 1 & -2 \end{array}, \quad \longrightarrow \quad \begin{bmatrix} 0 & 1 & 0 & 1 \\ 1 & -3 & 0 & 2 \\ 0 & 0 & 1 & -2 \end{bmatrix}, \tag{6.2.16}$$

whose rank is 3. Thus, there exists a single dimensionless product given by

$$\Pi = \frac{r}{(E/\rho_0)^{1/5} t^{2/5}} = K_3, \tag{6.2.17}$$

where K_3 is a constant. Taking logarithmic operation of this equation yields

$$\ln r = \ln K_3 + \frac{1}{5} \ln \left(\frac{E}{\rho_0} \right) + \frac{2}{5} \ln t. \tag{6.2.18}$$

This equation can be displayed graphically in a double logarithmic diagram spanned by $x = \ln t$ and $y = \ln r$, in which a straight line passing point ($x = 0$, $y = \ln K + \ln(E/\rho_0)/5$) with slope of 2/5 is drawn. The value of K_3 is approximated by $K_3 \sim 1$ from the theory of gas dynamics. Thus, if one can measure the radius of shock front at different times, then E can be estimated.[7]

The above discussions are based on a three-dimensional shock front. It is readily to verify that similar expressions for the two- and one-dimensional approximations of the shock front can be given respectively by

$$\frac{r}{(E/\rho_0)^{1/4} t^{1/2}} = K_2, \qquad \frac{r}{(E/\rho_0)^{1/3} t^{2/3}} = K_1, \tag{6.2.19}$$

where K_2 and K_1 are the constants in the two- and one-dimensional approximations, respectively. It is seen that the speed of shock front changes with the dimension of space where the shock wave travels. The derivations of Eq. (6.2.19) are left as an exercise.

---

[7]This was done by Taylor by using a movie film of the nuclear test when Americans were testing their atomic bombs in the Manhattan Project during World War II, although the strength of bomb was kept secret, while the movie film was not classified. Sir Geoffrey Ingram Taylor, 1886–1975, a British physicist and mathematician, who was a major figure in fluid dynamics and wave theory and was described as one of the most notable scientists in the twentieth century.

6.3 Mathematical Foundation of Dimensional Analysis

6.3.1 Transformation of Basic Units

Let m be the number of the fundamental dimensions associated with the basic units denoted by

$$[G]_j, \qquad j = 1, 2, \ldots, m, \tag{6.3.1}$$

where $[G]$ denotes the basic unit of the fundamental dimension G. The units of the dimensions of quantities A_j derived by using the power products of fundamental dimensions can be expressed as

$$[A]_j = \prod_{i=1}^{m} [G]_i^{a_{ij}}, \qquad j = 1, 2, \ldots, n, \tag{6.3.2}$$

in which it is assumed that there exist n derived quantities, and the symbol "$\prod$" represents the multiplication summation.

It is supposed that there exist two basic unit systems, denoted respectively by $[G]_k^o$ and $[G]_k^n$ standing for the old and new basic units, respectively. The derived quantities A_j thus possess different values in the old and new basic units. Let the values of A_j in terms of $[G]_k^o$ and $[G]_k^n$ be denoted by x_j and $\bar{x}_j$, respectively, and

$$[G]_k^o = \alpha_k [G]_k^n, \tag{6.3.3}$$

represents a conversion between the old and new basic units, where α_k is the conversion factor. With these, the values of A_j in terms of the old basic unit system $[G]_k^o$, i.e., x_j, can be transformed to

$$\begin{aligned} x_j \left\{ \left[G_1^o\right]^{a_{1j}} \left[G_2^o\right]^{a_{2j}} \cdots \left[G_m^o\right]^{a_{mj}} \right\} &= x_j \left\{ \alpha_1^{a_{1j}} \left[G_1^n\right]^{a_{1j}} \alpha_2^{a_{2j}} \left[G_2^n\right]^{a_{2j}} \cdots \alpha_m^{mj} \left[G_m^n\right]^{a_{mj}} \right\} \\ &= \bar{x}_j \left\{ \left[G_1^n\right]^{a_{1j}} \left[G_2^n\right]^{a_{2j}} \cdots \left[G_m^n\right]^{a_{mj}} \right\}, \end{aligned} \tag{6.3.4}$$

giving rise to

$$\bar{x}_j = x_j \prod_{k=1}^{m} \alpha_k^{a_{kj}}. \tag{6.3.5}$$

This equation is used to compute the value of a derived quantity in its dimensional units, if the basic units of fundamental dimensions have been changed.

6.3.2 Definition of Dimensional Homogeneity

Let y be a function of n variables given by $y = f(x_1, x_2, \ldots x_n)$, whose value changes to $\bar{y} = f(\bar{x}_1, \bar{x}_2, \ldots \bar{x}_n)$ if the basic units are changed in expressing the values of $(x_1, x_2, \ldots x_n)$. An equation is said to be *dimensionally homogeneous* if $y = f(x_1, x_2, \ldots x_n)$ can be brought to

$$\bar{y} = f(\bar{x}_1, \bar{x}_2, \cdots \bar{x}_n), \tag{6.3.6}$$

with the *same functional* f. This means that the equation is indifferent under the group of transformation which is generated by all possible changes of the fundamental dimensions in terms of the basic units.

In the group transformation given in Eq. (6.3.5), the terms α_ks may be arbitrarily positive constants. Applying the group transformation to Eq. (6.3.6) yields

$$\bar{y} = K_0 y, \qquad \bar{x}_j = K_j x_j, \tag{6.3.7}$$

where

$$K_0 = \prod_{k=1}^{m} \alpha_k^{a_{k0}}, \qquad K_j = \prod_{k=1}^{m} \alpha_k^{a_{kj}}. \tag{6.3.8}$$

The term a_{kj} represents the number appearing in the dimensional matrix of $(y, x_1, x_2, \ldots x_n)$ at the kth row and the jth column. If $y = f(x_1, x_2, \ldots x_n)$ is dimensionally homogeneous, it follows that the expression

$$\bar{y} = K_0 y = K_0 f(x_1, x_2, \cdots x_n) = f(\bar{x}_1, \bar{x}_2, \cdots \bar{x}_n) = f(K_1 x_1, K_2 x_2, \cdots K_n x_n), \tag{6.3.9}$$

must be satisfied for all variables $(x_1, x_2, \ldots x_n)$ and $(\alpha_1, \alpha_2, \ldots, \alpha_m)$. In this expression, all Ks are determined if all $\alpha_k s$ and the dimensional matrix of $(y, x_1, x_2, \ldots x_n)$ in terms of the basic units $[G]_j$ are known.

For example, consider the drag force F acting on a sphere submerged in a moving fluid, which is described mathematically by

$$F = f(V, d, \rho, \mu), \tag{6.3.10}$$

where d is the diameter of sphere, and $\{V, \rho, \mu\}$ are respectively the velocity, density, and dynamic viscosity of moving fluid. If Eq. (6.3.10) is dimensionally homogeneous, it follows from Eq. (6.3.9) that

$$K_0 F = f(K_1 V, K_2 d, K_3 \rho, K_4 \mu). \tag{6.3.11}$$

By choosing the MKS-System, the dimensional matrix of Eq. (6.3.10) is obtained as

	F	V	d	ρ	μ
M	1	0	0	1	1
L	1	1	1	−3	−1
t	−2	−1	0	0	−1

$$\longrightarrow \begin{bmatrix} 1 & 0 & 0 & 1 & 1 \\ 1 & 1 & 1 & -3 & -1 \\ -2 & -1 & 0 & 0 & -1 \end{bmatrix}, \tag{6.3.12}$$

by which it is found that

$$\begin{aligned} &K_0 = \alpha_1^1 \alpha_2^1 \alpha_3^{-2}, \quad K_1 = \alpha_1^0 \alpha_2^1 \alpha_3^{-1}, \quad K_2 = \alpha_1^0 \alpha_2^1 \alpha_3^0, \\ &K_3 = \alpha_1^1 \alpha_2^{-3} \alpha_3^0, \quad K_4 = \alpha_1^1 \alpha_2^{-1} \alpha_3^{-1}. \end{aligned} \tag{6.3.13}$$

Substituting these expressions into Eq. (6.3.11) yields

$$\alpha_1^1 \alpha_2^1 \alpha_3^{-2} F = f(\alpha_1^0 \alpha_2^1 \alpha_3^{-1} V, \alpha_1^0 \alpha_2^1 \alpha_3^0 d, \alpha_1^1 \alpha_2^{-3} \alpha_3^0 \rho, \alpha_1^1 \alpha_2^{-1} \alpha_3^{-1} \mu), \tag{6.3.14}$$

which can be fulfilled for all values of α_ks if

$$F = \rho V^2 d^2 f'\left(\frac{\rho V d}{\mu}\right), \tag{6.3.15}$$

where f' denotes another functional. Thus, Eq. (6.3.10) is a dimensionally homogeneous equation.

Another example is to consider an equation given by

$$y = x_1 x_2 x_3. \tag{6.3.16}$$

Applying Eq. (6.3.9) to this equation yields

$$K_0 y = K_1 x_1 K_2 x_2 K_3 x_3, \qquad K_0 = \prod_{j=1}^{m} \alpha_j^{a_{j0}}, \; K_i = \prod_{j=1}^{m} \alpha_j^{a_{ji}}. \tag{6.3.17}$$

With these, it follows that

$$\left(\prod_{j=1}^{m} \alpha_j^{a_{j0}}\right) x_1 x_2 x_3 = \left(\prod_{j=1}^{m} \alpha_j^{a_{j1}}\right) x_1 \left(\prod_{j=1}^{m} \alpha_j^{a_{j2}}\right) x_2 \left(\prod_{j=1}^{m} \alpha_j^{a_{j3}}\right) x_3, \tag{6.3.18}$$

giving rise to

$$\alpha_{j0} = \alpha_{j1} + \alpha_{j2} + \alpha_{j3}, \tag{6.3.19}$$

which is the condition of dimensional homogeneity of Eq. (6.3.16).

6.3.3 Two Special Forms of Dimensionally Homogeneous Equations

To further explore the concept of dimensional homogeneity, consider a special case of $y = f(x_1, x_2, \ldots x_n)$ given by

$$y = \sum_{i=1}^{n} x_i = x_1 + x_2 + \cdots + x_n, \tag{6.3.20}$$

where y is considered a dependent variable, and $(x_1, x_2, \ldots x_n)$ are independent variables. Applying Eq. (6.3.9) to the above equation yields

$$K_0(x_1 + x_2 + \cdots + x_n) = K_1 x_1 + K_2 x_2 + \cdots + K_n x_n, \tag{6.3.21}$$

which must be satisfied. It follows immediately that

$$K_0 = K_1 = K_2 = \cdots = K_n, \tag{6.3.22}$$

with which Eq. (6.3.8) becomes

$$a_{k_0} = a_{k_1} = a_{k_2} \cdots = a_{k_n}, \qquad k = 1, 2, \cdots, m. \tag{6.3.23}$$

This result indicates that Eq. (6.3.20) is dimensionally homogeneous *if and only if* all its members, i.e., y and $(x_1, x_2, \ldots x_n)$ have the same dimensions.

Consider another special case of $y = f(x_1, x_2, \ldots x_n)$ given by

$$y = \prod_{j=1}^{n} x_j^{k_j} = x_1^{k_1} x_2^{k_2} \cdots x_n^{k_n}, \tag{6.3.24}$$

where k_js are arbitrary. It is first assumed that this equation is dimensionally homogeneous, with which Eq. (6.3.9) must be fulfilled. Thus,

$$K_0\left(\prod_{j=1}^{n} x_j^{k_j}\right) = \prod_{j=1}^{n} (K_j x_j)^{k_j} = \left(\prod_{j=1}^{n} K_j^{k_j}\right)\left(\prod_{j=1}^{n} x_j^{k_j}\right), \tag{6.3.25}$$

which gives

$$K_0 = \left(\prod_{j=1}^{n} K_j^{k_j}\right). \tag{6.3.26}$$

Combining this equation with Eq. (6.3.8) results in

$$\prod_{k=1}^{m} \alpha_k^{a_{k0}} = \left(\prod_{k=1}^{m} \alpha_k^{a_{k1}}\right)^{k_1} \left(\prod_{k=1}^{m} \alpha_k^{a_{k2}}\right)^{k_2} \cdots \left(\prod_{k=1}^{m} \alpha_k^{a_{kn}}\right)^{k_n} = \prod_{k=1}^{m} \alpha_k^{\sum_{j=1}^{n} a_{kj} k_j}, \tag{6.3.27}$$

which yields the expression

$$a_{k0} = \sum_{j=1}^{n} a_{kj} k_j, \qquad k = 1, 2, \ldots, m. \tag{6.3.28}$$

Thus, if Eq. (6.3.24) is dimensionally homogeneous, Eq. (6.3.28) must be satisfied. On the other hand, expressing Eq. (6.3.24) in terms of the old and new basic units leads to

$$y = \prod_{j=1}^{n} x_j^{k_j}, \qquad \bar{y} = \prod_{j=1}^{n} \bar{x}_j^{k_j}. \tag{6.3.29}$$

Applying Eqs. (6.3.7) and (6.3.8) to Eq. (6.3.29) results in

$$\left(\prod_{k=1}^{m} \alpha_k^{a_{k0}}\right) y = \prod_{j=1}^{n}\left(\prod_{k=1}^{m} \alpha_k^{a_{kj}} x_j\right)^{k_j} = \prod_{j=1}^{n}\left(\prod_{k=1}^{m} \alpha_k^{a_{kj}k_j}\right) x_j^{k_j} = \left(\prod_{k=1}^{m} \alpha_k^{\sum_{j=1}^{n} a_{kj} k_j}\right) \prod_{j=1}^{n} x_j^{k_j}, \tag{6.3.30}$$

in which the last $\prod$-operation is nothing else than y itself. It is found that if

$$\prod_{k=1}^{m} \alpha_k^{a_{k0}} = \prod_{k=1}^{m} \alpha_k^{\sum_{j=1}^{n} a_{kj} k_j}, \qquad \longrightarrow \qquad a_{k0} = \sum_{j=1}^{n} a_{kj} k_j, \tag{6.3.31}$$

holds, then Eq. (6.3.24) is dimensionally homogeneous. It is concluded that Eq. (6.3.24) is dimensionally homogeneous *if and only if* Eq. (6.3.28) or $(6.3.31)_2$ is fulfilled.

6.3.4 Determination of Dimensionless Products

It is assumed that the power products among the variable $(x_1, x_2, \ldots, x_n)$ given by

$$\begin{aligned}
\Pi_{(1)} &= x_1^{k_1^{(1)}} x_2^{k_2^{(1)}} \cdots x_n^{k_n^{(1)}}, \\
\Pi_{(2)} &= x_1^{k_1^{(2)}} x_2^{k_2^{(2)}} \cdots x_n^{k_n^{(2)}}, \\
&\vdots \\
\Pi_{(p)} &= x_1^{k_1^{(p)}} x_2^{k_2^{(p)}} \cdots x_n^{k_n^{(p)}},
\end{aligned} \tag{6.3.32}$$

are dimensionless. If any two of the products $\Pi_{(i)}$ and $\Pi_{(j)}$ are linearly dependent, then a certain power of $\Pi_{(i)}$ must be equal to $\Pi_{(j)}$, or in general $\Pi_{(i)}^{h_i}\Pi_{(j)}^{h_j} = 1$ holds for some non-vanishing values of h_i and h_j. This implies that if $\Pi_{(1)}, \Pi_{(2)}, \ldots \Pi_{(p)}$ are linearly dependent, there should exist constants $h_1, h_2, \ldots h_p$ which do not vanish simultaneously, so that

$$\Pi_{(1)}^{h_1}\Pi_{(2)}^{h_2}\cdots\Pi_{(p)}^{h_p} = 1. \tag{6.3.33}$$

On the other hand, the sufficient and necessary condition of products $\Pi_{(1)}, \Pi_{(2)}, \ldots \Pi_{(p)}$ which are linearly independent of one another is that the rows of the matrix given by

$$\begin{bmatrix} k_1^{(1)} & k_2^{(1)} & \cdots & k_n^{(1)} \\ k_1^{(2)} & k_2^{(2)} & \cdots & k_n^{(2)} \\ \vdots & \vdots & & \vdots \\ k_1^{(p)} & k_2^{(p)} & \cdots & k_n^{(p)} \end{bmatrix}, \tag{6.3.34}$$

are linearly independent. The proof is summarized in the following.

To demonstrate the necessity, it is assumed that the rows in the matrix given in Eq. (6.3.34) are linearly dependent but the dimensionless products given in Eq. (6.3.32) are linearly independent. It follows immediately that there should exist constants $(h_1, h_2, \ldots, h_n)$, which do not vanish simultaneously, so that

$$h_1 k_i^{(1)} + h_2 k_i^{(2)} + \cdots + h_p k_i^{(p)} = 0, \qquad i = 1, 2, \ldots, n. \tag{6.3.35}$$

By using Eq. (6.3.32), it follows that

$$\Pi_{(1)}^{h_1}\Pi_{(2)}^{h_2}\cdots\Pi_{(p)}^{h_p} = x_1^{\sum_{j=1}^{p} h_j k_1^{(j)}} x_2^{\sum_{j=1}^{p} h_j k_2^{(j)}} \cdots x_n^{\sum_{j=1}^{p} h_j k_n^{(j)}}, \tag{6.3.36}$$

which, by using Eq. (6.3.35), is simplified to

$$\Pi_{(1)}^{h_1}\Pi_{(2)}^{h_2}\cdots\Pi_{(p)}^{h_p} = x_1^0 x_2^2 \cdots x_n^0 = 1. \tag{6.3.37}$$

This result contradicts to the previous assumption that the dimensionless products given in Eq. (6.3.32) are linearly independent. Thus, the rows of the matrix in Eq. (6.3.34) must be linearly independent.

Conversely, it is assumed that the rows in the matrix given in Eq. (6.3.34) are linearly independent but the dimensionless products given in Eq. (6.3.32) are linearly dependent. It follows immediately that the expression

$$\Pi_{(1)}^{h_1}\Pi_{(2)}^{h_2}\cdots\Pi_{(p)}^{h_p} = 1, \tag{6.3.38}$$

can be satisfied for a set of $(h_1, h_2, \ldots, h_p)$, which do not vanish simultaneously. By using Eq. (6.3.37), the above equation implies that

$$x_1^{\sum_{j=1}^{p} h_j k_1^{(j)}} x_2^{\sum_{j=1}^{p} h_j k_2^{(j)}} \cdots x_n^{\sum_{j=1}^{p} h_j k_n^{(j)}} = 1, \tag{6.3.39}$$

which can only be fulfilled if all the powers vanish. This result, in view of Eq. (6.3.35), contradicts to the assumption that the rows of the matrix in Eq. (6.3.34) are linearly independent. Thus, the dimensionless products given in Eq. (6.3.32) must be linearly independent.

Based on the previously discussions, it is concluded that a power product of x_j, e.g. Eq. (6.3.24), is dimensionless if and only if the conditions

$$\sum_{j=1}^{n} a_{kj} k_j = 0, \qquad k = 1, 2, \cdots, m, \tag{6.3.40}$$

are satisfied. These conditions possess $(n - r)$ linearly independent solutions to k_js, which are denoted by

$$k_j^{(1)}, k_j^{(2)}, \cdots k_j^{(n-r)}, \qquad j = 1, 2, \ldots, n, \tag{6.3.41}$$

where r is the rank of the matrix $[a_{kj}]$. Combining this expression with Eq. (6.3.34) yields that the solutions $k_j^{(1)}, k_j^{(2)}, \ldots k_j^{(n-r)}$ furnish the exponents for all dimensionless products. There are no additional ones, so the number of independent products in a complete set of dimensionless products of given variable $(x_1, x_2, \ldots, x_n)$ is simply $n - r$, where r is the rank of the dimensional matrix of $(x_1, x_2, \ldots, x_n)$.

Now going back to Eqs. (6.3.24)–(6.3.28). If y is not dimensionless, there should exist a product in the form given by

$$y = \prod_{j=1}^{n} x_j^{k_j}, \tag{6.3.42}$$

if and only if the dimensional matrix of $(x_1, x_2, \ldots, x_n)$ has the same rank of that of $(y, x_1, x_2, \ldots, x_n)$. The proof is given here. Consider the linear equation system given in Eq. (6.3.28), and let the reduced form of matrix $[a_{k_j}]$ be denoted by $[a_{k_j}]_R$, which is obtained by using any sequence of elementary row operations, and the augmented form of matrices $[a_{k_j}]$ and $[a_{k0}]$ be denoted by $[a_{k_j} : a_{k0}]$. If the rank of matrix $[a_{k_j}]$ is less than that of the augmented matrix $[a_{k_j} : a_{k0}]$, the reduced matrix $[a_{k_j}]_R$ must have at least one row of zero, while the corresponding row in the reduced augmented matrix $[a_{k_j} : a_{k0}]_R = [a_{k_j}]_R : [\beta_{k0}]$ has a nonzero element in this row of $[\beta_{k0}]$. This corresponds to an equation of the form

$$0k_1 + 0k_2 + \cdots + 0k_n = \beta_{k0} \neq 0, \tag{6.3.43}$$

which shows that no solutions to k_js can be found. It follows that the solutions to k_js of Eq. (6.3.28) exist only if the ranks of dimensional matrices of $(x_1, x_2, \ldots, x_n)$ and $(y, x_1, x_2, \ldots, x_n)$ are the same. In this case, Eq. (6.3.40) is reproduced, ensuring in turn that Eq. (6.3.42) is valid.

On the contrary, if $y = f(x_1, x_2, \ldots, x_n)$ is a dimensionally homogeneous equation and if y is dimensional, there should exist a product of powers of x_j which has the same dimension as y. The proof of this statement is left as an exercise.

6.3.5 Proof of the Buckingham Theorem

Let $(x_1, x_2, \ldots, x_n)$ be the independent variables involved in a physical process. They may be regarded as the Cartesian coordinates of an Euclidean space $\mathcal{E}$. As motivated by Eqs. $(6.3.7)_2$ and $(6.3.8)_2$, define

$$K_j \equiv \prod_{i=1}^{m} \alpha_i^{a_{ij}}, \qquad x_j = K_j x'_j, \qquad j = 1, 2, \ldots, n, \tag{6.3.44}$$

where α_is are positive constants and $[a_{ij}]$ is the dimensional matrix of $(x'_1, x'_2, \ldots, x'_n)$. The expression $x_j = K_j x'_j$ assigns to each point x'_j the coordinate x_j and vice versa. Thus, the space spanned by all values of x_js are called the $\mathcal{E}$-space, and the entities $(K_1, K_2, \ldots, K_n)$ generated by applying this expression to $(x_1, x_2, \ldots, x_n)$ may be regarded as the coordinates of a point in a n-dimensional space, called the $\mathcal{K}$-*space*. The expression (6.3.44) represents then a point transformation between the $\mathcal{E}$- and $\mathcal{K}$-spaces, which is called a K-*transformation*. All K-transformations satisfy the following properties:

- If $x_j = K_j^* x'_j$ and $x'_j = K_j^{**} x''_j$, it follows that

$$x_j = K_j^* K_j^{**} x''_j = K_j x''_j, \qquad K_j = \left(\prod_{i=1}^{m} \alpha_i^{*a_{ij}}\right)\left(\prod_{i=1}^{m} \alpha_i^{**a_{ij}}\right) = \prod_{i=1}^{m} (\alpha_i^* \alpha_i^{**})^{a_{ij}}. \tag{6.3.45}$$

  Thus, a composition of any two K-transformations is also a K-transformation.
- There exists an identity K-transformation, with which $x_j = x'_j$.
- Since $x_j = K_j x'_j$ and $x'_j = K_j^{-1} x_j$, it follows that

$$x_j = K_j K_j^{-1} x_j, \qquad K_j K_j^{-1} = 1, \qquad K_j^{-1} = \prod_{i=1}^{m} \left(\frac{1}{\alpha_i}\right)^{a_{ij}}. \tag{6.3.46}$$

  Thus, for every K_j there exists an inverse transformation.

Consider now a dimensionally homogeneous dimensionless function given by

$$\Pi = f(x_1, x_2, \cdots, x_n). \tag{6.3.47}$$

Since Π is dimensionless, its exponents of the fundamental dimensions, i.e., a_{i0} in view of Eq. (6.3.40), must all vanish. It follows form Eq. (6.3.8)$_1$ that $K_0 = 1$, with which Eq. (6.3.7) reduces to

$$\Pi = f(K_1 x_1, K_2 x_2, \cdots, K_n x_n). \tag{6.3.48}$$

Comparing this equation with Eq. (6.3.47) shows that the value of Π must be a constant irrespective of the values of $(K_1, K_2, \ldots, K_n)$ in the $\mathcal{K}$-space. Now let $(\Pi'_1, \Pi'_2, \ldots, \Pi'_p)$ be a set of the values of dimensionless products $(\Pi_1, \Pi_2, \ldots, \Pi_p)$, which are constructed by using the power products of x'_j and x''_j being two points in the $\mathcal{K}$-space. It follows that

$$\Pi'_\upsilon = (x'_1)^{k_1^\upsilon}(x'_2)^{k_2^\upsilon}\cdots(x'_n)^{k_n^\upsilon} = (x''_1)^{k_1^\upsilon}(x''_2)^{k_2^\upsilon}\cdots(x''_n)^{k_n^\upsilon}, \qquad \upsilon = 1, 2, \cdots, p. \tag{6.3.49}$$

Taking logarithm of this equation yields

$$r_1 k_1^\upsilon + r_2 k_2^\upsilon + \cdots + r_n k_n^\upsilon = r_j k_j^\upsilon = 0, \qquad r_j = \ln\left(\frac{x'_j}{x''_j}\right), \tag{6.3.50}$$

for all x'_js and x''_js are assumed to be positive. This is the crucial condition, without which the Buckingham theorem cannot be proved. The set of dimensionless products $(\Pi_1, \Pi_2, \ldots, \Pi_p)$ is complete, thus k_j^v should be the solutions to the linear equation system given by

$$\sum_{j=1}^{m} a_{ij} k_j^v = 0, \qquad i = 1, 2, \ldots, m, \quad v = 1, 2, \ldots, p. \tag{6.3.51}$$

Since both Eqs. (6.3.50) and (6.3.51) have the same solutions to k_j^v, the coefficients in the former equation must depend linearly on those in the latter equation. Consequently, there must exist non-vanishing numbers α_j^*, $j = 1, 2, \ldots, m$, so that

$$\sum_{j=1}^{m} \alpha_j^* a_{ji} = r_i = \ln\left(\frac{x'_i}{x''_i}\right), \tag{6.3.52}$$

implying that

$$x'_i = x''_i \exp\left(\sum_{j=1}^{m} \alpha_j^* a_{ji}\right) = x''_i \prod_{j=1}^{m} \exp\left(\alpha_j^* a_{ji}\right). \tag{6.3.53}$$

For simplicity, let $\alpha_j = e^{\alpha_j^*}$, $j = 1, 2, \ldots, m$, with which the above equation is simplified to

$$x'_i = \left(\prod_{j=1}^{m} \alpha_j^{a_{ji}}\right) x''_i = K_i x''_i, \qquad i = 1, 2, \ldots, n, \tag{6.3.54}$$

showing that both x'_i and x''_i belong to the same $\mathcal{K}$-space.

So, the proof of the Buckingham theorem is as follows: if $y = f(x_1, x_2, \ldots, x_n)$ is a dimensionally homogeneous equation and if y is dimensional, there should exist a product of powers of x_j which has the same dimension of y. It follows subsequently from the discussions in Sect. 6.3.3 that $y = f(x_1, x_2, \ldots, x_n)$ can be brought into the form of $\Pi = F(x_1, x_2, \ldots, x_n)$, where Π is dimensionless. Let $(\Pi_1, \Pi_2, \ldots, \Pi_p)$ be a complete set of dimensionless products belonging to $(x_1, x_2, \ldots, x_n)$. It follows from Eq. (6.3.48) that there is only one single $\mathcal{K}$-space to every set of the values of $(\Pi_1, \Pi_2, \ldots, \Pi_p)$. Equally, to every $\mathcal{K}$-space there is only one single value of Π, and consequently there is only one single value of Π to every set of the values of $(\Pi_1, \Pi_2, \ldots, \Pi_p)$, so that Π is a unique function of $(\Pi_1, \Pi_2, \ldots, \Pi_p)$. Thus, any arbitrarily dimensionally homogeneous equation $y = f(x_1, x_2, \ldots, x_n)$ can be brought to the form of $\Pi = F(\Pi_1, \Pi_2, \ldots, \Pi_p)$, and in view of Eqs. (6.3.40) and (6.3.41), $p = n - r$, where r is the rank of the dimensional matrices $(x_1, x_2, \ldots, x_n)$ (or $(y, x_1, x_2, \ldots, x_n)$). It is readily verified that the reverse statement, i.e., an equation of dimensionless products is dimensionally homogeneous, holds true. With all these, the proof of the Buckingham theorem is complete. Q.E.D.[8]

[8]The term "Q.E.D." is an initialism of the Latin phrase, which reads: "quod erat demonstrandum," with the translation as "this is what to be proved."

6.4 Theory of Physical Models

6.4.1 Model and Prototype

A *physical model* is a projection of nature or at least of a subprocess that occurs in nature in our world of experience to small scales. On the contrary, nature or subprocess occurring in nature is called a *prototype*. Since a model is only a projection of a real process, some information of the real process is lost during the projection, although a model is used for executing experiments and transporting answers to the corresponding prototype. There exists a point-to-point correlation between the model and prototype, and the corresponding points between the model and prototype are called the *homologous points*. Several homologous points form an agglomeration, and a set of agglomerations leads eventually to a homologous region or domain. If time-dependent processes are analyzed, the notation of homologous time must be introduced, which is accomplished by using Newton's second law of motion. That is, differences in times are declared to be homologous if a material point passes two homologous points on homologous trajectories.

Consider an Euclidean space with the Cartesian coordinates and time (x_1, x_2, x_3, t), and let the coordinates and time used in the model and prototype be denoted respectively by $(x_1^m, x_2^m, x_3^m, t^m)$ and $(x_1^p, x_2^p, x_3^p, t^p)$. Since a models is either an enlargement or a reduction in size of the prototype, it is plausible to define

$$x_1^m \equiv k_{x_1} x_1^p, \qquad x_2^m \equiv k_{x_2} x_2^p, \qquad x_3^m \equiv k_{x_3} x_3^p, \qquad t^m \equiv k_t t^p, \tag{6.4.1}$$

where $\{k_{x_1}, k_{x_2}, k_{x_3}\}$ are the *geometric (scale) factors* in the spatial coordinates $\{x_1, x_2, x_3\}$, and k_t is the *timescale*. A model is said to be *geometrically similar* to the prototype if $k_{x_1} = k_{x_2} = k_{x_3}$; otherwise, the model is said to be *distorted*. The timescale can be chosen as the ratio of times that elapse when a material point tracts the distance between two homologous points in the model and prototype. In a more general circumstance, let $y^p = f^p(x_1^p, x_2^p, \ldots, x_n^p)$ be an equation describing a process in the prototype, and the projection of this process in the model be described by the equation $y^m = f^m(x_1^m, x_2^m, \ldots, x_n^m)$. The function f^p is said to be *similar* to the function f^m, if the ratio f^m/f^p is a constant, provided that for the arguments $(x_1^p, x_2^p, \ldots, x_n^p)$ and $(x_1^m, x_2^m, \ldots, x_n^m)$, definite homologous points and times are chosen. The ratio f^m/f^p, denoted by k_f, is called the *scale of f*.

In additional to the geometrical similarity, a model is said to be *kinematically similar* to the prototype, if their motions are similar, namely, if homologous particles are to be found at homologous times in homologous points. Specifically, the kinematic similarity requires that the velocities and accelerations at the corresponding points are similar. Since in the model and prototype the velocities are given by

$$u_1^m = \frac{\mathrm{d}x_1^m}{\mathrm{d}t^m}, \quad u_2^m = \frac{\mathrm{d}x_2^m}{\mathrm{d}t^m}, \quad u_3^m = \frac{\mathrm{d}x_3^m}{\mathrm{d}t^m}; \qquad u_1^p = \frac{\mathrm{d}x_1^p}{\mathrm{d}t^p}, \quad u_2^p = \frac{\mathrm{d}x_2^p}{\mathrm{d}t^p}, \quad u_3^p = \frac{\mathrm{d}x_3^p}{\mathrm{d}t^p}, \tag{6.4.2}$$

it follows that

$$u_1^m = \frac{k_{x1}}{k_t} u_1^p, \qquad u_2^m = \frac{k_{x2}}{k_t} u_2^p, \qquad u_3^m = \frac{k_{x3}}{k_t} u_3^p, \tag{6.4.3}$$

for $\mathrm{d}x_i^m = k_{x_i}\mathrm{d}x_i^p$ (no summation) and $\mathrm{d}t^m = k_t\mathrm{d}t^p$. Thus, the *scale factors for velocity*, or *velocity factors* are given by

$$k_{u_1} = \frac{k_{x1}}{k_t}, \qquad k_{u_2} = \frac{k_{x2}}{k_t}, \qquad k_{u_3} = \frac{k_{x3}}{k_t}. \tag{6.4.4}$$

The *scale factors for acceleration*, or *acceleration factors* are obtained in an analogous manner, viz.,

$$k_{a_1} = \frac{k_{x1}}{k_t^2}, \qquad k_{a_2} = \frac{k_{x2}}{k_t^2}, \qquad k_{a_3} = \frac{k_{x3}}{k_t^2}. \tag{6.4.5}$$

Furthermore, a model is said to be *dynamically similar* to the prototype, if homologous points of the system are subject to similar forces, i.e., the *force factors* are invariant. To explore the idea, let the masses in the model and prototype be denoted respectively by m^m and m^p, and $k_m = m^m/m^p$ is defined as the *scale factor for mass*, or *mass factor*.[9] It follows from Newton's second law of motion that

$$\begin{aligned} F_1^m &= m^m a_1^m, \quad F_2^m = m^m a_2^m, \quad F_3^m = m^m a_3^m, \\ F_1^p &= m^p a_1^p, \quad F_2^p = m^p a_2^p, \quad F_3^p = m^p a_3^p, \end{aligned} \tag{6.4.6}$$

and the force ratios between the model and prototype are then obtained as

$$\frac{F_1^m}{F_1^p} = \frac{m^m a_1^m}{m^p a_1^p} = k_m\frac{k_{x1}}{k_t^2}, \qquad \frac{F_2^m}{F_2^p} = \frac{m^m a_2^m}{m^p a_2^p} = k_m\frac{k_{x2}}{k_t^2}, \qquad \frac{F_3^m}{F_3^p} = \frac{m^m a_3^m}{m^p a_3^p} = k_m\frac{k_{x3}}{k_t^2}, \tag{6.4.7}$$

with which the *scale factors for force* or *force factors* are given by

$$k_{F_1} = k_m\frac{k_{x1}}{k_t^2}, \qquad k_{F_2} = k_m\frac{k_{x2}}{k_t^2}, \qquad k_{F_3} = k_m\frac{k_{x3}}{k_t^2}. \tag{6.4.8}$$

It follows from Eqs. (6.4.4) and (6.4.5) that the scale factors for velocity and acceleration are not freely assignable, but must be computed from the scale factors of geometry and time. Analogously, for dynamic similarity the force factors are obtained automatically from the scale factors for geometry, mass, and time, as implied by Eq. (6.4.8).

The requirement of dynamic similarity is the most restrictive. The kinematic similarity requires the geometric similarity. On the other hand, the kinematical similarity is a necessary, but not a sufficient requirement to the dynamic similarity. In studying fluid motions experimentally, a model, in most cases, should be dynamically similar to the prototype to provide useful information.

As an illustration of the scale factors, consider an explosion at a point in an infinite compressible gas. The explosion generates a spherical pressure wave with pressure p, which depends on the radius r of the front of spherical pressure wave, the mass m of explosive substance, the initial pressure p_0, density ρ and bulk modulus E_v of gas. These six physical variables are described by an equation

$$f(p, r, m, p_0, \rho, E_v) = 0, \tag{6.4.9}$$

[9]Other scale factors can be defined in a similar manner.

which is assumed to be dimensionally homogeneous. By using the dimensional analysis, this equation is brought to the dimensionless form given by

$$f\left(\frac{p}{p_0}, \frac{p_0}{E_v}, \frac{m}{\rho r^3}\right) = 0. \tag{6.4.10}$$

If one requires the invariance of all the dimensionless products in the model and prototype, the scale factors for six physical variables are identified to be

$$k_{p_0} = \frac{p_0^m}{p_0^p} = k_p = \frac{p^m}{p^p} = k_{E_v} = \frac{E_v^m}{E_v^p}; \qquad k_m = \frac{m^m}{m^p} = k_\rho k_r^3 = \frac{\rho^m}{\rho^p}\left(\frac{r^m}{r^p}\right)^3. \tag{6.4.11}$$

If the explosion takes place in water in both the model and prototype, it follows that $k_{p_0} = k_p = k_{E_v} = k_\rho = 1$, and

$$k_m = k_r^3, \tag{6.4.12}$$

indicating that the mass factor of explosive substance must be the third power of geometric scale factor.

6.4.2 Modeling Law

For a problem, it is advantageous to first contemplate about which variables might have influence on the processes to be studied before a model test. For simplicity, let the process be described by a dimensionally homogeneous equation $f(y, x_1, x_2, \ldots, x_n) = 0$, where y is the object variable depending on the independent physical arguments $(x_1, x_2, \ldots, x_n)$. It follows from the dimensional analysis that this equation can be brought to a dimensionless homogeneous equation given by

$$f(\Pi, \Pi_1, \Pi_2, \cdots, \Pi_p) = 0, \tag{6.4.13}$$

which gives the dimensionless variable Π to be analyzed, corresponding to y in the dimensional form, as a function of other dimensionless variables $(\Pi_1, \Pi_2, \ldots, \Pi_p)$, corresponding to $(x_1, x_2, \ldots, x_n)$. If a model has to reproduce the process arising in the prototype correctly, the values of $(\Pi_1, \Pi_2, \ldots, \Pi_p)$ are not allowed to change freely when going from the prototype conditions to those of the model, if the same result for Π is to be delivered. The model is said to be *completely similar* to the prototype if specific conditions are fulfilled. This gives rise to the requirements which need to be satisfied in establishing a completely similar model, which are summarized in the following law:

6.2 (Modeling Law) *A model is capable to reproduce a process in a prototype with complete similarity, if all the dimensionless products describing the process have the same values in the model and prototype.*

This law is also called alternatively the *model design condition* or *similarity requirement*.

The complete similarity between a model and its prototype requires that the geometric, kinematic, and dynamic similarities all hold simultaneously. However, in practice, it is hardly possible to require all the dimensionless products to be the same in both the model and prototype, and one is regularly forced to hold only a reduced number of Π-products constant while the others are allowed to vary as dictated by the modeling law. In such a case, an *incomplete similarity* between the model and prototype is established and it is hoped that the Π-products which do not remain invariant in the projection will not, at least not much, influence the physical process that is studied. For such a circumstance, the model is said to have *scale effects*. On the contrary, if a process depends only on the Π-products which remain invariant in a model projection, this process is called *scale invariant*.

For example, consider the drag force acting on a ship traveling with a constant velocity in a still water. The drag force F possibly depends on the density ρ and dynamic viscosity μ of fluid, the gravitational acceleration g, the characteristic length L, and velocity V of ship. The considered physical process is described by

$$f(F, \rho, \mu, g, V, L) = 0, \tag{6.4.14}$$

which is assumed to be a dimensionally homogeneous equation and all physical variables influencing the process are included. By using the dimensional analysis, this equation is brought to

$$\frac{F}{\rho V^2 L^2} = f\left(\frac{\rho V L}{\mu}, \frac{V^2}{gL}\right), \tag{6.4.15}$$

which can be expressed alternatively as

$$\mathcal{C}_D = f\left(\mathcal{R}_e, \mathcal{F}_r\right), \qquad \mathcal{C}_D \equiv \frac{F}{\rho V^2 L^2}, \quad \mathcal{R}_e \equiv \frac{\rho V L}{\mu}, \quad \mathcal{F}_r \equiv \frac{V^2}{gL}. \tag{6.4.16}$$

The dimensionless products $\mathcal{C}_D$, $\mathcal{R}_e$, $\mathcal{F}_r$ are termed the *drag coefficient*, the *Reynolds number*, and the *Froude number*,[10] respectively, which should assume the same values in both the model and prototype. By requiring the same Froude number, it follows that

$$\frac{(V^m)^2}{gL^m} = \frac{(V^p)^2}{gL^p}, \qquad \longrightarrow \qquad \frac{V^m}{V^p} = \sqrt{\frac{L^m}{L^p}} = \sqrt{k_L}, \tag{6.4.17}$$

where k_L is the geometric (length) scale.[11] Equation (6.4.17)$_2$ delivers the velocity that needs to be assigned to the model ship in the model experiment to maintain the same Froude number. The term *Froude similitude* is used to denote a model experiment in which the Froude number remains invariant, with the model called a *Froude model*. Since $t = L/V$, it follows that $t^m = L^m/V^m$ in the model and $t^p = L^p/V^p$ in the prototype, and the scale factor for time is obtained as

$$\frac{t^m}{t^p} = \frac{L^m/V^p}{L^p/V^p} = \sqrt{k_L}, \tag{6.4.18}$$

[10]William Froude, 1810–1879, a British hydrodynamicist and engineer, who first formulated reliable laws for the resistance that water offers to ships and for predicting their stability.

[11]In deriving Eq. (6.4.17), the gravitational constant assumes the same value in both the model and prototype, for the experiments are normally conducted on the earth's surface.

with which the scale factor for acceleration is given by

$$\frac{a^m}{a^p} = \frac{V^m/t^m}{V^p/t^p} = 1. \tag{6.4.19}$$

Furthermore, requiring the same Reynolds number yields

$$\frac{V^m L^m}{\nu^m} = \frac{V^p L^m}{\nu^p}, \quad \longrightarrow \quad k_\nu = \frac{\nu^m}{\nu^p} = k_L^{3/2}, \tag{6.4.20}$$

in which the kinematic viscosity ν is used to replace μ/ρ, and k_ν is the scale factor for kinematic viscosity. This equation delivers the kinematic viscosity of the fluid that should be used in the model to maintain the same Reynolds number. The term *Reynolds similitude* is used to denote a model experiment in which the Reynolds number remains invariant, and the model is called a *Reynolds model*. Similarly, the scale factors for time and acceleration are obtained as

$$\frac{t^m}{t^p} = \frac{k_L^2}{k_\nu} = \sqrt{k_L}, \qquad \frac{a^m}{a^p} = \frac{k_\nu^2}{k_L^3} = 1, \tag{6.4.21}$$

which coincide to Eqs. (6.4.18) and (6.4.19).

For a complete similarity between the model and prototype, Eqs. (6.4.17)–(6.4.21) must be satisfied simultaneously. As a demonstration, consider a model ship which is constructed in a scale of $1:100$ of the prototype ship, with which $k_L = 1/100$, yielding $V^m = V^p/10$. This condition, however, can be fulfilled in experiments. Further, it follows from the Reynolds similitude that $\nu^m = \nu^p/1000$. This condition, however, can never be reached in practice, for in the prototype the fluid is water, and in reality mercury is the only fluid whose kinematic viscosity is less than that of water. Unfortunately, it is only about an order of magnitude less, so the kinematic viscosity ratio required to duplicate the Reynolds number cannot be reached, not to mention that water is almost the only fluid for most model tests of free surface flows. It is concluded that it is impossible in practice for this model/prototype scale of $1/100$ to reach a complete similarity. To obtain a complete similarity, one would require a *full-scale test*, which is, however, not meaningful in a model test.

For the considered problem, an incomplete similarity between the model and prototype needs to be developed. This is accomplished by requiring only the Froude similitude, and the experiments of the total drag in relation to the Froude number are conducted in the model test. Since the total drag consists of the wave resistance depending on the Froude number and frictional resistance depending on the Reynolds number, the boundary-layer theory is used to calculate the frictional resistance of model ship, by which the wave resistance can be extracted from the total drags measured in the experiments. This gives then the wave resistance of model ship as a function of the Froude number, which is also valid for the wave resistance of prototype ship due to the Froude similitude. The frictional resistance of prototype ship is determined again by using the boundary-layer theory. Combing the wave and frictional resistances yields the total drag in relation with the Froude number for the prototype ship. So, the incomplete similarity is overcome by using the analytical computations, and the model experiments are only conducted for the Froude similitude, not the Reynolds similitude.

However, there exists a question: Is there any rule that can be followed for the selection of dimensionless numbers which should be kept invariant in a practical situation to reach at least an approximate similarity? The answer depends on which physical influence dominates the process and requires the study of different dimensionless products in the process, which will be discussed in the next section. At the meantime, it is sufficient to introduce two basic rules as a guidance of model experiment for density-preserving fluids, which are given in the following:

- Rule 1: In the regions with *fixed boundaries* and geometrically similar boundary values, the Reynolds similitude is required.
- Rule 2: In the regions with *free boundaries* and geometrically similar boundary values, the Reynolds, Froude, and sometimes the Weber similitudes are required.

For more complicated flow circumstances, more dimensionless products should be introduced, and a complete or an incomplete model similitude can be established by using different dimensionless products.

6.5 Dimensionless Products in Fluid Mechanics

6.5.1 Non-dimensionalization of Differential Equations

To attain the requirements of model similitude, differential equations governing the flow behavior must be brought to dimensionless forms, yielding different dimensionless products known as the *dimensionless numbers* in fluid mechanics. This can be achieved by introducing

$$f = [f]\bar{f}, \tag{6.5.1}$$

for every variable f. The term $[f]$ is called the *scaling variable*, which assumes the same dimension as f and has a constant magnitude so that the dimensionless variable $\bar{f}$ assumes a value which is of an order of unity. The dimensionally homogeneous local balance equations summarized in Table 5.6, specifically the local balances of mass, linear momentum, and internal energy, are brought to dimensionless forms by using the concept of Eq. (6.5.1) to generate a certain dimensionless numbers as a demonstration.

The local mass balance reads

$$\frac{\partial \rho}{\partial t} + \operatorname{div}(\rho \boldsymbol{u}) = 0. \tag{6.5.2}$$

Defining the scaling variables

$$\rho = [\rho]\bar{\rho}, \qquad \boldsymbol{u} = [u]\bar{\boldsymbol{u}}, \qquad t = [\tau]\bar{t}, \qquad \boldsymbol{x} = [L]\bar{\boldsymbol{x}}, \tag{6.5.3}$$

and substituting these scaling variables into Eq. (6.5.2) results in

$$\frac{[L]}{[u][\tau]}\frac{\partial \bar{\rho}}{\partial \bar{t}} + \operatorname{div}(\bar{\rho}\bar{\boldsymbol{u}}) = 0, \tag{6.5.4}$$

which is the dimensionless form of local mass balance. The local balance of linear momentum in an inertia reference frame reads

$$\rho\left(\frac{\partial \boldsymbol{u}}{\partial t}+(\operatorname{grad}\boldsymbol{u})\boldsymbol{u}\right)=-\operatorname{grad}p+\rho\,\boldsymbol{b}+\operatorname{grad}(\lambda\operatorname{div}\boldsymbol{u})+2\operatorname{div}(\mu\boldsymbol{E}), \tag{6.5.5}$$

with $\boldsymbol{E}=\boldsymbol{D}-(\operatorname{div}\boldsymbol{u}/3)\boldsymbol{I}$. Defining the additional scaling variables given by

$$\lambda=[\lambda]\bar{\lambda},\qquad \mu=[\mu]\bar{\mu},\qquad p=[p]\bar{p},\qquad \boldsymbol{b}=[b]\bar{\boldsymbol{b}},\qquad \boldsymbol{E}=[E]\bar{\boldsymbol{E}}, \tag{6.5.6}$$

and substituting these scaling variables and those in Eq. (6.5.3) into Eq. (6.5.5) yields

$$\begin{aligned}\bar{\rho}\left(\frac{[L]}{[u][\tau]}\frac{\partial\bar{\boldsymbol{u}}}{\partial\bar{t}}+(\operatorname{grad}\bar{\boldsymbol{u}})\bar{\boldsymbol{u}}\right)&=-\frac{[p]}{[\rho][u]^2}\operatorname{grad}\bar{p}+\frac{[g][L]}{[u]^2}\bar{\rho}\,\bar{\boldsymbol{b}}\\&\quad+\frac{[\mu]}{[\rho][u][L]}\left\{\frac{[\lambda]}{[\mu]}\operatorname{grad}(\bar{\lambda}\operatorname{div}\bar{\boldsymbol{u}})+2\operatorname{div}(\bar{\mu}\bar{\boldsymbol{E}})\right\},\end{aligned} \tag{6.5.7}$$

which is the dimensionless local balance of linear momentum. The local balance of internal energy for the Fourier fluids, in which the specific internal energy is expressed by using the specific enthalpy given in Eq. (5.6.50), reads

$$\rho c_p\left(\frac{\partial T}{\partial t}+(\operatorname{grad}T)\cdot\boldsymbol{u}\right)=\operatorname{div}(k\operatorname{grad}T)+\lambda(\operatorname{div}\boldsymbol{u})^2+2\mu\operatorname{tr}\boldsymbol{E}^2+\rho\,\zeta. \tag{6.5.8}$$

Defining the additional scaling variables given by

$$\zeta=[\zeta]\bar{\zeta},\qquad c_p=[c_p]\bar{c}_p,\qquad k=[k]\bar{k},\qquad T=T_0+[\Delta T]\theta, \tag{6.5.9}$$

and substituting these scaling variables and those in Eqs. (6.5.3) and (6.5.6) into Eq. (6.5.8) gives

$$\begin{aligned}\bar{\rho}\bar{c}_p\left(\frac{[L]}{[u][\tau]}\frac{\partial\theta}{\partial\bar{t}}+(\operatorname{grad}\theta)\cdot\bar{\boldsymbol{u}}\right)&=\frac{[k]}{[\rho][c_p][u][L]}\operatorname{div}(\bar{k}\operatorname{grad}T)+\frac{[L][\zeta]}{[c_p][\Delta T][u]}\bar{\rho}\,\bar{\zeta}\\&\quad+\frac{[u]^2[\mu]}{[\rho][c_p][\Delta T][u][L]}\left\{\frac{[\lambda]}{[\mu]}\bar{\lambda}(\operatorname{div}\bar{\boldsymbol{u}})^2+2\bar{\mu}\operatorname{tr}\bar{\boldsymbol{E}}^2\right\},\end{aligned} \tag{6.5.10}$$

which is the dimensionless balance of internal energy for the Fourier fluids. In Eqs. (6.5.4), (6.5.7) and (6.5.10), the divergence, curl, gradient, and trace operations are referred to the dimensionless variables.

6.5.2 Dimensionless Numbers

There exist various combinations of the scaling variables in the dimensionless differential equations derived previously. These combinations define the dimensionless numbers corresponding to the Π-terms in the dimensional analysis, which are given by

$$\begin{aligned}&\mathcal{S}_t\equiv\frac{[L]}{[u][\tau]},\qquad \mathcal{E}_u\equiv\frac{[p]}{[\rho][u]^2},\qquad \mathcal{F}_r\equiv\frac{[u]^2}{[g][L]},\qquad \mathcal{R}_e\equiv\frac{[\rho][u][L]}{[\mu]},\\&\mathcal{P}_e\equiv\frac{[\rho][c_p][u][L]}{[k]},\quad \mathcal{E}_d\equiv\frac{[c_p][\Delta T]}{[u]^2}\frac{[u][L]}{[\mu]/[\rho]},\quad \mathcal{R}_a\equiv\frac{[c_p][\Delta T][u]}{[L][\zeta]},\end{aligned} \tag{6.5.11}$$

Table 6.3 Scaling expressions of physical and virtual forces in isothermal fluid flows

Physical and virtual forces	Expressions of scaling variables
Inertia force	$[\rho][u]^2[L]^2$
Local inertia force	$[\rho][L]^3[u]/[t]$
Convective inertia force	$[\rho][L]^3[u]^2/[L]$
Viscous force	$[\mu][u][L]$
Pressure force	$[p][L]^2$
Gravity force	$[\rho][L]^3[g]$
Surface tension force	$[\sigma][L]$
Compressibility force	$[E_v][L]^2$

which are called respectively the *Strouhal number*, *Euler number*, *Froude number*, *Reynolds number*, *Péclet number*, *dissipation number*, and *radiation number*.[12] The inverse of the Froude number is called the *Richardson number*,[13] and it is conventionally to introduce

$$\mathcal{P}_e = \mathcal{R}_e \mathcal{P}_r, \qquad \mathcal{E}_d = 2\mathcal{R}_e \mathcal{T}_h, \tag{6.5.12}$$

where $\mathcal{P}_r$ and $\mathcal{T}_h$ are respectively the *Prandtl number* and *temperature number* defined by

$$\mathcal{P}_r \equiv \frac{[\mu]/[\rho]}{[k]/([\rho][c_p])} = \frac{[\nu]}{[d_{th}]}, \qquad \mathcal{T}_h \equiv \frac{[c_p][\Delta T]}{[u]^2}, \tag{6.5.13}$$

with d_{th} the *thermal diffusivity* of fluid. More dimensionless numbers emerge in other differential equations subject to the similar non-dimensionalization procedures. For example, for atmospheric or ocean fluid flows in a rotating reference frame, one can derive the *Rossby number* and *Ekman number* in a similar manner.[14]

For the Newtonian fluids with constant density, dynamic viscosity, and heat conductivity, the dimensionless local balances of mass, linear momentum, and internal energy reduce respectively to

$$\begin{aligned}
&\operatorname{div}\bar{\boldsymbol{u}} = 0,\\
&\bar{\rho}\left(\mathcal{S}_t\frac{\partial\bar{\boldsymbol{u}}}{\partial\bar{t}} + (\operatorname{grad}\bar{\boldsymbol{u}})\bar{\boldsymbol{u}}\right) = -\mathcal{E}_u \operatorname{grad}\bar{p} + \frac{1}{\mathcal{F}_r}\bar{\rho}\bar{\boldsymbol{b}} + \frac{1}{\mathcal{R}_e}\{2\bar{u}\operatorname{lap}\bar{\boldsymbol{u}}\},\\
&\bar{\rho}\bar{c}_p\left(\mathcal{S}_t\frac{\partial\theta}{\partial\bar{t}} + (\operatorname{grad}\theta)\cdot\bar{\boldsymbol{u}}\right) = \frac{1}{\mathcal{P}_e}\bar{k}\operatorname{lap}T + \frac{1}{\mathcal{R}_a}\bar{\rho}\bar{\zeta} + \frac{1}{\mathcal{E}_d}\left\{2\bar{\mu}\operatorname{tr}\bar{\boldsymbol{D}}^2\right\}.
\end{aligned} \tag{6.5.14}$$

[12] Vincenc Strouhal, 1850–1922, a Czech physicist specializing in experimental physics. Jean Claude Eugène Péclet, 1793–1857, a French physicist.

[13] Lewis Fry Richardson, 1881–1953, a British mathematician, physicist and meteorologist, who contributed to the mathematical techniques of weather forecasting.

[14] Carl-Gustaf Arvid Rossby, 1898–1957, a Swedish-born American meteorologist, who first explained the large-scale motions of atmosphere in terms of fluid mechanics and identified and characterized both the jet stream and long waves in the westerlies that were later named the *Rossby waves*. Vagn Walfrid Ekman, 1874–1954, a Swedish oceanographer, who proposed the *Ekman spiral* to explain the moving trajectory of a moving object in a rotating environment.

Table 6.4 Conventional dimensionless numbers for flows of the isothermal Newtonian fluids with constant density and dynamic viscosity

Dimensionless numbers	Physical interpretations	Application fields
$\mathcal{C}_a = \dfrac{[\rho][u]^2}{[E_v]}$	$\dfrac{\text{inertia force}}{\text{compressibility force}}$	Compressible flows
$\mathcal{E}_u \dfrac{[p]}{[\rho][u]^2}$	$\dfrac{\text{pressure force}}{\text{inertia force}}$	Flows in which pressure difference is of interest
$\mathcal{F}_r = \dfrac{[u]^2}{[g][L]}$	$\dfrac{\text{inertia force}}{\text{gravitational force}}$	Flows with free surfaces
$\mathcal{M}_a = \dfrac{[u]}{[c]}$	$\dfrac{\text{inertia force}}{\text{compressibility force}}$	Compressible flows
$\mathcal{R}_e = \dfrac{[\rho][u][L]}{[\mu]}$	$\dfrac{\text{inertia force}}{\text{viscous force}}$	Important for all types of fluid flows
$\mathcal{S}_t = \dfrac{[\omega][L]}{[u]}$	$\dfrac{\text{local inertia force}}{\text{convective inertia force}}$	Unsteady flows with oscillation frequency
$\mathcal{W}_e = \dfrac{[\rho][u]^2[L]}{[\sigma]}$	$\dfrac{\text{inertia force}}{\text{surface tension force}}$	Flows in which surface tension is important

These equations indicate that to reach a complete model similarity of a prototype, the Strouhal number, Euler number, Froude number, Reynolds number, Péclet number, radiation number, and dissipation numbers must be invariant. Nevertheless, it is impossible to accomplish this requirement in a model test. An incomplete model similarity needs to be conducted.

Each dimensionless number has a physical interpretation. In fact, each dimensionless number represents a relative significance between any two physical influences, mostly forces, in a fluid motion. By using the previously introduced scaling variables, one can introduce the physical and virtual forces appearing in isothermal fluid flows, as those summarized in Table 6.3. With these, the Strouhal number is a measure to estimate the relative significance between the local and convective inertia forces. For larger values of $\mathcal{S}_t$, Eq. (6.5.14)$_2$ may be simplified to

$$\bar{\rho}\mathcal{S}_t \frac{\partial \bar{\boldsymbol{u}}}{\partial \bar{t}} = -\mathcal{E}_u \operatorname{grad} \bar{p} + \frac{1}{\mathcal{F}_r}\bar{\rho}\,\bar{\boldsymbol{b}} + \frac{1}{\mathcal{R}_e}\{2\bar{u}\,\mathrm{lap}\,\bar{\boldsymbol{u}}\}. \tag{6.5.15}$$

For solid-fluid interactions such as those in wind-structure systems, the Strouhal number is conventionally expressed as

$$\mathcal{S}_t = \frac{[\omega][L]}{[u]}, \qquad [\omega] = \frac{1}{[\tau]}. \tag{6.5.16}$$

The Euler number is a measure to estimate the relative significance between the pressure and inertia forces. It is closely related to the *pressure coefficient* C_p and *cavitation number* $\mathcal{C}_{av}$ defined respectively as

$$C_p \equiv \frac{[\Delta p]}{[\rho][u]^2/2}, \qquad \mathcal{C}_{av} \equiv \frac{[p - p_v]}{[\rho][u]^2}, \tag{6.5.17}$$

where p_v is the vapor pressure of fluid. The Froude number is understood as a measure to estimate the relative significance between the inertia and gravity forces, while the Reynolds number denotes the relative significance between the inertia and viscous forces. Depending on the relative values of $\mathcal{E}_u$, $\mathcal{F}_r$, and $\mathcal{R}_e$, Eq. (6.5.14)$_2$ can be simplified to a certain extent. For example, in the circumstance in which $\mathcal{R}_e \sim \mathcal{F}_r \gg 1$, the inertial force dominates the flow behavior when compared to the influence of viscous and gravity forces, with which Eq. (6.5.14)$_2$ is simplified to

$$\bar{\rho}\left(\mathcal{S}_t \frac{\partial \bar{\boldsymbol{u}}}{\partial \bar{t}} + (\operatorname{grad} \bar{\boldsymbol{u}})\bar{\boldsymbol{u}}\right) = -\mathcal{E}_u \operatorname{grad} \bar{p}. \tag{6.5.18}$$

In this case the flow behavior is dominated by the inertia and pressure forces.

Thus, non-dimensionalization of differential equations of fluid mechanics provides not only the definitions of dimensionless numbers, but also a systematic way to evaluate the relative contributions of each term appearing in the dimensionless equations. It is, however, more difficult to deduce the evaluation if the differential equations are in dimensional forms. For the Newtonian fluids in which heat transfer processes involve, similar interpretations can be found for the Péclet number, dissipation number, and radiation number, as will be shown in Sect. 8.5.

Based on the given physical and virtual forces in terms of the scaling variables, it is possible to define the *Cauchy number*, $\mathcal{C}_a$, as the ratio of inertia force divided by compressibility force, the *Mach number*, $\mathcal{M}_a$, as the square root of the Cauchy number, and the *Weber number*, $\mathcal{W}_e$, as the ratio of inertia force divided by surface tension force, which are given by

$$\begin{aligned} \mathcal{C}_a &\equiv \frac{[\rho][u]^2[L]^2}{[E_v][L]^2} = \frac{[\rho][u]^2}{[E_v]}, \qquad \mathcal{M}_a \equiv \frac{[u]}{\sqrt{[E_v]/[\rho]}} = \frac{[u]}{[c]}, \\ \mathcal{W}_e &\equiv \frac{[\rho][u]^2[L]^2}{[\sigma][L]} = \frac{[\rho][u]^2[L]}{[\sigma]}. \end{aligned} \tag{6.5.19}$$

Table 6.4 summarizes the dimensionless numbers frequently used for flows of the isothermal Newtonian fluids with constant density and dynamic viscosity. Of particular importance are the Reynolds number and Mach number. While the former is used to distinguish a flow to be laminar or turbulent, the latter is used to indicate the influence of fluid compressibility. The related discussions will be provided in the forthcoming chapters.

6.6 Exercises

6.1 Use the Buckingham theorem and dimensional analysis to derive Eq. (6.2.19), namely, the one- and two-dimensional approximations of the radius of shock front induced by an explosion of a bomb on the earth's surface.

6.2 The pressure drop Δp for a steady, incompressible viscous flow through a straight horizontal pipe depends on the pipe length ℓ, the average flow velocity V, the fluid dynamic viscosity μ, the pipe diameter d, the fluid density ρ, and the average pipe roughness e. Determine a set of dimensionless products that can be used to correlate the experiment data.

6.3 Consider a two-dimensional basin filled with an incompressible liquid shown in the figure. The basin has the cross-sectional area A_1 and is connected to a pipe with the cross-sectional area $A_2 \ll A_1$ and length L. Initially, the basin is filled with a liquid to the height h. Derive a dimensionless formula for the average velocity V over the cross-section at point 2 that is established shortly after the opening of valve as a function of h, L, A_1, A_2, gravitational acceleration, and time t.

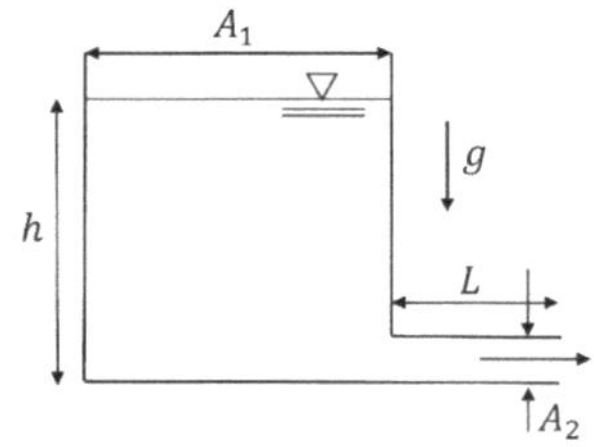

6.4 The figure shows a vertically discharging air jet. Experiments show that a ball placed in the jet is suspended in a stable position. The equilibrium height of ball h is found to depend on the diameter D and weight W of ball, the diameter d of jet-discharging hole, the density ρ and dynamic viscosity μ of air, and the velocity V of air jet. Find the dimensionless products that characterize this physical process.

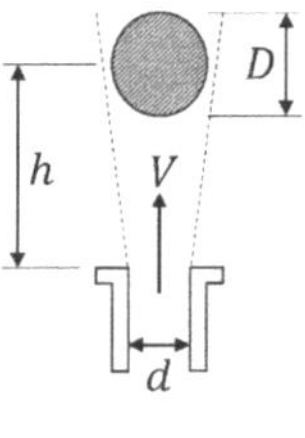

6.5 Derive the dimensionless formulas for the steady-flow rate Q through a *Thompson* and a *Poincelet overfall weirs*, as shown in the figure. These two weirs are used frequently to estimate the flow rate of an *open-channel flow*.

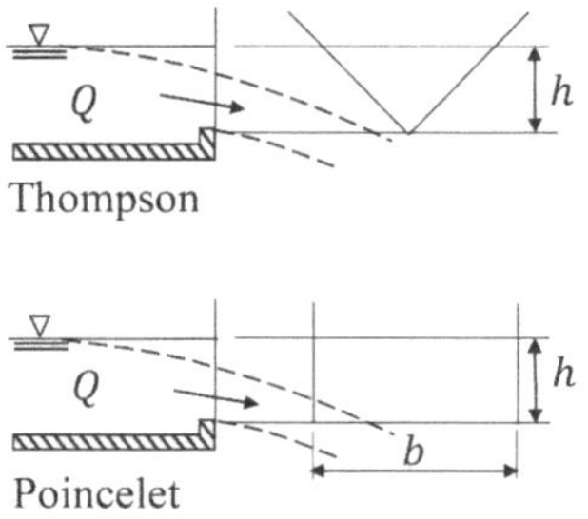

6.6 Show that if $y = f(x_1, x_2, \ldots, x_n)$ is a dimensionally homogeneous equation and if y is dimensional, there exists a product of powers of x_j which has the same dimension of y.

6.7 Let $y = f(x_1, x_2, x_3)$ be a dimensionally homogeneous function with the dimensions of the variables given by

	y	x_1	x_2	x_3
M	1	1	2	− 1
L	3	− 1	0	2
t	− 2	− 3	− 2	2

It is assumed that in a physical model the quantities (x_1, x_2, x_3) are to be reduced in magnitude, specifically, x_1 to a fifth, x_2 to a tenth and x_3 to a fourth of their values in nature. What is the change in scale for the variable y?

6.8 The drag of a sonar transducer is to be predicted by a model test in a wind tunnel. The prototype, which is a sphere with 0.3 m diameter, is to be towed at 10 km/h in water at 20 °C. The model sphere is with 0.15 m in diameter. Determine the required test speed of air in the wind tunnel. If the measured drag in the model is 30 N, estimate the drag in the prototype.

6.9 The equation describing the motion of a fluid in a pipe due to an applied pressure gradient, if the flow starts from rest, is given by

$$\frac{\partial u}{\partial t} = -\frac{\partial p}{\partial x} + \frac{\mu}{\rho}\left(\frac{\partial^2 u}{\partial r^2} + \frac{1}{r}\frac{\partial u}{\partial r}\right).$$

Use the average velocity V, pressure drop Δp, pipe length L, and pipe diameter d as the scaling variables to non-dimensionalize the equation. Obtain the dimensionless products that characterize the flow problem.

6.10 In atmospheric studies, the motion of earth's atmosphere can sometimes be approximated by the equation

$$\frac{\mathrm{D}\boldsymbol{u}}{\mathrm{D}t} + 2\boldsymbol{\omega} \times \boldsymbol{u} = -\frac{1}{\rho}\nabla p,$$

where $\boldsymbol{u}$ is the large-scale velocity of atmosphere across the earth's surface, ∇p denotes the climate pressure gradient and $\boldsymbol{\omega}$ represents the earth's angular velocity. Use the pressure difference Δp and typical length scale L to non-dimensionalize this equation. Obtain the dimensionless products that characterize the flow problem.

Further Reading

E. Buckingham, On physically similar system: illustrations of the use of dimensional equations. Phys. Rev. **4**(4), 345–376 (1914)

R.W. Fox, P.J. Pritchard, A.T. McDonald, *Introduction to Fluid Mechanics*, 7th edn. (Wiley, New York, 2009)

K. Hutter, K. Jönk, *Continuum Methods of Physical Modeling* (Springer, Berlin, 2004)

K. Hutter, Y. Wang, *Fluid and Thermodynamics. Volume 1: Basic Fluid Mechanics* (Springer, Berlin, 2016)

D.C. Ispen, *Units, Dimensions, and Dimensionless Numbers* (McGraw-Hill, New York, 1960)
S.J. Kline, *Similitude and Approximation Theory* (McGraw-Hill, New York, 1965)
B.S. Massey, *Units, Dimensional Analysis and Physical Similarity* (Van Nostrand Reinhold Company, London, 1971)
B.R. Munson, D.F. Young, T.H. Okiishi, *Fundamentals of Fluid Mechanics*, 3rd edn. (Wiley, New York, 1990)
R.H.F. Pao, *Fluid Mechanics* (Wiley, New York, 1961)
K.I. Sedov, *Similarity and Dimensional Methods in Mechanics* (Academic Press, New York, 1959)
E.S. Taylor, *Dimensional Analysis for Engineers* (Clarendon Press, Oxford, 1974)
G.I. Taylor, The formation of a blast wave by a very intensive explosion. Part I: Theoretical discussion, Part II: The atomic explosion of 1945, in *Proceeding of Royal Society London A*, vol. 201, pp. 159–186, 1945
M.S. Yalin, *Theory of Hydraulic Models* (Macmillan, London, 1971)

Ideal-Fluid Flows

7

Ideal fluids are a special fluid class, in which the density is constant and the frictional effect is neglected. Any phenomenon which is predicted by the theory of ideal fluid is due to the inertia effects. This chapter is devoted to the discussions on the characteristics of ideal-fluid flows in two- and three-dimensional circumstances. Nevertheless, for real fluids even liquids, the densities still experience variation under extremely high pressures, and the viscous effect plays a very significant role in the flow characteristics. Instead of interpreting the theory of ideal fluid as the discipline far away from practical reality, it does deliver insights into the flow features and in most cases provide the limiting situations, to which the results obtained from the theory of viscous flows must approach. This becomes more obvious if a moving fluid is in contact with a solid boundary, on which a very thin boundary layer exists. The theory of boundary-layer flows should deliver the results which coincide with those of ideal fluids on the edge of boundary layer.

Started with the discussions on ideal fluids and their features, the Euler and Bernoulli equations are introduced to study the important physical characteristics of ideal-fluid flows, followed by Kelvin's theorem to show the relation between the circulation and vorticity in an ideal-fluid flow. Specifically, incompressible and irrotational flows, i.e., potential flows in two- and three-dimensional circumstances, are discussed intensively. For two-dimensional potential flows, the focus is on the application of the principle of superposition to obtain complex flow patterns from simple ones. Typical outcomes are the theory of two-dimensional airfoils. For three-dimensional circumstances, Stokes' stream function is introduced, and d'Alembert's paradox is derived to show the limitations of potential-flow theory. Wave motions on the free surface of a liquid, or at the interface between two dissimilar fluids, are discussed by using a two-dimensional approximation to the potential-flow theory at the end.

© Springer International Publishing AG 2019

C. Fang, *An Introduction to Fluid Mechanics*, Springer Textbooks in Earth Sciences, Geography and Environment, https://doi.org/10.1007/978-3-319-91821-1_7

7.1 Ideal Fluids

Incompressible or density-preserving fluids without frictional effect are referred to as *ideal fluids*. Analysis of ideal fluids delivers the results limited to the flow fields in which the viscous and compressible effects are unimportant. The mathematical simplification which results from the assumption of ideal fluid is great, and consequently the topics of ideal-fluid flows are mathematically best understood. The frictional force per unit volume in the Navier-Stokes equation reads $\boldsymbol{f}_v = \mu\nabla^2\boldsymbol{u}$ if the dynamic viscosity is a constant. A *frictionless flow* can be achieved by either a vanishing dynamic viscosity of a fluid, termed an *inviscid fluid*, or by an *irrotational flow*, for $\nabla \times \boldsymbol{f}_v = \mu\nabla^2(\nabla \times \boldsymbol{u}) = \boldsymbol{0}$. Both circumstances lead to frictionless flows.

The balances of mass and linear momentum of ideal-fluid flows are given respectively by

$$\nabla \cdot \boldsymbol{u} = 0, \qquad \frac{\partial \boldsymbol{u}}{\partial t} + (\boldsymbol{u} \cdot \nabla)\boldsymbol{u} = -\frac{1}{\rho}\nabla p + \rho \boldsymbol{b}, \tag{7.1.1}$$

where the second equation is termed the *Euler equation*, which is nothing else than Newton's second law of motion per unit volume without viscous force or the Navier-Stokes equation with vanishing viscous effect. Theoretically, ideal-fluid flows in isothermal conditions can be studied by using these two equations with the appropriately formulated boundary conditions to obtain the pressure and velocity fields. The study of ideal-fluid flows is frequently referred to as *hydrodynamics*, and the two equations are called the *equations of hydrodynamics*. Obviously, the no-slip boundary condition is not appropriate for ideal-fluid flows, for the Euler equation is one order lower than the Navier-Stokes equation because the viscous term is dropped. Thus, the boundary condition must be relaxed under the approximation of negligible viscous effect. It may be achieved by requiring that the normal velocity on a solid boundary is retained but the tangential velocity is dropped, i.e.,

$$\boldsymbol{u} \cdot \boldsymbol{n} = \boldsymbol{u}_w \cdot \boldsymbol{n}, \tag{7.1.2}$$

where $\boldsymbol{u}_w$ is the velocity of solid boundary and $\boldsymbol{n}$ is the unit normal to the solid surface. Physically, this boundary condition implies that a solid boundary must be a streamline. Any boundary condition which is to be satisfied far away from the body is unaffected by the frictionless approximation.

For ideal fluids, if the frictionless assumption is accomplished by an irrotational flow,[1] the condition of irrotationality implies that

$$\nabla \times \boldsymbol{u} = 0, \qquad \longrightarrow \qquad \boldsymbol{u} = \nabla\phi, \tag{7.1.3}$$

where ϕ is termed the *velocity potential function*. The velocity field $\boldsymbol{u}$ can thus be obtained directly from Eq. (7.1.3) without solving Eq. (7.1.1), provided that ϕ is known. For ideal-fluid flows, the formulation of ϕ is rather trivial and will be discussed in Sect. 7.5.1. The pressure field, instead of using the Euler equation, can

[1] The flow field is initially irrotational and remains still irrotational even near the body, as indicated by *Kelvin's theorem*, to be discussed in Sect. 7.4.

equally be determined in a simpler manner. The *Bernoulli equation*,[2] to be discussed in Sect. 7.3, is an integration form of the Euler equation give by

$$\frac{\partial \phi}{\partial t} + \frac{p}{\rho} + \frac{1}{2}\nabla\phi \cdot \nabla\phi - G = F(t), \tag{7.1.4}$$

which is shown here to demonstrate the solving procedure, where G is the potential function of the conservative body forces and $F(t)$ represents the *unsteady Bernoulli constant*. The pressure field can be determined by using this equation if ϕ is determined. Consequently, for ideal-fluid flows, instead of solving the equations of hydrodynamics directly, the pressure and velocity fields can be obtained in a simpler manner by using Eqs. (7.1.3) and (7.1.4). The study of ideal-fluid flows by using this simpler solving procedure is termed the *potential-flow theory*. When compared to the hydrodynamic equations, the features of potential-flow theory are twofold: First, Eq. (7.1.4) is linear, whereas Eq. $(7.1.1)_2$ is nonlinear.[3] Second, the *principle of superposition* can be used for linear equations to superimpose simple solutions to obtain solutions to complex circumstances. This latter feature will be used extensively in the analyses of two- and three-dimensional potential flows.

7.2 The Euler Equation in Streamline Coordinates

Consider a two-dimensional ideal-fluid flow on the (y, z)-plane shown in Fig. 7.1a, in which the solid lines with arrows represent streamlines. At a specific point of a stream line, the direction $\boldsymbol{s}$ is defined as the tangential direction of the streamline at that point. The direction $\boldsymbol{n}$ is perpendicular to $\boldsymbol{s}$ and points outward. The coordinate system spanned by $\{\boldsymbol{s}, \boldsymbol{n}\}$ is termed the *streamline coordinate system*. Taking inner product of the Euler equation with two infinitesimal vectors d$\boldsymbol{s}$ and d$\boldsymbol{n}$ yields

$$\begin{aligned} \mathrm{d}\boldsymbol{s} \cdot \left(\frac{\partial \boldsymbol{u}}{\partial t} + (\boldsymbol{u}\cdot\nabla)\boldsymbol{u} = -\frac{1}{\rho}\nabla p + \rho\boldsymbol{b}\right), \quad &\longrightarrow \quad u_s\frac{\partial u_s}{\partial s} = -\frac{1}{\rho}\frac{\partial p}{\partial s} - g\frac{\partial z}{\partial s}, \\ \mathrm{d}\boldsymbol{n} \cdot \left(\frac{\partial \boldsymbol{u}}{\partial t} + (\boldsymbol{u}\cdot\nabla)\boldsymbol{u} = -\frac{1}{\rho}\nabla p + \rho\boldsymbol{b}\right), \quad &\longrightarrow \quad \frac{u_s^2}{R} = \frac{1}{\rho}\frac{\partial p}{\partial n} + g\frac{\partial z}{\partial n}, \end{aligned} \tag{7.2.1}$$

if the gravitational acceleration g is the only body force, and the steady-flow assumption is used, where u_s and u_n are respectively the velocity components in the s- and n-directions, and R denotes the radius of curvature at the evaluation point on the streamline. Equation $(7.2.1)_2$ is obtained by the fact that the acceleration a_n that is experienced by a fluid element at the point is the inverse of centrifugal acceleration given by $a_n = -u_s^2/R$.

[2]Daniel Bernoulli, 1700–1782, a Swiss mathematician and physicist, who was one of the many prominent mathematicians in the *Bernoulli family*, with his main contributions in mathematics, mechanics, fluid mechanics, probability, and statistics.

[3]Although the term $\nabla\phi \cdot \nabla\phi$ in Eq. (7.1.4) is nonlinear, it places no difficulty in the analysis.

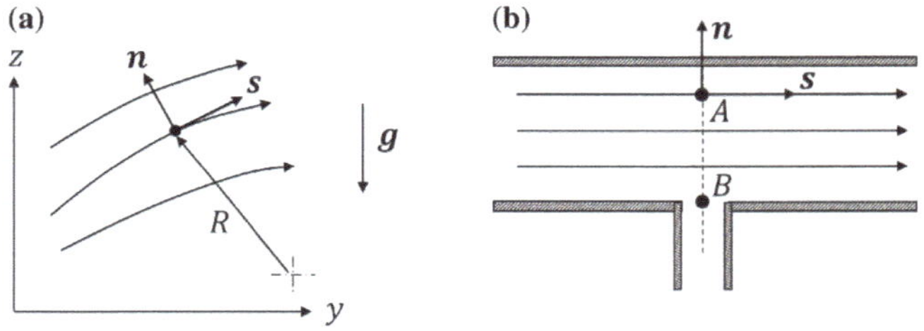

Fig. 7.1 Euler equation in a streamline coordinate system. **a** Illustration of the streamline coordinates. **b** Equivalence between the pressures at point A and point B in view of the Euler equation

The implications of Eq. (7.2.1) are straightforward. For example, consider a steady ideal-fluid flow along a horizontal streamline, for which the Euler equation in the s-direction reduces to

$$u_s \frac{\partial u_s}{\partial s} = -\frac{1}{\rho}\frac{\partial p}{\partial s}. \tag{7.2.2}$$

This equation indicates that a negative pressure gradient in the s-direction is required to have a positive velocity increase and vice versa. On the other hand, if the pressure is maintained as a constant along a streamline which has an elevation difference (i.e., z is not a constant), Eq. $(7.2.1)_1$ reduces to

$$u_s \frac{\partial u_s}{\partial s} = -\frac{\partial z}{\partial s}. \tag{7.2.3}$$

Similarly, in order to have a positive velocity increase, a negative elevation gradient in the s-direction is required. Without solving the Euler equation directly, Eqs. (7.2.2) and (7.2.3) deliver the important physical features of ideal-fluid flows which are summarized as follows:

- A fluid tends to flow locally from a high-pressure region to a low-pressure region.
- A fluid tends to flow locally from a high-elevation region to a low-elevation region.

For circumstances in which both elevation and pressure gradients present, the flow direction is determined by the relative significance between the pressure and gravitation forces. Although the above conclusions are obtained for ideal-fluid flows, they can equally be extended *qualitatively* for viscous fluid flows, except that the influence of viscous force needs to be taken into account.

For a steady horizontal streamline, the Euler equation in the n-direction reduces to

$$\frac{\partial p}{\partial n} = -g\frac{\partial z}{\partial n}, \tag{7.2.4}$$

which corresponds exactly to the hydrostatic equation given in Sect. 3.1. If the gravitational acceleration is further assumed to point in the x-direction (i.e., the direction perpendicular to the page), this equation is simplified to

$$\frac{\partial p}{\partial n} = 0, \tag{7.2.5}$$

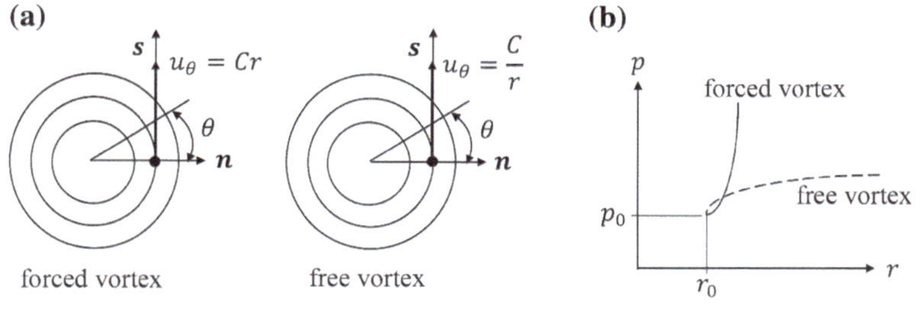

Fig. 7.2 Applications of the Euler equation in a streamline coordinate system. **a** Two-dimensional forced and free vortices. **b** The radial pressure distributions of two vortices

indicating that there exists no pressure variation in the n-direction. This equation provides the theoretical foundation of pressure measurement. For example, consider a two-dimensional flow shown in Fig. 7.1b, in which the gravitational acceleration points perpendicularly to the page. The pressure at point A is exactly the same as that at point B, which can be measured by using e.g. a manometer connected to the wall. However, caution must be made to ensure that the tube of manometer should strictly be perpendicular to the wall with completely flat connecting surfaces. If it is not the case, then at the connection region there exists a non-vanishing R, giving rise to a non-vanishing value of u_s^2/R on the left-hand-side of Eq. $(7.2.1)_2$. In this case, the pressure at point A is no longer the same as the pressure at point B.

Applications of the Euler equation in a streamline coordinate system are demonstrated in Fig. 7.2a by studying e.g. the pressure distributions of a two-dimensional forced and a free vortices, in which the gravitational acceleration points perpendicularly to the page. The pressure variations in the forced and free vortices are described by using the Euler equation in the radial direction given respectively by

$$\frac{\partial p}{\partial r} = \rho C^2 r, \qquad \frac{\partial p}{\partial r} = \frac{\rho C^2}{r^3}, \tag{7.2.6}$$

showing that the pressure variation in a forced vortex is proportional to r, while that in a free vortex is proportional to r^{-3}. Integrating these equations yields

$$p - p_0 = \frac{1}{2}\rho C^2 (r - r_0^2), \qquad p - p_0 = \frac{1}{2}\rho C^2 \left(\frac{1}{r_0^2} - \frac{1}{r^2}\right), \tag{7.2.7}$$

where the reference point is taken at $r = r_0$ with $p = p_0$, i.e., the atmospheric pressure. These results are illustrated graphically in Fig. 7.2b. Applying Eq. (7.2.7) to the free surfaces of two vortices gives rise respectively to

$$\frac{1}{2}\rho C^2 (r - r_0^2) = 0, \qquad \frac{1}{2}\rho C^2 \left(\frac{1}{r_0^2} - \frac{1}{r^2}\right) = 0, \tag{7.2.8}$$

which are the equations of free surfaces. The first equation corresponds to the free surface of a fluid in rigid rotational motion described in Sect. 3.5. Although these two equations have singular points, they serve as the simplest models to demonstrate qualitatively the features of forced and free vortices.[4]

[4]The singular point of a forced vortex occurs at $r \to \infty$, while that of a free vortex occurs at $r \to 0$.

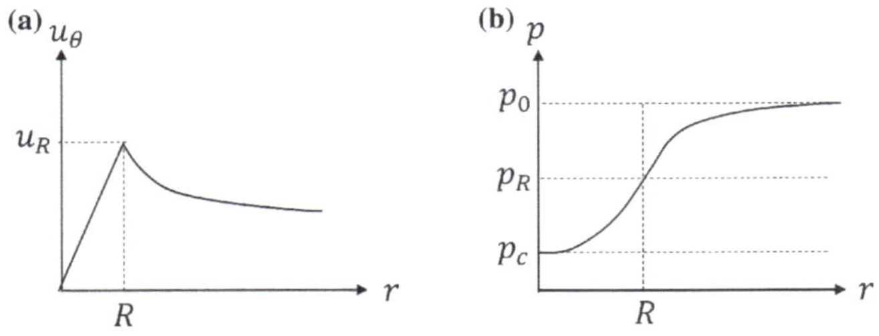

Fig. 7.3 Characteristics of a Rankine vortex in the radial direction. **a** Distribution of the tangential velocity. **b** Distribution of the thermodynamic pressure

A *Rankine vortex* is a combination of a forced vortex in the inner part and a free vortex in the outer part,[5] with the distributions of tangential velocity and pressure shown in Fig. 7.3, where p_c is the pressure at the vortex center, u_R represents the maximum tangential velocity with pressure p_R, and p_0 denotes the surrounding (atmospheric) pressure. The Rankine vortex is the simplest model to describe the features of a typhoon and can be used as a first engineering approximation to estimate the wind loads on structures due to the occurrence of a typhoon. In this circumstance, u_R is used to estimate the typhoon intensity, and R corresponds nearly to the edge of typhoon eye. It follows from the Euler equation that moist air flows from the surrounding toward the center of a typhoon, with the flow direction deflected to the right by the Coriolis force in the Northern Hemisphere, causing a typhoon to rotate counterclockwise. On the contrary, the flow direction is deflected to the left in the Southern Hemisphere, and a typhoon rotates clockwise. During the inward motion, the moist air experiences equally an upward motion to reach the top of a typhoon. Since the atmospheric temperature in high elevation is lower than that near the sea surface, it is likely possible that condensation of water vapor contained in the moist air takes place. The latent heat released by the condensation process provides an energy supply to maintain or even enhance the rotational motion of a typhoon, until all supplied energies are dissipated by the viscous and other effects.

7.3 The Bernoulli Equation

7.3.1 General Formulation

For the Newtonian fluids with constant density and dynamic viscosity, the Navier-Stokes equation with vanishing viscous force reads

$$\rho\frac{\partial \boldsymbol{u}}{\partial t} + \rho(\boldsymbol{u}\cdot\nabla)\boldsymbol{u} = -\nabla p + \rho\nabla G, \tag{7.3.1}$$

[5] William John Macquorn Rankine, 1820–1872, a Scottish mechanical engineer, who made contributions to various fields, and together with Rudolf Clausius and William Thomson, founded first law of thermodynamics.

where the body force per unit mass is considered a *conservative field* with its corresponding scalar potential given by G, i.e., $\boldsymbol{b} = \nabla G$. For example, the gravitational acceleration is a conservative force field, which can be determined by the gradient of gravitational potential energy per unit mass. Since

$$(\boldsymbol{u} \cdot \nabla)\boldsymbol{u} = \nabla \left(\frac{1}{2}\boldsymbol{u} \cdot \boldsymbol{u} \right) - \boldsymbol{u} \times \boldsymbol{\omega}, \tag{7.3.2}$$

where $\boldsymbol{\omega} = \nabla \times \boldsymbol{u}$, substituting this expression into Eq. (7.3.1) yields

$$\frac{\partial \boldsymbol{u}}{\partial t} + \nabla \left(\frac{1}{2}\boldsymbol{u} \cdot \boldsymbol{u} \right) - \boldsymbol{u} \times \boldsymbol{\omega} = -\frac{1}{\rho}\nabla p + \nabla G. \tag{7.3.3}$$

Taking inner product of this equation with a line element of a space curve, $\mathrm{d}\boldsymbol{\ell}$, gives

$$\frac{\partial \boldsymbol{u}}{\partial t} \cdot \mathrm{d}\boldsymbol{\ell} + \mathrm{d}\left(\frac{1}{2}\boldsymbol{u} \cdot \boldsymbol{u} \right) + \frac{\mathrm{d}p}{\rho} - \mathrm{d}G = (\boldsymbol{u} \times \boldsymbol{\omega}) \cdot \mathrm{d}\boldsymbol{\ell}, \tag{7.3.4}$$

along the tangential direction at a specific point on the space curve ℓ, where $\mathrm{d}\alpha$ denotes the total derivative of any quantity α. This equation is referred to as the *differential Bernoulli equation* or the *differential Bernoulli integral* for frictionless flow. Although Eq. (7.3.4) originates from the Euler equation, it is in fact a scalar equation denoting an energy balance, for each term represents a kind of energy. Integrating Eq. (7.3.4) results in

$$\int \frac{\partial \boldsymbol{u}}{\partial t} \cdot \mathrm{d}\boldsymbol{\ell} + \frac{1}{2}\boldsymbol{u} \cdot \boldsymbol{u} + \int \frac{\mathrm{d}p}{\rho} - G = \int (\boldsymbol{u} \times \boldsymbol{\omega}) \cdot \mathrm{d}\boldsymbol{\ell}, \tag{7.3.5}$$

which is termed the *Bernoulli equation* or the *Bernoulli integral* for frictionless flows along an arbitrary curve in space.

Several important simplifications to Eq. (7.3.5) can be made, which are explored in the following.

- For steady and incompressible flows along a streamline, Eq. (7.3.5) reduces to

$$\frac{u^2}{2} + \frac{p}{\rho} - G = \mathcal{C}, \qquad u = \|\boldsymbol{u}\|, \tag{7.3.6}$$

  for $\boldsymbol{u}$ is in parallel with $\mathrm{d}\boldsymbol{\ell}$, giving rise to a vanishing value of $(\boldsymbol{u} \times \boldsymbol{\omega}) \cdot \mathrm{d}\boldsymbol{\ell}$. The term $\mathcal{C}$ is an integration constant, called the *Bernoulli constant*, which is different in different streamlines. Obviously, the right-hand-side of Eq. (7.3.5) vanishes equally if the flow is irrotational. For such a circumstance, $\mathrm{d}\boldsymbol{\ell}$ represents then a line element of any curve in space, which is not necessary a streamline. If the flow experiences both the gravitational acceleration $g = \|g\|$ and a concentric acceleration $r\omega^2$ with r the radius and ω the angular speed, the scalar potential G is then given by

$$G = -gz + \frac{r^2\omega^2}{2}, \tag{7.3.7}$$

  with which Eq. (7.3.6) becomes

$$\frac{u^2 - r^2\omega^2}{2} + \frac{p}{\rho} + gz = \mathcal{C}, \tag{7.3.8}$$

where z is the elevation. If the fluid experiences no rotational motion, the above equation is simplified to

$$\frac{u^2}{2}+\frac{p}{\rho}+gz=\mathcal{C}, \tag{7.3.9}$$

which is the most common form of the Bernoulli equation, being valid for every point on a streamline. This equation indicates that the total mechanical energy, including the kinetic energy $u^2/2$, pressure energy p/ρ, and potential energy gz per unit mass, should be a constant through the entire streamline. From this perspective, Eq. (7.3.9) represents the simplest conservation of energy of fluid flows and provides a transformation rule for mechanical energy of steady, ideal-fluid flows along a streamline. The equivalent expressions of Eq. (7.3.9) are given by

$$\frac{u^2}{2g}+\frac{p}{\gamma}+z=\mathcal{C}, \qquad \frac{\rho u^2}{2}+p+\rho gz=\mathcal{C}. \tag{7.3.10}$$

- For steady and compressible flows along a streamline, Eq. (7.3.5) reduces to

$$\frac{u^2-r^2\omega^2}{2}+\int\frac{\mathrm{d}p}{\rho}+gz=\mathcal{C}. \tag{7.3.11}$$

Further simplifications to this equation become possible if the relations between ρ and p are prescribed. For example, consider an ideal gas characterized by $p=C\rho^\gamma$, where C is a constant and γ denotes the specific-heat ratio. Substituting this expression into the above equation yields

$$\frac{u^2-r^2\omega^2}{2}+\frac{\gamma}{\gamma-1}\frac{p}{\rho}+gz=\mathcal{C}, \tag{7.3.12}$$

which is different from Eq. (7.3.8) for incompressible flows due to the amount of energy stored in pressure form. The difference becomes more obvious if Eqs. (7.3.8) and (7.3.12) are expressed in terms of the Mach number $\mathcal{M}_a$ given respectively by[6]

$$\frac{p_2-p_1}{p_1}=\frac{\gamma(\mathcal{M}_a)_1^2}{2};\quad \frac{p_2-p_1}{p_1}=\frac{\gamma(\mathcal{M}_a)_1^2}{2}\left(1+\frac{1}{4}(\mathcal{M}_a)_1^2+\frac{2-\gamma}{24}(\mathcal{M}_a)_1^4+\cdots\right), \tag{7.3.13}$$

between any two points 1 and 2 on a streamline without the contributions of concentric acceleration, where $(\mathcal{M}_a)_1$ represents the Mach number at point 1. For air, $\gamma=1.4$, and it is ready to verify that the pressure energy stored in a compressible flow is larger than that in an incompressible flow. However, the difference is nearly 2% when $(\mathcal{M}_a)_1=0.3$. For larger values of $(\mathcal{M}_a)_1$, the disagreement between Eqs. $(7.3.13)_1$ and $(7.3.13)_2$ becomes obvious. These results imply that compressible flows can be approximated by using the theory of incompressible flows if the Mach number does not exceed 0.3.

[6] A more detailed discussion will be provided in Sect. 9.3.4.

Table 7.1 Different forms of the Bernoulli equation with the corresponding restrictions under conservative body forces

Form	Restrictions
$\int \frac{\partial \boldsymbol{u}}{\partial t} \cdot \mathrm{d}\boldsymbol{\ell} + \frac{1}{2}\boldsymbol{u} \cdot \boldsymbol{u} + \int \frac{\mathrm{d}p}{\rho} - G = \int (\boldsymbol{u} \times \boldsymbol{\omega}) \cdot \mathrm{d}\boldsymbol{\ell}$	F
$\frac{u^2 - r^2\omega^2}{2} + \frac{p}{\rho} + gz = \mathcal{C}$	F + I + S + L
$\frac{u^2}{2} + \frac{p}{\rho} + gz = \mathcal{C}$	F + I + S + L (gravitational field)
$\frac{u^2 - r^2\omega^2}{2} + \int \frac{\mathrm{d}p}{\rho} + gz = \mathcal{C}$	F + S + L
$\int \frac{\partial u}{\partial t} \mathrm{d}\ell + \frac{u^2 - r^2\omega^2}{2} + \frac{p}{\rho} + gz = \mathcal{C}$	F + I + L
$\frac{\partial \phi}{\partial t} + \int \frac{\mathrm{d}p}{\rho} + \frac{1}{2}\nabla\phi \cdot \nabla\phi - G = F(t)$	F + IR

F: frictionless flows, I: incompressible flows, IR: irrotational flows, S: steady flows, L: along a streamline

- For unsteady and incompressible flows along a stream line, Eq. (7.3.5) reduces to

$$\begin{aligned} &\int \frac{\partial u}{\partial t} \mathrm{d}\ell + \frac{u^2 - r^2\omega^2}{2} + \frac{p}{\rho} + gz = \mathcal{C}, \\ &\int_1^2 \frac{\partial u}{\partial t} \mathrm{d}\ell + \frac{(u_2^2 - r_2^2\omega^2) - (u_1^2 - r_1^2\omega^2)}{2} + \frac{p_2 - p_1}{\rho} + g(z_2 - z_1) = 0, \end{aligned} \tag{7.3.14}$$

  which should be evaluated between any two points 1 and 2 on a streamline.
- For unsteady and irrotational flows, the right-hand-side of Eq. (7.3.5) vanishes identically, and the equation reduces to

$$\frac{\partial \phi}{\partial t} + \int \frac{\mathrm{d}p}{\rho} + \frac{1}{2}\nabla\phi \cdot \nabla\phi - G = F(t), \tag{7.3.15}$$

  where $\boldsymbol{u} = \nabla\phi$ is used for irrotational flows with ϕ the velocity potential function, and $F(t)$ is termed the unsteady Bernoulli constant, even though it is not strictly a constant.

Table 7.1 summarizes different forms of the Bernoulli equation with the corresponding restrictions. The restriction "along a streamline" can be removed if the flow is irrotational. In that circumstance, the Bernoulli equation can be used between any two points on any curve in space.

The Bernoulli equation is frequently used to estimate the energy loss between any two points of a flow. The difference in total mechanical energies between any two points is defined as the energy loss in-between. For example, consider water flowing through a valve. Let point 1 be located before the valve and point 2 be located after the valve. The energy loss of water passing the valve per unit mass, ΔE, is then obtained as

$$\Delta E = \left(\frac{p_1}{\rho} + \frac{u_1^2}{2} + gz_1\right) - \left(\frac{p_2}{\rho} + \frac{u_2^2}{2} + gz_2\right), \tag{7.3.16}$$

for a steady, incompressible flow along a streamline with the gravitational field. It is noted that this equation only provides the general concept of energy loss. For viscous flows, it should be revised to take into account the influence of non-uniform velocity distributions in laminar and turbulent flows on the estimations on the kinetic energy. The topic will be discussed in Sect. 8.6.8.

7.3.2 Static, Dynamic, and Stagnation Pressures

The pressure in the Bernoulli equation is termed the thermodynamic pressure or static pressure. Instead of expressing the pressure as a gage one in determining the hydrostatic force on a submerged surface described in Sect. 3.2, in applying the Bernoulli equation the pressure needs to be expressed as an absolute value, for it represents a kind of energy stored in the fluid. The term $\rho u^2/2$ is called the *dynamic pressure*, which is an equivalent pressure due to the presence of fluid velocity. The sum of static and dynamic pressures is called the *stagnation pressure* p_s given by

$$p_s = p + \frac{1}{2}\rho u^2. \tag{7.3.17}$$

The stagnation pressure is interpreted as the pressure that a fluid element, initially associated with a velocity in an ideal fluid, experiences if it is brought to rest *isentropically*, in which the entropy of fluid remains unchanged. In such a process, the kinetic energy of a fluid element is converted completely to a form of pressure. For viscous-fluid flows, a moving fluid element experiences equally a pressure which is larger than the static pressure when it is brought to rest. However, this larger pressure is not the stagnation pressure, for the process is not isentropic, and only a part of the kinetic energy is converted to a form of pressure, while the other part is converted to heat due to the dissipative viscous effects.

The static and stagnation pressures of a fluid can be measured by e.g. a *Pitot tube* shown in Fig. 7.4a, which is a combination of two concentric circular tubes with the inner tube open in the front and outer tube having small open holes on the sides. The static pressure is measured at point B, while the stagnation pressure is measured at point A. It is possible to connect these two points with a regular manometer, with which the fluid velocity can be determined by

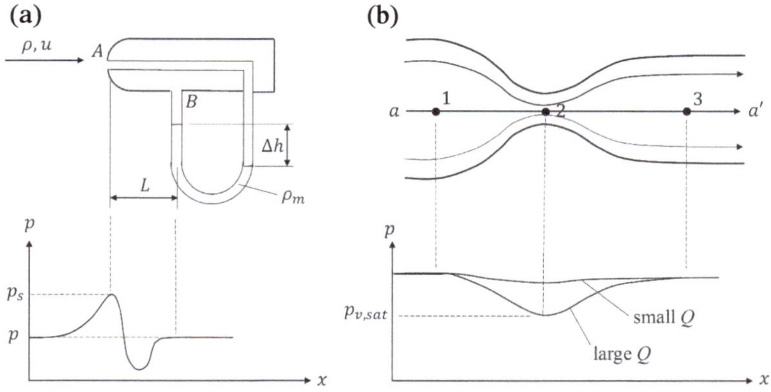

Fig. 7.4 Mechanical energy conversion in the Bernoulli equation. **a** The Pitot tube for flow velocity measurement. **b** The Venturi nozzle and cavitation

$$u = \sqrt{\frac{2\rho_m g \Delta h}{\rho}}, \tag{7.3.18}$$

where ρ_m is the density of fluid in the manometer, ρ represents the density of fluid whose velocity is to be measured, and Δh denotes the elevation difference in the manometer. In practice, the distance L of a Pitot tube needs to be chosen carefully to minimize the temporary pressure variation due to the presence of the Pitot tube in the flow field.

The most important implication of the Bernoulli equation is that the pressure, kinetic, and potential energies of an ideal fluid can be converted into one another along a streamline. The solid boundary which is in contact with a fluid can be so shaped to accomplish such an energy conversion. For example, consider a water flow through a *Venturi nozzle* or alternatively a *convergent-divergent nozzle* shown in Fig. 7.4b.[7] It is assumed that water is incompressible and the water flow is steady. It follows from the continuity equation that the flow velocity u_2 at cross-section A_2 is larger than the velocity u_1 at cross-section A_1, since $A_2 < A_1$. By using the Bernoulli equation along the streamline aa', the fluid pressure p_2 is smaller than the pressure p_1. When water flows subsequently from point 2 to point 3, the velocity is reduced due to a larger cross-sectional area A_3, giving rise to a larger pressure p_3, for part of the kinetic energy is converted to the pressure energy, i.e., $u_3 < u_2$ and $p_3 > p_2$. If the Venturi nozzle is not well constructed, the pressure p_2 at the throat region will be lower than the saturated vapor pressure of water $p_{v,sat}$ for larger flow rates Q. In such a circumstance, tiny water vapor bubbles form at the throat region, which flow subsequently downstream, where they are compressed significantly to create high vapor pressure p_v due to the larger ambient fluid pressure. When the water bubbles

[7]Giovanni Battista Venturi, 1746–1822, an Italian physicist, who discovered the *Venturi effect*.

occasionally are in contact with the solid boundary, the high vapor pressure exerts a temporarily strong impact to the solid wall and may cause a failure of the boundary material in a long-term operation. This phenomenon is known as the *cavitation*, in which the failure of boundary material is caused by mechanical impact, in contrast to the *erosion*, where the boundary material is eroded mainly by chemical reactions.

7.3.3 Illustrations of the Bernoulli Equation

Consider an incompressible inviscid liquid in a vertical U-tube with constant diameter shown in Fig. 7.5a. Initially, the two sides of U-tube are exposed to the atmospheric pressure, yielding the liquid free surface at the equilibrium position $z = 0$. A small pressure difference is applied on the two sides and creates an initial elevation difference in the liquid free surface. The two sides of U-tube are exposed again to the atmospheric pressure. The liquid column will then oscillate at a specific frequency ω, which needs to be determined.

Construct the coordinates shown in the figure, and identify the streamline connecting points 1 and 2 on the free surfaces on the two sides. It follows from the unsteady Bernoulli equation that

$$\int_1^2 \frac{\partial u}{\partial t} \mathrm{d}\ell + \frac{u_2^2}{2} + \frac{p_2}{\rho} + gz_2 = \frac{u_1^2}{2} + \frac{p_1}{\rho} + gz_1, \tag{7.3.19}$$

which reduces to

$$\int_1^2 \frac{\partial u}{\partial t} \mathrm{d}\ell + \frac{u_2^2}{2} + gz = \frac{u_1^2}{2} - gz, \tag{7.3.20}$$

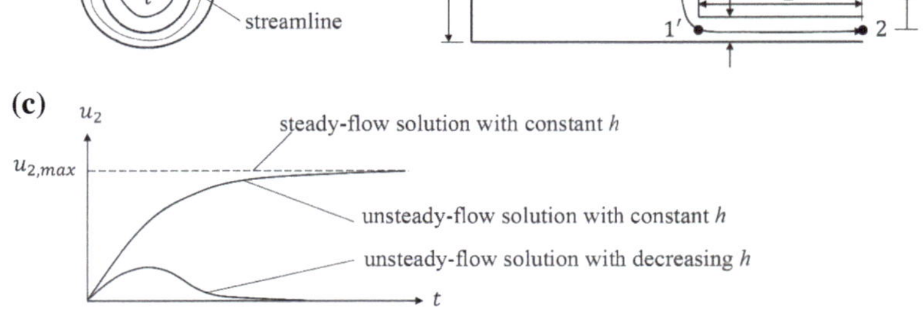

Fig. 7.5 Illustrations of the Bernoulli equation. **a** Oscillation of the liquid column in a U-tube. **b** Discharge flow of water from a large cylindrical tank through a horizontal circular pipe. **c** Time sequence of the exit velocity at point 2 of the problem in **b**

because $z_2 = z$, $z_1 = -z$, and $p_1 = p_2 = p_{atm}$. Since the diameter of U-tube is constant, it follows from the continuity equation that

$$u_2 = \frac{dz}{dt}, \qquad u_1 = -\frac{dz}{dt}, \qquad \int_1^2 \frac{\partial u}{\partial t} d\ell = \ell \frac{d^2 z}{dt^2}, \tag{7.3.21}$$

for u remains unchanged along the streamline inside the liquid, although $u = u(t)$. Substituting these expressions into Eq. (7.3.20) yields

$$\frac{d^2 z}{dt^2} + \frac{2g}{\ell} z = 0, \tag{7.3.22}$$

showing that the free surface elevation z experiences a simple harmonic oscillation, and the oscillating frequency ω is obtained as

$$\omega = \sqrt{\frac{2g}{\ell}}. \tag{7.3.23}$$

Another example is shown in Fig. 7.5b, in which a large cylindrical tank with diameter D is connected to a horizontal circular pipe having diameter d and length L. The tank is initially filled with water to the height h, and the valve at the exit of circular pipe is closed. At $t = 0$, the valve is opened, and water flows through the circular pipe with increasing velocity. It is required to determine (a) the steady-flow solution of exit velocity u_2 with constant h, (b) the unsteady-flow solution of u_2 with constant h, and (c) the unsteady-flow solution of u_2 with decreasing h.

Construct the streamline $1 - 1' - 2$, and locate the datum of elevation z shown in the figure, where point 1 locates on the water free surface in the tank, point $1'$ is at the connection region between the circular pipe and tank, while point 2 is at the exit of circular pipe. For a steady flow with constant h, the Bernoulli equation along the streamline reads

$$\frac{p_1}{\rho} + \frac{u_1^2}{2} + g z_1 = \frac{p_2}{\rho} + \frac{u_2^2}{2} + g z_2, \tag{7.3.24}$$

which reduces to

$$u_2 = u_{2,max} = \sqrt{2gh}, \tag{7.3.25}$$

for $z_1 = h$, $z_2 = 0$, $p_1 = p_2 = p_{atm}$, and $u_1 = dh/dt = 0$. This equation is known as the *Torricelli equation*,[8] which represents the maximum velocity that a free water jet from a container with constant water depth h can assume.

Alternatively, after the valve is opened, the velocity u_2 increases continuously with time. It needs a certain time duration to reach its maximum value if the height h remains unchanged. For this circumstance, the Bernoulli equation reads

$$\int_1^2 \frac{\partial u}{\partial t} d\ell + \frac{u_2^2}{2} = gh. \tag{7.3.26}$$

[8] Evangelista Torricelli, 1608–1647, an Italian physicist and mathematician, who is best known for his invention of the barometer, and his advances in optics and work on the *method of indivisibles*.

The line integral in the above equation is decomposed into two parts, viz.,

$$\int_1^2 \frac{\partial u}{\partial t}\mathrm{d}\ell = \int_1^{1'} \frac{\partial u}{\partial t}\mathrm{d}\ell + \int_{1'}^2 \frac{\partial u}{\partial t}\mathrm{d}\ell \sim \int_{1'}^2 \frac{\partial u}{\partial t}\mathrm{d}\ell. \tag{7.3.27}$$

Essentially, the first line integral does not vanish. However, if the height h is kept constant, the order of magnitude of the first line integral is much smaller than that of the second line integral and can be neglected as an engineering approximation. It follows from the continuity equation that the water velocity assumes the same value at different points inside the circular pipe at a specific time, although the value varies with time. With this, the second line integral is obtained as

$$\int_{1'}^2 \frac{\partial u}{\partial t}\mathrm{d}\ell = L\frac{\mathrm{d}u_2}{\mathrm{d}t}. \tag{7.3.28}$$

Substituting this equation into Eq. (7.3.26) yields

$$L\frac{\mathrm{d}u_2}{\mathrm{d}t} + \frac{u_2^2}{2} - gh = 0, \tag{7.3.29}$$

which is a nonlinear first-order ordinary differential equation of u_2, to which the solution is given by

$$\frac{u_2}{\sqrt{2gh}} = \tanh\left(\frac{\sqrt{2gh}}{2L}t\right). \tag{7.3.30}$$

As $t \to \infty$, $u_2 \to \sqrt{2gh}$, which corresponds to Eq. (7.3.25).

The most realistic circumstance is that h decreases gradually with time as water is discharged at point 2, for which the continuity equation reads

$$u_1\frac{\pi D^2}{4} = u_2\frac{\pi d^2}{4}, \qquad u_2 = \lambda^2\frac{\mathrm{d}h}{\mathrm{d}t}, \qquad \lambda = \frac{D}{d}, \qquad u_1 = \frac{\mathrm{d}h}{\mathrm{d}t}, \tag{7.3.31}$$

with which the Bernoulli equation reads

$$\int_1^{1'} \frac{\partial u}{\partial t}\mathrm{d}\ell + \int_{1'}^2 \frac{\partial u}{\partial t}\mathrm{d}\ell + \frac{1}{2}\left(\lambda^2\frac{\mathrm{d}h}{\mathrm{d}t}\right)^2 = \frac{1}{2}\left(\frac{\mathrm{d}h}{\mathrm{d}t}\right)^2 + gh. \tag{7.3.32}$$

When compared with the second line integral, the first line integral is even less significant, for h decreases during the flow. Thus, it is plausible to assume that the first line integral vanishes, while the second line integral is determined as

$$\int_1^{1'} \frac{\partial u}{\partial t}\mathrm{d}\ell + \int_{1'}^2 \frac{\partial u}{\partial t}\mathrm{d}\ell \sim \int_{1'}^2 \frac{\partial u}{\partial t}\mathrm{d}\ell = \lambda^2 L\frac{\mathrm{d}^2h}{\mathrm{d}t^2}. \tag{7.3.33}$$

Substituting this expression into Eq. (7.3.32) gives

$$2\lambda^2 L\frac{\mathrm{d}^2h}{\mathrm{d}t^2} + (\lambda^4 - 1)\left(\frac{\mathrm{d}h}{\mathrm{d}t}\right)^2 - 2gh = 0, \tag{7.3.34}$$

which is a nonlinear second-order ordinary differential equation of h, whose solution must be determined by using numerical integration. Once $h(t)$ is obtained, the velocity u_2 is determined by using Eq. (7.3.31)$_2$. For comparison, the obtained three expressions of u_2 are illustrated graphically in Fig. 7.5c. This problem demonstrates the applications of the disciplines of fluid mechanics to a realistic circumstance by studying the simplest case at the beginning, with additional considerations taken into account to approach the final real situation.

7.4 Kelvin's Theorem

It is assumed that the body force per unit mass $\boldsymbol{b}$ experienced by an ideal fluid is conservative, which can be expressed as the gradient of its corresponding scalar function G. With these, the Euler equation reads

$$\dot{\boldsymbol{u}} = -\frac{1}{\rho}\nabla p + \nabla G. \tag{7.4.1}$$

It follows from the definition of circulation Γ given in Sect. 4.2 that

$$\dot{\Gamma} = \frac{\mathrm{D}}{\mathrm{D}t}\oint \boldsymbol{u}\cdot \mathrm{d}\boldsymbol{\ell} = \oint \left[\dot{\boldsymbol{u}}\cdot \mathrm{d}\boldsymbol{\ell} + \boldsymbol{u}\cdot (\mathrm{d}\boldsymbol{\ell})^{\cdot}\right] = \oint \left[\dot{\boldsymbol{u}}\cdot \mathrm{d}\boldsymbol{\ell} + \boldsymbol{u}\cdot \mathrm{d}\boldsymbol{u}\right], \tag{7.4.2}$$

where $\mathrm{d}\boldsymbol{\ell}$ is an infinitesimal line segment, and $(\mathrm{d}\boldsymbol{\ell})^{\cdot} = \mathrm{D}(\mathrm{d}x_i)/\mathrm{D}t = \mathrm{d}(\mathrm{D}x_i)/\mathrm{D}t = \mathrm{d}\boldsymbol{u}$. Substituting Eq. (7.4.1) into the above equation leads to

$$\dot{\Gamma} = \oint \left[-\frac{\mathrm{d}p}{\rho} + \mathrm{d}G + \frac{1}{2}\mathrm{d}(\boldsymbol{u}\cdot\boldsymbol{u})\right] = -\oint \frac{\mathrm{d}p}{\rho}, \tag{7.4.3}$$

for the line integration is carried out in a closed contour. Since the flow is incompressible, the above equation yields that

$$\dot{\Gamma} = 0. \tag{7.4.4}$$

This result is equally valid for fluids whose pressures depend only on density, which are termed the *barotropic fluids*. If $p = p(\rho)$, then

$$\dot{\Gamma} = -\oint \frac{\mathrm{d}p}{\rho} = -\oint \frac{\mathrm{d}p}{\mathrm{d}\rho}\frac{\mathrm{d}\rho}{\rho} = 0. \tag{7.4.5}$$

The results given in Eqs. (7.4.4) and (7.4.5) are referred to as *Kelvin's theorem*, which is summarized in the following:

7.1 (Kelvin's Theorem) *The vorticity of each fluid element in a frictionless flow field is preserved, when subject to conservative body force fields with pressures is constant or depends only on density.*

Since the circulation on a closed contour is related to the vorticity of the area spanned by the contour, Kelvin's theorem states that the vorticity inside the contour will not change if a given contour is followed. However, in applying Kelvin's theorem, its restrictions should strictly be followed. It may be deduced that the vorticity may be changed in the presence of viscous forces, non-conservative body forces or density variations which are not simply related to pressure variations. Obviously, the closed contour should be in a *simply connected region*. Thus, for any closed contour in the fluid there exists some definite value of Γ, and Kelvin's theorem asserts that Γ will not change around the contour even though the contour itself may be deformed by the flow. If the closed contour initially contains no body, it cannot at any subsequent time include a body.

It is evident that the total vorticity associated with a vortex filament introduced in Sect. 4.4 is fixed and will not change as the vortex filament flows with the fluid, as implied by Kelvin's theorem. Distortion of the vortex filament may take place, but the total vorticity associated with it remains the same. However, the vortex filament should always consist of the same fluid points as it flows. If the vortex filament is elongated during the flow, the vorticity should decrease correspondingly, so that the total vorticity associated with the vortex filament remains fixed. The principal use of Kelvin's theorem is in the interpretation of lift force acting on a body in a flow field, which will be discussed in Sect. 8.4.9.

7.5 Two-Dimensional Potential Flows

7.5.1 Velocity Potential and Stream Functions

For an irrotational flow, the velocity potential function ϕ is so defined that the condition of irrotationality is automatically satisfied, despite whether the flow is compressible or incompressible. Similarly, for an incompressible flow there exists also a scalar function ψ, called the *stream function*, with which the continuity equation satisfies identically, despite whether the flow is rotational or irrotational. In two-dimensional circumstance, ψ is defined by

$$\frac{\partial \psi}{\partial y} \equiv u, \qquad \frac{\partial \psi}{\partial x} \equiv -v, \tag{7.5.1}$$

in the rectangular Cartesian coordinate system, where $\{u, v\}$ are the velocity components in the x- and y-directions, respectively. The definitions of ψ in the cylindrical and spherical coordinate systems can be given in a similar manner. For an incompressible and irrotational flow, both ϕ and ψ exist. To find the velocity potential or stream function, applying the incompressibility condition to ϕ and the irrotationality condition to ψ yields respectively

$$\nabla^2 \phi = 0, \qquad \nabla^2 \psi = 0, \tag{7.5.2}$$

showing that both ϕ and ψ satisfy the Laplace equation. The velocity field $\boldsymbol{u}$ can then be determined, once ϕ or ψ is determined by solving the Laplace equation subject to appropriately formulated boundary conditions.

In two-dimensional circumstances, the expression of $\phi = C$ with C an arbitrary constant represents a family of curves in the (x, y)-plane. The curves are called the *equipotential lines*. A specific equipotential line is obtained by assigning a definite value to the constant C. Similarly, $\psi = C$ denotes a family of curves, whose properties are discussed in the following.

- Consider a specific curve denoted by $\psi = C$, where C is a constant assuming a definite value. Taking total differential of $\psi = C$ gives

$$0 = \mathrm{d}\psi = \frac{\partial \psi}{\partial x}\mathrm{d}x + \frac{\partial \psi}{\partial y}\mathrm{d}y = -v\,\mathrm{d}x + u\,\mathrm{d}y, \qquad \longrightarrow \qquad \left(\frac{\mathrm{d}y}{\mathrm{d}x}\right)_\psi = \frac{v}{u}, \quad (7.5.3)$$

  which corresponds to the definition of streamline in the (x, y)-plane. Thus, the expression $\psi = C$ represents a family of *streamlines* in the (x, y)-plane, and a specific streamline is prescribed by assigning a definite value to the constant C and vice versa.
- Consider two adjacent streamlines in the (x, y)-plane described by $\psi = C_1$ and $\psi = C_2$ with $C_1 \neq C_2$. The flow rate Q passing between two streamlines is obtained as

$$Q = \int_A \boldsymbol{u} \cdot \mathrm{d}\boldsymbol{a} = \int_A^B u\,\mathrm{d}y - \int_A^B v\,\mathrm{d}x, \qquad (7.5.4)$$

  where A is a point lying on $\psi = C_1$ and B is a point lying on $\psi = C_2$, and the above integration is carried out along a line connecting points A and B with a positive slope for simplicity. Since $\mathrm{d}\psi = -v\mathrm{d}x + u\mathrm{d}y$, it follows that

$$Q = \int_A^B u\,\mathrm{d}y - \int_A^B v\,\mathrm{d}x = \int_{C_1}^{C_2} \mathrm{d}\psi = C_2 - C_1. \qquad (7.5.5)$$

  Thus, the difference in the values of stream function between two streamlines gives the flow rate between two streamlines.
- For an incompressible and irrotational flow, consider a stream and an equipotential lines in the (x, y)-plane described by $\psi = C_1$ and $\phi = C_2$. The slopes of two curves are given by

$$\left(\frac{\mathrm{d}y}{\mathrm{d}x}\right)_\psi = \frac{v}{u}, \quad \left(\frac{\mathrm{d}y}{\mathrm{d}x}\right)_\phi = -\frac{u}{v}, \qquad \longrightarrow \qquad \left(\frac{\mathrm{d}y}{\mathrm{d}x}\right)_\psi \left(\frac{\mathrm{d}y}{\mathrm{d}x}\right)_\phi = -1, \quad (7.5.6)$$

  showing that the streamlines $\psi =$ constant and equipotential lines $\phi =$ constant are *mutually orthogonal*. This property delivers the foundation of *flow-net analysis* in solving two-dimensional potential-flow problems.

7.5.2 Complex Potential and Complex Velocity

For an incompressible and irrotational flow in a two-dimensional rectangular coordinate system, it follows from the definitions of velocity potential and stream functions that

$$u = \frac{\partial \phi}{\partial x} = \frac{\partial \psi}{\partial y}, \qquad v = \frac{\partial \phi}{\partial y} = -\frac{\partial \psi}{\partial x}, \qquad (7.5.7)$$

indicating that ϕ and ψ satisfy the Cauchy-Riemann equations. This motivates the *complex potential* $F(z)$ defined by

$$F(z) \equiv \phi + i\psi. \tag{7.5.8}$$

If $F(z)$ is an analytical function, ϕ and ψ will satisfy the Cauchy-Riemann equations identically, and for every $F(z)$ the real part is a valid velocity potential function and the imaginary part is a valid stream function.

Since $F(z)$ is supposed to be analytic, its derivative with respect to z is a point function whose value is independent of the direction along which it is evaluated. The derivative of $F(z)$ with respect to z is denoted by $W(z)$, which is given by

$$W(z) = \frac{\mathrm{d}F}{\mathrm{d}z} = \frac{\partial F}{\partial x} = u - iv, \tag{7.5.9}$$

where $W(z)$ is called the *complex velocity*, although its imaginary part equals to $-iv$. It follows that

$$W\overline{W} = u^2 + v^2, \tag{7.5.10}$$

where $\overline{W}$ represents the *complex conjugate* or simply *conjugate* of W. This equation can be applied to evaluate the kinetic energy in the Bernoulli equation, i.e., $\boldsymbol{u} \cdot \boldsymbol{u} = \nabla\phi \cdot \nabla\phi = u^2 + v^2 = W\overline{W}$. In the study of two-dimensional potential flows, it is convenient to use the concepts of complex potential and complex velocity, and such a procedure is termed the *complex analysis*. The advantage of complex analysis is that by equating the real part of a given analytic function to ϕ and the imaginary part to ψ, the theory of complex variables guarantees that the Laplace equations of ϕ and ψ hold identically. The velocity is determined once ϕ or ψ are obtained. However, the complex analysis validates only for two-dimensional potential flows and cannot be generalized to three-dimensional potential flows. Despite these, the complex analysis avails itself of the powerful results of complex variable theory, avoids the difficulties in solving the partial differential equations, and will be used in the forthcoming discussions.

For the two-dimensional polar coordinates $\{r, \theta\}$ generated by rotating the rectangular coordinates $\{x, y\}$ counterclockwise along the z-axis by an angle θ, the velocity components u_r and u_θ are given by

$$\begin{bmatrix} u_r \\ u_\theta \end{bmatrix} = [\boldsymbol{Q}]^{\mathrm{T}} \begin{bmatrix} u \\ v \end{bmatrix}, \quad \begin{bmatrix} u \\ v \end{bmatrix} = [\boldsymbol{Q}] \begin{bmatrix} u_r \\ u_\theta \end{bmatrix} = \begin{bmatrix} \cos\theta & -\sin\theta \\ \sin\theta & \cos\theta \end{bmatrix} \begin{bmatrix} u_r \\ u_\theta \end{bmatrix}. \tag{7.5.11}$$

With these, the complex velocity in terms of the two-dimensional polar coordinates is obtained as

$$W = (u_r - iu_\theta)e^{-i\theta}. \tag{7.5.12}$$

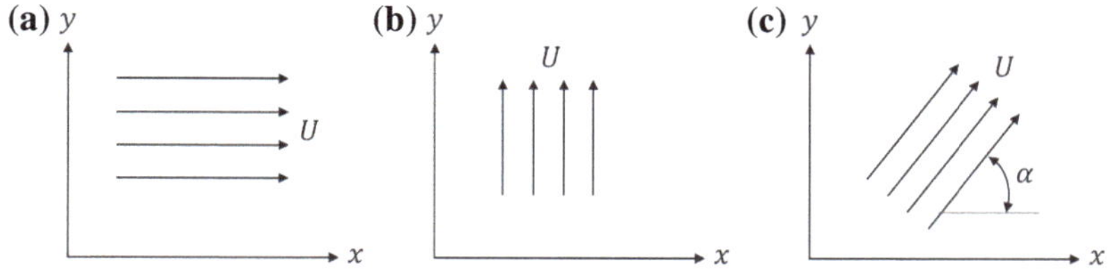

Fig. 7.6 Two-dimensional uniform flows in the (xy)-plane. **a** A rectilinear uniform flow in the x-direction, with $F(z) = Uz$. **b** A rectilinear uniform flow in the y-direction, with $F(z) = -iUz$. **c** An inclined uniform flow with an angle α with respect to the x-axis, with $F(z) = Ue^{-i\alpha}z$

7.5.3 Elementary Solutions

In this section, some elementary two-dimensional potential-flow solutions are discussed. Solutions to more complicated flow circumstances may be obtained by using the superposition principle to these elementary solutions.

Uniform flows. The simplest complex potential is that it is proportional to z, and the corresponding flow fields are uniform flows. First, let $F(z)$ be proportional to z given by

$$F(z) = Uz, \quad \longrightarrow \quad W(z) = U = u - iv, \quad \longrightarrow \quad u = U, \; v = 0, \quad (7.5.13)$$

where U is a real constant. The complex potential $F(z) = Uz$ thus represents a rectilinear uniform flow with constant velocity U in the positive x-direction, as shown in Fig. 7.6a. Next, let $F(z)$ be proportional to z given by

$$F(z) = -iUz, \quad \longrightarrow \quad W(z) = -iU = u - iv, \quad \longrightarrow \quad u = 0, \; v = U, \quad (7.5.14)$$

indicating that this complex potential represents a rectilinear uniform flow field with constant velocity U in the positive y-axis, as shown in Fig. 7.6b. Finally, let $F(z)$ be given by

$$F(z) = Ue^{-i\alpha}z, \quad (7.5.15)$$

by which the complex velocity is obtained as

$$\begin{aligned} &W(z) = Ue^{-i\alpha} = U\cos\alpha - iU\sin\alpha = u - iv, \\ \longrightarrow\; &u = U\cos\alpha, \; v = U\sin\alpha. \end{aligned} \quad (7.5.16)$$

This complex potential represents an inclined uniform flow field with constant velocity U by the angle α with respect to the x-axis, as shown in Fig. 7.6c. The term α is called the *angle of attack*, and for $\alpha = 0$ and $\alpha = \pi/2$, Eq. (7.5.15) reduces to Eqs. (7.5.13) and (7.5.14), respectively.

Source, sink, and vortex flows. Let $F(z)$ be proportional to $(\ln z)$ given by

$$F(z) = c\ln z = c\ln\left(re^{i\theta}\right) = c\ln r + ic\theta, \quad (7.5.17)$$

where c is a real constant, and $0 < \theta < 2\pi$ is considered in the two-dimensional polar coordinate system. It follows that

$$\phi = c \ln r, \qquad \psi = c\theta, \tag{7.5.18}$$

indicating that the equipotential lines are the circles with $r =$ constant, and the streamlines are the radial lines with $\theta =$ constant. With Eq. (7.5.17), the complex velocity is determined as[9]

$$W(z) = \frac{c}{r} e^{-i\theta}, \qquad \longrightarrow \qquad u_r = \frac{c}{r}, \ u_\theta = 0. \tag{7.5.19}$$

A *source* is obtained if $c > 0$, in which the flow velocity is purely radially outward with its magnitude decreasing as the fluid leaves the origin. The velocity decreases in such a way that the fluid volume crossing each circle should be the same, as implied by the continuity equation. If $c < 0$, the flow field is termed a *sink*, with purely radial velocity with increasing magnitude toward the origin. The *strength* m of a source is defined by the fluid volume leaving the source origin per unit time per unit depth, i.e.,

$$m \equiv \int_0^{2\pi} u_r (r\mathrm{d}\theta) = 2\pi c, \tag{7.5.20}$$

with which the complex potential of a source is recast alternatively as

$$F(z) = \frac{m}{2\pi} \ln z, \qquad F(z) = \frac{m}{2\pi} \ln(z - z_0), \tag{7.5.21}$$

for the circumstances in which the singular points of $F(z)$ locate at $z = 0$ and $z = z_0$, respectively. For a sink, the strength m is simply replaced by $-m$.

Alternatively, $F(z)$ can be given by

$$\begin{aligned} &F(z) = -ic \ln z = -ic \ln\left(re^{i\theta}\right) = c\theta - ic \ln r, \\ \longrightarrow\ &\phi = c\theta, \ \psi = -c \ln r, \end{aligned} \tag{7.5.22}$$

where c is a real constant. It follows that the equipotential lines are the radial lines with $\theta =$ constant, while the streamlines are the circles with $r =$ constant, with the directions determined by the complex velocity obtained as

$$W(z) = -i\frac{c}{r} e^{-i\theta}, \qquad \longrightarrow \qquad u_r = 0, \ u_\theta = \frac{c}{r}. \tag{7.5.23}$$

For $c > 0$, the direction of flow is counterclockwise and vice versa. The flow described by Eqs. (7.5.22) and (7.5.23) with a positive value of c is called a *vortex* in counterclockwise rotation, more specifically, a *free vortex*. The strength of a vortex is characterized by the circulation Γ associated with it, which is determined as

$$\Gamma = \oint \boldsymbol{u} \cdot \mathrm{d}\boldsymbol{\ell} = \int_0^{2\pi} u_\theta (r\mathrm{d}\theta) = 2\pi c, \tag{7.5.24}$$

[9] In Eq. (7.5.19), the origin is an isolated singular point of a source or a sink.

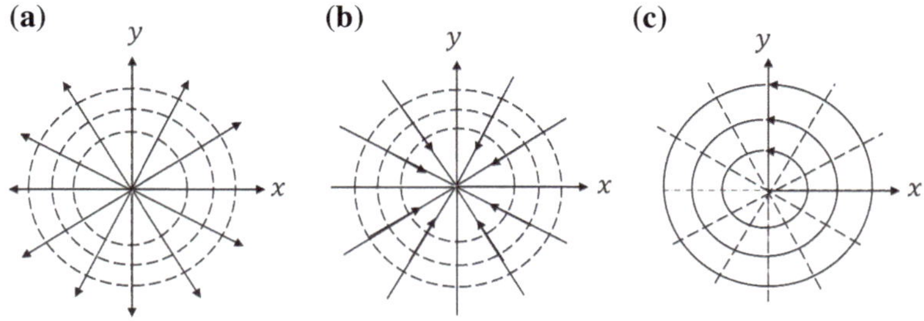

Fig. 7.7 Flow fields of two-dimensional source, sink and vortex in the (xy)-plane. **a** A source locating at the origin. **b** A sink locating at the origin. **c** A free vortex locating at the origin. The solid and dashed lines are respectively the streamlines and equipotential lines

where Γ is associated with the singularity at the origin.[10] With these, the complex potential of a free vortex becomes

$$F(z) = -i\frac{\Gamma}{2\pi}\ln z, \qquad F(z) = -i\frac{\Gamma}{2\pi}\ln(z - z_0), \tag{7.5.25}$$

with the singularities locating respectively at $z = 0$ and $z = z_0$. For a clockwise-free vortex, Γ is simply replaced by $-\Gamma$. The flow fields of a source, a sink, and a free vortex are shown graphically in Fig. 7.7.

Flows in a sector. Flows in sharp bends or sectors are represented by the complex potential which is proportional to z^n with $n \geq 1$, viz.,

$$\begin{aligned} &F(z) = Uz^n = Ur^n\cos(n\theta) + iUr^n\sin(n\theta), \\ \longrightarrow\ &\phi = Ur^n\cos(n\theta),\ \psi = Ur^n\sin(n\theta), \end{aligned} \tag{7.5.26}$$

where U is a real constant. This equation indicates that $\psi = 0$ at $\theta = 0$ and $\theta = \pi/n$. Thus, $\psi = 0$ represents two streamlines which are the radial lines of $\theta = 0$ and $\theta = \pi/n$. The directions of other streamlines described by $Ur^n\sin(n\theta) =$ constant are determined by the complex velocity given by

$$\begin{aligned} &W(z) = nUr^{n-1}\left[\cos(n\theta) + i\sin(n\theta)\right]e^{-i\theta}, \\ &u_r = nUr^{n-1}\cos(n\theta), \quad u_\theta = -nUr^{n-1}\sin(n\theta). \end{aligned} \tag{7.5.27}$$

For $0 < \theta < (\pi/2n)$, u_r assumes positive values, while u_θ is negative. For $(\pi/2n) < \theta < (\pi/n)$, both u_r and u_θ are negative. The obtained streamlines and equipotential lines in a sector between $\theta = 0$ and $\theta = \pi/n$ are shown in Fig. 7.8a. For $n = 1$, Eq. (7.5.26) yields a rectilinear uniform flow. For $n = 2$, it gives the complex potential for the flow in a right-angled corner.

Flows around a sharp edge. The complex potential for the flow around a sharp edge, e.g. the edge of a flat plate, is obtained by letting $F(z)$ be proportional to $z^{1/2}$, viz.,

$$F(z) = cz^{1/2} = cr^{1/2}e^{i\theta/2}, \quad \longrightarrow \quad \phi = cr^{1/2}\cos\left(\frac{\theta}{2}\right), \quad \psi = cr^{1/2}\sin\left(\frac{\theta}{2}\right), \tag{7.5.28}$$

[10]It is readily to show that the circulation with any closed contour which does not include the singularity vanishes, and thus the flow is irrotational.

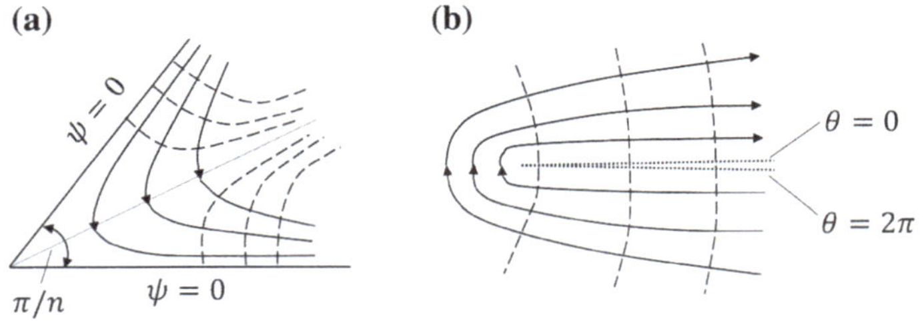

Fig. 7.8 Two-dimensional flows in a sector and around a sharp edge in the (*xy*)-plane. **a** A flow in a sector. **b** A flow around a sharp edge. The solid and dashed lines are respectively the streamlines and equipotential lines

where c is a real constant and $0 < \theta < 2\pi$. Thus, the radial lines of $\theta = 0$ and $\theta = 2\pi$ are the streamlines corresponding to $\psi = 0$, and the other streamlines are described by $cr^{1/2}\sin(\theta/2) = \text{constant}$. The direction of flow is determined by the complex velocity given by

$$\begin{aligned} W(z) &= \frac{c}{2r^{1/2}}\left[\cos\left(\frac{\theta}{2}\right) + i\sin\left(\frac{\theta}{2}\right)\right]e^{-i\theta}, \\ \longrightarrow u_r &= \frac{c}{2r^{1/2}}\cos\left(\frac{\theta}{2}\right), \quad u_\theta = -\frac{c}{2r^{1/2}}\sin\left(\frac{\theta}{2}\right). \end{aligned} \tag{7.5.29}$$

For $0 < \theta < \pi$, $u_r > 0$, $u_\theta < 0$, and for $\pi < \theta < 2\pi$, both u_r and u_θ assume negative values. The singular point of Eq. (7.5.28) is at the corner ($r = 0$), where the velocity components approach infinite. The flow field is shown in Fig. 7.8b.

Flows due to a doublet. Consider a source and a sink with same strength which locate on the real axis in a small distance ε from the origin, as shown in Fig. 7.9a. By using the principle of superposition, the complex potential is given by

$$F(z) = \frac{m}{2\pi}\ln(z+\varepsilon) - \frac{m}{2\pi}\ln(z-\varepsilon) = \frac{m}{2\pi}\ln\left(\frac{1+\varepsilon/z}{1-\varepsilon/z}\right). \tag{7.5.30}$$

It is assumed that $\varepsilon/z \sim 0$, so that the above expression can be approximated as

$$F(z) \sim \frac{m}{2\pi}\ln\left\{\left(1+\frac{\varepsilon}{z}\right)\left[1+\frac{\varepsilon}{z}+\mathcal{O}\left(\frac{\varepsilon^2}{z^2}\right)\right]\right\} = \frac{m}{2\pi}\ln\left[1+2\frac{\varepsilon}{z}+\mathcal{O}\left(\frac{\varepsilon^2}{z^2}\right)\right], \tag{7.5.31}$$

where the notation $\mathcal{O}(\varepsilon^2/z^2)$ denotes the terms of order (ε^2/z^2) or smaller. Since the sum of the second and third terms inside the bracket in the right-hand-side is much smaller than unity, Eq. (7.5.31) can be approximated by

$$F(z) = \frac{m}{2\pi}\left[2\frac{\varepsilon}{z}+\mathcal{O}\left(\frac{\varepsilon^2}{z^2}\right)\right]. \tag{7.5.32}$$

It is further assumed that $m \to \infty$ and $\varepsilon \to 0$ in such a way that $\lim_{\varepsilon\to 0}(m\varepsilon) = \pi\mu$, where μ is a constant. With these, Eq. (7.5.32) is simplified to

$$F(z) = \frac{\mu}{z} = \frac{\mu}{r}e^{-i\theta}. \tag{7.5.33}$$

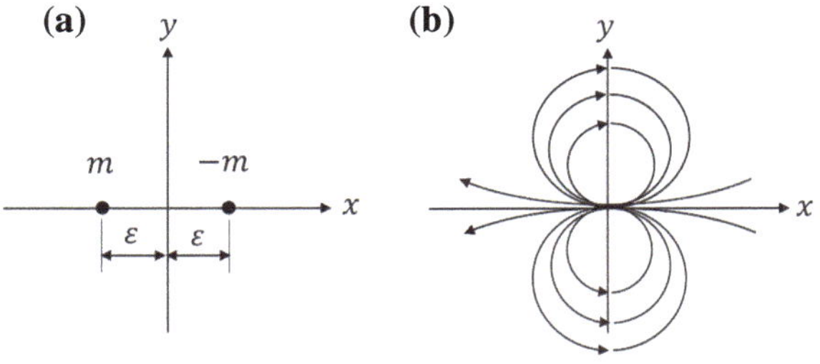

Fig. 7.9 Two-dimensional doublet flows in the (xy)-plane. **a** Superposition of a source and a sink. **b** The streamlines of a doublet flow

This equation is an equivalence of the superposition of a very strong source and a very strong sink which are very close together. It follows immediately that

$$F(z) = \frac{\mu \bar{z}}{z\bar{z}} = \frac{\mu(x - iy)}{x^2 + y^2}, \quad \longrightarrow \quad \phi = \frac{\mu x}{x^2 + y^2}, \quad \psi = -\frac{\mu y}{x^2 + y^2}. \tag{7.5.34}$$

The equation of streamlines is thus given by

$$x^2 + \left(y + \frac{\mu}{2\psi}\right)^2 = \left(\frac{\mu}{2\psi}\right)^2, \tag{7.5.35}$$

indicating a family of circles with radius $\mu/2\psi$ locating at $y = -\mu/2\psi$, as shown in Fig. 7.9b. The alternative expressions of velocity potential and stream functions are

Table 7.2 Velocity potential and stream functions, and velocity components of the elementary two-dimensional potential flows in terms of the polar coordinates

Flow field	ϕ	ψ	u_r	u_θ
Inclined uniform flows	$Ur\cos(\theta - \alpha)$	$Ur\sin(\theta - \alpha)$	$U\cos(\theta - \alpha)$	$U\sin(\theta - \alpha)$
Source flows	$\frac{m}{2\pi}\ln r$	$\frac{m}{2\pi}\theta$	$\frac{m}{2\pi r}$	0
Free vortex flows in counterclockwise rotation	$\frac{\Gamma}{2\pi}\theta$	$-\frac{\Gamma}{2\pi}\ln r$	0	$\frac{\Gamma}{2\pi r}$
Flows in sector	$Ur^n\cos(n\theta)$	$Ur^n\sin(n\theta)$	$nUr^{n-1}\cos(n\theta)$	$-nUr^{n-1}\sin(n\theta)$
Flows around sharp edge	$cr^{1/2}\cos\left(\frac{\theta}{2}\right)$	$cr^{1/2}\sin\left(\frac{\theta}{2}\right)$	$\frac{c}{2r^{1/2}}\cos\left(\frac{\theta}{2}\right)$	$-\frac{c}{2r^{1/2}}\sin\left(\frac{\theta}{2}\right)$
Doublet flows	$\frac{\mu\cos\theta}{r}$	$-\frac{\mu\sin\theta}{r}$	$-\frac{\mu\cos\theta}{r^2}$	$-\frac{\mu\sin\theta}{r^2}$

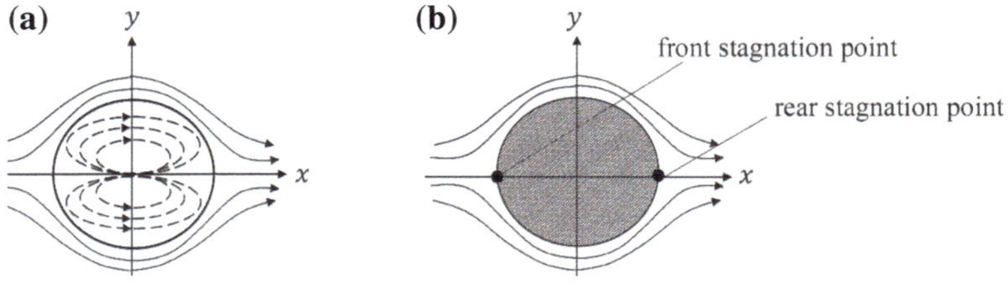

Fig. 7.10 Flow fields generated by the superposition of elementary potential-flow solutions in the (xy)-plane. **a** Superposition of a uniform and a doublet flows. **b** A uniform flow past a circular cylinder

given by $\phi = \mu \cos\theta/r$ and $\psi = -\mu \sin\theta/r$. The complex velocity in terms of the polar coordinates is then identified to be

$$\begin{aligned} W(z) &= -\frac{\mu}{r^2}(\cos\theta - i\sin\theta)e^{-i\theta}, \\ \longrightarrow u_r &= -\frac{\mu}{r^2}\cos\theta, \quad u_\theta = -\frac{\mu}{r^2}\sin\theta. \end{aligned} \tag{7.5.36}$$

The directions of streamlines in Fig. 7.9b are then determined by using the above equation. The flow field described by Eq. (7.5.33) is called a *doublet flow* with μ the *doublet strength*, whose single singularity locates at the origin, which is termed a *doublet*. The complex potential of a doublet flow locating at $z = z_0$ is obtained directly from Eq. (7.5.33) by changing z to $(z - z_0)$.

Table 7.2 summarizes the velocity potential and stream functions, and the velocity components of the elementary two-dimensional potential flows in the polar coordinate system.

7.5.4 Flows Around Circular Cylinder

The complex potential of a uniform flow along the positive x-axis around a doublet flow locating at the origin, by using the principle of superposition, is given by

$$F(z) = Uz + \frac{\mu}{z} = \left(Ua + \frac{\mu}{a}\right)\cos\theta + i\left(Ua - \frac{\mu}{a}\right)\sin\theta, \tag{7.5.37}$$

for a circle denoted by $z = a^{i\theta}$ with radius a. The doublet strength μ should be so chosen that this circle may become a streamline. To achieve this, the stream function corresponding to the circle $r = a$ is identified to be

$$\psi = \left(Ua - \frac{\mu}{a}\right)\sin\theta = 0, \qquad \longrightarrow \qquad \mu = Ua^2. \tag{7.5.38}$$

Thus, with $\mu = Ua^2$, the circle is identified as $\psi = 0$, which may become a streamline. As shown in Fig. 7.10a, the entire doublet flow field is inside the circle, while the uniform flow field is deflected by the doublet in such a way that it is entirely outside the circle. The circle is itself common to the two flow fields.

If a thin metal cylinder with radius a is placed perpendicularly into the uniform flow field in such a way that it coincides exactly with the streamline $\psi = 0$, the flow fields inside and outside the cylinder are not disturbed. Having done this, the flow field due to the doublet could be removed and the outer flow field would remain unchanged, as shown in Fig. 7.10b. Thus, for $r > a$, the flow field due to the doublet strength μ and uniform rectilinear flow of magnitude U gives the same flow field of a uniform flow of magnitude U past a circular cylinder with radius a, whose complex potential is then given by

$$F(z) = U\left(z + \frac{a^2}{z}\right). \tag{7.5.39}$$

There exist two *stagnation points* locating on the x-axis, where the kinetic energy of fluid is converted completely into the pressure. The upstream and downstream stagnation points are referred to as the *front* and *rear* stagnation points, respectively. However, Eq. (7.5.39) predicts no *drag* and *lift forces* due to the symmetries of flow field with respect to both x- and y-axes. This results from the fact that the viscous effect is neglected in the potential-flow theory. It will be shown in Sect. 8.4 that a thin boundary layer on the surface of cylinder is generated due to the viscous effect, and the resulting flow field is no longer symmetric with respect to the x-axis, giving rise to non-vanishing drag forces. Despite these, Eq. (7.5.39) still gives a valid solution outside the thin boundary layer and upstream of the vicinity of *separation point*. The solution also delivers the idealized flow situation which would be approached if the viscous effect is minimized.

Consider further the circumstance in which the established flow field is superposed by a clockwise-free vortex locating at the center of cylinder. Since the inclusion of a free vortex does not change the fact that the circle $r = a$ is a streamline, the complex potential is then given by

$$F(z) = U\left(z + \frac{a^2}{z}\right) + \frac{i\Gamma}{2\pi}\ln z + c, \tag{7.5.40}$$

where c is a constant used to maintain the conventional denotation that $\psi = 0$ on $r = a$. To evaluate the value of c, this equation is expressed in terms of the polar coordinates, which is subsequently applied to the circle $r = a$ to obtain

$$F(z) = 2Ua\cos\theta - \frac{\Gamma}{2\pi}\theta + \frac{i\Gamma}{2\pi}\ln a + c, \quad \longrightarrow \quad c = -\frac{i\Gamma}{2\pi}\ln a. \tag{7.5.41}$$

With this, the complex potential becomes

$$F(z) = U\left(z + \frac{a^2}{z}\right) + \frac{i\Gamma}{2\pi}\ln\left(\frac{z}{a}\right), \tag{7.5.42}$$

which describes a uniform rectilinear flow of magnitude U approaching a circular cylinder of radius a having a clockwise vortex with strength Γ around it.

The complex velocity in terms of the polar coordinates is obtained as

$$W(z) = \left\{U\left(1 - \frac{a^2}{r^2}\right)\cos\theta + i\left[U\left(1 + \frac{a^2}{r^2}\right)\sin\theta + \frac{\Gamma}{2\pi r}\right]\right\}e^{-i\theta}, \tag{7.5.43}$$

by which the velocity components are given by

$$u_r = U\left(1 - \frac{a^2}{r^2}\right)\cos\theta, \qquad u_\theta = -U\left(1 + \frac{a^2}{r^2}\right)\sin\theta - \frac{\Gamma}{2\pi r}. \tag{7.5.44}$$

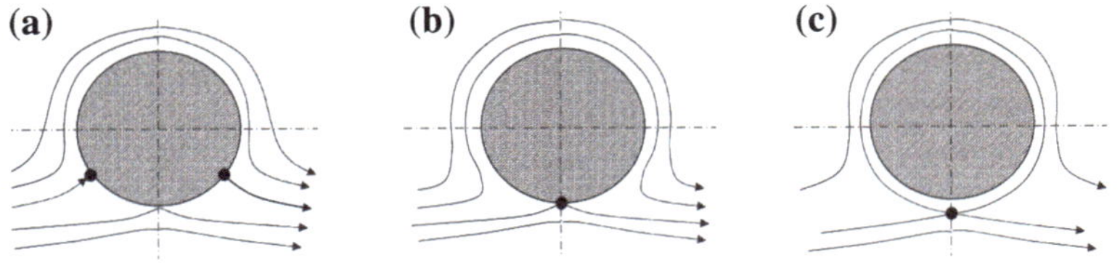

Fig. 7.11 Uniform flows with velocity U around a circular cylinder of radius a with clockwise circulation. **a** $0 < \Gamma/(4\pi Ua) < 1$. **b** $\Gamma/(4\pi Ua) = 1$. **c** $\Gamma/(4\pi Ua) > 1$

Applying these equations to the surface of cylinder yields

$$u_r = 0, \qquad u_\theta = -2U\sin\theta - \frac{\Gamma}{2\pi a}, \tag{7.5.45}$$

which indicates that $u_r = 0$ at $r = a$, as expected, since the circle represents the boundary condition with vanishing velocity component normal to the solid surface. The locations of the stagnation point at which all velocity components vanish are identified to be

$$\sin\theta_s = -\frac{\Gamma}{4\pi Ua}, \tag{7.5.46}$$

with θ_s denoting the value of θ corresponding to the stagnation point. For $\Gamma = 0$, $\theta_s = 0$ and $\theta_s = \pi$, which agree with the stagnation points of a uniform flow past a circular cylinder without circulation.

For non-vanishing values of Γ, the values of θ_s are determined as follows: For $0 < \Gamma/(4\pi Ua) < 1$, $\sin\theta_s < 0$, leading to that θ_s must locate in the third and fourth quadrants of the two-dimensional coordinate plane, as shown in Fig. 7.11a. There exist two stagnation points, and the points locating in the third and fourth quadrants correspond to the stagnation points $\theta = \pi$ and $\theta = 0$ of the non-circulating case, respectively. Furthermore, these two stagnation points are symmetric with respect to the y-axis, for $\sin\theta_s$ assumes a negative constant value. The physical interpretations of these outcomes are that since the circulation is clockwise, the flows due to the vortex and doublet are reinforced in the first and second quadrants, while two flow fields oppose each other in the third and fourth quadrants, so that at some points in these regions the net velocity is null. It follows that a negative circulation around the cylinder makes the front and real stagnation points approach each other in the lower surface of cylinder and vice versa.

For $\Gamma/(4\pi Ua) = 1$, $\sin\theta_s = -1$, and hence $\theta_s = 3\pi/2$. Thus, there exists a single stagnation point, with the corresponding flow field shown in Fig. 7.11b. In this circumstance, the front and rear stagnation points are brought together by the enhanced strength of bounded vortex such that they coincide to form a single stagnation point at the bottom surface of cylinder.

If $\Gamma > 4\pi Ua$, it is not possible to maintain a single stagnation point on the cylinder surface, and the stagnation point will move off into the fluid as either a single or two stagnation points. For this circumstance, the velocity components given in Eq. (7.5.44) must be satisfied by the coordinates (r_s, θ_s) of stagnation point. Since $r_s \neq a$,

Eq. $(7.5.44)_1$ yields $\theta_s = \pi/2$ or $\theta_s = 3\pi/2$. Substituting these values of θ_s into Eq. $(7.5.44)_2$ gives rise to

$$U\left(1+\frac{a^2}{r_s^2}\right)\sin\theta_s = \mp\frac{\Gamma}{2\pi r_s}, \tag{7.5.47}$$

where the minus and positive signs correspond to the cases of $\theta_s = \pi/2$ and $\theta_s = 3\pi/2$, respectively. To maintain the dimensional homogeneity for positive values of U and Γ, the minus sign must be rejected.[11] With this, Eq. (7.5.47) is recast alternatively as

$$\frac{r_s}{a} = \frac{\Gamma}{4\pi Ua}\left[1 \pm \sqrt{1-\left(\frac{4\pi Ua}{\Gamma}\right)^2}\,\right], \tag{7.5.48}$$

which is expanded to

$$\frac{r_s}{a} = \frac{\Gamma}{4\pi Ua}\left\{1 \pm \left[1-\frac{1}{2}\left(\frac{4\pi Ua}{\Gamma}\right)^2 + \cdots\right]\right\}. \tag{7.5.49}$$

The minus sign of this equation, however, leads to the result that $r_s \to 0$ as $4\pi Ua/\Gamma \to 0$, yielding the stagnation point locating inside the cylinder, which contradicts to the physical situation. Thus, the minus sign in Eq. (7.5.48) or (7.5.49) must be rejected. The coordinates of stagnation point in the fluid outside the cylinder are then identified to be

$$\theta_s = \frac{3\pi}{2}, \qquad \frac{r_s}{a} = \frac{\Gamma}{4\pi Ua}\left[1 + \sqrt{1-\left(\frac{4\pi Ua}{\Gamma}\right)^2}\,\right], \tag{7.5.50}$$

giving rise to a single stagnation point, with the corresponding flow field shown in Fig. 7.11c. It is seen that there is a portion of the fluid which perpetually encircles the cylinder.

In the previous discussions, the flow fields are symmetric to the y-axis, yielding no drag force, as the same in the case of no circulation around the cylinder. However, the flow fields are not symmetric with respect to the x-axis, implying that there exists a non-vanishing lift force acting on the cylinder, which will be explored in the next section.

7.5.5 Blasius' Integral Laws

Consider an arbitrarily shaped body in contact with a fluid in two-dimensional circumstance shown in Fig. 7.12, in which the body surface is denoted by C_1, and C_0 represents any closed surface embracing the entire body. The force components in the x- and y-directions and moment acting on the body by the surrounding fluid are

[11] This makes sense, for in the previous case it has been demonstrated that $\theta_s = 3\pi/2$ at $\Gamma/(4\pi Ua) = 1$. If the minus sign is used, it will lead to a large jump of θ_s for a small change of Γ.

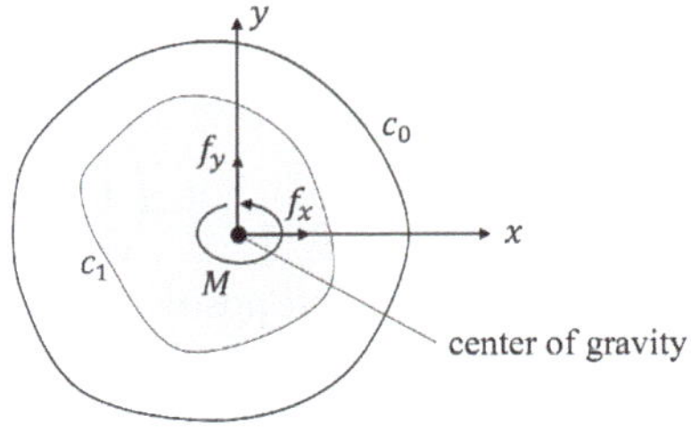

Fig. 7.12 Illustration of Blasius' integral laws for an arbitrarily shaped body with surface C_1 surrounded by a fluid, and an arbitrary surface C_0 embracing the entire body

denoted by f_x, f_y, and M, respectively. Choosing the region between C_0 and C_1 as the finite control-volume and applying the integral balance of linear momentum to the control-volume yield

$$-f_x - \int_{C_0} p\,\mathrm{d}y = \int_{C_0} \rho u(u\,\mathrm{d}y - v\,\mathrm{d}x), \quad -f_y + \int_{C_0} p\,\mathrm{d}x = \int_{C_0} \rho v(u\,\mathrm{d}y - v\,\mathrm{d}x),$$
$$-M + \int_{C_0} \big[px\,\mathrm{d}x + py\,\mathrm{d}y + \rho uy(u\,\mathrm{d}y - v\,\mathrm{d}x) - \rho vx(u\,\mathrm{d}y - v\,\mathrm{d}x)\big] = 0, \tag{7.5.51}$$

where the origin of coordinate system locates at the center of gravity of the body, and a line element of C_0 with a positive slope and no linear momentum transfer across C_1 are assumed for simplicity. By using the Bernoulli equation given by

$$p + \frac{1}{2}\rho\left(u^2 + v^2\right) = B, \tag{7.5.52}$$

Equation (7.5.51) can be expressed alternatively by eliminating its pressure, viz.,

$$f_x = \rho \int_{C_0} \left[uv\,\mathrm{d}x - \frac{1}{2}\left(u^2 - v^2\right)\mathrm{d}y\right], \quad f_y = -\rho \int_{C_0} \left[uv\,\mathrm{d}y + \frac{1}{2}\left(u^2 - v^2\right)\mathrm{d}x\right],$$
$$M = -\frac{\rho}{2} \int_{C_0} \left[\left(u^2 - v^2\right)(x\,\mathrm{d}x - y\,\mathrm{d}y) + 2uv(x\,\mathrm{d}y + y\,\mathrm{d}x)\right], \tag{7.5.53}$$

where it is noted that $\int_{C_0} B\mathrm{d}x = \int_{C_0} B\mathrm{d}y = 0$ around any closed contour C_0.

Equation (7.5.53) can be further simplified by using the complex velocity. Conducting the following two complex integrals

$$i\frac{\rho}{2} \int_{C_0} W^2\mathrm{d}z = i\frac{\rho}{2} \int_{C_0} (u - iv)^2(\mathrm{d}x + i\,\mathrm{d}y) = f_x - if_y,$$
$$Re\left(\frac{\rho}{2} \int_{C_0} zW^2\mathrm{d}z\right) = Re\left[\frac{\rho}{2} \int_{C_0} (x + iy)(u - iv)^2(\mathrm{d}x + i\,\mathrm{d}y)\right] = -M, \tag{7.5.54}$$

indicates that the force components and moment acting on the body can be evaluated by using the complex integrals given by

$$f_x - if_y = i\frac{\rho}{2} \int_{C_0} W^2\mathrm{d}z, \qquad M = -\frac{\rho}{2} Re\left(\int_{C_0} zW^2\mathrm{d}z\right), \tag{7.5.55}$$

where M is positive if it acts in the clockwise direction. These results are known as *Blasius' integral laws.*[12] The contour integrals in determining the force components

[12]Paul Richard Heinrich Blasius, 1883–1970, a German fluid dynamics physicist, who was one of the first students of Prandtl and contributed to a mathematical base for boundary-layer drag.

and moment are usually evaluated by using the residue theorem in the complex analysis.

As an illustration of Blasius' laws, consider the rectilinear flow around a circular cylinder with circulation in the last section. It follows from the complex velocity given in Eq. (7.5.43) that

$$W^2(z) = U^2 - \frac{2U^2a^2}{z^2} + \frac{U^2a^4}{z^4} + \frac{iU\Gamma}{\pi z} - \frac{iU\Gamma a^2}{\pi z^3} - \frac{\Gamma^2}{4\pi^2 z^2}. \tag{7.5.56}$$

Substituting this expression into Blasius' laws yields

$$f_x - if_y = i\frac{\rho}{2}\int_{C_0} W^2 dz = i\frac{\rho}{2}\left(2\pi i \sum \alpha\right), \tag{7.5.57}$$

where α is the residue of $W^2(z)$ inside C_0. Since Eq. (7.5.56) has only a single singular point at $z = 0$, it is already the Laurent series of $W^2(z)$ at $z = 0$, and the single residue is the coefficient of the term $1/z$. With these, Eq. (7.5.57) becomes

$$f_x - if_y = -i\rho U\Gamma, \qquad \longrightarrow \qquad f_x = 0, \;\; f_y = \rho U\Gamma. \tag{7.5.58}$$

This equation is known as the *Kutta-Joukowski law*,[13] and the phenomenon of non-vanishing lifting force acting on a rotating circular cylinder is called the *Magnus effect*.[14] For clockwise and counterclockwise circulations, f_y assumes positive and negative values, pointing consequently in the y- and negative y-directions, respectively. No lift force is generated if the cylinder has no circulation acting on it.

Physically, the flow direction of approaching flow and that induced by the cylinder with clockwise circulation shown in Fig. 7.11a are nearly in parallel with the upper region near the cylinder, while they are opposed in the regions below the cylinder. It follows from the Bernoulli equation that the pressure below the cylinder is larger than that above the cylinder, resulting in a vertical force acting in the y-direction. A reverse circumstance takes place if the cylinder is associated with a counterclockwise circulation, giving rise to a vertical force acting in the negative y-direction.

To evaluate the moment, the quantity zW^2 is determined to be

$$zW^2(z) = U^2 z - \frac{2U^2a^2}{z} + \frac{U^2a^4}{z^3} + \frac{iU\Gamma}{\pi} - \frac{iU\Gamma a^2}{\pi z^2} - \frac{\Gamma^2}{4\pi^2 z}, \tag{7.5.59}$$

[13]Martin Wilhelm Kutta, 1867–1944, a German mathematician, who codeveloped the *Runge-Kutta method* in the numerical analysis of ordinary differential equations. Nikolay Yegorovich Zhukovsky (or Joukowski), 1847–1921, a Russian scientist and mathematician, who was a founding father of modern aero- and hydrodynamics and was often called "Father of Russian Aviation".

[14]Heinrich Gustav Magnus, 1802–1870, a German experimental scientist, who discovered the first *platino-ammonium class of compounds*, which is also called the *Magnus green salt*.

by which the moment is obtained as

$$M = -\frac{\rho}{2} Re \left(\int_{C_0} zW^2 dz \right) = -\frac{\rho}{2} Re \left[2\pi i \left(-2U^2a^2 - \frac{\Gamma}{4\pi^2} \right) \right] = 0. \quad (7.5.60)$$

This result is justified, for the pressure distribution on the cylinder surface is symmetric with respect to the y-axis.

7.5.6 The Joukowski Transformation

The *Joukowski transformation* is one of the conformal transformations between the z- and ζ-planes, which is given by

$$z = \zeta + \frac{c^2}{\zeta}, \quad (7.5.61)$$

where c^2 is a constant, frequently be taken to be real. Three properties are associated with the Joukowski transformation. The first property is that for large values of $|\zeta|$, $z \to \zeta$, as implied by the equation. This means that in the region far from the origin, the transformation becomes an identity mapping; namely, the complex velocity becomes indifferent. For example, if a uniform flow with certain magnitude approaches a body at some angle of attack in the ζ-plane, a uniform flow with the same magnitude and angle of attack approaches the corresponding body in the z-plane far from the origin.

The second property is that there exists a single singular point locating at $\zeta = 0$, which is verified by the derivative of z with respect to ζ given by

$$\frac{dz}{d\zeta} = 1 - \frac{c^2}{\zeta^2}. \quad (7.5.62)$$

Since flows around some bodies are normally dealt with, this singularity locates generally within the body and is of no consequence. As implied by the above equation, $dz/d\zeta$ vanishes at $\zeta = c$ and $\zeta = -c$. This marks two *critical points*, and it is possible that smooth curves passing two critical points in the ζ-plane may become corners in the z-plane. For example, consider an arbitrary point z in the z-plane and its corresponding mapping ζ in the ζ-plane, as shown in Fig. 7.13a. It follows from Eq. (7.5.61) that the corresponding points of $\zeta = \pm c$ on the ζ-plane locate at $z = \pm 2c$ on the z-plane, and

$$\frac{z - 2c}{z + 2c} = \left(\frac{\zeta - c}{\zeta + c} \right)^2, \quad \longrightarrow \quad \frac{r_1}{r_2} e^{i(\theta_1 - \theta_2)} = \left(\frac{\rho_1}{\rho_2} \right)^2 e^{i2(\nu_1 - \nu_2)}, \quad (7.5.63)$$

where θ and ν are the angles measured respectively in the z- and ζ-planes. It follows that

$$\frac{r_1}{r_2} = \left(\frac{\rho_1}{\rho_2} \right)^2, \quad \theta_1 - \theta_2 = 2(\nu_1 - \nu_2), \quad (7.5.64)$$

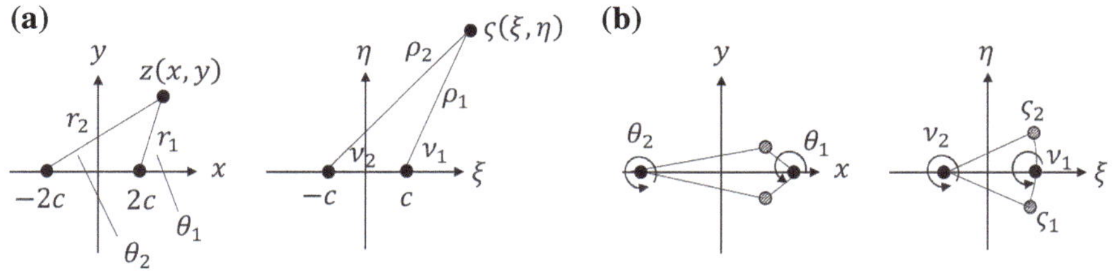

Fig. 7.13 Joukowski transformation. **a** The coordinates of critical points in the z- and ζ-planes. **b** A coordinate change corresponding to a smooth curve passing through $\zeta = c$

showing that if a smooth curve passes through point $\zeta = c$ in the ζ-plane, its corresponding curve in the z-plane will form a knife-edge or cusp. Let an infinitesimal smooth curve pass through point ζ_1 toward point ζ_2, as shown in Fig. 7.13b. The angle ν_1 changes from $3\pi/2$ to $\pi/2$, while the angle ν_2 changes from 2π to 0. These yield that the value of $(\nu_1 - \nu_2)$ changes from $-\pi/2$ to $\pi/2$, giving rise to an angle difference of π. The corresponding angle change in the z-plane thus becomes $(\theta_1 - \theta_2) = 2\pi$, implying a knife-edge or cusp. Consequently, if a smooth curve passes through either of the critical points $\zeta = \pm c$ in the ζ-plane, its corresponding curve in the z-plane will contain a knife-edge at the corresponding points of $z = \pm 2c$.[15]

The third property follows directly from the definition. For any value of ζ, Eq. (7.5.61) yields the same value of z for that value of ζ and also for c^2/ζ. Consequently, the Joukowski transformation is not a one-to-one mapping, but is a double-valued transformation.[16] In fluid mechanics, this double-valued property does not usually arise, for the mapping of points $|\zeta| < c$ usually lies inside some body about which the flow is to be studied, so that these points are not in the flow field in the z-plane.

As an application of the Joukowski transformation, consider a circle with radius $a > c$ in the ζ-plane which is to be centered at the origin and surrounded by an inclined uniform flow, as shown in Fig. 7.14b. By using Eq. (7.5.61), the corresponding curve in the z-plane is obtained as

$$z = ae^{i\nu} + \frac{c^2}{a}e^{-i\nu} = \left(a + \frac{c^2}{a}\right)\cos\nu + i\left(a - \frac{c^2}{a}\right)\sin\nu = x + iy, \qquad (7.5.65)$$

[15]For example, consider a circle with radius c locating in the origin of ζ-plane. The circle passes through two critical points $\zeta = \pm c$. By using the Joukowski transformation, the points on the circle are described by

$$z = ce^{i\nu} + ce^{-i\nu} = 2c\cos\nu,$$

indicating that these points form the strip of $y = 0$, $x = 2\nu\cos\nu$ in the z-plane. It is readily verified that the points outside the circle $|\zeta| = c$ in the ζ-plane cover the entire z-plane, so behave the points inside the circle $|\zeta| = c$. That is, the Joukowski transformation is a double-valued mapping.

[16]Mathematically, this is resolved by connecting two critical points $\zeta = \pm c$ via a branch cut along the x-axis in the z-plane, creating two Riemann sheets.

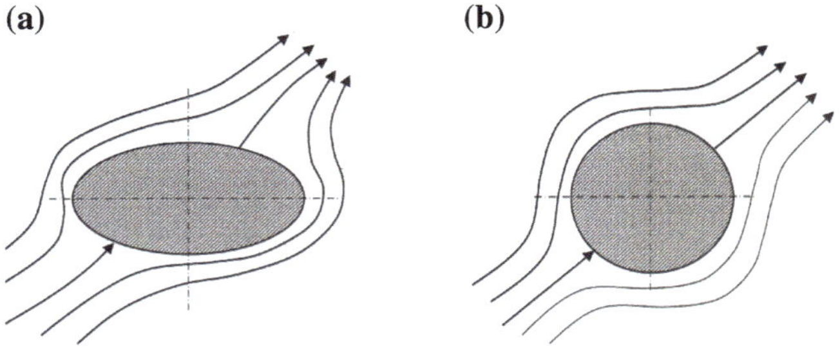

Fig. 7.14 Flows around an ellipse as an illustration of the Joukowski transformation. **a** An inclined uniform flow passing an ellipse in the z-plane. **b** An inclined uniform flow passing a circular cylinder in the ζ-plane

giving rise to

$$\left(\frac{x}{a+c^2/a}\right)^2+\left(\frac{y}{a-c^2/a}\right)^2=1, \tag{7.5.66}$$

which is the equation of an ellipse whose major semi-axis is of length $a+c^2/a$ aligned along the x-axis and minor semi-axis of length $a-c^2/a$ aligned along the y-axis. Thus, the ellipse is the corresponding curve in the z-plane of the circle with radius a in the ζ-plane.

Since the complex potential of an inclined uniform flow with magnitude U approaching a circular cylinder with radius a by an attack angle α in the ζ-plane is given by

$$F(\zeta)=U\left(\zeta e^{-i\alpha}+\frac{a^2}{\zeta}e^{i\alpha}\right), \tag{7.5.67}$$

its corresponding expression in the z-plane is obtained as

$$F(z)=U\left\{\left[z-\frac{z}{2}+\sqrt{\left(\frac{z}{2}\right)^2-c^2}\,\right]e^{-i\alpha}+\frac{a^2}{c^2}\left[\frac{z}{2}-\sqrt{\left(\frac{z}{2}\right)^2-c^2}\,\right]e^{i\alpha}\right\}, \tag{7.5.68}$$

in which ζ is replaced by $\zeta=z/2+\sqrt{(z/2)^2-c^2}$, as indicated by Eq. (7.5.65), for $\zeta\to z$ for large values of z. By writing $z/2$ as $z-z/2$ in the first term on the right-hand-side, the above equation can be further simplified to

$$F(z)=U\left[ze^{-i\alpha}+\left(\frac{a^2}{c^2}e^{i\alpha}-e^{-i\alpha}\right)\left(\frac{z}{2}-\sqrt{\left(\frac{z}{2}\right)^2-c^2}\right)\right], \tag{7.5.69}$$

which is the complex potential of an inclined uniform flow with attack angle α and magnitude U around an ellipse on the z-plane, as shown in Fig. 7.14a. The complex potential consists of two parts: that corresponding to an inclined uniform flow with attack angle α to the reference x-axis and that due to the perturbation which is larger near the ellipse but vanishes for large values of z. Since in the ζ-plane there exist

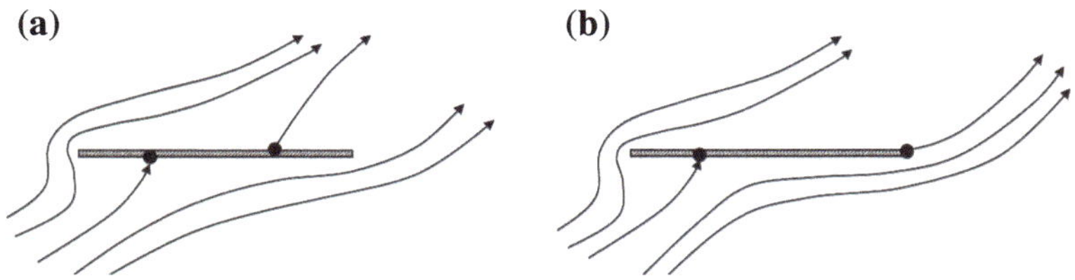

Fig. 7.15 Stagnation points on a flat-plate airfoil. **a** A flat plate without circulation. **b** A flat plate with circulation by the Kutta condition

two stagnation points locating at $\zeta \pm ae^{i\alpha}$, the corresponding stagnation points in the z-plane are identified to be

$$z = \pm ae^{i\alpha} \pm \frac{c^2}{a}e^{-i\alpha} = \pm\left(a + \frac{c^2}{a}\right)\cos\alpha \pm i\left(a - \frac{c^2}{a}\right)\sin\alpha, \quad (7.5.70)$$

which gives

$$x = \pm\left(a + \frac{c^2}{a}\right)\cos\alpha, \qquad y = \pm\left(a - \frac{c^2}{a}\right)\sin\alpha. \quad (7.5.71)$$

For $\alpha = 0$, Eq. (7.5.69) describes a uniform rectilinear flow approaching a horizontally oriented ellipse, with two stagnation points locating on the x-axis with the coordinates $(a + c^2/a, 0)$ and $(-a - c^2/a, 0)$ by using Eq. (7.5.71), corresponding to the physical observation. For $\alpha = \pi/2$, Eq. (7.5.69) describes a uniform vertical flow approaching the same oriented ellipse. In this case, two stagnation points locate on the y-axis with the coordinates $(0, a - c^2/a)$ and $(0, -a + c^2/a)$.

The Joukowski transformation is one of the most important transformations in the study of fluid mechanics. By means of this transformation and the elementary flow the solutions discussed previously, it is possible to obtain the solutions to more complex flow fields, e.g. the flows around a family of airfoils, to be discussed in the next section.

7.5.7 Theory of Airfoils

Flat-plate airfoil and the Kutta condition. In the previous case of a flow around an ellipse, if the constant c approaches the radius of circle a in the ζ-plane, the resulting ellipse in the z-plane degenerates to a flat plate defined by the strip $-2a \le x \le 2a$. It follows from Eq. (7.5.71) that two stagnation points on the z-plane locate at $x = \pm 2a\cos\alpha$, which are shown in Fig. 7.15a with the corresponding flow field. Since the flat plate has an attack angle with respect to the approaching flow, the upstream stagnation point locates on the lower surface, while the downstream stagnation point locates on the upper surface. The flows around the leading and trailing edges, however, are associated with a sharp edge having vanishing radius of curvature. This results in the infinite velocity components in these regions, which is physically impossible.

In practice, real air foils have finite thickness, and thus a finite radius of curvature possibly exists at the leading edge. However, the trailing edge is usually quite sharp, so infinite velocity components still exist there. This inconsistency would be overcome if the downstream stagnation point was actually at the trailing edge. This would be accomplished if a circulation exists around the flat plate with the required strength to rotate the rear stagnation point to the trailing edge. This condition is referred to as the *Kutta condition*, which reads: "*for bodies with sharp trailing edges which are at small attack angles to the free stream, the flow will adjust itself in such a way that the rear stagnation point coincides with the trailing edge*".

In the ζ-plane, the rear stagnation point locates at $\zeta = ae^{i\alpha}$. However, in view of the Kutta condition, it should be located at point $z = 2a$, corresponding to $\zeta = a$ in the ζ-plane. It follows immediately that the downstream stagnation point of circular cylinder in the ζ-plane should be rotated clockwise through the angle α. The strength of clockwise circulation which can accomplish this rotation, in view of Eq. (7.5.46), is given by

$$\Gamma = 4\pi Ua \sin\alpha, \tag{7.5.72}$$

by which the complex potential in the ζ-plane becomes

$$F(\zeta) = U\left(\zeta e^{-i\alpha} + \frac{a^2}{\zeta}e^{i\alpha}\right) + i2Ua \sin\alpha \ln\frac{\zeta}{a}. \tag{7.5.73}$$

Since the Joukowski transformation in the considered circumstance is given by

$$z = \zeta + \frac{a^2}{\zeta}, \qquad \longrightarrow \qquad \zeta = \frac{z}{2} + \sqrt{\left(\frac{z}{2}\right)^2 - a^2}, \tag{7.5.74}$$

where the second equation is given to meet the condition that $\zeta \to z$ as $z \to \infty$, substituting it into Eq. (7.5.73) yields

$$F(z) = U\left\{\left[\chi + \sqrt{\chi^2 - a^2}\right]e^{-i\alpha} + \frac{a^2 e^{i\alpha}}{\chi + \sqrt{\chi^2 - a^2}} + i2a \sin\alpha \ln\left[\frac{1}{a}\left(\chi + \sqrt{\chi^2 - a^2}\right)\right]\right\}, \tag{7.5.75}$$

where $\chi = z/2$. The flow field corresponding to the obtained complex potential in the z-plane is shown in Fig. 7.15b. Although the flow at the trailing edge becomes regular, the singularity at the leading edge still exists. In reality, an actual flow configuration indicates that the fluid would separate at the leading edge and reattach again on the top surface of airfoil. The streamline $\psi = 0$ would then correspond to a finite curvature, and the velocity components would remain finite at the leading edge.

The lift force f_y generated by the flat-plate airfoil with circulation may be determined by using the *Kutta-Joukowski law* given by

$$f_y = 4\pi\rho U^2 a \sin\alpha, \tag{7.5.76}$$

which is recast alternatively in terms of the dimensionless *lift coefficient* C_L as

$$C_L = \frac{2f_y}{\rho U^2 \ell}, \tag{7.5.77}$$

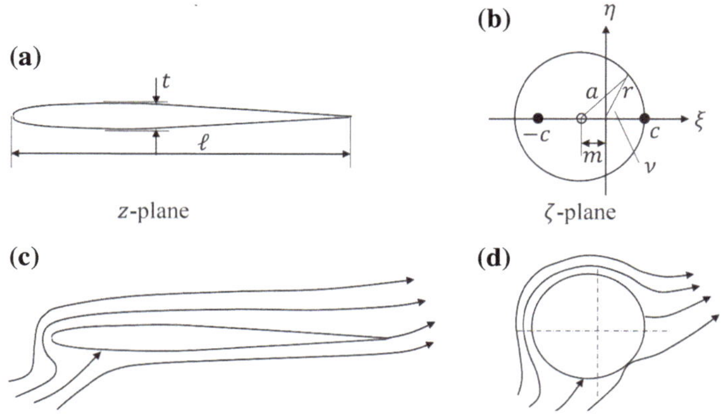

Fig. 7.16 The symmetric Joukowski airfoil. **a** The mapping of the offset circle in the z-plane. **b** The offset circle in the ζ-plane. **c** The flow field around a symmetric Joukowski airfoil in the z-plane. **d** The flow field around an offset circular cylinder with circulation in the ζ-plane

where ℓ is the length or *chord* of airfoil, given by $\ell = 4a$ in the z-plane, which yields $C_L = 2\pi \sin\alpha$. For small attack angles, $\sin\alpha \sim \alpha$, and hence the lift coefficient increases proportionally with α. This result is very close to the experimental outcomes and justifies the Kutta condition. If the Kutta condition were not valid, there would be no circulation around the flat plate, and no lift force would be generated.

The Symmetric Joukowski airfoil. The Joukowski transformation in conjunction with a series of circles in the ζ-plane whose centers are slightly displaced from the origin generates a family of airfoils, which are referred to as the *Joukowski family of airfoils*. Consider a circle on the ζ-plane whose center locates on the ξ-axis with an offset $-m$ from the origin, where m is a real constant, as shown in Fig. 7.16b. With this, the radius a of circle is chosen to be

$$a = c + m = c(1+\varepsilon), \qquad 0 \le \varepsilon = \frac{m}{c} \ll 1, \tag{7.5.78}$$

where c is the *Joukowski constant*. This is so chosen in order to let the critical point $\zeta = -c$ be contained inside the body to have a finite radius of curvature at the leading edge in the z-plane to avoid infinite velocity components. Equally, the radius a is so determined to let the circle pass through the critical point $\zeta = c$ to have a sharp trailing edge in the z-plane. With these, a *symmetric Joukowski airfoil* is established in the z-plane, as shown in Fig. 7.16a. A symmetric Joukowski airfoil is characterized by its chord ℓ and *maximum thickness t*. By choosing different values of ε, the Joukowski airfoils with different ℓ and t can be generated.

Substituting $\zeta = c$ and $\zeta = -(c + 2m)$ in to the Joukowski transformation yields

$$z = 2c, \qquad z = -c(1+2\varepsilon) - \frac{c}{1+2\varepsilon}, \tag{7.5.79}$$

with which the chord ℓ is obtained as

$$\ell = 4c, \tag{7.5.80}$$

in which a linearization has been conducted for the variations in ε, for it is assumed that $\varepsilon \ll 1$. The above equation indicates that within a first-order approximation of ε, the length of airfoil in the z-plane is unchanged by the shifting of the center of circle in the ζ-plane. It follows from Fig. 7.16b that

$$a^2 = r^2 + m^2 - 2rm\,[\cos(\pi - \nu)] = r^2 + m^2 + 2rm\cos\nu, \tag{7.5.81}$$

which is rewritten as

$$(c+m)^2 = r^2\left(1 + \frac{m^2}{r^2} + 2\frac{m}{r}\cos\nu\right), \tag{7.5.82}$$

for $a = c + m$. Since $r \geq c$, it follows that $m/r \leq m/c$, so that with a first-order approximation of ε, the term m^2/r^2 in the above equation may be neglected. With this, the equation of circle in the ζ-plane becomes

$$r = c\,[1 + \varepsilon(1 - \cos\nu)]\,. \tag{7.5.83}$$

Substituting this expression into the Joukowski transformation gives

$$z = c\,[1 + \varepsilon(1 - \cos\nu)]\,e^{i\nu} + \frac{ce^{-i\nu}}{1 + \varepsilon(1 - \cos\nu)}, \tag{7.5.84}$$

which describes the surface of a symmetric Joukowski airfoil in the z-plane. Again, with a linearization of ε, Eq. (7.5.84) is simplified to

$$z = c\,[2\cos\nu + i2\varepsilon(1 - \cos\nu)\sin\nu]\,, \tag{7.5.85}$$

yielding the parametric equations of airfoil surface, viz.,

$$x = 2c\cos\nu, \qquad y = 2c\varepsilon(1 - \cos\nu)\sin\nu. \tag{7.5.86}$$

Combining these two equations yields the conventional expression of airfoil surface in the form

$$y = \pm 2c\varepsilon\left(1 - \frac{x}{2c}\right)\sqrt{1 - \left(\frac{x}{2c}\right)^2}. \tag{7.5.87}$$

The location where y assumes an extreme value is determined by $\mathrm{d}y/\mathrm{d}\nu = 0$, yielding $\cos(2\nu) = \cos\nu$, and hence $\nu = 0$, $\nu = 2\pi/3$, and $4\pi/3$. Since $\nu = 0$ corresponds to the trailing edge, the values of $\nu = 2\pi/3$ and $4\pi/3$ are chosen. Thus, the points of maximum y locate at

$$x = -c, \qquad y = \pm\frac{3\sqrt{3}}{2}c\varepsilon \qquad \longrightarrow \qquad t = 3\sqrt{3}c\varepsilon, \tag{7.5.88}$$

from which the maximum thickness is expressed alternatively as

$$\frac{t}{\ell} = \frac{3\sqrt{3}}{4}\varepsilon. \tag{7.5.89}$$

This result indicates that the thickness-to-chord ratio of an airfoil is proportional to ε, which is the ratio of the offset of the center of circle in the ζ-plane, m, to the Joukowski constant c. Since the airfoil thickness is thought of as being specified, it is conventionally to express Eq. (7.5.89) in the form

$$\varepsilon = \frac{4}{3\sqrt{3}}\frac{t}{\ell} = 0.77\frac{t}{\ell}, \tag{7.5.90}$$

so that Eq. (7.5.87) becomes

$$\frac{y}{t} = \pm 0.385\left(1 - 2\frac{x}{\ell}\right)\sqrt{1 - \left(2\frac{x}{\ell}\right)^2}, \tag{7.5.91}$$

where the maximum and minimum values of y/t are 0.5 and -0.5, respectively, which take place at $x = -c$.

In order to satisfy the Kutta condition, it follows from Eq. (7.5.72) that

$$\Gamma = \pi U \ell \left(1 + 0.77\frac{t}{\ell}\right)\sin\alpha, \tag{7.5.92}$$

which is the required circulation strength to rotate the rear stagnation point to the trailing edge. The lift force, in view of the Kutta-Joukowski law, is then obtained as

$$f_y = \pi\rho U^2 \ell \left(1 + 0.77\frac{t}{\ell}\right)\sin\alpha, \quad \longrightarrow \quad C_L = 2\pi\left(1 + 0.77\frac{t}{\ell}\right)\sin\alpha. \tag{7.5.93}$$

As $t/\ell \to 0$, Eq. $(7.5.93)_2$ coincides exactly to that of a flat-plate airfoil, and it is seen that the finite thickness of an airfoil tends to increase the lift coefficient. However, this result cannot be used to produce high lift coefficients via thicker airfoils, for the flow tends to separate from bluff bodies much more readily than it does from streamlined bodies. This separation goes back to the viscous effect, which will be discussed in Sect. 8.4.8.

Since the center of circle in the ζ-plane locates at $\zeta = -m$, the complex potential of an inclined uniform flow passing a displaced circular cylinder with clockwise circulation in the ζ-plane is given by

$$F(\zeta) = U\left[(\zeta + m)e^{-i\alpha} + \frac{a^2}{\zeta + m}e^{i\alpha}\right] + \frac{i\Gamma}{2\pi}\ln\left(\frac{\zeta + m}{a}\right), \tag{7.5.94}$$

with

$$a = \frac{\ell}{4} + 0.77\frac{tc}{\ell}, \qquad m = 0.77\frac{tc}{\ell}, \tag{7.5.95}$$

where Γ is given in Eq. (7.5.92) and $c = \ell/4$. The flow fields in the z- and ζ-planes are shown respectively in Figs. 7.16c and d.

Circular-arc airfoil. Consider a circle with radius $a > c$ in the ζ-plane which locates on the η-axis with a distance m from the origin, as shown in Fig. 7.17b. In order to have a sharp trailing edge of the airfoil in the z-plane, the circle should pass through the critical point $\zeta = c$. However, in the considered circumstance, the circle also passes the critical point $\zeta = -c$, so that the leading edge of airfoil is also sharp, as shown in Fig. 7.17a. Substituting $\zeta = re^{i\nu}$ into the Joukowski transformation gives

$$z = \left(r + \frac{c^2}{r}\right)\cos\nu + i\left(r - \frac{c^2}{r}\right)\sin\nu, \tag{7.5.96}$$

where r is now a function of ν. This equation yields the parametric representations of airfoil in the z-plane given by

$$x = \left(r + \frac{c^2}{r}\right)\cos\nu, \qquad y = \left(r - \frac{c^2}{r}\right)\sin\nu, \tag{7.5.97}$$

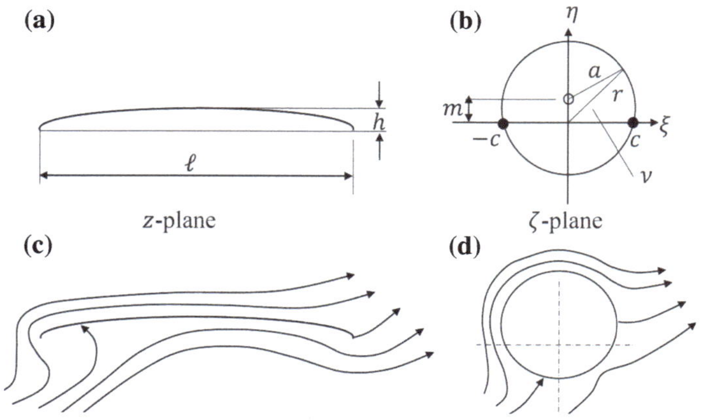

Fig. 7.17 A circular-arc airfoil. **a** The mapping of the offset circle in the z-plane. **b** The offset circle in the ζ-plane. **c** The flow field around the circular-arc airfoil in the z-plane. **d** The flow field around the offset circular cylinder with circulation in the ζ-plane

or alternatively as

$$x^2 \sin^2 \nu - y^2 \cos^2 \nu = 4c^2 \sin^2 \nu \cos^2 \nu. \tag{7.5.98}$$

It follows from Fig. 7.17b, by using the cosine rule, that

$$a^2 = r^2 + m^2 - 2rm \cos\left(\frac{\pi}{2} - \nu\right), \qquad \longrightarrow \qquad c^2 + m^2 = r^2 + m^2 - 2rm \sin \nu, \tag{7.5.99}$$

in which $a^2 = c^2 + m^2$ has been used. With Eq. (7.5.99)$_2$, it is ready to obtain

$$\sin^2 \nu = \frac{y}{2m}, \qquad \cos^2 \nu = 1 - \frac{y}{2m}, \tag{7.5.100}$$

so that Eq. (7.5.98) becomes

$$\frac{x^2 y}{2m} - y^2 \left(1 - \frac{y}{2m}\right) = \frac{4c^2 y}{2m}\left(1 - \frac{y}{2m}\right), \tag{7.5.101}$$

which can be expressed alternatively as

$$x^2 + \left[y + c\left(\frac{c}{m} - \frac{m}{c}\right)\right]^2 = c^2\left[4 + \left(\frac{c}{m} - \frac{m}{c}\right)^2\right]. \tag{7.5.102}$$

This equation describes a circle in the z-plane. Applying a linearized approximation of $\varepsilon = m/c \ll 1$ to the equation results in

$$x^2 + \left(y + \frac{c^2}{m}\right)^2 = c^2\left(4 + \frac{c^2}{m^2}\right), \tag{7.5.103}$$

which is the equation of a circle whose center locates at $y = -c^2/m$ in the z-plane with radius $c\sqrt{4 + c^2/m^2}$.

The circular-arc airfoil in Fig. 7.17a is characterized by the chord ℓ and *camber height h*. Since two ends of the airfoil lie on the x-axis with the locations $x = \pm 2c$, it is found that

$$\ell = 4c, \tag{7.5.104}$$

which is the same as those in the previous two cases. Since the center of circle in the ζ-plane locates on the η-axis, corresponding to $\nu = \pi/2$, it follows from Eq. (7.5.100)$_2$ that $y = 2m$, with which the camber height is obtained as

$$h = 2m. \tag{7.5.105}$$

With $c = \ell/4$ and $m = h/2$, Eq. (7.5.103) is simplified to

$$x^2 + \left(y + \frac{\ell^2}{8h}\right)^2 = \frac{\ell^2}{4}\left(1 + \frac{\ell^2}{16h^2}\right). \tag{7.5.106}$$

In order to satisfy the Kutta condition, the rear stagnation point must be rotated through an angle greater than the angle of attack of the inclined uniform flow. Since the rear stagnation point locates at $\zeta = c$, the additional angle necessary to this rotation is identified to be

$$\tan^{-1}\frac{m}{c} = \tan^{-1}\varepsilon \sim \varepsilon, \tag{7.5.107}$$

in which a linearized approximation of ε has been used. The total angle of rotation is thus $\alpha + \varepsilon$, which may be accomplished by the circulation strength given by

$$\Gamma = 4\pi U a \sin\left(\alpha + \frac{m}{c}\right), \tag{7.5.108}$$

which reduces to

$$\Gamma = 4\pi U c \sin\left(\alpha + \frac{m}{c}\right), \tag{7.5.109}$$

for $a = \sqrt{c^2 + m^2} \sim c$ under a linearized approximation of ε. The lift force, by using the Kutta-Joukowski law, is determined to be

$$f_y = 4\pi\rho U^2 c \sin\left(\alpha + \frac{m}{c}\right), \quad \longrightarrow \quad C_L = 8\pi\frac{c}{\ell}\sin\left(\alpha + \frac{m}{c}\right). \tag{7.5.110}$$

Using the facts that $c = \ell/4$ and $m = h/2$ in Eq. (7.5.110)$_2$ shows

$$C_L = 2\pi \sin\left(\alpha + \frac{2h}{\ell}\right), \tag{7.5.111}$$

indicating that the lift coefficient can be increased by a positive camber height, when compared to that of a flat-plate airfoil. For a circular-arc airfoil without camber height surrounded by a uniform flow without angle of attack, Eq. (7.5.111) delivers no lift force, corresponding to the previous results.

Since the center of circle in the ζ-plane locates at $\zeta = im$, the complex potential is given by

$$F(\zeta) = U\left[(\zeta - im)e^{-i\alpha} + \frac{a^2}{\zeta - im}e^{i\alpha}\right] + \frac{i\Gamma}{2\pi}\ln\left(\frac{\zeta - im}{a}\right), \tag{7.5.112}$$

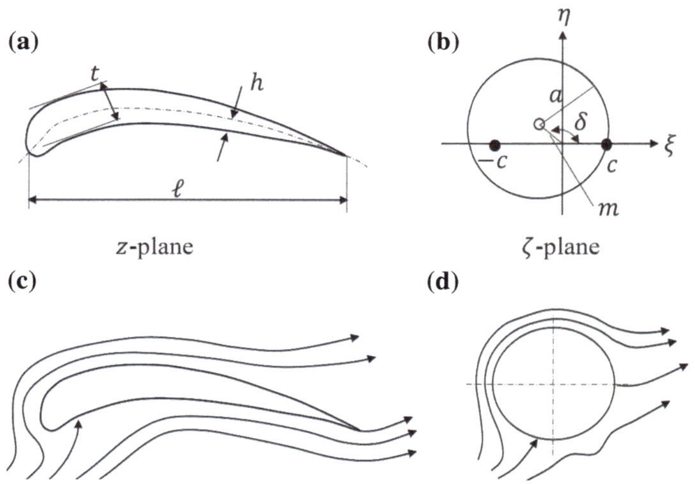

Fig. 7.18 The unsymmetric Joukowski airfoil. **a** The mapping of the offset circle in the z-plane. **b** The offset circle in the ζ-plane. **c** The flow field around the airfoil in the z-plane. **d** The flow field around the offset circular cylinder with circulation in the ζ-plane

with

$$a = \frac{\ell}{4}, \qquad m = \frac{h}{2}. \tag{7.5.113}$$

The strength of circulation is determined as

$$\Gamma = \pi U \ell \sin\left(\alpha + \frac{2h}{\ell}\right), \qquad c = \frac{\ell}{4}. \tag{7.5.114}$$

The flow fields described by the above complex potential are shown respectively in Figs. 7.17c and d in the z- and ζ-planes. Although there exists a singularity at the leading edge of airfoil in the z-plane, it would not exist for airfoils of finite nose radius and would not exist even for sharp leading edges, for flow separation occurs at the nose. Despite this local inaccuracy, the results are still representative for flows around thin chambered airfoils.

The Joukowski airfoil. Since the offset of the origin of circle in the ζ-plane along the negative ξ-axis generates a symmetric Joukowski airfoil with finite thickness in the z-plane, and the offset along the positive η-axis generates a circular-arc airfoil with finite camber height, it becomes possible to combine two offsets to generate an unsymmetric Joukowski airfoil with finite camber height in the z-plane, as shown respectively in Figs. 7.18a and b. The center of circle in the ζ-plane is displaced by a distance m from the origin at an angle δ, and the circle should pass the critical point $\zeta = c$ in order to have a sharp trailing edge. The mapped airfoil in Fig. 7.18a is referred to as the *Joukowski airfoil*, which is characterized by the chord ℓ, maximum thickness t, and maximum camber height h.

It follows from Eqs. (7.5.91) and (7.5.106) that the surface of the Joukowski airfoil in the z-plane, by using the principle of superposition, is described by

$$y = \sqrt{\frac{\ell^2}{4}\left(1+\frac{\ell^2}{16h^2}\right) - x^2} - \frac{\ell^2}{8h} \pm 0.385t\left(1-2\frac{x}{\ell}\right)\sqrt{1-\left(2\frac{x}{\ell}\right)^2}, \tag{7.5.115}$$

with the plus and minus signs assigned to the upper and lower surfaces, respectively. Since the thickness of an air foil increases the lift coefficient by an amount of $0.77t/\ell$, and the camber height increases the effective angle of attack to an amount of $2h/\ell$, the lift coefficient of the Joukowski airfoil is obtained as

$$C_L = 2\pi\left(1+0.77\frac{t}{\ell}\right)\sin\left(\alpha+\frac{2h}{\ell}\right). \tag{7.5.116}$$

The complex potential of the flow field around a Joukowski airfoil in the ζ-plane is obtained by using the principle of superposition, which is given by

$$F(\zeta) = U\left[(\zeta - me^{i\delta})e^{-i\alpha} + \frac{a^2 e^{i\alpha}}{\zeta - me^{i\delta}}\right] + \frac{i\Gamma}{2\pi}\ln\left(\frac{\zeta - me^{i\delta}}{a}\right), \tag{7.5.117}$$

with

$$m\cos\delta = -0.77\frac{tc}{\ell}, \qquad m\sin\delta = \frac{h}{2}, \qquad a = \frac{\ell}{4} + 0.77\frac{tc}{\ell}. \tag{7.5.118}$$

The strength of circulation Γ consists of the contributions of thickness and camber height of the airfoil. It follows that

$$\Gamma = \pi U\ell\left(1+0.77\frac{tc}{\ell}\right)\sin\left(\alpha+\frac{2h}{\ell}\right). \tag{7.5.119}$$

The flow fields in the ζ- and z-planes are shown respectively in Figs. 7.18c and d. As similar to the previous cases, as t and/or h increases, the body departs more and more from a streamlined airfoil and approaches a bluff body, causing flow separation near the nose. The flow separation induces dramatic decrease in the lift force, which is known as the *stall*, to be discussed later in a detailed manner in Sects. 8.4.8 and 8.4.9.

Table 7.3 Lift coefficients of different airfoils and the corresponding circulations required to satisfy the Kutta condition

Airfoil	C_L	Γ
Flat plate	$2\pi\sin\alpha$	$4\pi Ua\sin\alpha$
Sym. Joukowski	$2\pi\left(1+0.77\frac{t}{\ell}\right)\sin\alpha$	$\pi U\ell\left(1+0.77\frac{t}{\ell}\right)\sin\alpha$
Circular-arc	$2\pi\sin\left(\alpha+\frac{2h}{\ell}\right)$	$\pi U\ell\sin\left(\alpha+\frac{2h}{\ell}\right)$
Joukowski	$2\pi\left(1+0.77\frac{t}{\ell}\right)\sin\left(\alpha+\frac{2h}{\ell}\right)$	$\pi U\ell\left(1+0.77\frac{t}{\ell}\right)\sin\left(\alpha+\frac{2h}{\ell}\right)$

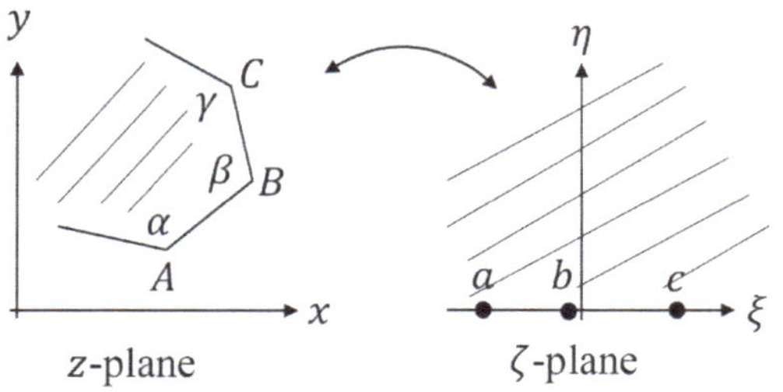

Fig. 7.19 The Schwarz-Christoffel transformation between two complex planes

Table 7.3 summarizes the lift coefficients of different airfoils in the z-plane and the corresponding circulations required to satisfy the Kutta condition in the ζ-plane derived previously. As a summary, in the context of two-dimensional potential-flow theory, the lift coefficient of an airfoil can be increased by increasing the angle of attack, the airfoil thickness, or the airfoil camber height.

7.5.8 The Schwarz-Christoffel Transformation

The *Schwarz-Christoffel transformation* is one of the conformal transformations which maps the interior of a closed polygon in the z-plane onto the upper half of the ζ-plane, while the boundary of polygon is mapped onto the ξ-axis, as shown in Fig. 7.19. The transformation is given by the solution to the equation

$$\frac{\mathrm{d}z}{\mathrm{d}\zeta} = K(\zeta - a)^{\alpha/\pi - 1}(\zeta - b)^{\beta/\pi - 1}(\zeta - c)^{\gamma/\pi - 1} \cdots , \tag{7.5.120}$$

where K is an arbitrary constant and $\{A, B, C, \cdots\}$ in the z-plane are the vertices subtended the interior angles $\{\alpha, \beta, \gamma, \cdots\}$, whose corresponding points in the ζ-plane are $\{a, b, c, \cdots\}$ lying on the ξ-axis. Since the polygon in the z-plane is closed, it is seen that

$$\alpha + \beta + \gamma + \cdots = (n - 2)\pi, \tag{7.5.121}$$

where n is the number of vertices of the polygon. Any three of the constants $a, b, c, \cdots$ are chosen arbitrarily (conventionally $-1, 0, 1$), and any remaining ones can be determined by the shape of polygon, i.e., the value of constant K.

The Schwarz-Christoffel transformation is of prime interest in the study of potential flows. As an illustration, consider a uniform rectilinear flow around a vertical flat plate with finite length shown in Fig. 7.20c. The considered flow may be approached by using the flow around a vertically oriented ellipse with angle of attack $\pi/2$ by a limiting procedure. Since the plate length in the z-plane becomes $4a$, and the flow field and plate are symmetric with respect to the stagnation streamline, only a half of the flow field and plate, e.g. the upper half, is chosen for simplicity. The plate is considered to be made up of line ABC which folds back on itself, as shown in Fig. 7.20a. The locations of vertices A, B, C are mapped on to points a, b, c in the ζ-plane, whose values are chosen to be $\zeta = -1, 0, 1$, respectively, as displayed in Fig. 7.20b.

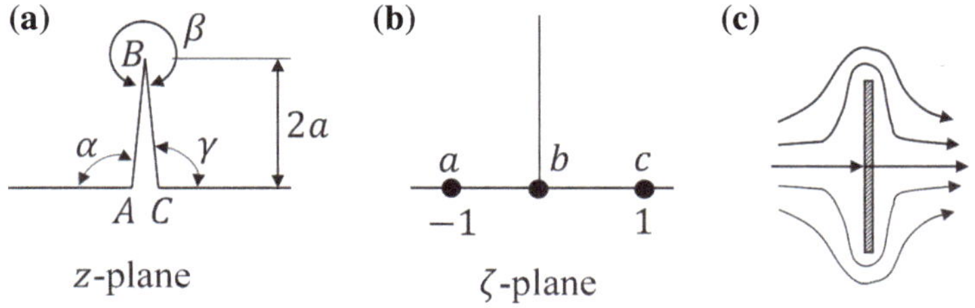

Fig. 7.20 Flows around a vertical plate with no separations. **a** The mapping of the Schwarz-Christoffel transformation. **b** Points in the ζ-plane. **c** The corresponding flow field in the z-plane

With these, the Schwarz-Christoffel transformation reads

$$\frac{dz}{d\zeta} = K(\zeta+1)^{-1/2}(\zeta-0)^{1}(\zeta-1)^{-1/2} = \frac{K\zeta}{\sqrt{\zeta^2-1}}, \tag{7.5.122}$$

for $\alpha = \pi/2$, $\beta = 2\pi$, and $\gamma = \pi/2$. Integrating this equation yields

$$z = K\sqrt{\zeta^2-1} + D, \tag{7.5.123}$$

where D is an integration constant, which is in general a complex number. Since the coordinates of points $\{A, B, C\}$ in the z-plane are given by $\{z = 0, z = i2a, z = 0\}$, $D = 0$ and $K = 2a$ are obtained, and Eq. (7.5.123) becomes

$$z = 2a\sqrt{\zeta^2-1}, \tag{7.5.124}$$

which is the required mapping function. It is found that as $\zeta \to \infty$, $z \to 2a\zeta$, and the complex velocity $W(\zeta) \to 2aW(z)$, as implicated by the property of conformal transformation described in Sect. 1.6.6. In order to obtain a uniform rectilinear flow with magnitude U in the z-plane, the magnitude of uniform flow in the ζ-plane should be $2aU$, with which the complex potential in the ζ-plane becomes

$$F(\zeta) = 2aU\zeta. \tag{7.5.125}$$

However, the inverse mapping of Eq. (7.5.124) is given by $\zeta = \pm\sqrt{(z/2a)^2+1}$. To fulfill the condition that $\zeta \to \infty$ as $z \to \infty$, the minus sign must be rejected. Substituting this into the above equation yields the complex potential in the z-plane given by

$$F(z) = U\sqrt{z^2+4a^2}. \tag{7.5.126}$$

The flow in the z-plane described by this complex potential is shown in Fig. 7.20c. Obviously, infinite velocity components exist at $y = \pm 2a$, for which the Kutta condition cannot be applied. So, the fluid must separate from two edges of the plate, and the complex potential derived previously is not able to represent this separation phenomenon, because the fluid does not remain in contact with the plate as was implicitly assumed in Eq. (7.5.125). A more representative flow configuration for this problem will be analyzed in Sect. 8.4.8 by using the theory of boundary layer. In the following, the Schwarz-Christoffel transformation is used to study three typical problems of two-dimensional potential flows, specifically the source in a channel, the flow through an aperture, and the flow past a vertical plate.

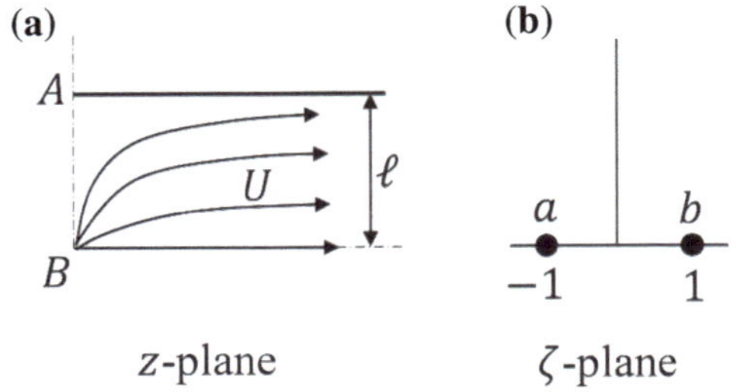

Fig. 7.21 Flows of a source in an infinitely long two-dimensional channel. **a** The flow field in the first quadrant in the z-plane. **b** The mapping of the Schwarz-Christoffel transformation in the ζ-plane

Source in a channel. Consider a two-dimensional channel of width 2ℓ and of infinite length, in which a source is located midway between the channel walls. The origin of coordinate system in the z-plane is taken to be at the location of source, so that the resulting flow field is symmetric with respect to both x- and y-axes. The entire x- and y-axes are the streamlines, and only the first quadrant of flow field is used for convenience, where $0 \leq x$ and $0 \leq y \leq \ell$, as shown in Fig. 7.21a. This region is considered to be bounded by the polygon which is to be mapped, and vertices A and B are chosen to correspond to points $\zeta = -1$ and $\zeta = 1$ in the ζ-plane, as shown in Fig. 7.21b. With these, the Schwarz-Christoffel transformation reads

$$\frac{\mathrm{d}z}{\mathrm{d}\zeta} = K(\zeta+1)^{-1/2}(\zeta-1)^{-1/2} = \frac{K}{\sqrt{\zeta^2-1}}, \qquad (7.5.127)$$
$$\longrightarrow z = K\cosh^{-1}\zeta + D,$$

where D is an integration constant. Since the coordinates of points $\{A, B\}$ are respectively $\{z = i\ell, z = 0\}$, corresponding to the coordinates of points $\{a, b\}$ given by $\{\zeta = -1, \zeta = 1\}$, it follows that $D = 0$ and $K = \ell/\pi$, with which the mapping function reduces to

$$z = \frac{\ell}{\pi}\cosh^{-1}\zeta, \qquad \longrightarrow \qquad \zeta = \cosh\frac{\pi z}{\ell}. \qquad (7.5.128)$$

The flow field in the ζ-plane now corresponds to a source located at point $\zeta = 1$, and the complex potential is then given by

$$F(\zeta) = \frac{m}{2\pi}\ln(\zeta - 1), \qquad (7.5.129)$$

by which the complex potential in the z-plane is obtained as

$$F(z) = \frac{m}{2\pi}\ln\left(\cosh\frac{\pi z}{\ell} - 1\right). \qquad (7.5.130)$$

This result can be simplified by using the identity that $\cosh(X+Y) - \cosh(X-Y) = 2\sinh X \sinh Y$, where $X = Y = \pi z/(2\ell)$. With this, Eq. (7.5.130) is simplified to

$$F(z) = \frac{m}{\pi}\ln\left(\sinh\frac{\pi z}{2\ell}\right) + \frac{m}{2\pi}\ln 2 = \frac{m}{\pi}\ln\left(\sinh\frac{\pi z}{2\ell}\right), \qquad (7.5.131)$$

for the constant term has no influence on the complex velocity. The flow field corresponding to this complex potential in the first quadrant in the z-plane is shown in Fig. 7.21a. It is readily verified that the total quantity of fluid leaving the source is

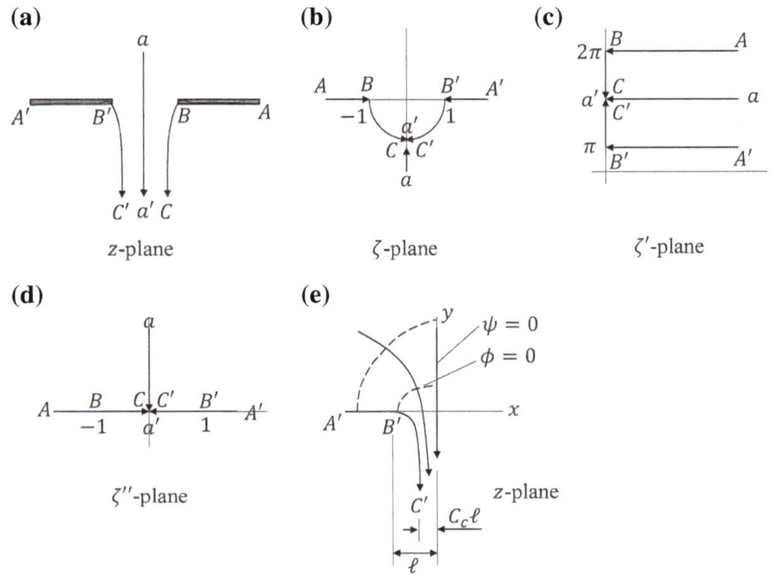

Fig. 7.22 Flows through a horizontal aperture. **a** The configuration in the z-plane. **b** The mapping in the ζ-plane. **c** The mapping in the ζ'-plane. **d** The mapping in the ζ''-plane. **e** Geometry of the free streamlines in the z-plane. Solid lines: streamlines; dashed lines: equipotential lines

$4\ell U$, so that the source strength should be $m = 4\ell U$, and the velocity in the channel is U.

Flows through an aperture. One of the most impressive applications of the Schwarz-Christoffel transformation, in the context of fluid mechanics, is the study of streaming motions which involve free streamlines. It is not frequently known where these free streamlines locate, and this information must come from the solution. The key to solving such problems is the so-called *hodograph plane*, which uses the fact that along such free streamlines the pressure remains unchanged. The idea is illustrated by considering a flow through a two-dimensional horizontal slit or aperture in the z-plane shown in Fig. 7.22a. The plate contains a semi-infinite expanse of a fluid above it, which is draining through the aperture defined by section BB' on the x-axis. At corners B and B', the flow will locally behave like that around a sharp edge, and thus the bounding streamlines along the horizontal plate will curve toward the vertical direction to allow the fluid to separate from the corners in order to avoid infinite velocity components. The magnitude of velocity in the resulting jet will reach a constant value U downstream of all edge effects.

To study this flow field, the transformation given by

$$\zeta = U\frac{\mathrm{d}z}{\mathrm{d}F} = \frac{U}{W} = \frac{U}{\sqrt{u^2+v^2}}e^{i\theta}, \tag{7.5.132}$$

is introduced, where θ is the angle subtended by the velocity vector in the z-plane, by which the flow field in the z-plane is transformed to the first hodograph plan, i.e., the ζ-plane. With this, the free streamlines whose positions are unknown in the z-plane are mapped onto a unit-circle in the ζ-plane, as shown in Fig. 7.22b. Along streamlines BC and $B'C'$, the pressure corresponds to the atmospheric one, so that $u^2 + v^2 = U^2 =$ constant, as indicated by the Bernoulli equation. Then, Eq. (7.5.132) reduces to $\zeta = e^{i\theta}$, representing a unit-circle in the ζ-plane. Since θ along streamline $A'B'$ is either 0 or 2π, that along streamline AB is π and that along streamline aa' is $3\pi/2$, it follows that the lower half of unit-circle in the ζ-plane represents streamlines BC and $B'C'$. In addition, since along streamline $A'B'$ the angle θ is either 0 or 2π, it leads to that $u^2 + v^2$ varies from 0 at point A' and to U^2 at point B', and hence $|\zeta|$ varies from infinite to unity correspondingly. Equally, along streamline AB, $|\zeta|$ varies from infinite at point A to unity at point B, while $\theta = \pi$. Moreover, along streamline aa', $\theta = 3\pi/2$, thus $u^2 + v^2$ varies from zero at point a to unity at point a', making $|\zeta|$ infinite and unit correspondingly. With these, the mapping in the ζ-plane shown in Fig. 7.22b is established, in which points a', C and C' mark a sink.

Consider a second mapping described by

$$\zeta' = \ln \zeta, \tag{7.5.133}$$

which maps the configuration in the ζ-plane to the second hodograph plan, i.e., the ζ'-plane. A point in the ζ-plane denoted by $\zeta = re^{i\theta}$ is mapped onto the ζ'-plane in the form

$$\zeta' = \ln r + i\theta, \tag{7.5.134}$$

where $r = U/(u^2 + v^2)^{1/2}$. The radial lines in the ζ-plane thus become the horizontal lines in the ζ'-plane. The unit-circle with $r = 1$ in the ζ-plane is mapped to the vertical line given by $\zeta' = i\theta$ in the ζ'-plane, as shown in Fig. 7.22c, in which the mappings of streamlines AB, $A'B'$, and aa' are displayed. The flow field in the ζ'-plane corresponds to that of a sink locating at the center at the end of a semi-infinite channel. The flow field in the ζ'-plane is mapped again onto the third hodograph plane, the ζ''-plane, viz.,[17]

$$\zeta'' = \cosh(\zeta' - i\pi) = -\cosh \zeta', \tag{7.5.135}$$

whose flow field is shown in Fig. 7.22d.

The flow field shown in Fig. 7.22d corresponds to that of a sink locating at $\zeta'' = 0$, and thus the complex potential is given by

$$F(\zeta'') = -\frac{m}{2\pi} \ln \zeta'' + K, \tag{7.5.136}$$

where K is a constant, which is used to permit the streamline $\psi = 0$ and equipotential line $\phi = 0$ to correspond to a chosen streamline and equipotential line in the z-plane, respectively. In view of Fig. 7.22e, it is chosen that $\psi = 0$ corresponds to streamline

[17] It is done so, for rectangle $ABCC'B'A'$ in the ζ'-plane has been taken as the equivalence of the half channel with width ℓ. In this circumstance, ℓ appearing in the transformation is π, and the quantity $\zeta' - i\pi$ is used to replace ζ' in order to bring corner B to the origin in the ζ'-plane.

aa' and $\phi = 0$ corresponds to that passes through points B' and B. Considering the flow between streamlines aa' and $A'B'C'$, it follows that

$$\psi_{aa'} - \psi_{A'B'C'} = 0 - \psi_{A'B'C'} = C_c \ell U, \tag{7.5.137}$$

where C_c is the *contraction coefficient* of fluid jet in the z-plane. This result is derived by using the property of stream function that the fluid volume flow rate between two streamlines equals the difference in the values of stream functions. Equally, by considering streamlines aa' and ABC,

$$\psi_{ABC} - \psi_{aa'} = \psi_{ABC} - 0 = C_c \ell U, \tag{7.5.138}$$

is obtained. As a result, at point B', $\phi = 0$ and $\psi = -C_c \ell U$, yielding the complex potential there to be $-iC_c \ell U$ and the complex potential at point B assumes the value of $iC_c \ell U$. Since at points B' and B, $\zeta'' = -1$ and $\zeta'' = 1$, respectively, applying these conditions to Eq. (7.5.136) gives

$$K = iC_c \ell U, \qquad m = 4C_c \ell U, \tag{7.5.139}$$

with which the equation becomes

$$F(\zeta'') = -\frac{2C_c \ell U}{\pi} \ln \zeta'' + iC_c \ell U. \tag{7.5.140}$$

Substituting Eqs. (7.5.132), (7.5.133) and (7.5.135) into the above equation results in

$$F(z) = -\frac{2C_c \ell U}{\pi} \ln \left\{ \cosh \left[\ln \left(U \frac{\mathrm{d}z}{\mathrm{d}F} - i\pi \right) \right] \right\} + iC_c \ell U, \tag{7.5.141}$$

which is an implicit differential equation of the complex potential in the z-plane, and the flow problem has in principle been solved. However, the value of C_c needs to be identified; otherwise, it is impossible to integrate Eq. (7.5.141) numerically to obtain $F(z)$.

To this end, construct a coordinate s, whose value is zero at point B' and increases along $B'C'$, and let a small element of streamline $B'C'$ be denoted by $\mathrm{d}s$ with positive slope. It follows from Fig. 7.22a that

$$\frac{\mathrm{d}x}{\mathrm{d}s} = \cos\theta, \qquad \longrightarrow \qquad x = x_0 + \int_0^s \cos\theta \, \mathrm{d}s, \tag{7.5.142}$$

denoting the x-coordinate of any point on $B'C'$, and x_0 is included to permit that $x = -\ell$ at $s = 0$. Substituting Eq. (7.5.132) into the above equation yields

$$x = x_0 + \int_{2\pi}^{0} \cos\theta \frac{\mathrm{d}s}{\mathrm{d}\zeta''} \frac{\mathrm{d}\zeta''}{\mathrm{d}\theta} \mathrm{d}\theta, \tag{7.5.143}$$

where the terms $\mathrm{d}s/\mathrm{d}\zeta''$ and $\mathrm{d}\zeta''\mathrm{d}\theta$ must be expressed in terms of θ. First, it follows from Eq. (7.5.132) that on streamline $B'C'$,

$$1 = |\zeta| = \left| \frac{U}{W} \right| = \left| U \frac{\mathrm{d}z}{\mathrm{d}F} \right| = \left| U \frac{\mathrm{d}z}{\mathrm{d}\zeta''} \frac{\mathrm{d}\zeta''}{\mathrm{d}F} \right|. \tag{7.5.144}$$

Taking derivative of Eq. (7.5.140) with respect to ζ'' gives

$$\frac{\mathrm{d}F}{\mathrm{d}\zeta''} = -\frac{2C_c \ell U}{\pi}\frac{1}{\zeta''}, \tag{7.5.145}$$

which is substituted into Eq. (7.5.144) to obtain

$$1 = \left| U\frac{\mathrm{d}z}{\mathrm{d}\zeta''}\frac{\pi}{2C_c \ell U}\zeta'' \right|, \quad \longrightarrow \quad \left|\frac{\mathrm{d}z}{\mathrm{d}\zeta''}\right| = \frac{2C_c\ell}{\pi}\frac{1}{\zeta''}, \tag{7.5.146}$$

because $\zeta'' > 0$ on streamline $B'C'$. Second, replacing $\mathrm{d}z$ on streamline $B'C'$ by $\mathrm{d}z = (\mathrm{d}s)e^{i\theta}$ yields

$$\frac{\mathrm{d}s}{\mathrm{d}\zeta''} = -\frac{2C_c\ell}{\pi}\frac{1}{\zeta''}, \tag{7.5.147}$$

for $\mathrm{d}\zeta'' < 0$ along streamline $B'C'$. Since on $B'C'$ the value of ζ is given by $\zeta = e^{i\theta}$, it follows from Eqs. (7.5.134) and (7.5.135) that $\zeta' = i\theta$ and $\zeta'' = -\cos\theta$. Substituting these into the above equation gives

$$\frac{\mathrm{d}s}{\mathrm{d}\zeta''} = \frac{2C_c\ell}{\pi}\frac{1}{\cos\theta}. \tag{7.5.148}$$

Now, turn back to Eq. (7.5.143), which, by using the above equation and $\mathrm{d}\zeta''/\mathrm{d}\theta = \sin\theta$, can be integrated to obtain

$$x = x_0 + \frac{2C_c\ell}{\pi}(1-\cos\theta) = -\ell + \frac{2C_c\ell}{\pi}(1-\cos\theta), \tag{7.5.149}$$

for at point B' $\theta = 2\pi$ and $x = x_0 = -\ell$. Since at point C', $x = -C_c\ell$ with $\theta = 3\pi/2$, it is found that

$$C_c = \frac{\pi}{\pi+2} \sim 0.611. \tag{7.5.150}$$

This result predicts that a free jet which emerges from an aperture will assume a width which is nearly 0.611 of the slit width. This theoretical result has been confirmed experimentally for openings under deep liquids. The contraction of the width of a liquid jet is called the *vena contracta*.

Flows past a vertical plate. Another example is a rectilinear uniform flow past a vertical plate, as shown in Fig. 7.23a. This problem can equally be solved by using the concept of hodograph plane. In the z-plane, the magnitude of uniform flow is U and the height of vertical plate is 2ℓ. The stagnation streamline aa' splits upon reaching the plate and forms two bounding streamlines ABC and $A'B'C'$, where BC and $B'C'$ are the free streamlines. The region downstream the plate between two free streamlines is interpreted as a cavity which has a uniform pressure p_c throughout.[18]

Consider the first mapping given by

$$\zeta = U\frac{\mathrm{d}z}{\mathrm{d}F} = \frac{U}{W} = \frac{U}{\sqrt{u^2+v^2}}e^{i\theta}, \tag{7.5.151}$$

[18] In real flows, this region is called a *wake*.

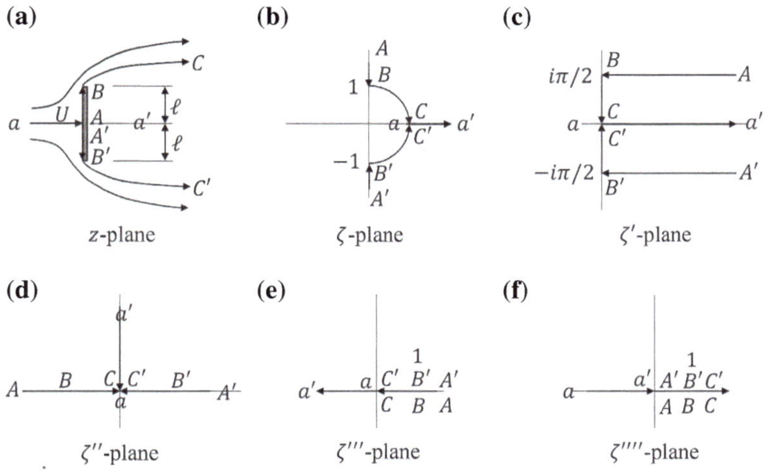

Fig. 7.23 Flows through a vertical plate. **a** The configuration in the z-plane. **b** The mapping in the ζ-plane. **c** The mapping in the ζ'-plane. **d** The mapping in the ζ''-plane. **e** The mapping in the ζ'''-plane. **f** The mapping in the ζ''''-plane

which maps the flow region in the z-plane to a unit-circle in the ζ-plane, as shown in Fig. 7.23b. The free streamlines BC and $B'C'$ become parts of the unit-circle, and θ along ABC lies between $\pi/2$ and 0, while θ along $A'B'C'$ lies between $-\pi/2$ and 0, so that two free streamlines construct the right half part of unit-circle. Since the flow boundary crosses the positive portion of real axis in the ζ-plane, it results in a multi-valued function for $0 \le \theta \le 2\pi$. A branch cut lying on the negative real axis in the ζ-plane is used to overcome this difficulty, so that the principal values of multi-valued functions will correspond to $-\pi \le \theta \le \pi$.

The radial lines and circular contour in the ζ-plane are further mapped onto the ζ'-plane with the second mapping given by

$$\zeta' = \ln \zeta, \tag{7.5.152}$$

which maps the flow boundary into that of a rectangular channel, as shown in Fig. 7.23c. Since $-\pi \le \theta \le \pi$, the lower and upper walls of this channel correspond to $\zeta' = -\pi/2$ and $\zeta' = \pi/2$, respectively, and the centerline coincides with the real axis in the ζ'-plane. The flow field shown in Fig. 7.23c corresponds to that of a sink in a channel. Thus, points B' and B locate at $\zeta' = -i\pi/2$ and $\zeta' = i\pi/2$, respectively, and the channel half width becomes π. With these, the third mapping ζ'' is proposed as

$$\zeta'' = \cosh\left(\zeta' + i\frac{\pi}{2}\right), \tag{7.5.153}$$

and the principal flow lines in the ζ''-plane can be made collinear by means of the fourth mapping, viz.,

$$\zeta''' = (\zeta'')^2. \tag{7.5.154}$$

The flow fields in the ζ''- and ζ'''-planes are displayed respectively in Figs. 7.23d and e. By using the above mapping, the angles subtended by the principal streamlines are doubled, so that the flow in the ζ'''-plane is unidirectional along the principal streamline. Since this flow field is still not that of a rectilinear uniform flow as the principal streamlines might suggest, an additional mapping needs to be introduced. Consider the flow field in the z-plane, which can be approximated by a source locating at point a with fluid flowing toward a sink locating at CC'. However, in the ζ'''-plane point a and plane CC' are at the same location, so that the flow there is probably a doublet one. With these, the last mapping is proposed as

$$\zeta'''' = \frac{1}{\zeta'''}, \tag{7.5.155}$$

in order to map the origin in the ζ'''-plane to infinite and vice versa, as illustrated in Fig. 7.23f. Thus, in the ζ''''-plane the fluid emanates from point a and flows toward CC', as was the case in the z-plane; i.e., the flow in the ζ''''-plane becomes a rectilinear uniform flow.

The complex potential in the ζ''''-plane is then given by

$$F(\zeta'''') = K\zeta'''', \tag{7.5.156}$$

where K is a constant, representing the magnitude of uniform flow. In order to determine its value, Eq. (7.5.151) is recast alternatively as

$$\zeta = U\frac{\mathrm{d}z}{\mathrm{d}\zeta''}\frac{\mathrm{d}\zeta''}{\mathrm{d}F}, \tag{7.5.157}$$

and it follows from Eqs. (7.5.154) and (7.5.156) that

$$F(\zeta''') = \frac{K}{\zeta'''}, \qquad F(\zeta'') = \frac{K}{(\zeta'')^2}, \tag{7.5.158}$$

with which

$$\frac{\mathrm{d}F}{\mathrm{d}\zeta''} = -\frac{2K}{(\zeta'')^3}. \tag{7.5.159}$$

In view of Eq. (7.5.152), ζ is expressed as $\zeta = e^{\zeta'}$, which can be further simplified to

$$\zeta = \exp\zeta' = \exp\left(\cosh^{-1}\zeta'' - i\frac{\pi}{2}\right) = -i\exp\left(\cosh^{-1}\zeta''\right) = -i\left(\zeta'' + \sqrt{(\zeta'')^2 - 1}\right), \tag{7.5.160}$$

in which Eq. (7.5.153) has been used, and it is noted that $\cosh^{-1}x = \ln(x + \sqrt{x^2 - 1})$ for any x. Substituting the last two equations into Eq. (7.5.157) results in

$$-U\frac{\mathrm{d}z}{\mathrm{d}\zeta''}\frac{(\zeta'')^3}{2K} = -i\left(\zeta'' + \sqrt{(\zeta'')^2 - 1}\right), \quad \longrightarrow \quad U\mathrm{d}z = i2K\frac{\zeta'' + \sqrt{(\zeta'')^2 - 1}}{(\zeta'')^3}\mathrm{d}\zeta'', \tag{7.5.161}$$

which is to be integrated into the region from B' to A' to obtain

$$\int_{-i\ell}^{0} U\mathrm{d}z = i2K\int_{1}^{\infty}\frac{\zeta'' + \sqrt{(\zeta'')^2 - 1}}{(\zeta'')^3}\mathrm{d}\zeta'', \tag{7.5.162}$$

where the upper and lower limits of integration correspond to points A' and B' in the ζ''-plane, respectively. With $\zeta'' = 1/\sin\nu$, this integration is recast alternatively as

$$U\int_{-i\ell}^{0} \mathrm{d}z = -i2K\int_{\pi/2}^{0}(1+\cos\nu)\cos\nu\,\mathrm{d}\nu, \quad\longrightarrow\quad K = \frac{2U\ell}{\pi+4}. \tag{7.5.163}$$

Substituting the obtained value of K into Eq. (7.5.156) yields

$$F(\zeta'''') = \frac{2U\ell}{\pi+4}\zeta'''', \quad\longrightarrow\quad F(z) = -\frac{2U\ell}{\pi+4}\frac{1}{\sinh^2\{\ln[U(\mathrm{d}z/\mathrm{d}F)]\}}, \tag{7.5.164}$$

in which all the mappings defined previously have been used. Again, this is an implicit solution to $F(z)$ for the flow field in the z-plane.

The drag force f_x acting on the plate results from the pressure distribution on the plate surface, which is given by

$$f_x = 2\int_{-\ell}^{0}(p-p_c)\,\mathrm{d}y, \tag{7.5.165}$$

where f_x is assumed to point in the positive x-direction, and the symmetry of flow field with respect to the x-axis is used. This equation is recast alternatively by using the Bernoulli equation, viz.,

$$f_x = 2\int_{-\ell}^{0}\frac{\rho}{2}\left[U^2-(u^2+v^2)\right]\mathrm{d}y = \rho U^2\int_{-\ell}^{0}\mathrm{d}y - \rho\int_{-\ell}^{0}(u^2+v^2)\,\mathrm{d}y, \tag{7.5.166}$$

in which the Bernoulli equation has been formulated on a point in the upstream region far away from the plate and a point well downstream the plate on a free streamline. By using the complex velocity $W(z)$, by which $v^2 = -W^2$, since $u = 0$ on the surface of plate, the drag force is obtained as

$$f_x = \rho U^2\ell + \rho\int_{-\ell}^{0}\left(\frac{\mathrm{d}F}{\mathrm{d}z}\right)^2\mathrm{d}y = \rho U^2\ell - i\rho\int_{-i\ell}^{0}\left(\frac{\mathrm{d}F}{\mathrm{d}z}\right)^2\mathrm{d}z, \tag{7.5.167}$$

because at $x = 0$ on the plate surface, $\mathrm{d}z = i\mathrm{d}y$. To determine f_x, it would be better to evaluate $F(\zeta'')$ and to conduct the required integration in the ζ''-plane rather than in the z-plane. This can be accomplished by using

$$f_x = \rho U^2\ell - i\rho\int_{1}^{\infty}\left(\frac{\mathrm{d}F}{\mathrm{d}\zeta''}\right)^2\frac{\mathrm{d}\zeta''}{\mathrm{d}z}\mathrm{d}\zeta'', \tag{7.5.168}$$

for in the ζ''-plane ζ'' varies from unity to infinite as z varies from $-i\ell$ to 0. Substituting Eqs. (7.5.159) and $(7.5.163)_2$ into the above equation results in

$$f_x = \rho U^2\ell - \frac{4\rho U^2\ell}{\pi+4}\int_{1}^{\infty}\frac{\mathrm{d}\zeta''}{(\zeta'')^3\left[\zeta''+\sqrt{(\zeta'')^2-1}\right]}. \tag{7.5.169}$$

By using $\zeta'' = 1/\sin\nu$, the drag force is given by

$$f_x = \rho U^2\ell\left[1+\frac{4}{\pi+4}\int_{\pi/2}^{0}(1-\cos\nu)\cos\nu\,\mathrm{d}\nu\right] = \frac{2\pi}{\pi+4}\rho U^2\ell. \tag{7.5.170}$$

Physically, it follows from the symmetry of flow field in the z-plane with respect to the x-axis that there exists no lift force acting on the plate. On the other hand, the unsymmetric flow field with respect to the y-axis implies the existence of a drag force, which is determined by Eq. (7.5.170).

7.6 Three-Dimensional Potential Flows

For three-dimensional potential flows, a spherical coordinate system (r, θ, ω) is constructed, as shown in Fig. 7.24a, where θ is the angle between the reference axis and the radius position r at point P, while the angle ω is subtended by the perpendicular to the reference axis which passes through point P. For practical interest, only the three-dimensional bodies which are axis-symmetric are considered, which, by definition, implies that $\partial\alpha/\partial\omega = 0$ for any quantity α.

7.6.1 Velocity Potential and Stokes' Stream Functions

The velocity potential function ϕ for irrotational flows should satisfy the Laplace equation, which, in terms of the spherical coordinates defined in Fig. 7.24a, reads

$$\frac{1}{r^2}\frac{\partial}{\partial r}\left(r^2\frac{\partial\phi}{\partial r}\right) + \frac{1}{r^2\sin\theta}\frac{\partial}{\partial\theta}\left(\sin\theta\frac{\partial\phi}{\partial\theta}\right) = 0. \tag{7.6.1}$$

Once ϕ is determined, the velocity components are directly given by

$$u_r = \frac{\partial\phi}{\partial r}, \qquad u_\theta = \frac{1}{r}\frac{\partial\phi}{\partial\theta}. \tag{7.6.2}$$

The stream function ψ is so defined that the continuity equation is satisfied identically, i.e.,

$$\frac{1}{r^2}\frac{\partial}{\partial r}\left(r^2 u_r\right) + \frac{1}{r\sin\theta}\frac{\partial}{\partial\theta}\left(u_\theta\sin\theta\right) = 0, \tag{7.6.3}$$

for the considered circumstance, giving rise to the definition of ψ, viz.,

$$u_r = \frac{1}{r^2\sin\theta}\frac{\partial\psi}{\partial\theta}, \qquad u_\theta = -\frac{1}{r\sin\theta}\frac{\partial\psi}{\partial r}. \tag{7.6.4}$$

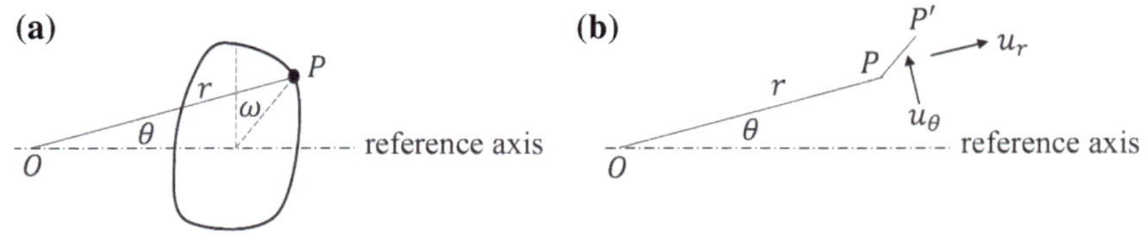

Fig. 7.24 Three-dimensional axis-symmetric potential flows. **a** Setup of the spherical coordinate system. **b** Illustration for the fluid volume crossing the revolution surface of OP with respect to the reference axis

The defined stream function is called *Stokes's stream function* for axis-symmetric incompressible flows. Consider a line segment PP' shown in Fig. 7.24b. The infinitesimal fluid volume $\mathrm{d}v$ crossing the unit-area generated by the revolution of PP' with respect to the reference axis is given by

$$\mathrm{d}v = 2\pi r \sin\theta \left[u_r(r\mathrm{d}\theta) - u_\theta \mathrm{d}r\right] = 2\pi\left(\frac{\partial\psi}{\partial\theta}\mathrm{d}\theta + \frac{\partial\psi}{\partial r}\mathrm{d}r\right) = 2\pi\mathrm{d}\psi, \quad (7.6.5)$$

with which the total fluid volume V crossing the surface of revolution which is formed by rotating OP around the reference axis is obtained as

$$V = \int \mathrm{d}v = 2\pi\psi. \quad (7.6.6)$$

Essentially, the velocity components of a three-dimensional potential flow can be determined by using either ϕ or ψ.[19] In the forthcoming discussions, the fundamental solutions will be established by solving the Laplace equation for ϕ by separation of variables, unless stated otherwise. It is assumed that $\phi(r, \theta)$ can be decomposed into

$$\phi(r, \theta) = R(r)\Theta(\theta). \quad (7.6.7)$$

Substituting this expression into Eq. (7.6.1) results in a regular *Sturm-Liouville problem* for $R(r)$ and a *Legendre's equation* for $\Theta(\theta)$,[20] whose solutions may be obtained by using the method of *eigenfunction expansion*. With these, the general solution to ϕ is obtained as

$$\phi(r, \theta) = \sum_{\iota=0}^{\infty}\left(A_\iota r^\iota + \frac{B_\iota}{r^{\iota+1}}\right)P_\iota(\cos\theta), \quad (7.6.8)$$

where P_ι represents *Legendre's function of the first kind*, which is defined by

$$P_\iota(x) \equiv \frac{1}{2^\iota \iota!}\frac{\mathrm{d}^\iota}{\mathrm{d}x^\iota}\left(x^2-1\right)^\iota, \qquad \forall x. \quad (7.6.9)$$

This expression is also known as *Legendre's polynomial of order* ι. For convenience, the first three terms in Legendre's polynomial are given here:

$$P_0(x) = 1, \qquad P_1(x) = x, \qquad P_2(x) = \frac{1}{2}\left(3x^2-1\right). \quad (7.6.10)$$

Equation (7.6.8) contains a number of fundamental solutions, which can be used to establish the solutions to more complicated circumstances by using the principle of superposition. These fundamental solutions are discussed in the next section.

[19] However, for rotational flows, velocity potential function does not exist, and the stream-function approach offers the only way for reducing the vector equations of motion to scalar equations.
[20] Jacques Charles Francois Sturm, 1803–1855; Joseph Liouville, 1809–1882; Adrien-Marie Legendre, 1752–1833, all are French mathematicians.

7.6.2 Fundamental Solutions

Uniform flows. One of the fundamental solutions contained in Eq. (7.6.8) is obtained by setting

$$B_\iota = 0, \quad \forall \iota; \qquad A_\iota = \begin{Bmatrix} 0, & \iota \neq 1 \\ U, & \iota = 1 \end{Bmatrix}, \tag{7.6.11}$$

corresponding to a uniform flow. With these, the velocity potential function becomes

$$\phi(r, \theta) = Ur \cos\theta. \tag{7.6.12}$$

By using Eqs. (7.6.2) and (7.6.4), it follows that

$$\frac{\partial \psi}{\partial \theta} = Ur^2 \sin\theta \cos\theta, \qquad \frac{\partial \psi}{\partial r} = Ur \sin^2\theta, \tag{7.6.13}$$

which are integrated to obtain

$$\psi(r, \theta) = \frac{1}{2} Ur^2 \sin^2\theta. \tag{7.6.14}$$

This stream function can also be derived by using Eq. (7.6.6), for $2\pi\psi = U\pi (r \sin\theta)^2$, giving rise to the same result of ψ given in Eq. (7.6.14).

Source and sink flows. The velocity potential function corresponding to a three-dimensional source or sink is obtained from Eq. (7.6.8) by letting

$$A_\iota = 0, \quad \forall \iota; \qquad B_\iota = \begin{Bmatrix} 0, & \iota \neq 0 \\ B_0, & \iota = 0 \end{Bmatrix}, \tag{7.6.15}$$

with which $\phi(r, \theta)$ is identified to be

$$\phi(r, \theta) = \frac{B_0}{r}, \qquad \longrightarrow \qquad u_r = -\frac{B_0}{r^2}, \quad u_\theta = 0, \tag{7.6.16}$$

showing that the velocity is purely radial, and its magnitude increases as the origin is approached. A singularity locates at $r = 0$, where there exists a source or a sink. The strength Q of a source or a sink is defined as the fluid volume leaving or entering a specific surface per unit time, which is given viz.,

$$Q = \int_A \boldsymbol{u} \cdot \boldsymbol{n} \mathrm{d}a = \int_0^{2\pi} \mathrm{d}\omega \int_0^{\pi} \left(\frac{B_0}{r^2} \right) r^2 \sin\theta \, \mathrm{d}\theta = -4\pi B_0, \tag{7.6.17}$$

in which a spherical surface has been used. With this, Eq. $(7.6.16)_1$ becomes

$$\phi(r, \theta) = -\frac{Q}{4\pi r}, \tag{7.6.18}$$

where the minus sign is associated with a source. For a sink, $-Q$ is simply replaced by Q.

As referred to Fig. 7.24b, it is assumed that a source is located slightly to the right of origin O, so that the fluid volume crossing the revolution surface generated by OP will be $2\pi\psi + Q$. It follows that

$$2\pi\psi + Q = \int_0^\theta u_r \cos\theta (2\pi r \sin\theta) \frac{r\mathrm{d}\theta}{\cos\theta}, \qquad \longrightarrow \qquad \psi(r, \theta) = -\frac{Q}{4\pi} (1 + \cos\theta). \tag{7.6.19}$$

If a source with strength Q is located slightly to the left of origin, the constant term contained in this equation would have been different. However, the velocity components would remain unchanged.

7.6.3 Solutions of Superimposing Flows

Flows due to a doublet. Consider a source with strength Q locating at the origin and a sink with same strength locating at a distance δx from the origin, as shown in Fig. 7.25a. The distance from the source to some point P in the fluid is denoted by r, so its distance to the sink is $r - \delta r$. The velocity potential function corresponding to the considered circumstance is obtained by using the principle of superposition, viz.,

$$\phi(r,\theta) = -\frac{Q}{4\pi r} + \frac{Q}{4\pi(r-\delta r)} = -\frac{Q}{4\pi r}\left(1 - \frac{1}{1-\delta r/r}\right). \tag{7.6.20}$$

For small values of $\delta r/r$, i.e., the source and sink are very close to each other, and this equation can be approximated by

$$\phi(r,\theta) = -\frac{Q}{4\pi r}\left\{1 - \left[1 + \frac{\delta r}{r} + \mathcal{O}\left(\frac{\delta r}{r}\right)^2\right]\right\} = \frac{Q}{4\pi r}\left[\frac{\delta r}{r} + \mathcal{O}\left(\frac{\delta r}{r}\right)^2\right]. \tag{7.6.21}$$

In view of Fig. 7.25a, applying the cosine rule to the triangle yields

$$(r-\delta r)^2 = r^2 + (\delta x)^2 - 2r\delta x\cos\theta, \longrightarrow \cos\theta = \frac{\delta r}{\delta x}\left[1 + \mathcal{O}\left(\frac{\delta r}{r}\right)\right]. \tag{7.6.22}$$

Substituting this result into Eq. (7.6.21) gives

$$\phi(r,\theta) = \frac{Q}{4\pi r}\left\{\frac{\delta x}{r}\cos\theta\left[1 + \mathcal{O}\left(\frac{\delta r}{r}\right)\right]\right\} = \frac{\mu}{4\pi r^2}\cos\theta, \tag{7.6.23}$$

in which it is assumed that as $\delta x \to 0$ and $Q \to \infty$, the product $(Q\delta x) \to \mu$, which defines the strength of a doublet denoted by μ. It is seen that a doublet expels fluids along the negative portion of reference axis and absorbs fluids along the positive portion.

The stream function is obtained by integrating the equations

$$u_r = \frac{\partial\phi}{\partial r} = -\frac{\mu}{2\pi r^3}\cos\theta = \frac{1}{r^2\sin\theta}\frac{\partial\psi}{\partial\theta}, \quad u_\theta = \frac{1}{r}\frac{\partial\phi}{\partial\theta} = -\frac{\mu}{4\pi r^3}\sin\theta = -\frac{1}{r\sin\theta}\frac{\partial\psi}{\partial r}, \tag{7.6.24}$$

to yield

$$\psi(r,\theta) = -\frac{\mu}{4\pi r}\sin^2\theta, \tag{7.6.25}$$

showing again that a doublet discharges and attracts fluids along the negative and positive portions of reference axis, respectively.

Flows near a blunt nose. By superimposing the solution of a uniform flow with that of a source, the flow around a long cylinder with a blunt nose is obtained. Combining the stream functions for a uniform flow with magnitude U and a source with strength Q locating at the origin yields

$$\psi(r,\theta) = \frac{1}{2}Ur^2\sin^2\theta - \frac{Q}{4\pi}(1+\cos\theta). \tag{7.6.26}$$

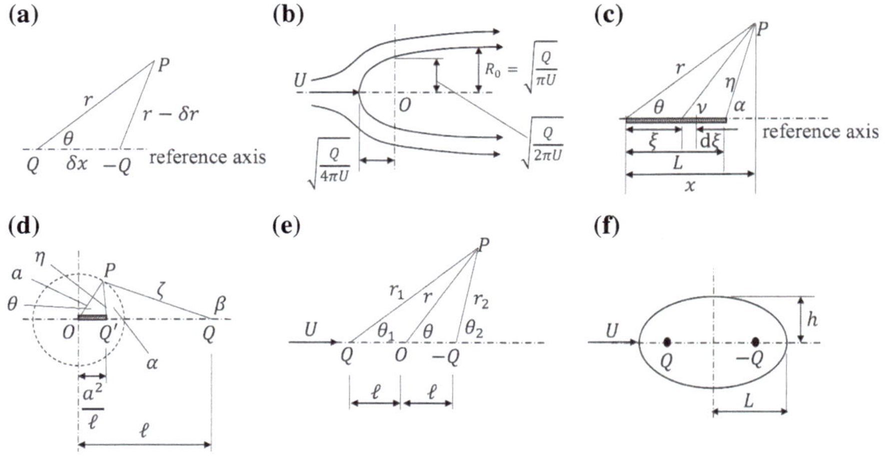

Fig. 7.25 Flows by using the principle of superposition. **a** A source and a sink in a close distance apart to form a doublet as $\delta x \to 0$. **b** Flows around an axis-symmetric body generated by a source in a uniform flow. **c** A line-distributed source along the reference axis. **d** A line-distributed sink with two sources. **e** A uniform flow with a source and a sink. **f** A uniform flow approaching a Rankine solid

With the assumption that $\psi =$ constant, it is found that

$$r = \sqrt{\frac{2\psi}{U\sin^2\theta} + \frac{Q}{4\pi U \sin^2(\theta/2)}}, \quad \longrightarrow \quad r_0 = \sqrt{\frac{Q}{4\pi U}}\frac{1}{\sin(\theta/2)}, \tag{7.6.27}$$

where $r = r_0$ represents the radius for $\psi = 0$, corresponding to the principal values of θ given by

$$\theta = 0, \ \ r_0 \to \infty; \quad \theta = \frac{\pi}{2}, \ \ r_0 = \sqrt{\frac{Q}{2\pi U}}; \quad \theta = \pi, \ \ r_0 = \sqrt{\frac{Q}{4\pi U}}, \tag{7.6.28}$$

which defines the stream surface $\psi = 0$ shown in Fig. 7.25b.

Although $r_0 \to \infty$ at $\theta = 0$, the cylindrical radius R_0 is, however, finite. Since $R = r\sin\theta$ (thus $R_0 = r_0 \sin\theta$), it follows from Eq. $(7.6.27)_2$ that

$$R_0 = \sqrt{\frac{Q}{4\pi U}}\frac{\sin\theta}{\sin(\theta/2)} = \sqrt{\frac{Q}{\pi U}}, \tag{7.6.29}$$

as $\theta \to 0$, for $\sin\theta/\sin(\theta/2) \to 2$. Hence, the fluid which emanates from the source locating at the origin does not mix with that of the uniform flow. A shell could be fitted to the shape of surface corresponding to $\psi = 0$, and the source could be removed without disturbing the outer flow, so that the stream function given in Eq. (7.6.26) corresponds exactly to that of a uniform flow approaching a semi-infinite body with a blunt nose. The corresponding velocity potential function is determined directly from Eqs. (7.6.12) and (7.6.18), viz.,

$$\phi(r,\theta) = Ur\cos\theta - \frac{Q}{4\pi r}. \tag{7.6.30}$$

Flows around a sphere. The stream function for a uniform flow past a sphere may be obtained by superimposing the solution of a uniform flow and that of a doublet, which is given by

$$\psi(r,\theta) = \frac{1}{2}Ur^2\sin^2\theta - \frac{\mu}{4\pi r}\sin^2\theta, \tag{7.6.31}$$

by which the stream surface corresponding to $\psi = 0$ is described by

$$r = r_0 = \left(\frac{\mu}{2\pi U}\right)^{1/3}, \tag{7.6.32}$$

which is a constant. Thus, the surface which corresponds to $\psi = 0$ is the surface of a sphere. If the strength of doublet is chosen to be $\mu = 2\pi U a^3$, then the radius of sphere will be $r_0 = a$. For such a case, the stream and velocity potential functions of a uniform flow through a sphere with radius a are obtained respectively as

$$\psi(r,\theta) = \frac{1}{2}U\left(r^2 - \frac{a^3}{r}\right)\sin^2\theta, \qquad \phi(r,\theta) = U\left(r + \frac{1}{2}\frac{a^3}{r^2}\right)\cos\theta, \tag{7.6.33}$$

which are symmetric with respect to both the horizontal and vertical axes with the coordinate origin locating at the center of sphere.

Flows by a line-distributed source. Consider a source which is distributed uniformly over the section $0 \le x \le L$ of reference axis, as shown in Fig. 7.25c. The source strength per unit length is denoted by $q = Q/L$, so that qL represents the total fluid volume emanating from the source per unit time. The origin of coordinate system locates at the left end of line source, by which the coordinates of an arbitrary point P are denoted by (r,θ). The distance of this point from the other end of line source is η with angle α with respect to the x-axis. Let an element of the line-distributed source in a distance ξ from the origin be denoted by $\mathrm{d}\xi$, and the line connecting it and point P subtends an angle ν to the x-axis. The stream function for the line source is then given by the summation of that line element $\mathrm{d}\xi$ over the entire L, viz.,

$$\psi(r,\theta) = -\int_0^L \frac{q\,\mathrm{d}\xi}{4\pi}(1+\cos\nu). \tag{7.6.34}$$

In view of Fig. 7.25c, it is seen that

$$x - \xi = R\cot\nu, \qquad \longrightarrow \qquad -\mathrm{d}\xi = -R\csc^2\nu\,\mathrm{d}\nu, \tag{7.6.35}$$

where $R = r\sin\theta$. Substituting these expressions into Eq. (7.6.34) yields

$$\psi(r,\theta) = -\frac{qR}{4\pi}\left(\cot\theta - \cot\alpha + \frac{1}{\sin\theta} - \frac{1}{\sin\alpha}\right) = -\frac{q}{4\pi}(L + r - \eta), \tag{7.6.36}$$

in which the identities

$$x = R\cot\theta, \qquad x - L = R\cot\alpha, \qquad r = \frac{R}{\sin\theta}, \qquad \eta = \frac{R}{\sin\alpha}, \tag{7.6.37}$$

have been used.

The velocity potential function corresponding to Eq. (7.6.36) is obtained in an analogous manner, and is given by

$$\phi(r, \theta) = -\int_0^L \frac{q\,\mathrm{d}\xi}{4\pi(R/\sin\nu)}, \tag{7.6.38}$$

which, by using Eq. (7.6.35), is recast alternatively as

$$\phi(r, \theta) = -\frac{q}{4\pi}\int_\theta^\alpha \sin\nu\csc^2\nu\,\mathrm{d}\nu = -\frac{q}{4\pi}\ln\left[\frac{\tan(\alpha/2)}{\tan(\theta/2)}\right]. \tag{7.6.39}$$

Sphere in a source flow. Figure 7.25d shows a source with strength Q locating on the reference axis with distance ℓ from the origin and a source with strength Q' locating also on the reference axis at the distance a^2/ℓ from the origin. Between points O and Q', there exists a line-distributed sink with strength q per unit length. It is assumed that $qa^2/\ell = Q'$ to make the spherical surface $r = a$ be a stream surface. With these, the stream function is obtained as

$$\psi(r, \theta) = -\frac{Q}{4\pi}(1 + \cos\beta) - \frac{Q'}{4\pi}(1 + \cos\alpha) + \frac{Q'}{4\pi}\frac{\ell}{a^2}\left(\frac{a^2}{\ell} + r - \eta\right), \tag{7.6.40}$$

which reduces to

$$\psi(a, \theta) = -\frac{Q}{4\pi}(1 + \cos\beta) - \frac{Q'}{4\pi}(1 + \cos\alpha) + \frac{Q'}{4\pi}\left(1 + \frac{\ell}{a} - \frac{\ell\eta}{a^2}\right), \tag{7.6.41}$$

for the sphere surface with $r = a$. Let point P be located on this sphere. In view of Fig. 7.25d, since

$$\frac{a^2/\ell}{a} = \frac{a}{\ell}, \qquad \longrightarrow \qquad \frac{OQ'}{OP} = \frac{OP}{OQ}, \tag{7.6.42}$$

which denotes the propositions between line segments OQ', OP, and OQ, it follows that two triangles OPQ' and OQP are similar to the common angle θ, and the angle η is identified to be

$$\eta = \frac{a^2}{\ell}\cos(\pi - \alpha) + a\cos(\pi - \beta) = -\frac{a^2}{\ell}\cos\alpha - a\cos\beta. \tag{7.6.43}$$

Substituting this expression into Eq. (7.6.41) gives

$$\psi(a, \theta) = (1 + \cos\beta)\left(-\frac{Q}{4\pi} + \frac{Q'}{4\pi}\frac{\ell}{a}\right) = 0, \quad \longrightarrow \quad Q = Q'\frac{a}{\ell}, \tag{7.6.44}$$

in which the source strength Q' is chosen to be equal to aQ/ℓ, so that the surface $r = a$ corresponds to the stream function $\psi = 0$. The stream function for a sphere of radius a whose center is at the origin and which is exposed to a point source of strength Q locating a distance along the positive part of reference axis is then given by

$$\psi(r,\theta) = -\frac{Q}{4\pi}(1+\cos\beta) - \frac{Q}{4\pi}\frac{a}{\ell}(1+\cos\alpha) + \frac{Q}{4\pi}\left(\frac{a}{\ell}+\frac{r}{a}-\frac{\eta}{a}\right). \quad (7.6.45)$$

Similarly, the velocity potential function is obtained as

$$\phi(r,\theta) = -\frac{Q}{4\pi\zeta} - \frac{Q'}{4\pi\eta} + \frac{q}{4\pi}\ln\left[\frac{\tan(\alpha/2)}{\tan(\theta/2)}\right], \quad (7.6.46)$$

which reduces to

$$\phi(r,\theta) = -\frac{Q}{4\pi\zeta} - \frac{Qa}{4\pi\eta\ell} + \frac{Q}{4\pi a}\ln\left[\frac{\tan(\alpha/2)}{\tan(\theta/2)}\right], \quad (7.6.47)$$

where ζ is the distance between points P and Q.

Rankine solids. By superimposing a source and a sink of equal strength in a uniform flow field, a family of bodies, known as the *Rankine solids*, can be obtained. Let the magnitude of uniform flow be denoted by U, the strength of source and sink be denoted by Q, and the source and sink be located with equal distance ℓ from the origin, as shown in Fig. 7.25e. It follows from Eqs. (7.6.14) and $(7.6.19)_2$ that

$$\psi(r,\theta) = \frac{1}{2}Ur^2\sin^2\theta - \frac{Q}{4\pi}(\cos\theta_1 - \cos\theta_2). \quad (7.6.48)$$

It is assumed that $\psi = 0$ on the surface $r = r_0$, and then r_0 must satisfy

$$0 = \frac{1}{2}Ur_0^2\sin^2\theta - \frac{Q}{4\pi}(\cos\theta_1 - \cos\theta_2), \quad (7.6.49)$$

which reduces to

$$R_0^2 = \frac{Q}{2\pi U}(\cos\theta_1 - \cos\theta_2), \quad (7.6.50)$$

where $R_0 = r_0\sin\theta$. It is found that $R_0 = 0$ as $\theta_1 = \theta_2 = 0$ or $\theta_1 = \theta_2 = \pi$, and R_0 assumes the maximum value at $\cos\theta_1 = -\cos\theta_2$, corresponding to $\theta = \pi/2$ or $\theta = 3\pi/2$. Thus, the stream surface $\psi = 0$ represents the surface of body. The principal dimensions of this body are the half width L and half height h, as shown in Fig. 7.25f; both depend on the free stream velocity U, the strength Q of source and sink, and the distance ℓ.

Since the velocity at the downstream stagnation point must vanish, it follows that

$$U + \frac{Q}{4\pi(L+\ell)^2} - \frac{Q}{4\pi(L-\ell)^2} = 0, \quad (7.6.51)$$

which reduces to

$$(L^2-\ell^2)^2 - \frac{Q\ell}{\pi U}L = 0. \quad (7.6.52)$$

This equation must be satisfied by L. It is noted that $R_0 = h$ if $\cos\theta_1 = -\cos\theta_2$, where $\tan\theta_1 = h/\ell$. Hence,

$$h^2 = \frac{Q}{2\pi U}\frac{2\ell}{\sqrt{h^2+\ell^2}}, \quad \longrightarrow \quad h^2\sqrt{h^2+\ell^2} - \frac{Q\ell}{\pi U} = 0. \quad (7.6.53)$$

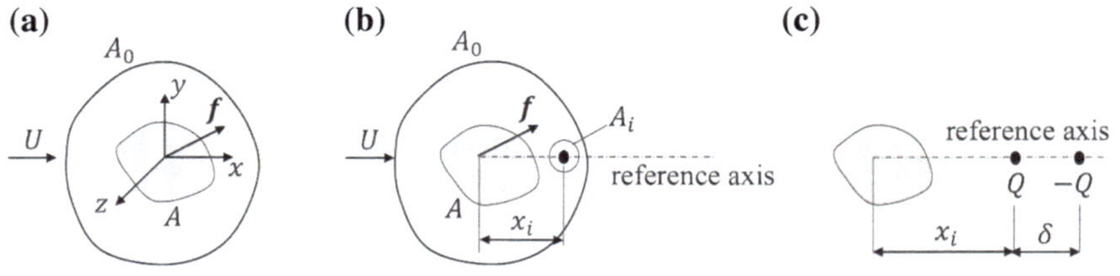

Fig. 7.26 Hydrodynamic force acting on a three-dimensional body immersed in a fluid. **a** The configuration of d'Alembert's paradox. **b** The configuration of a force induced by a singularity. **c** A source and a sink which are close together near the body to form a singularity

For various values of U, Q, and ℓ, Eqs. (7.6.52) and (7.6.53)$_2$ define a family of bodies of revolution for which the stream function is given by Eq. (7.6.48). The corresponding velocity potential function is obtained as

$$\phi(r,\theta) = Ur\cos\theta - \frac{Q}{4\pi r_1} + \frac{Q}{4\pi r_2}. \tag{7.6.54}$$

7.6.4 D'Alembert's Paradox

Consider an arbitrarily three-dimensional body with surface A and unit outward normal $\boldsymbol{n}$, which is completely immersed in a moving fluid with velocity U, as shown in Fig. 7.26a. A spherical surface A_0, with unit outward normal $\boldsymbol{n}_0$, embraces the whole body, where $\boldsymbol{n}_0 = \boldsymbol{e}_r$, which is the unit radial vector. Let the hydrodynamic force acting on the body by the surrounding fluid be denoted by $\boldsymbol{f}$. It is required to determine this hydrodynamic force.

The region between surfaces A and A_0 is taken as the finite control-volume, and it is assumed that there exists no transfer of linear momentum across surface A, for it is a stream surface. Applying the integral balance of linear momentum to the fluid contained inside the control-volume yields

$$-\boldsymbol{f} - \int_{A_0} p\boldsymbol{n}_0 \mathrm{d}a = \int_{A_0} [\boldsymbol{u}(\rho\boldsymbol{u}\cdot\boldsymbol{n}_0)]\,\mathrm{d}a, \tag{7.6.55}$$

in which the steady-flow assumption has been used. If the flow is further assumed to be irrotational, then the Bernoulli equation is formulated as

$$p + \frac{1}{2}\boldsymbol{u}\cdot\boldsymbol{u} = B, \tag{7.6.56}$$

where B is the Bernoulli constant, by which Eq. (7.6.55) is recast alternatively as

$$\boldsymbol{f} = \rho\int_{A_0}\left[\frac{1}{2}(\boldsymbol{u}\cdot\boldsymbol{u})\boldsymbol{n}_0 - \boldsymbol{u}(\boldsymbol{u}\cdot\boldsymbol{n}_0)\right]\mathrm{d}a, \tag{7.6.57}$$

for $\int_{A_0} B\boldsymbol{n}_0\mathrm{d}a = \boldsymbol{0}$. For convenience, let the velocity $\boldsymbol{u}$ be decomposed into

$$\boldsymbol{u} = \boldsymbol{U} + \boldsymbol{u}', \tag{7.6.58}$$

where $\boldsymbol{U}$ is the free stream velocity and $\boldsymbol{u}'$ is the perturbation velocity. In the free stream region, $\boldsymbol{u}' = \boldsymbol{0}$, while it becomes larger in the region near the body. Substituting this expression into Eq. (7.6.57) gives

$$\boldsymbol{f} = \rho \int_{A_0} \left\{ \left(\frac{1}{2}U^2 + \boldsymbol{U} \cdot \boldsymbol{u}' + \frac{1}{2}\boldsymbol{u}' \cdot \boldsymbol{u}' \right) \boldsymbol{n}_0 - (\boldsymbol{U} + \boldsymbol{u}') \left[(\boldsymbol{U} + \boldsymbol{u}') \cdot \boldsymbol{n}_0 \right] \right\} da. \tag{7.6.59}$$

To evaluate this integration, it is noted that

$$\int_{A_0} U^2 \boldsymbol{n}_0 \mathrm{d}a = \boldsymbol{0}, \quad \int_{A_0} \boldsymbol{U} \cdot \boldsymbol{n}_0 \, \mathrm{d}a = 0, \quad \boldsymbol{U} \times \left(\boldsymbol{u}' \times \boldsymbol{n}_0 \right) = \boldsymbol{u}' \left(\boldsymbol{U} \cdot \boldsymbol{n}_0 \right) - \boldsymbol{n}_0 \left(\boldsymbol{U} \cdot \boldsymbol{u}' \right), \tag{7.6.60}$$

because U^2 is a constant and $\boldsymbol{U}$ is a constant vector. Substituting these expressions into Eq. (7.6.59) gives rise to

$$\boldsymbol{f} = \rho \int_{A_0} \left[-\boldsymbol{U} \times \left(\boldsymbol{u}' \times \boldsymbol{n}_0 \right) + \frac{1}{2} \left(\boldsymbol{u}' \cdot \boldsymbol{u}' \right) \boldsymbol{n}_0 - \boldsymbol{u}' \left(\boldsymbol{u}' \cdot \boldsymbol{n}_0 \right) \right] da. \tag{7.6.61}$$

Let ϕ' be the velocity potential function corresponding to $\boldsymbol{u}'$. It follows from Eq. (7.6.8) that ϕ' must be of the form

$$\phi' = \frac{Q}{4\pi r} + \frac{\mu \cos \theta}{4\pi r^2} + \mathcal{O}\left(\frac{1}{r^3} \right), \tag{7.6.62}$$

where the first two terms on the right-hand-side are the contributions of a source and a doublet, respectively, while the last term denotes the contributions depending on $1/r^3$. Since $\boldsymbol{u}' = \nabla \phi'$, the magnitude of $\boldsymbol{u}'$ is determined as

$$\left\| \boldsymbol{u}' \right\| = \mathcal{O}\left(\frac{1}{r^2} \right). \tag{7.6.63}$$

That is, the perturbation velocity varies at most with $1/r^2$. It is also found that

$$\boldsymbol{u}' \times \boldsymbol{n}_0 = \left[-\frac{Q}{4\pi r^2} \boldsymbol{e}_r + \mathcal{O}\left(\frac{1}{r^3} \right) \right] \times \boldsymbol{e}_r, \longrightarrow \left\| \boldsymbol{u}' \times \boldsymbol{n}_0 \right\| = \mathcal{O}\left(\frac{1}{r^3} \right). \tag{7.6.64}$$

Moreover, a surface element $\mathrm{d}a$ on the surface A_0 is given by $\mathrm{d}a = r^2 \sin \theta \, \mathrm{d}\omega$, and its magnitude depends on $\mathcal{O}(r^2)$. With these estimations, the orders of magnitude of the terms on the right-hand-side Eq. (7.6.61) are identified to be

$$\begin{aligned} &\int_{A_0} \boldsymbol{U} \times \left(\boldsymbol{u}' \times \boldsymbol{n}_0 \right) \mathrm{d}a = \mathcal{O}\left(\frac{1}{r} \right), \quad \int_{A_0} \left(\boldsymbol{u}' \cdot \boldsymbol{u}' \right) \boldsymbol{n}_0 \mathrm{d}a = \mathcal{O}\left(\frac{1}{r^2} \right), \\ &\int_{A_0} \boldsymbol{u}' \left(\boldsymbol{u}' \cdot \boldsymbol{n}_0 \right) \mathrm{d}a = \mathcal{O}\left(\frac{1}{r^2} \right). \end{aligned} \tag{7.6.65}$$

As the radius r of surface A_0 approaches infinite, all the terms in the above equation vanish, so that

$$\boldsymbol{f} = \boldsymbol{0}, \tag{7.6.66}$$

indicating a vanishing hydrodynamic force acting on a body which is immersed completely in a moving fluid.

Since it is a well-known fact that any body submerged in a flow field experiences a drag force, Eq. (7.6.66) becomes a paradox which is known as *d'Alembert's paradox*. The resolution of paradox lies in the fact that the viscous effect is omitted in deriving Eq. (7.6.66). It will be seen in Sect. 8.4 that there is a thin fluid layer around a body in which the viscous effect cannot be neglected, which is called the boundary layer, exerting a shear stress on the body to give rise to a drag force. Occasionally, the boundary layer may separate from the body surface, and creates a low-pressure wake, inducing an additional drag, called the *form drag*, resulted from the pressure differential around the surface of body.

On the contrary, a force does exist for a three-dimensional body if it is exposed to a point singularity in the fluid, as shown in Fig. 7.26b, in which the singularity locates at point $x = x_i$ on the reference axis. A small spherical surface A_i with radius ε and unit outward normal $\boldsymbol{n}_i$ embraces the singularity. The region between A_0, A_i, and the surface of body A with unit outward normal $\boldsymbol{n}$ is taken as the finite control-volume. Applying the integral balance of linear momentum to the control-volume yields

$$-\boldsymbol{f} - \int_{A_0} p\boldsymbol{n}_0 \,\mathrm{d}a + \int_{A_i} p\boldsymbol{n}_i \,\mathrm{d}a = \int_{A_0} [\boldsymbol{u}(\rho\boldsymbol{u}\cdot\boldsymbol{n}_0)]\,\mathrm{d}a - \int_{A_i} [\boldsymbol{u}(\rho\boldsymbol{u}\cdot\boldsymbol{n}_i)]\,\mathrm{d}a, \tag{7.6.67}$$

which is simplified to

$$\boldsymbol{f} = \int_{A_i} \big[p\boldsymbol{n}_i + \rho\boldsymbol{u}(\boldsymbol{u}\cdot\boldsymbol{n}_i)\big]\,\mathrm{d}a, \tag{7.6.68}$$

in which Eqs. (7.6.55) and (7.6.66) have been used. With the Bernoulli equation given by $p + \boldsymbol{u}\cdot\boldsymbol{u}/2 = B$, Eq. (7.6.68) is expressed alternatively as

$$\boldsymbol{f} = \rho \int_{A_i} \left[-\frac{1}{2}(\boldsymbol{u}\cdot\boldsymbol{u})\boldsymbol{n}_i + \boldsymbol{u}(\boldsymbol{u}\cdot\boldsymbol{n}_i)\right]\mathrm{d}a. \tag{7.6.69}$$

Consider first the singularity at $x = x_i$ to be a source with strength Q, with which the velocity on the surface A_i is obtained as

$$\boldsymbol{u} = \frac{Q}{4\pi\varepsilon^2}\boldsymbol{e}_\varepsilon + \boldsymbol{u}_i, \tag{7.6.70}$$

where the first term is the contribution of source, with $\boldsymbol{e}_\varepsilon$ representing the unit vector radial from point $x = x_i$, while $\boldsymbol{u}_i$ is the velocity induced by all means other than the source. It follows immediately that

$$\boldsymbol{u}\cdot\boldsymbol{u} = \frac{Q^2}{16\pi^2\varepsilon^4} + \frac{Q}{2\pi\varepsilon^2}\boldsymbol{e}_\varepsilon\cdot\boldsymbol{u}_i + \boldsymbol{u}_i\cdot\boldsymbol{u}_i, \quad \boldsymbol{u}\cdot\boldsymbol{n}_i = \frac{Q}{4\pi\varepsilon^2} + \boldsymbol{u}_i\cdot\boldsymbol{e}_\varepsilon, \tag{7.6.71}$$

with which Eq. (7.6.69) becomes

$$\boldsymbol{f} = \rho \int_{A_i} \left[\frac{Q^2}{32\pi^2\varepsilon^4}\boldsymbol{e}_\varepsilon - \frac{1}{2}(\boldsymbol{u}_i\cdot\boldsymbol{u}_i)\boldsymbol{e}_\varepsilon + \frac{Q}{4\pi\varepsilon^2}\boldsymbol{u}_i + (\boldsymbol{u}_i\cdot\boldsymbol{e}_\varepsilon)\boldsymbol{u}_i\right]\mathrm{d}a, \tag{7.6.72}$$

which is simplified to

$$f = \rho \int_{A_i} \frac{Q}{4\pi\varepsilon^2} \boldsymbol{u}_i \, \mathrm{d}a. \tag{7.6.73}$$

This result is so obtained that the first integral in Eq. (7.6.72) vanishes since it involves constant times $\boldsymbol{e}_\varepsilon$ around a closed surface. The second integral vanishes equally, because as $\varepsilon \to 0$ the term $\boldsymbol{u}_i \cdot \boldsymbol{u}_i$ may be considered to be constant over the surface A_i. The last integral vanishes due to the fact that $\boldsymbol{u}_i$ is constant, and hence the integration of $(\boldsymbol{u}_i \cdot \boldsymbol{e}_\varepsilon)\boldsymbol{u}_i$ over A_i will be zero, although $\boldsymbol{e}_\varepsilon$ changes its direction around A_i. By using the approximation that $\boldsymbol{u}_i$ is constant as $\varepsilon \to 0$, Eq. (7.6.73) becomes

$$\boldsymbol{f} = \frac{\rho Q}{4\pi} \boldsymbol{u}_i \int_0^{2\pi} \mathrm{d}\omega \int_o^{\pi} \sin\theta \, \mathrm{d}\theta = \rho Q \boldsymbol{u}_i. \tag{7.6.74}$$

That is, the force acting on the body and on the source is proportional to the source strength and the magnitude of velocity $\boldsymbol{u}_i$ induced at the location of source by all mechanisms other than the source itself. For a sink, the term Q is simply replaced by $-Q$.

Consider now the singularity be accomplished by a doublet, which is generated by superimposing a source and a sink of equal strength. Let the source with strength Q be located at $x = x_i$ and the sink with same strength be located at $x = x_i + \delta$, as shown in Fig. 7.26c, in which δ is a nearly vanishing small distance. The velocities at $x = x_i$ and $x = x_i + \delta$ are given respectively by

$$\boldsymbol{u}^{x_i} = \frac{Q}{4\pi\delta^2}\boldsymbol{e}_x + \boldsymbol{u}_i, \qquad \boldsymbol{u}^{x_i+\delta} = \frac{Q}{4\pi\delta^2}\boldsymbol{e}_x + \boldsymbol{u}_i + \delta\frac{\partial \boldsymbol{u}_i}{\partial x} + \cdots, \tag{7.6.75}$$

in which the Taylor series expansion has been used. As the same in the previous case, $\boldsymbol{u}_i$ is the fluid velocity due to all components of the flow except that induced by the considered source and sink. It follows from Eq. (7.6.73) that the forces acting on the body due to the source and sink are given respectively by

$$\boldsymbol{f}_{source} = \rho Q\left(\frac{Q}{4\pi\delta^2}\boldsymbol{e}_x + \boldsymbol{u}_i\right), \quad \boldsymbol{f}_{sink} = -\rho Q\left(\frac{Q}{4\pi\delta^2}\boldsymbol{e}_x + \boldsymbol{u}_i + \delta\frac{\partial \boldsymbol{u}_i}{\partial x} + \cdots\right), \tag{7.6.76}$$

by which the net force acting on the body is obtained as

$$\boldsymbol{f} = \boldsymbol{f}_{source} + \boldsymbol{f}_{sink} = -\rho Q\delta\frac{\partial \boldsymbol{u}_i}{\partial x}. \tag{7.6.77}$$

It is assumed that as $\delta \to 0$, $Q \to \infty$, so that the product $(Q\delta) \to \mu$, where μ is the doublet strength. With this, Eq. (7.6.77) becomes

$$\boldsymbol{f} = -\rho\mu\frac{\partial \boldsymbol{u}_i}{\partial x}, \tag{7.6.78}$$

which is the net force acting on the body due to a doublet with strength μ locating at $x = x_i$.

As an example of the derived results, consider a sphere in the presence of a source, as already discussed in Sect. 7.6.3, in which a source with strength Q locates at $x = \ell$, a source with strength Qa/ℓ locates at $x = a^2/\ell$, and a line-distributed sink with strength Q/a distributes over the region $0 \le x \le a^2/\ell$. With these, the velocity $\boldsymbol{u}_i$ at $x = \ell$ due to all causes except the source is obtained as

$$\boldsymbol{u}_i = \frac{Qa}{4\pi\ell} \frac{1}{(\ell - a^2/\ell)^2}\boldsymbol{e}_x - \int_0^{a^2/\ell} \frac{Q}{4\pi a} \frac{\boldsymbol{e}_x}{(\ell - x)^2}\mathrm{d}x = \frac{Qa^3}{4\pi\ell(\ell^2 - a^2)^2}\boldsymbol{e}_x, \quad (7.6.79)$$

with which the force acting on the sphere due to the source is given by

$$\boldsymbol{f} = \frac{\rho Q^2 a^3}{4\pi\ell(\ell^2 - a^2)^2}\boldsymbol{e}_x, \quad (7.6.80)$$

showing that the sphere is attracted to the source with a force being proportional to Q^2.

7.6.5 Kinetic Energy of Moving Fluid and Apparent Mass

The kinetic energy associated with a fluid in a uniform flow around a stationary body will be infinite if the flow field is infinite in extent. On the contrary, the kinetic energy induced in a quiescent fluid by the passing of a body through it will be finite, even if the flow field is infinite in extent. Based on this, the discussions on the kinetic energy are on a reference frame in which the fluid far away from the body is at rest and the body is moving. As shown in Fig. 7.27, an arbitrary body with surface A and unit outward normal $\boldsymbol{n}$ is moving with velocity $\boldsymbol{U}$ through a stationary fluid, and the body is embraced by an arbitrarily shaped surface A_0 with the same unit outward normal $\boldsymbol{n}$ for simplicity. The region between A and A_0 is taken as the finite control-volume, and the kinetic energy T of the fluid contained inside this control-volume is given by

$$T = \int_V \frac{1}{2}\rho\,(\boldsymbol{u} \cdot \boldsymbol{u})\,\mathrm{d}v = \frac{1}{2}\rho \int_V \nabla\phi \cdot \nabla\phi\,\mathrm{d}v = \frac{1}{2}\rho \int_\Sigma \phi\frac{\partial\phi}{\partial n}\mathrm{d}a, \quad (7.6.81)$$

in which the Green theorem has been used, where ϕ is the velocity potential function corresponding to the fluid motion induced by the moving body and Σ is the control-surface, which consists of surfaces A and A_0. Expanding the above equation yields

$$T = \frac{1}{2}\rho \int_{A_0} \phi\frac{\partial\phi}{\partial n}\mathrm{d}a - \frac{1}{2}\rho \int_A \phi\frac{\partial\phi}{\partial n}\mathrm{d}a. \quad (7.6.82)$$

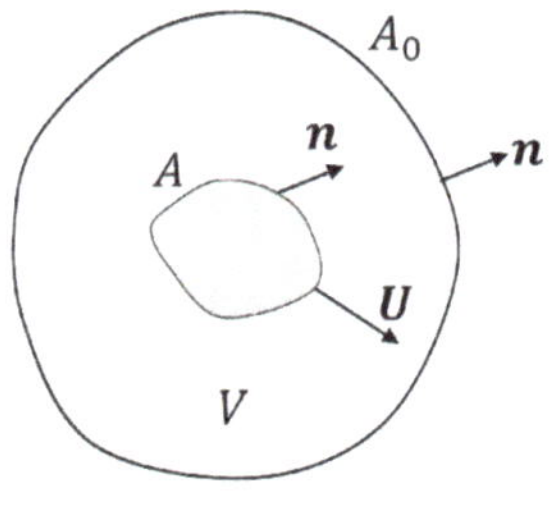

Fig. 7.27 Control-surfaces for an arbitrary body moving through a quiescent fluid

It follows from the continuity equation that

$$\int_V (\nabla \cdot \boldsymbol{u})\, \mathrm{d}v = 0, \qquad \longrightarrow \qquad \int_{A_0} \boldsymbol{u} \cdot \boldsymbol{n}\, \mathrm{d}a - \int_A \boldsymbol{u} \cdot \boldsymbol{n}\, \mathrm{d}a = 0, \tag{7.6.83}$$

in which the Gauss theorem has been used. However, it is noted that $\boldsymbol{u} \cdot \boldsymbol{n} = \partial\phi/\partial n$ and $\boldsymbol{u} = \boldsymbol{U}$ on A, with which Eq. (7.6.83)$_2$ is recast alternatively as

$$\int_{A_0} \frac{\partial \phi}{\partial n} \mathrm{d}a - \int_A \boldsymbol{U} \cdot \boldsymbol{n}\, \mathrm{d}a = 0, \qquad \longrightarrow \qquad \int_{A_0} C \frac{\partial \phi}{\partial n} \mathrm{d}a = 0, \tag{7.6.84}$$

in which the second integral on the left-hand-side of first equation is null, for $\boldsymbol{U}$ is a constant vector, and the inclusion of a constant C does not alter the resulting equation. Substituting this result into Eq. (7.6.82) gives

$$T = \frac{1}{2}\rho \int_{A_0} (\phi - C) \frac{\partial \phi}{\partial n} \mathrm{d}a - \frac{1}{2}\rho \int_A \phi \frac{\partial \phi}{\partial n} \mathrm{d}a. \tag{7.6.85}$$

Since in the region far away from the body the fluid velocity is zero, the velocity potential function there can at most be a constant. Let A_0 be so large and the value of C be the value of ϕ, and the first integral on the right-hand-side vanishes, so that the kinetic energy induced in the fluid by the motion of body is obtained as

$$T = -\frac{1}{2}\rho \int_A \phi \frac{\partial \phi}{\partial n} \mathrm{d}a, \tag{7.6.86}$$

where ϕ is the velocity potential function corresponding to the body moving through a stationary fluid.

When a body moves through a stationary fluid, a certain mass of the fluid is driven to move to some greater or lesser extent. The *apparent mass* of a fluid, m', is then defined as the mass of fluid which, if it were moving with the same velocity of body, would have the same kinetic energy as the entire fluid, i.e.,

$$\frac{1}{2} m' U^2 \equiv -\frac{1}{2}\rho \int_A \phi \frac{\partial \phi}{\partial n} \mathrm{d}a, \qquad \longrightarrow \qquad m' = -\frac{\rho}{U^2} \int_A \phi \frac{\partial \phi}{\partial n} \mathrm{d}a. \tag{7.6.87}$$

Since for arbitrarily shaped bodies ϕ depends on the direction of flow, the apparent mass of fluid associated with a given body becomes a property of body shape. As similar to the inertia, there exist in general three principal axes of the apparent mass. For axis-symmetric bodies, there exist two principal values of m', while for spherical bodies there exists only one.

As an illustration of the concept of apparent mass, consider a sphere which is moving in a stationary fluid. The velocity potential function given in Eq. (7.6.33)$_2$ corresponds to a stationary sphere with radius a in a uniform flow of magnitude U. To meet the configuration of apparent mass, a velocity potential function of a uniform flow of magnitude U in the negative x-axis is superimposed to Eq. (7.6.33)$_2$, so that

$$\phi(r, \theta) = U \left(r + \frac{1}{2} \frac{a^3}{r^2} \right) \cos\theta - U r \cos\theta = \frac{1}{2} U \frac{a^3}{r^2} \cos\theta, \tag{7.6.88}$$

which is the required velocity potential function. It follows immediately that

$$\frac{\partial \phi}{\partial n} = \frac{\partial \phi}{\partial r} = -U \frac{a^3}{r^3} \cos\theta, \qquad \longrightarrow \qquad \left(\phi \frac{\partial \phi}{\partial n} \right)_A = -\frac{1}{2} U^2 a \cos^2\theta. \tag{7.6.89}$$

Substituting these results into Eq. (7.6.87) results in

$$m' = -\frac{\rho}{U^2}\int_0^{2\pi} \mathrm{d}\omega \int_0^{\pi}\left(-\frac{1}{2}U^2 a\cos^2\theta\right)a^2\sin\theta\,\mathrm{d}\theta = \frac{2\pi}{3}\rho a^3. \quad (7.6.90)$$

That is, m' for a sphere is one-half of the mass of the same volume of fluid. This apparent mass may be added to the actual mass of sphere, and the total mass may be used in the dynamic equations of sphere. In other words, the existence of fluid may be ignored if its apparent mass is incorporated into the actual mass of body.

7.7 Surface Waves

The effect of gravity on liquid surfaces is discussed in this section. Flows associated with surface waves are assumed to be potential in nature, which is a valid approximation for many free surface phenomena. Most of the discussions in the following are conducted for two-dimensional circumstances, unless stated otherwise.

7.7.1 General Formulation

When a quiescent liquid body experiences gravity waves on its free surface, the motion induced by the surface waves may be considered to be irrotational in most cases. This implies that the governing equations of surface wave problems are the same as those in potential flows, except that the boundary conditions need to be formulated accordingly.

Consider a liquid body in which waves exist on its free surface, as shown in Fig. 7.28a, in which the free surface is described by $y = \eta(x, z, t)$ with mean liquid depth h, on which the coordinate x locates. Three boundary conditions on the free surface and bottom bed must be allocated. The first boundary condition imposed on $y = \eta$ is called the *kinematic boundary condition*, which states that a liquid particle which is at some time on the free surface will always remain on the free surface at subsequent times. Mathematically, it is described by

$$\frac{\mathrm{D}}{\mathrm{D}t}(y-\eta) = \frac{\partial}{\partial t}(y-\eta) + \boldsymbol{u}\cdot\nabla(y-\eta) = 0, \longrightarrow -\frac{\partial\eta}{\partial t} - u\frac{\partial\eta}{\partial x} + v - w\frac{\partial\eta}{\partial z} = 0, \quad (7.7.1)$$

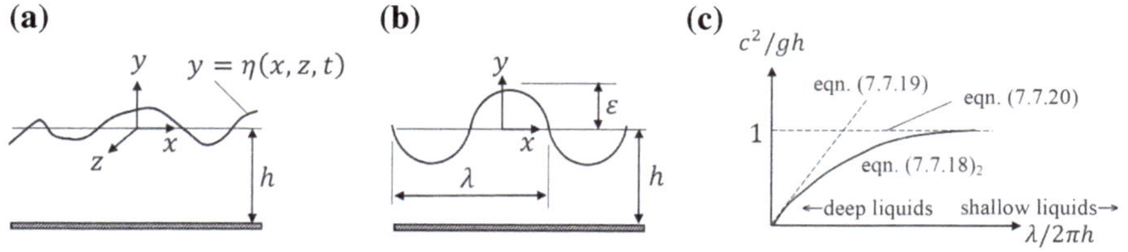

Fig. 7.28 Configuration of surface waves. **a** The coordinate system for two-dimensional surface wave problems. **b** A two-dimensional small-amplitude plane wave in purely sinusoidal form. **c** The propagation speeds of small-amplitude surface waves in sinusoidal form

which is recast alternatively as

$$\frac{\partial \eta}{\partial t} + \frac{\partial \phi}{\partial x}\frac{\partial \eta}{\partial x} + \frac{\partial \phi}{\partial z}\frac{\partial \eta}{\partial z} = \frac{\partial \phi}{\partial y}, \tag{7.7.2}$$

in which $\boldsymbol{u} = \nabla\phi$ has been used. The second boundary condition imposed on $y = \eta$, termed the *dynamic boundary condition*, is that the pressure p on the free surface should satisfy $p = P(x, z, t)$, where P comes from the pressure of the Bernoulli equation, which is given by

$$\frac{\partial \phi}{\partial t} + \frac{P}{\rho} + \frac{1}{2}\nabla\phi \cdot \nabla\phi + g\eta = F(t), \tag{7.7.3}$$

in which only the gravitational field is taken into account. The third boundary condition should be imposed at the bottom bed. It is required that the velocity component normal to the bed should vanish. For the flat bed shown in Fig. 7.28a, this corresponds simply to $\partial\phi/\partial y = 0$ at $y = -h$, which is called the *bed boundary condition*.

As a summary, the governing equation in terms of the velocity potential function ϕ for surface wave problems is given by

$$\nabla^2\phi = \frac{\partial^2 \phi}{\partial x^2} + \frac{\partial^2 \phi}{\partial y^2} + \frac{\partial^2 \phi}{\partial z^2} = 0, \tag{7.7.4}$$

which is associated respectively with the kinematic, dynamic, and bed boundary conditions given by

$$\begin{aligned} y = \eta: &\quad \frac{\partial \eta}{\partial t} + \frac{\partial \phi}{\partial x}\frac{\partial \eta}{\partial x} + \frac{\partial \phi}{\partial z}\frac{\partial \eta}{\partial z} = \frac{\partial \phi}{\partial y}; \quad \frac{\partial \phi}{\partial t} + \frac{P}{\rho} + \frac{1}{2}\nabla\phi \cdot \nabla\phi + g\eta = F(t), \\ y = -h: &\quad \frac{\partial \phi}{\partial y} = 0. \end{aligned} \tag{7.7.5}$$

As an illustration of the formulation, consider a two-dimensional flow field in the (xy)-plane with waves on the surface, for which Eq. (7.7.4) reduces to

$$\frac{\partial^2 \phi}{\partial x^2} + \frac{\partial^2 \phi}{\partial y^2} = 0. \tag{7.7.6}$$

For simplicity, it is assumed that the wave amplitude is small compared with other characteristic lengths such as the mean liquid height h and wavelength of the waves, which leads to that the value of η is small compared with the wavelength. This implies that $\partial\eta/\partial x$ and $\partial\phi/\partial x$ are both small, for $\partial\eta/\partial x$ is the slope of free surface, and $\partial\phi/\partial x$ represents a velocity component which is small for waves with low frequencies. With these, the kinematic boundary equation is simplified to

$$\frac{\partial \eta}{\partial t}(x, t) = \frac{\partial \phi}{\partial y}(x, \eta, t) = \frac{\partial \phi}{\partial y}(x, 0, t) + \eta\frac{\partial^2 \phi}{\partial y^2}(x, 0, t) + \mathcal{O}(\eta^2), \tag{7.7.7}$$

in which a Taylor series expansion has been made for $\partial\phi/\partial y$ at $y = \eta$ about the line $y = 0$. Applying a first-order approximation to the above equation gives

$$\frac{\partial \eta}{\partial t}(x, t) = \frac{\partial \phi}{\partial y}(x, 0, t). \tag{7.7.8}$$

Since the liquid is essentially quiescent and any liquid motion is induced by the waves, the nonlinear term $\nabla\phi \cdot \nabla\phi$ in Eq. (7.7.3) may be neglected as being quadratically small, so that the dynamic boundary condition may be simplified to

$$\frac{\partial \phi}{\partial t}(x, \eta, t) + \frac{P(x, t)}{\rho} + g\eta(x, t) = F(t). \tag{7.7.9}$$

Similarly, the first term on the left-hand-side in this equation may be expanded in a Taylor series about $y = 0$, and only the first-order term in the resulting expansion needs to be retained. This yields

$$\frac{\partial \phi}{\partial t}(x, 0, t) + \frac{P(x, t)}{\rho} + g\eta(x, t) = F(t), \tag{7.7.10}$$

which is further simplified to

$$\begin{aligned} &\frac{\partial}{\partial t}\left(\phi - \int F(t)\mathrm{d}t\right)(x, 0, t) + \frac{P(x, t)}{\rho} + g\eta(x, t) = 0, \\ \longrightarrow\ &\frac{\partial \phi}{\partial t}(x, 0, t) + \frac{P(x, t)}{\rho} + g\eta(x, t) = 0, \end{aligned} \tag{7.7.11}$$

by introducing a new velocity potential function given by $\left(\phi - \int F(t)\mathrm{d}t\right)$ without changing the symbol for simplicity. Taking time derivative of this equation gives

$$\frac{\partial^2 \phi}{\partial t^2}(x, 0, t) + \frac{1}{\rho}\frac{\partial P(x, t)}{\partial t} + g\frac{\partial \phi}{\partial y}(x, 0, t) = 0, \tag{7.7.12}$$

in which Eq. (7.7.8) has been used. This equation is the preferred form of dynamic boundary condition at $y = \eta$. The bed boundary equation given in Eq. $(7.7.5)_2$ is unaffected by the approximation of small-amplitude waves, i.e.,

$$\frac{\partial \phi}{\partial y}(x, -h, t) = 0. \tag{7.7.13}$$

Equations (7.7.6), (7.7.8), (7.7.12) and (7.7.13) are the two-dimensional approximation to Eqs. (7.7.4) and (7.7.5) with small-amplitude surface waves.

To apply the formulation, consider a small-amplitude sinusoidal wave with amplitude ε and wavelength λ traveling along the surface of liquid with velocity c shown in Fig. 7.28b. The free surface is described by

$$\eta(x, t) = \varepsilon \sin\frac{2\pi}{\lambda}(x - ct), \tag{7.7.14}$$

corresponding to a wave traveling in the positive x-direction with velocity c, which is an unknown and needs to be determined for given values of ε, λ, and h. For simplicity, the effect of surface tension is assumed to be negligible at the present stage, so that the pressure on the free surface of liquid is constant and equals the atmospheric pressure, i.e., $P(x, t) = p_0 =$ constant. With this, Eqs. (7.7.6), (7.7.8), (7.7.12) and (7.7.13) become

$$\begin{aligned} &0 = \frac{\partial^2 \phi}{\partial x^2} + \frac{\partial^2 \phi}{\partial y^2}, &&0 = \frac{\partial \phi}{\partial y}(x, 0, t) + \varepsilon\frac{2\pi c}{\lambda}\cos\frac{2\pi}{\lambda}(x - ct), \\ &0 = \frac{\partial^2 \phi}{\partial t^2}(x, 0, t) + g\frac{\partial \phi}{\partial y}(x, 0, t), &&0 = \frac{\partial \phi}{\partial y}(x, -h, t). \end{aligned} \tag{7.7.15}$$

The solution to the Laplace equation by using separation of variables will be trigonometric in x, and hence it will be exponential or hyperbolic in y. Inspecting the kinematic boundary condition at $y = 0$ implies that ϕ must vary as $\cos[2\pi(x - ct)/\lambda]$. In addition, the separation constants in the x- and y-directions must be $2\pi/\lambda$. Hence, an appropriate form of the solution to the Laplace equation is given by

$$\phi(x, y, t) = \cos\frac{2\pi}{\lambda}(x - ct)\left(C_1 \sinh\frac{2\pi y}{\lambda} + C_2 \cosh\frac{2\pi y}{\lambda}\right), \tag{7.7.16}$$

where C_1 and C_2 are constants. Substituting the bed boundary condition at $y = -h$ into this solution yields

$$\cos\frac{2\pi}{\lambda}(x - ct)\left(\frac{2\pi}{\lambda}C_1 \cosh\frac{2\pi h}{\lambda} - \frac{2\pi}{\lambda}C_2 \sinh\frac{2\pi h}{\lambda}\right) = 0, \quad \longrightarrow \quad C_1 = C_2 \tanh\frac{2\pi h}{\lambda}, \tag{7.7.17}$$

for Eq. $(7.7.17)_1$ must be satisfied for all values of x and t. Substituting Eq. $(7.7.17)_2$ and the dynamic boundary condition on the free surface into Eq. (7.7.16) gives rise to

$$C_2 \cos\frac{2\pi}{\lambda}(x - ct)\left[-\left(\frac{2\pi c}{\lambda}\right)^2 + \frac{2\pi g}{\lambda}\tanh\frac{2\pi h}{\lambda}\right] = 0, \quad \longrightarrow \quad \frac{c^2}{gh} = \frac{\lambda}{2\pi h}\tanh\frac{2\pi h}{\lambda}, \tag{7.7.18}$$

for Eq. $(7.7.18)_1$ must be satisfied equally for all values of x and t. It is noted that the obtained result is valid only for $\varepsilon \ll \lambda$ and $\varepsilon \ll h$.

Depending on the relative magnitudes between λ and h, Eq. $(7.7.18)_2$ can be further simplified. First, consider the liquid to be deep, i.e., $h \gg \lambda$, with which $2\pi h/\lambda$ becomes large, so that $\tanh(2\pi h/\lambda) \to 1$. With this, Eq. $(7.7.18)_2$ yields

$$\frac{c^2}{gh} = \frac{\lambda}{2\pi h}, \tag{7.7.19}$$

which is valid for $\varepsilon \ll \lambda \ll h$. On the contrary, consider the liquid to be shallow, in which $h \ll \lambda$. In this case, $2\pi h/\lambda$ will be small, so that $\tanh(2\pi h/\lambda) \sim 2\pi h/\lambda$, and Eq. $(7.7.18)_2$ gives

$$\frac{c^2}{gh} = 1, \tag{7.7.20}$$

which is valid for $\varepsilon \ll h \ll \lambda$. Figure 7.28c shows the propagation speeds for small-amplitude waves of sinusoidal form with different liquid depths.

An arbitrarily shaped wave train may be considered to be a superposition of sinusoidal waves with different amplitudes and wavelengths, so that it can be Fourier analyzed and decomposed into a number of purely sinusoidal waves. Such a wave will not in general propagate in an undisturbed way, because the propagation speed c depends on the wavelength, as indicated by Eq. $(7.7.18)_2$. Unless the *shallow-liquid conditions* are applied, the different Fourier components of an arbitrarily shaped wave will all travel at different speeds, so that the waveform will change continuously. This process is frequently referred to as a *dispersion*.

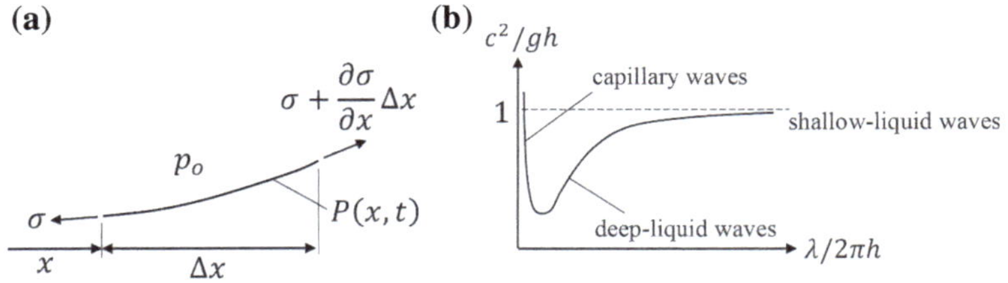

Fig. 7.29 Effect of surface tension on the propagation speed of surface waves. **a** A line element on the free surface of a liquid. **b** Propagation speeds of surface waves in sinusoidal form under the influence of surface tension

7.7.2 Effect of Surface Tension

To evaluate the influence of surface tension on the propagation speed of surface waves, consider a line element Δx locating at x on the liquid surface, as shown in Fig. 7.29a, in which the pressure above the liquid surface is denoted by p_0, the pressure on the liquid surface is denoted by $P(x, t)$, and the surface tension intensity is denoted by σ. In a static equilibrium state, the forces in the y-direction must be balanced, namely

$$(P - p_0)\Delta x + \left(\sigma + \frac{\partial \sigma}{\partial x}\Delta x\right)\left(\frac{\partial \eta}{\partial x} + \frac{\partial^2 \eta}{\partial x^2}\Delta x\right) - \sigma \frac{\partial \eta}{\partial x} = 0, \qquad (7.7.21)$$

which, by neglecting the terms in which $(\Delta x)^2$ involves, reduces to

$$(P - p_0) + \sigma \frac{\partial^2 \eta}{\partial x^2} + \frac{\partial \sigma}{\partial x}\frac{\partial \eta}{\partial x} = 0. \qquad (7.7.22)$$

Applying a first-order approximation of small values of σ in the above equation yields

$$P(x, t) = p_0 - \sigma \frac{\partial^2 \eta}{\partial x^2}, \quad \longrightarrow \quad \frac{\partial P}{\partial t} = -\sigma \frac{\partial^2}{\partial x^2}\left(\frac{\partial \eta}{\partial t}\right) = -\sigma \frac{\partial^3 \phi}{\partial x^2 \partial y}(x, 0, t), \qquad (7.7.23)$$

in which Eq. (7.7.8) has been used. Substituting the second equation into Eq. $(7.7.11)_2$ results in

$$\frac{\partial^2 \phi}{\partial t^2}(x, 0, t) - \frac{\sigma}{\rho}\frac{\partial^3 \phi}{\partial x^2 \partial y}(x, 0, t) + g\frac{\partial \phi}{\partial y}(x, 0, t) = 0, \qquad (7.7.24)$$

which is the revised form of dynamic boundary condition.

Since the surface tension has influence only on the dynamic boundary condition, it follows that the velocity potential function is still given by

$$\phi(x, y, t) = C_2 \cos \frac{2\pi}{\lambda}(x - ct)\left(\tanh \frac{2\pi h}{\lambda} \sinh \frac{2\pi y}{\lambda} + \cosh \frac{2\pi y}{\lambda}\right). \qquad (7.7.25)$$

Applying Eq. (7.7.24) to this solution yields

$$C_2 \cos\frac{2\pi}{\lambda}(x-ct)\left[-\left(\frac{2\pi c}{\lambda}\right)^2+\frac{\sigma}{\rho}\left(\frac{2\pi}{\lambda}\right)^3\tanh\frac{2\pi h}{\lambda}+\frac{2\pi g}{\lambda}\tanh\frac{2\pi h}{\lambda}\right]=0, \tag{7.7.26}$$

which is valid for all values of x and t. It is concluded that

$$\frac{c^2}{gh}=\frac{\lambda}{2\pi h}\left[1+\frac{\sigma}{\rho g}\left(\frac{2\pi}{\lambda}\right)^2\right]\tanh\frac{2\pi h}{\lambda}, \tag{7.7.27}$$

where the influence of surface tension is indicated by the terms inside the bracket. It is readily verified that c increases as σ increases. However, if σ is negligibly small, Eq. (7.7.27) coincides exactly with Eq. $(7.7.18)_2$.

For deep liquids in which $2\pi h/\lambda$ is large, Eq. (7.7.27) becomes

$$\frac{c^2}{gh}=\frac{\lambda}{2\pi h}\left[1+\frac{\sigma}{\rho g}\left(\frac{2\pi}{\lambda}\right)^2\right], \tag{7.7.28}$$

which reduces subsequently to

$$\frac{c^2}{gh}=\frac{2\pi\sigma}{\rho g\lambda h}, \qquad \forall\ \frac{\sigma}{\rho g}\left(\frac{2\pi}{\lambda}\right)^2\gg 1. \tag{7.7.29}$$

The waves satisfying this condition and so traveling at the speed defined by the above equation are called the *capillary waves*. The propagation speed of capillary waves depends on the wavelength λ, so that an arbitrarily shaped wave will disperse because of different propagation speeds of its Fourier components. The propagation speeds of sinusoidal waves are shown in Fig. 7.29b for deep and shallow liquids as functions of the parameter $\lambda/(2\pi h)$, in which the surface tension only plays a significant role in deep liquids. This is because the condition

$$\frac{\sigma}{\rho g}\left(\frac{2\pi}{\lambda}\right)^2\gg 1, \tag{7.7.30}$$

can only be accomplished for small values of λ, corresponding to deep liquid waves.

7.7.3 Shallow-Liquid Waves of Arbitrary Form

It was deduced previously that waves of arbitrary form will disperse unless the liquid is shallow. Such a deduction is verified in this section. To achieve this, consider a two-dimensional plane wave shown in Fig. 7.30a, in which a surface wave of arbitrary form is assumed, and the smallest wavelength of its Fourier components is large compared with the mean depth h, so that a one-dimensional approximation may be applied. That is, the x-component of velocity will be assumed to be constant over the entire liquid depth, while the y-component will be neglected as being small.

An element of length Δx of the liquid which extends from the bottom to the free surface is shown in Fig. 7.30b, which is considered a control-volume. On the left surface of this line element, there exists an entering mass flow rate, on the right surface

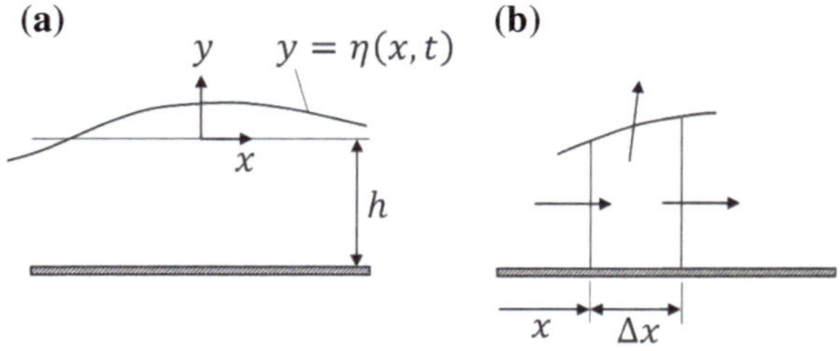

Fig. 7.30 Shallow-liquid waves of arbitrary form. **a** The configuration and coordinate system. **b** The control-volume and control-surface of a line element over the entire liquid depth under a one-dimensional approximation

a mass flow rate leaving the line element exists, while a mass flow rate leaves the line element at the top surface. Applying the balance of mass to the control-volume yields

$$\rho u(h+\eta) = \rho u(h+\eta) + \frac{\partial}{\partial x}\left[\rho u(h+\eta)\right]\Delta x + \rho\frac{\partial \eta}{\partial t}\Delta x, \tag{7.7.31}$$

which reduces to

$$\frac{\partial \eta}{\partial t} + \frac{\partial}{\partial x}\left[u(h+\eta)\right] = 0, \tag{7.7.32}$$

as $\Delta x \to 0$. Since in this equation the product $(u\eta)$ is of a second order smaller than the other terms, it can be neglected in the context of linear approximation. Hence, the linearized form of mass balance is obtained as

$$\frac{\partial \eta}{\partial t} + h\frac{\partial u}{\partial x} = 0. \tag{7.7.33}$$

Equally, applying the balance of linear momentum in the x-direction to the control-volume gives

$$\frac{\partial}{\partial t}\left[\rho u(h+\eta)\Delta x\right] + \frac{\partial}{\partial x}\left[\rho u^2(h+\eta)\right]\Delta x + \rho u\frac{\partial \eta}{\partial t}\Delta x = -\frac{\partial}{\partial x}\left[\frac{1}{2}\rho g(h+\eta)^2\right]\Delta x, \tag{7.7.34}$$

where the first term on the left-hand-side represents the time increase of the linear momentum of liquid contained inside the control-volume, the second term represents the net change of the linear momentum of liquid entering and leaving the control-volume in the x-direction, while the third term represents the net change of the linear momentum of liquid leaving the control-volume at the top surface of control-volume. The right-hand-side is the net external force acting on the control-volume, resulted from the hydrostatic pressure distributions on the left and right control-surfaces. Dividing this equation by $\rho\Delta x$ yields

$$\frac{\partial}{\partial t}\left[u(h+\eta)\right] + \frac{\partial}{\partial x}\left[u^2(h+\eta)\right] + u\frac{\partial \eta}{\partial t} = -g(h+\eta)\frac{\partial \eta}{\partial x}, \tag{7.7.35}$$

which, by using a linearization approximation to small values of u and η, is simplified to

$$\frac{\partial u}{\partial t} + g\frac{\partial \eta}{\partial x} = 0, \tag{7.7.36}$$

for the product $(u\eta)$ and all the terms in which u^2 involves are of second order smaller and are all neglected. This result is the linearized form of balance of linear momentum.

Two unknowns u and η are solved by using linearized Eqs. (7.7.33) and (7.7.36). Taking derivatives of the first equation with respect to x and the second equation with respect to t and eliminating the common terms in the resulting equations give

$$\frac{\partial^2 u}{\partial t^2} - gh\frac{\partial^2 u}{\partial x^2} = 0, \qquad \frac{\partial^2 \eta}{\partial t^2} - gh\frac{\partial^2 \eta}{\partial x^2} = 0, \tag{7.7.37}$$

showing that both u and η satisfy one-dimensional wave equations. Their general solutions are given by

$$u(x,t) = f_1(x - \sqrt{gh}t) + g_1(x + \sqrt{gh}t), \quad \eta(x,t) = f_2(x - \sqrt{gh}t) + g_2(x + \sqrt{gh}t), \tag{7.7.38}$$

where $\{f_1, g_1\}$ and $\{f_2, g_2\}$ are any differentiable functions. The first term in each of two sets, i.e., f_1 or f_2, represents a wave traveling in the positive x-direction with velocity $\sqrt{gh}$, while the second term denotes a wave traveling in the reverse direction with same velocity. Consequently, if an arbitrary wave is traveling along the surface of a shallow liquid, it will continue to travel with velocity $\sqrt{gh}$. Since this result confirms the propagation speed derived previously for a sinusoidal wave form in a shallow liquid, it is concluded that the wave shape does not change the wave speed as it travels along the surface of a shallow liquid. If the shape of wave is known as a function of x at some time, it will be known for all values of x and t.

7.7.4 Particle Trajectories in Traveling Waves

Traveling waves are waves which move along the free liquid surface. It is intended to determine the trajectories of particles in traveling waves. In the context of small-amplitude surface waves in a liquid of arbitrary depth, a sinusoidal wave is described by

$$\eta(x,t) = \varepsilon \sin\frac{2\pi}{\lambda}(x - ct), \tag{7.7.39}$$

for which the velocity potential function is given via Eq. (7.7.25), i.e.,

$$\phi(x,y,t) = C_2 \cos\frac{2\pi}{\lambda}(x - ct)\left(\tanh\frac{2\pi h}{\lambda}\sinh\frac{2\pi y}{\lambda} + \cosh\frac{2\pi y}{\lambda}\right). \tag{7.7.40}$$

Applying the kinematic boundary condition on the free surface, i.e., Eq. (7.7.8), to this solution yields

$$C_2\frac{2\pi}{\lambda}\cos\frac{2\pi}{\lambda}(x - ct)\tanh\frac{2\pi h}{\lambda} = -\varepsilon\frac{2\pi c}{\lambda}\cos\frac{2\pi}{\lambda}(x - ct), \tag{7.7.41}$$

giving rise to

$$C_2 = -\frac{c\varepsilon}{\tanh(2\pi h/\lambda)}, \tag{7.7.42}$$

with which Eq. (7.7.40) becomes

$$\phi(x, y, t) = -c\varepsilon \cos \frac{2\pi}{\lambda}(x - ct) \left(\sinh \frac{2\pi y}{\lambda} + \coth \frac{2\pi h}{\lambda} \cosh \frac{2\pi y}{\lambda} \right). \quad (7.7.43)$$

This is the velocity potential function for a *traveling sinusoidal wave*, where the propagation speed c is given by Eq. $(7.7.18)_2$.

It follows from the Cauchy-Riemann equations that

$$\begin{aligned} \frac{\partial \psi}{\partial y} &= \frac{2\pi c\varepsilon}{\lambda} \sin \frac{2\pi}{\lambda}(x - ct) \left(\sinh \frac{2\pi y}{\lambda} + \coth \frac{2\pi h}{\lambda} \cosh \frac{2\pi y}{\lambda} \right), \\ \frac{\partial \psi}{\partial x} &= \frac{2\pi c\varepsilon}{\lambda} \cos \frac{2\pi}{\lambda}(x - ct) \left(\cosh \frac{2\pi y}{\lambda} + \coth \frac{2\pi h}{\lambda} \sinh \frac{2\pi y}{\lambda} \right). \end{aligned} \quad (7.7.44)$$

Integrating these equations gives

$$\psi(x, y, t) = c\varepsilon \sin \frac{2\pi}{\lambda}(x - ct) \left(\cosh \frac{2\pi y}{\lambda} + \coth \frac{2\pi h}{\lambda} \sinh \frac{2\pi y}{\lambda} \right), \quad (7.7.45)$$

which is the corresponding stream function. With Eqs. (7.7.43) and (7.7.45), the complex potential $F(z, t)$ of a traveling sinusoidal wave is obtained as

$$F(z, t) = -\frac{c\varepsilon}{\sinh(2\pi h/\lambda)} \left[\cosh \frac{2\pi h}{\lambda} \cos \frac{2\pi}{\lambda}(z - ct) - i \sinh \frac{2\pi h}{\lambda} \sin \frac{2\pi}{\lambda}(z - ct) \right], \quad (7.7.46)$$

in which it is noted that $\sin(i\alpha) = i \sinh \alpha$ and $\cos(i\alpha) = \cosh \alpha$ for any quantity α. Since $x - ct + iy = z - ct$, Eq. (7.7.46) can be recast alternatively as

$$F(z, t) = -\frac{c\varepsilon}{\sinh(2\pi h/\lambda)} \cos \frac{2\pi}{\lambda}(z - ct + ih), \quad (7.7.47)$$

which is the complex potential of a traveling sinusoidal wave for the determination of particle trajectories.

As a wave train travels across the surface of an otherwise quiescent liquid, an individual particle of the liquid undergoes a small cyclic motion. To identify this motion, consider a specific particle P in the liquid shown in Fig. 7.31a, whose instantaneous position is indicated by using a fixed position z_0 and an additional position z_1 which varies with time. Taking time derivative of the complex conjugate of z_1 yields

$$\frac{\mathrm{d}\bar{z}_1}{\mathrm{d}t} = \frac{\mathrm{d}x_1}{\mathrm{d}t} - i\frac{\mathrm{d}y_1}{\mathrm{d}t} = u - iv = W = \frac{\mathrm{d}F}{\mathrm{d}z}. \quad (7.7.48)$$

Substituting Eq. (7.7.47) into this equation results in

$$\frac{\mathrm{d}\bar{z}_1}{\mathrm{d}t} = \frac{(2\pi/\lambda)c\varepsilon}{\sinh(2\pi h/\lambda)} \sin \frac{2\pi}{\lambda}(z - ct + ih), \quad (7.7.49)$$

which is to be integrated with respect to t to obtain

$$\bar{z}_1 = \frac{\varepsilon}{\sinh(2\pi h/\lambda)} \cos \frac{2\pi}{\lambda}(z - ct + ih), \quad (7.7.50)$$

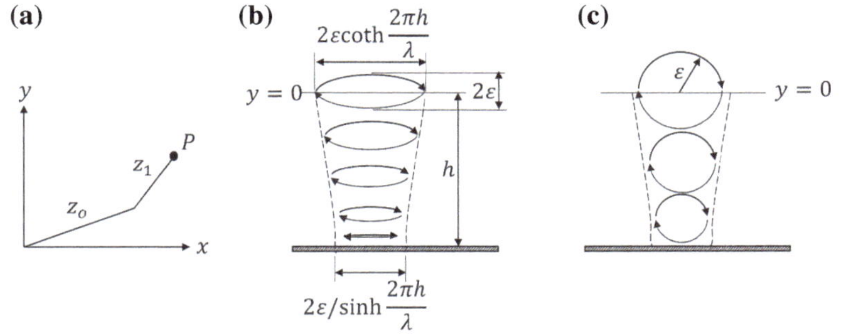

Fig. 7.31 Particle trajectories in traveling waves. **a** The configuration setup and coordinate system. **b** Particle trajectories in a sinusoidal wave. **c** Particle trajectories in deep liquids

in which the integration constant is chosen to be zero without loss of generality, for it does not affect the trajectories of liquid particles. Comparing this equation with Eq. (7.7.47) shows that

$$\bar{z}_1 = -\frac{F(z,t)}{c}, \tag{7.7.51}$$

which indicates that

$$\begin{aligned} x_1 &= -\frac{\phi(x,y,t)}{c} = \varepsilon\cos\frac{2\pi}{\lambda}(x-ct)\left(\sinh\frac{2\pi y}{\lambda}+\coth\frac{2\pi h}{\lambda}\cosh\frac{2\pi y}{\lambda}\right), \\ y_1 &= \frac{\psi(x,y,t)}{c} = \varepsilon\sin\frac{2\pi}{\lambda}(x-ct)\left(\cosh\frac{2\pi y}{\lambda}+\coth\frac{2\pi h}{\lambda}\sinh\frac{2\pi y}{\lambda}\right). \end{aligned} \tag{7.7.52}$$

Thus, the instantaneous coordinates of the trajectory of a liquid particle depend on its x- and y-coordinates and time. Eliminating time t in the above equations gives rise to

$$\begin{aligned} &\frac{x_1^2}{\varepsilon^2[\sinh(2\pi y/\lambda)+\coth(2\pi h/\lambda)\cosh(2\pi y/\lambda)]^2} \\ &\quad+\frac{y_1^2}{\varepsilon^2[\cosh(2\pi y/\lambda)+\coth(2\pi h/\lambda)\sinh(2\pi y/\lambda)]^2} = 1, \end{aligned} \tag{7.7.53}$$

which shows that the trajectory of a liquid particle depends only on its depth of submergence. It follows that each particle of the liquid experiences the same waves passing above it, irrespective of its x- coordinate. Thus, the motion experienced by any two particles which are separated in the x-direction will be the same, only the phasing will be different. Since Eq. (7.7.53) describes an ellipse, the trajectories of liquid particles will be an ellipse whose dimensions are determined by the value of y of the particles. For those particles lying on the free surface, at which $y = 0$, Eq. (7.7.53) reduces to

$$\frac{x_1^2}{[\varepsilon\coth(2\pi h/\lambda)]^2}+\frac{y_1^2}{\varepsilon^2} = 1, \tag{7.7.54}$$

showing that the particle trajectories are ellipses whose semi-axes are ε and $\varepsilon\coth(2\pi h/\lambda)$ in the y- and x-directions, respectively. For the particles at the bottom, the semi-axis in the y-direction becomes null, but the semi-axis in the x-direction becomes $\varepsilon/\sinh(2\pi h/\lambda)$, so that the ellipse degenerates to a line described by $-\varepsilon/\sinh(2\pi h/\lambda) \le x_1 \le \varepsilon/\sinh(2\pi h/\lambda)$. For $-h < y < 0$, the trajectories of particles are ellipses, as indicated by Eq. (7.7.53). These results are displayed in Fig. 7.31b.

For shallow liquids, the ellipses determined previously become elongated in the x-direction, while for deep liquids they become circles, as shown in Fig. 7.31c. This is verified by that for deep liquids, and the term $2\pi h/\lambda$ becomes very large, so that $\coth(2\pi h/\lambda)$ approaches unity. With this, Eq. (7.7.53) becomes

$$\frac{x_1^2}{\varepsilon^2[\sinh(2\pi y/\lambda)+\cosh(2\pi y/\lambda)]^2}+\frac{y_1^2}{\varepsilon^2[\sinh(2\pi y/\lambda)+\cosh(2\pi y/\lambda)]^2}=1, \tag{7.7.55}$$

which represents a circle with radius $\varepsilon|\sinh(2\pi y/\lambda)+\cosh(2\pi y/\lambda)|$. In other words, at the free surface the radius is ε, which decreases as y becomes more and more negative.

7.7.5 Particle Trajectories in Standing Waves

Standing waves are waves which remain stationary; namely, the free surface moves only vertically. It is intended to determine the trajectories of liquid particles in standing waves. Let η_1 and η_2 denote the free surfaces of two identical traveling waves which are moving in opposite direction, which are given by

$$\eta_1(x,t)=\frac{1}{2}\varepsilon\sin\frac{2\pi}{\lambda}(x-ct), \qquad \eta_2(x,t)=\frac{1}{2}\varepsilon\sin\frac{2\pi}{\lambda}(x+ct), \tag{7.7.56}$$

by which the free surface of a standing wave, η, is obtained by superimposing η_1 and η_2, i.e.,

$$\eta(x,t)=\eta_1(x,t)+\eta_2(x,t)=\varepsilon\sin\frac{2\pi x}{\lambda}\cos\frac{2\pi ct}{\lambda}. \tag{7.7.57}$$

It is readily verified that the free surface of a standing wave is a single function in x, which, for any value of x, oscillates vertically in time. It is assumed that the standing wave possesses a sinusoidal shape. The complex potential is then obtained by superimposing the complex potentials of two traveling waves moving in opposite direction, which, by using Eq. (7.7.47), is given by

$$F(z,t)=\frac{c\varepsilon}{2\sinh(2\pi h/\lambda)}\left[-\cos\frac{2\pi}{\lambda}(z-ct+ih)+\cos\frac{2\pi}{\lambda}(z+ct+ih)\right], \tag{7.7.58}$$

where the wave amplitude is $\varepsilon/2$ and wavelength is λ. By expanding the cosine function of $z+ih$ as one element and ct as the other, this equation is recast alternatively as

$$F(z,t)=-\frac{c\varepsilon}{\sinh(2\pi h/\lambda)}\sin\frac{2\pi}{\lambda}(z+ih)\sin\frac{2\pi ct}{\lambda}, \tag{7.7.59}$$

which is the complex potential for a standing sinusoidal wave with wavelength λ and oscillating frequency $2\pi c/\lambda$.

Substituting the obtained complex potential into Eq. (7.7.48) yields

$$\frac{d\bar{z}_1}{dt} = -\frac{(2\pi/\lambda)c\varepsilon}{\sinh(2\pi h/\lambda)} \cos\frac{2\pi}{\lambda}(z+ih)\sin\frac{2\pi ct}{\lambda}, \tag{7.7.60}$$

which is integrated with respect to t to obtain

$$\bar{z}_1 = \frac{\varepsilon}{\sinh(2\pi h/\lambda)} \cos\frac{2\pi}{\lambda}(z+ih)\cos\frac{2\pi ct}{\lambda}. \tag{7.7.61}$$

Expressing $z + ih = x + i(y + h)$ and expanding the trigonometric function of this argument give

$$\begin{aligned}\bar{z}_1 &= \frac{\varepsilon}{\sinh(2\pi h/\lambda)}\cos\frac{2\pi ct}{\lambda}\left[\cos\frac{2\pi x}{\lambda}\cosh\frac{2\pi}{\lambda}(y+h) - i\sin\frac{2\pi x}{\lambda}\sinh\frac{2\pi}{\lambda}(y+h)\right] \\ &= r_1 e^{-i\theta_1},\end{aligned} \tag{7.7.62}$$

leading to

$$\begin{aligned}r_1 &= \frac{\varepsilon}{\sinh(2\pi h/\lambda)}\cos\frac{2\pi ct}{\lambda}\left[\cos^2\frac{2\pi x}{\lambda}\cosh^2\frac{2\pi}{\lambda}(y+h)\right. \\ &\qquad \left. + \sin^2\frac{2\pi x}{\lambda}\sinh^2\frac{2\pi}{\lambda}(y+h)\right]^{1/2}, \\ \theta_1 &= \tan^{-1}\left[\tan\frac{2\pi x}{\lambda}\tanh\frac{2\pi}{\lambda}(y+h)\right].\end{aligned} \tag{7.7.63}$$

These two equations show that for given values of x and y, the value of θ_1 of particle trajectory is constant whereas the value of r_1 oscillates in time. This implies that the particle trajectories are straight lines whose inclinations depend on the locations of particles under consideration. Specifically, if $x = n\lambda/2$, it is seen that

$$r_1 = \varepsilon\cos\frac{2\pi ct}{\lambda}\frac{\cosh(2\pi/\lambda)(y+h)}{\sinh(2\pi h/\lambda)}, \qquad \theta_1 = 0,\ \pi, \tag{7.7.64}$$

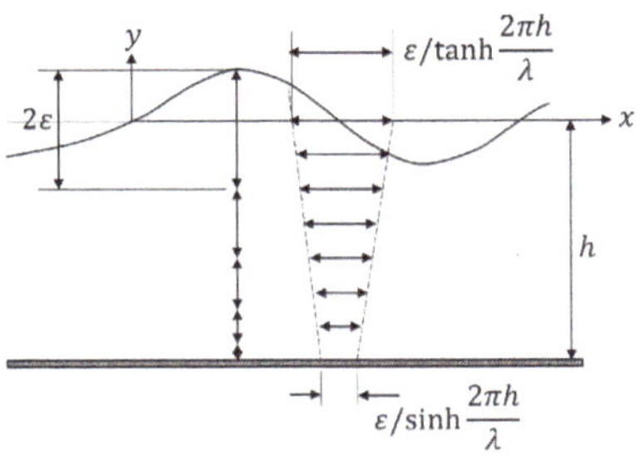

Fig. 7.32 Particle trajectories in a standing waves having sinusoidal form with amplitude ε and wavelength λ

which describes a family of horizontal lines whose length r_1 decreases with the depth of submergence. The location $x = n\lambda/2$ corresponds to the *nodes* of free surface, i.e., the points of free surface where no vertical motion takes place. The horizontal motion of these points shown in Fig. 7.32 should satisfy the continuity equation as the maximum amplitude of wave shifts from one side of the node to the other as the surface oscillations take place.

In the regions between the nodes, e.g. at $x = (2n+1)\lambda/4$, Eq. (7.7.63) reduces to

$$r_1 = \varepsilon \cos \frac{2\pi ct}{\lambda} \frac{\sinh(2\pi/\lambda)(y+h)}{\sinh(2\pi h/\lambda)}, \qquad \theta_1 = \frac{\pi}{2}, \frac{3\pi}{2}, \tag{7.7.65}$$

which defines a family of vertical lines whose r_1 decreases as the submergence increases and reaches zero on the bottom, as shown in the figure. Obviously, the vertical motion ceases at $y = -h$.

7.7.6 Waves in Rectangular and Cylindrical Containers

Consider a two-dimensional rectangular container of width 2ℓ which is filled by a liquid to depth h, as shown in Fig. 7.33a. It is assumed that if standing waves exist on the free surface of liquid, they must satisfy the Laplace equation for the velocity potential function and the associated boundary conditions, which are given by

$$\begin{aligned} 0 &= \frac{\partial^2 \phi}{\partial x^2} + \frac{\partial^2 \phi}{\partial y^2}, \quad 0 = \frac{\partial^2 \phi}{\partial t^2}(x, h, t) + g\frac{\partial \phi}{\partial y}(x, h, t), \\ 0 &= \frac{\partial \phi}{\partial y}(x, 0, t), \quad 0 = \frac{\partial \phi}{\partial x}(\pm\ell, y, t), \end{aligned} \tag{7.7.66}$$

in which the second equation is the pressure condition at the free surface in which the kinematic boundary condition is involved, and the third and fourth equations prevent normal velocity components on the bottom and sidewalls of container, respectively.

By assuming that the standing wave is steady, ϕ should have trigonometric time dependence. This is so, because the sidewall eliminates the possibility of traveling waves. Thus, the time dependence should be of the standing wave type chosen to be $\sin(2\pi ct/\lambda)$ without loss of generality, because any phase change merely corresponds to a shifting of the time domain. With these, the velocity potential function is given by

$$\phi(x, y, t) = \left(A_1 \sin \frac{2\pi x}{\lambda} + A_2 \cos \frac{2\pi x}{\lambda}\right)\left(B_1 \sinh \frac{2\pi y}{\lambda} + B_2 \cosh \frac{2\pi y}{\lambda}\right) \sin \frac{2\pi ct}{\lambda}, \tag{7.7.67}$$

where $\{A_1, A_2, B_1, B_2\}$ are constants. This is done so, for the trigonometric functions of x are used to meet the homogeneous boundary conditions at $x = \pm\ell$, while the dependency in y is chosen to be in the hyperbolic form in order to satisfy the Laplace equation and boundary condition at $y = 0$. Substituting Eq. (7.7.66)$_3$ into this solution yields immediately that $B_1 = 0$, with which Eq. (7.7.67) becomes

$$\phi(x, y, t) = \left(D_1 \sin \frac{2\pi x}{\lambda} + D_2 \cos \frac{2\pi x}{\lambda}\right) \cosh \frac{2\pi y}{\lambda} \sin \frac{2\pi ct}{\lambda}, \tag{7.7.68}$$

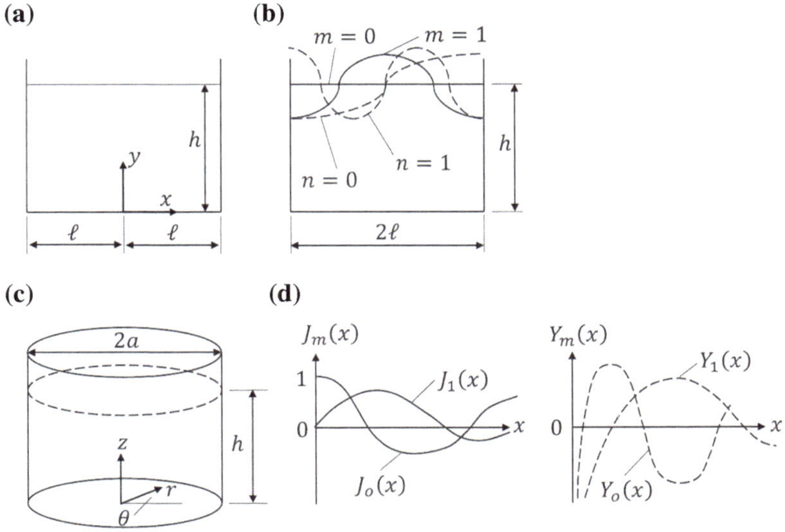

Fig. 7.33 Standing waves in rigid containers. **a** The geometry of a liquid in a rectangular container. **b** The first two fundamental modes of surface oscillation in **a**. **c** The geometry of a liquid in a cylindrical container. **d** The first two terms of the Bessel functions of the first and second kinds

with D_1 and D_2 constants. Applying Eq. $(7.7.66)_2$ to this solution yields

$$\left[-\left(\frac{2\pi c}{\lambda}\right)^2 \cosh\frac{2\pi h}{\lambda}+\frac{2\pi g}{\lambda}\sinh\frac{2\pi h}{\lambda}\right]\left(D_1\sin\frac{2\pi x}{\lambda}+D_2\cos\frac{2\pi x}{\lambda}\right)\sin\frac{2\pi ct}{\lambda}=0, \tag{7.7.69}$$

which should be satisfied for all values of x and t. It follows that

$$\frac{c^2}{gh}=\frac{\lambda}{2\pi h}\tanh\frac{2\pi h}{\lambda}, \tag{7.7.70}$$

which establishes the frequencies of wave motion. As implied by this equation, each Fourier component of the waveform has a different frequency of motion. Substituting Eq. $(7.7.66)_4$ into Eq. (7.7.68) results in

$$\frac{2\pi}{\lambda}\left(D_1\cos\frac{2\pi\ell}{\lambda}\mp D_2\sin\frac{2\pi\ell}{\lambda}\right)\cosh\frac{2\pi y}{\lambda}\sin\frac{2\pi ct}{\lambda}=0, \tag{7.7.71}$$

giving rise to

$$D_1\cos\frac{2\pi\ell}{\lambda}=\pm D_2\sin\frac{2\pi\ell}{\lambda}, \tag{7.7.72}$$

since Eq. (7.7.71) must be fulfilled for all values of x and t. Although Eq. (7.7.72) can be fulfilled by $D_1=D_2=0$, this yields a trivial solution $\phi=0$, which does not suit the considered problem.

Suppose first that $D_1\neq 0$ and $D_2=0$, it follows from Eq. (7.7.72) that

$$\cos\frac{2\pi\ell}{\lambda_n}=0, \qquad \longrightarrow \qquad \lambda_n=\frac{4\ell}{2n+1}, \tag{7.7.73}$$

which brings Eq. (7.7.68) to the form

$$\phi_n(x, y, t) = D_{1n} \sin \frac{(2n+1)\pi x}{2\ell} \cosh \frac{(2n+1)\pi y}{2\ell} \sin \frac{(2n+1)\pi c_n t}{2\ell}, \tag{7.7.74}$$

where c_n and λ_n are related to each other by using Eq. (7.7.70). Next, suppose $D_1 = 0$ and $D_2 \neq 0$, Eq. (7.7.72) then gives

$$\sin \frac{2\pi\ell}{\lambda_m} = 0, \quad \longrightarrow \quad \lambda_m = \frac{2\ell}{m}, \tag{7.7.75}$$

with which Eq. (7.7.68) becomes

$$\phi_m(x, y, t) = D_{2m} \cos \frac{m\pi x}{\ell} \cosh \frac{m\pi y}{\ell} \sin \frac{m\pi c_m t}{\ell}, \tag{7.7.76}$$

where the relation between c_m and λ_m is equally given by Eq. (7.7.70). Equations (7.7.74) and (7.7.76) provide respectively the solutions to different modes of ϕ_n and ϕ_m, whose first two surface modes are shown in Fig. 7.33b. It is verified that out of the continuous spectrum of wavelengths, only those waves whose particle paths are vertical at $x = \pm\ell$ are permissible solutions. This leads to an even spectrum of modes, corresponding to $D_1 = 0$, and an odd spectrum of modes, corresponding to $D_2 = 0$. In other words, there is a discrete spectrum of wavelengths whose particle paths are vertical at $x = \pm\ell$ in order to satisfy the boundary conditions at the sidewalls.

A more general solution to ϕ may be obtained by superimposing all the ϕ_n- and ϕ_m-solutions, which is given by

$$\begin{aligned}\phi(x, y, t) = &\sum_{n=0}^{\infty} D_{1n} \sin \frac{(2n+1)\pi x}{2\ell} \cosh \frac{(2n+1)\pi y}{2\ell} \sin \frac{(2n+1)\pi c_n t}{2\ell} \\ &+ \sum_{m=0}^{\infty} D_{2m} \cos \frac{m\pi x}{\ell} \cosh \frac{m\pi y}{\ell} \sin \frac{m\pi c_m t}{\ell},\end{aligned} \tag{7.7.77}$$

with

$$\frac{c_n^2}{gh} = \frac{2\ell}{(2n+1)\pi h} \tanh \frac{(2n+1)\pi h}{2\ell}, \quad \frac{c_m^2}{gh} = \frac{1}{m\pi h} \tanh \frac{m\pi h}{\ell}, \tag{7.7.78}$$

where the coefficients D_{1n} and D_{2m} remain undetermined, unless other conditions are provided. An example of the analysis is the establishment of the response of a water body subject to an earthquake. The water body may be in an artificial reservoir or a lake whose shape can be approximated by a rectangular tank. Seismographic records for the area would indicate the magnitude and frequency of the expected acceleration, which are then analyzed by using the Fourier analysis to establish the surface waveform and oscillating frequency at the end of an earthquake event. This provides the initiation of standing waves, and the coefficients D_{1n} and D_{2m} may be used to fit the data. The subsequent motion of standing waves is then described by Eqs. (7.7.77) and (7.7.78).

A similar analysis can be made to a cylindrical container shown in Fig. 7.33c, in which the radius of container is a and the height of liquid is h. By using the cylindrical coordinate system shown in the figure, the velocity potential function ϕ needs to satisfy the equations given by

$$0=\frac{1}{r}\frac{\partial}{\partial r}\left(r\frac{\partial\phi}{\partial r}\right)+\frac{1}{r^2}\frac{\partial^2\phi}{\partial\theta^2}+\frac{\partial^2\phi}{\partial z^2},\quad 0=\frac{\partial^2\phi}{\partial t^2}(r,\theta,h,t)+g\frac{\partial\phi}{\partial z}(r,\theta,h,t),$$
$$0=\frac{\partial\phi}{\partial z}(r,\theta,0,t),\qquad 0=\frac{\partial\phi}{\partial r}(a,\theta,z,t). \tag{7.7.79}$$

Let the solution to ϕ be given by

$$\phi(r,\theta,z,t)=R(r)\Theta(\theta)Z(z)\sin\omega t, \tag{7.7.80}$$

in which the time dependence is taken to be sinusoidal, which corresponds to standing waves. Substituting this expression into Eq. $(7.7.79)_1$ yields

$$\frac{r}{R}\frac{\mathrm{d}}{\mathrm{d}r}\left(r\frac{\mathrm{d}R}{\mathrm{d}r}\right)+\frac{1}{\Theta}\frac{\mathrm{d}^2\Theta}{\mathrm{d}\theta^2}+\frac{r^2}{Z}\frac{\mathrm{d}^2Z}{\mathrm{d}z^2}=0. \tag{7.7.81}$$

To solve this equation, let

$$\frac{1}{\Theta}\frac{\mathrm{d}^2\Theta}{\mathrm{d}\theta^2}=-m^2,\qquad\longrightarrow\qquad \Theta(\theta)=A_1\sin(m\theta)+A_2\cos(m\theta), \tag{7.7.82}$$

where m is an integer. With these, Eq. (7.7.81) is recast alternatively as

$$\frac{1}{rR}\frac{\mathrm{d}}{\mathrm{d}r}\left(r\frac{\mathrm{d}R}{\mathrm{d}r}\right)+\frac{m^2}{r^2}+\frac{1}{Z}\frac{\mathrm{d}^2Z}{\mathrm{d}z^2}=0. \tag{7.7.83}$$

Next, let

$$\frac{1}{Z}\frac{\mathrm{d}^2Z}{\mathrm{d}z^2}=k^2,\qquad\longrightarrow\qquad Z(z)=B_1\sinh(kz)+B_2\cosh(kz), \tag{7.7.84}$$

where k is also an integer. The hyperbolic form has been chosen to meet the finite extent in the z-direction. With these, Eq. (7.7.81) is expressed alternatively as

$$r\frac{\mathrm{d}}{\mathrm{d}r}\left(r\frac{\mathrm{d}R}{\mathrm{d}r}\right)+\left(k^2r^2-m^2\right)R=0, \tag{7.7.85}$$

which is *Bessel's equation*,[21] to which the solution is given by

$$R(r)=D_{1m}J_m(kr)+D_{2m}Y_m(kr), \tag{7.7.86}$$

where J_m and Y_m are respectively *Bessel's function of the first kind* and *Bessel's function of the second kind*. The first two terms of J_m and Y_m are shown in Fig. 7.33d.

Since Y_m diverges at $x=0$ for all values of m, D_{2m} must be zero. It follows that for any integer m, the solution ϕ_m should be in the form

$$\phi_m(r,\theta,z,t)=\left[A_{1m}\sin(m\theta)+A_{2m}\cos(m\theta)\right]\cdot \left[B_{1m}\sinh(kz)+B_{2m}\cosh(kz)\right]J_m(kr)\sin\omega t. \tag{7.7.87}$$

[21] Friedrich Wilhelm Bessel, 1784–1846, a German mathematician and physicist, who was the first astronomer to determine reliable values of the distance from the sun to another star by the *method of parallax*.

It is verified that $B_{1m} = 0$, when Eq. (7.7.79)$_3$ is applied to the above solution, and the oscillating frequency is determined to be

$$\omega^2 = gk \tanh(kh), \tag{7.7.88}$$

if Eq. (7.7.79)$_2$ is used. With these, Eq. (7.7.87) reduces to

$$\phi_m(r, \theta, z, t) = [K_{1m} \sin(m\theta) + K_{2m} \cos(m\theta)] \cosh(kz) J_m(kr) \sin \omega t. \tag{7.7.89}$$

Applying Eq. (7.7.79)$_4$ to this solution shows that

$$J'_m(ka) = 0, \quad \longrightarrow \quad J'_m(k_{mn}a) = 0, \tag{7.7.90}$$

where the first equation must be fulfilled to have a non-trivial solution, with the prime denoting differentiation. Since this can be satisfied by an infinite number of the discrete values of k, a specific value of k, denoted by k_{mn}, which is the nth root of the J_m Bessel function, as indicated in the second equation. Consequently, one solution to ϕ of the considered problem is obtained as

$$\phi_{mn}(r, \theta, z, t) = [K_{1mn} \sin(m\theta) + K_{2mn} \cos(m\theta)] \cosh(k_{mn}z) J_m(k_{mn}r) \sin \omega_{mn} t. \tag{7.7.91}$$

A more general solution may be obtained by superimposing all possible forms of ϕ_{mn}, viz.,

$$\phi(r, \theta, z, t) = \sum_{m=0}^{\infty} \sum_{n=0}^{\infty} [K_{1mn} \sin(m\theta) + K_{2mn} \cos(m\theta)] \cosh(k_{mn}z) J_m(k_{mn}r) \sin \omega_{mn} t, \tag{7.7.92}$$

with

$$\omega_{mn}^2 = gk_{mn} \tanh(k_{mn}h), \qquad J'_m(k_{mn}a) = 0. \tag{7.7.93}$$

The coefficients K_{1mn} and K_{2mn} remain undetermined, unless other conditions are provided. An example of the obtained results may be seen by a cup of coffee. If this cup of coffee is jarred slightly by striking it squarely on a flat table, the liquid may be excited to vibrate in a purely radial mode, so that the fundamental mode at which the surface $r = a$ vibrates in and out may be induced. This causes the surface waves which will have no θ dependence. Letting $m = 0$ in Eq. (7.7.92) shows that ϕ is proportional to $J_0(k_{0n}r)$, indicating that the surface assumes the shape predicted by the J_0 Bessel function, which can be observed in experiments.

7.7.7 Interfacial Wave Propagations

In the previous sections, the surface liquid waves in contact with the atmospheric air have been discussed. In this section, the focus is on a propagating surface wave between two dissimilar fluids, which is shown in Fig. 7.34, in which the wavy surface is described by $y = \eta(x, t)$, below which a fluid with density ρ_1 flows with mean velocity U_1 in the x-direction. Above the interface, a fluid with density ρ_2 moves with mean velocity U_2 also in the x-direction. To simplify the analysis, the wave at the interface is assumed to have a sinusoidal waveform, which is expressed by

$$\eta(x, t) = \varepsilon \exp\left\{ i \frac{2\pi}{\lambda}(x - \sigma t) \right\}, \tag{7.7.94}$$

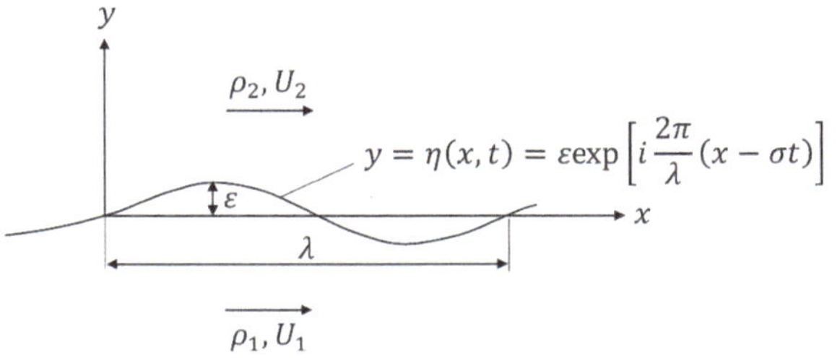

Fig. 7.34 A wavy interface between two dissimilar fluids traveling at different mean speeds

where the wave amplitude and wavelength are ε and λ, respectively. The term σ represents the propagation speed, with real values indicating that the wave is traveling in the x-direction, while the wave is decaying (for $\sigma/i < 0$) or growing (for $\sigma/i > 0$) if σ assumes imaginary values. The third circumstance denotes an unstable interface.

The velocities of two fluids are rewritten as

$$\boldsymbol{u}_i = \boldsymbol{U}_i + \nabla\phi_i = U_i\boldsymbol{e}_x + \nabla\phi_i, \qquad i = 1, 2, \tag{7.7.95}$$

where ϕ_i is the velocity potential function for the perturbation to the uniform flow caused by the wave at the interface. With this, the material derivative becomes

$$\frac{\mathrm{D}\alpha}{\mathrm{D}t} = \frac{\partial\alpha}{\partial t} + (\boldsymbol{u}_i \cdot \nabla)\alpha = \frac{\partial\alpha}{\partial t} + U_i\frac{\partial\alpha}{\partial x} + \nabla\phi_i \cdot \nabla\alpha, \tag{7.7.96}$$

for any scalar quantity α. Substituting this expression into the kinematic boundary condition at the interface yields

$$\frac{\mathrm{D}(y-\eta)}{\mathrm{D}t} = 0 = -\frac{\partial\eta}{\partial t} - U_i\frac{\partial\eta}{\partial x} + \frac{\partial\phi_i}{\partial y} - \nabla\phi_i \cdot \nabla\eta. \tag{7.7.97}$$

For small-amplitude waves, the last term on the right-hand-side is quadratically small and can be neglected. Thus, the kinematic boundary condition is simplified to

$$\frac{\partial\eta}{\partial t}(x,t) + U_i\frac{\partial\eta}{\partial x}(x,t) = \frac{\partial\phi_i}{\partial y}(x,0,t). \tag{7.7.98}$$

Substituting Eq. (7.7.95) into the Bernoulli equation for a constant pressure surface in which $F(t)$ is incorporated into the velocity potential function and applying the resulting equation to the interface yield

$$\rho_i\frac{\partial\phi_i}{\partial t}(x,0,t) + \rho_i U_i\frac{\partial\phi_i}{\partial x}(x,0,t) + \rho_i g\eta(x,t) = \text{constant}, \tag{7.7.99}$$

in which the quadratic terms have been neglected in the context of a first-order approximation.

By using Eqs. (7.7.98) and (7.7.99), it becomes possible to define the conditions that should be satisfied by ϕ_1 and ϕ_2 for the considered problem, which are given in the following:

$$\begin{aligned} y < 0 &: \frac{\partial^2\phi_1}{\partial x^2} + \frac{\partial^2\phi_1}{\partial y^2} = 0, \quad \|\nabla\phi_1\| = \text{finite}, \\ y > 0 &: \frac{\partial^2\phi_2}{\partial x^2} + \frac{\partial^2\phi_2}{\partial y^2} = 0, \quad \|\nabla\phi_2\| = \text{finite}. \end{aligned} \tag{7.7.100}$$

Since at the interface, which is linearized to $y = 0$, the kinematic boundary condition must be fulfilled separately by ϕ_1 and ϕ_2, it follows from Eq. (7.7.98) that

$$\frac{\partial \eta}{\partial t}(x,t) + U_1 \frac{\partial \eta}{\partial x}(x,t) = \frac{\partial \phi_1}{\partial y}(x,0,t), \quad \frac{\partial \eta}{\partial t}(x,t) + U_2 \frac{\partial \eta}{\partial x}(x,t) = \frac{\partial \phi_2}{\partial y}(x,0,t). \tag{7.7.101}$$

Equally, the pressures determined by using the Bernoulli equations from two fluids at the interface must be the same. It follows from Eq. (7.7.99) that

$$\begin{aligned} &\rho_1 \frac{\partial \phi_1}{\partial t}(x,0,t) + \rho_1 U_1 \frac{\partial \phi_1}{\partial x}(x,0,t) + \rho_1 g \eta(x,t) \\ &= \rho_2 \frac{\partial \phi_2}{\partial t}(x,0,t) + \rho_2 U_2 \frac{\partial \phi_2}{\partial x}(x,0,t) + \rho_2 g \eta(x,t). \end{aligned} \tag{7.7.102}$$

Equations (7.7.100)–(7.7.102) represent the conditions to be satisfied by any perturbation to the considered problem.

Since the wave shape at the interface is prescribed by Eq. (7.7.94), the solution to Eq. (7.7.100) should be trigonometric in x, with the y dependence being exponential. For finite values of $\nabla\phi_1$ and $\nabla\phi_2$, the negative and positive exponentials in the solutions to ϕ_1 and ϕ_2 should be rejected, respectively. Hence, the solutions to ϕ_1 and ϕ_2 are given by

$$\begin{aligned} \phi_1(x,y,t) &= A_1 \exp\left(\frac{2\pi y}{\lambda}\right) \exp\left\{i\frac{2\pi}{\lambda}(x-ct)\right\}, \\ \phi_2(x,y,t) &= A_2 \exp\left(-\frac{2\pi y}{\lambda}\right) \exp\left\{i\frac{2\pi}{\lambda}(x-ct)\right\}, \end{aligned} \tag{7.7.103}$$

where A_1 and A_2 are constants. Applying Eq. (7.7.101) to these solutions yields

$$A_1 = i\varepsilon(-\sigma + U_1), \qquad A_2 = -i\varepsilon(-\sigma + U_2), \tag{7.7.104}$$

with which ϕ_1 and ϕ_2 become

$$\begin{aligned} \phi_1(x,y,t) &= -i\varepsilon(\sigma - U_1) \exp\left(\frac{2\pi y}{\lambda}\right) \exp\left\{i\frac{2\pi}{\lambda}(x-ct)\right\}, \\ \phi_2(x,y,t) &= i\varepsilon(\sigma - U_2) \exp\left(-\frac{2\pi y}{\lambda}\right) \exp\left\{i\frac{2\pi}{\lambda}(x-ct)\right\}. \end{aligned} \tag{7.7.105}$$

Applying Eq. (7.7.102) to Eq. (7.7.105) gives

$$\begin{aligned} &\rho_1 i(\sigma - U_1)\left(i\sigma\frac{2\pi}{\lambda}\right) - \rho_1 U_1 i(\sigma - U_1)\left(i\frac{2\pi}{\lambda}\right) + \rho_1 g \\ &= \rho_2 i(\sigma - U_2)\left(-i\sigma\frac{2\pi}{\lambda}\right) + \rho_2 U_2 i(\sigma - U_1)\left(-i\frac{2\pi}{\lambda}\right) + \rho_2 g, \end{aligned} \tag{7.7.106}$$

which reduces to

$$-\rho_1 \frac{2\pi}{\lambda}(\sigma - U_1)^2 + \rho_1 g = \rho_2 \frac{2\pi}{\lambda}(\sigma - U_2)^2 + \rho_2 g. \tag{7.7.107}$$

This result delivers a solution to σ given by

$$\sigma = \frac{\rho_1 U_1 + \rho_2 U_2}{\rho_1 + \rho_2} \pm \sqrt{\left(\frac{\rho_1 - \rho_2}{\rho_1 + \rho_2}\right)\frac{\lambda g}{2\pi} - \frac{\rho_1 \rho_2}{(\rho_1 + \rho_2)^2}(U_2 - U_1)^2}, \quad (7.7.108)$$

which shows that σ may be real, imaginary, or complex, depending on the parameters of two free streams.

Consider a special case, in which $U_1 = U_2 = 0$ and $\rho_2 = 0$, corresponding to two stationary fluids in which the density of upper fluid is much smaller than that of lower fluid, e.g. a gas over a stationary liquid. Equation (7.7.108) then shows that

$$\sigma = \pm\sqrt{\frac{\lambda g}{2\pi}}, \quad (7.7.109)$$

which delivers the propagation speed of surface waves in deep liquids and coincides with Eq. (7.7.19). The minus sign represents a wave traveling in the negative x-direction. Since σ assumes real values, the waves at the interface will propagate, so that the surface of separation will remain intact; i.e., the interface is stable. Consider another special case, in which only $\rho_2 = 0$, but other parameters remain non-vanishing. This corresponds to a gas blowing over a liquid surface, and σ is determined to be

$$\sigma = U_1 \pm \sqrt{\frac{\lambda g}{2\pi}}, \quad (7.7.110)$$

which can be regarded as a Galilean transformation of Eq. (7.7.109) by using a relative speed U_1. Thus, the waves move along the surface of liquid at the speed of liquid plus/minus the speed of waves on a quiescent liquid body. The interface remains intact and is stable.

Consider a third special case, in which $\rho_1 = \rho_2$. This circumstance corresponds to a discontinuity in the velocity (e.g. a shear layer) in a homogeneous fluid. With this, Eq. (7.7.108) gives

$$\sigma = \frac{U_2 + U_1}{2} \pm i\frac{U_2 - U_1}{2}, \quad (7.7.111)$$

which shows that unless $U_2 = U_1$, σ will have an imaginary part, resulting in that the interfacial waves will grow exponentially with time. That is, the interface at the shear layer is unstable. Such an instability is referred to as the *Helmholtz instability* or the *Rayleigh instability*. Finally, consider a special case in which both fluids are quiescent so that $U_1 = U_2 = 0$, but $\rho_1 \neq \rho_2$, for which Eq. (7.7.108) shows that

$$\sigma = \pm\sqrt{\frac{\lambda g}{2\pi}\left(\frac{\rho_1 - \rho_2}{\rho_1 + \rho_2}\right)}. \quad (7.7.112)$$

For $\rho_1 > \rho_2$, i.e., a heavier fluid is below a lighter fluid, σ assumes real values, so that the interface will be stable. On the contrary, if $\rho_1 < \rho_2$, σ will assume imaginary values, and hence the interface will be unstable, as implied by the physical fact that an unstable interface exists if a heavier fluid is above a lighter fluid. This form of instability is referred to as the *Taylor instability*.

7.8 Exercises

7.1 To model the velocity distribution in a curved inlet section of a wind tunnel shown in the figure, the radius of curvature of streamline is expressed as $r = Lr_0/2y$. As a first approximation, it is assumed that the air speed V along each streamline is a constant. Determine the pressure change from $y = 0$ to the tunnel wall at $y = L/2$.

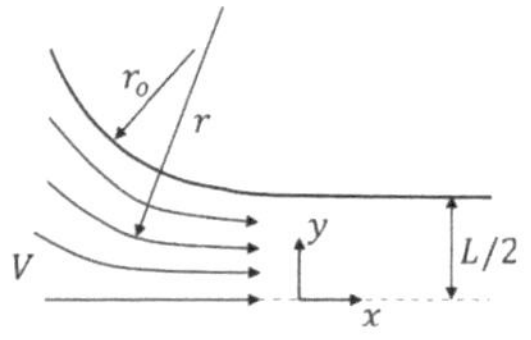

7.2 Water flowing through a pipe reducer is shown in the figure. The static pressure difference between points 1 and 2 is measured by using an inverted manometer containing a liquid with specific weight γ_0, which is smaller than that of water. Determine the manometer reading h if the water velocity at point 2 is V.

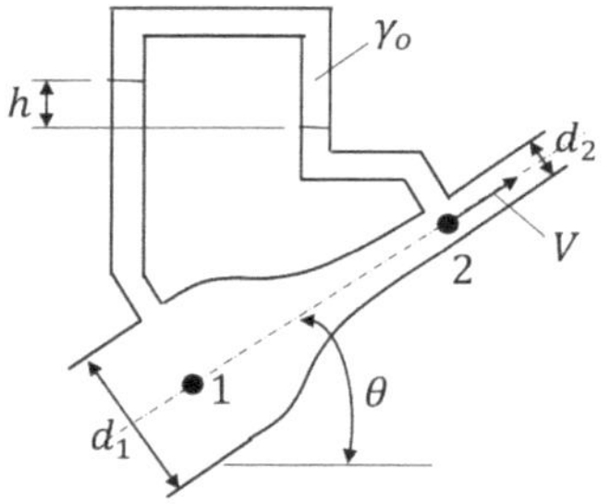

7.3 Water flows at low speed through a circular tube with inside diameter d shown in the figure. A smoothly contoured body of diameter $d_1 < d$ is held at the end of tube, where the water discharges to the atmosphere. It is assumed that the frictional effect is negligible and at each cross-section the velocity profile is uniform. Determine the gage pressure in the upstream region from the body and the force f required to hold the body.

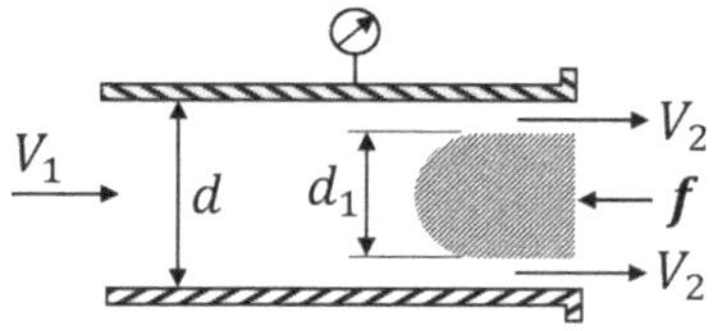

7.4 A tank associated with a *Borda mouthpiece* is shown in the figure. It is assumed that the fluid is incompressible and frictionless. The Borda mouthpiece essentially eliminates the flow along the tank wall, so the pressure there is nearly hydrostatic. Determine the contraction coefficient $C_c = A_j/A_0$.

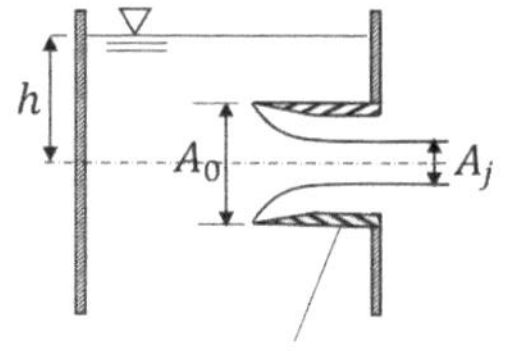

7.5 Water flows under an inclined sluice gate, as shown in the figure. The flow is assumed to be incompressible and frictionless. Derive an expression for the volume flow rate in terms of the parameters shown in the figure, if the gate width perpendicular to the page is b. Determine also the force acting on the inclined gate by the water.

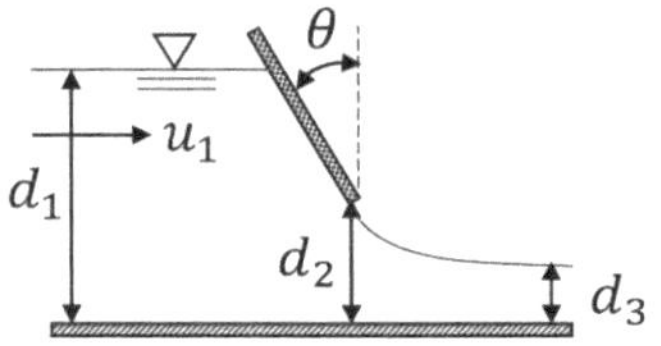

7.6 Consider a cylindrical container filled with a liquid which rotates about its axis with a constant rotational speed ω, as shown already in Fig. 3.9b. Use the Bernoulli equation to derive the expression of liquid free surface, i.e., $z = (\omega r)^2/2g$, if the origin of cylindrical coordinate system locates at the lowest point of liquid free surface.

7.7 A water jet with diameter d is used to support a cone-shaped object shown in the figure. Derive an expression for the combined mass of cone and water, M, that can be supported by the water jet in terms of the given parameters associated with a suitably chosen control-volume.

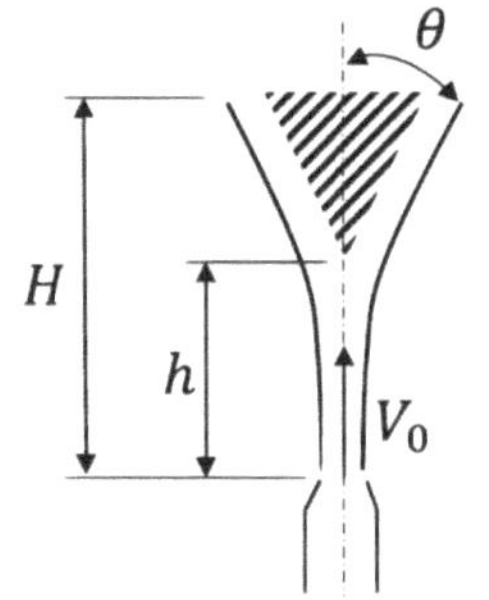

7.8 Consider the U-tube container shown in the figure. The left vertical tube has a constant diameter d_1 with length ℓ_1, while those for the right vertical tube are d_2 and $\ell_2 = \ell_1$, respectively. Two vertical tubes are connected by a linearly shrinking tube in diameter from the left tube to the right tube, whose length is L. Initially, a liquid is placed inside the U-tube container, with its equilibrium free surface at $z = 0$. A small pressure difference is applied to the openings of two vertical tubes to create a difference in the free surfaces. The pressure difference is then removed to let the openings be exposed again to the atmosphere, and the free surfaces experience an oscillation, whose frequency needs to be determined. The liquid is assumed to be incompressible and frictionless.

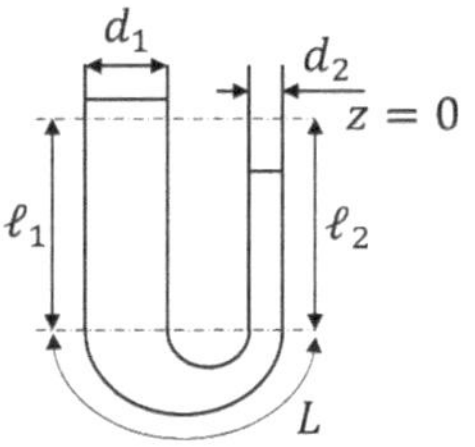

7.9 Consider the cylindrical container shown in the figure. The container is connected with a vertical tube with cross-sectional area A, and both the container and vertical tube are stationary. At the other end of vertical tube, two rotating horizontal tubes with cross-sectional area $A/2$ are associated, which rotate at a constant rotating speed ω. Determine (a) the steady-state expression of fluid velocity V at the exit of horizontal tube and (b) the expression of V as a function of time before it reaches its steady value. The fluid is assumed to be incompressible and frictionless, and the height h is a constant.

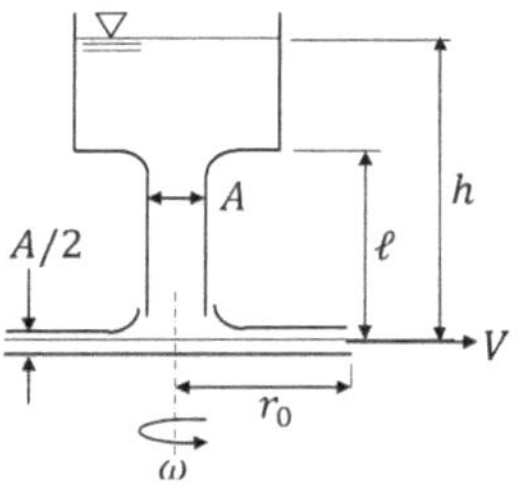

7.10 A fluid flows through a two-dimensional horizontal convergent-divergent channel shown in the figure. It is observed that vortices exist between cross-sections 3 and 4. If the flow velocity at the inlet and exit of channel are uniform with magnitude U, determine the pressure drop between cross-sections 1 and 4. The cross-sectional area between points 1 and 2, and 3 and 4 is A, while that between points 2 and 3 is A', with $A'/A = \mu$.

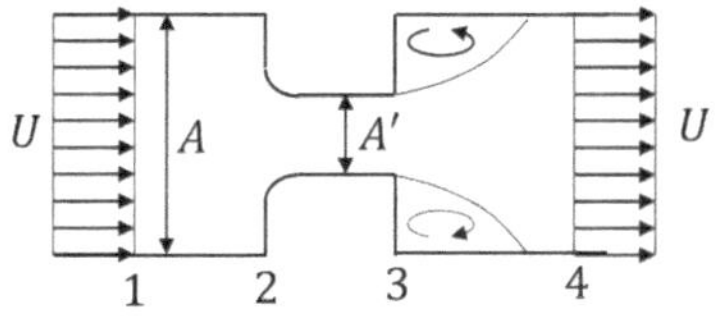

7.11 The stream function corresponding to a two-dimensional incompressible and irrotational flow in the vicinity of a right-angled corner is given by

$$\psi = 2r^2 \sin(2\theta),$$

which is expressed in terms of the polar coordinates. Determine the corresponding velocity potential function and velocity components.

7.12 A source with strength m is located at distance ℓ from a vertical plane, as shown in the figure. (a) Determine the velocity potential function of flow field. (b) Show that there is no flow through the wall. (c) Determine the velocity and pressure distributions along the wall surface. For simplicity, it is assumed that $p = p_0$ in the region far away from the source, and the gravity effect is neglected.

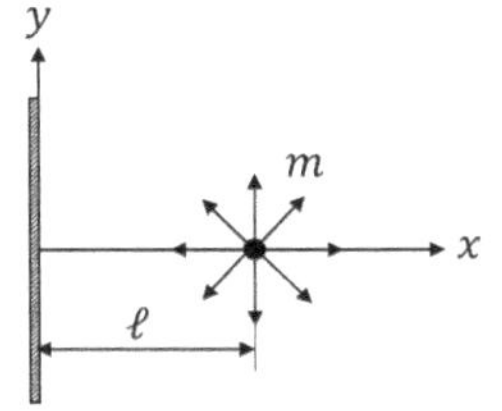

7.13 Use the Bernoulli equation to obtain an expression of the pressure distribution $p(a, \theta)$ on the cylinder surface with radius a in a uniform flow with magnitude U, in which the cylinder has a bound clockwise vortex with strength Γ. Integrate the product $-p(a, \theta)a \sin\theta$ around the cylinder surface to confirm the validity of the Kutta-Joukowski law.

7.14 Determine the complex potential for a circular cylinder with radius a in a flow field which is produced by a counterclockwise vortex with strength Γ locating in a distance ℓ from the axis of cylinder. Obtain the force acting on the cylinder by using Blasius' laws to a contour which includes the cylinder but excludes the vortex at $z = \ell$.

7.15 Find the transformation which maps the interior of sector $0 \le \theta \le \pi/n$ in a complex z-plane onto the upper half of another complex ζ-plane. Consider then a uniform flow in the ζ-plane to obtain the complex potential for the flow around the sector in the z-plane.

7.16 Use the definition of Stokes's stream function and the ω-component of the condition of irrotationality to show that the equation to be satisfied by $\psi(r, \theta)$ for axis-symmetric flows is given by

$$r^2 \frac{\partial^2 \psi}{\partial r^2} + \sin\theta \frac{\partial}{\partial \theta}\left(\frac{1}{\sin\theta}\frac{\partial \psi}{\partial \theta}\right) = 0.$$

7.17 Show that the force acting on a sphere with radius a owing to a doublet of strength μ locating a distance ℓ from the center of sphere along the x-axis is given by

$$\boldsymbol{f} = \frac{3\rho\mu^2 a^3 \ell}{2\pi(\ell^2 - a^2)^4}\boldsymbol{e}_x,$$

where $\boldsymbol{e}_x$ is the unit vector in the positive x-axis.

7.18 A sphere of radius a moves along the x-axis with velocity $U(t)$. A fixed-origin coordinate system is defined by the location of sphere at $t = 0$, as shown in the figure, so that the location of sphere at any subsequent time is obtained as

$$x_0(t) = \int_0^t U(\tau)\mathrm{d}\tau.$$

Let P be a fixed point, whose coordinates relative to the sphere, denoted by (r, θ), will change with time. Obtain the velocity potential function for the sphere in a stationary fluid in terms of r, θ, x, R, and x_0. If the undisturbed pressure is p_∞, find the force acting on the sphere. Compare the result with that obtained by using the concept of apparent mass in conjunction with Newton's second law of motion.

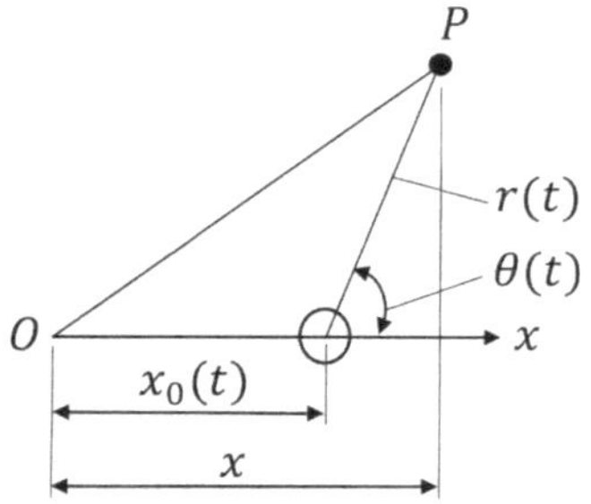

7.19 Use the complex potential for a traveling wave on a quiescent liquid surface to derive that the complex potential for a stationary wave on the surface of a liquid with mean velocity c along the negative x-axis is given by

$$F(z) = -cz - \frac{c\varepsilon}{\sinh(2\pi h/\lambda)} - \cos\frac{2\pi}{\lambda}(z + ih).$$

For deep liquids, show that this equation may be reduced to

$$F(z) = -cz - c\varepsilon \exp\left[-i\left(\frac{2\pi}{\lambda}\right)z\right].$$

Use this result to determine the stream function $\psi(x, y)$ for a stationary wave on the surface of a deep liquid layer, whose mean velocity is c. Show also that $\psi(x, \eta) = 0$ gives the equation of free surface, viz.,

$$\eta = \varepsilon \exp\left(\frac{2\pi}{\lambda}\eta\right) \sin\frac{2\pi x}{\lambda}.$$

7.20 The potential and kinetic energies per wavelength of a wave train are given respectively by

$$PE = \int_0^\lambda \frac{1}{2}\rho g\eta^2 \mathrm{d}x \qquad KE = \frac{1}{2}\rho \int_0^\lambda \phi \frac{\partial\phi}{\partial y}\bigg|_{y=0} \mathrm{d}x.$$

Use these expressions to show that the potential and kinetic energies per wavelength of the wave described by $\eta = \varepsilon \sin[2\pi(x - ct)/\lambda]$ are given by

$$PE = KE = \frac{1}{4}\rho g\varepsilon^2\lambda.$$

7.21 The work done on a vertical plane in a liquid layer is given by

$$W = \int_{-h}^0 p\frac{\partial\phi}{\partial x}\mathrm{d}y,$$

where p is the pressure and ϕ represents the velocity potential function. Use the linearized form of the Bernoulli equation and the velocity potential function of a traveling sinusoidal wave to show that the work done W across a vertical plane of liquid which has a traveling wave defined by $\eta = \varepsilon \sin[2\pi(x - ct)/\lambda]$ on its surface is given by

$$W = \frac{1}{2}\rho g c \varepsilon^2 \sin^2 \frac{2\pi}{\lambda}(x - ct)\left[1 + \frac{2\pi h/\lambda}{\sinh(2\pi h/\lambda)\cosh(2\pi h/\lambda)}\right].$$

Further, show that for deep liquids, the time average of work done is one-half of the sum of kinetic and potential energies per wavelength.

Further Reading

F. Charru, *Hydrodynamic Instabilities* (Cambridge University Press, Cambridge, 2011)

I.G. Currie, *Fundamental Mechanics of Fluids*, 2nd edn. (McGraw-Hill, Singapore, 1993)

E. Guyon, J.P. Hulin, L. Petit, C.D. Mitescu, *Physical Hydrodynamics*, 2nd edn. (Oxford University Press, Oxford, 2005)

K. Hutter, Y. Wang, *Fluid and Thermodynamics. Volume 1: Basic Fluid Mechanics* (Springer, Berlin, 2016)

H. Lamb, *Hydrodynamics*, 6th edn. (Dover, New York, 1932)

H. Liu, *Wind Engineering: A Handbook for Structural Engineers* (Prentice-Hall, New Jersey, 1991)

J. Lighthill, *Waves in Fluids* (Cambridge University Press, Cambridge, 1978)

C.C. Mei, *The Applied Dynamics of Ocean Surface Waves*, 2nd edn. (World Scientific Pub. Co., Inc, New York, 1989)

L.M. Milne-Thompson, *Theoretical Hydrodynamics*, 4th edn. (The Macmillan Company, New York, 1962)

J.M. Panton, *Hydrodynamics in Theory and Applications* (Prentice-Hall, New York, 1965)

R.H.F. Pao, *Fluid Mechanics* (Wiley, New York, 1961)

D.J. Tritton, *Physical Fluid Dynamics* (Oxford University Press, Oxford, 1988)

C.S. Yih, *Fluid Mechanics* (McGraw-Hill, New York, 1969)

Incompressible Viscous Flows

8

Flows of viscous fluids are discussed in this chapter, in which the fluid viscosity is intrinsically important. For simplicity, fluid density is considered constant, and the focus is on the characteristics of incompressible viscous flows. First, a general formulation of the field equations for viscous flows is presented, and the vorticity equation is derived, which provides a useful perspective in describing viscous flows. The exact solutions to the full Navier-Stokes equation for selected problems are presented. The approximate solutions to the Navier-Stokes equation for low-Reynolds-number flows, in the context of Stokes' approximation, are discussed for selected problems. Similarly, large-Reynolds-number flows are introduced in the context of boundary-layer theory and Prandtl's boundary-layer equations. These are considered equally an approximation to the Navier-Stokes equation, and some exact solutions to the obtained boundary-layer equations are presented by using similarity methods. On the other hand, the momentum integral and the Kármán-Pohlhausen method are introduced as the approximate methods in solving the boundary-layer equations, with a discussion on the stability of boundary layer. Buoyancy-driven flows, which are induced essentially by density variation, are discussed in the context of the Boussinesq approximation to the Navier-Stokes and thermal energy equations. The solutions to the resulting equations are presented for some problems with simple geometric configurations. The stability of a horizontal fluid layer is explored to study the conditions of the onset of thermal convection.

The obtained theories are valid for laminar flows. However, the most encountered flows in reality are turbulent. The last section deals with a fundamental concept of turbulence. A brief description of the characteristics of turbulence is provided, with the focus on the concepts of correlations, turbulent eddies and wave and energy spectra. The turbulence equations are derived by using the Reynolds-filter process to address the importance of energetic quantities resulted from the correlations of fluctuating field quantities, e.g. Reynolds' stress, for which turbulence closure models of different orders are required to arrive at a mathematically well-posed problem. Fully developed turbulent flows in circular pipes are discussed. The friction factors of

© Springer International Publishing AG 2019

C. Fang, *An Introduction to Fluid Mechanics*, Springer Textbooks in Earth Sciences, Geography and Environment, https://doi.org/10.1007/978-3-319-91821-1_8

fully developed laminar and turbulent pipe-flows with the applications of the Moody chart are given. With this information, the characteristics of viscous flows in a single circular pipe or in a multi-connected pipe system may be determined.

8.1 General Formulation and Vorticity Equation

For incompressible viscous flows of the Newtonian fluids, the governing equations, namely the local balances of mass and linear momentum, are given respectively by

$$\nabla \cdot \boldsymbol{u} = 0, \qquad \frac{\partial \boldsymbol{u}}{\partial t} + (\boldsymbol{u} \cdot \nabla)\boldsymbol{u} = -\frac{1}{\rho}\nabla p + \nu \nabla^2 \boldsymbol{u} + \boldsymbol{b}, \tag{8.1.1}$$

where ν is the kinematic viscosity which is a constant. These equations are used to determine the primitive fields of pressure and velocity. Since the formulation seems to be mathematically well-posed, one has the chance to obtain the values of primitive fields by integrating the equations simultaneously, provided that the boundary conditions are appropriately prescribed. This is accomplished by

$$\boldsymbol{u} = \boldsymbol{u}_w, \tag{8.1.2}$$

which is the no-slip boundary condition, where $\boldsymbol{u}_w$ is the velocity of solid boundary. The boundary condition for pressure is frequently taken from the pressure far away from the solid boundary, which must be the same as that of the free stream. Although the formulation is complete, it will be seen later that the solutions obtained by directly integrating Eqs. (8.1.1) and (8.1.2), called the *exact solutions*, are relatively few in number. These exact solutions are cherished and are used as the base for perturbation schemes to solve the problems which are close to the exact-solution configurations. They can also be used to test the accuracy of numerical techniques and to calibrate measuring instruments.

Frequently, the characteristics of incompressible viscous flows with constant density and dynamic viscosity are studied by using the vorticity $\boldsymbol{\omega}$, whose time evolution is described by the *vorticity equation*. One reason of the interest of vorticity equation is that it enables us to understand more about the physics of a given flow field. Also, in the analysis of some flow fields, it is frequently possible to make statements about the vorticity distribution which facility the analysis if the problem is posed in terms of vorticity. To obtain the vorticity equation, Eq. $(8.1.1)_2$ is expressed alternatively as

$$\frac{\partial \boldsymbol{u}}{\partial t} + \nabla\left(\frac{1}{2}\boldsymbol{u} \cdot \boldsymbol{u}\right) - \boldsymbol{u} \times (\nabla \times \boldsymbol{u}) = -\nabla\left(\frac{p}{\rho}\right) + \nu \nabla^2 \boldsymbol{u} + \boldsymbol{b}. \tag{8.1.3}$$

Taking curl of this equation yields

$$\frac{\partial \boldsymbol{\omega}}{\partial t} - \nabla \times (\boldsymbol{u} \times \boldsymbol{\omega}) = \nu \nabla^2 \boldsymbol{\omega} + \nabla \times \boldsymbol{b}, \qquad \boldsymbol{\omega} = \nabla \times \boldsymbol{u}, \tag{8.1.4}$$

where the second term on the left-hand-side can be expanded to

$$\nabla \times (\boldsymbol{u} \times \boldsymbol{\omega}) = \boldsymbol{u}\,(\nabla \cdot \boldsymbol{\omega}) - \boldsymbol{\omega}\,(\nabla \cdot \boldsymbol{u}) - (\boldsymbol{u} \cdot \nabla)\,\boldsymbol{\omega} + (\boldsymbol{\omega} \cdot \nabla)\,\boldsymbol{u}. \tag{8.1.5}$$

Substituting this into Eq. (8.1.4) gives

$$\frac{\partial \boldsymbol{\omega}}{\partial t} + (\boldsymbol{u} \cdot \nabla)\,\boldsymbol{\omega} = (\boldsymbol{\omega} \cdot \nabla)\,\boldsymbol{u} + \nu \nabla^2 \boldsymbol{\omega} + \nabla \times \boldsymbol{b}, \tag{8.1.6}$$

for $\nabla \cdot \boldsymbol{u} = 0$ and $\nabla \cdot \boldsymbol{\omega} = 0$. This equation is the so-called *vorticity equation*. If $\boldsymbol{b}$ is a conservative force field, then $\nabla \times \boldsymbol{b} = \nabla \times (\nabla G) = 0$. For two-dimensional circumstances, $\boldsymbol{\omega}$ is perpendicular to the coordinate plane, so that $(\boldsymbol{\omega} \cdot \nabla)\,\boldsymbol{u} = \boldsymbol{0}$, and the vorticity equation reduces to

$$\frac{\partial \boldsymbol{\omega}}{\partial t} + (\boldsymbol{u} \cdot \nabla)\,\boldsymbol{\omega} = \nu \nabla^2 \boldsymbol{\omega} + \nabla \times \boldsymbol{b}. \tag{8.1.7}$$

The advantage of vorticity equation is that the fluid pressure appears in neither Eq. (8.1.6) nor Eq. (8.1.7), so that the vorticity field may be obtained without any knowledge of the pressure field.

To determine the pressure field, taking divergence of the Navier-Stokes equation yields

$$\nabla^2 \left(\frac{p}{\rho}\right) = \boldsymbol{\omega} \cdot \boldsymbol{\omega} + \boldsymbol{u} \cdot \left(\nabla^2 \boldsymbol{u}\right) - \frac{1}{2} \nabla^2 (\boldsymbol{u} \cdot \boldsymbol{u}) + \nabla \cdot \boldsymbol{b}, \tag{8.1.8}$$

resulted from the facts that

$$\begin{aligned} &\nabla \cdot \frac{\partial \boldsymbol{u}}{\partial t} = \frac{\partial}{\partial t} (\nabla \cdot \boldsymbol{u}) = 0, \qquad \nabla \cdot \left(\nu \nabla^2 \boldsymbol{u}\right) = \nu \nabla^2 (\nabla \cdot \boldsymbol{u}) = 0, \\ &\nabla \cdot [(\boldsymbol{u} \cdot \nabla)\,\boldsymbol{u}] = \frac{1}{2} \nabla^2 (\boldsymbol{u} \cdot \boldsymbol{u}) - \boldsymbol{u} \cdot \left(\nabla^2 \boldsymbol{u}\right) - \boldsymbol{\omega} \cdot \boldsymbol{\omega}. \end{aligned} \tag{8.1.9}$$

Once $\boldsymbol{\omega}$ is determined, the pressure field can be determined by using Eq. (8.1.8). It is noted that Eq. (8.1.6) or (8.1.7) (and hence $\boldsymbol{\omega}$) satisfies the *diffusion equation*, while Eq. (8.1.8) (and hence p) fulfills the *Poisson equation*.[1]

Essentially, viscous flows may be classified into two categories: laminar and turbulent flows. The phenomena and treatments of turbulent flows are different from the other fundamental aspects of fluid flows. Up to Sect. 8.5, only laminar flows are discussed. In Sect. 8.6, a fundamental concept of turbulent flows is given, associated with the applications of turbulent flows in pipes.

8.2 Exact Solutions

In this section, a few number of exact solutions to the coupled local balances of mass and linear momentum of an isothermal, incompressible viscous Newtonian fluid will be established. So few exact solutions have been found, so that they are important in the theoretical, numerical and experimental analyses of fluid motion. The main difficulty in obtaining exact solutions to viscous-flow problems lies in the existence of nonlinear convection terms in the Navier-Stokes equation. In general,

[1] Siméon Denis Poisson, 1781–1840, a French mathematician and physicist, who obtained many important results in mathematics, statistics, and physics.

the obtained exact solutions can be classified into two categories. In the first category, the non-linear term $(\boldsymbol{u} \cdot \nabla)\boldsymbol{u}$ is identically null due to the simple geometric nature of a flow field. The second broad category is that the nonlinear convective term is not identically null, but the governed partial differential equations can be reduced to ordinary differential equations, although they need to be solved numerically. In the following, some exact solutions from two categories are presented to show the characteristics of incompressible viscous flows of the Newtonian fluids.

8.2.1 The Couette Flow

Consider the flow between two parallel plates shown in Fig. 8.1a, in which the z-direction is assumed to be very large compared with the distance h between two plates. The lower plate is stationary, while the upper plate is moving along the x-direction with constant velocity U, and there exists a pressure gradient along the x-direction. Based on the geometric configuration, the flow in the x-direction is assumed to be steady and *fully developed*, so that $u = u(y)$ only. For simplicity, the gravitational acceleration is assumed to point perpendicular to the page. With these, the local mass balance reads

$$\nabla \cdot \boldsymbol{u} = \frac{\partial u}{\partial x} + \frac{\partial v}{\partial y} = 0, \qquad \longrightarrow \qquad v = f(x), \tag{8.2.1}$$

where $f(x)$ is an undetermined function. Since at $y = 0$, $v = 0$, for the plate is not porous, it follows that $f(x) = 0$, giving rise to $v = 0$. The Navier-Stokes equation is given by

$$\begin{aligned} u\frac{\partial u}{\partial x} + v\frac{\partial u}{\partial y} &= -\frac{1}{\rho}\frac{\partial p}{\partial x} + \nu\left(\frac{\partial^2 u}{\partial x^2} + v\frac{\partial^2 u}{\partial y^2}\right), \\ u\frac{\partial v}{\partial x} + v\frac{\partial v}{\partial y} &= -\frac{1}{\rho}\frac{\partial p}{\partial y} + \nu\left(\frac{\partial^2 v}{\partial x^2} + v\frac{\partial^2 v}{\partial y^2}\right), \end{aligned} \tag{8.2.2}$$

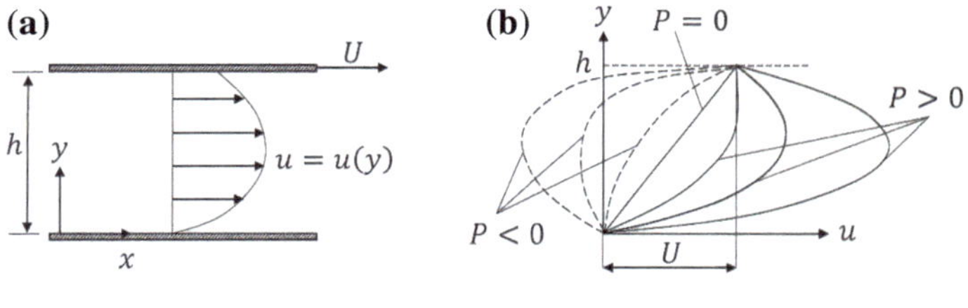

Fig. 8.1 A general two-dimensional Couette flow. **a** The geometric configurations and coordinate system. **b** The velocity profiles for variations in dimensionless pressure parameter P

in the x- and y-directions, respectively. Substituting $v = 0$ into the second equation yields $\partial p/\partial y = 0$, indicating that $p = p(x)$ only. With this, the first equation is simplified to

$$\frac{\mathrm{d}^2 u}{\mathrm{d}y^2} = \frac{1}{\mu}\left(\frac{\mathrm{d}p}{\mathrm{d}x}\right). \tag{8.2.3}$$

Since the right-hand-side of this equation depends only on x, the equation can be integrated directly with respect to y to obtain

$$u(y) = \frac{1}{\mu}\left(\frac{\mathrm{d}p}{\mathrm{d}x}\right)\left(\frac{y^2}{2} + C_1 y + C_2\right), \tag{8.2.4}$$

where C_1 and C_2 are integration constants. Applying the no-slip boundary conditions, i.e., $u = 0$ at $y = 0$, and $u = U$ at $y = h$, to the above solution yields

$$C_1 = \frac{\mu U}{h}\frac{1}{\mathrm{d}p/\mathrm{d}x} - \frac{1}{2}h, \qquad C_2 = 0, \tag{8.2.5}$$

with which the velocity distribution in the x-direction is obtained as

$$\frac{u(y)}{U} = \frac{y}{h} + \frac{h^2}{2\mu U}\left(\frac{\mathrm{d}p}{\mathrm{d}x}\right)\left[\left(\frac{y}{h}\right)^2 - \left(\frac{y}{h}\right)\right]. \tag{8.2.6}$$

Let the dimensionless pressure parameter P be defined by

$$P \equiv -\frac{h^2}{2\mu U}\left(\frac{\mathrm{d}p}{\mathrm{d}x}\right), \tag{8.2.7}$$

with which Eq. (8.2.6) is recast alternatively as

$$\frac{u(y)}{U} = \frac{y}{h} + P\left(\frac{y}{h}\right)\left[1 - \left(\frac{y}{h}\right)\right]. \tag{8.2.8}$$

The flow field described by this velocity profile or Eq. (8.2.6) is referred to as the *general Couette flow*,[2] and the velocity profiles for variations in P are shown in Fig. 8.1b. The fluid velocity consists of two contributions: the first contribution results from the motion of upper plate, as indicated by the first term on the right-hand side of Eq. (8.2.8), and the second contribution results from the influence of pressure gradient along the x-direction, as indicated by the second term. Specifically, the flow triggered by the motion of upper plate is referred to as the *plane Couette flow*, while that resulted from a non-vanishing pressure gradient with two stationary plates is referred to as the *plane Poiseuille flow*.[3] For $P = 0$, a plane Couette flow is recovered, while for $P \neq 0$, the pressure gradient will either assist or resist the viscous shear motion. For example, if $P > 0$ (i.e., $\mathrm{d}p/\mathrm{d}x < 0$), the pressure gradient assists the viscously induced motion to overcome the shear force at the lower plate. On the other hand, if $P < 0$ (i.e., $\mathrm{d}p/\mathrm{d}x > 0$), the pressure gradient resists the fluid motion which is induced by the motion of upper plate. In such a circumstance, a

---

[2]Maurice Marie Alfred Couette, 1858–1943, a French physicist, who is known for his studies of fluidity of matters.

[3]Jean Léonard Marie Poiseuille, 1797–1869, a French physicist and physiologist, who is best known for his work on laminar flow characteristics in circular pipes, which is referred to as *Poiseuille's law*.

region of *reverse flow* may occur near the lower plate, as also shown in Fig. 8.1b. Pressure gradient assisting fluid motion is termed *favorable pressure gradient*, whilst that resisting fluid motion is called *adverse pressure gradient*.

The shear stress τ_{yx}, by using Newton's law of viscosity, is obtained as

$$\tau_{yx} = \mu \frac{du}{dy} = \frac{\mu U}{h} + \frac{\mu U P}{h}\left[1 - 2\left(\frac{y}{h}\right)\right], \tag{8.2.9}$$

whose values at $y = 0$ and $y = h$ are given respectively by

$$\tau_{yx}|_{y=0} = \frac{\mu U}{h}(1 + P), \qquad \tau_{yx}|_{y=h} = \frac{\mu U}{h}(1 - P). \tag{8.2.10}$$

For a plane Couette flow, it follows that $\tau_{yx}|_{y=0} = \tau_{yx}|_{y=h} = \mu U/h$, which are both positive, while for a plan Poiseuille flow, $\tau_{yx}|_{y=0} = \mu U P/h$, which is a positive shear stress, and $\tau_{yx}|_{y=h} = -\mu U P/h$, which is a negative shear stress. These results are physically justified. The volume flow rate Q per unit depth perpendicular to the page and the corresponding average velocity u_{av} are determined as

$$Q = \int_0^h u(y)\mathrm{d}y = \frac{Uh}{2}\left(1 + \frac{P}{3}\right), \qquad u_{av} = \frac{Q}{h} = \frac{U}{2}\left(1 + \frac{P}{3}\right). \tag{8.2.11}$$

The location of maximum velocity is identified to be

$$\frac{du}{dy} = 0 = \frac{U}{h} + \frac{UP}{h}\left[1 - 2\left(\frac{y}{h}\right)\right], \quad \longrightarrow \quad y = \left(\frac{1 + P}{2P}\right)h, \tag{8.2.12}$$

with the maximum velocity u_{max} given by

$$\frac{u_{max}}{U} = \frac{(1 + P)^2}{4P}. \tag{8.2.13}$$

The Reynolds number corresponding to the considered flow is defined by

$$\mathcal{R}_e \equiv \frac{\rho u_{av} h}{\mu}. \tag{8.2.14}$$

Experiments show that the conducted analysis is only valid for the laminar flows characterized by $\mathcal{R}_e < 1400$ with vanishing pressure gradient. Not much information is available if a non-vanishing pressure gradient presents.

8.2.2 The Poiseuille Flow

A steady flow of a viscous fluid in a conduit of arbitrary but constant cross-section is referred to as a *Poiseuille flow*. Consider an arbitrary cross-sectional conduit in the (yz)-plane shown in Fig. 8.2a, in which the gravity is neglected for simplicity. It is assumed that the flow is fully developed, i.e., $u = u(y, z)$. It follows from the geometric configuration that the transverse velocity components v and w are null, and the pressure cannot vary in the transverse direction, so that $p = p(x)$ only. With

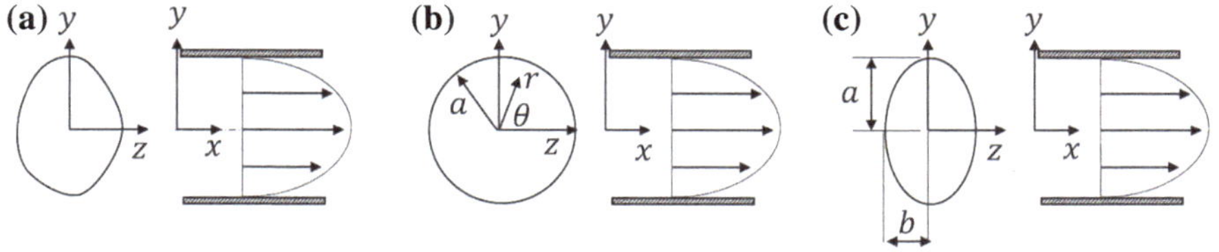

Fig. 8.2 Poiseuille flows along conduits of various cross-sections with the coordinate systems. **a** An arbitrary cross-section. **b** A circular cross-section. **c** An elliptic cross-section

these, the continuity equation holds identically, and the Navier-Stokes equation in the x-direction reduces to

$$\frac{\partial^2 u}{\mathrm{d}y^2} + \frac{\partial^2 u}{\partial z^2} = \frac{1}{\mu}\left(\frac{\mathrm{d}p}{\mathrm{d}x}\right), \tag{8.2.15}$$

which is a Poisson-type equation, in which the right-hand-side must be a constant at most. Although there exists no general solution to the above equation for arbitrary cross-section, exact solutions for a few specific sections are possible.

For the special case in which the cross-section is circular with radius a shown in Fig. 8.2b, the cylindrical coordinate system (r, θ, x) is preferred, so that the axial velocity component u is only a function of r, for the flow is axis-symmetric. Thus, Eq. (8.2.15) is expressed as

$$\frac{1}{r}\frac{\mathrm{d}}{\mathrm{d}r}\left(r\frac{\mathrm{d}u}{\mathrm{d}r}\right) = \frac{1}{\mu}\left(\frac{\mathrm{d}p}{\mathrm{d}x}\right). \tag{8.2.16}$$

Since the right-hand-side does not depend on r, integrating this equation respect to r twice yields

$$u(r) = \frac{1}{\mu}\left(\frac{\mathrm{d}p}{\mathrm{d}x}\right)\frac{r^2}{4} + C_1 \ln r + C_2, \tag{8.2.17}$$

where C_1 and C_2 are integration constants. Applying the boundary conditions that $u(r = a) = 0$ and $u(r = 0) =$ finite to this solution yields $C_1 = 0$ and $C_2 = -(\mathrm{d}p/\mathrm{d}x)a^2/(4\mu)$, with which the axial velocity profile is obtained as

$$u(r) = -\frac{a^2}{4\mu}\left(\frac{\mathrm{d}p}{\mathrm{d}x}\right)\left[1 - \left(\frac{r}{a}\right)^2\right]. \tag{8.2.18}$$

It is seen that the flow can be triggered by a non-vanishing pressure gradient along the x-direction, and the resulting axial velocity profile is parabolic. The shear stress τ_{rx} is determined as

$$\tau_{rx} = \mu\frac{\mathrm{d}u}{\mathrm{d}y} = -\mu\frac{\mathrm{d}u}{\mathrm{d}r} = \frac{r}{2}\left(\frac{\mathrm{d}p}{\mathrm{d}x}\right), \qquad y = a - r, \tag{8.2.19}$$

whose values on the conduit wall and at the centerline are given respectively by

$$\tau_{rx}|_{r=a} = -\frac{a}{2}\left(\frac{\mathrm{d}p}{\mathrm{d}x}\right), \qquad \tau_{rx}|_{r=0} = 0. \tag{8.2.20}$$

The first equation indicates that $\tau_{rx}|_{r=a} > 0$ for a negative pressure gradient along the x-direction, which is physically justified. The second equation shows that at $r = 0$ the shear stress vanishes, implying that there should be the location at which the axial velocity is maximum, as will be demonstrated later. The volume flow rate Q and average axial velocity u_{av} are obtained respectively as

$$Q = \int_0^a u(r)(2\pi r \mathrm{d}r) = -\frac{\pi a^4}{8\mu}\left(\frac{\mathrm{d}p}{\mathrm{d}x}\right), \qquad u_{av} = -\frac{a^2}{8\mu}\left(\frac{\mathrm{d}p}{\mathrm{d}x}\right). \tag{8.2.21}$$

If the pressure gradient is constant, it can be approximated by $\mathrm{d}p/\mathrm{d}x = (p_2 - p_1)/\ell = -\Delta p/\ell$, where ℓ is the pipe length between any two points 1 and 2 of the circular conduit with the corresponding *pressure drop* $\Delta p = p_1 - p_2$. With these, Q is frequently expressed as

$$Q = -\frac{\pi a^4}{8\mu}\left(-\frac{\Delta p}{\ell}\right) = \frac{\pi d^4 \Delta p}{128\mu\ell}, \qquad d = 2a, \tag{8.2.22}$$

where d is the diameter of circular conduit. This equation is called the *Hagen-Poiseuille equation*,[4] or simply *Poiseuille's law* for laminar flows in horizontal circular pipes driven by pressure gradient. The location at which the maximum axial velocity takes place is identified to be

$$\frac{\mathrm{d}u}{\mathrm{d}r} = \frac{r}{2\mu}\left(\frac{\mathrm{d}p}{\mathrm{d}x}\right) = 0, \qquad \longrightarrow \qquad r = 0, \tag{8.2.23}$$

and the maximum axial velocity is obtained as

$$u_{max} = -\frac{a^2}{4\mu}\left(\frac{\mathrm{d}p}{\mathrm{d}x}\right) = 2u_{av}, \qquad \longrightarrow \qquad \frac{u}{u_{max}} = 1 - \left(\frac{r}{a}\right)^2. \tag{8.2.24}$$

The Reynolds number in the considered flow is defined by

$$\mathcal{R}_e \equiv \frac{\rho u_{av} d}{\mu}. \tag{8.2.25}$$

Experiments show that the previously obtained results are only valid for laminar flows characterized by $\mathcal{R}_e < 2100$.

For the special case in which the cross-section is an ellipse, as shown in Fig. 8.2c, the condition $y^2/a^2 + z^2/b^2 - 1 = 0$ must hold on the conduit wall, so that the solution to the axial velocity may be proportional to this term. Thus, a solution to Eq. (8.2.15) is sought in the form

$$u(y, z) = \alpha\left(\frac{y^2}{a^2} + \frac{z^2}{b^2} - 1\right), \tag{8.2.26}$$

where α is an undetermined constant. Substituting this expression into Eq. (8.2.15) shows that

$$\alpha = \frac{1}{2\mu}\left(\frac{\mathrm{d}p}{\mathrm{d}x}\right)\frac{a^2 b^2}{a^2 + b^2}. \tag{8.2.27}$$

[4]Gotthilf Heinrich Ludwig Hagen, 1797–1884, a German civil engineer, who made contributions to fluid dynamics, hydraulic engineering and probability theory.

Hence, the axial velocity profile in a horizontal conduit with an elliptic cross-section is obtained as

$$u(y,z)=\frac{1}{2\mu}\left(\frac{\mathrm{d}p}{\mathrm{d}x}\right)\frac{a^2b^2}{a^2+b^2}\left(\frac{y^2}{a^2}+\frac{z^2}{b^2}-1\right). \tag{8.2.28}$$

8.2.3 Flows Between Two Concentric Cylinders

Consider a Newtonian fluid with constant density and dynamic viscosity contained in the annual region between two concentric cylinders shown in Fig. 8.3a, in which the outer cylinder has radius r_o with angular velocity ω_o, while those for the inner cylinder are r_i and ω_i, respectively. Both cylinders are assumed to be long compared with their diameters, so that the considered rotating flow will be two-dimensional. The origin of cylindrical coordinate system is located at the center of cylinders, with the x-direction pointing perpendicular to the page. It follows from the geometric configurations that the non-vanishing velocity component will be the tangential velocity u_θ, which depends only on r for fully developed laminar flows. With these, the continuity equation holds identically, and the Navier-Stokes equations in the r- and θ-directions reduce respectively to

$$0=\frac{u_\theta}{r^2}-\frac{1}{\rho}\left(\frac{\mathrm{d}p}{\mathrm{d}r}\right),\qquad 0=\frac{\mathrm{d}^2u_\theta}{\mathrm{d}r^2}+\frac{\mathrm{d}}{\mathrm{d}r}\left(\frac{u_\theta}{r}\right). \tag{8.2.29}$$

The first equation shows that there is a balance between the centrifugal force which acts on a fluid element and the force which is produced by the induced pressure field. The second equation indicates the tangential velocity distribution in the annual region.

Integrating the second equation yields

$$u_\theta(r)=C_1\frac{r}{2}+\frac{C_2}{r}, \tag{8.2.30}$$

where C_1 and C_2 are integration constants. Applying the no-slip boundary conditions, namely at $r=r_i$, $u_\theta=r_i\omega_i$, and at $r=r_o$, $u_\theta=r_o\omega_o$ to the above solution gives

$$C_1=\frac{2(\omega_or_o^2-\omega_ir_i^2)}{r_o^2-r_i^2},\qquad C_2=-r_i^2r_o^2\frac{\omega_o-\omega_i}{r_o^2-r_i^2}, \tag{8.2.31}$$

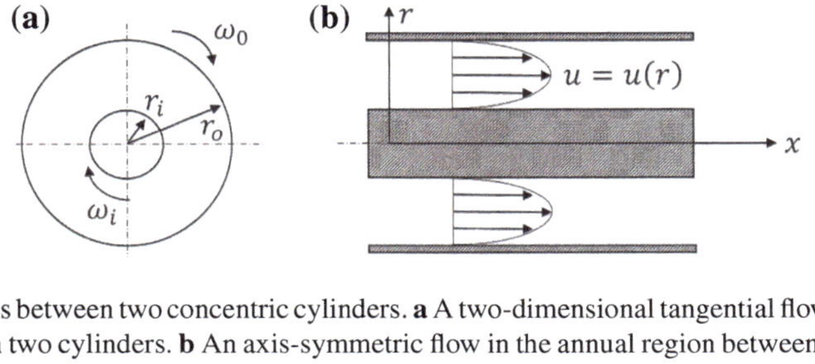

Fig. 8.3 Flows between two concentric cylinders. **a** A two-dimensional tangential flow in the annual space between two cylinders. **b** An axis-symmetric flow in the annual region between two cylinders

with which the profile of tangential velocity component is obtained as

$$u_\theta(r) = \frac{1}{r_o^2 - r_i^2}\left[\left(\omega_o r_o^2 - \omega_i r_i^2\right) r - (\omega_o - \omega_i)\frac{r_i^2 r_o^2}{r}\right]. \tag{8.2.32}$$

Substituting this solution into Eq. (8.2.29)$_1$ results in

$$\begin{aligned}\frac{\mathrm{d}p}{\mathrm{d}r} = \frac{\rho}{(r_o^2 - r_i^2)^2}&\left[\left(\omega_o r_o^2 - \omega_i r_i^2\right)^2 r - 2(\omega_o - \omega_i)\left(\omega_o r_o^2 - \omega_i r_i^2\right)\frac{r_i^2 r_o^2}{r}\right.\\ &\left. +(\omega_o - \omega_i)^2 \frac{r_i^4 r_o^4}{r^3}\right],\end{aligned} \tag{8.2.33}$$

which is integrated to obtain

$$\begin{aligned}p(r) = \frac{\rho}{(r_o^2 - r_i^2)^2}&\left[\left(\omega_o r_o^2 - \omega_i r_i^2\right)^2 \frac{r^2}{2} - 2(\omega_o - \omega_i)\left(\omega_o r_o^2 - \omega_i r_i^2\right) r_i^2 r_o^2 \ln r\right.\\ &\left. -(\omega_o - \omega_i)^2 \frac{r_i^4 r_o^4}{2r^2}\right] + C,\end{aligned} \tag{8.2.34}$$

which is the pressure distribution of flow field, with C an integration constant. It needs to be determined in any particular problem by specifying the value of p on $r = r_o$ or $r = r_i$. For the special case in which $\omega_i = 0$ and $\omega_o = \omega$, i.e., the inner cylinder is stationary while the outer cylinder rotates at a constant angular speed ω, Eq. (8.2.32) coincides exactly to Eq. (5.7.25), and the difference in pressures p_i and p_o obtained by using Eq. (8.2.34) corresponds exactly to that given in Eq. (5.7.28).

Now let two cylinders be stationary, and consider a fully developed, axial laminar flow along the x-direction, as shown in Fig. 8.3b. The equation describing the velocity component $u(r)$ along the x-direction is the same as Eq. (8.2.17), except that different no-slip boundary conditions should be allocated, which are given by $u = 0$ at $r = r_i$ and $r = r_o$. With these, the velocity profile of u is obtained as

$$u(r) = \frac{1}{4\mu}\left(\frac{\mathrm{d}p}{\mathrm{d}x}\right)\left[r^2 - r_o^2 + \frac{r_i^2 - r_o^2}{\ln(r_o/r_i)}\ln\left(\frac{r}{r_o}\right)\right]. \tag{8.2.35}$$

The volume flow rate is identified to be

$$\begin{aligned}Q &= -\frac{\pi}{8\mu}\left(\frac{\mathrm{d}p}{\mathrm{d}x}\right)\left[r_o^4 - r_i^4 - \frac{(r_o^2 - r_i^2)^2}{\ln(r_o/r_i)}\right],\\ \longrightarrow Q &= \frac{\pi\Delta p}{8\mu\ell}\left[r_o^4 - r_i^4 - \frac{(r_o^2 - r_i^2)^2}{\ln(r_o/r_i)}\right],\end{aligned} \tag{8.2.36}$$

where the second equation is obtained for a pipe with length ℓ and pressure drop Δp between two pipe ends, if the pressure gradient along the x-direction is constant. The location r_m at which the axial velocity is maximum is obtained by $\mathrm{d}u/\mathrm{d}r = 0$, which is given by

$$r_m = \sqrt{\frac{r_o^2 - r_i^2}{2\ln(r_o/r_i)}}. \tag{8.2.37}$$

It is readily verified that the maximum axial velocity does not occur at the midpoint of the annual region, and it occurs rather nearer the inner cylinder, with its specific location depending on the values of r_o and r_i.

The established results of an axial flow between two concentric cylinders are only valid for laminar flows. Since the geometry of flow field is annual instead of circular, the corresponding Reynolds number is revised as

$$\mathcal{R}_e \equiv \frac{\rho u_{av} d_h}{\mu}, \tag{8.2.38}$$

where u_{av} and d_h are the average velocity and *hydraulic diameter* defined by

$$u_{av} = \frac{Q}{A}, \qquad d_h \equiv \frac{4A}{L_w}. \tag{8.2.39}$$

The term L_w is called the *wetted perimeter*, which is the solid length in contact with the fluid. For the considered problem, $A = \pi(r_0^2 - r_i^2)$ and $L_w = 2\pi(r_0 + r_i)$. The validity of previous analysis is that the value of $\mathcal{R}_e$ should be smaller than 2100.

8.2.4 Stokes' First and Second Problems

Consider a fluid which is located on an initially stationary horizontal solid plate, as shown in Fig. 8.4a. At $t = 0$, the solid plate starts to move with a constant velocity U along the x-direction, which triggers a flow in the above fluid with its velocity depending on time. The response of fluid, i.e., the velocity distribution due to the sudden motion of solid boundary, needs to be determined. This two-dimensional problem is referred to as *Stokes' first problem*, which has counterparts in many branches of engineering and physics.

Since the motion of solid boundary is in the x-direction, it is plausible to assume that the motion of fluid will also be in the same direction. Thus, the non-vanishing

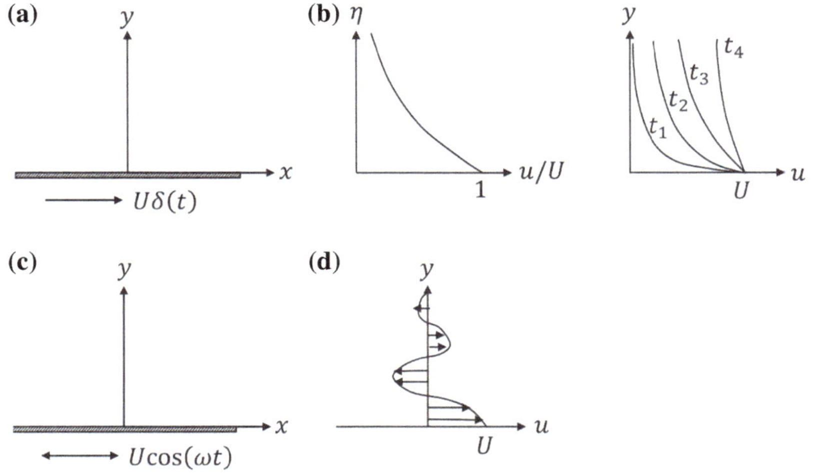

Fig. 8.4 Flows induced by moving boundaries. **a** Stokes' first problem. **b** Dimensionless and dimensional velocity profiles corresponding to **a** with respect to η and y, in which $t_4 > t_3 > t_2 > t_1$. **c** Stokes' second problem. **d** Dimensional velocity profile corresponding to **c** with respect to y

velocity component will be u, which is only a function of y and t. For simplicity, the gravitational acceleration is assumed to point perpendicular to the page. Then, the pressure will be independent on y, and subsequently independent on x, for u is independent of x, and becomes a constant everywhere in the fluid. With these, the continuity equation holds identically, and the Navier-Stokes equation along the x-direction reduces to

$$\frac{\partial u}{\partial t}=\nu\frac{\partial^2 u}{\partial y^2}, \tag{8.2.40}$$

which is subject to the boundary conditions given by

$$u(0,t)=U\delta(t), \qquad u(y\to\infty,t)=\text{finite}, \tag{8.2.41}$$

where $\delta(t)$ is the *unit-step function* having $\delta(t)=1$ for $t>0$, and $\delta(t)=0$ for $t\leq 0$. The above formulation lends itself to solution by the Laplace transform or by similarity methods. Since similarity solutions are the only ones which exist for some nonlinear problems arising in the boundary-layer theory and other situations, this approach will be applied to establish a base for the forthcoming discussions.

Similarity solutions are a special class of solutions which exist for problems which are governed by parabolic partial differential equations in two independent arguments, where there is no geometric length scale in the problem. The Stokes first problem meets these restrictions. It may be anticipated that the velocity u will reach some specific value u^* at different values of y which will depend on the values of t. At some time t_1, the velocity will have the value of u^* at some distance y_1, and at some later time t_2, the same velocity magnitude will exist at some different distance y_2, and so on. This suggests that there exists some combination of y and t, so that if this combined quantity is constant, the velocity will also be constant. Thus, a solution to the problem may exist in the form

$$\frac{u(y,t)}{U}=f(\eta), \qquad \eta(y,t)=\alpha\frac{y}{t^n}, \tag{8.2.42}$$

where α and n are constants, and η is called a *similarity variable*. This is done so, because if η is a constant, u is also a constant. Substituting these expressions into Eq. (8.2.40) yields

$$-Un\frac{\eta}{t}f'=\nu U\frac{\alpha^2}{t^{2n}}f'', \tag{8.2.43}$$

where the primes denote differentiation with respect to η. By choosing $\eta=1/2$, this equation can be simplified to

$$f''+\frac{\eta}{2\nu\alpha^2}f'=0. \tag{8.2.44}$$

Hence, the original partial differential equation has been reduced to an ordinary differential equation in the context of similarity methods. Since η must be dimensionless, the quantity α must be a function of $1/\sqrt{\nu}$, which is chosen to be $\alpha=1/(2\sqrt{\nu})$ for simplicity. With this, it follows that

$$\eta=\frac{y}{2\sqrt{\nu t}}, \qquad \longrightarrow \qquad f''+2\eta f'=0. \tag{8.2.45}$$

The solution to Eq. (8.2.45)$_2$ is given by

$$f(\eta) = C_1 \int_0^{\eta} e^{-\xi^2} \mathrm{d}\xi + C_2, \tag{8.2.46}$$

where C_1 and C_2 are integration constants. Using the boundary conditions to this solution yields $C_1 = -2/\sqrt{\pi}$ and $C_2 = 1$, with which the velocity component $u(y, t)$ is obtained as

$$\frac{u(y,t)}{U} = 1 - \frac{2}{\sqrt{\pi}} \int_0^{y/(2\sqrt{\nu t})} e^{-\xi^2} \mathrm{d}\xi = 1 - \mathrm{erf}\left(\frac{y}{2\sqrt{\nu t}}\right), \tag{8.2.47}$$

where "erf(x)" stands for the error function of x. Figure 8.4b illustrates the dimensionless velocity profiles in terms of η and the corresponding dimensional velocity profiles at different times based on Eq. (8.2.47).

An estimation on the fluid depth which is affected by the motion of moving boundary is obtained by requiring that $u/U \sim 0.04$, which gives rise to $\eta = 3/2$. Thus,

$$\eta = \frac{3}{2}, \qquad \longrightarrow \qquad \Delta = 3\sqrt{\nu t}, \tag{8.2.48}$$

where Δ represents the value of y at which $u/U \sim 0.04$, which denotes the thickness of fluid layer which is influenced significantly by the motion of boundary. It is proportional to the square root of time and the square root of kinematic viscosity. Outside this layer the fluid may be considered to be unaffected by the moving boundary. Equation (8.2.48) shows equally the role played by the kinematic viscosity in the diffusion of linear momentum from the moving boundary toward the fluid.

Now consider the same configuration again, except that the solid boundary experiences a harmonic oscillation with frequency ω, as shown in Fig. 8.4c. This problem is referred to as *Stokes' second problem*. The governing equation for this flow is the same as Eq. (8.2.40), but the allocated boundary conditions are changed to

$$u(0, t) = U\cos(\omega t), \qquad u(y \to \infty, t) = \text{finite}. \tag{8.2.49}$$

Since the boundary at $y = 0$ is oscillating in time, it is expected that the fluid will also oscillate in the x-direction in time with the same frequency but different amplitude and phase due the viscous effect. Hence, a steady-state solution is sought in the form

$$u(y, t) = Re\left[f(y)e^{i\omega t}\right], \tag{8.2.50}$$

which is substituted into Eq. (8.2.40) to yield

$$f''(y) - i\frac{\omega}{\nu} f = 0, \tag{8.2.51}$$

where the primes denote differentiation with respect to y. The solution to this ordinary differential equation is given by

$$f(y) = C_1 \exp\left[-(1+i)\sqrt{\frac{\omega}{2\nu}}y\right] + C_2 \exp\left[(1+i)\sqrt{\frac{\omega}{2\nu}}y\right], \tag{8.2.52}$$

in which $\sqrt{i} = \pm(1+i)/\sqrt{2}$ has been used, and C_1 and C_2 are integration constants. Applying the boundary conditions to this solution yields $C_2 = 0$ and $C_1 = U$, with which the velocity distribution is obtained as

$$\frac{u(y,t)}{U} = \exp\left(-\sqrt{\frac{\omega}{2\nu}}y\right)\cos\left(\omega t - \sqrt{\frac{\omega}{2\nu}}y\right). \tag{8.2.53}$$

It is seen that the velocity is oscillating in time with the same frequency as that of the boundary. The amplitude assumes the maximum value at $y = 0$ and decreases exponentially as y increases. There exists a phase shift in the motion of fluid, which is proportional to y and to the square root of the oscillating frequency of boundary. The results are shown graphically in Fig. 8.4d.

The distance Δ away from the oscillating boundary within which the fluid is influenced by the motion of boundary is obtained by requiring that the maximum amplitude of the oscillating velocity of fluid equals U/e^2. That is,

$$\frac{1}{e^2} = \exp\left(-\sqrt{\frac{\omega}{2\nu}}\Delta\right), \quad \longrightarrow \quad \Delta = 2\sqrt{\frac{2\nu}{\omega}}. \tag{8.2.54}$$

For $y > \Delta$, the fluid may be considered to be essentially unaffected by the motion of oscillating boundary. The viscous effect extends over a distance which is proportional to $\sqrt{\nu}$, and Δ varies inversely as the square root of the frequency of motion. The faster the motion is, the smaller will be the value of Δ.[5]

8.2.5 Pulsating Flows in Channels and Circular Conduits

An exact solution exists for a flow induced by an oscillating pressure gradient in a fluid layer which is bounded by two parallel planes. Consider the configuration shown in Fig. 8.1a again. Now let two fixed parallel planes be located at $y = \pm a$, between which a Newtonian fluid layer is placed. An oscillating pressure gradient with time exists along the x-direction. It follows from the geometric configurations that the velocity will be in the x-direction, which oscillates in time, i.e., $u = u(y, t)$. For simplicity, the gravity is assumed to point perpendicular to the page. For the considered circumstance, the mass balance holds identically, and the Navier-Stokes equation in the x-direction reduces to

$$\frac{\partial u}{\partial t} = -\frac{1}{\rho}\frac{\partial p}{\partial x} + \nu\frac{\partial^2 u}{\partial y^2}, \tag{8.2.55}$$

which is associated with the no-slip boundary conditions given by $u(-a, t) = u(a, t) = 0$. The pressure gradient is further expressed in the form

$$\frac{\partial p}{\partial x} = p_x \cos(\omega t) = Re\left[p_x e^{i\omega t}\right], \tag{8.2.56}$$

where p_x represents the amplitude of pressure oscillation, which is a constant.

[5]The responses of the non-Newtonian fluids which are subject to the boundary conditions of Stokes' first and second problems are completely different. For details, see e.g. Fang, C., Wang, Y., Hutter, K., A unified evolution equation for the Cauchy stress tensor of an isotropic elasto-visco-plastic material, I, On thermodynamically consistent evolution, *Continuum Mech. Thermodyn.*, 19(7), 423–440, 2008; and Fang, C., Lee, CH., A unified evolution equation for the Cauchy stress tensor of an isotropic elasto-visco-plastic material, II, Normal stress difference in a viscometric flow, and an unsteady flow with a moving boundary, *Continuum Mech. Thermodyn.*, 19(7), 441–455, 2008.

By virtue of the oscillating nature of pressure gradient, it is expected that the fluid velocity oscillates in time with the same frequency and a possible phase lag relative to the pressure oscillation, which is proposed as

$$u(y,t) = Re\left[f(y)e^{i\omega t}\right], \tag{8.2.57}$$

with which Eq. (8.2.55) becomes

$$f'' - i\frac{\omega}{\nu}f = \frac{p_x}{\rho\nu}, \tag{8.2.58}$$

which is a non-homogeneous ordinary differential equation of $f(y)$, where the primes denote differentiation with respect to y. The solution to $f(y)$ is given by

$$f(y) = i\frac{p_x}{\rho\omega} + C_1\cosh\left[(1+i)\sqrt{\frac{\omega}{2\nu}}y\right] + C_2\sinh\left[(1+i)\sqrt{\frac{\omega}{2\nu}}y\right], \tag{8.2.59}$$

where C_1 and C_2 are integration constants. This solution is so obtained that the term $(1+i)/\sqrt{2}$ has been used to replace $\sqrt{i}$, and the hyperbolic form has been chosen due to the finite extent of flow field in the y-direction. Applying the no-slip boundary conditions to this solution yields

$$C_1 = -\frac{ip_x}{\rho\omega\cosh[(1+i)\sqrt{(\omega/2\nu)}a]}, \qquad C_2 = 0, \tag{8.2.60}$$

with which $f(y)$ is given by

$$f(y) = i\frac{p_x}{\rho\omega}\left\{1 - \frac{\cosh[(1+i)\sqrt{(\omega/2\nu)}y]}{\cosh[(1+i)\sqrt{(\omega/2\nu)}a]}\right\}, \tag{8.2.61}$$

and the fluid velocity is then obtained as

$$u(y,t) = Re\left[i\frac{p_x}{\rho\omega}\left(1 - \frac{\cosh[(1+i)\sqrt{(\omega/2\nu)}y]}{\cosh[(1+i)\sqrt{(\omega/2\nu)}a]}\right)\right]. \tag{8.2.62}$$

Equation (8.2.62) can further be expanded to yield the real part explicitly. Although the concept is straightforward, the details are cumbersome. Hence, the implicit form given in Eq. (8.2.62) is considered the final expression of solution. It is readily verified that the velocity oscillates with the same frequency as the pressure gradient, but with a phase lag depending on y. The motion of fluid which is adjacent to the boundaries has a time-wise phase shift relative to the motion near the centerline of channel. The amplitude of motion near the boundaries is equally different from that near the centerline. This amplitude will approach null as the boundaries are approached, in order to satisfy the boundary conditions.

Now consider the horizontal circular conduit shown in Fig. 8.2b again. Let the pressure gradient along the x-direction oscillate in time, which is prescribed by

$$\frac{\mathrm{d}p}{\mathrm{d}x} = -\rho p_x e^{i\omega t}, \tag{8.2.63}$$

where ρp_x is the amplitude of pressure oscillation. The Navier-Stokes equation in the x-direction reads the form

$$\rho\frac{\partial u}{\partial t} = -\frac{\mathrm{d}p}{\mathrm{d}x} + \frac{\mu}{r}\frac{\mathrm{d}}{\mathrm{d}r}\left(r\frac{\mathrm{d}u}{\mathrm{d}r}\right). \tag{8.2.64}$$

Substituting Eq. (8.2.63) into the above equation results in an ordinary differential equation for $u(r, t)$. With the similar procedures described previously, the solution is obtained as

$$u(r,t) = R_e\left[\frac{p_x}{i\omega}e^{i\omega t}\left(1 - \frac{J_0\left(r\sqrt{-i\omega/\nu}\right)}{J_0\left(a\sqrt{-i\omega/\nu}\right)}\right)\right], \tag{8.2.65}$$

where J_0 is the Bessel function of the first kind, which depends on the complex argument z. For different values of z, J_0 can be approximated by the series given by

$$\begin{aligned} z < 2,\ J_0(z) &\sim 1 - \frac{z^2}{4} + \frac{z^4}{64} - \cdots, \\ z > 2,\ J_0(z) &\sim \sqrt{\frac{2}{\pi z}}\cos\left(z - \frac{\pi}{4}\right). \end{aligned} \tag{8.2.66}$$

With these approximations, the velocity profiles $u(r, t)$ at different oscillating frequencies are obtained as

$$\begin{aligned} \bar{\omega} < 4,\ \frac{u(r,t)}{u_{max}} &\sim \left(1 - \bar{r}^2\right)\cos(\bar{\omega}t) + \frac{\bar{\omega}}{16}\left(\bar{r}^4 + 4\bar{r}^2 - 5\right)\sin(\bar{\omega}t) + \mathcal{O}(\bar{\omega}^2), \\ \bar{\omega} > 4,\ \frac{u(r,t)}{u_{max}} &\sim \frac{4}{\bar{\omega}}\left[\sin(\bar{\omega}t) - \frac{e^{-A}}{\sqrt{\bar{r}}}\sin(\bar{\omega}t - A)\right] + \mathcal{O}(\bar{\omega}^{-2}), \end{aligned} \tag{8.2.67}$$

with the dimensionless variables defined by

$$\bar{r} = \frac{r}{a}, \qquad \bar{\omega} = \frac{\omega a^2}{\nu}, \qquad u_{max} = \frac{a p_x}{4\nu}, \qquad A = (1 - \bar{r}^2)\sqrt{\frac{\bar{\omega}}{2}}. \tag{8.2.68}$$

For very small values of $\bar{\omega}$, the flow is nearly a quasi-steady Poiseuille flow in phase with the slowly varying pressure gradient, and the second term of Eq. $(8.2.67)_1$ adds a lagging component which reduces the velocity at the centerline. For larger values of $\bar{\omega}$, it follows from Eq. $(8.2.67)_2$ that the flow lags approximately the pressure gradient by angle $\pi/2$, and again the velocity at the centerline is less than u_{max}. However, near the conduit wall there is a region of high velocity, as implied by Eq. $(8.2.67)_2$. Averaging this equation over one cycle yields the mean square velocity $\bar{u}_m^2$ given by

$$\frac{\bar{u}_m^2}{p_x^2/2\omega^2} = 1 - \frac{2}{\sqrt{\bar{r}}}e^{-A}\cos A + \frac{e^{-2A}}{\bar{r}}, \tag{8.2.69}$$

which shows an overshoot of $\bar{u}_m^2$ near the conduit wall.

8.2.6 The Hiemenz Flow

Consider a flow approaching vertically a two-dimensional stationary plate, as shown in Fig. 8.5a, in which the flow direction coincides with the negative y-axis, and the plate surface coincides with the x-axis. The plate boundary may be considered to be curved, e.g. the surface of a circular cylinder, provided that the region under consideration is small in extent compared with the radius of curvature of the surface.

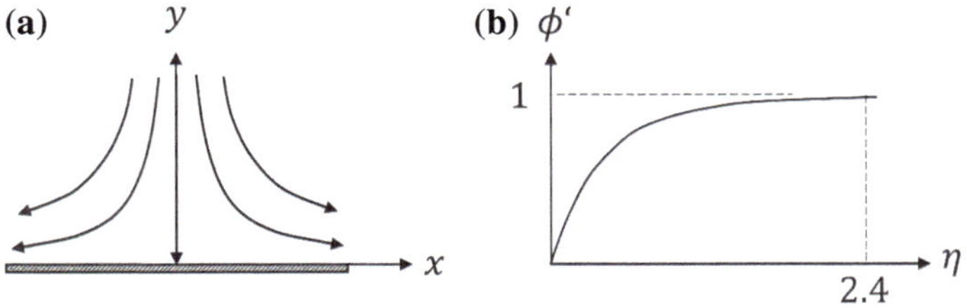

Fig. 8.5 A two-dimensional Hiemenz flow (stagnation-point flow). **a** The geometric configurations and coordinate system. **b** The functional form of the solution ϕ'

The flow field is frequently referred to as the *Hiemenz flow*,[6] or alternatively the *stagnation-point flow*. The solution to the problem is obtained by modifying the corresponding potential-flow solution in such a way that the Navier-Stokes equation and associated no-slip boundary conditions can be fulfilled.

By using Eq. (7.5.26) with $n = 2$, the velocity components of the corresponding potential flow are given by

$$u = 2Ux, \qquad v = -2Uy, \tag{8.2.70}$$

with which the pressure p is obtained by using the Bernoulli equation, viz.,

$$p = p_s - 2\rho U^2 \left(x^2 + y^2\right), \tag{8.2.71}$$

where p_s is the pressure at the stagnation point. The velocity components and pressure obtained by using the potential-flow theory also satisfy the Navier-Stokes equation, for the viscous term vanishes identically, i.e., $\mu\nabla^2\boldsymbol{u} = \mu\nabla^2(\nabla\phi) = \mu\nabla(\nabla^2\phi) = \boldsymbol{0}$. However, they do not satisfy the no-slip boundary conditions. To meet this requirement, the velocity component in the x-direction is revised to

$$u = 2Uxf'(y), \tag{8.2.72}$$

where f is an undetermined function with the prime denoting differentiation with respect to y. It follows form the continuity equation that

$$\frac{\partial v}{\partial y} = -\frac{\partial u}{\partial x} = -2Uf'(y), \qquad \longrightarrow \qquad v = -2Uf(y). \tag{8.2.73}$$

If $f(y)$ is stipulated that $f(y) \to y$ as $y \to \infty$, the potential-flow solutions can be recovered far away from the boundary. The Navier-Stokes equations in the x- and y-directions of the consider problem are given respectively by

$$\begin{aligned} u\frac{\partial u}{\partial x} + v\frac{\partial u}{\partial y} &= -\frac{1}{\rho}\frac{\partial p}{\partial x} + \nu\left(\frac{\partial^2 u}{\partial x^2} + \frac{\partial^2 u}{\partial y^2}\right), \\ u\frac{\partial v}{\partial x} + v\frac{\partial v}{\partial y} &= -\frac{1}{\rho}\frac{\partial p}{\partial y} + \nu\left(\frac{\partial^2 v}{\partial x^2} + \frac{\partial^2 v}{\partial y^2}\right), \end{aligned} \tag{8.2.74}$$

[6] Karl Hiemenz, 1885–1973, a German mathematician and physicist, who was one of Prandtl's' student and contributed to the theory of boundary layer.

which, by using Eqs. (8.2.72) and (8.2.73)$_2$, are recast alternatively as

$$\begin{aligned} 4U^2x\left(f'\right)^2 - 4U^2xff'' &= -\frac{1}{\rho}\frac{\partial p}{\partial x} + 2U\nu xf''', \\ 4U^2ff' &= -\frac{1}{\rho}\frac{\partial p}{\partial y} - 2U\nu f''. \end{aligned} \tag{8.2.75}$$

Integrating the second equation yields

$$\begin{aligned} p(x,y) &= -2\rho U^2f^2 - 2\rho U\nu f' + g(x) \\ &= p_s - 2\rho U^2f^2 + 2\rho U\nu(1-f') - 2\rho U^2x^2, \end{aligned} \tag{8.2.76}$$

where $g(x)$ is an undetermined function, which can be determined by using Eq. (8.2.71) at $y \to \infty$. Substituting this equation into Eq. (8.2.75)$_1$ gives

$$4U^2x\left(f'\right)^2 - 4U^2xff'' = 4U^2x + 2U\nu xf''', \quad \longrightarrow \quad \frac{\nu}{2U}f''' + ff'' - \left(f'\right)^2 + 1 = 0. \tag{8.2.77}$$

Three boundary conditions must be allocated to Eq. (8.2.77)$_2$, which are given by

$$f(0) = f'(0) = 0, \qquad f'(y \to \infty) \to 1, \tag{8.2.78}$$

resulted from the facts that $u(x,0) = v(x,0) = 0$, yielding $f'(0) = f(0) = 0$, respectively, and the potential-flow solution should be recovered at $y \to \infty$, yielding $f(y \to \infty) = y$ or $f'(y \to \infty) = 1$.

Defining the new variables given by

$$\phi(\eta) \equiv \sqrt{\frac{2U}{\nu}}f(y), \qquad \eta \equiv \sqrt{\frac{2U}{\nu}}y, \tag{8.2.79}$$

and substituting these expressions into Eqs. (8.2.77)$_2$ and (8.2.78) result respectively in

$$\phi''' + \phi\phi'' - \left(\phi'\right)^2 + 1 = 0, \qquad \phi(0) = \phi'(0) = 0, \;\; \phi'(\eta \to \infty) \to 1, \tag{8.2.80}$$

where the primes become differentiation with respect to η. The established nonlinear ordinary differential equation with the prescribed boundary conditions cannot be solve analytically, and the solution must be obtained numerically. Since it is much easier to obtain the solutions to this ordinary differential equation than those to the original partial differential equation, the formulation given in Eq. (8.2.80) is usually considered to be exact. Once ϕ is determined, f is subsequently determined, and the velocity components and pressure distribution are then obtained. The numerical solution to ϕ in terms of η is shown in Fig. 8.5b.

It follows from the numerical results that ϕ assumes unity value at $\eta = 2.4$. The thickness of viscous layer, Δ, in which the fluid characteristics are influenced by the solid boundary, is obtained as

$$\eta = \sqrt{\frac{2U}{\nu}}\Delta = 2.4, \qquad \longrightarrow \qquad \Delta = 2.4\sqrt{\frac{\nu}{2U}}. \tag{8.2.81}$$

In other words, the viscous effects are confined to a layer adjacent to the boundary, whose thickness varies as the square root of the kinematic viscosity of fluid, and inversely as the square root of the magnitude of approaching flow.

8.2.7 Flows in Convergent and Divergent Channels

Figure 8.6 shows a Newtonian fluid with constant density and dynamic viscosity flowing through a two-dimensional convergent and a divergent channels, in which the cylindrical coordinate system (r, θ, x) is chosen with x pointing perpendicular to the page. For simplicity, the gravitational acceleration is assumed to be in the x-direction. It follows from the geometric configurations that u_r is the only non-vanishing velocity component, which depends on r and θ. With these, the continuity equation reads

$$\frac{1}{r}\frac{\partial}{\partial r}(ru_r) = 0, \tag{8.2.82}$$

and the Navier-Stokes equations in the r- and θ-directions are given respectively by

$$\begin{aligned} u_r\frac{\partial u_r}{\partial r} &= -\frac{1}{\rho}\frac{\partial p}{\partial r} + \nu\left[\frac{1}{r}\frac{\partial}{\partial r}\left(r\frac{\partial u_r}{\partial r}\right) - \frac{u_r}{r^2} + \frac{1}{r^2}\frac{\partial^2 u_r}{\partial \theta^2}\right], \\ 0 &= -\frac{1}{\rho r}\frac{\partial p}{\partial \theta} + \nu\left(\frac{2}{r^2}\frac{\partial u_r}{\partial \theta}\right). \end{aligned} \tag{8.2.83}$$

By using the method of separation variables, a solution to u_r is decomposed into

$$u_r(r, \theta) = R(r)\Theta(\theta) = \frac{\nu}{r}\Theta(\theta), \tag{8.2.84}$$

resulted from the fact that u_r must be proportional to $1/r$, as implied by the continuity equation, with ν the kinematic viscosity as the proportional factor to render $\Theta(\theta)$ dimensionless.

Substituting the above expression into Eq. (8.2.83) yields respectively

$$-\frac{\nu^2}{r^3}\Theta^2 = -\frac{1}{\rho}\frac{\partial p}{\partial r} + \frac{\nu^2}{r^3}\Theta'', \qquad 0 = -\frac{1}{\rho r}\frac{\partial p}{\partial \theta} + 2\frac{\nu^2}{r^3}\Theta', \tag{8.2.85}$$

with the primes denoting differentiations with respect to θ. Taking partial derivative with respect to θ to the first equation, and the partial derivative with respect to r to the second equation, and eliminating the common term $\partial^2 p/(\partial r\partial\theta)$ of two resulting equations yields

$$\Theta''' + 4\Theta' + 2\Theta\Theta' = 0, \tag{8.2.86}$$

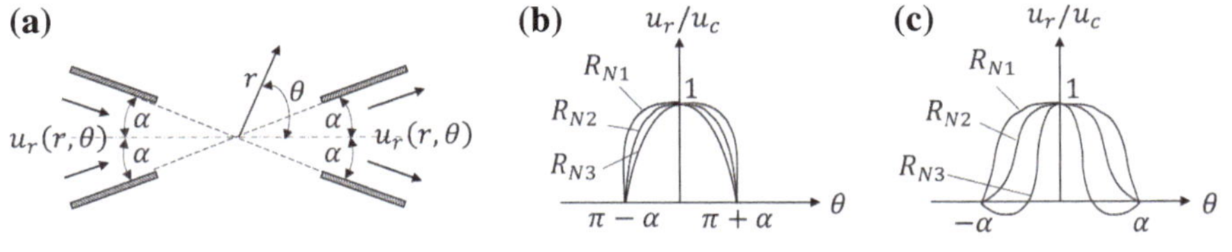

Fig. 8.6 Flows in two-dimensional convergent and divergent channels. **a** The geometric configurations and cylindrical coordinate system. **b** The dimensionless velocity profiles in the convergent channel. **c** The dimensionless velocity profiles in the divergent channel, with $\mathcal{R}_{N1} > \mathcal{R}_{N2} > \mathcal{R}_{N3}$

which is integrated once to obtain

$$\Theta'' + 4\Theta + \Theta^2 = K, \tag{8.2.87}$$

where K is an integration constant. Let $G(\Theta) = \Theta'$, with which Eq. (8.2.87) is recast as

$$G\frac{\mathrm{d}G}{\mathrm{d}\Theta} + 4\Theta + \Theta^2 = K, \qquad \longrightarrow \qquad \frac{\mathrm{d}}{\mathrm{d}\Theta}\left(\frac{G^2}{2}\right) = K - 4\Theta - \Theta^2, \tag{8.2.88}$$

for $\mathrm{d}G/\mathrm{d}\Theta = \Theta''/G$. Integrating the second equation gives

$$\frac{G^2}{2} = A + K\Theta - 2\Theta^2 - \frac{\Theta^3}{3}, \quad \longrightarrow \quad G(\Theta) = \frac{\mathrm{d}\Theta}{\mathrm{d}\theta} = \sqrt{2\left(A + K\Theta - 2\Theta^2 - \frac{\Theta^3}{3}\right)}, \tag{8.2.89}$$

where A is an integration constant. Although this equation does not deliver an explicit expression of $\Theta(\theta)$, the result may be put in the form of an integral expression for θ as a function of Θ given by an elliptic integral, viz.,

$$\theta = \int_0^{\Theta} \frac{\mathrm{d}\xi}{\sqrt{2(A + K\xi - 2\xi^2 - \xi^3/3)}} + B, \tag{8.2.90}$$

where B is an integration constant. Equations (8.2.84) and (8.2.90) define the velocity distribution of considered problem.

The no-slip boundary conditions on the channel walls require that

$$u_r(r, \alpha) = u_r(r, -\alpha) = 0; \qquad u_r(r, \pi + \alpha) = u_r(r, \pi - \alpha) = 0, \tag{8.2.91}$$

for the divergent and convergent channels, respectively. The velocity profiles in the divergent and convergent channels should also respectively satisfy

$$\frac{\partial u_r}{\partial \theta}(r, 0) = 0; \qquad \frac{\partial u_r}{\partial \theta}(r, \pi) = 0, \tag{8.2.92}$$

for the flow fields in both channels are symmetric with respect to the reference axis, as shown in the figure. With these, the conditions that should be fulfilled by Θ are obtained as

$$\Theta(\alpha) = \Theta(-\alpha) = \Theta'(0) = 0; \qquad \Theta(\pi + \alpha) = \Theta(\pi - \alpha) = \Theta'(\pi) = 0, \tag{8.2.93}$$

for the divergent and convergent channels, respectively. The Eq. (8.2.90) with the conditions given in Eq. (8.2.93) cannot be solved analytically to express Θ in terms of θ, and numerical integration must be used. Once $\Theta(\theta)$ is numerically determined, the numerical determinations of velocity given in Eq. (8.2.84) are accomplished, which are shown in Fig. 8.6b for the convergent channel, and in Fig. 8.6c for the divergent channel, in which $\mathcal{R}_N$ represents the Reynolds number defined by

$$\mathcal{R}_N \equiv \frac{u_c r}{\nu}, \tag{8.2.94}$$

where u_c is the fluid velocity along the centerline of channel. At low Reynolds numbers, the velocity profiles in the convergent channel are quite different from those in the divergent channel. This is due to the fact that an adverse pressure gradient in the divergent channel may overcome the inertia effect of fluid near the channel wall, where the viscous effects have reduced the velocity, giving rise to a reverse-flow configuration. The flow separation from the channel wall in a divergent channel has been well verified experimentally, in particular for large values of angle α.

8.2.8 Flows over Porous Boundary

In the previous discussions, the exact solutions were obtained for the flows in contact with solid boundaries, on which the tangential and normal components of fluid velocities were required to coincide with those of the solid boundaries. Exact solutions may equally exist if solid boundaries are allowed to permit non-vanishing normal velocity components on themselves. Boundaries satisfying this condition are termed *porous boundaries*. In Fig. 8.7, the plate is stationary and porous, above which a uniform flow with magnitude U along the x-direction exists, while a flow in the y-direction is induced near the porous plate. The flow is assumed to be steady and fully developed in the x-direction with the velocity component given by $u = u(y)$, while the velocity component along the y-direction is denoted by $v(x, y)$. The gravitational acceleration is assumed to point perpendicular to the page for simplicity. With these, the continuity equation and Navier-Stokes equations in the x- and y-directions reduce to

$$\frac{\partial v}{\partial y} = 0, \qquad v\frac{\mathrm{d}u}{\mathrm{d}y} = \nu\frac{\mathrm{d}^2u}{\mathrm{d}y^2}, \qquad u\frac{\partial v}{\partial x} + v\frac{\partial v}{\partial y} = \nu\left(\frac{\partial^2 v}{\partial x^2} + \frac{\partial^2 v}{\partial y^2}\right), \tag{8.2.95}$$

for the pressure is constant in the whole flow field. The associated boundary conditions are prescribed by

$$u(0) = 0, \qquad v(x, 0) = -V, \qquad u(y \to \infty) \to U, \tag{8.2.96}$$

where $V = \text{constant} > 0$, and $-V$ is termed the *suction velocity*. It follows immediately from Eqs. $(8.2.95)_1$ and $(8.2.96)_2$ that $v(x, y) = -V$.

With $v = -V$, Eq. $(8.2.95)_3$ is satisfied identically, while Eq. $(8.2.95)_2$ reduces to

$$-V\frac{\mathrm{d}u}{\mathrm{d}y} = \nu\frac{\mathrm{d}^2u}{\mathrm{d}y^2}, \tag{8.2.97}$$

to which the solution is given by

$$u(y) = C_1 + C_2 \exp\left(-\frac{V}{\nu}y\right), \tag{8.2.98}$$

where C_1 and C_2 are integration constants. Applying Eqs. $(8.2.96)_{1,3}$ to this solution yields $C_1 = U$ and $C_2 = -C_1$, with which the velocity component $u(y)$ is obtained as

$$u(y) = U\left[1 - \exp\left(-\frac{V}{\nu}y\right)\right]. \tag{8.2.99}$$

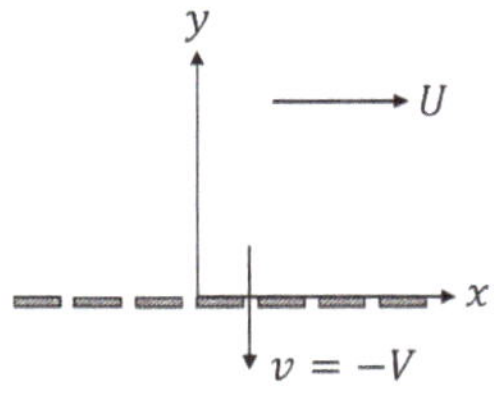

Fig. 8.7 A two-dimensional uniform flow over a horizontal porous plate with suction

To determine the thickness Δ of fluid layer, in which the fluid characteristics are affected by the viscous effect, let $u/U = 1 - 1/e^5$ at $y = \Delta$, it follows then

$$\Delta = 5\frac{\nu}{V}. \tag{8.2.100}$$

Thus, the distance away from the plate surface at which the uniform flow is essentially recovered is proportional to the kinematic viscosity of fluid and inversely proportional to the suction velocity.

If instead of a suction but a blowing is provided at $y = 0$, i.e., V assumes a negative value, the solution given in Eq. (8.2.99) diverges. The reason can be seen from the vorticity equation. It follows from Eq. (8.1.6) that for the considered circumstance, the vorticity equation reads the form

$$-V\frac{d\omega}{dy} = \nu\frac{d^2\omega}{dy^2}, \qquad \boldsymbol{\omega} = (0, 0, \omega), \tag{8.2.101}$$

which is integrated with respect to y to obtain

$$-V\omega = \nu\frac{d\omega}{dy}. \tag{8.2.102}$$

The left-hand-side represents the convection of vorticity toward the boundary along the negative y-direction in assistance with the suction velocity V, while the right-hand side represents the diffusion of vorticity along the positive y-direction via the kinematic viscosity of fluid. This equation shows that there is a balance between two transportation mechanisms of vorticity, so that the solution in the form of $u = u(y)$ prevails. If a blowing is provided (i.e., $V < 0$), these two transportation mechanisms will be along the same direction, and the assumed solution form of $u = u(y)$ is no longer valid.

8.3 Low-Reynolds-Number Solutions

For a flow problem in which an exact solution does not exist, it may be possible to obtain an approximate solution to the coupled local balances of mass and linear momentum. In this section, the full governing equations will be approximated for flows with low Reynolds numbers, and a certain exact solutions to the simplified equations, termed the low-Reynolds-number solutions, will be established.

8.3.1 Stokes' Approximation

The Reynolds number $\mathcal{R}_e$ is defined as the ratio of inertial force to viscous force of a fluid. For very small values of $\mathcal{R}_e$, the inertia force may be neglected in comparison with other presented forces. The essential feature of *Stokes' approximation* is that all the convective components of the inertia force are assumed to be small compared with the viscous force, so that the local mass balance and the Navier-Stokes equation reduce respectively to

$$\nabla \cdot \boldsymbol{u} = 0, \qquad \frac{\partial \boldsymbol{u}}{\partial t} = -\frac{1}{\rho}\nabla p + \nu\nabla^2 \boldsymbol{u}. \tag{8.3.1}$$

These equations are referred to as *Stokes' equations* for very slow motions of an incompressible viscous Newtonian fluid, in which Eq. $(8.3.1)_2$ is considered to be an asymptotic limit of the Navier-Stokes equation corresponding to vanishing values of $\mathcal{R}_e$, while the space coordinates remain of order of unity. For higher-order approximations of Stokes' equations for a problem, the velocity $\boldsymbol{u}$ and pressure p may be expanded in the ascending powers of the Reynolds number, so that sequences of differential equations would have to be solved by a limiting procedure to the Navier-Stokes equation, as will be shown later.

Taking double curl of Eq. $(8.3.1)_2$ yields

$$\frac{\partial}{\partial t}\left[\nabla(\nabla \cdot \boldsymbol{u}) - \nabla^2 \boldsymbol{u}\right] = \nu\nabla^2\left[\nabla(\nabla \cdot \boldsymbol{u}) - \nabla^2 \boldsymbol{u}\right], \tag{8.3.2}$$

in which $\nabla \times (\nabla \times \boldsymbol{u}) = \nabla(\nabla \cdot \boldsymbol{u}) - \nabla^2 \boldsymbol{u}$ and $\nabla \times \nabla p = 0$ have been used. It follows from this equation that

$$\nabla^2\left(\frac{\partial \boldsymbol{u}}{\partial t}\right) = \nu\nabla^4 \boldsymbol{u}, \qquad \nabla^2 p = 0. \tag{8.3.3}$$

To obtain the first equation, the continuity equation has been used, while the second equation has been derived by taking divergence of Eq. $(8.3.1)_2$. These two equations are the alternative form of the Stokes equations, with the advantage that the pressure field has been separated mathematically from the velocity field by the cost of highest differentials changed to fourth order instead of second order. Solutions to the Stokes' equations may be obtained by two different approaches. By using directly the equations subject to appropriately formulated boundary conditions, the solutions to the formulated boundary-value problems for geometry of interest may be obtained. Or the fundamental solutions may be established first for simple problems, then the solutions to complex problems may be obtained by superimposing the fundamental solutions. The latter approach is used in the forthcoming discussions for the benefit that a clear understanding of which elements in a solution are responsible for producing forces and torques.

8.3.2 Fundamental Solutions

Uniform flows. The simplest solution to Stokes' equations is that of a uniform flow. For a uniform flow with constant velocity $\boldsymbol{U}$ and pressure p, Eq. (8.3.1) holds identically. Thus, a solution to a uniform flow is given by

$$\boldsymbol{u} = U\boldsymbol{e}_x, \qquad p = \text{constant}, \tag{8.3.4}$$

where $\boldsymbol{e}_x$ is the unit vector with its direction parallel to $\boldsymbol{U}$. With these velocity and pressure distributions, no force or turning moment acting on the fluid exists.

Doublet. Any potential flow is an exact solution to the Navier-Stokes equation, for the viscous term vanishes identically. Thus, any steady potential flow is also a solution to Stokes' equations, provided that the pressure gradient vanishes, yielding a constant pressure field. By using the results derived in Sect. 7.6.3, the velocity potential function $\phi(r, \theta)$ of a doublet flow is given by

$$\phi(r, \theta) = A\frac{\cos\theta}{r^2} = A\frac{x}{r^3}, \qquad x = r\cos\theta, \tag{8.3.5}$$

with the coordinate system defined in Fig. 7.24a. The fluid velocity is obtained as

$$\boldsymbol{u} = \nabla\phi = A\left(\frac{1}{r^3}\boldsymbol{e}_x - \frac{3x}{r^4}\boldsymbol{e}_r\right), \qquad p = \text{constant}, \tag{8.3.6}$$

where $\boldsymbol{e}_x$ and $\boldsymbol{e}_r$ are respectively the unit vectors along the x- and radial directions. The above solution to the fluid velocity is only valid for a viscous fluid and cannot be proved to be valid from the upstream irrotational conditions. In addition, in order to satisfy the linear momentum equation, the pressure must be a constant field. Although there exists a singularity in the flow field described by Eq. (8.3.6), it does not exert a force or a moment on the surrounding fluid, for $p =$ constant.

Rotlet. Consider a steady flow field described by

$$\boldsymbol{u} = \boldsymbol{r} \times \nabla\chi, \qquad u_i = \varepsilon_{ijk} x_j \frac{\partial \chi}{\partial x_k}, \tag{8.3.7}$$

where $\boldsymbol{r}$ is the position vector, and χ represents a scalar quantity. Taking divergence of this equation yields

$$\nabla\cdot\boldsymbol{u} = \frac{\partial u_i}{\partial x_i} = \varepsilon_{ijk}\left(\frac{\partial x_j}{\partial x_i}\frac{\partial \chi}{\partial x_k} + x_j \frac{\partial^2 \chi}{\partial x_i \partial x_k}\right) = 0, \tag{8.3.8}$$

for the first term inside the paragraph vanishes identically, and the second term is a symmetric tensor. This equation indicates that the proposed flow field satisfies the continuity equation. Further, it is assumed that the pressure field is constant, with which Eq. $(8.3.1)_2$ reduces to

$$\nabla^2\boldsymbol{u} = \boldsymbol{0}, \quad \longrightarrow \quad \nabla^2 u_i = \varepsilon_{ijk}\left(\frac{\partial^2 x_j}{\partial x_m \partial x_m}\frac{\partial \chi}{\partial x_k} + x_j \frac{\partial}{\partial x_k}\frac{\partial^2 \chi}{\partial x_m \partial x_m}\right) = \nabla^2\chi = 0, \tag{8.3.9}$$

in which Eq. (8.3.7) has been used. Equation (8.3.9) shows that the proposed velocity field also satisfies Stokes' equations under a constant pressure field, provided that the scalar function χ satisfies the Laplace equation. Thus, the problem reduces to the determination of an axis-symmetric solution to the three-dimensional Laplace equation. It follows from the results in Sect. 7.6.3 that the solution corresponding to a doublet is in the form

$$\chi = B\frac{\cos\theta}{r^2} = B\frac{x}{r^3}, \tag{8.3.10}$$

with which the velocity is identified as

$$\boldsymbol{u} = B\boldsymbol{r} \times \nabla\left(\frac{x}{r^3}\right) = B\frac{\boldsymbol{e}_r \times \boldsymbol{e}_x}{r^2}, \qquad p = \text{constant}, \tag{8.3.11}$$

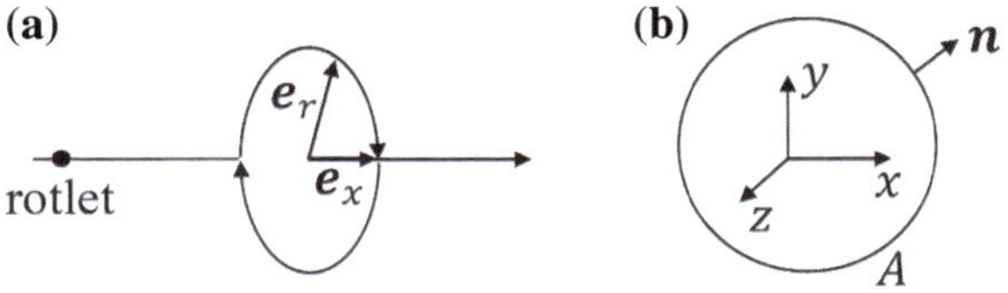

Fig. 8.8 A rotlet in a viscous fluid. **a** Typical streamlines. **b** A spherical control-surface embracing a rotlet with the coordinate system

since $\boldsymbol{r} = r\boldsymbol{e}_r$. The streamlines corresponding to the established velocity field with $B > 0$ are shown in Fig. 8.8a, which must be perpendicular to both $\boldsymbol{e}_r$ and $\boldsymbol{e}_x$, so that they form circles whose centers lie on the x-axis. The singularity of flow field locates at $r = 0$, which is termed a *rotlet*.

To identify the force and torque acting on the surrounding fluid by a rotlet, construct a spherical control-surface embracing the rotlet, as shown in Fig. 8.8b. Let the force acting on the fluid contained inside the control-surface be denoted by $\boldsymbol{f}$, it follow that

$$f_i = -\int_A t_{ij} n_j \mathrm{d}a, \tag{8.3.12}$$

where $\boldsymbol{t}$ is the stress tensor, A denotes the area of control-surface, and $\boldsymbol{n}$ represents the unit outward normal vector of A. For the Newtonian fluids with constant density and dynamic viscosity, the stress tensor is given in Eq. (5.6.33) with vanishing value of $\lambda(\nabla \cdot \boldsymbol{u})$. Substituting this into Eq. (8.3.12) yields

$$f_i = -\int_A \left[-p\delta_{ij} + \mu \left(\frac{\partial u_i}{\partial x_j} + \frac{\partial u_j}{\partial x_i} \right) \right] n_j \mathrm{d}a \sim \frac{1}{r}, \tag{8.3.13}$$

which results from that the first integration vanishes for a constant pressure, $u_i \sim r^{-2}$, as indicated by Eq. (8.3.11), and $\mathrm{d}a \sim r^2$. If the control-surface is assumed to be very large, then $f_i = 0$, as $r \to \infty$. Hence, there is no net force acting on the fluid due to a rotlet. Similarly, the torque $\boldsymbol{M}$ acting on the fluid contained inside the control-surface by a rotlet is given by

$$\boldsymbol{M} = \int_A \boldsymbol{r} \times \boldsymbol{t}\boldsymbol{n}\, \mathrm{d}a, \qquad M_i = \int_A \varepsilon_{ijk} x_j t_{km} n_m \mathrm{d}a. \tag{8.3.14}$$

Substituting the expression of the stress tensor into this expression gives

$$\begin{aligned} M_i &= \int_A \varepsilon_{ijk} x_j \left[-p\delta_{km} + \mu \left(\frac{\partial u_k}{\partial x_m} + \frac{\partial u_m}{\partial x_k} \right) \right] n_m \mathrm{d}a \\ &= \frac{\mu}{r} \int_A \varepsilon_{ijk} x_j x_m \left(\frac{\partial u_k}{\partial x_m} + \frac{\partial u_m}{\partial x_k} \right) \mathrm{d}a, \end{aligned} \tag{8.3.15}$$

for the first integration vanishes due to $p =$ constant, and $n_m = x_m/r$. The obtained expression is valid for any velocity distribution whatsoever. Since the velocity given in Eq. (8.3.11) is a homogeneous function of degree 2, it follows that[7]

$$x_m \frac{\partial u_k}{\partial x_m} = -2u_k, \qquad x_m \frac{\partial u_m}{\partial x_k} = \frac{\partial}{\partial x_k}(x_m u_m) - u_m \frac{\partial x_m}{\partial x_k} = -u_k, \tag{8.3.16}$$

[7]In three-dimensional circumstances, a *homogeneous function of order n* is one which satisfies

$$f\left(\frac{x}{\lambda}, \frac{y}{\lambda}, \frac{z}{\lambda}\right) = \lambda^n f(x, y, z), \qquad \forall\ \lambda.$$

resulted from the fact that the first term on the right-hand-side of the second equation vanishes because it corresponds to $\nabla(\boldsymbol{r}\cdot\boldsymbol{u})=\boldsymbol{0}$, for $\boldsymbol{u}$ is perpendicular to $\boldsymbol{r}$, as implied by Eq. (8.3.11). With these, Eq. (8.3.15) is simplified to

$$M_i=-3\frac{\mu}{r}\int_A \varepsilon_{ijk}x_j u_k \mathrm{d}a,\qquad \boldsymbol{M}=-3\frac{\mu}{r}\int_A \boldsymbol{r}\times\boldsymbol{u}\,\mathrm{d}a, \tag{8.3.17}$$

which is expressed alternatively as

$$\boldsymbol{M}=-3B\mu\int_A\left(\frac{x}{r}\boldsymbol{e}_r-\boldsymbol{e}_x\right)\frac{\mathrm{d}a}{r^2}, \tag{8.3.18}$$

in which Eq. (8.3.11) has been used. With the transformation relations between the rectangular and spherical coordinate systems given in Sect. 1.4, Eq. (8.3.18) is identified to be

$$\begin{aligned}\boldsymbol{M}=-3B\mu\int_0^{2\pi}\mathrm{d}\psi\int_0^{\pi}&\left[\left(\cos^2\theta-1\right)\boldsymbol{e}_x+\sin\theta\cos\theta\cos\psi\boldsymbol{e}_y\right.\\&\left.+\sin\theta\cos\theta\sin\psi\boldsymbol{e}_z\right]\sin\theta\,\mathrm{d}\theta=8\pi B\mu\boldsymbol{e}_x.\end{aligned} \tag{8.3.19}$$

Thus, the singularity exerts no force but a turning moment on the surrounding fluid. The magnitude of turning moment is proportional to the magnitude of fluid velocity and acts along the positive x-direction.

Stokeslet. Since the pressure must satisfy the three-dimensional Laplace equation, as indicated by Eq. $(8.3.3)_2$, it follows form the discussions in Sect. 7.6.3 that the pressure solution to the doublet-type, i.e., $p\sim\cos\theta/r^2$, may meet the requirement. It is assumed that the pressure is given by

$$p=2c\mu\frac{x}{r^3},\qquad x=r\cos\theta. \tag{8.3.20}$$

The flow is assumed to be steady, with which Eq. $(8.3.1)_2$ reduces to

$$\nabla^2\boldsymbol{u}=\frac{1}{\mu}\nabla p, \tag{8.3.21}$$

which must be satisfied by the fluid velocity $\boldsymbol{u}$. Substituting Eq. (8.3.20) into this equation for the y- and z-components yields

$$\nabla^2 v=-6c\frac{xy}{r^5},\qquad \nabla^2 w=-6c\frac{xz}{r^5}, \tag{8.3.22}$$

where v and w are the velocity components of $\boldsymbol{u}$ in the y- and z-directions, respectively. By using the properties of harmonic functions, the solutions to Eq. (8.3.22) are given by

$$v=\frac{cxy}{r^3},\qquad w=\frac{cxz}{r^3}, \tag{8.3.23}$$

For homogeneous functions, *Euler's theorem* states that

$$x\frac{\partial f}{\partial x}+y\frac{\partial f}{\partial y}+z\frac{\partial f}{\partial z}=-nf.$$

with which the equation to be satisfied by the velocity component u reduces to

$$\nabla^2 u = c\left(\frac{2}{r^3} - 6\frac{x^2}{r^5}\right). \tag{8.3.24}$$

In view of the solutions to v and w, u may be expected to be in the form

$$u = c\frac{x^2}{r^3}. \tag{8.3.25}$$

With Eqs. (8.3.22) and (8.3.25), the complete expression of $\boldsymbol{u}$ corresponding to the prescribed pressure field given in Eq. (8.3.20) is obtained as

$$\boldsymbol{u} = c\left(\frac{x^2}{r^3}\boldsymbol{e}_x + \frac{xy}{r^3}\boldsymbol{e}_y + \frac{xz}{r^3}\boldsymbol{e}_z\right) + \boldsymbol{u}' = c\frac{x}{r^2}\boldsymbol{e}_r + \boldsymbol{u}', \qquad \boldsymbol{u}' = (u', v', w'), \tag{8.3.26}$$

where $\boldsymbol{u}'$ is another solution corresponding to $\nabla^2\boldsymbol{u}' = \boldsymbol{0}$, for $u + u'$, $v + v'$, and $w + w'$ are also solutions to the equations satisfied by u, v, and w, respectively. Taking divergence of Eq. (8.3.26) yields

$$\nabla \cdot \boldsymbol{u} = c\frac{x}{r^3} + \nabla \cdot \boldsymbol{u}', \tag{8.3.27}$$

showing that $\boldsymbol{u}' = c\boldsymbol{e}_x/r$ must be chosen in order to satisfy the continuity equation. It is noted that the form of $\boldsymbol{u}'$ also fulfills $\nabla^2\boldsymbol{u}' = \boldsymbol{0}$. Consequently, the solution to Stokes' equations corresponding to a doublet type of the pressure field is summarized in the following:

$$p = 2c\mu\frac{x}{r^3}, \qquad \boldsymbol{u} = c\left(\frac{x}{r^2}\boldsymbol{e}_r + \frac{1}{r}\boldsymbol{e}_x\right), \tag{8.3.28}$$

with the singularity locating at the origin, which is called a *Stokeslet*.

By substituting the above expressions into Eq. $(8.3.13)_1$, the force acting on the surrounding fluid due to the presence of a Stokeslet is given by

$$f_i = -\int_A \left[-2c\mu\frac{x}{r^3}\delta_{ij} + \mu\left(\frac{\partial u_i}{\partial x_j} + \frac{\partial u_j}{\partial x_i}\right)\right] n_j \mathrm{d}a, \tag{8.3.29}$$

which reduces to

$$f_i = -\int_A \left[-2c\mu\frac{x}{r^3}\frac{x_i}{r} - \frac{\mu}{r}x_j\left(\frac{\partial u_i}{\partial x_j} + \frac{\partial u_j}{\partial x_i}\right)\right] \mathrm{d}a, \tag{8.3.30}$$

if A is chosen to be a spherical control-surface embracing the Stokeslet, with radius r and $n_j = x_j/r$ as the unit outward normal. Since the velocity distribution given in Eq. $(8.3.28)_2$ is homogeneous of order 1, it follows from Euler's theorem that

$$\begin{aligned} x_j\frac{\partial u_i}{\partial x_j} &= -u_i = -c\left(\frac{x}{r^3}x_i + \frac{\delta_{i1}}{r}\right), \\ x_j\frac{\partial u_j}{\partial x_i} &= \frac{\partial}{\partial x_i}(x_j u_j) - u_j\frac{\partial x_j}{\partial x_i} = \frac{\partial}{\partial x_i}(\boldsymbol{r}\cdot\boldsymbol{u}) - u_i = c\left(\frac{\delta_{i1}}{r} - 3\frac{xx_i}{r^3}\right). \end{aligned} \tag{8.3.31}$$

With these, Eq. (8.3.30) is simplified to

$$\begin{aligned} f_i &= -\int_A \left[-2c\mu\frac{xx_i}{r^4} - \frac{c\mu}{r}\left(\frac{xx_i}{r^3} + \frac{\delta_{i1}}{r}\right) + \frac{c\mu}{r}\left(\frac{\delta_{i1}}{r} - 3\frac{xx_i}{r^3}\right)\right] \mathrm{d}a \\ &= 6c\mu \int_A \frac{xx_i}{r^4} \mathrm{d}a, \end{aligned} \tag{8.3.32}$$

which, in vector notation, is expressed as

$$\boldsymbol{f} = 6c\mu \int_A \frac{x}{r^3} \boldsymbol{e}_r \mathrm{d}a. \tag{8.3.33}$$

Again, with the relations between the rectangular and spherical coordinate systems, Eq. (8.3.33) is further identified to be

$$\boldsymbol{f} = 6c\mu \int_0^{2\pi} \mathrm{d}\psi \int_0^{\pi} \cos\theta \left(\cos\theta \boldsymbol{e}_x + \sin\theta\cos\psi \boldsymbol{e}_y + \sin\theta\sin\psi \boldsymbol{e}_z\right) \mathrm{d}\theta = 8\pi c\mu \boldsymbol{e}_x. \tag{8.3.34}$$

In other words, a Stokeslet exerts a force on the surrounding fluid along the positive x-axis with the strength proportional to the pressure parameter c, if $c > 0$. However, a Stokeslet does not exert a torque $\boldsymbol{M}$ on the surrounding fluid. The derivation of this result is left as an exercise.

8.3.3 Interactions Between a Sphere and a Viscous Fluid

Two fundamental interactions between a sphere and a viscous fluid are discussed. First, consider a sphere with radius a rotating with constant angular speed ω about the x-axis in an otherwise quiescent fluid. The induced flow field is similar to that of a rotlet. Thus, the velocity distribution is given by Eq. (8.3.11). Since on the surface $r = a$, the fluid velocity is given by $\boldsymbol{u} = a\omega \boldsymbol{e}_r \times \boldsymbol{e}_x$, the constant B is determined as $B = \omega a^3$, so that the fluid velocity becomes

$$\boldsymbol{u} = \frac{\omega a^3}{r^2} \boldsymbol{e}_r \times \boldsymbol{e}_x, \tag{8.3.35}$$

which also satisfies the condition that $\|\boldsymbol{u}(r \to \infty)\| =$ finite. Although the singularity of a rotlet locates at $r = 0$, it has no influence on the flow field around a rotating sphere, for the singularity is now embraced by the spherical surface. But the rotlet exerts a turning moment on the surrounding fluid. It follows from Newton's third law of motion that there exists equally a turning moment with same magnitude but reverse direction on the sphere given by

$$\boldsymbol{M} = -8\pi\mu\omega a^3 \boldsymbol{e}_x. \tag{8.3.36}$$

Next, consider a uniform flow past a sphere, whose solution to the velocity field is obtained by superimposing the flow fields of a uniform flow, a doublet and a Stokeslet. It follows from Eqs. (8.3.4), (8.3.6) and (8.3.28) that

$$\boldsymbol{u} = U\boldsymbol{e}_x + A\left(\frac{1}{r^3}\boldsymbol{e}_x - \frac{3x}{r^4}\boldsymbol{e}_r\right) + c\left(\frac{x}{r^2}\boldsymbol{e}_r + \frac{1}{r}\boldsymbol{e}_x\right), \qquad p = 2c\mu\frac{x}{r^3}, \tag{8.3.37}$$

for the velocity and pressure fields, respectively. Since the Reynolds number of the considered circumstance is very small, the velocity at the rear stagnation point must vanish. Substituting the condition $\boldsymbol{u}(r=x=a)=\mathbf{0}$ into Eq. (8.3.37)$_1$ yields

$$\mathbf{0}=U\boldsymbol{e}_x+A\left(\frac{1}{a^3}\boldsymbol{e}_x-\frac{3}{a^3}\boldsymbol{e}_r\right)+\frac{c}{a}\left(\boldsymbol{e}_r+\boldsymbol{e}_x\right), \tag{8.3.38}$$

which gives rise to a pair of equations given by

$$0=U+\frac{A}{a^3}+\frac{c}{a},\quad 0=-\frac{3A}{a^3}+\frac{c}{a},\qquad\longrightarrow\qquad A=-\frac{Ua^3}{4},\quad c=-\frac{3Ua}{4}, \tag{8.3.39}$$

with which the velocity and pressure distributions are obtained as

$$\boldsymbol{u}=U\left\{\left[1-\frac{a}{4r}\left(\frac{a^2}{r^2}+3\right)\right]\boldsymbol{e}_x+\frac{3ax}{4r^2}\left(\frac{a^2}{r^2}-1\right)\boldsymbol{e}_r\right\},\qquad p=-\frac{3}{2}\mu U\frac{ax}{r^3}. \tag{8.3.40}$$

It is seen that $\boldsymbol{u}=\mathbf{0}$ over the entire surface of sphere.

Since only the Stokeslet exerts a force on the surrounding fluid, and it is inside the spherical surface $r=a$, the surrounding fluid exerts an equal but opposite force on the sphere, which, by using Eq. (8.3.34), is given by

$$\boldsymbol{f}=6\pi\mu Ua\boldsymbol{e}_x, \tag{8.3.41}$$

which is referred to as *Stokes' drag law* for the drag force experienced by a stationary sphere in a uniform flow, and is valid for flows with low Reynolds numbers. Since the direction of this force is in the direction of uniform flow, the drag force is frequently expressed in terms of the drag coefficient C_D defined by[8]

$$C_D\equiv\frac{2\|\boldsymbol{f}\|}{\rho U^2 A},\qquad A=\pi a^2, \tag{8.3.42}$$

where A is the frontal area of sphere. Combining Eqs. (8.3.41) and (8.3.42) results in

$$C_D=\frac{24}{\mathcal{R}_e},\qquad \mathcal{R}_e\equiv\frac{2\rho Ua}{\mu}, \tag{8.3.43}$$

in which the diameter of sphere is chosen as the characteristic length of the Reynolds number. The drag coefficient of a sphere in a uniform flow in terms of the Reynolds number is shown in Fig. 8.9. For the entire range of $\mathcal{R}_e$, Eq. (8.3.43) is the only closed-form analytic solution which exists. It is valid for $\mathcal{R}_e<1$, in which the viscous force dominates.

[8]The drag coefficient and the related lift coefficient will be discussed in Sect. 8.4.9.

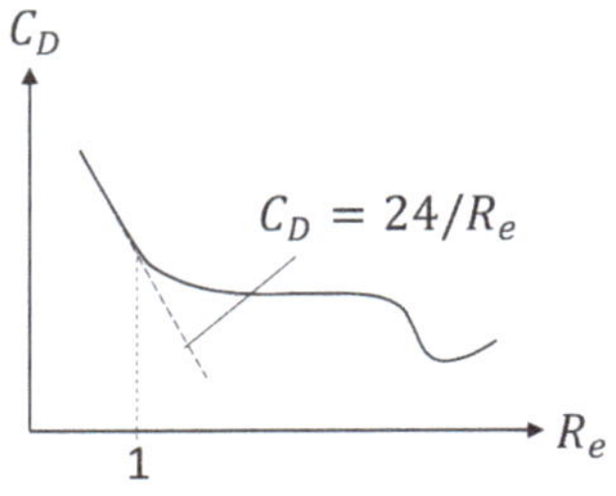

Fig. 8.9 The drag coefficient as a function of the Reynolds number for a stationary sphere in a uniform flow

8.3.4 Stokes' Paradox and the Oseen Approximation

Consider a uniform flow past a two-dimensional circular cylinder. The flow is assumed to be steady, for which Eq. (8.3.1)$_2$ reduces to

$$\mathbf{0} = -\frac{1}{\rho}\nabla p + \nu\nabla^2 \boldsymbol{u}. \tag{8.3.44}$$

Taking curl of this equation yields

$$\mathbf{0} = \nabla^2 \boldsymbol{\omega}, \qquad 0 = \nabla^2 \omega, \tag{8.3.45}$$

for $\nabla \times \nabla p = \mathbf{0}$, and in two-dimensional circumstances, the vorticity vector is expressed as $\boldsymbol{\omega} = (0, 0, \omega)$. In the two-dimensional rectangular coordinate system, ω is given by

$$\omega = \frac{\partial v}{\partial x} - \frac{\partial u}{\partial y} = -\left(\frac{\partial^2 \psi}{\partial x^2} + \frac{\partial^2 \psi}{\partial y^2}\right) = -\nabla^2 \psi, \tag{8.3.46}$$

in which ψ is the stream function. With this, the vorticity component ω must satisfy the *biharmonic equation* given by

$$\nabla^4 \psi = 0, \tag{8.3.47}$$

which is expressed in terms of the cylindrical coordinates (r, θ), viz.,

$$\left(\frac{\partial^2}{\partial r^2} + \frac{1}{r}\frac{\partial}{\partial r} + \frac{1}{r^2}\frac{\partial^2}{\partial \theta^2}\right)^2 \psi = 0. \tag{8.3.48}$$

Since the stream function of a uniform flow is given by $\psi = Uy = Ur\sin\theta$, it is plausible to assume that the solution to Eq. (8.3.48) may be in the form

$$\psi(r, \theta) = R(r)\sin\theta, \tag{8.3.49}$$

with which Eq. (8.3.48) reduces to

$$\left(\frac{\mathrm{d}^2}{\mathrm{d}r^2} + \frac{1}{r}\frac{\mathrm{d}}{\mathrm{d}r} - \frac{1}{r^2}\right)^2 R = 0, \tag{8.3.50}$$

which is an equi-dimensional equation. Integrating this equation gives

$$\begin{aligned} R(r) &= C_1 r^3 + C_2 r \ln r + C_3 r + \frac{C_4}{r}, \\ \longrightarrow \quad \psi(r, \theta) &= \left(C_1 r^3 + C_2 r \ln r + C_3 r + \frac{C_4}{r}\right)\sin\theta, \end{aligned} \tag{8.3.51}$$

where C_1–C_4 are integration constants. Since a uniform flow far away from the cylinder must be recovered by the obtained solution by requiring that $\psi(r \to \infty, \theta) = Ur \sin\theta$, it follows immediately that $C_1 = C_2 = 0$ and $C_3 = U$, with which Eq. $(8.3.51)_2$ becomes

$$\psi(r, \theta) = \left(Ur + \frac{C_4}{r} \right) \sin\theta. \tag{8.3.52}$$

In addition, on the surface of cylinder with radius a, the tangential and normal velocity components must vanish to satisfy the no-slip boundary conditions, which are given by

$$\psi(a, \theta) = 0, \qquad \frac{\partial \psi}{\partial r}(a, \theta) = 0. \tag{8.3.53}$$

The first condition is so obtained that since both partial derivatives of ψ with respect to r and θ must vanish, and $\partial\psi/\partial\theta = 0$ for all values of θ, the condition of vanishing tangential velocity component is equivalent to $\psi(a, \theta) =$ constant, with the constant chosen to be null without loss of generality. It is found that there is no choice of C_4 which satisfies the two conditions given in Eq. (8.3.53). If the first condition is satisfied by the solution, the second condition can never be fulfilled. It is concluded that there is no solution to the two-dimensional Stokes' equations which can satisfy both the near and far boundary conditions. Such a conclusion is referred to as *Stokes' paradox*.

The difference between two- and three-dimensional Stokes' equations is recognized by using the dimensionless Navier-Stokes equation given by

$$\bar{\rho} \left(\mathcal{R}_e \mathcal{S}_t \frac{\partial \bar{\boldsymbol{u}}}{\partial \bar{t}} + \mathcal{R}_e (\operatorname{grad} \bar{\boldsymbol{u}}) \bar{\boldsymbol{u}} \right) = -\mathcal{R}_e \mathcal{E}_u \operatorname{grad} \bar{p} + 2\bar{u} \operatorname{lap} \bar{\boldsymbol{u}}, \tag{8.3.54}$$

quoted from Eq. $(6.5.14)_2$, in which the external body force is omitted for simplicity. This equation is recast alternatively as

$$\frac{\partial \bar{\boldsymbol{u}}}{\partial \bar{t}} + \mathcal{R}_e (\bar{\boldsymbol{u}} \cdot \overline{\nabla}) \bar{\boldsymbol{u}} = -\overline{\nabla} \bar{p} + \overline{\nabla}^2 \bar{\boldsymbol{u}}, \tag{8.3.55}$$

with the scaling variables newly defined as

$$\boldsymbol{u} = U \bar{\boldsymbol{u}}, \qquad p = \frac{\rho \nu U}{\ell} \bar{p}, \qquad \boldsymbol{x} = \ell \bar{\boldsymbol{x}}, \qquad t = \frac{\ell^2}{\nu} \bar{t}. \tag{8.3.56}$$

Since Stokes' equations correspond to $\mathcal{R}_e \to 0$, a more accurate solution to the stream function for low-Reynolds-number flows could be sought in the form

$$\psi = \psi_0 + \mathcal{R}_e \psi_1 + \mathcal{O}(\mathcal{R}_e^2), \tag{8.3.57}$$

which represents an asymptotic expansion of ψ. Thus, a solution corresponding to ψ_0 exists for a sphere but not for a cylinder. On the contrary, it has been found that a solution to ψ_1 does not exist for a sphere. Such a situation is called the *Whitehead's paradox*.[9] The paradox occurs in the first-order problems for two-dimensional circumstances and in the second-order problems for three-dimensional situations.

[9] Alfred North Whitehead, 1861–1947, a British mathematician and philosopher, who is best known as the defining figure of the philosophical school known as the process philosophy, which has found application to a wide variety of disciplines, including ecology and physics.

Mathematically, the emerging difficulty is referred to as a *singular perturbation.*[10] Stokes' approximation is in fact a first-order problem arising out of a perturbation type of solution to the Navier-Stokes equation, with the instability rendering the perturbation singular to match the required boundary conditions. For two-dimensional circumstances, the difficulty associated with this singular perturbation appears immediately, while for three-dimensional circumstances, the difficulty is postponed to the second-order term in the expansion. Physically, the difficulty results from the neglecting of convective linear momentum of the fluid, an assumption which is invalid far from the body. Assuming $\mathcal{R}_e \to 0$ is equivalent to completely neglect the convection in comparison with the viscous diffusion in the fluid. Due to the nature of viscous boundary conditions near the body, the viscous diffusion is larger near the body, whereas the convection is small for the retardation of velocity by the body. On the contrary, the velocity gradient far away from the body nearly vanishes, so that the viscous diffusion is reduced, where the fluid velocity is close to that of free stream. The convection in the fluid becomes more and more important while the viscous diffusion exhibits a reverse tendency when leaving the body. This means that the conditions which are required to satisfy Stokes' approximation are violated. Hence, Stokes' approximation is valid close to the body, but losses its validity far away from the body.

This difficulty may be overcome by linearizing the Navier-Stokes equation, so that the linear momentum is transported not with the local velocity (as in the exact cases) or with zero velocity (as in Stokes' approximation), but with the free stream velocity. Back to the considered uniform flow with magnitude U along the x-axis past a two-dimensional circular cylinder, the formulations now become

$$\nabla \cdot \boldsymbol{u} = 0, \qquad \frac{\partial \boldsymbol{u}}{\partial t} + U\frac{\partial \boldsymbol{u}}{\partial x} = -\frac{1}{\rho}\nabla p + \nu\nabla^2\boldsymbol{u}, \tag{8.3.58}$$

which is known as the *Oseen approximation.*[11] Solutions to the above equations can be obtained in a similar manner to those introduced to obtain the solutions to Stokes' equations. Unfortunately, the obtained results are valid far from the body but fail close to the body. This is exactly the opposite of the solutions to Stokes' equations. By matching two solutions of the same problem, a uniformly valid expression will result which is valid for small Reynolds numbers. The method of overcoming the difficulties of singular perturbation is called the *method of matched asymptotic expansion.*

8.4 Boundary-Layer Flows

This section deals with large-Reynolds-number flows. Specifically, Prandtl's boundary-layer approximation to the full Navier-Stokes equation is explored. The exact solution

[10] Singular perturbation is sometimes called *non-uniform expansion.*

[11] Carl Wilhelm Oseen, 1879–1944, a Swedish theoretical physicist and the Director of the Nobel Institute for Theoretical Physics in Stockholm, who also contributed to the fundamentals of elasticity theory for liquid crystals, known as the *Oseen elasticity theory.*

to the established boundary-layer equations may be obtained via the similarity methods. The Kármán-Pohlhausen method is discussed as an example of the approximate solution to the boundary-layer equations, which is known as the approach of momentum integral. The stability of boundary layer is then introduced, followed by the drag and lift forces experienced by an object immersed in a viscous fluid, which are closely related to the boundary-layer separation.

8.4.1 Concept of Boundary-Layer

When a flowing viscous fluid with uniform velocity U is in contact with a solid surface, the viscous effect ensures that the fluid velocity on the solid surface vanishes, yielding the so-called no-slip boundary condition. The viscous effect is transmitted from the solid surface toward the fluid to retard the velocities of fluid subsequently, until the viscous effect becomes insignificant when compared with other forces taking place in the fluid. This forms a very thin fluid layer adjacent to the body surface, in which strong viscous effect exists, and the layer is referred to as the *boundary layer*. Typical boundary layer on a flat plate is shown in Fig. 8.10a, in which the boundary layer originates at the leading edge and moves downstream near the surface of flat plate, with the boundary layer edge displayed by the dashed line. Outside the boundary layer the velocity gradients are not large, and so the viscous effect is negligible. If the compressible effect may be ignored further, the fluid may be considered to be ideal, and the results of ideal-fluid flows in Chap. 7 may be employed. Consequently, if the flow field far upstream is uniform and irrotational, the flow outside the boundary layer is equally everywhere irrotational, as implied by Kelvin's

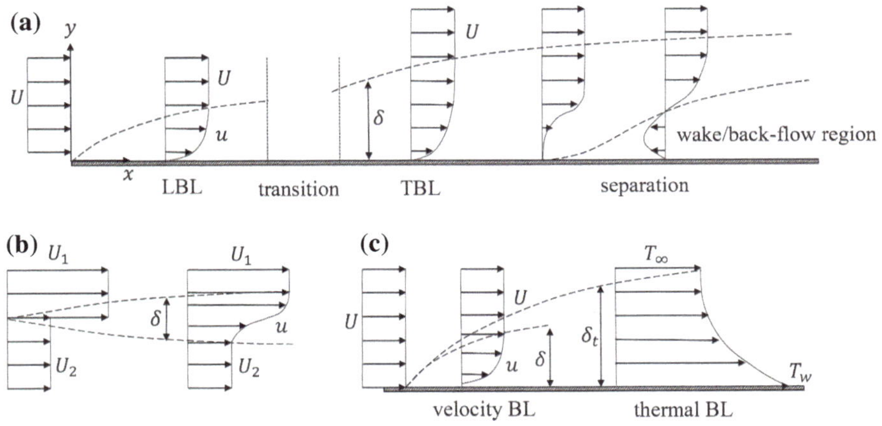

Fig. 8.10 Boundary layer over a horizontal flat plate. **a** Laminar and turbulent boundary layers, flow separation and wake region. **b** Formation of the boundary layer near the interface between two uniform flows at different velocities. **c** Velocity boundary layer δ and thermal boundary layer δ_t of a uniform flow with temperature T_∞ over a flat plate with temperature $T_w > T_\infty$

theorem. The potential-flow field outside the boundary layer is frequently referred to as the *outer flow*.

Inside the boundary layer, strong viscous effect takes place due to the significant velocity gradients, as induced by the no-slip boundary condition on the flat plate reducing the uniform velocity U in the outer flow to null on the surface. The flow inside the boundary layer is referred to as the *inner flow*. Here, the vorticity does not vanish. It is generated along the surface of flat plate, and diffused and convected along the boundary layer by the mean flow. The flow inside the boundary layer may be laminar or turbulent, which are respectively referred to as the *laminar boundary layer* (LBL) or *turbulent boundary layer* (TBL). The boundary layer is laminar in a short distance downstream from the leading edge of flat plate; transition occurs over a short region of the plate rather than at a single line across the plate. The transition region extends downstream to the locations where the boundary-layer flow becomes completely turbulent.[12] Toward the rear of flat plate, the boundary layer may encounter an adverse pressure gradient, causing the boundary layer to separate from the flat plate to form a so-called *wake region* or a *back-flow region*. The velocity gradients in the wake region are not large, so that the viscous effect is not too significant. However, all the vorticities existing in the boundary layer are convected to the wake, so that the flow in the wake is not irrotational. If the boundary layer remains still laminar at the separation point, a shear layer of the type discussed in Sect. 7.7.6 may exist. Such shear layers were found to be unstable, and over a wide range of the Reynolds number this instability manifests itself in the form of a periodic wake, which is the well-known *von Kármán vortex street*.

Boundary layers may also form when two fluid layers with different uniform velocities are in contact, as shown in Fig. 8.10b, in which the boundary layer originates from the interface between two fluid layers, and grows gradually downstream. Boundary layers can also take place when the body and fluid have different temperatures, as shown in Fig. 8.10c, in which a uniform flow passes a flat plate whose temperature is higher than that of fluid. In addition to the boundary layer caused by the no-slip boundary condition on the plate, another boundary layer presents due to the temperature difference between the fluid and plate. Boundary layers caused by the no-slip boundary condition for velocity are referred to as the *velocity boundary layers* (Velocity BL), while those caused by the temperature difference are termed

---

[12]For an incompressible flow over a smooth plate with vanishing pressure gradient along the x-direction and without heat transfer between the plate and fluid, the transition from laminar to turbulent boundary-layer flows are characterized by the critical Reynolds number given by

$$\mathcal{R}_{e_{cr}} = \frac{\rho U x_{cr}}{\mu},$$

where x_{cr} marks the location in the x-direction with $\mathcal{R}_{e_{cr}} > 10^6$, if all external disturbances are minimized. For practical calculation, the critical Reynolds number is chosen to be 5×10^5. For example, for air at standard conditions and with $U = 30$ m/s, the critical Reynolds number corresponds to $x_{cr} = 0.24$ m. The thickness of boundary layer grows as x increases. It will be seen later that the thickness of turbulent boundary layer grows faster than that of laminar boundary layer.

the *thermal boundary layers* (Thermal BL). Boundary layers also exist in the atmospheric environment. The atmosphere of earth is semi-transparent to incoming solar radiation. It obtains nearly 20% of its energy strictly by absorption, and about 30% energy is reflected or scattered to space. The rest of energy passes through the atmosphere, which is absorbed by the surface of earth. Later this energy is transferred back, primarily to the lowest kilometer of atmosphere. This lowest portion of atmosphere, which intensively exchanges heat as well as mass and momentum with the earth surface, is referred to as the *atmospheric boundary layer* (ABL). It has great practical and scientific importance. Almost all human and biological activities take place in this layer. The mass and energy transfer within the atmospheric boundary layer regulates a broad variety of processes in the entire atmosphere. Obviously, the atmospheric boundary layer is a combined phenomenon of velocity and thermal boundary layers and is caused by the velocity and temperature differences between air and earth surface.

Velocity gradients exist in both laminar and turbulent boundary layers. However, they approach asymptotically null when approaching the edges of boundary layers. Hence, it is difficult to determine the boundary-layer thickness, which cannot be defined simply as the location where the velocity of inner flow equals that of outer flow. Several measures of the boundary-layer thickness are proposed. Consider a boundary-layer flow over a flat plate shown in Fig. 8.10a again. Let U and u be the velocities of outer and inner flows along the x-direction, respectively. The most straightforward measure of boundary-layer thickness is the *disturbance thickness* δ, which is defined as the distance from the solid surface at which $u \sim 0.99U$, as shown in Fig. 8.11a, which is given by

$$y = \delta, \qquad u \sim 0.99U. \tag{8.4.1}$$

The second measure is the *displacement thickness* δ^*, which is defined as the distance from the solid surface where the undisturbed outer flow is displaced from the solid boundary by a stagnant layer which removes the same mass flux from the flow field as the actual boundary layer, as shown in Fig. 8.11b. In other words, δ^* is the thickness of a zero-velocity layer which has the same mass flux defect as the actual boundary-layer flow, so that

$$\rho U\delta^* b = \int_0^\infty \rho(U-u)b\mathrm{d}y, \quad \longrightarrow \quad \delta^* = \int_0^\infty \left(1-\frac{u}{U}\right)\mathrm{d}y = \int_0^\delta \left(1-\frac{u}{U}\right)\mathrm{d}y, \tag{8.4.2}$$

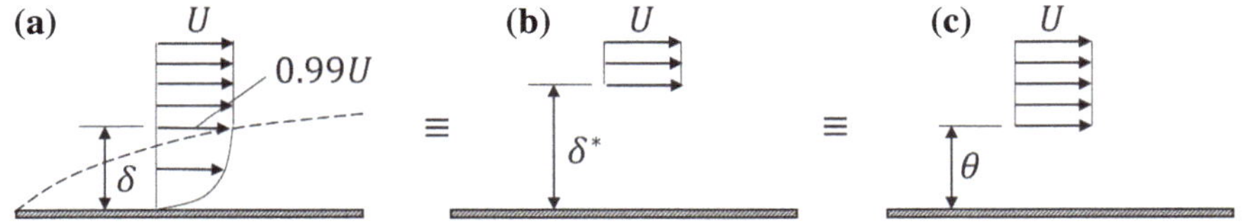

Fig. 8.11 Three measures of boundary-layer thickness. **a** The disturbance thickness δ. **b** The displacement thickness δ^*. **c** The momentum thickness θ

where b is the width of flow field. The third measure is called the *momentum thickness* θ, which is similar to the displacement thickness, except that the momentum-flux defect is taken into account, which is given by

$$\rho U^2\theta b = \int_0^\infty \rho u(U-u)b\mathrm{d}y, \longrightarrow \theta = \int_0^\infty \frac{u}{U}\left(1-\frac{u}{U}\right)\mathrm{d}y = \int_0^\delta \frac{u}{U}\left(1-\frac{u}{U}\right)\mathrm{d}y. \tag{8.4.3}$$

Although the integrations in Eqs. (8.4.2) and (8.4.3) are defined to be taken from $y = 0$ to $y \to \infty$, they are taken from $y = 0$ to $y = \delta$ in practice, for the integrands are essentially null for $y \geq \delta$. Since δ^* and θ are defined in terms of integrals, they are called the *integral thicknesses*, which are appreciably easier to be evaluated accurately from experimental outcomes than δ. This accounts for their common use in specifying the boundary-layer thickness when coupled with their physical significance. The various thicknesses defined previously are to some extent an indication of the distance over which the viscous effect extends. Conventionally, the disturbance thickness is larger than the displacement thickness, which is in turn usually larger than the momentum thickness, i.e., $\delta > \delta^* > \theta$.

The importance of boundary layer lies in the fact that it provides a link that had been missing between the theory and practice of fluid mechanics. Since the establishment of the Euler equation in 1755, the science of fluid mechanics had been developing in rather two different directions: the *theoretical hydrodynamics*, evolving from the Euler equation for frictionless flows. Although mathematically elegant, the obtained results contradicted to many experimental observations, e.g. a body experiences no drag under the assumption of inviscid flow discussed in Sect. 7.6.4. On the other hand, practical needs in engineering applications called an empirical art of *hydraulics*, which was based on experimental data and differed significantly from the purely mathematical approach of theoretical hydrodynamics. Although the Navier-Stokes equation describing the whole picture of the motion of a viscous fluid had been developed in 1827 by Navier and independently by Stokes in 1845, the mathematical difficulty in solving the coupled balances of mass and linear momentum still prohibited a theoretical advance of viscous flows, except for a few simple circumstances, and two diverse developments of fluid mechanics continued, until Prandtl proposed the well-known concept of boundary layer in 1904.

Prandtl realized that many viscous flows may be analyzed by dividing the flow into two regions: one is adjacent to solid boundaries, and the other covering the rest of flow. Only in the region adjacent to a solid boundary, namely the boundary layer, is the effect of viscosity important. Outside the boundary layer, the effect of viscosity is negligible and the fluid may be treated as inviscid. Prandtl's contribution was a historical breakthrough. The concept of boundary layer delivered not only the estimations on drags on objects theoretically, but also permitted the solutions to viscous-flow problems that would have been impossible through the applications of the full Navier-Stokes equation. Hence, the introduction of boundary layer marked the beginning of modern era of fluid mechanics.

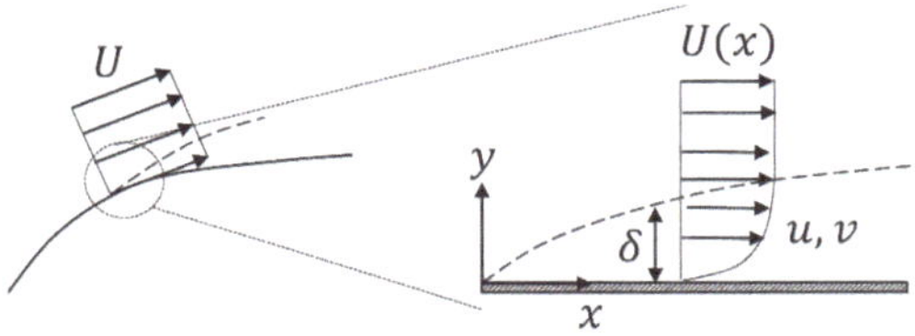

Fig. 8.12 A viscous flow with uniform velocity over a curved surface approximated by a boundary-layer flow over a flat plate, in which the boundary-layer thickness is much smaller than the radius of curvature of the curved surface

8.4.2 Boundary-Layer Equations

The boundary-layer equations are derived by using the physical arguments proposed by Prandtl. Consider a uniform flow with velocity $U(x)$ over a curved surface, as shown in Fig. 8.12, in which δ marks the thickness of boundary layer, and the gravitational acceleration is assumed to point perpendicular to the page for simplicity. If the order of magnitude of δ is much smaller than the radius of curvature of the curved surface, the flow field may be approximated as that over a flat plate. It follows that in all points in the boundary layer, $\delta/x \ll 1$ is satisfied, except near the leading edge of plate. It is further assumed that the order of magnitude of u, which is the fluid velocity inside the boundary layer along the x-direction, is similar to that of U in the outer flow, and $\partial/\partial x$ inside the boundary layer is of order $1/x$. Hence, $\partial u/\partial x \sim U/x$, and so is the same for $\partial v/\partial y \sim U/x$, as implied by the continuity equation. Since $\delta/x \ll 1$, this implies that v is much smaller than u, but $\partial/\partial y$ is much larger than $\partial/\partial x$. These conditions can be fulfilled by choosing

$$u \sim U, \quad v \sim U\frac{\delta}{x}; \qquad \frac{\partial}{\partial x} \sim \frac{1}{x}, \quad \frac{\partial}{\partial y} \sim \frac{1}{\delta}. \tag{8.4.4}$$

For the consider two-dimensional steady flow, the Navier-Stokes equations in the x- and y-directions read respectively

$$\begin{aligned} u\frac{\partial u}{\partial x} + v\frac{\partial u}{\partial y} &= -\frac{1}{\rho}\frac{\partial p}{\partial x} + \nu\left(\frac{\partial^2 u}{\partial x^2} + \frac{\partial^2 u}{\partial y^2}\right), \\ u\frac{\partial v}{\partial x} + v\frac{\partial v}{\partial y} &= -\frac{1}{\rho}\frac{\partial p}{\partial y} + \nu\left(\frac{\partial^2 v}{\partial x^2} + \frac{\partial^2 v}{\partial y^2}\right), \end{aligned} \tag{8.4.5}$$

whose orders of magnitude, by using Eq. (8.4.4), are estimated as

$$\frac{U^2}{x} + \frac{U^2}{x} = -\frac{1}{\rho}\frac{\partial p}{\partial x} + \nu\left(\frac{U}{x^2} + \frac{U}{\delta^2}\right), \quad \frac{\delta U^2}{x^2} + \frac{\delta U^2}{x^2} = -\frac{1}{\rho}\frac{\partial p}{\partial y} + \nu\left(\frac{\delta U}{x^3} + \frac{U}{x\delta}\right), \tag{8.4.6}$$

in which the orders of magnitude of pressure gradients remain at the moment un-estimated. The inertia terms in Eq. $(8.4.5)_1$ are of the same order, but the viscous term $\partial^2 u/\partial y^2$ is much larger than its counter part $\partial^2 u/\partial x^2$, so that the latter can be neglected in the boundary layer. Since the viscous and inertia terms along the

x-direction are assumed to be of the same order of magnitude in the boundary layer, for fluid particles there may be accelerated by a comparable inertia force, it follows from the analysis of order of magnitude that

$$\frac{U^2}{x} \sim \nu \frac{U}{\delta^2}, \quad \longrightarrow \quad \delta \sim \sqrt{\frac{\nu x}{U}}, \tag{8.4.7}$$

indicating that the disturbance thickness δ is proportional to $\sqrt{x}$. In addition, since $\delta/x \ll 1$, it is seen that

$$\frac{x^2}{\delta^2} \sim \frac{Ux}{\nu} \gg 1, \quad \longrightarrow \quad \mathcal{R}_{e_x} = \frac{Ux}{\nu} \gg 1, \tag{8.4.8}$$

showing that the assumption $\delta/x \ll 1$ corresponds to $\mathcal{R}_{e_x} \gg 1$. Now turn to the Navier-Stokes equation in the y-direction, i.e., Eqs. $(8.4.5)_2$ and $(8.4.6)_2$. Applying the analysis of order of magnitude to two equations yields that the inertial terms assume an order of δ/x, which are much smaller than their counterparts in the x-direction and can be neglected. Equally, the viscous terms are of order δ/x, and are much smaller than their counterparts in the x-direction, which can equally be neglected. With these, Eq. $(8.4.5)_2$ is simplified to

$$0 = -\frac{1}{\rho}\frac{\partial p}{\partial y}, \quad \longrightarrow \quad p = p(x). \tag{8.4.9}$$

Based on the established results, the continuity and Navier-Stokes equations for flows inside the boundary layer are given respectively by

$$\frac{\partial u}{\partial x} + \frac{\partial v}{\partial y} = 0, \qquad u\frac{\partial u}{\partial x} + v\frac{\partial u}{\partial y} = -\frac{1}{\rho}\frac{\mathrm{d}p}{\mathrm{d}x} + \nu\frac{\partial^2 u}{\partial y^2}, \tag{8.4.10}$$

which are devoted to the two velocity components $u(x, y)$ and $v(x, y)$. These equations are referred to as *Prandtl's boundary-layer equations*, or simply the *boundary-layer equations* for steady, incompressible, isothermal, two-dimensional boundary-layer flows. When compared with the original Navier-Stokes equation which is elliptic, Eq. $(8.4.10)_2$ is parabolic due to the neglecting of highest derivatives in the x-direction. Since $p = p(x)$, it follows that $\mathrm{d}p/\mathrm{d}x$ is the same in both the inner and outer flows, and the pressure gradient in the x-direction can be estimated by using the Bernoulli equation $p + \rho U^2/2 =$ constant, for the outer flow is incompressible and frictionless. With these, Eq. $(8.4.10)_2$ is further simplified to

$$u\frac{\partial u}{\partial x} + v\frac{\partial u}{\partial y} = U\frac{\mathrm{d}U}{\mathrm{d}x} + \nu\frac{\partial^2 u}{\partial y^2}. \tag{8.4.11}$$

The boundary conditions associated with the boundary-layer equations result from two physical observations: the no-slip boundary condition on the plate, and the condition that the outer flow should be recovered far from the plate surface, which are given respectively by

$$u(x, y = 0) = 0, \quad v(x, y = 0) = 0, \qquad u(x, y \to \infty) \to U(x). \tag{8.4.12}$$

The last condition effectively matches the inner flow to the outer flow, so that the corresponding potential-flow solution must be known before a boundary-layer problem can be solved.

An alternative way to derive the boundary-layer equations from the Navier-Stokes equation involves a limiting procedure similar to that used to extract Stokes' equations from the full Navier-Stokes equation. The Navier-Stokes equation is first expressed in dimensionless form, which results in a coefficient $1/\mathcal{R}_e$ in front of the viscous terms. The stretched coordinates $X = x$, $Y = \sqrt{\mathcal{R}_e}y$ are then introduced, which remove the coefficient $1/\mathcal{R}_e$ from one of the viscous terms. A limiting procedure with $\mathcal{R}_e \to \infty$ is conducted under fixed values of X and Y, with which the boundary-layer equations can be derived. The detailed derivation is left as an exercise. This derivation is useful if higher-order approximations to the boundary-layer theory are required, namely if an expansion of solution is sought. However, the nature of coordinate stretching is not obvious without appealing to the physical approach, as demonstrated previously.

8.4.3 Blasius' Solution

An exact solution to the boundary-layer equations was obtained by Blasius by considering the velocity of outer flow to be a constant, i.e., $U(x) =$ constant, with $\delta = \delta(x)$. With these, Eqs. $(8.4.10)_1$ and (8.4.11) reduce respectively to

$$\frac{\partial u}{\partial x} + \frac{\partial v}{\partial y} = 0, \qquad u\frac{\partial u}{\partial x} + v\frac{\partial u}{\partial y} = \nu\frac{\partial^2 u}{\partial y^2}. \tag{8.4.13}$$

Replacing the velocity components u and v by using the stream function ψ yields that the first equation is satisfied identically, while the second equation becomes

$$\frac{\partial \psi}{\partial y}\frac{\partial^2 \psi}{\partial x \partial y} - \frac{\partial \psi}{\partial x}\frac{\partial^2 \psi}{\partial y^2} = \nu\frac{\partial^3 \psi}{\partial y^3}, \tag{8.4.14}$$

which is a parabolic partial differential equation of ψ. Since there exists no geometric length scale in the problem, it is possible to use the similarity transformation given by

$$\psi(x, y) \sim f(\eta), \qquad \eta \sim \frac{y}{x^n} = \frac{y}{\sqrt{\nu x/U}}, \tag{8.4.15}$$

where the power $n = 1/2$ is chosen for a flat plate, and the quantities ν and U are incorporated to make η dimensionless. With these expressions, the velocity component u is given by

$$u = \frac{\partial \psi}{\partial y} \sim \sqrt{\frac{U}{\nu x}}f'(\eta), \tag{8.4.16}$$

where the primes denote differentiations with respect to η. It is seen that if η is constant, u is also constant, so that the proportional factor between ψ and $f(\eta)$ should include the term $\sqrt{x}$. Since ψ assumes the unit of length square divided by time, this proportional factor should also include the term $\sqrt{\nu U}$ to become dimensionally consistent. Consequently, a similarity solution to the problem is obtained as

$$\psi(x, y) = \sqrt{\nu U x}\, f\left(\frac{y}{\sqrt{\nu x/U}}\right). \tag{8.4.17}$$

Table 8.1 Numerical integrations of the Blasius solution for a laminar boundary-layer flow with constant velocity over a flat plate

η	$f'(\eta)$	η	$f'(\eta)$	η	$f'(\eta)$
0	0	0.4	0.1328	0.8	0.2647
1.2	0.3938	1.6	0.5168	2.0	0.6298
2.4	0.7290	2.8	0.8115	3.2	0.8761
3.6	0.9233	4.0	0.9555	4.4	0.9759
4.8	0.9878	5.0	0.9916	5.2	0.9943
5.6	0.9975	6.0	0.9990	∞	1.0000

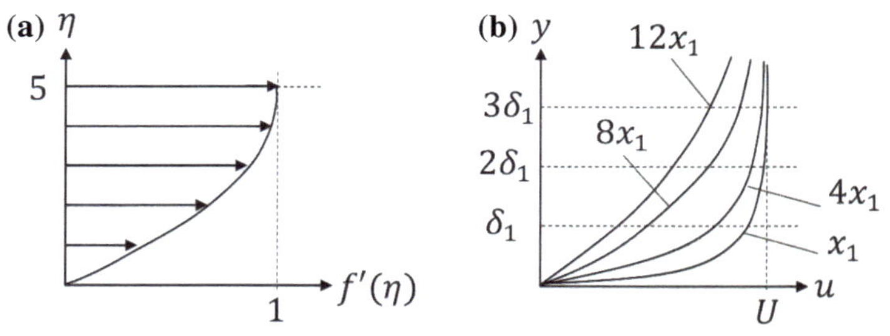

Fig. 8.13 The velocity profiles from the Blasius solution. **a** The dimensionless profile in terms of η. **b** The dimensional profiles at different locations along the plate, where δ_1 is the disturbance thickness at $x = x_1$

Substituting these expressions into Eq. (8.4.14) yields

$$-\frac{U^2}{2x} f f'' = \frac{U^2}{x} f''', \qquad \longrightarrow \qquad f''' + \frac{1}{2} f f'' = 0, \tag{8.4.18}$$

which is subject to the boundary conditions given by

$$f(0) = f'(0) = 0, \qquad f'(\eta \to \infty) \to 1, \tag{8.4.19}$$

as implied by Eq. (8.4.12). Equations (8.4.18) and (8.4.19) construct a mathematically well-posed problem, but the solution demands, however, numerical integration. Despite this, the Blasius solution is still considered an exact solution, for the original partial differential equations have been brought to ordinary differential equations. The results of the numerical integrations of $f(\eta)$ are summarized in Table 8.1, and the dimensionless and dimensional velocity profiles are shown graphically in Figs. 8.13a and b, respectively.

From Table 8.1, it is seen that $u/U \sim 0.99$ at $\eta = 5.0$; thus, the disturbance thickness δ is identified as

$$\delta = 5.0\sqrt{\frac{\nu x}{U}}, \qquad \frac{\delta}{x} = \frac{5.0}{\sqrt{\mathcal{R}_{e_x}}}, \tag{8.4.20}$$

with the displacement and momentum thicknesses obtained respectively as

$$\frac{\delta^*}{x} = \frac{1.721}{\sqrt{\mathcal{R}_{e_x}}}, \qquad \frac{\theta}{x} = \frac{0.664}{\sqrt{\mathcal{R}_{e_x}}}. \tag{8.4.21}$$

These results show that all three thicknesses are very thin at large values of $\mathcal{R}_{e_x}$ and grow as $\sqrt{x}$ increases, in which the inequality $\theta < \delta^* < \delta$ holds. The shear stress $\tau_w(x)$ on the plate is determined to be

$$\tau_w(x) = \mu\frac{\partial u}{\partial y}(x,0) = \mu\sqrt{\frac{U^3}{\nu x}}f''(0), \qquad \longrightarrow \qquad \frac{\tau_w(x)}{\frac{1}{2}\rho U^2} = \frac{0.664}{\sqrt{\mathcal{R}_{e_x}}}, \tag{8.4.22}$$

showing that τ_w falls off as $\sqrt{x}$ along the surface of flat plate. The drag force $F_D(x)$ per unit width due to the skin friction is evaluated by integrating the shear stress to a specific point x, which is given by

$$F_D(x) = \int_0^x \tau_w(\xi)\,\mathrm{d}\xi, \tag{8.4.23}$$

showing that F_D increases proportionally with $\sqrt{x}$. With this, the drag coefficient C_D is identified as[13]

$$C_D(x) = \frac{F_D/x}{\frac{1}{2}\rho U^2} = \frac{1}{x}\int_0^x \frac{\tau_w(\xi)}{\frac{1}{2}\rho U^2}\mathrm{d}\xi = \frac{1.328}{\sqrt{\mathcal{R}_{e_x}}}, \tag{8.4.24}$$

in which Eq. $(8.4.22)_2$ has been used. In fact, the obtained result of $\tau_w(x)$ should not be applied near the leading edge of flat plate in order to maintain the assumptions of boundary-layer flows. Fortunately, any difference between the actual shear stress and that predicted by Eq. $(8.4.22)_2$ is not significant, for relatively short distance involves. Although $x = 0$ is a singularity of $\tau_w(x)$, it can be integrable, so that F_D and C_D are not singular.

8.4.4 The Falkner-Skan Solutions

A whole family of similarity solutions to the boundary-layer equations were obtained by Falkner and Skan by seeking a general formulation of similarity-type solutions. An interpretation of each solution is then given for a specific flow field.[14] It is assumed that a general similarity-type solution is in the form

$$u(x,y) = U(x)f'(\eta), \qquad \eta = \frac{y}{\xi(x)}, \tag{8.4.25}$$

where $U(x)$ is the velocity of outer flow with $\xi(x)$ an undetermined function of x. The corresponding stream function is then given by

$$\psi(x,y) = U(x)\xi(x)f(\eta). \tag{8.4.26}$$

Substituting this expression into Eq. (8.4.11) yields

$$U\frac{\mathrm{d}U}{\mathrm{d}x}\left(f'\right)^2 - U\frac{\mathrm{d}U}{\mathrm{d}x}ff'' - U^2\frac{1}{\xi}\frac{\mathrm{d}\xi}{\mathrm{d}x}ff'' = U\frac{\mathrm{d}U}{\mathrm{d}x} + \nu\frac{U}{\xi^2}f''', \tag{8.4.27}$$

[13] Essentially, this drag coefficient consists only the contribution of skin friction, which is referred to as the *skin friction coefficient*.

[14] For more details, see Falkner, V.M., Skan, S.W., Some approximate solutions of the boundary layer equations, *Phil. Mag.* 12, 865–896, 1931; *ARC RM*, 1314, 1930.

where the primes denote differentiations with respect to η. Combining the second and third terms on the left-hand-side gives

$$f''' + \left[\frac{\xi}{\nu}\frac{\mathrm{d}}{\mathrm{d}x}(U\xi)\right] ff'' + \left[\frac{\xi^2}{\nu}\frac{\mathrm{d}U}{\mathrm{d}x}\right]\left[1-\left(f'\right)^2\right] = 0. \tag{8.4.28}$$

If a similarity solution $f(\eta)$ exists, this equation must be an ordinary differential equation of f. It follows that the terms inside two brackets must at most be constant, namely

$$\frac{\xi}{\nu}\frac{\mathrm{d}}{\mathrm{d}x}(U\xi) = \alpha_1, \qquad \frac{\xi^2}{\nu}\frac{\mathrm{d}U}{\mathrm{d}x} = \alpha_2, \tag{8.4.29}$$

where α_1 and α_2 are two constants. An alternative to one of these two equations is obtained by

$$\frac{\mathrm{d}}{\mathrm{d}x}\left(U\xi^2\right) = 2\xi\frac{\mathrm{d}}{\mathrm{d}x}(U\xi) - \xi^2\frac{\mathrm{d}U}{\mathrm{d}x} = \nu(2\alpha_1 - \alpha_2). \tag{8.4.30}$$

Any two of the equations given in Eqs. (8.4.29) and (8.4.30) are sufficient to relate U and ξ to the undetermined constants α_1 and α_2.

In terms of α_1 and α_2, Eq. (8.4.28) is expressed as

$$f''' + \alpha_1 ff'' + \alpha_2\left[1-\left(f'\right)^2\right] = 0, \tag{8.4.31}$$

which is subject to the same boundary conditions given in Eq. (8.4.19), e.g. a boundary-layer flow over a flat plate. If a formulated problem is solvable, then an exact solution to the boundary-layer equations may be found. The crucial step in obtaining a solution is to choose the values of α_1 and α_2. Once it is done, a particular flow configuration is then considered, and the values of $U(x)$ and $\xi(x)$ may be determined by using Eqs. (8.4.29) and (8.4.30), where $U(x)$ is the velocity of the corresponding potential flow for the geometry under consideration. Then, a solution of $f(\eta)$ is sought by solving Eq. (8.4.31) subject to the boundary conditions given in Eq. (8.4.19). With these, the stream function of flow field is obtained by Eq. (8.4.26), and all properties of the flow field may be known. It is noted that for $\alpha_1 = 1$, numerical solutions of the described procedure show that $f''(0) \to 0$ as α_2 is decreased. The value of $f''(0) = 0$ corresponds to $\alpha_2 = -0.1988$. Any value of α_2 smaller than this value yields $f'(\eta) > 1$ at some location, corresponding to $u > U$, which is physically unjustified. Thus, for $\alpha_1 = 1$, the value of α_2 must be greater than -0.1988. The applications of the Falkner-Skan solutions to the boundary-layer equations are demonstrated for selected problems in the following.

Flows over a flat plate. The solution is obtained by choosing $\alpha_1 = 1/2$ and $\alpha_2 = 0$. It follows from Eqs. $(8.4.29)_2$ and (8.4.30) that

$$\frac{\xi^2}{\nu}\frac{\mathrm{d}U}{\mathrm{d}x} = 0, \quad \frac{\mathrm{d}}{\mathrm{d}x}\left(U\xi^2\right) = \nu, \qquad \longrightarrow \qquad U(x) = c, \ \ \xi(x) = \sqrt{\frac{\nu x}{c}}, \tag{8.4.32}$$

where c is a constant, and $\xi(x)$ does not vanish in general. Since $U(x) = c$, a flat surface rather than a curved one is considered. With the chosen values of α_1 and α_2, Eqs. (8.4.31) and (8.4.19) reduce respectively to

$$f''' + \frac{1}{2}ff'' = 0, \qquad f(0) = f'(0) = 0, \ \ f'(\eta \to \infty) \to 1, \tag{8.4.33}$$

and the corresponding stream function is obtained as

$$\psi(x, y) = \sqrt{c\nu x}\, f\left(\frac{y}{\sqrt{\nu x/c}}\right), \tag{8.4.34}$$

which agrees identically with that of the Blasius solution.

Flows over a wedge. The solution is obtained by letting $\alpha_1 = 1$ with arbitrary value of α_2. In doing so, the equations that need to be satisfied by $U(x)$ and $\xi(x)$ become

$$\frac{\mathrm{d}}{\mathrm{d}x}\left(U\xi^2\right) = \nu(2 - \alpha_2), \qquad \frac{\xi^2}{\nu}\frac{\mathrm{d}U}{\mathrm{d}x} = \alpha_2. \tag{8.4.35}$$

Integrating the first equation yields

$$\xi^2 U = \nu(2 - \alpha_2)x, \tag{8.4.36}$$

which is used to divide the second equation to obtain

$$\frac{1}{U}\frac{\mathrm{d}U}{\mathrm{d}x} = \frac{\alpha_2}{2 - \alpha_2}\frac{1}{x}. \tag{8.4.37}$$

Integrating this equation gives

$$\ln U = \frac{\alpha_2}{2 - \alpha_2}\ln x + \ln c, \qquad \longrightarrow \qquad U(x) = c x^{\alpha_2/(2-\alpha_2)}, \tag{8.4.38}$$

where c is an integration constant. By using Eq. $(8.4.35)_2$, it is seen that

$$\xi(x) = \sqrt{\frac{\nu(2 - \alpha_2)}{c}}\, x^{(1-\alpha_2)/(2-\alpha_2)}. \tag{8.4.39}$$

By comparing the results of two-dimensional potential flows in Sect. 7.5, Eq. $(8.4.38)_2$ indicates that the outer flow corresponds to that over a wedge of angle $\pi\alpha_2$, and has the same forms of velocity components u and v near the flow boundary in a sector of angle π/n. The value of α_2 is then determined to be

$$n - 1 = \frac{\alpha_2}{2 - \alpha_2}, \tag{8.4.40}$$

which gives the angle of an half wedge measured in the fluid. Since the potential-flow field is symmetric, the angle of wedge must be $2(\pi - \pi/n)$, and hence corresponds to $\pi\alpha_2$. The obtained flow field is shown in Fig. 8.14a.

With the chosen values of α_1 and α_2, Eqs. (8.4.31) and (8.4.19) reduce to

$$f''' + ff'' + \alpha_2\left[1 - \left(f'\right)^2\right] = 0, \qquad f(0) = f'(0) = 0, \;\; f'(\eta \to \infty) \to 1, \tag{8.4.41}$$

where the solution to $f(\eta)$ needs to be determined numerically. Having done this, the stream function is then given by

$$\psi(x, y) = \sqrt{c(2 - \alpha_2)\nu x}^{\,1/(2-\alpha_2)} f\left(\frac{y}{\sqrt{(2 - \alpha_2)\nu/c}} x^{-(1-\alpha_2)(2-\alpha_2)}\right). \tag{8.4.42}$$

Stagnation-point flows. The solution is obtained by letting $\alpha_1 = \alpha_2 = 1$, which corresponds to the solution to a flow over a wedge with angle π, and is equivalent

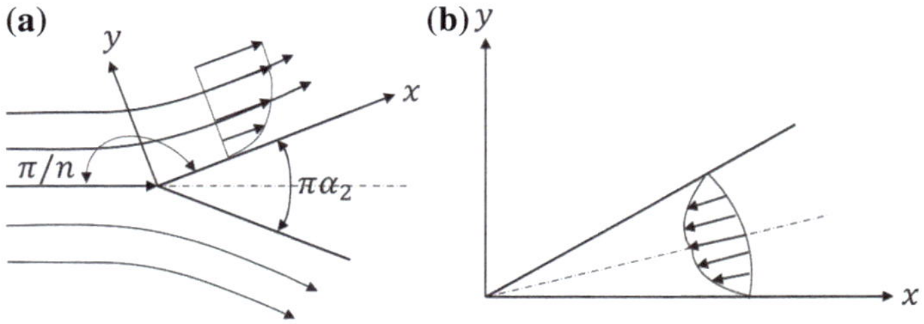

Fig. 8.14 Applications of the Falkner-Skan solutions to the boundary-layer equations. **a** A boundary-layer flow over a wedge. **b** A boundary-layer flow near the wall of a two-dimensional convergent channel

to a flow impinging on a flat surface, yielding a plane stagnation point. By letting $\alpha_2 = 1$, Eqs. (8.4.38)$_2$, (8.4.39) and (8.4.41) reduce respectively to

$$U(x) = cx, \quad \xi(x) = \sqrt{\frac{\nu}{c}}, \quad f''' + ff'' + 1 - \left(f'\right)^2 = 0, \qquad (8.4.43)$$
$$f(0) = f'(0) = 0, \, f'(\eta \to \infty) \to 1.$$

Solving $f(\eta)$ numerically gives the stream function in the form

$$\psi(x, y) = \sqrt{c\nu} x f\left(\frac{y}{\sqrt{\nu/c}}\right). \qquad (8.4.44)$$

It is verified that this stream function coincides to the exact solution to the full Navier-Stokes equation of the Hiemenz flow. Thus, an exact solution to the boundary-layer equations is also an exact solution to the full Navier-Stokes equation in this case.

Flows in a convergent channel. The solution is obtained by choosing $\alpha_1 = 0$ and $\alpha_2 = 1$. With these, Eqs. (8.4.29)$_2$ and (8.4.30) become respectively

$$\frac{\xi^2}{\nu}\frac{\mathrm{d}U}{\mathrm{d}x} = 1, \qquad \frac{\mathrm{d}}{\mathrm{d}x}\left(U\xi^2\right) = -\nu. \qquad (8.4.45)$$

Integrating the second equation and dividing the first equation by the integrated equation yield

$$\frac{1}{U}\frac{\mathrm{d}U}{\mathrm{d}x} = -\frac{1}{x}, \qquad \longrightarrow \qquad U(x) = -\frac{c}{x}, \qquad (8.4.46)$$

for the velocity of outer flow, where c is an integration constant. It is found that

$$\xi(x) = \sqrt{\frac{\nu}{c}} x. \qquad (8.4.47)$$

Equation (8.4.46)$_2$ is the velocity of a potential flow toward the apex of channel walls in a two-dimensional convergent channel. Hence, the obtained solution corresponds to a boundary-layer flow in the same geometric configuration, as shown in Fig. 8.14b. It follows from Eq. (8.4.47) that for $c < 0$, which corresponds to a flow in a divergent channel, no solution exists. This is due to the fact that the flow in a divergent channel experiences an adverse pressure gradient, causing the boundary layer to separate from the channel wall to induce a reverse flow. For $\alpha_1 = 0$ and $\alpha_2 = 1$, Eq. (8.4.31) and the associated boundary conditions become

$$f''' + 1 - (f')^2 = 0, \qquad f(0) = f'(0) = 0, \;\; f'(\eta \to \infty) \to 1, \tag{8.4.48}$$

which needs to be solved numerically. Once it is done, the values of $f(\eta)$ can be determined, and the stream function is then obtained as

$$\psi(x, y) = -\sqrt{c\nu}\, f\left(\frac{y}{\sqrt{\nu/cx}}\right). \tag{8.4.49}$$

8.4.5 Momentum Integral for a Flat Plate

When an exact solution to the boundary-layer equations does not exist, an approximate solution may be sought. One of the classical approximate methods is introduced by von Kármán and refined by Pohlhausen.[15] Consider a boundary-layer flow over a flat plate again. The basic idea is that if the boundary-layer equations are integrated across the boundary-layer thickness, the resulting equation will represent a balance between the averaged inertia and viscous forces for each x-location. Then, a velocity profile may be obtained which fulfills the averaged force balance. The outcomes are found to be reasonable accurate in most circumstances.

Equation (8.4.11) is recast in the form

$$\frac{\partial}{\partial x}\left(u^2\right) + \frac{\partial}{\partial y}(uv) = \nu \frac{\partial^2 u}{\partial y^2}, \tag{8.4.50}$$

in which $\partial u/\partial x$ has been replaced by $-\partial v/\partial y$ from the continuity equation, and the velocity of outer flow, U, is considered a constant. Integrating this equation with respect to y across the boundary-layer thickness yields

$$\int_0^\delta \frac{\partial}{\partial x}\left(u^2\right) \mathrm{d}y + (uv)\Big|_0^\delta = \nu \left[\frac{\partial u}{\partial y}\right]\Big|_0^\delta, \quad \longrightarrow \quad \int_0^\delta \frac{\partial}{\partial x}\left(u^2\right) \mathrm{d}y + Uv(x, \delta) = -\frac{\tau_w}{\rho}, \tag{8.4.51}$$

where τ_w is the shear stress on the plate. This is so obtained that $u(x, 0) = v(x, 0) = 0$, as implied by the no-slip boundary condition, and $u(x, \delta) = U$, $\mu \partial u/\partial y = \tau_w$ at $y = 0$, and $\partial u/\partial y = 0$ at $y = \delta$, since the velocity profile should blend smoothly into the outer-flow at the edge of boundary layer. Integrating the continuity equation across the boundary layer gives

$$\int_0^\delta \frac{\partial u}{\partial x} \mathrm{d}y + [v]\|_0^\delta = 0, \qquad \longrightarrow \qquad Uv(x, \delta) = -U \int_0^\delta \frac{\partial u}{\partial x} \mathrm{d}y, \tag{8.4.52}$$

with which Eq. $(8.4.51)_2$ is recast alternatively as

$$\int_0^\delta \frac{\partial}{\partial x}\left(u^2\right) \mathrm{d}y - U \int_0^\delta \frac{\partial u}{\partial x} \mathrm{d}y = -\frac{\tau_w}{\rho}. \tag{8.4.53}$$

[15] Ernst Pohlhausen, 1890–1964, a German mathematician.

This equation, by using the rule of Leibnitz, can be further simplified to[16]

$$\frac{\mathrm{d}}{\mathrm{d}x}\int_0^{\delta} u(U-u)\,\mathrm{d}y = \frac{\tau_w}{\rho}, \tag{8.4.54}$$

which is known as the *momentum integral*. It is valid for a boundary-layer flow over a flat plate with a constant velocity of the outer flow and states that the rate of change of momentum in the entire boundary layer at any x-location equals the force produced by the shear stress at the plate surface at the same location.

To use the momentum integral, the velocity profile, conventionally a polynomial in y, should be first assumed. The arbitrary constants in the assumed velocity profile are calibrated to match the required boundary conditions given by

$$u(x,0)=0, \quad u(x,\delta)=U, \quad \frac{\partial u}{\partial y}(x,\delta)=0, \tag{8.4.55}$$

where the first equation is the no-slip boundary condition, the second is the requirement that at the edge of boundary layer the velocity is the same as that of outer flow, while the last is used to ensure that the matching at $y=\delta$ is smooth.[17] For demonstration, the velocity profile is proposed as

$$\frac{u}{U} = a + b\left(\frac{y}{\delta}\right) + c\left(\frac{y}{\delta}\right)^2, \qquad a,b,c \in \mathcal{R}^1. \tag{8.4.56}$$

Applying Eq. (8.4.55) to the assumed velocity profile yields

$$a=0, \qquad a+b+c=1, \qquad b+2c=0, \tag{8.4.57}$$

giving rise to $a=0, b=2$ and $c=-1$, so that the assumed velocity profile becomes

$$\frac{u}{U} = 2\left(\frac{y}{\delta}\right) - \left(\frac{y}{\delta}\right)^2. \tag{8.4.58}$$

Substituting the obtained velocity profile into the momentum integral gives

$$\frac{\mathrm{d}}{\mathrm{d}x}\left(\frac{2}{15}\delta U^2\right) = 2\frac{\nu U}{\delta}, \tag{8.4.59}$$

---

[16]For any function $f(x,y)$, the rule of Leibnitz reads

$$\int_{\alpha(x)}^{\beta(x)} \frac{\partial f}{\partial x}(x,y)\mathrm{d}y = \frac{\mathrm{d}}{\mathrm{d}x}\int_{\alpha(x)}^{\beta(x)} f(x,y)\mathrm{d}y - f(x,\beta(x))\frac{\mathrm{d}\beta(x)}{\mathrm{d}x} + f(x,\alpha(x))\frac{\mathrm{d}\alpha(x)}{\mathrm{d}x}.$$

[17]All higher derivatives of velocity should equally vanish at $y=\delta$, since the transition from the inner to the outer flows is assumed to be smooth. The number of conditions which should be satisfied depends on the number of free parameters in the assumed velocity profile. Similarly, a series of boundary conditions should also be imposed at $y=0$. This follows from that the boundary-layer equations and no-slip boundary condition result automatically in that the higher derivatives of velocity on the surface of plate should vanish, if the velocity profile is assumed correctly. Since the assumed velocity profile may not be correct, these boundary conditions must be imposed separately. Likewise, by differentiating the boundary-layer equations, the conditions for higher derivatives of velocity are obtained, which should be imposed additionally in the approximate solution. Normally, the three conditions given in Eq. (8.4.55) are included in the order of priority in which they appear, then the condition $\partial^2 u/\partial y^2=0$ at $y=0$ is imposed, then $\partial^2 u/\partial y^2=0$ at $y=\delta$, and so on.

which is integrated with $\delta = 0$ at $x = 0$ (i.e., the condition at the leading edge of plate) to obtain

$$\delta = \sqrt{30}\sqrt{\frac{\nu x}{U}}, \qquad \longrightarrow \qquad \frac{\delta}{x} = \frac{5.48}{\sqrt{\mathcal{R}_{e_x}}}, \tag{8.4.60}$$

which compares favorable with the Blasius solution. The shear stress on the plate surface is obtained as

$$\frac{\tau_w}{\frac{1}{2}\rho U^2} = \frac{0.73}{\sqrt{\mathcal{R}_{e_x}}}, \tag{8.4.61}$$

which again compares favorable with Eq. $(8.4.22)_2$ in the Blasius solution.

The momentum integral is capable to produce meaningful results, even if it is used in conjunction with a rather crude approximation to the form of velocity profile. More accurate results can be obtained if third- or higher-order polynomials are used, for which more boundary conditions at $y = 0$ and $y = \delta$ need to be included to determine the free parameters in the assumed velocity profiles. On the other hand, if a straight line is used as the assumed velocity profile, only the first two conditions in Eq. (8.4.55) are necessary. Table 8.2 summarizes the obtained values of δ, τ_w and C_D by using different velocity profiles in the momentum integral for a laminar boundary-layer flow over a flat plate.

8.4.6 General Momentum Integral

With $U = U(x)$, the boundary-layer equations read the form

$$\frac{\partial u}{\partial x} + \frac{\partial v}{\partial y} = 0, \qquad \frac{\partial}{\partial x}\left(u^2\right) + \frac{\partial}{\partial y}(uv) = U\frac{\mathrm{d}U}{\mathrm{d}x} + \nu\frac{\partial^2 u}{\partial y^2}. \tag{8.4.62}$$

Integrating the second equation across the boundary layer yields

$$\int_0^\delta \frac{\partial}{\partial x}\left(u^2\right)\mathrm{d}y + Uv(x,\delta) = \frac{\mathrm{d}U}{\mathrm{d}x}\int_0^\delta U\mathrm{d}y - \frac{\tau_w}{\rho}, \tag{8.4.63}$$

Table 8.2 Results from the momentum integral for a laminar boundary-layer flow over a flat plate

Velocity profile	$\delta(x)\sqrt{\mathcal{R}_{e_x}}/x$	$2\tau_w(x)\sqrt{\mathcal{R}_{e_x}}/(\rho U^2)$	$C_D(x)\sqrt{\mathcal{R}_{e_x}}$
Blasius solution	5.00	0.664	1.328
Linear: $u/U = y/\delta$	3.46	0.578	1.156
Parabolic: $u/U = 2y/\delta - (y/\delta)^2$	5.48	0.730	1.460
Cubic: $u/U = 3y/2\delta - (y/\delta)^3/2$	4.64	0.646	1.292
Sine wave: $u/U = \sin(\pi y/2\delta)$	4.79	0.655	1.310

in which $\mathrm{d}U/\mathrm{d}x$ depends only on x, and $u(x,0)=0, u(x,\delta)=U, \mu\,\partial u/\partial y(x,0)=\tau_w$ and $\partial u/\partial y(x,\delta)=0$. Integrating the continuity equation gives

$$v(x,\delta)=-\int_0^\delta \frac{\partial u}{\partial x}\mathrm{d}y, \tag{8.4.64}$$

which is substituted into Eq. (8.4.63) to obtain

$$\int_0^\delta \frac{\partial}{\partial x}\left(u^2\right)\mathrm{d}y-U\int_0^\delta \frac{\partial u}{\partial x}\mathrm{d}y=\frac{\mathrm{d}U}{\mathrm{d}x}\int_0^\delta U\mathrm{d}y-\frac{\tau_w}{\rho}. \tag{8.4.65}$$

By using the rule of Leibnitz, this equation is recast alternatively as

$$\frac{\mathrm{d}}{\mathrm{d}x}\int_0^\delta u^2\,\mathrm{d}y-\frac{\mathrm{d}}{\mathrm{d}x}\int_0^\delta Uu\,\mathrm{d}y+\frac{\mathrm{d}U}{\mathrm{d}x}\int_0^\delta u\,\mathrm{d}y=\frac{\mathrm{d}U}{\mathrm{d}x}\int_0^\delta U\mathrm{d}y-\frac{\tau_w}{\rho}, \tag{8.4.66}$$

in which the identity

$$U\frac{\mathrm{d}}{\mathrm{d}x}\int_0^\delta u\,\mathrm{d}y=\frac{\mathrm{d}}{\mathrm{d}x}\int_0^\delta Uu\,\mathrm{d}y-\frac{\mathrm{d}U}{\mathrm{d}x}\int_0^\delta u\,\mathrm{d}y, \tag{8.4.67}$$

has been used. Combining the first with the second integrals and the third with the fourth integrals in Eq. (8.4.66) yields

$$\frac{\mathrm{d}}{\mathrm{d}x}\int_0^\delta u(U-u)\,\mathrm{d}y+\frac{\mathrm{d}U}{\mathrm{d}x}\int_0^\delta (U-u)\,\mathrm{d}y=\frac{\tau_w}{\rho}. \tag{8.4.68}$$

Since the integrands of two integrals vanish essentially for $y>\delta$, it is possible to express the above equation as

$$\frac{\mathrm{d}}{\mathrm{d}x}\left[U^2\int_0^\infty \frac{u}{U}\left(1-\frac{u}{U}\right)\mathrm{d}y\right]+U\frac{\mathrm{d}U}{\mathrm{d}x}\int_0^\infty\left(1-\frac{u}{U}\right)\mathrm{d}y=\frac{\tau_w}{\rho}, \tag{8.4.69}$$

where the first integral is the momentum thickness θ, while the second integral corresponds to the displacement thickness δ^*. Thus, the above equation may be expressed alternatively as

$$\frac{\mathrm{d}}{\mathrm{d}x}\left(U^2\theta\right)+\frac{\mathrm{d}U}{\mathrm{d}x}U\delta^*=\frac{\tau_w}{\rho}, \qquad \longrightarrow \qquad \frac{\mathrm{d}\theta}{\mathrm{d}x}+(2\theta+\delta^*)\frac{1}{U}\frac{\mathrm{d}U}{\mathrm{d}x}=\frac{\tau_w}{\rho U^2}, \tag{8.4.70}$$

which is referred to as the *general momentum integral* for a uniform flow with velocity $U(x)$ over a flat plate.

For any assumed velocity profile across the boundary layer, the values of θ, δ^* and τ_w can be evaluated from their definitions, by which Eq. (8.4.70) provides an ordinary differential equation for the boundary-layer thickness δ. To demonstrate the idea, the velocity profile is assumed to be in the form

$$\frac{u}{U}=a_1+a_2\eta+a_3\eta^2+a_4\eta^3+a_5\eta^4, \qquad \eta(x,y)=\frac{y}{\delta(x)}, \tag{8.4.71}$$

which is a fourth-order polynomial and is referred to as the *Kármán-Pohlhausen method*. The coefficients $\{a_1,a_2,a_3,a_4,a_5\}$ are functions of x in general. The associated boundary conditions are given by

$$\begin{aligned} &u(x,0)=0, \quad u(x,\delta)=U(x), \quad \frac{\partial u}{\partial y}(x,\delta)=0,\\ &\frac{\partial^2 u}{\partial y^2}(x,0)=-\frac{U(x)}{\nu}\frac{\mathrm{d}U(x)}{\mathrm{d}x}, \quad \frac{\partial^2 u}{\partial y^2}(x,\delta)=0, \end{aligned} \tag{8.4.72}$$

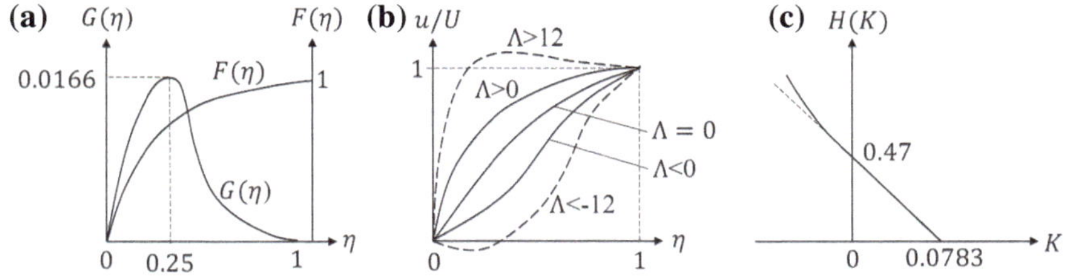

Fig. 8.15 Numerical integrations of the solutions by the Kármán-Pohlhausen method. **a** The distributions of $F(\eta)$ and $G(\eta)$. **b** The velocity profiles for variations in Λ. **c** The distribution of the function $H(k)$ (solid line) with the straight-line approximation (dashed line)

in which the fourth condition results from the boundary-layer equations together with the no-slip boundary condition. These boundary conditions are recast in dimensionless forms given by

$$\begin{aligned} \eta = 0: &\quad \frac{u}{U} = 0, \quad \frac{\partial^2}{\partial \eta^2}\left(\frac{u}{U}\right) = -\frac{\delta^2}{\nu}\frac{\mathrm{d}U(x)}{\mathrm{d}x} = -\Lambda(x), \\ \eta = 1: &\quad \frac{u}{U} = 1, \quad \frac{\partial}{\partial \eta}\left(\frac{u}{U}\right) = \frac{\partial^2}{\partial \eta^2}\left(\frac{u}{U}\right) = 0, \end{aligned} \tag{8.4.73}$$

in which $\Lambda(x)$ is introduced as a dimensionless parameter which is a measure of the pressure gradient in the outer flow. Applying Eq. (8.4.73) to Eq. (8.4.71) gives

$$a_1 = 0, \quad a_2 = 2 + \frac{\Lambda}{6}, \quad a_3 = -\frac{\Lambda}{2}, \quad a_4 = -2 + \frac{\Lambda}{2}, \quad a_5 = 1 - \frac{\Lambda}{6}, \tag{8.4.74}$$

with which the velocity profile becomes

$$\frac{u}{U} = \left(2 + \frac{\Lambda}{6}\right)\eta - \frac{\Lambda}{2}\eta^2 - \left(2 - \frac{\Lambda}{2}\right)\eta^3 + \left(1 - \frac{\Lambda}{6}\right)\eta^4, \tag{8.4.75}$$

which is recast alternatively as

$$\frac{u}{U} = 1 - (1+\eta)(1-\eta)^3 + \frac{\Lambda}{6}\eta(1-\eta)^3 = F(\eta) + \Lambda G(\eta), \tag{8.4.76}$$

where $F(\eta)$ and $G(\eta)$ are shown graphically in Fig. 8.15a for variations in η. The function $F(\eta)$ increases monotonically from 0 to 1 as η goes from 0 to 1. On the other hand, $G(\eta)$ increases from 0 at $\eta = 0$ to its maximum value of 0.0166 at $\eta = 0.25$, after that it drops off to null at $\eta = 1$. With these, the calculated values of u/U for variations in Λ are displayed in Fig. 8.15b.

For $\Lambda = 0$, the velocity profile corresponds to that of a flat plate in which the assumed velocity profile is a fourth-order polynomial. For $\Lambda > 12$, the value of u/U is greater than 1.0 for larger values of η, which is not physically justified. Equally, for $\Lambda < -12$, there exists a reverse flow for smaller values of η. Although reverse flows take place physically, they cannot be captured by the assumptions used in the analysis. It follows that to reach a physically justified result, the value of Λ should be restricted by

$$-12 < \Lambda(x) < 12. \tag{8.4.77}$$

Within this restriction, the displacement and momentum thicknesses, and the shear stress on the plate surface are determined to be

$$\delta^* = \delta\left(\frac{3}{10} - \frac{\Lambda}{120}\right), \quad \theta = \delta\left(\frac{37}{315} - \frac{\Lambda}{945} - \frac{\Lambda^2}{9072}\right), \quad \tau_w = \frac{\mu U}{\delta}\left(2 + \frac{\Lambda}{6}\right), \tag{8.4.78}$$

which are functions of the disturbance thickness δ. Since δ is yet known, these relations follow up purely from the assumed velocity profile. The additional relation which is required to determine the absolute values of these quantities should be provided by the momentum integral.

Multiplying Eq. $(8.4.70)_2$ by $U\theta/\nu$ yields

$$\frac{U}{2}\frac{\mathrm{d}}{\mathrm{d}x}\left(\frac{\theta^2}{\nu}\right) + \left(2 + \frac{\delta^*}{\theta}\right)\frac{\theta^2}{\nu}\frac{\mathrm{d}U}{\mathrm{d}x} = \frac{\tau_w\theta}{\mu U}, \tag{8.4.79}$$

which is recast alternatively as

$$\frac{\theta^2}{\nu}\frac{\mathrm{d}U}{\mathrm{d}x} = K(x), \qquad K(x) = \left(\frac{37}{315} - \frac{\Lambda}{945} - \frac{\Lambda^2}{9072}\right)^2 \Lambda, \tag{8.4.80}$$

in which the definition of Λ and Eq. $(8.4.78)_2$ has been used. Similarly, δ^*/θ and $\tau_w\theta/(\mu U)$ are obtained as

$$\begin{aligned} \frac{\delta^*}{\theta} &= f(K), \quad f(K) = \left(\frac{3}{10} - \frac{\Lambda}{120}\right) \Big/ \left(\frac{37}{315} - \frac{\Lambda}{945} - \frac{\Lambda^2}{9072}\right), \\ \frac{\tau_w\theta}{\mu U} &= g(K), \; g(K) = \left(2 + \frac{\Lambda}{6}\right)\left(\frac{37}{315} - \frac{\Lambda}{945} - \frac{\Lambda^2}{9072}\right), \end{aligned} \tag{8.4.81}$$

in which the functions f and g depend on $\Lambda(x)$ and hence on x. Since K depends also on x, f and g may be considered functions of K. Substituting Eqs. (8.4.80) and (8.4.81) into the momentum integral gives

$$\frac{U}{2}\frac{\mathrm{d}}{\mathrm{d}x}\left(\frac{\theta^2}{\nu}\right) + [2 + f(K)]\,K = g(K), \tag{8.4.82}$$

which, by letting $Z = \theta^2/\nu$, is expressed as

$$U\frac{\mathrm{d}Z}{\mathrm{d}x} = H(K), \qquad H(K) = 2\left\{g(K) - [2 + f(K)]\,K\right\}, \tag{8.4.83}$$

where $H(K)$ and K are functions of Λ, which may be determined once the value of Λ is prescribed. A curve of $H(K)$ in relation with K is shown graphically in Fig. 8.15c. It is seen that $H(K)$ is approximately linear in K over the range of interest. Thus, the function $H(K)$ may be approximated by the linear equation

$$H(K) = 0.47 - 6K, \tag{8.4.84}$$

with which the momentum integral becomes

$$\frac{1}{U^5}\frac{\mathrm{d}}{\mathrm{d}x}\left(ZU^6\right) = 0.47. \tag{8.4.85}$$

Integrating this equation results in

$$Z(x) = \frac{0.47}{U^6(x)}\int_0^x U^5(\xi)\,\mathrm{d}\xi, \longrightarrow \theta^2(x) = \frac{0.47\nu}{U^6(x)}\int_0^x U^5(\xi)\,\mathrm{d}\xi. \tag{8.4.86}$$

For any given geometric shape in practice, the specific potential-flow problem should be solved first to obtain $U(x)$ of the outer flow. This $U(x)$ is then substituted into Eq. (8.4.86) to evaluate the momentum thickness $\theta(x)$. The pressure parameter $\Lambda(x)$ is subsequently obtained by using Eq. (8.4.80). Having done these, the disturbance thickness δ is determined by using Eq. $(8.4.78)_2$, which is substituted into Eqs. $(8.4.78)_{1,3}$ to obtain the values of δ^* and τ_w, respectively. Finally, the velocity profile is determined by using Eq. (8.4.76). Although straightforward, it is in fact difficult to evaluate $\Lambda(x)$ directly from Eq. (8.4.80) in practice, unless it is a constant. It is hence much simpler to prescribe specific functions of $\Lambda(x)$ and use the forgoing equations to determine the velocity of outer-flow field and the nature of geometric shape.

To explore the idea, consider the Kármán-Pohlhausen approximation for a boundary-layer flow over a flat surface. Since $U(x)$ is a constant, it follows from Eq. $(8.4.86)_2$ that

$$\theta^2 = 0.47\frac{\nu x}{U}, \qquad \longrightarrow \qquad \frac{\theta}{x} = \frac{0.686}{\sqrt{\mathcal{R}_{e_x}}}. \tag{8.4.87}$$

On the other hand, it follows from Eq. (8.4.80) that $\Lambda = 0$, and

$$\theta = \frac{37}{315}\delta, \qquad \longrightarrow \qquad \frac{\delta}{x} = \frac{5.84}{\sqrt{\mathcal{R}_{e_x}}}, \tag{8.4.88}$$

as implied by Eq. $(8.4.78)_2$. Equally, by using Eqs. $(8.4.78)_{1,3}$, the displacement thickness and shear stress on the plate surface are obtained as

$$\frac{\delta^*}{x} = \frac{1.75}{\sqrt{\mathcal{R}_{e_x}}}, \qquad \frac{\tau_w}{\frac{1}{2}\rho U^2} = \frac{0.686}{\sqrt{\mathcal{R}_{e_x}}}. \tag{8.4.89}$$

The obtained results compare favorably with those of the Blasius solution. For example, the obtained value of τ_w by using the Kármán-Pohlhausen approximation is within 3.5% of the exact solution.

8.4.7 Transition from Laminar to Turbulent Boundary-Layer Flows

The previous discussions are restricted only to laminar boundary-layer flows over a flat plate. They agree quite well with the experimental outcomes up to the points where the boundary-layer flows become turbulent, which is characterized by the value of critical Reynolds number $\mathcal{R}_{e_{cr}}$. The occurrence of turbulent boundary-layer flows takes place for any free stream velocity and any fluid, provided that the flat plate is sufficiently long. The value of $\mathcal{R}_{e_{cr}}$ at the transition location is a rather complex function of various parameters, e.g. the roughness and curvature of plate surface, and some measures of the disturbances in the outer flow. On a flat plate with a sharp leading edge in a standard atmospheric air, the transition takes place at a distance from the leading edge, where $\mathcal{R}_{e_{cr}} = 2 \times 10^5 \sim 3 \times 10^6$. For engineering application, $\mathcal{R}_{e_{cr}} = 5 \times 10^5$ is frequently chosen.

The actual transition from laminar to turbulent boundary-layer flows may not occur at a specific location, but over a region of the plate. This is partly due to the *spottiness* of the transition. Typically, the transition begins at some random locations on the plate with the values of $\mathcal{R}_{e_x}$ approaching $\mathcal{R}_{e_{cr}}$. The spots grow rapidly as they are convected downstream until the entire plate width is covered by the turbulent boundary-layer flows. The complex transition process involves the stability of flow field. Small disturbances imposed on the boundary-layer flows will either grow or decay, corresponding to instability and stability, respectively, which depend on where the disturbances are introduced. If the disturbances occur at a location with $\mathcal{R}_{e_x} < \mathcal{R}_{e_{cr}}$, they will die out, and the boundary-layer flow will return to laminar at that location. A reverse circumstance takes place if $\mathcal{R}_{e_x} > \mathcal{R}_{e_{cr}}$. The analysis of the stability of boundary-layer flows will be discussed in the next section.

When changing to the turbulent boundary-layer flows, the velocity profiles involve a noticeable change in the shape. The profiles obtained in the neighborhood of transition location are shown graphically in Fig. 8.16a. The turbulent velocity profile is flatter, having a larger velocity gradient at the wall, and produces a larger boundary-layer thickness than its laminar counterpart. The structure of a turbulent-boundary-layer flow is very complex, random and irregular, but it shares many of the characteristics of turbulent pipe-flows, which will be discussed in Sect. 8.6. Specifically, the velocity at any given location in the flow is unsteady in a random manner. The flow may be thought of as a jumbled mix of intertwined eddies of different sizes, and various quantities such as mass, momentum, and energy are convected downstream not only in the direction which is parallel to that of the outer flow, but also convected across the boundary layer in the direction perpendicular to the plate by the random transport of finite-sized fluid particles associated with the turbulent eddies. There exists a considerable mixing with these finite turbulent eddies, which are more considerable than that associated with laminar boundary-layer flows, where it is confined to the molecular scale. Despite of the intensive mixing across the boundary layer, the largest mass transportation still takes place in the direction parallel to the plate. However, there exists a significant x-component momentum transfer across the boundary layer due to the random motions of fluid particles. Fluid particles moving toward the plate

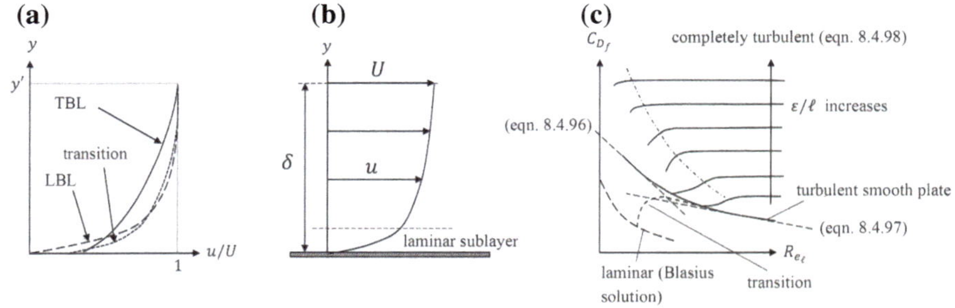

Fig. 8.16 Characteristics of turbulent boundary-layer flows over a flat plate with vanishing pressure gradient. **a** Typical velocity profiles in laminar, transition, and turbulent regions. **b** Turbulent pipe flow as a model for turbulent boundary-layer flow. **c** The frictional drag coefficient in relation with the Reynolds number and relative surface roughness

have some of their excess momentum removed by the plate, while those leaving the plate gain momentum from the fluid. This gives rise to the circumstance that the plate acts as a momentum sink, which continuously extracts momentum from the fluid. For laminar boundary-layer flows, such a cross-stream momentum transfer is, however, restricted to the molecular level. Since in turbulent boundary-layer flows the randomness is associated with fluid-particle mixing, the shear stress on the plate surface is considerably larger than its laminar counterpart.

Unfortunately, there exists no "exact" solution to a turbulent boundary-layer flow, since there is no precise expression for the shear stress on the plate surface. Approximate solutions may be obtained by using the momentum integral, which is valid for both laminar and turbulent boundary-layer flows. The required information is the reasonable approximations to the velocity profiles and a functional relation describing the wall shear stress. However, the details of velocity gradient at the wall are not well understood. It is thus necessary to use some empirical relations for the wall shear stress. To demonstrate the idea, consider a turbulent boundary-layer flow over a flat plate. The velocity profile within the boundary layer is found to be the same as that of a turbulent flow in a circular pipe, as will be discussed in Sect. 8.6. At the section where the turbulent flow is *fully developed*, the velocity profile is shown in Fig. 8.16b. As an approximation, Blasius' one-seventh-power law for the velocity distribution given by

$$\frac{u}{U} = \left(\frac{y}{\delta}\right)^{1/7}, \tag{8.4.90}$$

is used for the velocity profile over a smooth flat plate. However, the shear stress at the plate surface, τ_w, is no longer determined by Newton's law of viscosity, for it approaches infinity at $y \to 0$. There exists thus a thin *laminar sublayer*, or *viscous sublayer* in the immediate vicinity of plate surface, and the above power-law equation applies only to the region outside the laminar sublayer. The shear stress outside the laminar sublayer is transmitted to the plate surface through the viscous action in the laminar sublayer. With the power law velocity profile, an expression for the shear stress in a turbulent boundary-layer flow on a smooth plate surface can be given by

$$\tau_w = 0.0233\rho U^2 \left(\frac{\mu}{\rho U \delta}\right)^{1/4}, \tag{8.4.91}$$

where U is the velocity of outer flow. This expression is motivated by the wall shear stress evaluated for turbulent pipe-flows.

Substituting Eqs. (8.4.90) and (8.4.91) into the momentum integral equation yields

$$\delta^{1/4}\frac{\mathrm{d}\delta}{\mathrm{d}x} = 0.239\left(\frac{\mu}{\rho U}\right)^{1/4}, \quad \longrightarrow \quad \frac{4}{5}\delta^{5/4} = 0.239\left(\frac{\mu}{\rho U}\right)^{1/4} x, \tag{8.4.92}$$

in which the integration has been conducted between 0 and x, for the turbulent boundary layer is assumed to occur over the entire length of plate with the initial condition $\delta(x = 0) = 0$. It follows that

$$\delta(x) = 0.379x\left(\frac{\mu}{\rho U x}\right)^{1/5} = \frac{0.379}{\mathcal{R}_{ex}^{1/5}}x, \tag{8.4.93}$$

indicating that the thickness of a turbulent boundary layer on a smooth plate increases as $x^{4/5}$, whereas that in a laminar boundary layer varies as $\sqrt{x}$. Thus, the turbulent boundary layer grows at a more rapid rate along the length of plate than the laminar boundary layer. With this, the shear stress τ_w is obtained as

$$\tau_w = 0.0588\frac{\rho U^2}{2}\left(\frac{\mu}{\rho U x}\right)^{1/5}, \tag{8.4.94}$$

which indicates that τ_w decreases as $x^{1/5}$ along the length of plate. The total frictional drag F_{D_f} and the corresponding frictional drag coefficient C_{D_f} on one side of the smooth plate with length ℓ per unit width produced by a turbulent boundary-layer flow are given respectively by[18]

$$F_{D_f} = \int_0^\ell \tau_w \mathrm{d}x = 0.072\frac{\rho U^2}{2}\ell\left(\frac{\mu}{\rho U \ell}\right)^{1/5}, \qquad C_{D_f} = \frac{0.072}{\mathcal{R}_{e_\ell}^{1/5}}. \tag{8.4.95}$$

However, the estimated frictional drag coefficient is found to agree better with the experimental outcomes if the numerical constant 0.072 is changed to 0.074, with which Eq. (8.4.95)$_2$ becomes

$$C_{D_f} = \frac{0.074}{\mathcal{R}_{e_\ell}^{1/5}}, \tag{8.4.96}$$

which is valid for the range $5 \times 10^5 < \mathcal{R}_{e_\ell} < 10^7$.

If the logarithmic universal velocity profile for a turbulent flow in a smooth pipe is used in conjunction with the momentum integral equation, the expression of frictional drag coefficient, as calibrated by experiments, is obtained as

$$C_{D_f} = \frac{0.455}{(\log \mathcal{R}_{e_\ell})^{2.58}}, \tag{8.4.97}$$

which is valid for the range $5 \times 10^5 < \mathcal{R}_{e_\ell} < 10^9$. By using the logarithmic universal velocity distribution for turbulent flows is rough pipes, Schlichting has derived an empirical expression for the frictional drag coefficient caused by a completely turbulent boundary-layer flow over a rough plate of length ℓ, viz.,[19]

$$C_{D_f} = \frac{1}{\left(1.89 + 1.62\log\dfrac{\ell}{\varepsilon}\right)^{2.5}}, \tag{8.4.98}$$

where ε is the surface roughness. This expression is valid for $10^2 < \ell/\varepsilon < 10^6$.

Essentially, the frictional drag coefficient C_{D_f} for a flat plate with length ℓ is a function of the Reynolds number $\mathcal{R}_{e_\ell}$, and the relative roughness ε/ℓ. The results of numerous experiments covering a wide range of the parameters of interest are shown

---

[18] In reality, the boundary-layer flow is laminar over the forward portion of plate, which will become turbulent farther on downstream of the plate.

[19] Hermann Schlichting, 1907–1982, a German fluid dynamics engineer, who contributed to the theory of boundary-layer transitioning from laminar to turbulent states, which is known as the *Tollmien-Schlichting waves*.

Table 8.3 Semi-empirical equations for the frictional drag coefficient C_{D_f} of a boundary-layer flow over a flat plate with vanishing pressure gradient

Frictional drag coefficient	Flow circumstances	Restrictions
$1.328/(\mathcal{R}_{e_\ell})^{1/2}$	Laminar flow	$\mathcal{R}_{e_\ell} < 10^5$
$0.455/(\log \mathcal{R}_{e_\ell})^{2.58} - 1700/\mathcal{R}_{e_\ell}$	Transitional region	$\mathcal{R}_{e_\ell} = 5 \times 10^5$
$0.074/(\mathcal{R}_{e_\ell})^{1/5}$	Turbulent flow, smooth plate	$5 \times 10^5 < \mathcal{R}_{e_\ell} < 10^7$
$0.455/(\log \mathcal{R}_{e_\ell})^{2.58}$	Turbulent flow, smooth plate	$5 \times 10^5 < \mathcal{R}_{e_\ell} < 10^9$
$[1.89 + 1.62 \log(\ell/\varepsilon)]^{-2.5}$	Completely turbulent	$5 \times 10^5 < \mathcal{R}_{e_\ell} < 10^9$ $10^2 < \ell/\varepsilon < 10^6$

graphically in Fig. 8.16c. For laminar boundary-layer flows, C_{D_f} depends solely on $\mathcal{R}_{e_\ell}$, while for completely turbulent boundary-layer flows, the surface roughness does affect the shear stress, and hence influences the values of C_{D_f}. These characteristics are very similar to those of laminar and turbulent pipe flows, which will be discussed in Sect. 8.6.[20] Table 8.3 summarizes some semi-empirical equations for the frictional drag coefficient of boundary-layer flows over a flat plate with vanishing pressure gradient.[21]

8.4.8 Separation and Stability of Boundary Layers

It is known from experiments that boundary layers have a tendency to separate from the surface over which they form a wake behind a body, which is known as *boundary-layer separation*. The separation of boundary layer results from the influence of pressure gradient, which may be determined by the outer flow. If a region with an adverse pressure gradient exists in the outer flow, this pressure gradient will equally exert itself along the body surface near which the fluid velocity assumes smaller values. The momentum contained in the fluid layers adjacent to the body surface will be insufficient to overcome the force exerted by the pressure gradient, so that at a specific location a region of reverse flow takes place. That is, at some point the adverse pressure gradient will cause the fluid layers adjacent to the body surface to flow in a direction opposite to that of outer flow. This marks that the boundary layer has separated from the body surface and is deflected over the reverse-flow region.

As shown in Fig. 8.10a, the velocity gradient at the body surface prior to separation assumes a positive value, so that the shear stress there opposes the outer-flow field. After the separation, the velocity gradient is negative, so that the shear stress has changed its sign and direction. These observations motivate that separation may be

[20] However, the underlying physics are quite different, as will be discussed later.

[21] Data quoted from Schlichting, H., *Boundary Layer Theory*, 7th ed., McGraw-Hill, New York, 1979.

estimated by the location where the velocity gradient vanishes, i.e.,

$$\frac{\partial u}{\partial y}(x, 0) = 0. \tag{8.4.99}$$

Along the surface $y = 0$, Eq. $(8.4.10)_2$ reduces to

$$-\frac{\mathrm{d}p}{\mathrm{d}x} + \mu\frac{\partial^2 u}{\partial y^2} = 0, \tag{8.4.100}$$

in which the no-slip boundary condition has been used. It is seen that the curvature of velocity profile is proportional to the pressure gradient along the surface. It follows that the curvature of velocity profile is negative if $\mathrm{d}p/\mathrm{d}x < 0$, and will remain negative at the surface just as it is at the edge of boundary layer. On the contrary, the curvature of velocity profile will be positive if $\mathrm{d}p/\mathrm{d}x > 0$. Since $\partial^2 u/\partial y$ may still be negative at the edge of boundary layer, the velocity profile must experience an inflection point somewhere between $y = 0$ and $y = \delta$. Such a velocity profile may lead to separation if the curvature at $y = 0$ is sufficiently positive to yield a reverse-flow. Separation will not occur in a region in which $\mathrm{d}p/\mathrm{d}x < 0$, and such a pressure gradient is termed a favorable pressure gradient. Separation can occur in a region with $\mathrm{d}p/\mathrm{d}x > 0$, which is called an adverse pressure gradient.

A precise determination of the location of separation point is not an easy task. It may be obtained by solving the potential-flow problem for the body under consideration first. The obtained pressure gradient is then substituted into the boundary-layer equations, which are solved by using either an analytical or a numerical approach. From the solutions to the boundary-layer equations, the location of vanishing shear stress on the body surface may be determined. The difficulty lies in that as soon as the boundary layer separates, the pressure distribution will be different from that obtained by the potential-flow theory, for the latter applies to a different streamline configuration. Although two principal approaches have been proposed to overcome the difficulty in determining the location of separation point,[22] the subject of boundary-layer separation is the one which is not well understood analytically. What is known is that boundary layers will separate in adverse pressure gradients, whose magnitude and extent should be minimized. In other words, bodies should be streamlined rather than bluff with small angles of attack. Also, sharp corners which bend away from the fluid become separation points, and should be avoided if separation is to be delayed as far as possible.

Now turn to the stability of boundary layer. Like any fluid-flow circumstance, boundary layers may become unstable. Usually, the instabilities of boundary layers manifest themselves in turbulence, and a laminar boundary layer which becomes

[22]The first approach was used by Hiemenz, in which the determination of pressure distribution around the body was accomplished experimentally. The drawback of approach lies in the fact that the pressure distribution must be established experimentally for each body shape and for each Reynolds number of interest, which is not only time-consuming, but also difficult in measuring data. The second approach is to revise the potential-flow model, from which the pressure distribution is obtained. The difficulty with this approach is that some empirical constants exist in the potential-flow model and experimental calibrations need to be conducted to obtain their values.

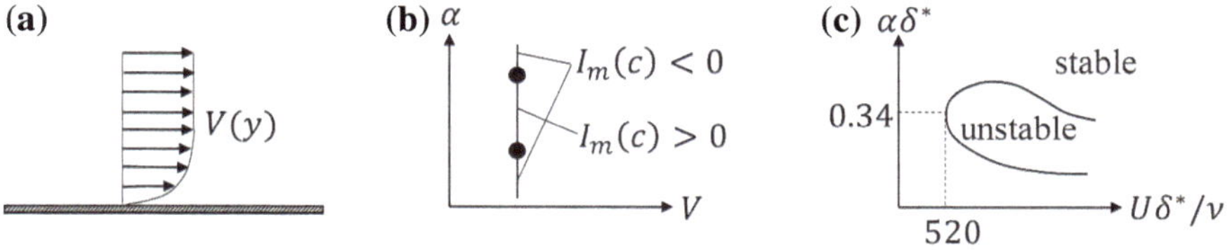

Fig. 8.17 Stability of boundary layer. **a** A undisturbed boundary-layer velocity profile. **b** The determination of stability for fixed values of V. **c** The stability diagram in terms of the Reynolds number

unstable usually becomes a turbulent boundary layer. The properties of laminar and turbulent boundary layers are quite different. An important difference, for example, is that the location of separation on a circular cylinder starts at angles of 82° and of 108° in a laminar and a turbulent boundary-layer flows, respectively, where the angle is measured from the upstream stagnation point. This leads to an appreciable drop in the drag coefficient, as already shown in Fig. 8.9.

Consider the velocity profile in a narrow strip of a boundary layer shown in Fig. 8.17a, in which the velocity component in the x-direction, V, is considered only a function of y, with vanishing velocity component in the y-direction. Based on the classical stability analysis, a disturbance is introduced to this boundary-layer velocity, so that $u(x,y,t)$, $v(x,y,t)$, and $p(x,y,t)$ become

$$\begin{aligned} &u(x,y,t) = V(y) + u'(x,y,t), \qquad v(x,y,t) = v'(x,y,t),\\ &p(x,y,t) = p_0(x) + p'(x,y,t), \end{aligned} \tag{8.4.101}$$

in which the primed quantities represent disturbances (perturbations), and it is assumed that $|u'/V|$, $|v'/V|$, and $|p'/p_0|$ are all small compared with unity. Substituting these expressions into the continuity and Navier-Stokes equations in the x- and y-directions yields respectively

$$\begin{aligned} &\frac{\partial u'}{\partial x} + \frac{\partial v'}{\partial y} = 0,\\ &\frac{\partial u'}{\partial t} + (V+u')\frac{\partial u'}{\partial x} + v'\left(\frac{\mathrm{d}V}{\mathrm{d}y} + \frac{\partial u'}{\partial y}\right) = -\frac{1}{\rho}\left(\frac{\mathrm{d}p_0}{\mathrm{d}x} + \frac{\partial p'}{\partial x}\right)\\ &\qquad\qquad + \nu\left(\frac{\partial^2 u'}{\partial x^2} + \frac{\mathrm{d}^2 V}{\mathrm{d}y^2} + \frac{\partial^2 u'}{\partial y^2}\right),\\ &\frac{\partial v'}{\partial t} + (V+u')\frac{\partial v'}{\partial x} + v'\frac{\partial v'}{\partial y} = -\frac{1}{\rho}\frac{\partial p'}{\partial y} + \nu\left(\frac{\partial^2 v'}{\partial x^2} + \frac{\partial^2 v'}{\partial y^2}\right). \end{aligned} \tag{8.4.102}$$

If the perturbations are null, the above equations reduce to

$$-\frac{1}{\rho}\frac{\mathrm{d}p_0}{\mathrm{d}x} + \nu\frac{\mathrm{d}^2 V}{\mathrm{d}y^2} = 0, \tag{8.4.103}$$

which can be extracted from Eq. $(8.4.102)_2$. Since the perturbations are assumed to be small, the products of all primed quantities can be neglected, so that a linearized form of Eq. (8.4.102) is obtained as

$$\begin{aligned} &\frac{\partial u'}{\partial x}+\frac{\partial v'}{\partial y}=0,\\ &\frac{\partial u'}{\partial t}+V\frac{\partial u'}{\partial x}+v'\frac{\mathrm{d}V}{\mathrm{d}y}=-\frac{1}{\rho}\frac{\partial p'}{\partial x}+\nu\left(\frac{\partial^2 u'}{\partial x^2}+\frac{\partial^2 u'}{\partial y^2}\right),\\ &\frac{\partial v'}{\partial t}+V\frac{\partial v'}{\partial x}=-\frac{1}{\rho}\frac{\partial p'}{\partial y}+\nu\left(\frac{\partial^2 v'}{\partial x^2}+\frac{\partial^2 v'}{\partial y^2}\right). \end{aligned} \tag{8.4.104}$$

Introducing the perturbation stream function $\psi_p(x, y, t)$ defined by

$$u'\equiv\frac{\partial\psi_p}{\partial y},\qquad v'\equiv-\frac{\partial\psi_p}{\partial x}, \tag{8.4.105}$$

and substituting these expressions into Eqs. $(8.4.104)_{2,3}$ give respectively

$$\begin{aligned} &\frac{\partial^2\psi_p}{\partial y\partial t}+V\frac{\partial^2\psi_p}{\partial x\partial y}-\frac{\partial\psi_p}{\partial x}\frac{\mathrm{d}V}{\mathrm{d}y}=-\frac{1}{\rho}\frac{\partial p'}{\partial x}+\nu\left(\frac{\partial^3\psi_p}{\partial x^2\partial y}+\frac{\partial^3\psi_p}{\partial y^3}\right),\\ &-\frac{\partial^2\psi_p}{\partial x\partial t}-V\frac{\partial^2\psi_p}{\partial x^2}=-\frac{1}{\rho}\frac{\partial p'}{\partial y}-\nu\left(\frac{\partial^3\psi_p}{\partial x^3}+\frac{\partial^3\psi_p}{\partial x\partial y^2}\right), \end{aligned} \tag{8.4.106}$$

which, by forming the mixed partial derivatives $\partial^2 p'/(\partial x\partial y)$, can be combined into a single equation given by

$$\left(\frac{\partial}{\partial t}+V\frac{\partial}{\partial x}\right)\left(\frac{\partial^2\psi_p}{\partial x^2}+\frac{\partial^2\psi_p}{\partial y^2}\right)-\frac{\mathrm{d}^2V}{\mathrm{d}y^2}\frac{\partial\psi_p}{\partial x}=\nu\left(\frac{\partial^4\psi_p}{\partial x^4}+2\frac{\partial^4\psi_p}{\partial x^2\partial y^2}+\frac{\partial^4\psi_p}{\partial y^4}\right). \tag{8.4.107}$$

This equation must be satisfied by the introduced perturbation stream function ψ_p.

Since the introduced disturbance is arbitrary in form, it may be Fourier analyzed in the x-direction. That is, ψ_p can be expressed in terms of the Fourier integral, viz.,

$$\psi_p(x, y, t)=\int_0^\infty \Psi_p(y)e^{i\alpha(x-ct)}\mathrm{d}\alpha, \tag{8.4.108}$$

where $\alpha\in\mathcal{R}^1>0$. In this expression, the time variation has been taken to be $e^{-i\alpha ct}$, so that if $Im(c)>0$, the disturbance will grow, and vice versa. For $c=0$, the disturbance will introduce a neutrally stable state. Substituting this expression into Eq. (8.4.107) results in

$$\begin{aligned} &\int_0^\infty\left[(-i\alpha c+i\alpha V)\left(\Psi_p''-\alpha^2\Psi_p\right)-i\alpha\Psi_pV''\right]e^{i\alpha(x-ct)}\mathrm{d}\alpha\\ &=\int_0^\infty\left[\nu(\Psi_p''''-2\alpha^2\Psi_p''+\alpha^4\Psi_p)\right]e^{i\alpha(x-ct)}\mathrm{d}\alpha, \end{aligned} \tag{8.4.109}$$

which is an integral-differential equation, where the primes denote differentiations with respect to y. This equation should be valid for any arbitrary disturbance, so that

$$(V-c)(\Psi_p''-\alpha^2\Psi_p)-V''\Psi_p=\frac{\nu}{i\alpha}(\Psi_p''''-2\alpha^2\Psi_p''+\alpha^4\Psi_p), \tag{8.4.110}$$

which is known as the *Orr-Sommerfeld equation.*[23] The associated boundary conditions are derived from the conditions that the disturbances should vanish at $y = 0$ and $y \to \infty$, which are given by

$$u'(x, 0, t) = v'(x, 0, t) = 0, \qquad u'(x, y \to \infty, t) = v'(x, y \to \infty, t) \to 0, \tag{8.4.111}$$

corresponding respectively to

$$\Psi_p(0) = \Psi_p'(0) = 0, \qquad \Psi_p(y \to \infty) = \Psi_p'(y \to \infty) \to 0. \tag{8.4.112}$$

To obtain the solutions to the Orr–Sommerfeld equation, both $V(y)$ and α must be evaluated for a given undisturbed velocity profile and a disturbance wavelength. Having done these, Eqs. (8.4.110) and (8.4.112) form an eigenvalue problem for the time coefficient c. If each possible wavelength is considered, the regions which are stable (i.e., $Im(c) < 0$) and unstable (i.e., $Im(c) > 0$) may be identified. Typical results are shown in Fig. 8.17b. By considering all possible values of the undisturbed boundary-layer velocities which are less than that of the outer flow, a stable diagram can be constructed. In other words, for all possible values of $V(y)$ in the range $0 \leq V(y) \leq U(x)$, the stable boundaries for a specific x-location can be established. Typical results for a boundary-layer flow over a flat plate are shown in Fig. 8.17c, in which the Reynolds number is based on the displacement thickness, and the inverse wavelength $\alpha\delta^*$ is non-dimensionalized by the same quantity. It is found that the lower limit of the Reynolds number for which an instability may occur is 520, which gives

$$\frac{U\delta_{cr}^*}{\nu} = 520. \tag{8.4.113}$$

Hence, an arbitrary disturbance which has a Fourier component whose wavelength is such that $\alpha\delta^* = 0.34$ will lie on the stability region. It is expected that for the Reynolds numbers greater than 520, arbitrary disturbances will lead to unstable states. Such instabilities will manifest themselves in the form of turbulence at the Reynolds number slightly larger than its critical value.

8.4.9 Drag and Lift

When a viscous fluid passes around a solid body, as shown in Fig. 8.18a, it exerts a force $\boldsymbol{F}$ acting on the body. The force may be resolved into the components parallel and perpendicular to the flow direction of undisturbed upstream fluid. The force component which is parallel to the flow direction, F_D, is called the *drag*, while the perpendicular force component, F_L, is termed the *lift*. Conventionally, two force components are expressed in terms of the dimensionless parameters given by

$$C_D = \frac{F_D}{\frac{1}{2}\rho U^2 A_J}, \qquad C_L = \frac{F_L}{\frac{1}{2}\rho U^2 A_J}, \tag{8.4.114}$$

[23] William McFadden Orr, 1866–1934, a British and Irish mathematician. Arnold Johannes Wilhelm Sommerfeld, 1868–1951, a German theoretical physicist, who pioneered developments in atomic and quantum physics.

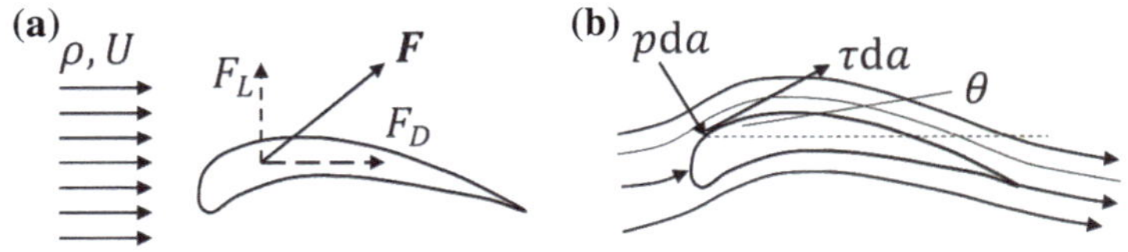

Fig. 8.18 Force $\boldsymbol{F}$ acting on an immersed airfoil. **a** The drag force F_D and lift force F_L. **b** The normal and shear forces acting on a surface element da of an airfoil

which are referred to as the *drag* and *lift coefficients*, respectively, where A_J is the characteristic area, either the largest projected area of immersed body or the projected area perpendicular to the flow direction. The term $\rho U^2/2$ is the dynamic pressure of fluid with density ρ and velocity U. Except for a few simple cases, both C_D and C_L should be determined experimentally.

Careful studies reveal that $\boldsymbol{F}$ acting on a completely immersed body depends on (a) the geometric configuration of body, (b) the fluid properties such as density, dynamic viscosity and elastic property, etc., and (c) the velocity of flow. Dimensional analysis shows that both C_D and C_L are functions of the geometric configuration, the Reynolds number, and the Mach number. The relative significance between the Reynolds and Mach numbers depends on whether the flow is considered incompressible or compressible. For incompressible flows, only the Reynolds number plays a significant role. The effect of the Mach number becomes important when the flow velocity approaches or exceeds the sonic velocity, which will be discussed in Sect. 9.5. Physically, the force $\boldsymbol{F}$ results from the shear stress and normal stress (the pressure) distributions on the entire body surface, as shown in Fig. 8.18b. The total force on each surface element can be resolved into a normal and a tangential components. The normal component is the pressure force, and the resultant in the direction of fluid motion is the *pressure drag*, F_{D_p}, which is given by

$$F_{D_p} = \int_A (p\mathrm{d}a) \sin\theta. \tag{8.4.115}$$

The tangential component is the frictional resistance, and the resultant in the direction of flow is called the *friction drag*, or alternatively the *skin friction*, F_{D_f}, which is given viz.,

$$F_{D_f} = \int_A (\tau\mathrm{d}a) \cos\theta. \tag{8.4.116}$$

The relative magnitude between two drag components depends to a great extent on the shape and orientation of immersed body. In view of Eq. (8.4.114), the pressure and skin friction may equally be expressed in terms of their drag coefficients.

While the friction drag, for a few simple cases, can be estimated by using the theory of boundary-layer flows, the pressure drag depends significantly on the separation of boundary layer and the wake region, as shown in Fig. 8.19 for a viscous flow past a circular cylinder as an example. In the front portion of cylinder where the flow is accelerated, the boundary-layer flow "adheres" to the cylinder surface. The

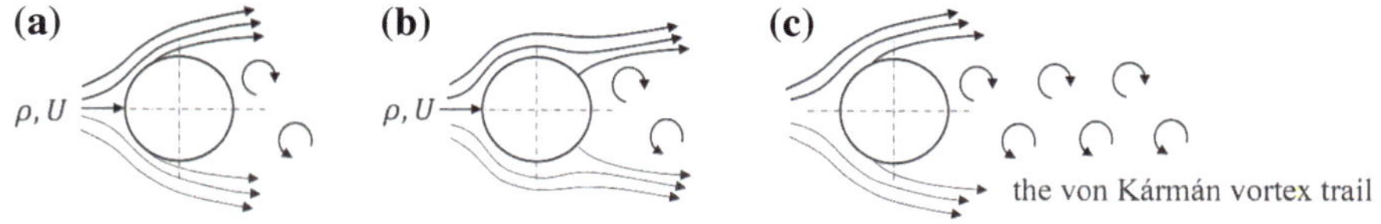

Fig. 8.19 Viscous flows past a two-dimensional circular cylinder. **a** A laminar boundary-layer flow with advanced separation points. **b** A turbulent boundary-layer flow with delayed points of separation. **c** The von Kármán vortex trail

pressure distribution is therefore nearly the same as that of an irrotational flow, since the accelerative action caused by the favorable pressure gradient along the surface is somewhat balanced by the decelerative action of viscous shear in the boundary layer. As soon as fluid moves into the region of deceleration in the rear portion of cylinder, an adverse pressure gradient causes the fluid streams to separate from the cylinder surface, with the flow patterns shown in Figs. 8.19a and b for a laminar and a turbulent boundary-layer flows, respectively. The difference between two patterns of streamlines lies in the locations of separation point. In Fig. 8.19a, the flow remains laminar up to the point of separation, while that in Fig. 8.19b is turbulent in the front portion of cylinder. Fluids in turbulent motion possess more momentum, so that they can move farther along the cylinder surface to make the way into the regions of higher pressure. The point of separation is therefore farther downstream toward the rear of cylinder than its counterpart in the laminar boundary-layer flow.

Downstream the separation point, the flow is characterized by the formation of turbulent eddies and vortices which persist for some distance until they are dissipated by the viscous action of fluid. The entire disturbed downstream region is called the *turbulent wakes*, and the main flow is diverted to the outside of wakes. Experiments show that the eddies formed at the separation points will be shed regularly in an alternating manner from the cylinder, so that at a greater distance downstream from the cylinder a regular pattern of vortices will be observed to move alternatively clockwise and counterclockwise, as shown in Fig. 8.19c. These vortices are usually referred to as the von Kármán vortex trail . The alternating shedding of eddying vortices produces periodic transverse forces on the cylinder which may cause transverse oscillation. If the natural vibration frequency of cylinder is in resonance with that of eddy formation, severe deflection and damage can result. It is this resonant phenomenon which gives rise to the aerodynamic instability of suspension bridges and tall chimneys.

Within the turbulent wakes downstream from the separation points, the mean fluid pressure is approximately the same as that at the points of separation. Since these points are usually in the region of high velocity and low pressure, the pressure behind the zone of separation is invariably lower than that at the front, as shown in Figs. 8.20a and b.[24] Thus, flow separation produces a net force in the direction of

[24]So that the wake region is called the "Totwassergebiet" in German Language.

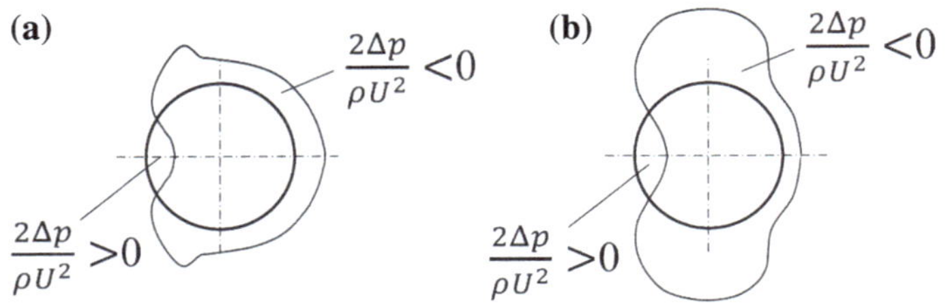

Fig. 8.20 Pressure distributions on the surface of a two-dimensional circular cylinder. **a** For a laminar boundary-layer flow. **b** For a turbulent boundary-layer flow

flow due to the pressure difference between the front and rear of cylinder. This force is commonly known as the *pressure* or *form drag*. The sum of form drag and friction drag which is estimated before the fluid reaches the separation points gives the total drag.

Since the form drag depends to a great extent on the geometric configuration of body and the location of separation point, a body may be so well-streamlined to reduce the zone of flow separation. Experimental studies reveal that at high Reynolds numbers the flow separation is limited to a very small region at the tail of streamlined body, so that the pressure distribution over the streamlined body surface is in good agreement with that determined by using an irrotational flow. Consequently, fairly accurate estimations on the friction and pressure drags for streamlined shapes can be accomplished separately by means of boundary-layer theory and theory of potential flows, respectively.

For many complex body shapes, flow separation occurs at an early section of the surface, and the total drag is the sum of the friction drag on the forward portion and the form drag due to the pressure difference caused by the flow separation and the subsequent turbulent wakes. Theoretical estimations on drag become difficult and often require detailed empirical data concerning the distributions of pressure and shear stress over the entire body surface. The data can be quoted from carefully experimental measurements. Under such circumstances, it is more feasible to measure the total drag on a scale model of the form in question in a wind tunnel, water tunnel, or towing tank.

Now turn to the lift. Since airfoils have been investigated very thoroughly, they are used to illustrate the general principles of dynamic lift. From the mathematical point of view, the theory of lift is intimately related to the circulation around the body.[25] The lift force experienced by an airfoil in a uniform stream equals the product of fluid density, stream velocity and circulation, and has a direction perpendicular to the stream velocity. Experiments show that the establishment of a circulation around an airfoil is accomplished by the formation of vortices of definite strength at the trailing edge of airfoil.

[25] It has been shown in Sect. 7.5.7 that the lift acting on a two-dimensional circular cylinder is given by the Kutta-Joukowski law.

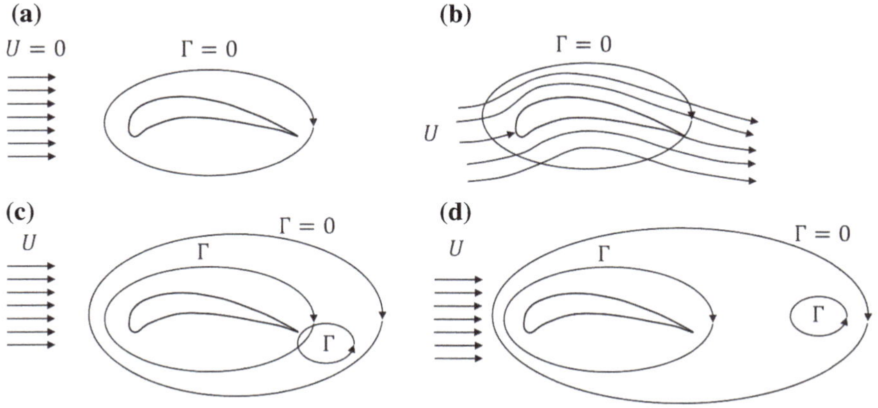

Fig. 8.21 Development of the body circulation Γ around an airfoil. **a** The first stage, in which no circulation exists. **b** The second stage, in which the flow is essentially irrotational, with vanishing value of Γ. **c** The third stage, in which the starting vortex at the trail develops, initiating a body circulation around the airfoil. **d** The fourth stage, in which the starting vortex leaves the airfoil, leaving the body circulation around the airfoil

Consider an airfoil shown in Fig. 8.21a, in which the flow motion just starts, and the circulation around the airfoil is simply null. As the uniform motion of fluid begins, the flow pattern is at first seen to be essentially irrotational, as shown in Fig. 8.21b, and there can be no circulation around the airfoil, yielding vanishing lift. This pattern of irrotational flow cannot persist too long, for the fluid layers that pass over the upper and lower surfaces of airfoil meet at the trailing edge with slightly different velocities, which results in the formation of a *surface of discontinuity*, across which there is a sharp velocity gradient. The fluid viscosity immediately causes the formation of a counterclockwise *starting vortex* which is shed from the trailing edge, as shown in Fig. 8.21c. In order to counterbalance the starting vortex with definite strength, a clockwise circulation with same strength must be set around the airfoil, as implied by Kelvin's theorem, for the initial circulation is null. This clockwise circulation around the airfoil is frequently referred to as the *body circulation*. As the strength of boundary circulation around the airfoil increases, the flow pattern changes until a steady state is eventually reached, in which the strengths of starting vortex at the trailing edge and boundary circulation around the airfoil attain a constant limiting value. The starting vortex then breaks away from the airfoil and moves downstream with the general fluid, leaving behind only the boundary circulation around the airfoil. A constant lift is thus set up on the airfoil, as indicated by the Kutta-Joukowski law, as shown in Fig. 8.21d. The starting vortex is instrumental in inducing a boundary circulation around the airfoil. Its subsequent history and eventual dissipation have no influence on the already existing boundary circulation.

Typical coefficients of drag (friction drag + pressure drag) and lift in relation with the angle of attack for a low-drag airfoil of infinite span is shown in Fig. 8.22a. The nearly linear relationship between C_L and α is representative for all normal airfoils

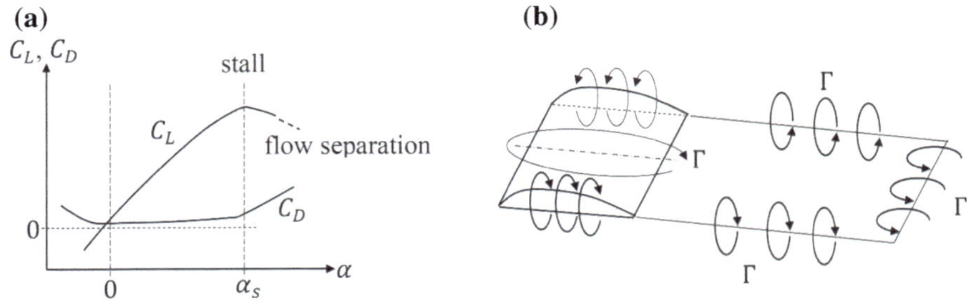

Fig. 8.22 Aerodynamic performance of a low-drag airfoil. **a** The drag coefficient C_D and lift coefficient C_L of an airfoil with infinite span in relation with the angle of attack α. **b** The vortices in the vicinity of an airfoil with finite span

at subsonic speeds. As the angle of attack increases, a condition known as *stall* will be reached owing to the separation of flow starting at the leading edge of airfoil. When the airfoil is stalled, the lift curve departs from the straight line, which is also accompanied by a rapid rise in the drag resulted from the boundary-layer separation and the subsequent large increase in the pressure drag. The previous discussions and results are based on two-dimensional airfoils, i.e., the span perpendicular to the page is infinitely long. For airfoils with finite span, the pressure on the underside is greater than that on the upper side, and fluids tend to escape around two ends in a direction from the below toward the top. There is thus a general flow outward from the center to the two ends along the underside of airfoil, turning upward around the ends and then inward from the two ends toward the center along the upper side. As a result, there are two so-called *tip vortex filaments trailing* from two ends of an airfoil with finite span, as shown in Fig. 8.22b.

According to the discussions in Sect. 4.4, the axis of boundary circulation around an airfoil with finite span cannot terminate at two ends, but must either extend to infinity or form a closed path. The closed path is a large *vortex ring* consisting of the finite airfoil, the axes of two tip vortices and the starting vortices, as shown in the figure. However, Kelvin's theorem still holds for this large vortex ring, since two tip vortices are of equal strength and opposite in direction. The total circulation around this large vortex ring still adds up to zero.

8.5 Buoyancy-Driven Flows

There exists a large class of fluid flows which is triggered by buoyancy. Buoyant force may result from a density variation in the presence of gravitational field. This section is devoted to the discussions on buoyancy-driven flows. The Boussinesq approximation to the full Navier-Stokes and thermal energy equations is introduced. The solutions to the approximate equations are presented for selected problems. The stability of a horizontal fluid layer is discussed to study the conditions required for the onset of thermal convection.

8.5.1 The Boussinesq Approximation

For incompressible viscous fluids, in which the gravity provides the only significant body force, the continuity and Navier-Stokes equations respectively read

$$\nabla \cdot \boldsymbol{u} = 0, \qquad \rho \frac{\partial \boldsymbol{u}}{\partial t} + \rho(\boldsymbol{u} \cdot \nabla)\boldsymbol{u} = -\nabla p + \mu \nabla^2 \boldsymbol{u} - \rho g \boldsymbol{e}_z, \tag{8.5.1}$$

in which the gravity is assumed to point along the negative z-axis with unit vector $\boldsymbol{e}_z$. For a static circumstance, the first equation holds identically, while the second equation reduces to

$$-\nabla p_0 - \rho_0 g \boldsymbol{e}_z = \boldsymbol{0}, \tag{8.5.2}$$

where p_0 and ρ_0 are the presenting pressure and density distributions under static equilibrium. It is assumed that the fluid motion is triggered by the buoyant force; hence, the pressure, density, and velocity during the convective motion are given respectively by

$$p = p_0 + p^*, \qquad \rho = \rho_0 + \rho^*, \qquad \boldsymbol{u} = \boldsymbol{0} + \boldsymbol{u}^*, \tag{8.5.3}$$

where p^* and ρ^* are respectively the pressure and density relative to their static values, and $\boldsymbol{u}^*$ is the fluid velocity triggered by the convective motion. Substituting these expressions into Eq. (8.5.1) yields

$$\nabla \cdot \boldsymbol{u}^* = 0, \quad (\rho_0 + \rho^*)\frac{\partial \boldsymbol{u}^*}{\partial t} + (\rho_0 + \rho^*)(\boldsymbol{u}^* \cdot \nabla)\boldsymbol{u}^* = -\nabla p^* + \mu \nabla^2 \boldsymbol{u}^* - \rho^* g \boldsymbol{e}_z, \tag{8.5.4}$$

in which Eq. (8.5.2) has been used. These equations are the local balances of mass and linear momentum for incompressible fluids with density variation, or alternatively for *stratified fluids* in which stratification in density takes place.

The *Boussinesq approximation* is accomplished by neglecting any density variation in the equations, except that in the gravitational term.[26] It is done so, for the density variation is assumed to play only the significant role in the body force, while it has a minor effect on the inertia force. This may be considered to be justified if a relatively small density difference exists over moderate distances. Hence, by considering ρ to be constant, the Boussinesq approximation to Eq. (8.5.4) is obtained as

$$\nabla \cdot \boldsymbol{u} = 0, \qquad \rho \frac{\partial \boldsymbol{u}}{\partial t} + \rho(\boldsymbol{u} \cdot \nabla)\boldsymbol{u} = -\nabla p + \mu \nabla^2 \boldsymbol{u} - (\Delta\rho) g \boldsymbol{e}_z, \tag{8.5.5}$$

in which the superscript $*$ has been removed for simplicity, and $\Delta\rho$ represents the density difference relative to the static density distribution, which is positive if the

[26] Joseph Valentin Boussinesq, 1842–1929, a French mathematician and physicist, who made contributions to the fields of hydrodynamics, vibration, light, and heat.

density is greater than its static value. Although the Boussinesq approximation is valid for a fluid with density variation, the fluid itself still remains incompressible. The same concept can equally be extended to compressible fluids, provided that the variation in density is small, and has negligible effects in all terms in the governing equations, except the gravitational term.

For demonstrations of the Boussinesq approximation, consider a density variation be caused by a temperature variation in a fluid, which is termed the *thermal convection*, with the density variation given by

$$\rho = \rho_0 \left[1 - \beta(T - T_0)\right], \tag{8.5.6}$$

where β is the *coefficient of thermal expansion* of fluid, which is a fluid property to be determined experimentally, and T_0 is the fluid temperature which presents at static equilibrium. The above expression is valid for a moderate departure of temperature T from its static value T_0 for an incompressible fluid.[27] Substituting this expression into Eq. (8.5.5) gives rise respectively to

$$\nabla \cdot \boldsymbol{u} = 0, \qquad \rho \frac{\partial \boldsymbol{u}}{\partial t} + \rho(\boldsymbol{u} \cdot \nabla)\boldsymbol{u} = -\nabla p + \mu \nabla^2 \boldsymbol{u} + \rho g \beta (T - T_0)\boldsymbol{e}_z, \tag{8.5.7}$$

in which $\Delta\rho = -\rho_0\beta(T - T_0) = -\rho\beta(T - T_0)$ has been used, and the density ρ is assumed to be constant, which equals its static value. These two equations govern the motion of a fluid in a thermal convection circumstance. They consist of four scalar equations for five unknowns, i.e., the velocity $\boldsymbol{u}$, pressure p, and temperature T. To arrive at a mathematically well-posed problem, an additional equation, namely the thermal energy equation, must be provided. Thus, the problem is a coupled thermo-mechanical system. It follows from the results in Sect. 5.6.4 that the appropriate form of the conservation of energy is given by

$$\rho \frac{\partial h}{\partial t} + \rho(\boldsymbol{u} \cdot \nabla)h = \frac{\partial p}{\partial t} + (\boldsymbol{u} \cdot \nabla)p + \nabla \cdot (k \nabla T) + \Phi, \qquad h = h(\rho, T), \tag{8.5.8}$$

where h is the specific enthalpy, Φ denotes the dissipation function, and ρ is a constant in the context of the Boussinesq approximation.

Although $h = h(\rho, T)$ in general, it can be demonstrated that $h = h(T) = c_p T$ for ideal gases, where c_p is the specific heat at constant pressure.[28] For the fluids which can be approximated as ideal gases, Eq. (8.5.8) is simplified to

$$\rho c_p \frac{\partial T}{\partial t} + \rho c_p (\boldsymbol{u} \cdot \nabla)T = \frac{\partial p}{\partial t} + (\boldsymbol{u} \cdot \nabla)p + \nabla \cdot (k \nabla T) + \Phi, \qquad h = c_p T, \tag{8.5.9}$$

[27] In general, the thermal equation of state of a Newtonian fluid is written as $\rho = \rho(p, T)$, which can be expanded as

$$\rho = \rho_0 + (p - p_0)\frac{\partial \rho}{\partial p}(T_0, p_0) + (T - T_0)\frac{\partial \rho}{\partial T}(p_0, T_0) + \cdots,$$

under a linear approximation in the context of the Taylor series expansion. For incompressible flows, the second term on the right-hand-side can be neglected, leading to that ρ is only a function of temperature.

[28] The derivations are provided in Sect. 11.8.5.

with the pressure p measured relative to its static value p_0. Equations (8.5.7) and (8.5.8) are the general Boussinesq approximation for incompressible fluids, while Eqs. (8.5.7) and (8.5.9) are the simplified formulation for the circumstances where the fluid density is only a function of temperature.

8.5.2 Boundary-Layer Approximation

Consider a vertical surface in contact with an incompressible fluid shown in Fig. 8.23, in which the fluid motion is driven by buoyancy. The circumstance is similar to a boundary-layer flow over a flat plate, and there exist two boundary layers: the velocity boundary layer with thickness δ, and the thermal boundary layer with thickness δ_t, which is assumed to be of the same order of magnitude as δ. The boundary-layer approximation to Eq. (8.5.7) is the same as that described in the last section, except that the buoyancy term acts along the x-direction, while that for Eq. (8.5.9) needs to be formulated. For two-dimensional steady flows with constant ρ, μ and k, Eq. (8.5.9) reads

$$\begin{aligned}\rho c_p\left(u\frac{\partial T}{\partial x}+v\frac{\partial T}{\partial y}\right)=&\,u\frac{\partial p}{\partial x}+v\frac{\partial p}{\partial y}+k\left(\frac{\partial^2 T}{\partial x^2}+\frac{\partial^2 T}{\partial y^2}\right)\\&+2\mu\left[\left(\frac{\partial u}{\partial x}\right)^2+\left(\frac{\partial v}{\partial y}\right)^2\right]+\mu\left(\frac{\partial u}{\partial y}+\frac{\partial v}{\partial x}\right)^2.\end{aligned} \tag{8.5.10}$$

It is observed that

- the two convective terms on the left-hand-side are of the same order of magnitude, as was the case for the convection of linear momentum in the boundary layer;
- on the right-hand-side the pressure gradient across the boundary layer is negligible small, as implied by the momentum transportation in the boundary layer along the y-direction;
- the heat conduction component with the second derivative of y is considerably larger than that with respect to x, as similar to the viscous terms in the boundary layer; and
- the dissipation function Φ may be neglected for moderate velocities induced by thermal convection.

With these, Eq. (8.5.10) is simplified to

$$\rho c_p\left(u\frac{\partial T}{\partial x}+v\frac{\partial T}{\partial y}\right)=u\frac{\partial p}{\partial x}+k\frac{\partial^2 T}{\partial y^2}, \tag{8.5.11}$$

so that a set of equations for buoyancy-driven thermal convections in the context of boundary-layer approximation consists of the continuity equation, momentum equation along the x-direction, and the simplified thermal energy equation. These

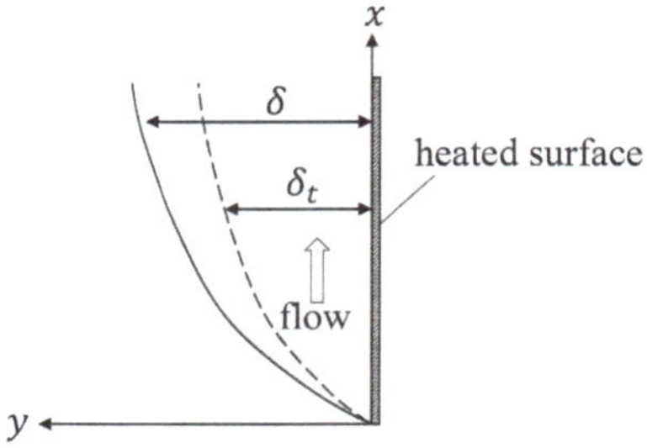

Fig. 8.23 Velocity and thermal boundary layers on a vertical heated surface

equations are summarized in the following for convenience:

$$\begin{aligned}
&\frac{\partial u}{\partial x}+\frac{\partial v}{\partial x}=0,\\
&u\frac{\partial u}{\partial x}+v\frac{\partial u}{\partial y}=-\frac{1}{\rho}\frac{\mathrm{d}p}{\mathrm{d}x}+\nu\frac{\partial^2 u}{\partial y^2}+g\beta(T-T_0),\\
&u\frac{\partial T}{\partial x}+v\frac{\partial T}{\partial y}=\frac{1}{\rho c_p}\left(u\frac{\mathrm{d}p}{\mathrm{d}x}\right)+\kappa\frac{\partial^2 T}{\partial y^2},
\end{aligned} \tag{8.5.12}$$

where $\kappa = k/(\rho c_p)$, called the *thermal diffusivity* of fluid. These equations are to be solved subject to the no-slip boundary condition on the surface $y = 0$, and to the condition that the velocity should vanish far from the heated surface. In addition, either the temperature or heat flux on the heated surface needs to be prescribed as an additional boundary condition.

8.5.3 Flows by Isothermal Vertical Surface

Consider a vertical flat plate shown in Fig. 8.23 again, in which the plate is assumed to have a constant temperature T_s, which is larger than the ambient constant fluid temperature T_0. Since the flow is induced by buoyancy, the pressure gradient in the x-direction is negligible, so that Eq. (8.5.12) reduces to

$$\frac{\partial u}{\partial x}+\frac{\partial v}{\partial y}=0,\quad u\frac{\partial u}{\partial x}+v\frac{\partial v}{\partial y}=\nu\frac{\partial^2 u}{\partial y^2}+g\beta(T-T_0),\quad u\frac{\partial T}{\partial x}+v\frac{\partial T}{\partial y}=\kappa\frac{\partial^2 T}{\partial y^2}, \tag{8.5.13}$$

which are subject to the boundary conditions given by

$$\begin{aligned}
&u(x,0)=0,\quad v(x,0)=0,\qquad u(x,y\to\infty)\to 0;\\
&T(x,0)=T_s,\, T(x,y\to\infty)\to T_0.
\end{aligned} \tag{8.5.14}$$

Introducing the stream function $\psi(x, y)$ to replace the velocity components u and v, and the dimensionless temperature difference $\theta(x, y)$ given by

$$\theta(x,y)=\frac{T(x,y)-T_0}{T_s-T_0}, \tag{8.5.15}$$

and substituting these into Eqs. $(8.5.13)_{2,3}$ yields respectively

$$\frac{\partial \psi}{\partial y}\frac{\partial^2 \psi}{\partial x \partial y} - \frac{\partial \psi}{\partial x}\frac{\partial^2 \psi}{\partial y^2} = \nu \frac{\partial^3 \psi}{\partial y^3} + g\beta(T_s - T_0)\theta, \qquad \frac{\partial \psi}{\partial y}\frac{\partial \theta}{\partial x} - \frac{\partial \psi}{\partial x}\frac{\partial \theta}{\partial y} = \kappa \frac{\partial^2 \theta}{\partial y^2}. \tag{8.5.16}$$

By using the methods used in the last section, a similarity solution is sought in the form

$$\psi(x, y) = C_1 x^m f(\eta), \qquad \theta(x, y) = F(\eta), \qquad \eta(x, y) = C_2 \frac{y}{x^n}, \tag{8.5.17}$$

where C_1 and C_2 are constants, whose values should render the functions f, F, and η dimensionless, and m and n are undetermined exponents. Substituting these expressions into Eq. (8.5.16) gives

$$C_1^2 C_2^2 x^{2(m-n)-1}\left[(m-n)\left(f'\right)^2 - mff''\right] = \nu C_1 C_2^3 x^{m-3n} f''' + g\beta(T_s - T_0)F, \tag{8.5.18}$$
$$-mC_1 x^{m-n-1} fF' = \kappa C_2 x^{-2n} F'',$$

where the primes denote differentiations with respect to η. If two equations are to be reduced to a pair of ordinary differential equations, the powers of x in the first equation must be zero, while those on the two sides of second equation must be the same. It follows that

$$m = \frac{3}{4}, \qquad n = \frac{1}{4}, \tag{8.5.19}$$

with which Eq. (8.5.18) becomes

$$f''' + \frac{C_1}{4\nu C_2}\left[3ff'' - 2\left(f'\right)^2\right] + \frac{g\beta(T_s - T_0)}{\nu C_1 C_2^3}F = 0, \qquad F'' + \frac{3}{4}\frac{C_1}{\kappa C_2} fF' = 0. \tag{8.5.20}$$

The constants C_1 and C_2 should be so chosen that the functions f, F, and η are dimensionless. By using dimensionality consideration, they are prescribed as

$$C_1 = \frac{\nu}{4}\left[\frac{4g\beta(T_s - T_0)}{\nu^2}\right]^{1/4}, \qquad C_2 = \left[\frac{4g\beta(T_s - T_0)}{\nu^2}\right]^{1/4}, \tag{8.5.21}$$

with which Eq. (8.5.20) is simplified to

$$f''' + 3ff'' - 2\left(f'\right)^2 + F = 0, \qquad F'' + 3\mathcal{P}_r fF' = 0, \tag{8.5.22}$$

where $\mathcal{P}_r$ is the Prandtl number given by $\mathcal{P}_r = \nu/\kappa$, which assumes the values of about 0.7 and about 7.0 for air and water, respectively. The boundary conditions associated with Eq. (8.5.22) are given by

$$f(0) = f'(0) = 0, \;\; f'(\eta \to \infty) \to 0; \qquad F(0) = 1, \;\; F(\eta \to \infty) \to 0. \tag{8.5.23}$$

Once the solutions to the formulations given in Eqs. (8.5.22) and (8.5.23) are obtained numerically, the solutions to the stream function and dimensionless temperature difference can be determined as

$$\psi(x, y) = \frac{\nu}{4}\left[\frac{4g\beta(T_s - T_0)}{\nu^2}\right]^{1/4} x^{3/4} f(\eta), \quad \theta(x, y) = F(\eta),$$
$$\eta = \left[\frac{4g\beta(T_s - T_0)}{\nu^2}\right]^{1/4} \frac{y}{x^{1/4}}. \tag{8.5.24}$$

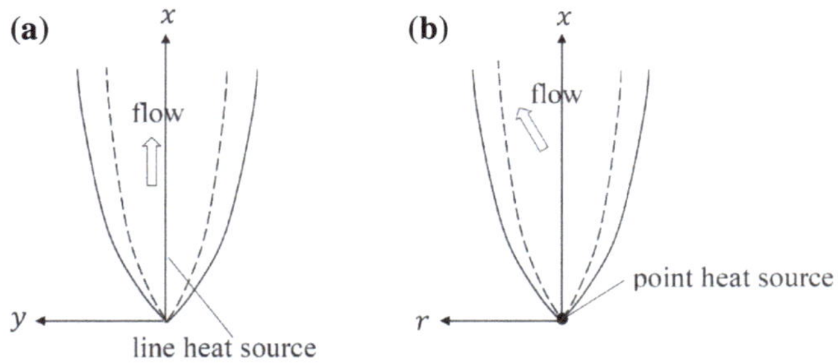

Fig. 8.24 Two-dimensional buoyancy-driven flows. **a** A flow by a line source of heat. **b** A flow by a point source of heat. Solid lines: velocity boundary layers; dashed lines: thermal boundary layers

The formulated problem has been solved by Pohlhausen for $\mathcal{P}_r = 0.733$. It has been found that the rate at which the convective heat transfer takes place between the vertical surface and ambient fluid is determined by the dimensionless number

$$\mathcal{N}_u = \frac{h\ell}{k} = 0.359\,(\mathcal{G}_r)^{1/4}, \qquad \mathcal{G}_r = \frac{g\ell^3(T_s - T_0)}{\nu^2 T_0}, \tag{8.5.25}$$

where $\mathcal{N}_u$ and $\mathcal{G}_r$ are respectively the Nusselt and Grashof numbers,[29] h is the rate of heat transfer per unit area per unit time between the plate and fluid, and ℓ denotes the surface length of plate, over which the heat transfer takes place. The Nusselt number is interpreted as the dimensionless heat transfer, while the Grashof number is the dimensionless temperature differential which triggers the thermal convection.

8.5.4 Flows by Line and Point Sources of Heat

Consider a line source of heat shown in Fig. 8.24a, which is immersed in an otherwise stationary fluid. The governing equations for the considered buoyancy-driven flow are the same as those in the last subsection, except that the boundary conditions are different, which are given by

$$\begin{aligned}&\frac{\partial u}{\partial y}(x,0) = 0, \quad v(x,0) = 0, \quad \int_{-\infty}^{\infty} \rho u c_p (T - T_0)\mathrm{d}y = Q, \quad \frac{\partial T}{\partial y}(x,0) = 0,\\ &T(x, y \to \pm\infty) \to T_0,\end{aligned} \tag{8.5.26}$$

in which Q is the total amount of heat leaving the line heat source per unit time per unit length. The first and fourth conditions result from that the x-axis is a symmetric line, the third condition ensures that the total heat rising from the heat source is the same at all streamwise locations, while the other conditions are the same as the previous case. Since there is no real physical surface in the considered problem, the

[29]Ernst Kraft Wilhelm Nusselt, 1882–1957; Franz Grashof, 1826–1893, both are German engineers.

surface temperature needs not to be normalized, and the appropriate dimensionless temperature is proposed as

$$\theta(x, y) = \beta[T(x, y) - T_0], \tag{8.5.27}$$

with which Eq. (8.5.16) becomes

$$\frac{\partial \psi}{\partial y}\frac{\partial^2 \psi}{\partial x \partial y} - \frac{\partial \psi}{\partial x}\frac{\partial^2 \psi}{\partial y^2} = \nu \frac{\partial^3 \psi}{\partial y^3} + g\theta, \qquad \frac{\partial \psi}{\partial y}\frac{\partial \theta}{\partial x} - \frac{\partial \psi}{\partial x}\frac{\partial \theta}{\partial y} = \kappa \frac{\partial^2 \theta}{\partial y^2}. \tag{8.5.28}$$

A similarity solution is sought in the form

$$\psi(x, y) = C_1 x^m f(\eta), \qquad \eta(x, y) = C_2 \frac{y}{x^n}, \qquad \theta(x, y) = C_3 x^r F(\eta), \tag{8.5.29}$$

where m, n, and r are undermined exponents, with C_1-C_3 being constants rendering the functions f, F, and η dimensionless. Substituting these expressions into Eq. (8.5.28) yields respectively

$$\begin{aligned} &C_1^2 C_2^2 x^{2(m-n)-1}\left[(m-n)\left(f'\right)^2 - m f f''\right] = \nu C_1 C_2^3 x^{m-3n} f''' + g C_3 x^r F, \\ &C_1 C_2 C_3 x^{m-n+r-1}\left(r f' F - m f F'\right) = \kappa C_2^2 C_3 x^{r-2n} F''. \end{aligned} \tag{8.5.30}$$

If a similarity solution exists, the x-dependence in two equations must cancel, giving rise to

$$m = \frac{3+r}{4}, \qquad n = \frac{1-r}{4}. \tag{8.5.31}$$

It is verified that for the special case $r = 0$, the solution obtained in Sect. 8.5.3 is recovered. On the other hand, substituting Eq. (8.5.29) into Eq. (8.5.26)$_3$ yields

$$Q = \int_{-\infty}^{\infty} \rho c_p \frac{\theta}{\beta}\frac{x^n}{C_2}\frac{\partial \psi}{\partial y} \mathrm{d}\eta = C_1 C_3 \frac{\rho c_p}{\beta} x^{m+r} \int_{-\infty}^{\infty} f' F \mathrm{d}\eta, \tag{8.5.32}$$

for the integration is carried out in a plane with $x =$ constant, so that $\mathrm{d}y$ is proportional to $x^n \mathrm{d}\eta$. Since the quantity Q should be independent of x, it follows that $m + r$ should be null. This condition, together with Eq. (8.5.31), determines the values of m, n, and r given by

$$m = \frac{3}{5}, \qquad n = \frac{2}{5}, \qquad r = -\frac{3}{5}, \tag{8.5.33}$$

with which Eq. (8.5.30) becomes

$$f''' + \frac{C_1}{5\nu C_2}\left[3 f f'' - \left(f'\right)^2\right] + \frac{g C_3}{\nu C_1 C_2^3} F = 0, \qquad F'' + \frac{3C_1}{5\kappa C_2}\frac{\mathrm{d}}{\mathrm{d}\eta}(f F) = 0. \tag{8.5.34}$$

In order to render f, F, and η dimensionless, the constants C_1-C_3 are chosen to be

$$C_1 = \nu\left(\frac{\beta Q}{\rho \nu c_p}\frac{g}{\nu^2}\right)^{1/5}, \quad C_2 = \frac{1}{5}\left(\frac{\beta Q}{\rho \nu c_p}\frac{g}{\nu^2}\right)^{1/5}, \quad C_3 = \nu\left(\frac{\rho^4 \nu^4 c_p^4}{\beta^4 Q^4}\frac{g}{\nu^2}\right)^{-1/5}, \tag{8.5.35}$$

with which Eq. (8.5.34) is simplified to

$$f''' + 3 f f'' - \left(f'\right)^2 + F = 0, \qquad F'' + 3\mathcal{P}_r \frac{\mathrm{d}}{\mathrm{d}\eta}(f F) = 0, \tag{8.5.36}$$

which are subject to the boundary conditions given by

$$f(0) = f''(0) = 0, \quad \int_{-\infty}^{\infty} f' F \mathrm{d}\eta = 1, \quad F'(0) = 0, \quad F(\eta \to \pm\infty) \to 0. \tag{8.5.37}$$

The solutions to Eq. (8.5.36) are of the forms

$$f(\eta) = A \tanh(\alpha\eta), \qquad F(\eta) = B \operatorname{sech}^2(\alpha\eta), \tag{8.5.38}$$

where A and B are constants. Substituting these expressions into Eqs. (8.5.36) and (8.5.37) gives

$$\alpha = \frac{5}{6}A, \quad B = \frac{50}{27}A^4; \qquad \alpha = 3\mathcal{P}_r A, \quad B = \frac{3}{4}A, \tag{8.5.39}$$

which cannot be fulfilled simultaneously by the constants α, A, and B alone; a specific value of $\mathcal{P}_r$ is required. For example, the solutions to the above equations may be obtained as

$$\mathcal{P}_r = \frac{5}{18}, \quad \longrightarrow \quad A = \left(\frac{81}{200}\right)^{1/5}, \quad B = \frac{3}{4}\left(\frac{200}{81}\right)^{1/5}, \quad \alpha = \frac{5}{6}\left(\frac{81}{200}\right)^{1/5}. \tag{8.5.40}$$

Hence, a similarity solution can be found for a specific value of $\mathcal{P}_r$, and the corresponding stream function and dimensionless temperature differential are obtained as

$$\begin{aligned} \psi(x, y) &= \frac{6\alpha\nu}{5}\left(\frac{\beta Q}{\rho\nu c_p}\frac{g}{\nu^2}\right)^{1/5} x^{3/5} \tanh(\alpha\eta), \\ \theta(x, y) &= \frac{5}{8\alpha}\left(\frac{\beta^4 Q^4}{\rho^4\nu^4 c_p^4}\frac{\nu^2}{g}\right)^{1/5} x^{-3/5} \operatorname{sech}^2(\alpha\eta), \end{aligned} \tag{8.5.41}$$

with

$$\eta(x, y) = \frac{1}{5}\left(\frac{\beta Q}{\rho\nu c_p}\frac{g}{\nu^2}\right)^{1/5} \frac{y}{x^{2/5}}, \qquad \alpha = \frac{5}{6}\left(\frac{81}{200}\right)^{1/5}. \tag{8.5.42}$$

It follows that along the line source of heat, the temperature differential $[T(x, 0) - T_0]$ varies as $x^{-3/5}$.

Now consider a buoyancy-driven flow induced by a point source of heat shown in Fig. 8.24b, in which there exists an angular symmetry abut the x-axis. So, it is more convenient to use the cylindrical coordinate system (r, θ, x) to describe the fluid motion, with which the corresponding forms of Eq. (8.5.13) are given by

$$\begin{aligned} &\frac{\partial}{\partial x}(ru) + \frac{\partial}{\partial r}(ru_r) = 0, \\ &u\frac{\partial u}{\partial x} + u_r\frac{\partial u}{\partial r} = \frac{\nu}{r}\frac{\partial}{\partial r}\left(r\frac{\partial u}{\partial r}\right) + g\beta(T - T_0), \\ &u\frac{\partial T}{\partial x} + u_r\frac{\partial T}{\partial r} = \frac{\kappa}{r}\frac{\partial}{\partial r}\left(r\frac{\partial T}{\partial r}\right), \end{aligned} \tag{8.5.43}$$

where u and u_r are the velocity components in the x- and r-directions, respectively, and the r-coordinate is perpendicular to the x-coordinate. To obtain a solution to these equations, the Stokes stream function $\psi(r, x)$ and dimensionless temperature $\theta(r, x)$ are proposed as

$$ru = \frac{\partial \psi}{\partial r}, \quad ru_r = -\frac{\partial \psi}{\partial x}; \qquad \theta = \beta(T - T_0). \tag{8.5.44}$$

Substituting these expressions into Eqs. $(8.5.43)_{2,3}$ yields respectively

$$\begin{aligned} &\frac{1}{r}\frac{\partial \psi}{\partial r}\frac{\partial}{\partial x}\left(\frac{1}{r}\frac{\partial \psi}{\partial r}\right) - \frac{1}{r}\frac{\partial \psi}{\partial x}\frac{\partial}{\partial r}\left(\frac{1}{r}\frac{\partial \psi}{\partial r}\right) = \frac{\nu}{r}\frac{\partial}{\partial r}\left[r\frac{\partial}{\partial r}\left(\frac{1}{r}\frac{\partial \psi}{\partial r}\right)\right] + g\beta(T - T_0), \\ &\frac{\partial \psi}{\partial r}\frac{\partial T}{\partial x} - \frac{\partial \psi}{\partial x}\frac{\partial T}{\partial r} = \kappa\frac{\partial}{\partial r}\left(r\frac{\partial T}{\partial r}\right), \end{aligned} \tag{8.5.45}$$

where Eq. $(8.5.43)_1$ holds identically. The associated boundary conditions are given by

$$\begin{aligned} &\frac{\partial u}{\partial r}(x, 0) = \frac{\partial}{\partial r}\left(\frac{1}{r}\frac{\partial \psi}{\partial r}\right)\Big|_{r=0} = 0, \quad u_r(x, 0) = -\frac{1}{r}\left(\frac{\partial \psi}{\partial x}\right)\Big|_{r=0} = 0, \\ &\int_0^\infty \rho u c_p (T - T_0) 2\pi r \,\mathrm{d}r = \int_0^\infty 2\pi\rho c_p \frac{\theta}{\beta}\frac{\partial \psi}{\partial r}\mathrm{d}r = Q, \\ &\frac{\partial T}{\partial r}(x, 0) = \frac{1}{\beta}\left(\frac{\partial \theta}{\partial r}\right)\Big|_{r=0} = 0, \quad T(x, r \to \pm\infty) \to T_0, \quad \theta(x, r \to \pm\infty) \to 0, \end{aligned} \tag{8.5.46}$$

which result essentially from the symmetric configuration of flow field.

The solutions to the system of differential equations are sought of the forms

$$\psi(x, r) = C_1 x^m f(\eta), \qquad \eta(x, r) = C_2 \frac{r}{x^n}, \qquad \theta(x, r) = C_3 x^r F(\eta), \tag{8.5.47}$$

where m, n, r are undetermined exponents, and C_1-C_3 are constants, which render the functions f, η, and F dimensionless. Substituting these expressions into Eq. (8.5.45) leads to

$$m = 1, \qquad 4n + r = 1, \tag{8.5.48}$$

while Eq. $(8.5.46)_3$ requires that

$$2\pi\rho C_1 C_3 \frac{c_p}{\beta} x^{m+r} \int_0^\infty f' F \mathrm{d}\eta = Q. \tag{8.5.49}$$

Since the quantity Q must be independent of x, it follows that $(m + r) = 0$. Combining this with Eq. (8.5.48) gives

$$m = 1, \qquad n = \frac{1}{2}, \qquad r = -1, \tag{8.5.50}$$

with which Eq. (8.5.45) becomes

$$f''' - \left(1 - \frac{C_1}{\nu} f\right)\frac{\mathrm{d}}{\mathrm{d}\eta}\left(\frac{f'}{\eta}\right) + \frac{g C_3}{\nu C_1 C_2^4}\eta F = 0, \qquad F' + \frac{C_1}{\kappa\eta} f F = 0. \tag{8.5.51}$$

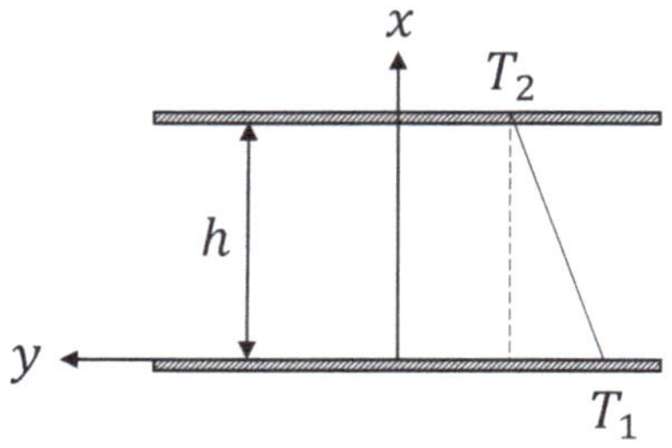

Fig. 8.25 A horizontal fluid layer between two parallel surfaces with different temperatures

In order to simplify the coefficients in these differential equations and Eq. (8.5.49), the constants C_1-C_3 are chosen to be

$$C_1 = \nu, \qquad C_2 = \left(\frac{\beta Q}{\rho \nu c_p}\right)^{1/4} \left(\frac{g}{\nu^2}\right)^{1/4}, \qquad C_3 = \frac{\beta Q}{\rho \nu c_p}, \tag{8.5.52}$$

with which Eq. (8.5.51) is simplified to

$$f''' - (1-f)\frac{\mathrm{d}}{\mathrm{d}\eta}\left(\frac{f'}{\eta}\right) + \eta F = 0, \qquad F' + \mathcal{P}_r \frac{fF}{\eta} = 0, \tag{8.5.53}$$

in which Eq. $(8.5.51)_2$ has been integrated once and Eq. $(8.5.46)_4$ has been used. The boundary conditions associated with two ordinary differential equations are thus given by

$$f(0) = f'(0) = F'(0) = 0, \qquad \int_0^\infty f' F \mathrm{d}\eta = \frac{1}{2\pi}. \tag{8.5.54}$$

Equations (8.5.53) and (8.5.54) define a boundary-value problem. The solutions to the formulated problem, with $\mathcal{P}_r = 1$, are of the forms

$$f(\eta) = A\frac{\eta^2}{a+\eta^2} \qquad F(\eta) = B\frac{1}{(a+\eta^2)^3}, \tag{8.5.55}$$

where $\{a, A, B\}$ are undetermined constants. By using the solution procedure described previously, these constants are determined to be

$$A = 6, \qquad a = 12\sqrt{2\pi}, \qquad B = \frac{(12\sqrt{2\pi})^3}{3\pi}. \tag{8.5.56}$$

The subsequent expressions of $\psi(x,r)$, $\theta(x,r)$, and $\eta(x,r)$ can be obtained by substituting the above expressions and Eq. (8.5.52) into Eq. (8.5.47). The derivations of Eq. (8.5.56) and the resulting expressions of $f(\eta)$ and $F(\eta)$ are left as an exercise.

8.5.5 Stability of a Horizontal Layer

Consider a horizontal fluid layer between two parallel plates, as shown in Fig. 8.25. The fluid layer is initially at rest, and two plates assume different constant temperatures, e.g. T_1 for the lower plate and T_2 for the upper plate with $T_2 < T_1$. The fluid is then either heated from below, or cooled from above, and a buoyant force takes place which results in a convective motion of the fluid layer. There exists a heat flux from the lower plate through the fluid layer toward the upper plate. Suppose that while the fluid is still at rest, a small-amplitude disturbance is introduced to the fluid. If the viscous force acting on the disturbing motion exceeds the buoyant force, the disturbance will decay and the motion will cease, and vice versa. These observations suggest that a stability analysis could identify the minimum value of buoyant force, below which no fluid motion can be triggered.

The equations governing the depicted circumstance are unsteady and three-dimensional. By following the Boussinesq approximation, the variation in density is considered to be important only in the gravitational term. It is further assumed that the fluid properties are constant, and the dissipation function and pressure variations in the thermal energy equation are neglected for simplicity. With these, the local balances of mass, linear momentum, and energy are given respectively by

$$\nabla\cdot\boldsymbol{u}=0,\quad \frac{\partial\boldsymbol{u}}{\partial t}+(\boldsymbol{u}\cdot\nabla)\boldsymbol{u}=-\frac{1}{\rho}\nabla p+\nu\nabla^2\boldsymbol{u}-g\,[1-\beta(T-T_0)]\,\boldsymbol{e}_x,$$
$$\frac{\partial T}{\partial t}+(\boldsymbol{u}\cdot\nabla)T=\kappa\nabla^2 T, \tag{8.5.57}$$

where $\boldsymbol{e}_x$ is the unit vector along the x-direction. In parallel, the geometric configuration implies that the static temperature distribution, $T_s(x)$, may be expressed as

$$T_s(x)=T_1-(T_1-T_2)\,\frac{x}{h}. \tag{8.5.58}$$

Substituting this expression into Eq. $(8.5.57)_2$ yields

$$0=-\frac{1}{\rho}\frac{\mathrm{d}p_0}{\mathrm{d}x}-g\left\{1-\beta\left[(T_1-T_0)-(T_1-T_2)\frac{x}{h}\right]\right\}, \tag{8.5.59}$$

for the fluid is initially at rest before the introduction of disturbance, where p_0 is the pressure distribution in the stationary state, and the density is evaluated at the reference temperature T_0. Now, let the disturbance be introduced, and the field quantities are assumed to be perturbed as

$$\boldsymbol{u}(x,y,z,t)=\boldsymbol{0}+\boldsymbol{u}'(x,y,z,t),\qquad p(x,y,z,t)=p_0(x)+p'(x,y,z,t),$$
$$T(x,y,z,t)=T_s(x)+T'(x,y,z,t), \tag{8.5.60}$$

where the primes represent perturbations, whose magnitudes are assumed to be small. Substituting these expressions and Eq. (8.5.58) into Eq. (8.5.57) yields respectively

$$\nabla\cdot\boldsymbol{u}'=0,\quad \frac{\partial\boldsymbol{u}'}{\partial t}=-\frac{1}{\rho}\nabla p'+\nu\nabla^2\boldsymbol{u}'+g\beta T'\boldsymbol{e}_x,\quad \frac{\partial T'}{\partial t}-u'\frac{(T_1-T_2)}{h}=\kappa\nabla^2 T', \tag{8.5.61}$$

in which the products of primed quantities are neglected in the context of linear approximation. Taking curl of Eq. $(8.5.61)_2$ yields

$$\left(\nabla^2-\frac{1}{\nu}\frac{\partial}{\partial t}\right)\nabla^2\boldsymbol{u}'=-\frac{g\beta}{\nu}\left(\boldsymbol{e}_x\nabla^2-\nabla\frac{\partial}{\partial x}\right)T', \tag{8.5.62}$$

in which

$$\nabla\times(\nabla\times\boldsymbol{u}')=\nabla(\nabla\cdot\boldsymbol{u}')-\nabla^2\boldsymbol{u}'=-\nabla^2\boldsymbol{u}', \tag{8.5.63}$$

and Eq. $(8.5.61)_1$ have been used. The y- and z-components of Eq. (8.5.62) may be eliminated by taking dot product with $\boldsymbol{e}_x$. As a result, only the perturbed velocity component u' exists, for which Eqs. (8.5.62) and $(8.5.61)_3$ reduce respectively to

$$\begin{aligned} \left(\nabla^2 - \frac{1}{\nu}\frac{\partial}{\partial t}\right)\nabla^2 u' &= -\frac{g\beta}{\nu}\left(\nabla^2 - \nabla\frac{\partial}{\partial x^2}\right)T', \\ \left(\nabla^2 - \frac{1}{\kappa}\frac{\partial}{\partial t}\right)T' &= -\frac{(T_1 - T_2)}{\kappa h}u'. \end{aligned} \tag{8.5.64}$$

Since the disturbances u' and T' are arbitrary in form, they may be expressed by the Fourier integrals in the y- and z-directions. Specifically, the Fourier integrals of u' and T' are given respectively by

$$\begin{aligned} u'(x, y, z, t) &= \int_{-\infty}^{\infty}\int_{-\infty}^{\infty} U'(x, t)\exp\left[i\left(k_y y + k_z z\right)\right]\mathrm{d}k_y\mathrm{d}k_z, \\ T'(x, y, z, t) &= \int_{-\infty}^{\infty}\int_{-\infty}^{\infty} \theta'(x, t)\exp\left[i\left(k_y y + k_z z\right)\right]\mathrm{d}k_y\mathrm{d}k_z, \end{aligned} \tag{8.5.65}$$

where U' and θ' are simply the integrands. Substituting these expressions into Eq. (8.5.64) gives

$$\begin{aligned} \left(\frac{\partial^2}{\partial x^2} - k^2 - \frac{1}{\nu}\frac{\partial}{\partial t}\right)\left(\frac{\partial^2}{\partial x^2} - k^2\right)U' &= \frac{g\beta}{\nu}k^2\theta', \\ \left(\frac{\partial^2}{\partial x^2} - k^2 - \frac{1}{\kappa}\frac{\partial}{\partial t}\right)\theta' &= -\frac{(T_1 - T_2)}{\kappa h}U', \end{aligned} \tag{8.5.66}$$

for the resulting equations should be valid for all wavelengths of disturbance, where $k^2 = k_y^2 + k_z^2$. Since the coefficients in Eq. (8.5.66) are constant, it follows that the solutions to U' and θ' are of the forms

$$U'(x, t) = U(x)\exp\left(\frac{\sigma\kappa}{h^2}t\right), \qquad \theta'(x, t) = \theta(x)\exp\left(\frac{\sigma\kappa}{h^2}t\right), \tag{8.5.67}$$

where σ is a parameter, with $\sigma\kappa/h^2$ a dimensionless parameter representing the time required for heat to diffuse across the fluid layer, and $U(x)$ and $\theta(x)$ are two undetermined functions. Incorporating these expressions into Eq. (8.5.65) results in

$$\left(D^2 - \alpha^2 - \frac{\sigma}{\mathcal{P}_r}\right)\left(D^2 - \alpha^2\right)U = \frac{g\beta}{\nu h^2}\alpha^2\theta, \quad \left(D^2 - \alpha^2 - \sigma\right)\theta = -\frac{(T_1 - T_2)}{\kappa h}U, \tag{8.5.68}$$

with

$$\alpha = hk, \qquad D = h\frac{\mathrm{d}}{\mathrm{d}x}, \tag{8.5.69}$$

where α is the dimensionless wave number and D represents the dimensionless derivative with respect to x. Eliminating θ from two equations gives rise to

$$\begin{aligned} &\left[\left(D^2 - \alpha^2\right)\left(D^2 - \alpha^2 - \sigma\right)\left(D^2 - \alpha^2 - \frac{\sigma}{\mathcal{P}_r}\right) + \alpha^2\mathcal{R}_a\right]U = 0, \\ &\mathcal{R}_a = \frac{gh^3\beta(T_1 - T_2)}{\kappa\nu}, \end{aligned} \tag{8.5.70}$$

which is the stability equation of considered circumstance, where $\mathcal{R}_a$ is the Rayleigh number with $\mathcal{R}_a = \mathcal{P}_r \mathcal{G}_r$. It is a measure of the strength of buoyant force which initiates a convective motion. The boundary conditions associated with Eq. (8.5.70), in view of Fig. 8.25, are given by

$$U|_{x/h=0,1} = DU|_{x/h=0,1} = D^2\left(D^2 - 2\alpha^2 - \frac{\sigma}{\mathcal{P}_r}\right)U|_{x/h=0,1} = 0. \tag{8.5.71}$$

These conditions result from that the perturbed velocity vanishes at $x = 0$ and $x = h$, which can be fulfilled by requiring $DU = 0$, as implied by the continuity equation, and are also based on the fact that the perturbed temperature should vanish at the same locations of x, as implied by Eq. $(8.5.68)_1$. Equations (8.5.70) and (8.5.71) construct an eigenvalue problem. For given values of $\mathcal{R}_a$, α, and $\mathcal{P}_r$, the eigenvalues will be the time coefficient σ, which satisfy the conditions described above. If the value of α changes, different values of σ will be obtained, whose largest real value will define the Fourier component of the disturbance which is the fast growing.

The minimum value of buoyant force for the onset of thermal convection corresponds to the wavelength of the fastest-growing component with $\sigma = 0$, and all other components will be decaying. Thus, at the onset of instability, the time coefficient in Eqs. (8.5.70) and (8.5.71) will be null, so that

$$\begin{aligned} &\left[\left(D^2 - \alpha^2\right)^3 + \alpha^2\mathcal{R}_a\right]U = 0, \\ &U|_{x/h=0,1} = DU|_{x/h=0,1} = \left(D^2 - \alpha^2\right)^2 U|_{x/h=0,1} = 0, \end{aligned} \tag{8.5.72}$$

must be fulfilled, which shows that the eigenvalue becomes the Rayleigh number. The minimum value of $\mathcal{R}_a$, with respect to α, is referred to as the *critical Rayleigh number*, which corresponds to the magnitude of smallest temperature gradient by which all disturbances (i.e., all possible wave numbers) will decay rather than grow in time to produce convective motion. For the problem described by Eq. (8.5.72), a solution yields a value of 1707.8 for the critical Rayleigh number. If one of the boundaries is free, this value is identified to be 1100.7, and changes to 657.5 for two free boundaries.

8.6 Turbulent Pipe-Flows

A brief description of the characteristics of turbulent flows is dealt with in this section. In turbulent flows, all physical quantities experience fluctuations in the values. The fluctuating quantities may combine with each other, or even with themselves to produce various ergodic terms. These ergodic terms have significant influence on the mean flow characteristics, which may be estimated by using the correlation coefficients. The Navier-Stokes equation is averaged with respect to time to show the most important velocity correlation, namely the Reynolds stress, for which different turbulence closure models are required and discussed. Fully developed turbulent flows in circular pipes are studied to show the application of turbulence theory.

8.6.1 Brief Description of Turbulent Flows

As discussed in Sect. 2.8.3, fluid properties experience random fluctuations in turbulent flows, which result from the flow instabilities. This feature is best understood e.g. by using the von Kármán vortex trail in the wake of an obstacle. The velocity at a fixed point relative to the obstacle varies periodically and roughly sinusoidally. The phase of this fluctuation is arbitrary, which depends on the small disturbances at the time the flow commenced. Thus, a prediction of the instantaneous velocity cannot be given within certain limits. This lack of predictability arises from the instability producing the vortex trail. Hence, a brief definition of turbulence may be given as that *turbulence is a state of continuous instability*. Each time a flow changes as a result of instability, and the ability to predict the details of motion is reduced. When successive instabilities have reduced the level of predictability so much, it becomes appropriate to describe a flow *statistically* rather than in every detail, and the flow is then referred to as turbulent. This implies that the random features of flow are dominant. However, a turbulent flow is not completely random. All turbulent flows involve more or less organized structures, for which theoretical and experimental studies are possible.

The statistical description of a turbulent flow starts by decomposing any property α into its mean (average) and fluctuating parts, denoted respectively by $\overline{\alpha}$ and α'. For theoretical purpose, it is convenient to think of the average as an *ensemble average*, i.e., a large number of identical systems is considered, and the average of any quantity at corresponding instant over all these systems is taken. However, in practice, the average is usually a *time average*. The value of any quantity at a point over a long period is observed and averaged. The period should be sufficiently long for separate measurements to give effectively the same result, so that the time average of a quantity α may be given by

$$\overline{\alpha} = \frac{1}{2s}\int_{-s}^{s} \alpha\,\mathrm{d}t, \qquad \alpha' = \alpha - \overline{\alpha}, \qquad \overline{\alpha'} = 0, \tag{8.6.1}$$

where s is large compared with any timescale involved in the variations of α, and the above expression is known as the *Reynolds-filter process*. To demonstrate the concept of time average and its application, consider a rectilinear turbulent flow with velocity $u = U + u'$, where $U = \overline{u}$, representing the mean motion of fluid. Information about the structure of velocity fluctuations is given by some average quantities. The first one is the mean square fluctuations $\overline{u'^2}^{1/2}$, which is called the *intensity of turbulence component*. The second one is the *intensity of turbulence* $\overline{q^2}^{1/2}$, which is given by

$$\sqrt{\overline{q^2}} = \sqrt{\overline{u'^2} + \overline{v'^2} + \overline{w'^2}}, \tag{8.6.2}$$

which is related directly to the *turbulent kinetic energy* per unit volume associated with the velocity fluctuations, k, viz.,

$$k = \frac{1}{2}\rho\overline{q^2}. \tag{8.6.3}$$

The mapping between the turbulence intensity and velocity fluctuations is not unique, for the same intensity can in principle be produced by different patterns of velocity fluctuations.

There exists an alternative statistical representation of the fluctuations of velocity components. The *probability distribution function* $P(u')$ of the fluctuating velocity component u' at one point is so defined that the probability of fluctuation velocity between u' and $u' + du'$ is $P(u')du'$. It follows that

$$\int_{-\infty}^{\infty} P(u')\mathrm{d}u' = 1, \qquad \longrightarrow \qquad \overline{u'^2} = \int_{-\infty}^{\infty} u'^2 P(u')\mathrm{d}u'. \tag{8.6.4}$$

The probability distribution function contains more information than the turbulence intensity. The relationships between the velocity fluctuations at different points (or times) are indicated by the *joint probability distribution functions*. For example, for a second-order function, $P(u_1', u_2')$ may be so defined that the probability of fluctuation velocity at one point between u_1' and $u_1' + \mathrm{d}u_1'$ and that at the other point simultaneously between u_2' and $u_2' + \mathrm{d}u_2'$ is $P(u_1', u_2')\mathrm{d}u_1'\mathrm{d}u_2'$. In principle, for a complete representation of a turbulence, the process needs to be continued to all orders of the fluctuating quantities.

The information about the velocity fluctuations at different points (or times) may also be expressed by the *correlation coefficients* c_r, which, for two velocity fluctuations u_1' and u_2', is defined by

$$c_r \equiv \overline{u_1' u_2'} / \left(\overline{u_1'^2}\,\overline{u_2'^2}\right)^{1/2}, \tag{8.6.5}$$

in which u_1' and u_2' represent general quantities. This expression can be extended e.g. for simultaneous values of the same fluctuating quantity at two different points, or two different fluctuating quantities at a single point. If u_1' and u_2' are independent of each another, $c_r = 0$. However, any turbulent flow is governed by the usual equations, which do not allow such a complete independence, in particular for fluctuations at points close to one another. As similar to the probability distribution function, the correlation coefficient can be extended to higher orders such as $\overline{u_1' u_2' u_3'}$. A complete specification of a turbulence may be accomplished by considering all orders of c_r up to infinity. In practice, it is usually confined to *double correlations* $\overline{u_1' u_2'}$ with a briefer study on *triple correlations*.

8.6.2 Interpretations of Correlations and Spectra

Correlation coefficients play an important role in both theoretical and experimental studies of turbulence. Consider a double correlation c_r given in Eq. (8.6.5). If u_1' and u_2' are the fluctuating velocity components at different points but at the same instant, it is called a *space correlation*, as shown in Fig. 8.26a. The correlations of the same fluctuating velocity component at points separated in a distance either parallel to that fluctuating velocity component, or perpendicular to it, as shown respectively in Figs. 8.26b and c, are called respectively the *longitudinal* and *lateral correlations*.

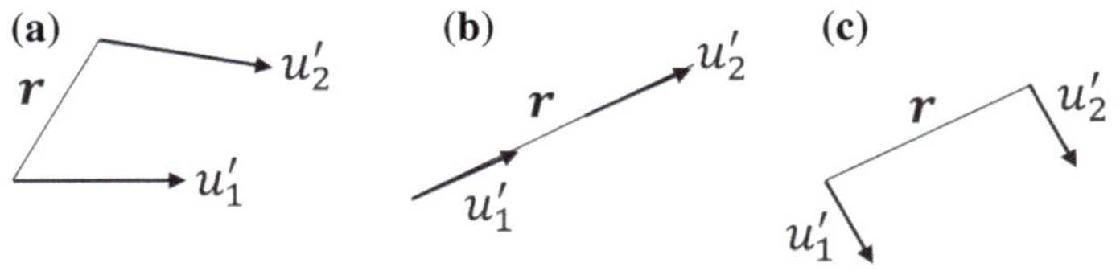

Fig. 8.26 Illustrations of double velocity correlations. **a** A general space correlation. **b** A longitudinal correlation. **c** A lateral correlation

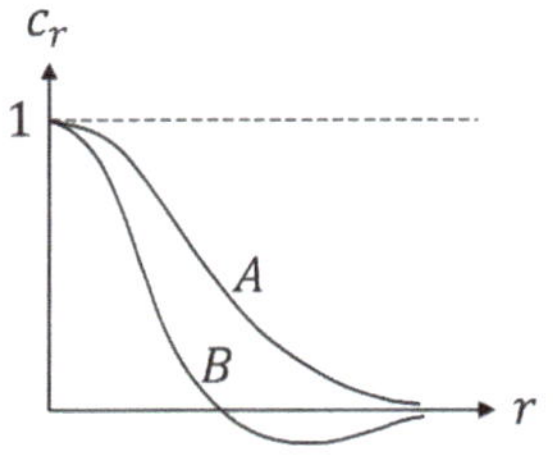

Fig. 8.27 Typical curves of a double correlation, in which curve A is representative for the longitudinal correlation, while curve B may be representative for the lateral correlation

The correlation depends on both the magnitude and direction of separation displacement $\boldsymbol{r}$. Different behaviors in different directions may provide information about the structure of turbulence. Let $r = \|\boldsymbol{r}\|$, and it follows from Eq. (8.6.5) that $c_r = 1$ if $r = 0$ and $u'_1 = u'_2$ with same direction. As r increases, u'_1 and u'_2 become independent of each another, so that c_r approaches null asymptotically. Typical relations between c_r and r are shown in Fig. 8.27, in which the curvature at $r = 0$ is usually large and the experimentally measured correlations often appear to have finite slope there, although the theoretical slope is identified to be null. A negative region in curve B implies that u'_1 and u'_2 tend to be in opposite direction more than in same direction. For longitudinal correlation, this implies dominant converging and/or diverging flow patterns. Since such patterns are not expected, the longitudinal correlation will usually behave as curve A. On the other hand, the lateral correlation may have a negative region, for the continuity equation requires the instantaneous transport of fluid across any plane by letting the turbulent fluctuations be null, although such responses are not always expected. In such a case, the curve itself may be informative about the structure of turbulence.

A correlation curve indicates the distance over which the motion at one point significantly affects that at another. It is used to describe a length scale in turbulence. This concept is extended to associate a variety of length scales with turbulence. Similarly, the correlation for the same fluctuation velocity component, i.e., $u'_1 = u'_2$, at a single point but at different times is known as an *autocorrelation*, which depends on the time separation in a similar manner to the dependence of a space coordinate. It can be used to define a typical timescale in turbulence. As an example, for a turbulent motion occurring in a flow with large mean velocity, the turbulence is advected past the point of observation more rapidly than the changing of fluctuating patterns, so that the autocorrelation is related directly to the corresponding space correlation with separation in the mean flow direction. With these, the curve of space correlation can be applied for the autocorrelation, provided that r/U is used as the time separation. Such a transformation is called *Taylor's hypothesis*. For complex circumstances, in which the fluctuating velocity components at different locations and times are considered, the emerging correlations are called the *space-time correlations*, which are useful to describe the trajectories of certain features such as the *turbulent eddies*. Essentially, the concept of correlation can be extended for the fluctuating pressure and velocity components, e.g. the term $\overline{p'v'}$ inside the parenthesis on the right-hand-side of Eq. (8.6.23), to be shown in Sect. 8.6.3. Since such a correlation is difficult to be measured, it has received less attention.

An alternative method to obtain various timescales associated with turbulence is the Fourier analysis. The turbulence signal is passed through a frequency filter before squaring and averaging. Let the fluctuating velocity component signal be denoted by $u'(t)$, its output from the filter is then given by

$$\chi(t) = \int_0^\infty u'(t - t_1)\Psi(t_1)\mathrm{d}t_1, \tag{8.6.6}$$

with t_1 a dummy variable, where $\Psi(t)$ is the response function of filter,[30] and $\chi(t)$ is a fluctuating function, whose mean square is obtained as

$$\overline{\chi^2} = \int_0^\infty \int_0^\infty \overline{u'(t - t_1)u'(t - t_2)}\Psi(t_1)\Psi(t_2)\mathrm{d}t_1\mathrm{d}t_2, \tag{8.6.7}$$

where the time average is taken over t, as defined in Eq. (8.6.1). Since

$$\overline{u'(t - t_1)u'(t - t_2)} = \overline{u'^2}c_r(t_1 - t_2), \tag{8.6.8}$$

where $c_r(s)$ is the autocorrelation for the time interval s, which is assumed to be an even function. Applying the Fourier transform to this expression gives

$$\overline{u'^2}c_r(s) = \int_0^\infty \phi(\omega)\exp(i\omega s)\,\mathrm{d}\omega. \tag{8.6.9}$$

Substituting this expression into Eq. (8.6.7) yields

$$\overline{\chi^2} = \int_0^\infty \int_0^\infty \int_0^\infty \phi(\omega)\Psi(t_1)\Psi(t_2)\exp\left[i\omega(t_1 - t_2)\right]\mathrm{d}\omega\,\mathrm{d}t_1\,\mathrm{d}t_2, \tag{8.6.10}$$

which is recast alternatively as

$$\overline{\chi^2} = \int_0^\infty \phi(\omega)\Lambda(\omega)\Lambda^*(\omega)\,\mathrm{d}\omega, \tag{8.6.11}$$

with

$$\Lambda(\omega) = \int_0^\infty \exp\left[i\omega t_1\right]\Psi(t_1)\,\mathrm{d}t_1, \qquad \Lambda^*(\omega) = \int_0^\infty \exp\left[i\omega t_2\right]\Psi(t_2)\,\mathrm{d}t_2. \tag{8.6.12}$$

The quantity $\Lambda(\omega)$ is the amplitude of output signal if the input signal is sinusoidal with angular frequency ω. Essentially, the product $\Lambda(\omega)\Lambda^*(\omega)$ is much larger over

[30]That is, it is the output at time t if the input is a delta function at $t = 0$.

a narrow frequency range centered on ω_0 than elsewhere. Hence, Eq. (8.6.11) may be simplified to

$$\overline{\chi^2} = C\phi(\omega_0), \tag{8.6.13}$$

where C is a calibration constant. This procedure determines the Fourier transform of the autocorrelation.

For the special case in which $s = 0$, Eq. (8.6.9) reduces to

$$\overline{u'^2} = \int_0^\infty \phi(\omega)\,\mathrm{d}\omega, \tag{8.6.14}$$

indicating that $\phi(\omega)$ may be interpreted as the contribution from the frequency ω to the energy of turbulence, which is known as the *energy spectrum*. Similarly, the *wave number spectrum*, namely the Fourier transforms of space correlations, can be defined. Since the topic is beyond the scope of the book, it is simply to state here that the distribution of energy over different length scales, $E(\xi)$, with ξ the magnitude of wave number, can be so defined that

$$\frac{k}{\rho} = \frac{1}{2}\overline{q^2} = \int_0^\infty E(\xi)\mathrm{d}\xi. \tag{8.6.15}$$

Although $E(\xi)$ is an important parameter in the theoretical study of turbulence, it cannot be measured experimentally, for a simultaneous information from every point of the flow is required. In practice, Taylor's hypothesis can be used to derive a spatial spectrum from an observed time spectrum. Even this is accomplished, the established spectrum is a one-dimensional one with respect to the component of wave number in the mean flow direction, which is in general not representative for the three-dimensional spectral characteristics.

8.6.3 Turbulence Equations

Applying the Reynolds-filter process to decompose the velocity and pressure fields into their mean and fluctuating parts and substituting the resulting expressions into the continuity and Navier-Stokes equations yields respectively

$$\begin{aligned}&\frac{\partial}{\partial x_i}\left(U_i + u'_i\right) = 0,\\ &\frac{\partial}{\partial t}\left(U_i + u'_i\right) + \left(U_j + u'_j\right)\frac{\partial}{\partial x_j}\left(U_i + u'_i\right) = -\frac{1}{\rho}\frac{\partial}{\partial x_i}\left(P + p'\right) + \nu\frac{\partial^2}{\partial x_j^2}\left(U_i + u'_i\right),\end{aligned} \tag{8.6.16}$$

where $\boldsymbol{U}$ and P are respectively the mean parts of velocity and pressure fields, with their fluctuating parts denoted by $\boldsymbol{u}'$ and p'. Taking time average of these equations gives

$$\begin{aligned}&\frac{\partial U_i}{\partial x_i} = 0, \quad \frac{\partial U_i}{\partial t} + U_j\frac{\partial U_i}{\partial x_j} = -\frac{1}{\rho}\frac{\partial P}{\partial x_i} + \nu\frac{\partial^2 U_i}{\partial x_j^2} - \frac{\partial}{\partial x_j}\left(\overline{u'_i u'_j}\right),\\ &\frac{\partial u'_i}{\partial x_i} = 0, \quad \frac{\partial u'_i}{\partial t} + U_j\frac{\partial u'_i}{\partial x_j} + u'_j\frac{\partial U_i}{\partial x_j} + u'_j\frac{\partial u'_i}{\partial x_j} - \overline{u'_j\frac{\partial u'_i}{\partial x_j}} = -\frac{1}{\rho}\frac{\partial p'}{\partial x_i} + \nu\frac{\partial^2 u'_i}{\partial x_j^2},\end{aligned} \tag{8.6.17}$$

in which the third equation has been used in deriving the second equation. Equation $(8.6.17)_2$ is referred to as the *Reynolds-Averaged-Navier-Stokes equation*, or *RANS* equation for the mean flows.

Equations $(8.6.17)_{1,3}$ indicate that the mean and fluctuating parts of velocity field separately satisfy the usual form of continuity equation, while Eq. $(8.6.17)_2$ differs from its laminar counterpart by the last term, which represents the action of velocity fluctuations on the mean flow arising from the nonlinearity of the Navier-Stokes equation. It is frequently large compared with the viscous term, with the result that the mean velocity distribution is very different from the corresponding laminar counterpart. To demonstrate the influence of this term, consider a stationary, two-dimensional boundary-layer flow in the (x, y)-plane, for which there is no variation of the mean quantities in the z-direction and the terms such as $\partial(\overline{u'w'})/\partial z$ vanish, although the turbulent fluctuations are essentially three-dimensional. With these, the boundary-layer equation for the mean motion in the x-direction is given by

$$\begin{aligned} U\frac{\partial U}{\partial x} + V\frac{\partial U}{\partial y} &= -\frac{1}{\rho}\frac{\partial P}{\partial x} + \nu\frac{\partial^2 U}{\partial y^2} - \frac{\partial}{\partial y}\left(\overline{u'v'}\right) \\ &= -\frac{1}{\rho}\frac{\partial P}{\partial x} + \frac{1}{\rho}\frac{\partial}{\partial y}\left[\mu\frac{\partial U}{\partial y} - \rho\left(\overline{u'v'}\right)\right], \end{aligned} \tag{8.6.18}$$

where V represents the mean velocity component in the y-direction. This equation shows that the velocity fluctuations produce a stress on the mean flow. Its gradient produces a net acceleration to the fluid in the same way as the gradient of viscous stress. The quantity $(-\rho\overline{u'v'})$, and more generally the quantity $(-\rho\overline{u'_i u'_j})$, is called the *Reynolds stress*, with its geometric illustrations shown in Fig. 8.28.

The Reynolds stress arises from the correlation of any two fluctuation velocity components at the same point. A non-vanishing correlation implies that any two fluctuating velocity components are not independent of one another. For example, if $\overline{u'v'} < 0$, then at the instant at which u' is positive, v' is more likely to be negative, and vice versa. At the coordinates with 45° counterclockwise to the x- and y-directions, the fluctuating velocity components are obtained as

$$u'_{45} = \frac{1}{\sqrt{2}}\left(u'_0 + v'_0\right), \qquad v'_{45} = \frac{1}{\sqrt{2}}\left(u'_0 - v'_0\right), \tag{8.6.19}$$

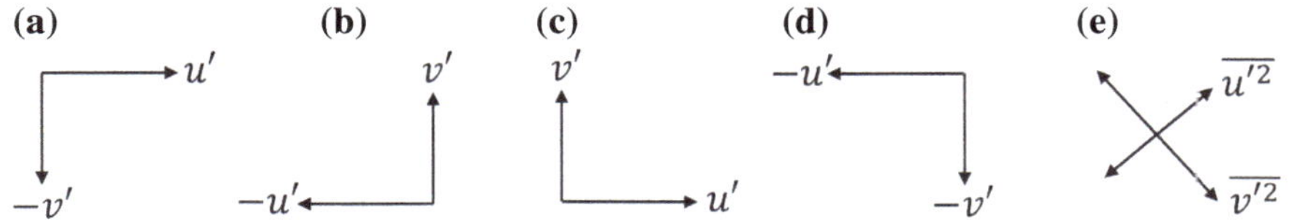

Fig. 8.28 Geometric illustrations of the Reynolds stress. The fluctuating velocity components with the patterns in **a** and **b** take place more frequently than those in **c** and **d**, giving rise to a negative $\overline{u'v'}$ than $\overline{v'^2} > \overline{u'^2}$ shown in **e**

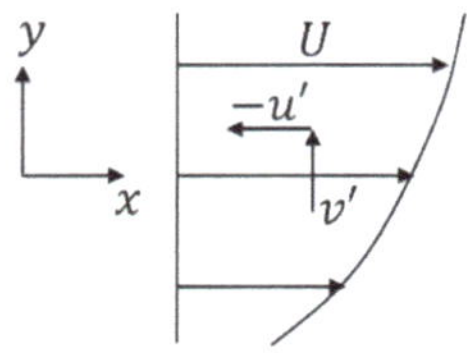

Fig. 8.29 Illustration of the generation of the Reynolds stress in a mean shear flow

where u'_0 and v'_0 are the fluctuating velocity components in the original (x, y)-coordinate system. With these, the velocity correlation becomes

$$\overline{u'_{45}v'_{45}} = \frac{1}{2}\left[\overline{(u'_0)^2} - \overline{(v'_0)^2}\right], \tag{8.6.20}$$

showing that turbulence is *anisotropic*, i.e., it has different intensities in different directions. Figure 8.28 shows the geometric significance of this anisotropic feature of turbulence. A correlation of this kind can arise in a mean shear flow, as shown in Fig. 8.29, in which $\partial U/\partial y > 0$. A fluid particle with positive v' is being carried out by the turbulent eddies in the positive y-direction. Since it comes from a region where the mean velocity is less, it moves downstream more slowly than its new environment, likely having negative u' than positive. A reverse circumstance takes place if v' is negative.[31] In both circumstances, an additional stress acts in the reverse direction of mean flow, which may be described by

$$-\overline{u'v'} = \nu_T \frac{\partial U}{\partial y}, \tag{8.6.21}$$

where ν_T is termed the *eddy viscosity*. Unlike its counterpart in laminar flows, namely the kinematic viscosity ν, ν_T is a representation of the action of turbulence on the mean flow, which is not a fluid property. Moreover, it is also a representation that simplifies the dynamics of that action, for the large-scale coherent motions yield that the Reynolds stress at any point depends on the whole velocity profile, not just the local gradient. Equation (8.6.21) should thus be regarded as the definition of ν_T rather than an equation for $\overline{u'v'}$. From this perspective, ν_T is not a constant, although for approximate calculations an empirical constant is conventionally used.

Further information on the interactions between the mean and fluctuating motions may be obtained by multiplying u'_i with Eq. (8.6.17)$_4$. Taking time average of the resulting equation yields

$$\begin{aligned}\frac{1}{2}\frac{\partial}{\partial t}\left(\overline{u_i'^2}\right) + \frac{1}{2}U_j\frac{\partial}{\partial x_j}\left(\overline{u_i'^2}\right) = -\overline{u'_iu'_j}\frac{\partial U_i}{\partial x_j} - \frac{1}{2}\frac{\partial}{\partial x_j}\left(\overline{u_i'^2u'_j}\right) \\ -\frac{1}{\rho}\frac{\partial}{\partial x_i}\left(\overline{p'u'_i}\right) + \nu\overline{u'_i\frac{\partial^2 u'_i}{\partial x_j^2}},\end{aligned} \tag{8.6.22}$$

[31] The process is essentially (not in detail) analogous to the *Brownian motion* of molecules giving rise to fluid viscosity. Robert Brown, 1773–1858, a Scottish botanist and paleobotanist.

in which Eq. (8.6.17)3 has been used. For the previously considered stationary, two-dimensional boundary-layer flow over a horizontal flat plate, this equation reduces to

$$\frac{1}{2}U\frac{\partial}{\partial x}\left(\overline{q^2}\right)+\frac{1}{2}V\frac{\partial}{\partial y}\left(\overline{q^2}\right)=-\overline{u'v'}\frac{\partial U}{\partial y}-\frac{\partial}{\partial y}\left(\frac{1}{2}\overline{q^2v'}+\frac{1}{\rho}\overline{p'v'}\right)+\nu\overline{u_i'\frac{\partial^2 u_i'}{\partial x_j^2}}, \tag{8.6.23}$$

in which Eq. (8.6.3) has been used. Equation (8.6.23) describes a balance statement of the energy induced by the fluctuating velocity components. The whole left-hand-side and the second term on the right-hand-side vanish when the equation is integrated over the whole flow layer. They represent the energy transfer from place to place by the mean motion and the turbulence itself.[32] The input of energy to compensate the dissipation must be provided by the only contribution, i.e., the first term on the right-hand-side, which is positive. This results from that the term $\overline{u'v'}$ are likely to be negative when $\partial U/\partial y > 0$. Although $\overline{u'v'} > 0$ may sometimes occur, they cannot occupy the majority of flow, or the turbulence cannot be maintained.

Similarly, a balance statement for the energy of mean flow can be obtained, in which the first term on the right-hand-side of Eq. (8.6.23) with negative values presented. Hence, this term represents an energy transfer from the mean flow to the turbulence. It may be concluded that the Reynolds stress works against the mean velocity gradient to remove the energy from the mean flow, just like the viscous stress works against the velocity gradient. The removed energy is directly dissipated, reappearing as heat, whereas the action of the Reynolds stress delivers the energy to the turbulence. This energy is ultimately dissipated by the action of viscosity on the turbulent fluctuations. Usually, the loss of mean flow energy to turbulence is large compared with that caused by the direct viscous dissipation.

8.6.4 Eddies in Turbulence

Since a turbulent flow is associated with various length scales, it is useful to divide a turbulent motion into the interacting submotions on various lengths, for different lengths play rather different roles in the dynamics of whole motion. This is frequently expressed as the *eddies* of different sizes. Although it is not a well-defined concept, a turbulent eddy is a very useful one for the description of turbulence. A turbulent eddy may not be necessary a circulatory motion, but for large eddies such a characteristic can often be identified, so that such eddies are called the *coherent structures* of turbulence. In contrast to the Fourier components, no matter how small their wavelengths (corresponding to how large the values of ξ) extending over the whole flow are, an eddy is rather localized. That is, the extent of an eddy in indicated by its length scale. Small eddies contribute to large wave number components of the

[32] For laminar flows, the viscous term can be divided into two parts: one is essentially negative, representing the viscous dissipation; the other integrates to zero and is another energy transfer process.

spectrum. The spectrum curve is often interpreted roughly in terms of the energy associated with the eddies of different sizes.

For a separation r between two points, the correlation coefficient is determined by all eddies larger than r. Only the largest eddies can thus be related directly to the correlation measurements. On the contrary, the observable spectrum functions possess a value at the wave number ξ which is influenced by all eddies smaller than $1/\xi$. This statement holds true for one-dimensional circumstance, for only one-dimensional spectra can be observed. It is usually most convenient to use the correlation measurements to provide information about the larger scales and spectrum measurements for the smaller scales.

An important physical observation of turbulence is that there exists an energy flow between turbulent eddies with different sizes, i.e., an energy transfer from eddies of a certain size to the next smaller eddies. This transfer is the result of a number of interactions between such eddies. However, as one progresses through this energy cascade, the memory how the turbulence might have been generated will be lost, that is, the energy spectrum for large wave numbers (small wavelengths) must be independent of its generation and hence must assume an universal form as $\xi \to \infty$. This universal law is referred to as *Kolmogorov's law*,[33] which reads "*the spectral energy density falls for large values of ξ as $\xi^{-5/3}$*", and the smallest scale corresponding to this condition is referred to as *Kolmogorov's scale*.

8.6.5 Turbulence Closure Models

The Reynolds stresses appearing in the RANS equation act as additional terms, which need to be prescribed as a function of the mean fields to arrive at a mathematically well-posed problem. Different prescriptions of the Reynolds stress and other ergodic terms lead to the *turbulence closure models of different orders*, which may be derived theoretically by using e.g. the variational or thermodynamic approach. The outcomes must be supplemented by experimental data.[34] Turbulence closure models of various orders are introduced in the following.

- *Closure models of zeroth order*. The double correlation coefficients of various fluctuating quantities, e.g. velocity, temperature, pressure, etc., are postulated as functions of mean quantities. These functions are further simplified or simply set equal to constants. The common procedure is to ignore further specifications of these correlations at this level of closure. For example, the eddy viscosity ν_T given

[33] Andrey Nikolaevich Kolmogorov, 1903–1987, a Russian mathematician, who contributed to the mathematics of probability theory, topology, turbulence, classical mechanics, etc.

[34] From the mathematical perspective, the formulations of turbulence closure models are in principle the same as the constitutive or material equations, for the purpose of formulation is to reach a mathematically well-posed problem. This similarity and the possible derivations by using the thermodynamic approach was pointed out by Rivlin.

in Eq. (8.6.21) may be assumed as a function of the mean velocity gradient, or even simply a constant.

- *Closure models of first order*. For the specific *turbulent kinetic energy* or another scalar quantity related to it, a *transport equation* is established, and the eddy viscosity is algebraically connected with this quantity that is evolving in time and space. In a more general sense, two scalar quantities, namely the specific turbulent kinetic energy and specific *turbulent dissipation*, also other combinations of scalar quantities, are described by using transport-like equations. For example, the well-known k-ε model belongs to this category. The eddy viscosity is again connected to these variables, whose description is often motivated by means of dimensional analysis.
- *Closure models of second order*. The double correlations are described by using transport-like equations, which contain new triple, even higher-order correlations. The higher-order correlations need to be parameterized by closure conditions of the gradient-type or other possible parameterizations, which are often motivated by the outcomes of dimensional analysis. Typical examples are the *Reynolds stress model* (RSM), the *algebraic Reynolds stress model* (ARSM), etc.
- *Closure models of mixed-type*. In addition to the above closure models, mixed-type relations are equally possible and often applied. For example, a closure scheme of second order may be applied for the Reynolds stress, while the turbulent heat flux may be parameterized by a closure model of zeroth order.

In parallel, there exist also other possibilities to describe the characteristics of turbulence, for example, the *Large Eddy Simulation* (LES), or the *Direct Numerical Simulation* (DNS), which belong essentially to the numerical approach of turbulence.

8.6.6 Entrance Length and Fully Developed Flows in Pipes

The theory of turbulence may better be demonstrated by considering a turbulent flow from a reservoir in a circular pipe, as shown in Fig. 8.30a. Typically, the fluid enters the pipe with nearly uniform velocity profile at section A. As the motion inside the pipe continues, the viscous effect causes the fluid to adhere to the pipe wall (i.e., the no-slip boundary condition), and the boundary layer starts to develop, so that the velocity profile at a later cross-section, e.g. at section B, is different from its initial uniform velocity profile. This circumstance continues until the edge of boundary layer reaches to the centerline of the pipe, or alternatively the edges of boundary layers from the pipe wall emerge at the centerline, e.g. at section C. After this location, the velocity profile does not vary with respect to the x-coordinate, and the flows are referred to as *fully developed*,[35] which can be either laminar or turbulent,

[35] Thus, fully developed flows may be interpreted as complete boundary-layer flows.

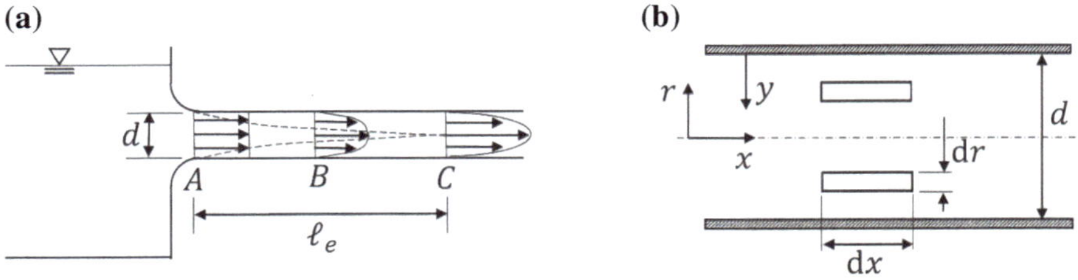

Fig. 8.30 Characteristics of pipe-flows. **a** The entrance length and fully developed flows, with dashed lines denoting the edges of boundary layers. **b** An annual differential control-volume in a horizontal pipe with diameter d

characterized by the Reynolds number given by

$$\mathcal{R}_e = \frac{\rho u_{av} d}{\mu}, \tag{8.6.24}$$

where ρ and μ are respectively the density and dynamic viscosity of fluid, u_{av} denotes the average velocity, and d is the diameter of circular pipe. Like flows in boundary layers, flows in circular pipes are completely laminar or turbulent for $\mathcal{R}_e < 2300$ or $\mathcal{R}_e > 4000$, respectively. In-between the flows are in the *transition region.*

The length between sections A and C is called the *entrance length* ℓ_e, whose value depends on the Reynolds number and flow characteristics. For laminar and turbulent flows, ℓ_e is identified respectively as

$$\frac{\ell_e}{d} \sim 0.06\mathcal{R}_e, \qquad \frac{\ell_e}{d} \sim 4.4\mathcal{R}_e^{1/6}. \tag{8.6.25}$$

For very low Reynolds numbers, ℓ_e can be quite short, e.g. $\ell_e = 0.6d$ for $\mathcal{R}_e = 10$. For very large Reynolds numbers, it may take a length equal to many pipe diameters before the end of entrance length is reached. For example, $\ell_e = 120d$ for $\mathcal{R}_e = 2000$, and $20d < \ell_e < 30d$ for $10^4 < \mathcal{R}_e < 10^5$. The entrance length does not take place only once. For every change in the geometric configuration, e.g. through a pipe bend or a pipe reduction, the boundary layers re-establish, yielding new entrance lengths.

Consider a horizontal pipe with diameter d shown in Fig. 8.30b. Applying the local balance of linear momentum to the annual differential control-volume in a steady flow yields

$$-\frac{\partial p}{\partial x}(2\pi r \mathrm{d}r)\mathrm{d}x + \tau_{rx}(2\pi \mathrm{d}r \mathrm{d}x) + \frac{\mathrm{d}\tau_{rx}}{\mathrm{d}r}(2\pi r \mathrm{d}r)\mathrm{d}x = 0, \tag{8.6.26}$$

which reduces to

$$\frac{\partial p}{\partial x} = \frac{\tau_{rx}}{r} + \frac{\mathrm{d}\tau_{rx}}{\mathrm{d}r} = \frac{1}{r}\frac{\mathrm{d}}{\mathrm{d}r}(r\tau_{rx}). \tag{8.6.27}$$

If the pressure is assumed to be uniform at each cross-section, the left-hand-side of this equation depends at most on x, while the right-hand-side is at most a function of r. It follows that

$$r\frac{\partial p}{\partial x} = \frac{\mathrm{d}}{\mathrm{d}r}(r\tau_{rx}) = C, \tag{8.6.28}$$

where C is a constant. This equation implies not only that the pressure drops uniformly along the pipe length in a constant-diameter pipe, but also that the pressure drop can be used as an estimation on the shear stress on the pipe wall. Integrating Eq. (8.6.28) with vanishing value of τ_{rx} at $r = 0$ gives

$$\tau_{rx} = \frac{r}{2}\frac{\partial p}{\partial x}, \tag{8.6.29}$$

indicating that the shear stress distributes linearly at the pipe section, provided that the pressure gradient along the x-direction is constant.[36] The results obtained in Eqs. (8.6.28) and (8.6.29) are valid for both fully developed laminar and turbulent pipe-flows.

8.6.7 Turbulent Velocity Profiles in Pipe-Flows

For fully developed laminar flows, the value of τ_{rx} at the pipe wall, i.e., $\tau_w = -\tau_{rx}|_{r=d/2}$, can immediately be determined by using Newton's law of viscosity, and substituting the determined expression of τ_{rx} into Eq. (8.6.29) yields the governing ordinary differential equation for the velocity profile $u(r)$, which can be integrated to obtain $u(r)$ with appropriately formulated boundary conditions, as accomplished in Sect. 8.2.2. For fully developed turbulent flows, it follows from the discussions in Sect. 8.6.3 that τ_w consists of two contributions given by

$$\tau_w = -\frac{d}{4}\frac{\partial p}{\partial x} = \mu\frac{\mathrm{d}\overline{u}}{\mathrm{d}y} - \overline{\rho u'v'} = \tau_{lam} + \tau_{turb}, \qquad y = \frac{d}{2} - r, \tag{8.6.30}$$

where y denotes the distance measured from the pipe wall for convenience, $\overline{u}$ is the time-averaged mean velocity, and $\{u', v'\}$ are the fluctuating velocity components in the x- and y-directions, respectively. In this equation, τ_{lam} denotes the viscous shear stress, while τ_{turb} is the Reynolds stress, which is the momentum transfer of fluid within the random turbulent eddies. The relative significance between two contributions is different at different locations on a fixed cross-section and is a complex function depending on the specific flow under consideration. Typical measures of two contributions are shown in Fig. 8.31a, in which the horizontal axis is in logarithmic scale.

In a very narrow region near the pipe wall, there exists a very thin layer, termed the *viscous sublayer*, or *laminar sublayer*, in which the shear stress τ_{lam} is dominant. Away from the wall is a relatively thick layer, termed the *outer layer*, in which τ_{turb} becomes dominant. The transition between two layers occurs in the so-called *overlap layer*, in which both τ_{lam} and τ_{turb} are of equal importance. A typical velocity

[36] A fully developed steady flow in a horizontal pipe of constant cross-section implies that there is a balance between the pressure and viscous forces, giving rise to a constant pressure gradient. In the entrance region, there exists a balance between the inertia, viscous, and pressure forces. Hence, the pressure gradient may not be constant.

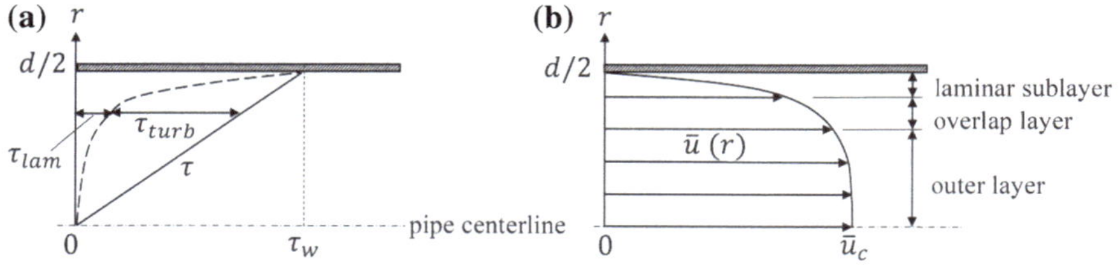

Fig. 8.31 Characteristics of fully developed pipe-flows. **a** The distribution of shear stress. **b** The distribution of mean velocity component $\overline{u}(r)$ in the axial direction, where $\overline{u}_c$ is the mean velocity at the pipe centerline

profile of a fully developed turbulent pipe-flow is shown in Fig. 8.31b,[37] for which turbulence closure models could involve for the prescriptions of the Reynolds stress. For example, by using the eddy viscosity, the Reynolds stress could be expressed as

$$\tau_{turb} = \rho \nu_T \frac{\mathrm{d}\overline{u}}{\mathrm{d}y}, \tag{8.6.31}$$

which, by using *Prandtl's mixing length theory*, can be further simplified to

$$\rho\nu_T = \rho \ell_m^2 \left|\frac{\mathrm{d}\overline{u}}{\mathrm{d}y}\right|, \quad \longrightarrow \quad \tau_{turb} = \rho \ell_m^2 \left|\frac{\mathrm{d}\overline{u}}{\mathrm{d}y}\right| \frac{\mathrm{d}\overline{u}}{\mathrm{d}y}, \tag{8.6.32}$$

where ℓ_m is called the *mixing length*, which represents a characteristic length, over which the momentum of a fluid bundle is transported randomly. In doing this, the problem is shifted to the determination of ℓ_m. Further studies indicate that the mixing length is not a constant through the flow field, and additional assumptions should be made regarding how ℓ_m varies through the flow. This example demonstrates the nature of turbulence closure model, and somewhere the closure modeling must be cut off by artificial assumptions calibrated by experimental data.

To obtain the velocity profile, let δ be the thickness of laminar sublayer. Experimental measurements show that the velocity distribution within δ is nonlinear. However, since δ is very small compared with the radius of pipe, the flows inside δ may be assumed to be laminar for simplicity with a linear approximation to the mean velocity gradient given by

$$\left.\frac{\mathrm{d}\overline{u}}{\mathrm{d}y}\right|_{0\le y\le\delta} = \frac{\overline{u}|_{y=\delta}}{\delta} = \frac{\overline{u}}{y}, \tag{8.6.33}$$

with which τ_w is determined to be

$$\tau_w = \mu \left.\frac{\mathrm{d}\overline{u}}{\mathrm{d}y}\right|_{y=0} = \mu \frac{\overline{u}|_{y=\delta}}{y} = \mu \frac{\overline{u}}{\delta}. \tag{8.6.34}$$

[37] These two figures are conceptual rather than realistic, for the horizontal and vertical scales are distorted. In reality, τ_{turb} is hundred to thousand times larger than τ_{lam} in the outer layer, with the reverse tendency in the viscous sublayer.

Dividing both sides of this equation by ρ yields

$$\frac{\overline{u}}{u^*} = \frac{yu^*}{\nu} = \text{constant}, \qquad u^* = \sqrt{\frac{\tau_w}{\rho}}, \tag{8.6.35}$$

where u^* is called the *friction velocity*, which is not an actual velocity of the fluid. It is only a quantity that has a dimension of velocity. Experimental data shows that the proposed linear mean velocity profile is valid in the range

$$0 \leq \frac{yu^*}{\nu} = \mathcal{R}_e^* \leq 5 \sim 7, \tag{8.6.36}$$

where $\mathcal{R}_e^*$ is called the *modified Reynolds number*, which represents a dimensionless distance from the pipe wall.[38] As y increases, the flow transits gradually from laminar to turbulent through the transition region, i.e., the overlap layer, which is also called the *buffer layer*, characterized by $5 \sim 7 \leq \mathcal{R}_e^* \leq 30$. For $\mathcal{R}_e^* > 30$, the turbulent core is reached, whose velocity profile is quite well represented by the semi-logarithmic curve-fit equation given by

$$\frac{\overline{u}}{u^*} = 2.5 \ln\left(\frac{yu^*}{\nu}\right) + 5.0, \tag{8.6.37}$$

where the constants 2.5 and 5.0 have been determined experimentally. Applying this equation to the pipe centerline yields[39]

$$\frac{\overline{u}_c - \overline{u}}{u^*} = 2.5 \ln\left(\frac{d}{2y}\right), \tag{8.6.38}$$

where $\overline{u}_c$ is the mean velocity at the pipe centerline. This equation is referred to as the *defect law*, indicating that the mean velocity defect (and hence the general shape

---

[38]It is found that for water at 20 °C flowing through a horizontal pipe with diameter of 0.1 m and $Q = 4 \times 10^{-2}$ m^3/s, the thickness δ is estimated as $\delta \sim 0.02$ mm under a pressure gradient of 2.59 kPa/m. Thus, the thickness of laminar sublayer is only 0.02% of the pipe diameter.

[39]Equation (8.6.38) can also be derived differently. Prandtl assumed that the mixing length ℓ_m should be a linear function of y for circular pipes, which is proposed as

$$\ell_m \propto y, \qquad \longrightarrow \qquad \ell_m = \kappa y,$$

where κ is the proportionality, which is called the *universal constant*. Substituting this expression into Eq. $(8.6.32)_2$ and subsequently the resulting equation into Eq. (8.6.34) gives

$$\frac{\mathrm{d}\overline{u}}{\mathrm{d}y} = \frac{u^*}{\kappa}\frac{1}{y}, \qquad \longrightarrow \qquad \overline{u} = \frac{u^*}{\kappa} \ln y + C,$$

where C is an integration constant. Applying this expression to the pipe centerline yields $C = \overline{u}_c - u^* \ln(d/2)/\kappa$, with which the mean velocity profile is obtained as

$$\frac{\overline{u}_c - \overline{u}}{u^*} = 2.5 \ln\left(\frac{d}{2y}\right).$$

This equation is called *Prandtl's universal velocity distribution equation/law*, for it has been confirmed that $\kappa \sim 0.4$ for the regions very close to the pipe wall, and this value is also applicable to the central region of a pipe.

of mean velocity profile in the neighborhood of centerline) is only a function of the distance ratio, and does not depend on the fluid viscosity. The characteristics of Eqs. (8.6.36) and (8.6.37) are shown graphically in Fig. 8.32a as two dashed lines.

In practice, it is more convenient to use the *power law equation* for the velocity profiles in fully developed turbulent pipe-flows, which is given by

$$\frac{\overline{u}}{\overline{u}_c} = \left(\frac{2y}{d}\right)^{1/n} = \left(1 - \frac{2r}{d}\right)^{1/n}, \tag{8.6.39}$$

where the exponent n varies with the Reynolds number defined by $\mathcal{R}_{ec} = \overline{u}_c d/\nu$. Since this equation gives an infinite mean velocity gradient at the pipe wall, it cannot be used to evaluate the wall shear stress τ_w. Specifically, it is not applicable in the region within $2y/d < 0.04$. As suggested by the experimental data, the variation in n with $\mathcal{R}_{ec}$ is given by

$$n = -1.7 + 1.8 \log \mathcal{R}_{ec}, \tag{8.6.40}$$

for $\mathcal{R}_{ec} > 2 \times 10^4$. With these, the ratio of the average mean velocity $\overline{u}_{av}$ to the mean velocity at the pipe centerline $\overline{u}_c$, by using $\overline{u}_{av} = Q/A$ with Q and A, respectively, the flow rate and cross-sectional area of the pipe, is obtained as

$$\frac{\overline{u}_{av}}{\overline{u}_c} = \frac{2n^2}{(n+1)(2n+1)}. \tag{8.6.41}$$

As n increases by virtue of increasing $\mathcal{R}_{ec}$, the ratio increases correspondingly. Thus, for large values of $\mathcal{R}_{ec}$, the velocity profile becomes blunter, as shown in Fig. 8.32b. As a representation, $n = 7$ is often used, which gives the *one-seventh-power profile* for fully developed turbulent pipe-flows. For comparison, the parabolic velocity profile for fully developed laminar flows and the one-seventh-power velocity profile for fully developed turbulent flows are shown in Fig. 8.33.

The obtained results are only valid for smooth pipes. For rough pipes, while the surface roughness of pipes, ε, plays no role for fully developed laminar flows, for the whole flow is inside the boundary layer, it has a significant role for fully developed turbulent flows. This results from the facts that there exists a viscous sublayer, and the

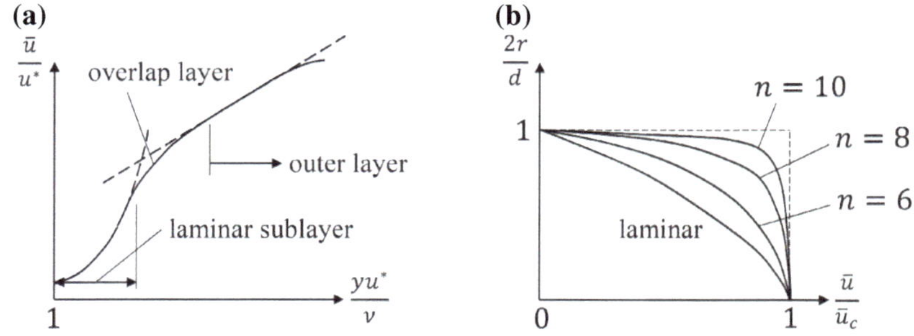

Fig. 8.32 The velocity profiles for fully developed turbulent flows in smooth pipes. **a** The velocity distributions in the laminar sublayer and outer layer. Solid line: experimental data, dashed lines: theoretical estimations. **b** The velocity distributions in terms of the power-law equation

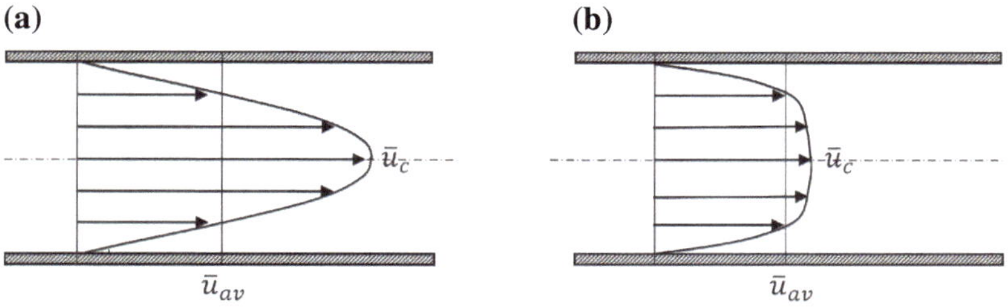

Fig. 8.33 Typical velocity profiles in fully developed pipe-flows. **a** The parabolic velocity distribution for laminar flows. **b** The velocity distribution for turbulent flows with one-seventh-power equation

influence of ε depends on the relative thickness between itself and δ. If the Reynolds number is of such a value that $\varepsilon < \delta$, the surface roughness is submerged within the viscous sublayer. Hence, ε does not interfere with the formation of viscous sublayer and overlap layer. In such a circumstance, the velocity profile is exactly the same as that for smooth pipes, and this rough pipe is referred to as *hydraulically smooth*. If $\varepsilon > \delta$, the surface roughness protrudes beyond the viscous sublayer, creating additional turbulence in the flow. The relative roughness ε/d may thus be regarded as a similarity parameter for rough pipes, for the geometrically similar rough surfaces will result in dynamically similar turbulent-flow patterns. Various theoretical and experimental studies have been devoted to the influence of relative surface roughness on the velocity profile as well as on other physical characteristics. In the context of the book, its influence will be considered macroscopically in the context of lump energy loss, which will be introduced in the next subsection.

8.6.8 Energy Loss, Friction Factor, and the Moody Chart

The three terms in the Bernoulli equation consist of the total mechanical energy of a fluid, which motivates the *loss of mechanical energy* h_{lt} between any two points 1 and 2 defined by

$$h_{lt} \equiv \left(\frac{p_1}{\rho g} + \frac{u_1^2}{2g} + z_1\right) - \left(\frac{p_2}{\rho g} + \frac{u_2^2}{2g} + z_2\right), \tag{8.6.42}$$

where h_{lt} represents the energy loss in terms of head. The mechanical energy loss between any two points by any means can be evaluated in principle by using this equation. Since the Bernoulli equation is devoted to ideal fluids, in which V represents a uniform velocity, Eq. (8.6.42) must be revised for viscous flows. By requiring that

$$\frac{1}{2}\alpha\dot{m}\overline{u}_{av}^2 = \int_A \frac{1}{2}\overline{u}^2(\rho\overline{u}\mathrm{d}a), \qquad \beta\dot{m}\overline{u}_{av} = \int_A \overline{u}(\rho\overline{u}\mathrm{d}a), \tag{8.6.43}$$

where $\dot{m} = \rho \overline{u}_{av} A$, representing the mass flow rate across a pipe section with area A, the *kinetic energy coefficient* α, and *momentum coefficient* β are defined by

$$\alpha \equiv \frac{1}{A\overline{u}_{av}^3} \int_A \overline{u}^3 \mathrm{d}a, \qquad \beta \equiv \frac{1}{A\overline{u}_{av}^2} \int_A \overline{u}^2 \mathrm{d}a. \tag{8.6.44}$$

By using the first equation, Eq. (8.6.42) is recast alternatively as

$$h_{lt} \equiv \left(\frac{p_1}{\rho g} + \frac{\alpha \overline{u}_{av1}^2}{2g} + z_1 \right) - \left(\frac{p_2}{\rho g} + \frac{\alpha \overline{u}_{av2}^2}{2g} + z_2 \right), \tag{8.6.45}$$

while the momentum coefficient is used in the global balance of linear momentum if viscous fluids are considered. For ideal fluids, $\alpha = \beta = 1$; for fully developed laminar flows, $\alpha = 2$ and $\beta = 4/3$, while for fully developed turbulent flows they become

$$\alpha = \frac{(n+1)^3(2n+1)^3}{4n^4(n+3)(2n+3)}, \qquad \beta = \frac{(n+1)^2(2n+1)^2}{2n^2(n+2)(2n+2)}, \tag{8.6.46}$$

if the power law equation is used for the velocity profile.[40] Specifically, $\alpha = 1.06$ and $\beta = 1.02$ when $n = 7$. The discrepancies in the estimated values of α and β for laminar and turbulent flows also reveal the characteristics of their velocity profiles.

Two physical mechanisms contribute to the total mechanical energy loss h_{lt}. The first one is the losses due to the frictional one, and the second one results from all other effects except the frictional effect, such as entrances, fittings, area changes. The frictional losses are termed the *major losses*, denoted by h_l, while the others are referred to as the *minor losses*, denoted by h_{lm}. That is, $h_{lt} = h_l + h_{lm}$. For horizontal pipes with constant cross-section, it follows from Eq. (8.6.45) that

$$h_{lt} = h_l = \frac{p_1 - p_2}{\rho g} = \frac{\Delta p}{\rho g}, \tag{8.6.47}$$

showing that the mechanical energy loss may be indicated by the pressure drop, resulted from the shear stress on the pipe wall. Since the head loss represents the energy converted by the frictional effect from the mechanical part to the thermal part, it depends only on the details of flow field through the conduit and is independent of the pipe orientation.

For fully developed laminar flows, it follows from the Hagen-Poiseuille equation that

$$\Delta p = \frac{128 \mu \ell Q}{\pi d^4} = 32 \frac{\ell}{d} \frac{\mu u_{av}}{d}, \tag{8.6.48}$$

so that Eq. (8.6.47) may be brought to the form

$$h_l = 32 \frac{\ell}{d} \frac{\mu u_{av}}{\rho g d} = f \frac{\ell}{d} \frac{u_{av}^2}{2g}, \qquad \longrightarrow \qquad f = \frac{64}{\mathcal{R}_e}, \quad \mathcal{R}_e = \frac{\rho u_{av} d}{\mu}, \tag{8.6.49}$$

[40] Since the velocity of a turbulent flow near the pipe wall is low, the error in calculating the integral quantities such as mass, momentum, and energy fluxes at a cross-section is relatively small.

where f is called the *friction factor*.[41] This equation indicates that the friction loss is proportional to the pipe length ℓ and the square of average velocity u_{av}, and is inversely proportional to the pipe diameter d, although f decreases as u_{av} increases by larger values of $\mathcal{R}_e$. For fully developed turbulent flows, the functional relation of the pressure drop is identified to be

$$\Delta p = F\left(\rho, \mu, \bar{u}_{av}, d, \ell, \varepsilon\right). \tag{8.6.50}$$

Applying the dimensional analysis to this functional relation yields

$$\frac{\Delta p}{\frac{1}{2}\rho\bar{u}_{av}^2} = F\left(\frac{\rho\bar{u}_{av}d}{\mu}, \frac{\ell}{d}, \frac{\varepsilon}{d}\right), \tag{8.6.51}$$

which differs from Eq. (8.6.48), for the influence of relative surface roughness ε/d has been taken into account. Experimental studies show that the dimensionless pressure drop is directly proportional to ℓ/d, with which Eq. (8.6.51) may be written as

$$\frac{\Delta p}{\frac{1}{2}\rho\bar{u}_{av}^2} = \frac{\ell}{d}F'\left(\frac{\rho\bar{u}_{av}d}{\mu}, \frac{\varepsilon}{d}\right) = \frac{\ell}{d}F'\left(\mathcal{R}_e, \frac{\varepsilon}{d}\right), \tag{8.6.52}$$

where F' represents a different functional relation from F. The friction factor for fully developed turbulent flows is thus defined viz.,

$$f \equiv F'\left(\mathcal{R}_e, \frac{\varepsilon}{d}\right), \tag{8.6.53}$$

so that the friction loss h_l is expressed as

$$h_l = f\frac{\ell}{d}\frac{\bar{u}_{av}^2}{2g}, \qquad f = \text{funct.}\left(\mathcal{R}_e, \frac{\varepsilon}{d}\right). \tag{8.6.54}$$

Although the friction losses for fully developed laminar and turbulent flows are expressed by the same equation, the friction factor for laminar flows is only a function of the Reynolds number, as indicated by Eq. $(8.6.49)_2$, while that for turbulent flows depends on the Reynolds number as well as on the relative surface roughness.

Various experiments have been conducted for the determination of Eq. $(8.6.54)_2$, with the results summarized in Fig. 8.34, which is known as the *Moody chart*, established by Moody in 1944.[42] The horizontal axis denotes the values of the Reynolds number, the left vertical axis represents the values of friction factor, while the right vertical axis expresses the values of relative surface roughness. For a specific value of $\mathcal{R}_e$ within the laminar flow region, the value of f is directly determined by the chart. Increasing u_{av} is to increase $\mathcal{R}_e$, until the critical Reynolds number is reached,

[41] It is also termed the *Darcy friction factor*. The less frequently used one is the *Fanning friction factor*, which is defined as

$$f_F \equiv \frac{\tau_w}{\frac{1}{2}\rho\bar{u}_{av}^2}.$$

It is readily verified that for fully developed pipe-flows, $f = 4f_F$. Henry Philibert Gaspard Darcy, 1803–1858, a French engineer, who made several important contributions to hydraulics.

[42] Lewis Ferry Moody, 1880–1953, an American engineer and professor, who is best known for the Moody chart.

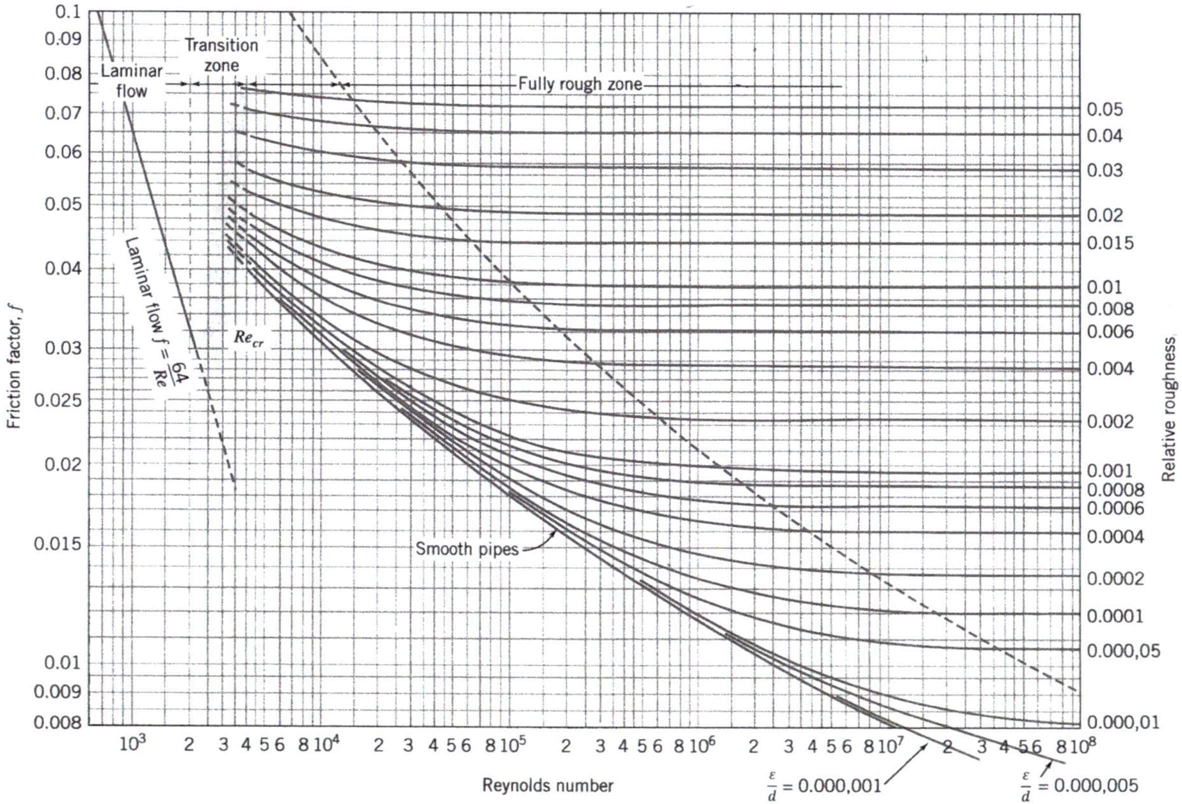

Fig. 8.34 Moody chart for the determination of friction factor for fully developed flows in circular pipes. Data quoted from Fox, R.W., Pritchard, P.J., McDonald, A.T., *Introduction to Fluid Mechanics*, 7th ed., John Wiley & Sons, New York, 2009. Used with permission. Original data quoted from Moody, L.F., Friction factors for pipe flows, *Transactions of the ASME*, 66, 8, 671–684, 1944

at which transition occurs, and laminar flows give way to turbulent flows. Since the velocity gradient at the pipe wall is much larger in turbulent flows than in laminar flows, the transition causes the wall shear stress to increase sharply, whose effect is reflected by a sharp increase in the friction factor. For larger values of $\mathcal{R}_e$ within the turbulent flow region, with $\varepsilon/d \leq 0.001$, the friction factor at first tends to follow the smooth-pipe curve, along which f is only a function of the Reynolds number. As even larger values of $\mathcal{R}_e$ present, the thickness of viscous sublayer decreases, so that as roughness elements begin to poke through the viscous sublayer, the effect of surface roughness becomes important. In such circumstances, additional information of ε/d is required to determine f. The dashed line marks the edge of *fully rough zone*. In the region to the right of this dashed line, most of the roughness elements on the pipe wall protrude through the viscous sublayer, so that the friction loss depends nearly only the size of roughness elements. For $\varepsilon/d \geq 0.001$, f is greater than the smooth-pipe value as $\mathcal{R}_e$ increases. The value of $\mathcal{R}_e$ at which the flow becomes fully rough decreases with increasing ε/d.

By and large, increasing $\mathcal{R}_e$ is to decrease the values of f, as long as the flow remains laminar. At the transition region, f increases sharply due to the sharp change of velocity gradient. In the turbulent region, f decreases gradually and finally levels out at a constant value for large values of $\mathcal{R}_e$. However, these do not imply that h_l decreases as $\mathcal{R}_e$ increases, for $h_l \propto u_{av}$ in the laminar flow region. In the transition region, there exists a sharp increase in h_l. In the fully rough zone, $h_l \propto \overline{u}_{av}^2$, and for the rest of turbulent region, h_l increases at a rate somewhere between $\overline{u}_{av}$ and $\overline{u}_{av}^2$. Thus, h_l always increases as the average flow velocity increases, and it increases more rapidly when the flow is turbulent.[43]

There exist some mathematical expressions for the determination of friction factor, which are calibrated by experimental data. For example, the most widely used formulation in implicit form is given by

$$\frac{1}{\sqrt{f}} = -2.0 \log \left(\frac{\varepsilon/d}{3.7} + \frac{2.51}{\mathcal{R}_e \sqrt{f}} \right), \tag{8.6.55}$$

which is referred to as the *Colebrook formula*. It is valid for the entire non-laminar range of the Moody chart. An alternative explicit formulation is given by

$$\frac{1}{\sqrt{f}} = -1.8 \log \left[\left(\frac{\varepsilon/d}{3.7} \right)^{1.11} + \frac{6.9}{\mathcal{R}_e} \right], \tag{8.6.56}$$

which was proposed by Haaland as an approximation to the Colebrook formula. The results obtained by using this equation is within 2% of the Colebrook formula for $\mathcal{R}_e > 3000$.

[43]The data in the Moody chart are the average values for new pipes with accuracy of nearly $\pm 10\%$. After a long period of service, corrosion and/or deposition take place, and the surface roughness may experience a dramatic change. In such circumstances, the relative roughness ε/d may be increased by factors of 5-10 for used pipes.

The minor losses h_{lm} are expressed in terms of either the *loss coefficient* K or *equivalent length* ℓ_e of a straight pipe given by

$$h_{lm} = K\frac{\overline{u}_{av}^2}{2g} = f\frac{\ell_e}{d}\frac{\overline{u}_{av}^2}{2g}. \tag{8.6.57}$$

In the expressions, h_{lm} is directly identified by the values of K, or transformed to a length of a straight pipe, whose friction loss (major loss) is equivalent to the minor loss. These two coefficients should be determined in principle by experiments. Experimental studies show that the loss coefficient varies with different configurations of pipe bends and fittings, while the equivalent length tends toward a constant, which is more convenient for practical application. The most encountered minor losses in practice are summarized in the following:

- inlets and exits,
- enlargements and contractions,
- pipe bends,
- valves and fittings, and
- pumps, fans and blowers, etc.

The values of loss coefficients and equivalent lengths for these minor losses can be found in any handbook of fluid engineering. However, the data are scattered among a variety of sources. Different sources may give different values of K and ℓ_e for the same flow configuration. Applications of the data must be conducted with care.

For non-circular conduits, Eqs. (8.6.49) and (8.6.54) can still be applied to estimate the friction losses, provided that the pipe diameter d is replaced by the hydraulic diameter d_h given in Eq. $(8.2.39)_2$. The corresponding Reynolds number is then given by Eq. (8.2.38), and the relative surface roughness becomes

$$\frac{\varepsilon}{d} = \frac{\varepsilon}{d_h}. \tag{8.6.58}$$

The validity of this approach is limited to turbulent flows in the conduits of rectangular, triangular, and elliptical cross-sections which do not depart significantly from a circular proportion, i.e., with the aspect ratio a_r smaller than 4, where a_r is defined by $a_r = h/b$, with h and b respectively the height and width of a rectangular conduit. For non-circular conduits, there exists a phenomenon that fluid particles flow away from the central portion and toward the corners of conduit at any flow section, as shown in Fig. 8.35. This phenomenon is called the *secondary flow*, which is superimposed on the longitudinal flow of fluid particles, and the secondary motion of fluid continuously transports momentum from the rest of the flow section toward the corners. As a result, comparatively large longitudinal velocities were measured at the corners. Energy losses caused by secondary flows increase rapidly in more extreme geometries. Experimental studies must be used if precise design information is required for specific problems.

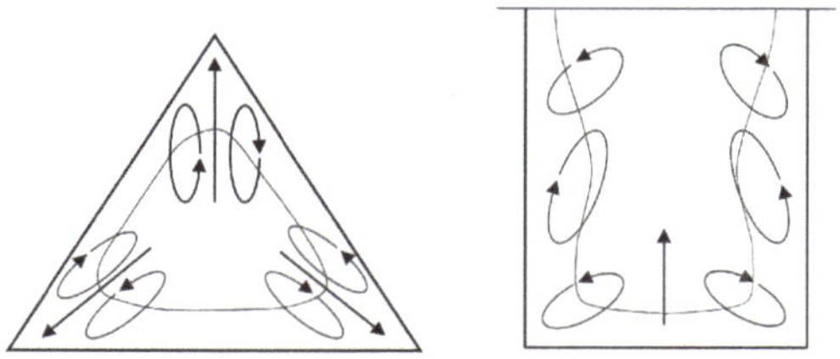

Fig. 8.35 Secondary flows in conduits with non-circular cross-sections

8.6.9 Pipe-Flow Problems

By using the obtained results, it becomes possible to deal with pipe-flow problems which are encountered frequently in practical engineering application. Specifically, pipe-flow problems are classified into two categories: the *single-pipe system* and the *multiple-pipe system*.

Single-pipe system. In this category, the system configuration such as pipe material, pipe surface roughness, devices contributing to minor losses, as well as fluid properties, e.g. ρ and μ, are usually known, and the goal is the determination of one of the following information:

- pressure drop and flow rate for a given pipe length and diameter;
- pipe length, if the pressure drop, pipe diameter, and flow rate are given;
- flow rate for given pipe length, pipe diameter, and pressure drop; or
- pipe diameter with given pipe length, pressure drop, and flow rate.

In solving the problems, the most important step is the determination of the Reynolds number, by which the flow state as laminar or turbulent can be identified. This information is necessary for the determination of friction factor to determine the energy loss in terms of the pressure drop. Occasionally, a try-and-error procedure needs to be conducted until an energy balance is reached.

Multiple-pipe system. Many practical pipe systems consist of a network of pipes of various diameters and lengths assembled in a complicated configuration that may contain parallel and serial pipe connections. The solution procedure is essentially similar to that to a single-pipe system, in which an energy balance should be formulated to each individual pipe and the whole system. Usually, a try-and-error procedure with the aid of numerical calculation should be accomplished to reach an energy balance between any two points in the multiple-pipe system.

To explore the idea, consider a single pipe connecting two tanks shown in Fig. 8.36a, in which the fluid flows from tank A to tank B through different constant-diameter pipes with total length ℓ and surface roughness ε, and six right-angled elbows. If the flow rate Q is given, it is required to determine the pipe diameter d. For the considered problem, the energy equation between points 1 and 2 reads

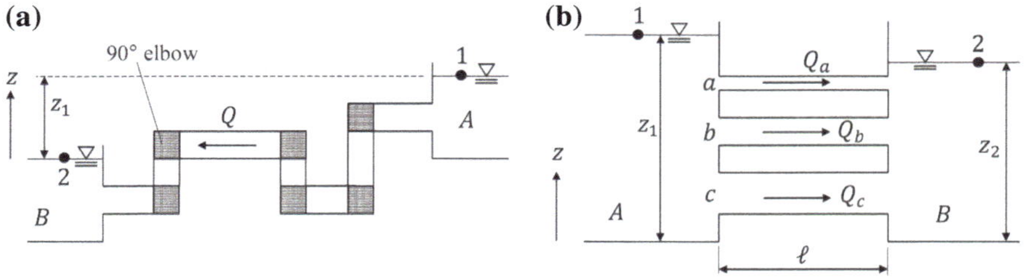

Fig. 8.36 Illustrations of pipe-flow problems. **a** A single-pipe system for the determination of pipe diameter. **b** A multiple-pipe system for the determination of mean average velocities is different pipes

$$\frac{p_1}{\gamma} + \alpha\frac{\bar{u}_{av1}^2}{2g} + z_1 = \frac{p_2}{\gamma} + \alpha\frac{\bar{u}_{av2}^2}{2g} + z_2 + h_{lt}. \tag{8.6.59}$$

Since $p_1 = p_2 = p_{atm}$, $\bar{u}_{av1} = \bar{u}_{av2} = 0$, for the two tanks are assumed to be sufficiently large that the free surfaces remain fixed during the flow, and $z_2 = 0$ if the elevation datum is set at point 2, this equation reduces to

$$z_1 = h_{lt} = \frac{\bar{u}_{av}^2}{2g}\left(f\frac{\ell}{d} + \sum K\right), \qquad \bar{u}_{av} = \frac{4Q}{\pi d^2}, \tag{8.6.60}$$

where $\bar{u}_{av}$ represents the average velocity in the pipe, which is an unknown, because d is yet determined, and K is the loss coefficient of minor losses. For convenience, let $\sum K$ be expressed as

$$\sum K = K_1 + K_2 + \cdots + K_8, \tag{8.6.61}$$

where K_1 is the loss coefficient of inlet loss in tank A, $K_2 = K_3 = \cdots = K_7$ are the same loss coefficient of a 90° elbow, and K_8 represents the loss coefficient of exit loss in tank B. All these eight values can be taken directly from a handbook of fluid engineering. If z_1 is known, Eq. (8.6.60) becomes an algebraic equation for the friction factor f and pipe diameter d. A try-and-error procedure is conducted as follows: First, a specific value of d is prescribed, with which $\bar{u}_{av}$ can be determined by using Eq. (8.6.60)$_2$, which is used subsequently to identify the value of the Reynolds number. Equally, the relative surface roughness ε/d is also obtained. With the information of $\mathcal{R}_e$ and ε/d, the value of friction factor can be obtained from the Moody chart. The value of f is then substituted into Eq. (8.6.60)$_1$ to check if this equation holds. If it is not the case, the procedure is repeated again by prescribing another value to d, until Eq. (8.6.60)$_1$ is satisfied.

For the multiple-pipe system shown in Fig. 8.36b, tanks A and B are connected with each other by three smooth straight pipes denoted by $\{a, b, c\}$ with different diameters but same length ℓ. If it is assumed that fluid flows from tank A to tank B, it follows that the total flow rate Q consists of the flow rates Q_a, Q_b, and Q_c in three pipes, viz.,

$$Q = Q_a + Q_b + Q_c, \tag{8.6.62}$$

with the total energy loss h_{lt} between points 1 and 2 given by

$$h_{lt} = h_{la} + h_{lb} + h_{lc}, \qquad h_{la} = h_{lb} = h_{lc}, \tag{8.6.63}$$

for only major losses in three straight pipes are considered for simplicity. Equations (8.6.62) and (8.6.63) are further recast alternatively as

$$Q=\frac{\pi}{4}\left(\bar{u}_{ava}d_a^2+\bar{u}_{avb}d_b^2+\bar{u}_{avc}d_c^2\right), z_1 - z_2=\frac{\ell}{2g}\left(f_a\frac{\bar{u}_{ava}^2}{d_a}+f_b\frac{\bar{u}_{avb}^2}{d_b}+f_c\frac{\bar{u}_{avc}^2}{d_c}\right),$$
$$f_a\frac{\bar{u}_{ava}^2}{d_a}=f_b\frac{\bar{u}_{avb}^2}{d_b}=f_c\frac{\bar{u}_{avc}^2}{d_c}, \tag{8.6.64}$$

where z_1 and z_2 are the elevations of points 1 and 2, respectively, and $\{f_a, f_b, f_c\}$ represent the friction factors in three straight pipes with diameters $\{d_a, d_b, d_c\}$ and mean average velocities $\{\bar{u}_{ava}, \bar{u}_{avb}, \bar{u}_{avc}\}$. If the total flow rate Q, the elevation difference $z_1 - z_2$ and the diameters of three straight pipes are known, then Eq. (8.6.64) becomes three equations for the friction factors and mean average velocities. A try-and-error procedure is now conducted as follows: First, a specific value of $\bar{u}_{ava}$ is prescribed to determine the value of f_a by using the Moody chart with the corresponding value of the Reynolds number. The values of $\bar{u}_{ava}$ and f_a are substituted into Eqs. $(8.6.64)_{2,3}$ to check if Eq. $(8.6.64)_2$ is satisfied. If it is not the case, the procedure is repeated again until the correct values of $\bar{u}_{ava}$ and f_a are found, so that Eqs. $(8.6.64)_{1,3}$ then provide a set of equations for $\bar{u}_{avb}$ and $\bar{u}_{avc}$, for which the Reynolds numbers in straight pipes b and c need to be determined. The value of $\bar{u}_{avb}$ is chosen, for which the value of $\bar{u}_{avc}$ is obtained by solving Eq. $(8.6.64)_1$. With the obtained values of $\bar{u}_{avb}$ and $\bar{u}_{avc}$, the corresponding Reynolds numbers are then determined, by which the values of f_b and f_c are obtained. The obtained values of $\bar{u}_{avb}, \bar{u}_{avc}, f_b$, and f_c are then substituted into Eq. $(8.6.64)_3$ to check if this equation holds. Again, an another try-and-error procedure should be initiated to achieve the goal.

In the above analysis, the minor losses of pipe inlets and exits were neglected for simplicity. A more complicated circumstance may be encountered if these and other minor losses are taken into account.

8.7 Exercises

8.1 Use the concept of an infinitesimal volume element, as that described in Fig. 5.10a, to derive the velocity distributions of a two-dimensional Couette flow in a horizontal channel and an axis-symmetric Poiseuille flow in a horizontal circular pipe.

8.2 Consider a fully developed laminar flow in a circular pipe which is titled by a counterclockwise angle θ with respect to the horizontal line. Derive the axial velocity distribution of flow.

8.3 Consider the configuration shown in Fig. 8.3a. Derive the profile of tangential velocity if the inner cylinder rotates clockwise with angular velocity ω, while the outer cylinder rotates counterclockwise with the same angular speed. Determine the location where the tangential velocity vanishes.

8.4 Use the solutions obtained in Sect. 8.2.3 to deduce the velocity distribution induced by a circular cylinder which is rotating with constant angular velocity ω_i in an infinite fluid which is otherwise at rest. Compare the result with that for a line vortex of strength $\Gamma = 2\pi r_i^2 \omega_i$ in a frictionless fluid which is at rest at infinity.

8.5 Derive Eqs. (8.2.35) and (8.2.37), namely, the expressions for the profile of axial velocity between two stationary concentric long cylinders, and the maximum axial velocity.

8.6 Consider the configuration shown in Fig. 8.3b. If both cylinders move axially along the x-direction with velocity U_i of the inner cylinder and U_o of the outer cylinder, derive the profile of axial velocity of the fluid contained in the annual region between two cylinders. For simplicity, the pressure gradient along the x-direction is assumed to vanish, and the fluid motion is only induced by the motions of two cylinders.

8.7 Consider the configuration shown in Fig. 8.1a. The upper plate is held stationary, while the lower plate is set to oscillate harmonically whose velocity is described by $U\cos(\omega t)$, where U is the amplitude and ω denotes the frequency. If the fluid contained between two plates is a Newtonian fluid with constant density and dynamic viscosity, determine its velocity profile in the x-direction. For simplicity, there exists no pressure gradient along the x-direction, and the gravity points perpendicular to the page. That is, the fluid motion is induced by the oscillating lower plate, while bounded by the upper plate.

8.8 A flow field is given by

$$u_r = -ar, \qquad u_\theta = \frac{K}{r}, \qquad u_z = 2az,$$

where a and K are constants. The given flow field satisfies the continuity equation everywhere, except at $r = 0$, where a singularity exists. Show that the flow field also satisfies the Navier-Stokes equation everywhere except at $r = 0$, and find the pressure distribution in the flow field. Modify the flow field as

$$u_r = -ar, \qquad u_\theta = \frac{K}{r} f(r), \qquad u_z = 2az,$$

where f is an undetermined function. Determine this function so that the modified flow field satisfies the governing equations for an incompressible viscous Newtonian fluid, and show that the original flow field can be recovered if $r \to \infty$.

8.9 Show that a Stokeslet in a low-Reynolds-number flow does not exert any torque on the surrounding viscous fluid.

8.10 A flow field is given by

$$\boldsymbol{u} = \nabla\chi \times \boldsymbol{\Omega}, \qquad p = 0,$$

where $\mathbf{\Omega}$ is a constant vector, and χ is a scalar function. Show that the given flow field is a solution to Stokes' equations, provided that χ must satisfy

$$\nabla^2\chi - \frac{1}{\nu}\frac{\partial \chi}{\partial t} = 0.$$

Solve this equation for χ, and find the velocity field generated by a sphere of radius a which is rotating with a periodic angular velocity $\|\mathbf{\Omega}\|e^{i\omega t}$.

8.11 Use the Oseen approximation to obtain the stream function of the flow induced by a uniform flow with magnitude U passing through a circular cylinder with radius a.

8.12 Derive the boundary-layer equations for a two-dimensional uniform flow with velocity $U(x)$ over a horizontal flat plate by using the limiting procedure to the full Navier-Stokes equation. The limiting procedure is similar to that use to derive Stokes' equations from the full Navier-Stokes equation.

8.13 A two-dimensional jet enters a reservoir which contains a stationary fluid, as shown in the figure. It is assumed that the jet is a laminar boundary-layer flow, and there is no pressure gradient along the jet (i.e., along the x-direction). If a similarity solution to the stream function of this jet is given by

$$\psi(x, y) = 6\alpha\nu x^{1/3} f(\eta), \qquad \eta = \alpha\frac{y}{x^{2/3}},$$

where α is a dimensional constant and ν represents the kinematic viscosity of fluid, obtain an expression for the function $f(\eta)$ and the corresponding boundary conditions. From the solution to $f(\eta)$, obtain the solution to the stream function.

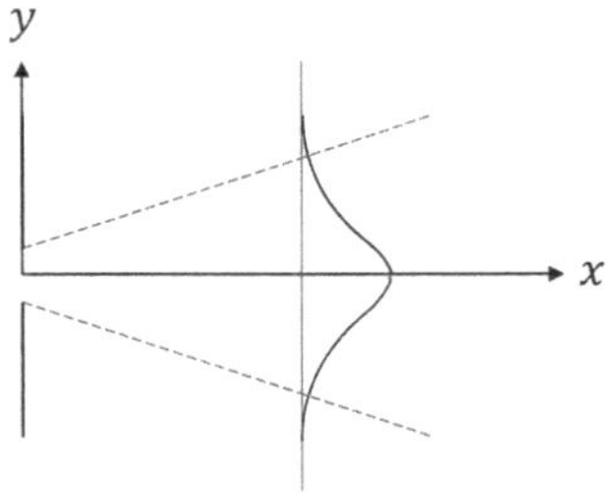

8.14 Use the momentum integral to verify the results summarized in Table 8.2.

8.15 Use the Kármán-Polhausen approximation to obtain a solution to the boundary layer which develops on a surface for which the outer flow velocity is given by

$$U(x) = Ax^{1/6},$$

where A is a constant. From the solution, determine the disturbance thickness δ, displacement thickness δ^*, momentum thickness θ, and shear stress τ_w on the surface.

8.16 Obtain the expressions of $\psi(x, r)$, $\theta(x, r)$ and $\eta(x, r)$ for a point source of heat in an otherwise quiescent fluid with $\mathcal{P}_r = 1$. The results given in Eq. (8.5.56) can be used as a beginning.

8.17 Show that for a point source of heat in a fluid for which $\mathcal{P}_r = 2$, a solution exists in the forms

$$f(\eta) = A\frac{\eta^2}{a+\eta^2}, \qquad F(\eta) = B\frac{1}{(a+\eta^2)^4}.$$

Determine the values of constants A, B, and a which satisfy Eqs. (8.5.22) and $(8.5.23)_1$.

8.18 Derive Eq. (8.6.41), i.e., the ratio of mean average velocity to mean velocity at the centerline of a fully developed turbulent flow in a smooth circular pipe, if the velocity profile is expressed by using the power law equation.

8.19 For fully developed laminar pipe-flows, show that the kinetic energy coefficient α and momentum coefficient β are given by $\alpha = 2$ and $\beta = 4/3$. For fully developed turbulent pipe-flows, if the velocity distribution is expressed by the power law equation, show that the expressions of α and β are given by Eq. (8.6.46).

8.20 Consider a nozzle installed inside a circular pipe, as shown in the figure. Apply the basic equations to the indicated control-volume (the volume enclosed by the dashed lines) to show that the permanent head loss across the nozzle can be expressed as the head loss coefficient given by

$$C_\ell = \frac{p_1 - p_3}{p_1 - p_2} = \frac{1 - A_2/A_1}{1 + A_2/A_1},$$

in which A_1 and A_2 represent respectively the cross-sectional areas at sections 1 and 2 in the figure.

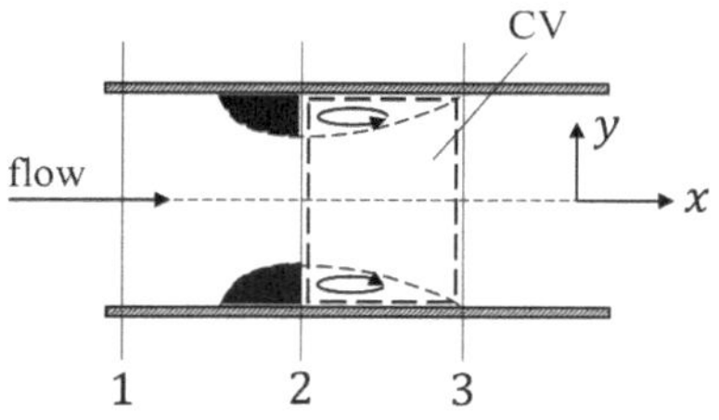

Further Reading

P. Bradshaw, *An Introduction to Turbulence and its Measurements* (Pergamon Press, New York, 1971)

I.G. Currie, *Fundamental Mechanics of Fluids*, 2nd edn. (McGraw-Hill, Singapore, 1993)

O. Darrigol, *Worlds of Flow: A History of Hydrodynamics from the Bernoulli to Prandtl* (Oxford University Press, Oxford, 2005)

R.W. Fox, P.J. Pritchard, A.T. McDonald, *Introduction to Fluid Mechanics*, 7th edn. (Wiley, New York, 2009)

R.J. Goldstein (ed.), *Fluid Mechanics Measurements*, 2nd edn. (Taylor & Francis, New York, 1996)
J. Happel, *Low Reynolds Number Hydrodynamics* (Prentice-Hill, New Jersey, 1965)
J.O. Hinze, *Turbulence*, 2nd edn. (McGraw-Hill, New York, 1975)
W.M. Kays, M.E. Crawford, *Convective Heat and Mass Transfer*, 3rd edn. (McGraw-Hill, Singapore, 1993)
B.R. Munson, D.F. Young, T.H. Okiishi, *Fundamentals of Fluid Mechanics*, 3rd edn. (Wiley, New York, 1990)
R.L. Panton, *Incompressible Flow*, 2nd edn. (Wiley, New York, 1996)
R.H.F. Pao, *Fluid Mechanics* (Wiley, New York, 1961)
L. Rosenhead, *Laminar Boundary Layers* (Dover, New York, 1963)
H. Schlichting, *Boundary Layer Theory*, 7th edn. (McGraw-Hill, New York, 1979)
F.S. Sherman, *Viscous Flow* (McGraw-Hill, New York, 1990)
Z. Sorbjan, *Structure of the Atmospheric Boundary Layer* (Prentice-Hall, New Jersey, 1989)
H. Tennkes, J.L. Lumley, *A First Course in Turbulence* (The MIT Press, Cambridge, 1972)
D.J. Tritton, *Physical Fluid Dynamics* (Oxford University Press, Oxford, 1988)
C. Tropea, A. Yarin, J.F. Foss (eds.), *Springer Handbook of Experimental Fluid Mechanics* (Springer, Berlin, 2007)
A. Tsinober, *An Informal Conceptual Introduction to Turbulence*, 2nd edn. (Springer, Berlin, 2009)
J.M. Wallace, P.V. Hobbs, *Atmospheric Science: An Introductory Survey*, 2nd edn. (Elsevier, New York, 2006)
F.M. White, *Viscous Fluid Flow*, 3rd edn. (McGraw-Hill, New York, 2006)
M. Van Dyke, *Perturbation Methods in Fluid Mechanics* (The Parabolic Press, Stanford, 1975)
M. Van Dyke, *An Album of Fluid Motion* (The Parabolic Press, Stanford, 1988)

Compressible Inviscid Flows

9

Selected phenomena associated with fluid compressibility, and the methods which are used to obtain quantitative descriptions of compressible flows, are discussed in this chapter. For simplicity, the viscous effect is neglected, while the compressible effect, which is a measure of the inertial effect, is taken into account due to its significant influence in high-speed flows. Hence, this chapter is devoted to the discussions on compressible inviscid flows.[1]

The first section deals with a general formulation of the governing equations for compressible inviscid fluids, and Crocco's equation is derived to show that irrotational flows of a compressible fluid correspond to isentropic flows. The second section is devoted to the propagation of disturbances with infinitesimal and finite amplitudes in compressible fluids, by which the propagation speed of sonic signal and the phenomenon of shock waves, including the normal and oblique ones, are considered, which are supplemented by the discussions on the Rankine-Hugoniot equations. The third section concerns with one-dimensional flows, in which how pressure signals reacting upon reaching the interfaces between different fluids and solid boundaries are treated. Non-adiabatic flows, specifically flows in which heat transfer and friction effect involve, are introduced, giving rise to the Fanno and Rayleigh lines to determine the flow conditions graphically. The fourth section deals with multi-dimensional flows in both subsonic and supersonic regions. The Prandtl-Glauert rule relating subsonic flows to incompressible flows and Ackeret's theory of supersonic flows are the main topics of the section. The chapter is ended by a qualitative description of the influence of fluid compressibility on the drag and lift coefficients of a solid body in a compressible flow.

[1]Compressible frictionless flow is a more appropriate terminology, for neglecting of the viscous effect can be accomplished by using either $\mu = 0$ or the assumption of irrotational flow, as discussed in Sect. 7.1.

© Springer International Publishing AG 2019

C. Fang, *An Introduction to Fluid Mechanics*, Springer Textbooks in Earth Sciences, Geography and Environment, https://doi.org/10.1007/978-3-319-91821-1_9

9.1 General Formulation and Crocco's Equation

For compressible flows with negligible viscous effect, the local balances of mass, linear momentum, and energy read respectively

$$\begin{aligned} &\frac{\partial \rho}{\partial t}+\nabla\cdot(\rho\boldsymbol{u})=0, \quad \rho\frac{\partial \boldsymbol{u}}{\partial t}+\rho(\boldsymbol{u}\cdot\nabla)\boldsymbol{u}=-\nabla p, \\ &\rho\frac{\partial e}{\partial t}+\rho(\boldsymbol{u}\cdot\nabla)e=-p\nabla\cdot\boldsymbol{u}+\nabla\cdot(k\nabla T), \end{aligned} \tag{9.1.1}$$

which are supplemented by the state equations given by

$$p=p(\rho,T), \qquad e=e(\rho,T), \tag{9.1.2}$$

in which the body force is assumed to vanish for simplicity, where e is the specific internal energy, T denotes the temperature, and k represents the thermal conductivity of fluids. The inclusion of thermal energy equation results from the fact that the fluid density becomes a field quantity, for which an additional independent equation must be supplied to arrive at a mathematically well-posed problem. The state equations, which can be considered a kind of material equations, are so proposed that the considered compressible fluids are assumed to be simple compressible substances, whose states are determined by the definite values of any two independent intensive properties. Equations (9.1.1) and (9.1.2) are to be solved for the unknown fields $\boldsymbol{u}$, ρ and T. By introducing the specific enthalpy $h=e+p/\rho$, Eq. (9.1.1)$_3$ can be recast alternatively as

$$\rho\frac{\partial h}{\partial t}+\rho(\boldsymbol{u}\cdot\nabla)h=\frac{\partial p}{\partial t}+(\boldsymbol{u}\cdot\nabla)p+\nabla\cdot(k\nabla T), \tag{9.1.3}$$

which is an alternative form of the energy equation.

For the special case in which heat conduction is negligible, Eq. (9.1.1)$_3$ is simplified to

$$\rho\frac{\mathrm{D}e}{\mathrm{D}t}=-p\nabla\cdot\boldsymbol{u}. \tag{9.1.4}$$

For ideal gases, it follows from thermodynamics that

$$e=e(T), \qquad de=c_v\mathrm{d}T, \tag{9.1.5}$$

where c_v is the specific heat at constant volume. Substituting these expressions and the ideal gas state equation into Eq. (9.1.4) yields

$$\rho c_v\frac{\mathrm{D}T}{\mathrm{D}t}=-p\nabla\cdot\boldsymbol{u}, \tag{9.1.6}$$

which is another form of the thermal energy equation. By using Eq. (9.1.1)$_1$ to replace the term $\nabla\cdot\boldsymbol{u}$ and the ideal gas state equation to replace T, this equation can be brought to the form

$$\frac{1}{p}\frac{\mathrm{D}p}{\mathrm{D}t}=\left(\frac{R+c_v}{c_v}\right)\frac{1}{\rho}\frac{\mathrm{D}\rho}{\mathrm{D}t}=\frac{\gamma}{\rho}\frac{\mathrm{D}\rho}{\mathrm{D}t}, \qquad c_p-c_v=R, \ \gamma=\frac{c_p}{c_v}, \tag{9.1.7}$$

where c_p is the specific heat at constant pressure, R represents the gas constant, and γ denotes the specific-heat ratio. Integrating this equation gives

$$\frac{p}{\rho^\gamma}=\text{constant}, \tag{9.1.8}$$

along each streamline, which is the *isentropic law* in thermodynamics. Thus, the assumption of inviscid fluid with negligible heat conduction is compatible with an isentropic flow.[2] Equation (9.1.8) shows that a constant value of p/ρ^γ along each streamline corresponds to a constant *entropy* along the same streamline. If a flow originates from a region where the entropy is constant everywhere, the constant in the equation remains the same for all streamlines, and hence p/ρ^γ will be constant everywhere. The boundary conditions associated with Eq. (9.1.1) may be given by prescribing the velocity and temperature or heat flux on the boundaries. Since the flows are assumed to be inviscid, instead of the conventional no-slip boundary condition, Eq. (7.1.2) will be used.

On the other hand, isentropic flows also correspond to irrotational flows, which can be justified by *Crocco's equation*.[3] Consider a flow of an inviscid fluid without any body force, for which Euler's equation reduces to

$$\frac{\partial \boldsymbol{u}}{\partial t} + (\boldsymbol{u} \cdot \nabla)\boldsymbol{u} = -\frac{1}{\rho}\nabla p, \tag{9.1.9}$$

which is expressed alternatively as

$$\frac{\partial \boldsymbol{u}}{\partial t} + \nabla\left(\frac{1}{2}\boldsymbol{u}\cdot\boldsymbol{u}\right) - \boldsymbol{u}\times\boldsymbol{\omega} = -\frac{1}{\rho}\nabla p, \qquad \boldsymbol{\omega} = \nabla\times\boldsymbol{u}, \tag{9.1.10}$$

in which the identity

$$(\boldsymbol{u}\cdot\nabla)\,\boldsymbol{u} = \nabla\left(\frac{1}{2}\boldsymbol{u}\cdot\boldsymbol{u}\right) - \boldsymbol{u}\times\boldsymbol{\omega}, \tag{9.1.11}$$

has been used. The $T\mathrm{d}S$ equations of thermodynamics are given by[4]

$$T\mathrm{d}s = \mathrm{d}e + p\mathrm{d}\left(\frac{1}{\rho}\right) = \mathrm{d}h - \frac{1}{\rho}\mathrm{d}p, \tag{9.1.12}$$

where s is the *specific entropy*. Since $\mathrm{d}\boldsymbol{\ell}\cdot\nabla\alpha = \mathrm{d}\alpha$, which represents the total derivative of any quantity α for any infinitesimal line segment $\mathrm{d}\boldsymbol{\ell}$, it follows that

$$T\nabla s = \nabla h - \frac{1}{\rho}\nabla p. \tag{9.1.13}$$

Substituting this equation into Eq. (9.1.10) results in

$$\boldsymbol{u}\times\boldsymbol{\omega} + T\nabla s = \nabla\left(h + \frac{1}{2}\boldsymbol{u}\cdot\boldsymbol{u}\right) + \frac{\partial \boldsymbol{u}}{\partial t}, \tag{9.1.14}$$

which is known as *Crocco's equation* for inviscid flows without any body force.

---

[2]The inviscid assumption eliminates any irreversible loss, while negligible heat conduction implies adiabatic. A reversible adiabatic process is an isentropic process, to be discussed in Sect. 11.5.2

[3]The equation was first enunciated by Friedmann in a paper in 1922. However, credit has been given to Crocco. Alexander Alexandrovich Friedmann, 1888–1925, a Russian physicist and mathematician, who is best known for his theory that the universe was expanding, known as the *Friedmann equations*. Luigi Crocco, 1909–1986, an Italo-American mathematician and space engineer.

[4]The equations will be discussed in Sect. 11.8.

For adiabatic flows of an inviscid fluid without any body force, the energy and Euler equations reduce respectively to

$$\rho\frac{\mathrm{D}h}{\mathrm{D}t} = \frac{\mathrm{D}p}{\mathrm{D}t}, \qquad \rho\frac{\mathrm{D}\boldsymbol{u}}{\mathrm{D}t} = -\nabla p. \tag{9.1.15}$$

Taking inner product of Eq. $(9.1.15)_2$ with $\boldsymbol{u}$ yields

$$\rho\frac{\mathrm{D}}{\mathrm{D}t}\left(\frac{1}{2}\boldsymbol{u}\cdot\boldsymbol{u}\right) = -\boldsymbol{u}\cdot\nabla p, \tag{9.1.16}$$

which is substituted into Eq. $(9.1.15)_1$ to obtain

$$\begin{aligned} &\rho\frac{\mathrm{D}}{\mathrm{D}t}\left(h+\frac{1}{2}\boldsymbol{u}\cdot\boldsymbol{u}\right) = \frac{\mathrm{D}p}{\mathrm{D}t} - \boldsymbol{u}\cdot\nabla p, \qquad \longrightarrow \qquad \rho\frac{\mathrm{D}h_s}{\mathrm{D}t} = \frac{\partial p}{\partial t}, \\ &h_s = h + \frac{1}{2}\boldsymbol{u}\cdot\boldsymbol{u}, \end{aligned} \tag{9.1.17}$$

in which h_s is called the *specific stagnation enthalpy*. For steady flows, the right-hand-side of Eq. $(9.1.17)_2$ vanishes, indicating that h_s is constant along each streamline. Substituting this result into Eq. (9.1.14) gives

$$\boldsymbol{u}\times\boldsymbol{\omega} + T\nabla s = \nabla h_s, \tag{9.1.18}$$

which is valid for steady, adiabatic flows of an inviscid fluid without any body force. It is seen that the term $T\nabla s$ must be perpendicular to streamlines, for the terms ∇h_s and $\boldsymbol{u}\times\boldsymbol{\omega}$ are also perpendicular to streamlines. As a result, the above equation can be reduced to a scalar one given by

$$\|\boldsymbol{u}\|\,\|\boldsymbol{\omega}\| + T\frac{\mathrm{d}s}{\mathrm{d}n} = \frac{\mathrm{d}h_s}{\mathrm{d}n}, \tag{9.1.19}$$

where n represents a local coordinate perpendicular to a specific streamline.

It occurs frequently that if h_s is constant along each streamline, it is constant everywhere. With this, Eq. (9.1.19) is simplified to

$$\|\boldsymbol{u}\|\,\|\boldsymbol{\omega}\| + T\frac{\mathrm{d}s}{\mathrm{d}n} = 0, \tag{9.1.20}$$

showing that if s = constant, then $\|\boldsymbol{\omega}\| = 0$. Conversely, if $\|\boldsymbol{\omega}\| = 0$, then $\mathrm{d}s/\mathrm{d}n$ must vanish, yielding a constant value of the specific entropy. It follows that isentropic flows are irrotational flows and vice versa, provided that the flows are steady, frictionless, and adiabatic without any body force.

9.2 Shock Waves

This section deals with the characteristics of shock waves occurring in supersonic flows. First, the propagation of infinitesimal internal waves (internal disturbances) is examined, resulting in the speed of sound in a gas. The obtained result is followed to study the propagation of finite-amplitude disturbances, and the features of steady flows in which standing shock waves involve. The Rankine-Hugoniot equations for normal shock waves are derived and discussed. The influence of boundary angle relative to the flow direction, which may induce oblique shock waves in supersonic flows, are also studied.

9.2.1 Propagation of Infinitesimal Disturbances

Consider a fluid as an ideal gas, which is initially at rest and through which an infinitesimal small one-dimensional (or plane) disturbance is traveling along the x-direction. The disturbance is assumed to travel sufficiently fast that the heat conduction occurring in the fluid may be neglected, yielding an adiabatic circumstance, for which Eqs. (9.1.1)$_{1,2}$ and (9.1.18) reduce respectively to

$$\frac{\partial \rho}{\partial t}+\frac{\partial}{\partial x}(\rho u)=0, \qquad \frac{\partial u}{\partial t}+u\frac{\partial u}{\partial x}=-\frac{1}{\rho}\frac{\partial p}{\partial x}, \qquad \frac{p}{\rho^{\gamma}}=\text{constant}, \tag{9.2.1}$$

for the undetermined fields p, ρ, and u along each streamline. Since Eq. (9.2.1)$_3$ implies that the flow is isentropic, $p=p(\rho,s)=p(\rho)$, i.e., the pressure field is only a function of density. It follows that

$$\frac{\partial p}{\partial x}=\frac{\mathrm{d}p}{\mathrm{d}\rho}\frac{\partial \rho}{\partial x}, \tag{9.2.2}$$

with which Eqs. (9.2.1)$_{1,2}$ become

$$\frac{\partial \rho}{\partial t}+u\frac{\partial \rho}{\partial x}+\rho\frac{\partial u}{\partial x}=0, \qquad \frac{\partial u}{\partial t}+u\frac{\partial u}{\partial x}+\frac{1}{\rho}\frac{\mathrm{d}p}{\mathrm{d}\rho}\frac{\partial \rho}{\partial x}=0. \tag{9.2.3}$$

Let the field quantities be decomposed as

$$p=p_0+p', \qquad \rho=\rho_0+\rho', \qquad u=0+u', \tag{9.2.4}$$

where p_0 and ρ_0 are the undisturbed values which are constants, and the primes denote the perturbations in the values caused by the passage of disturbance. Substituting these expressions into Eq. (9.2.3) yields

$$\frac{\partial \rho'}{\partial t}+u'\frac{\partial \rho'}{\partial x}+(\rho_0+\rho')\frac{\partial u'}{\partial x}=0, \qquad \frac{\partial u'}{\partial t}+u'\frac{\partial u'}{\partial x}+\frac{1}{\rho_0+\rho'}\frac{\mathrm{d}p}{\mathrm{d}\rho}\frac{\partial \rho'}{\partial x}=0. \tag{9.2.5}$$

Since the terms ρ'/ρ_0, p'/p_0, and u' are small for small-amplitude disturbances, neglecting the products of primed quantities and their quadratic terms gives

$$\frac{\partial \rho'}{\partial t}+\rho_0\frac{\partial u'}{\partial x}=0, \qquad \frac{\partial u'}{\partial t}+\frac{1}{\rho_0}\left(\frac{\mathrm{d}p}{\mathrm{d}\rho}\right)_0\frac{\partial \rho'}{\partial x}=0, \tag{9.2.6}$$

which is a linearized form of Eq. (9.2.5), where the term $\mathrm{d}p/\mathrm{d}\rho$ has been expanded in a Taylor series about the undisturbed state, with $(\mathrm{d}p/\mathrm{d}\rho)_0$ the first term, i.e., the value of $\mathrm{d}p/\mathrm{d}\rho$ in the undisturbed state. Combining two equations results in

$$\frac{\partial^2 \rho'}{\partial t^2}-\left(\frac{\mathrm{d}p}{\mathrm{d}\rho}\right)_0\frac{\partial^2 \rho'}{\partial x^2}=0, \qquad \frac{\partial^2 u'}{\partial t^2}-\left(\frac{\mathrm{d}p}{\mathrm{d}\rho}\right)_0\frac{\partial^2 u'}{\partial x^2}=0, \tag{9.2.7}$$

showing that both the density perturbation ρ' and velocity perturbation u' have the same functional form, and u' may be considered a function of ρ' only, although ρ' and u' are functions of x and t.

Since Eq. (9.2.7)$_1$ is a one-dimensional wave equation, its solution is given by

$$\rho'(x,t) = f_1\left(x - \sqrt{\left(\frac{\mathrm{d}p}{\mathrm{d}\rho}\right)_0} t\right) + f_2\left(x + \sqrt{\left(\frac{\mathrm{d}p}{\mathrm{d}\rho}\right)_0} t\right), \tag{9.2.8}$$

where f_1 and f_2 are any two differentiable functions, which represent respectively a wave traveling in the positive x-axis with velocity $\sqrt{(\mathrm{d}p/\mathrm{d}\rho)_0}$, and a wave traveling in the negative x-axis with the same velocity. The speed at which the density perturbation (and hence the velocity perturbation) travels is then obtained as[5]

$$a_0 = \sqrt{\left(\frac{\mathrm{d}p}{\mathrm{d}\rho}\right)_0}. \tag{9.2.9}$$

Since the disturbance was assumed to be small, and sound is also a small disturbance, this equation represents then the *speed of sound* in a quiescent ideal gas.

The obtained results were based on the assumption that u' should be small, which needs to be examined. Substituting Eq. (9.2.9) into Eq. (9.2.6)$_2$ gives

$$\frac{\partial u'}{\partial t} + \frac{a_0^2}{\rho_0}\frac{\partial \rho'}{\partial x} = 0. \tag{9.2.10}$$

Let the wave of u' traveling in the positive x-axis be denoted by $u' = f_1(x - a_0 t)$. It follows that

$$\frac{\partial u'}{\partial t} = -a_0 f_1'(x - a_0 t) = -a_0 \frac{\partial u'}{\partial x}, \tag{9.2.11}$$

where f_1' represents its derivative with respect to the arguments. Substituting Eq. (9.2.11) into Eq. (9.2.10) gives

$$\frac{\partial u'}{\partial x} = \frac{a_0}{\rho_0}\frac{\partial \rho'}{\partial x}, \tag{9.2.12}$$

[5]Another familiar form of Eq. (9.2.9) may be obtained from Eq. (9.2.1)$_3$. It is seen that

$$\frac{p}{\rho^\gamma} = \frac{p_0}{\rho_0^\gamma}, \qquad \longrightarrow \qquad \frac{\mathrm{d}p}{\mathrm{d}\rho} = \gamma\frac{p}{\rho}.$$

Substituting the ideal gas state equation into this equation yields

$$\frac{\mathrm{d}p}{\mathrm{d}\rho} = \gamma R T,$$

giving rise to

$$a_0 = \sqrt{\left(\frac{\mathrm{d}p}{\mathrm{d}\rho}\right)_0} = \sqrt{\gamma R T_0} = \sqrt{\gamma\frac{p_0}{\rho_0}},$$

where T_0 is the gas temperature in the undisturbed state. This equation indicates that the speed of sound is only a function of gas temperature and increases as $\sqrt{T}$.

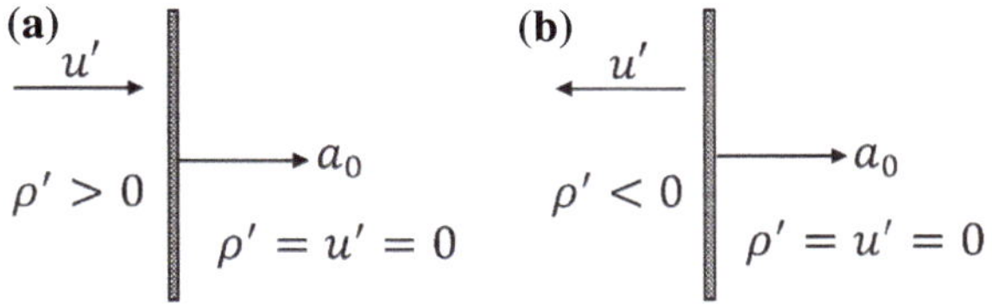

Fig. 9.1 The fluid velocities before and after an infinitesimal wave front traveling at the speed of sound a_0. **a** A compressive wave front. **b** An expansive wave front

which is integrated with respect to x to obtain

$$\frac{u'}{a_0} = \frac{\rho'}{\rho_0}, \tag{9.2.13}$$

with the condition that $u' = 0$ if $\rho' = 0$. This equation shows that $u'/a_0 \ll 1$, provided that $\rho'/\rho_0 \ll 1$, which has been used in the analysis. It exposes also a simple relation between u' and ρ', as implied by Eq. (9.2.7). The waves induced by infinitesimal disturbances may be compressive or expansive. For the former case ρ' is positive, so that Eq. (9.2.13) delivers that u' is also positive. In other words, the fluid velocity behind a compressive wave is such that the fluid particles tend to follow the wave, as shown in Fig. 9.1a. On the contrary, for expansive waves, ρ' is negative, so is u', and the fluid behind an expansive wave tends to move away from the wave front, as shown in Fig. 9.1b.

9.2.2 Propagation of Finite Disturbances

Consider the same circumstance in the last section, except that the one-dimensional disturbance assumes a finite amplitude. The local balances of mass and linear momentum are the same as Eqs. $(9.2.1)_{1,2}$, respectively. Although $p = p(\rho)$ and $u = u(\rho)$ were assumed for infinitesimal disturbances, they are used here again for simplicity. With these, it follows that

$$\frac{\partial \rho}{\partial t} = \frac{\mathrm{d}\rho}{\mathrm{d}u}\frac{\partial u}{\partial t}, \qquad \frac{\partial \rho}{\partial x} = \frac{\mathrm{d}\rho}{\mathrm{d}u}\frac{\partial u}{\partial x}, \qquad \frac{\partial p}{\partial x} = \frac{\mathrm{d}p}{\mathrm{d}\rho}\frac{\mathrm{d}\rho}{\mathrm{d}u}\frac{\partial u}{\partial x}, \tag{9.2.14}$$

so that Eqs. $(9.2.1)_{1,2}$ are expressed alternatively as

$$\frac{\mathrm{d}\rho}{\mathrm{d}u}\left[\frac{\partial u}{\partial t} + u\frac{\partial u}{\partial x}\right] + \rho\frac{\partial u}{\partial x} = 0, \qquad \left[\frac{\partial u}{\partial t} + u\frac{\partial u}{\partial x}\right] = -\frac{1}{\rho}\frac{\mathrm{d}p}{\mathrm{d}\rho}\frac{\mathrm{d}\rho}{\mathrm{d}u}\frac{\partial u}{\partial x}. \tag{9.2.15}$$

Combining two equations yields

$$\rho\frac{\mathrm{d}u}{\mathrm{d}\rho}\frac{\partial u}{\partial x} = \frac{1}{\rho}\frac{\mathrm{d}p}{\mathrm{d}\rho}\frac{\mathrm{d}\rho}{\mathrm{d}u}\frac{\partial u}{\partial x}, \qquad \longrightarrow \qquad \mathrm{d}u = \pm\sqrt{\frac{\mathrm{d}p}{\mathrm{d}\rho}\frac{\mathrm{d}\rho}{\rho}}, \tag{9.2.16}$$

which is recast alternatively as

$$\frac{\mathrm{d}u}{a} = \pm\frac{\mathrm{d}\rho}{\rho}, \qquad a = \sqrt{\frac{\mathrm{d}p}{\mathrm{d}\rho}}. \tag{9.2.17}$$

The physical interpretation of a is yet clear at the moment, although it is observed that $a \to a_0$ if the amplitude of disturbance is infinitesimal. Since Eq. (9.2.13) can also be expressed as $\mathrm{d}u/a_0 = \mathrm{d}\rho/\rho_0$, to which Eq. (9.2.17) must reduce under a linearized approximation, comparing two equations shows that the plus and minus signs in Eq. (9.2.17) are devoted respectively to forward-running and backward-running waves (i.e., for compression and expansion waves). This is done so, in order that the fluid-particle velocities following a compression wave or moving away from an expansion wave may be recovered.

Substituting the case of forward-running waves of Eq. (9.2.17) into Eq. $(9.2.1)_2$ gives

$$\frac{\partial u}{\partial t} + (u + a)\frac{\partial u}{\partial x} = 0, \tag{9.2.18}$$

to which the solution is given by

$$u(x, t) = f\left[x - (u + a)t\right], \tag{9.2.19}$$

where f represent any differentiable functional, in which both u and a are functions of x and t. This equation indicates a wave traveling in the positive x-direction with velocity $U = u + a$. Substituting Eq. $(9.2.1)_3$ into Eq. $(9.2.17)_2$ yields

$$a = \sqrt{\gamma \frac{p_0}{\rho_0}} \left(\frac{\rho}{\rho_0}\right)^{(\gamma-1)/2} = a_0 \left(\frac{\rho}{\rho_0}\right)^{(\gamma-1)/2}, \tag{9.2.20}$$

which is incorporated into Eq. $(9.2.17)_1$ to obtain

$$\mathrm{d}u = \frac{a_0}{\rho_0^{(\gamma-1)/2}} \rho^{(\gamma-3)/2} \mathrm{d}\rho. \tag{9.2.21}$$

Integrating this equation results in

$$u = \frac{2}{\gamma - 1}(a - a_0), \qquad \longrightarrow \qquad a = a_0 + \frac{\gamma - 1}{2}u, \tag{9.2.22}$$

in which the condition $u = 0$ at $\rho = \rho_0$ and Eq. (9.2.20) have been used. This result shows that $u > 0$, for $a > a_0$ in general if $\gamma > 1$, and the difference between a and a_0 is proportional to the local fluid velocity u. In view of these and under the assumption that $u > 0$, the propagation speed of a finite-amplitude disturbance is obtained as

$$U(x, t) = a + u = a_0 + \frac{\gamma + 1}{2}u, \tag{9.2.23}$$

showing that U is greater than the speed of sound a_0 for $u > 0$, and is no longer a constant but a function depending on the local fluid velocity. In addition, since U depends on x and t, it is not an equilibrium speed. In other words, the propagation speed of a finite-amplitude disturbance changes in space and time.

The distance L that is travelled by a finite-amplitude disturbance in a time duration τ is obtained as

$$L = \left(a_0 + \frac{\gamma + 1}{2}u\right)\tau, \tag{9.2.24}$$

whose Galilean transformation L^* to an observer moving at the speed a_0 is given by

$$L^* = \frac{\gamma + 1}{2}u\tau. \tag{9.2.25}$$

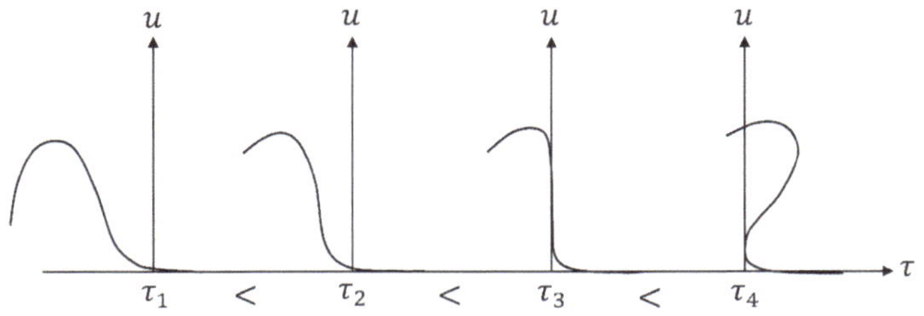

Fig. 9.2 The progression of a finite-amplitude disturbance in an otherwise quiescent fluid relative to an observer moving at the speed of sound

That is, relative to this observer the wave travels a distance which depends on the magnitude and sign of local fluid velocity u. The regions of high local fluid velocity will travel faster than the regions with low local fluid velocity, yielding a smooth disturbance in arbitrary form of the wave shown in Fig. 9.2, in which $\tau_1 < \tau_2 < \tau_3 < \tau_4$. At time τ_1 a smooth fluid velocity profile is assumed, which travels along the positive x-direction in subsequent times. At time τ_2, relative to an observer moving at speed a_0, the regions with high local fluid velocity advance farther than those with lower velocity. This circumstance continues, until at a specific instant, say time τ_3, the wave front becomes vertical as the high-velocity regions continue to advance faster than the slower regions. Finally, at time τ_4, the regions with high velocity have overtaken the portion of signal which is moving at the speed of sound a_0, which is an unjustified configuration, for three values of u exist at a fixed location. Hence, the wave front will steepen as described, until the circumstance at time τ_3 is reached. At this stage, a sharp discontinuity in the field quantities exists, which is called a *shock wave*. For $t > \tau_3$, the shock wave will propagate in an equilibrium configuration. The formation of shock wave is somewhat similar to the formation of tsunami wave with large amplitude near coastal regions, although the underlying physical mechanisms are different. By and large, a smooth, finite-amplitude compression wave propagates in a non-equilibrium configuration with its different parts traveling at different speeds in such a way that the wave front will steepen during the motion. Eventually, the steeping of wave front will reach to a state, at which a sharp change in the field quantities takes place over a very narrow region in space, yielding the formation of shock wave, which will continue to travel at an equilibrium speed.

The obtained results are valid for $u > 0$, corresponding to compression waves moving forward. For expansion waves, $u < 0$ for forward-moving waves, so that the wave front will move more slowly than the speed of sound, as indicated by Eq. (9.2.23). In parallel, the more intensive parts of wave move most slowly, resulting in the spreading of wave front. It is concluded that compression waves steepen as they propagate, while expansion waves spread out during the propagation.

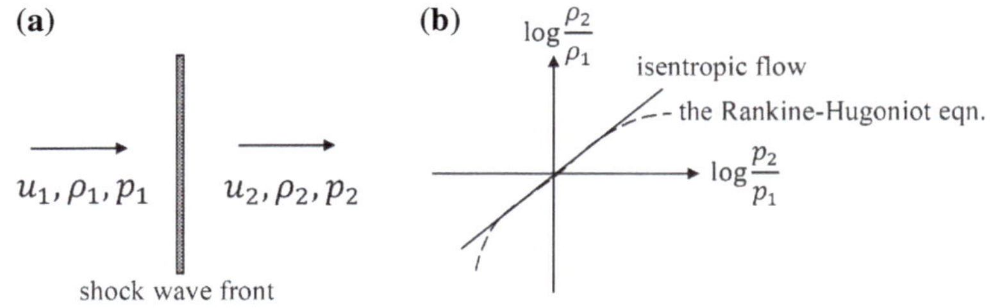

Fig. 9.3 Characteristics of normal shock waves. **a** The geometric configuration of a stationary shock wave. **b** The difference between an isentropic flow (solid line) and the Rankine-Hugoniot equations (dashed line)

9.2.3 The Rankine-Hugoniot Equations

Shock waves are very thin compared with most macroscopic length scales of flows, so that they are conventionally approximated as line discontinuities in the fluid properties. Although a shock wave is moving in a fluid, it becomes stationary by using the Galilean transformation, so that the fluid approaching it at one state and leaving it at another state. These two states are denoted by using subscripts 1 and 2, respectively, as shown in Fig. 9.3a. The velocity, pressure, and density of incoming flow are denoted by $\{u_1, p_1, \rho_1\}$, respectively, while those of leaving flow are given by $\{u_2, p_2, \rho_2\}$. Since the shock wave is oriented normal to the fluid velocity, it is referred to as a *normal shock wave*.

For a line discontinuity, differential equations cannot be used directly for the quantities across it. Rather, algebraic equations need to be formulated. The mass flow rates and linear momentums per unit area before and after the shock wave are given respectively by $\{\rho_1 u_1, \rho_1 u_1^2\}$ and $\{\rho_2 u_2, \rho_2 u_2^2\}$, and the conservations of mass and linear momentum require that

$$\rho_1 u_1 = \rho_2 u_2, \qquad p_1 + \rho_1 u_1^2 = p_2 + \rho_2 u_2^2, \tag{9.2.26}$$

for a steady flow through a shock wave. The conservation of energy between points 1 and 2 reads

$$h_1 + \frac{1}{2}u_1^2 = h_2 + \frac{1}{2}u_2^2, \quad \longrightarrow \quad \frac{\gamma}{\gamma - 1}\frac{p_1}{\rho_1} + \frac{1}{2}u_1^2 = \frac{\gamma}{\gamma - 1}\frac{p_2}{\rho_2} + \frac{1}{2}u_2^2, \tag{9.2.27}$$

in terms of the specific enthalpy, in which it follows from thermodynamics that

$$h = c_p T = c_p \frac{p}{\rho R} = \frac{\gamma}{\gamma - 1}\frac{p}{\rho}, \tag{9.2.28}$$

with the assumptions that the fluid is an ideal gas, and the flow is steady and adiabatic.

Equation (9.2.26) is further simplified to

$$u_2 - u_1 = \frac{p_1 - p_2}{\rho_1 u_1}, \tag{9.2.29}$$

which is multiplied by $(u_1 + u_2)$ to obtain

$$u_2^2 - u_1^2 = \frac{p_1 - p_2}{\rho_1}\left(1 + \frac{u_2}{u_1}\right) = (p_1 - p_2)\left(\frac{1}{\rho_1} + \frac{1}{\rho_2}\right), \tag{9.2.30}$$

in which Eq. (9.2.26)$_1$ has been used. Substituting Eq. (9.2.27)$_2$ into this equation yields

$$\frac{2\gamma}{\gamma - 1}\left(\frac{p_1}{\rho_1} - \frac{p_2}{\rho_2}\right) = (p_1 - p_2)\left(\frac{1}{\rho_1} + \frac{1}{\rho_2}\right), \tag{9.2.31}$$

which relates the pressures and densities across the shock wave. This equation can be rearranged to obtain an expression for the density ratio given viz.,

$$\frac{\rho_2}{\rho_1} = \frac{p_1 + mp_2}{mp_1 + p_2}, \qquad m = \frac{\gamma + 1}{\gamma - 1}, \tag{9.2.32}$$

which, by using Eq. (9.2.26)$_1$, is expressed alternatively as

$$\frac{\rho_2}{\rho_1} = \frac{1 + m(p_2/p_1)}{m + p_2/p_1} = \frac{u_1}{u_2}, \tag{9.2.33}$$

which is referred to as the *Rankine-Hugoniot equations*. Equation (9.2.33) relates the density ratio across a shock wave to the pressure and velocity ratios. The pressure, density, and velocity of a flow after a shock wave are immediately determined by using these equations, provided that their values before the shock wave are known.

Physically, the formation of a shock wave is not an isentropic process. If it were the case, it follows from Eq. (9.2.1)$_3$ that the density ratio across a shock wave would be given by

$$\frac{\rho_2}{\rho_1} = \left(\frac{p_2}{p_1}\right)^{1/\gamma}, \tag{9.2.34}$$

which does not coincide with Eq. (9.2.33). A graphic illustration of Eqs. (9.2.33) and (9.2.34) is given in Fig. 9.3b. It is seen that shock waves depart from isentropic flows, unless the pressure and density ratios are very close to unity. In such a circumstance, the shock waves become very weak, which may more appropriately be described as infinitesimal disturbances.

9.2.4 Normal Shock Waves

A normal shock wave can form under the definite conditions. For an ideal gas in a steady flow shown in Fig. 9.3a, the second law of thermodynamics reads the form

$$s_2 - s_1 = c_p \ln\left(\frac{T_2}{T_1}\right) - R\ln\left(\frac{p_2}{p_1}\right) = c_v \ln\left(\frac{p_2}{p_1}\right) - c_p \ln\left(\frac{\rho_2}{\rho_1}\right), \tag{9.2.35}$$

which is the difference in the specific entropy of fluid across a shock wave. Denoting $(s_2 - s_1)$ by Δs in this equation yields

$$\frac{\Delta s}{c_v} = \ln\left(\frac{p_2}{p_1}\right) - \gamma \ln\left(\frac{\rho_2}{\rho_1}\right). \tag{9.2.36}$$

Using this equation to evaluate the values of Δs for the process which satisfies the Rankine-Hugoniot equations and the process which is isentropic under the same pressure ratio gives respectively

$$\frac{\Delta s}{c_v}\bigg|_{\mathrm{RH}} = \ln\left(\frac{p_2}{p_1}\right) - \gamma\ln\left(\frac{\rho_2}{\rho_1}\right)\bigg|_{\mathrm{RH}}, \qquad \frac{\Delta s}{c_v}\bigg|_{\mathrm{is}} = 0 = \ln\left(\frac{p_2}{p_1}\right) - \gamma\ln\left(\frac{\rho_2}{\rho_1}\right)\bigg|_{\mathrm{is}}, \tag{9.2.37}$$

with the subscripts "RH" and "is" representing the Rankine-Hugoniot and isentropic processes, respectively. Combining two equations gives

$$\frac{\Delta s}{c_v}\bigg|_{\mathrm{RH}} = \gamma\left[\ln\left(\frac{\rho_2}{\rho_1}\right)\bigg|_{\mathrm{is}} - \ln\left(\frac{\rho_2}{\rho_1}\right)\bigg|_{\mathrm{RH}}\right]. \tag{9.2.38}$$

Since the second law of thermodynamics requires that $\Delta s \geq 0$, it follows that

$$\ln\left(\frac{\rho_2}{\rho_1}\right)\bigg|_{\mathrm{is}} \geq \ln\left(\frac{\rho_2}{\rho_1}\right)\bigg|_{\mathrm{RH}}, \tag{9.2.39}$$

which can only be fulfilled if $\ln(\rho_2/\rho_1) > 0$ and $\ln(p_2/p_1) > 0$, corresponding to the first quadrant in Fig. 9.3b. It is concluded that

$$\frac{\rho_2}{\rho_1} \geq 1, \qquad \frac{u_2}{u_1} \leq 1, \tag{9.2.40}$$

indicating that after the fluid has passed through a shock wave, it density is increased with the cost of a velocity drop.

Dividing Eq. $(9.2.26)_2$ by Eq. $(9.2.26)_1$ gives

$$u_1 + \frac{p_1}{\rho_1 u_1} = u_2 + \frac{p_2}{\rho_2 u_2}, \tag{9.2.41}$$

which is expressed alternatively as

$$u_1 + \frac{a_1^2}{\gamma u_1} = u_2 + \frac{a_2^2}{\gamma u_2}, \tag{9.2.42}$$

in which $a_1^2 = \gamma p_1/\rho_1$ and $a_2^2 = \gamma p_2/\rho_2$. Substituting Eq. $(9.2.27)_2$ into this equation yields

$$\frac{1}{2}u_1^2 + \frac{a_1^2}{\gamma - 1} = \frac{1}{2}u_2^2 + \frac{a_2^2}{\gamma - 1} = \frac{\gamma + 1}{2(\gamma - 1)}a_*^2, \qquad u_{1*} = u_{2*} = a_{1*} = a_{2*} = a_*, \tag{9.2.43}$$

where the subscript "$*$" is used to denote the local sonic velocity. With this, the velocity difference across a shock wave becomes

$$u_1 - u_2 = \frac{1}{\gamma u_2}\left(\frac{\gamma + 1}{2}a_*^2 - \frac{\gamma - 1}{2}u_2^2\right) - \frac{1}{\gamma u_1}\left(\frac{\gamma + 1}{2}a_*^2 - \frac{\gamma - 1}{2}u_1^2\right), \tag{9.2.44}$$

which reduces to

$$u_1 u_2 = a_*^2. \tag{9.2.45}$$

This equation is referred to as the *Prandtl* or *Meyer relation*,[6] which relates the fluid velocities before and after a shock wave to the local speed of sound.

Multiplying Eq. (9.2.40)$_2$ by u_1 and substituting the Meyer relation into the resulting equation gives

$$\frac{u_1^2}{a_*^2} \geq 1. \tag{9.2.46}$$

Equally, dividing Eq. (9.2.43) by u_1^2 leads to

$$\frac{u_1^2}{a_*^2} = \frac{(\gamma+1)\mathcal{M}_{a1}^2}{2+(\gamma-1)\mathcal{M}_{a1}^2}, \qquad \mathcal{M}_{a1} = \frac{u_1}{a_*}, \tag{9.2.47}$$

where $\mathcal{M}_{a1}$ is the Mach number of the fluid before a shock wave. This equation is substituted into Eq. (9.2.46) to obtain

$$\mathcal{M}_{a1} \geq 1, \tag{9.2.48}$$

showing that a shock wave can take place only if the incoming flow is *supersonic*; that is, its velocity exceeds the local sonic velocity. Substituting this result into the Meyer relation results in

$$\mathcal{M}_{a2} \leq 1, \qquad \mathcal{M}_{a2} = \frac{u_2}{a_*}, \tag{9.2.49}$$

where $\mathcal{M}_{a2}$ is the Mach number of the fluid after a shock wave. This equation shows that the fluid velocity after a shock wave is *subsonic*, i.e., the fluid velocity is smaller than the local sonic velocity. In other words, the fluid will be compressed as it passes through a shock wave. Equations (9.2.48) and (9.2.49) are the alternative expressions of Eq. (9.2.40). They are so derived to satisfy the fundamental physics, namely, the second law of thermodynamics, which is the necessary condition of the existence of a normal shock wave.

With the obtained results, it becomes possible to evaluate the downstream state of a shock wave in terms of the upstream state. Specifically, the upstream state is characterized by $\{u_1, p_1, \rho_1\}$, while those in the downstream region are $\{u_2, p_2, \rho_2\}$, as shown in Fig. 9.3a. For convenience, they are replaced respectively by $\{\mathcal{M}_{a1}, p_1, \rho_1\}$ and $\{\mathcal{M}_{a2}, p_2, \rho_2\}$. It follows from Eq. (9.2.43) that

$$\frac{a_*^2}{u_1^2} = 2\frac{\gamma-1}{\gamma+1}\left[\frac{1}{2}+\frac{1}{(\gamma-1)\mathcal{M}_{a1}^2}\right], \qquad \frac{a_*^2}{u_2^2} = 2\frac{\gamma-1}{\gamma+1}\left[\frac{1}{2}+\frac{1}{(\gamma-1)\mathcal{M}_{a2}^2}\right]. \tag{9.2.50}$$

Combining these equations with the Meyer relation yields

$$\mathcal{M}_{a2}^2 = \frac{1+[(\gamma-1)/2]\mathcal{M}_{a1}^2}{\gamma\mathcal{M}_{a1}^2-(\gamma-1)/2}. \tag{9.2.51}$$

[6]Theodor Meyer, 1882–1972, a German mathematician, who was a student of Prandtl, and contributed to the foundation of compressible flows or gas dynamics.

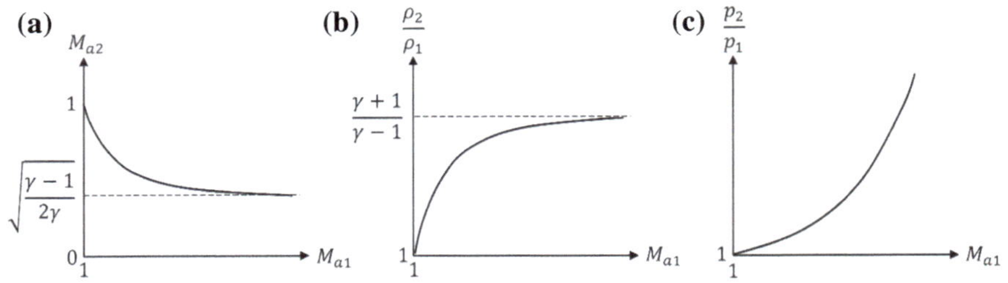

Fig. 9.4 The fluid state in the downstream region of a normal shock wave. **a** The Mach number. **b** The density ratio. **c** The pressure ratio, in which $\mathcal{M}_{a1}$ is the Mach number of the fluid in the upstream region

This equation shows that the downstream Mach number depends only on the upstream Mach number and the specific-heat ratio of gas. As $\mathcal{M}_{a1}$ increases, $\mathcal{M}_{a2}$ decreases with an asymptotic limiting value of $\mathcal{M}_{a2}^2 \to (\gamma-1)/(2\gamma)$ as $\mathcal{M}_{a1} \to \infty$, as shown in Fig. 9.4a.

On the other hand, it follows from the Meyer relation that

$$\frac{u_2}{u_1} = 2\frac{\gamma-1}{\gamma+1}\left[\frac{1}{2} + \frac{1}{(\gamma-1)\mathcal{M}_{a1}^2}\right] = \frac{(\gamma-1)\mathcal{M}_{a1}^2 + 2}{(\gamma+1)\mathcal{M}_{a1}^2}, \tag{9.2.52}$$

in which Eq. (9.2.50) has been used. Since the conservation of mass requires that $\rho_1 u_1 = \rho_2 u_2$, the density ratio is then obtained as

$$\frac{\rho_2}{\rho_1} = \frac{(\gamma+1)\mathcal{M}_{a1}^2}{(\gamma-1)\mathcal{M}_{a1}^2 + 2}, \tag{9.2.53}$$

which indicates that the density ratio is a function of the upstream Mach number and specific-heat ratio, with its characteristics shown in Fig. 9.4b. The density ratio increases monotonically as $\mathcal{M}_{a1}$ increases, until it reaches an asymptotic limiting value of $\rho_2/\rho_1 = (\gamma+1)/(\gamma-1)$. Finally, substituting the obtained expression of density ratio into the Rankine-Hugoniot equations gives rise to the pressure ratio given by

$$\frac{p_2}{p_1} = 1 + \frac{2\gamma}{\gamma+1}\left(\mathcal{M}_{a1}^2 - 1\right), \tag{9.2.54}$$

with its characteristics shown in Fig. 9.4c. The pressure ratio increases without any asymptotic limit as the upstream Mach number increases. It implies that the fluid density may be increased significantly in the downstream region, if the Mach number of incoming flow is extremely large.

9.2.5 Oblique Shock Waves

In contrast to normal shock waves which are perpendicular to the flow direction, oblique shock waves are inclined to the free stream at an angle which is different

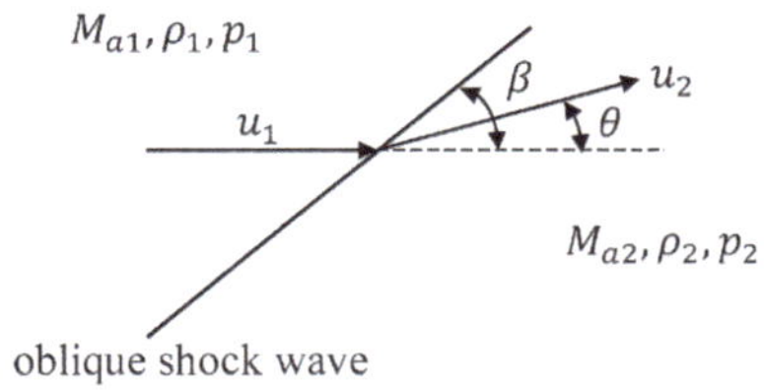

Fig. 9.5 The geometric configurations of an oblique shock wave

from $\pi/2$, as shown e.g. in Fig. 9.5, in which the shock wave assumes an angle β with respect to the incoming flow direction, and the fluid velocity is deflected through an angle θ by the wave front. Form the geometric configurations, the velocity components normal to the shock wave are given by $u_1 \sin\beta$ and $u_2 \sin(\beta - \theta)$, which need to satisfy

$$u_2 \sin(\beta - \theta) \leq u_1 \sin\beta, \tag{9.2.55}$$

as implied by Eq. (9.2.40)$_2$. On the contrary, the tangential velocity components must be the same, for there exist no pressure differentials or other forces acting along the tangential direction. The reducing in the normal velocity component and preservation of tangential velocity component give rise to the downstream fluid velocity u_2 which is bent toward the wave front, as shown in the figure.

For the normal velocity components, it follows from Eqs. (9.2.51), (9.2.53) and (9.2.54) that

$$\begin{aligned} \mathcal{M}_{a2}^2 \sin^2(\beta - \theta) &= \frac{1 + [(\gamma - 1)/2]\mathcal{M}_{a1}^2 \sin^2\beta}{\gamma \mathcal{M}_{a1}^2 \sin^2\beta - (\gamma - 1)/2}, \\ \frac{\rho_2}{\rho_1} &= \frac{(\gamma + 1)\mathcal{M}_{a1}^2 \sin^2\beta}{(\gamma - 1)\mathcal{M}_{a1}^2 \sin^2\beta + 2}, \\ \frac{p_2}{p_1} &= 1 + \frac{2\gamma}{\gamma + 1}\left(\mathcal{M}_{a1}^2 \sin^2\beta - 1\right), \end{aligned} \tag{9.2.56}$$

must hold. For the tangential velocity components, it is seen that

$$u_2 \cos(\beta - \theta) = u_1 \cos\beta, \qquad \longrightarrow \qquad \frac{u_1}{u_2} = \frac{\cos(\beta - \theta)}{\cos\beta}. \tag{9.2.57}$$

Applying the conservation of mass to the normal velocity components across the shock wave yields

$$\frac{u_1}{u_2} = \frac{\rho_2}{\rho_1} \frac{\sin(\beta - \theta)}{\sin\beta}, \tag{9.2.58}$$

which is substituted into Eq. (9.2.57) to obtain

$$\frac{\rho_2}{\rho_1} = \frac{\tan\beta}{\tan(\beta - \theta)}. \tag{9.2.59}$$

Incorporating this equation with Eq. (9.2.56)$_2$ leads to

$$\frac{(\gamma + 1)\mathcal{M}_{a1}^2 \sin^2\beta}{(\gamma - 1)\mathcal{M}_{a1}^2 \sin^2\beta + 2} = \frac{\tan\beta}{\tan(\beta - \theta)}. \tag{9.2.60}$$

The downstream conditions may essentially be determined by using Eq. (9.2.56), provided that the upstream conditions and two angles β and θ are prescribed. For the oblique shock wave generated by the leading edge of a body, the angle θ is normally known, for the downstream velocity must be tangent to the body surface. In view of these, Eq. (9.2.60) then delivers an implicit equation for the angle β, for the upstream Mach number $\mathcal{M}_{a1}$ is essentially known.

Solving Eq. (9.2.60) for $\mathcal{M}_{a1}$ yields

$$\mathcal{M}_{a1}^2 = \frac{2\tan\beta}{\sin^2\beta[(\gamma+1)\tan(\beta-\theta)-(\gamma-1)\tan\beta]}, \tag{9.2.61}$$

which is simplified to

$$\mathcal{M}_{a1}^2 = \frac{2\cos(\beta-\theta)}{\sin\beta[\sin(2\beta-\theta)-\gamma\sin\theta]}. \tag{9.2.62}$$

In this expression, the values of $\mathcal{M}_{a1}$ and θ are known, so that the value of β can be determined. Typical solutions to this equation are shown in Fig. 9.6a. For given values of $\mathcal{M}_{a1}$ and θ, there exist two values of β, corresponding to two shock waves, and the possible values of β are restricted by

$$\sin^{-1}\left(\frac{1}{\mathcal{M}_{a1}}\right) \le \beta \le \frac{\pi}{2}, \tag{9.2.63}$$

resulted from that $\mathcal{M}_{a1}\sin\beta \ge 1$. Thus, the upper limit in this equation corresponds to a normal shock wave, by which the maximum pressure and density ratios for a given approaching $\mathcal{M}_{a1}$ can be obtained. The lower limit is the angle of a *Mach wave*, which represents the sonic end of shock-wave spectrum, so that the pressure and density ratios across a Mach wave are unity. This angle is that to the leading edge of a sound wave which is being continuously emitted by a source of sound moving with $\mathcal{M}_{a1}$. Based on these, oblique shock waves are classified into two categories: the strong and weak ones, corresponding respectively to $\beta \sim \pi/2$ and $\beta \sim \sin^{-1}(1/\mathcal{M}_{a1})$.

The downstream Mach number $\mathcal{M}_{a2}$, by using Eq. $(9.2.56)_1$, is obtained as

$$\mathcal{M}_{a2}^2 = \frac{1+[(\gamma-1)/2]\mathcal{M}_{a1}^2\sin^2\beta}{\sin^2(\beta-\theta)[\gamma\mathcal{M}_{a1}^2\sin^2\beta-(\gamma-1)/2]}, \tag{9.2.64}$$

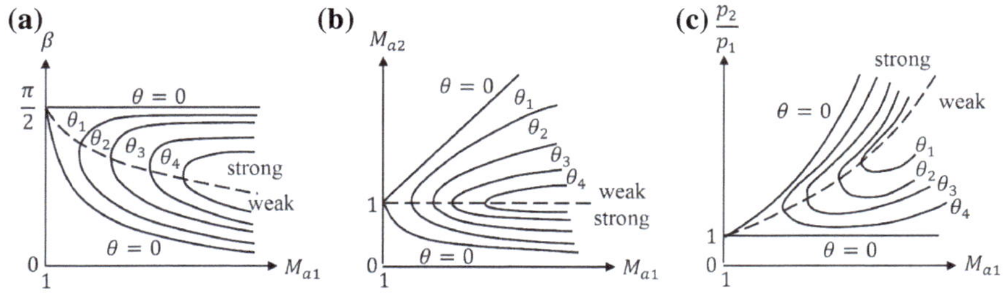

Fig. 9.6 Downstream conditions of oblique shock waves. **a** The inclined angle. **b** The Mach number. **c** The pressure ratio, in which $\mathcal{M}_{a1}$ is the Mach number of upstream flow, with the dashed lines marking the distinctions between strong and weak shock waves

whose characteristics are shown in Fig. 9.6b for given values of $\mathcal{M}_{a1}$ and θ, with the values of β determined by Eq. (9.2.62). It is seen that two possible downstream conditions may be obtained: the supersonic and subsonic flows. For a normal shock wave, the downstream flow is subsonic, while for an oblique shock wave with small values of β, the supersonic downstream flows may be established, resulted from the unaffected tangential velocity components. The pressure ratio across an oblique shock wave is given in Eq. $(9.2.56)_3$, whose characteristics are shown in Fig. 9.6c. The strength of a shock wave is defined by the dimensionless pressure coefficient $(p_2 - p_1)/p_1$, which is larger for strong shock waves than for weak shock waves. The density ratio across an oblique shock wave is given in Eq. $(9.2.56)_2$.

Consequently, with Eqs. $(9.2.56)_{2,3}$, (9.2.62) and (9.2.64), the downstream conditions of an oblique shock wave may be determined, provided that the type of shock wave, either strong or weak, is known. Unfortunately, there exists no mathematical criterion in determining whether a shock wave is strong or weak. The configuration which will be adopted by nature depends on the geometry of projectile or the boundary inducing a shock wave. For example, consider a blunt-nosed body in a supersonic flow, as shown in Fig. 9.7a. The boundary conditions on the blunt-nosed-body surface require that the velocity to be close to the vertical line in the vicinity of front stagnation point.[7] Since $\beta \sim \pi/2$, this shock wave will be strong, so that the Mach number after the shock wave will be smaller than unity, yielding a subsonic flow. Moving away from the front stagnation point along the body surface, the angle θ of downstream fluid velocity changes continuously, and this angle will at some point reach its critical value, so that the matching of boundary conditions by deflecting the flow through a weak shock wave becomes possible. The shock wave will then bent back with the flow far from the body, with which the downstream flow becomes supersonic. Thus, a subsonic-flow region exists in the vicinity of body nose, while the rest of flow field belongs to the supersonic-flow region. Figure 9.7b shows a sharp-nosed slender body in the same supersonic flow as that in Fig. 9.7a, in which an *attached shock wave* exists. In view of the geometric configurations, the velocity will be deflected by the shock wave through just the correct angle to fulfill

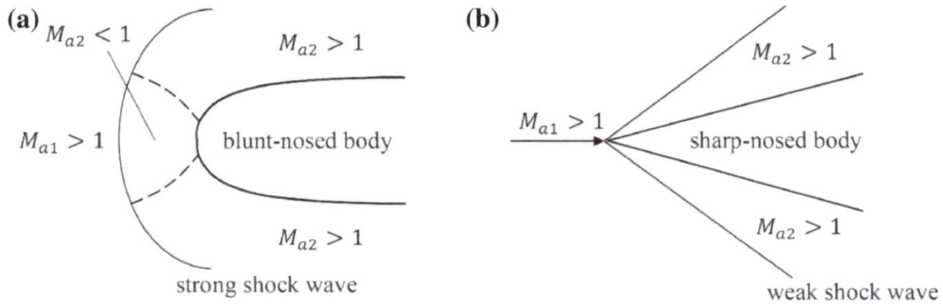

Fig. 9.7 Bodies in a supersonic-flow field and the corresponding possible shock waves. **a** For a blunt-nosed body. **b** For a sharp-nosed body

[7]This occurs only for a *detached* shock wave.

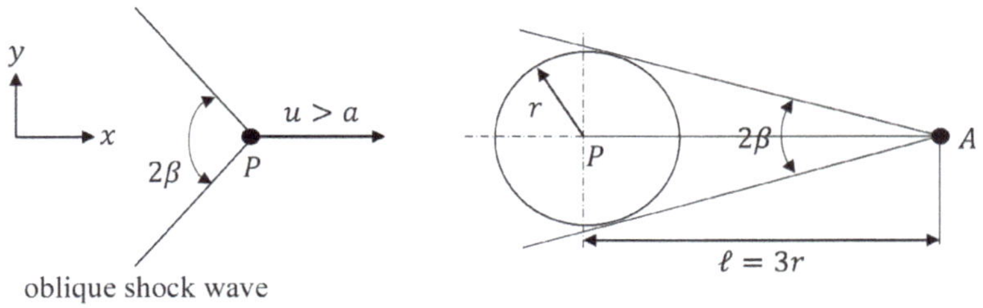

Fig. 9.8 A simple graphic method in determining the deflection angle of an attached oblique shock wave of a sharp-nosed body moving with supersonic velocity

the boundary conditions, so that the body surface becomes a streamline. In such a case, the shock wave is weak, and the downstream flow remains supersonic.

An insufficiently accurate but convenient method to determine the angle of an oblique shock wave of a sharp-nosed body moving with supersonic velocity is given graphically in Fig. 9.8. Let point P represents a moving body, whose location is used as the center of a circle with radius r representing the local sonic velocity a. The proportional factor between r and a is arbitrary, e.g. one may use $r = 1$ cm or $r = 5$ cm to denote the local sonic velocity graphically. It is supposed that the velocity of moving body is $u = 3a$ along the positive x-direction, then a horizontal line starting from point P to the positive x-direction with length $\ell = 3r$ is conducted to obtain point A. Making two tangent lines of the circle $r = a$ to pass through point A forms the angle 2β. In this case, the value of β is identified to be $\beta = \sin^{-1}(1/3)$.

9.3 One-Dimensional Flows

The features of one-dimensional compressible frictionless flows in subsonic and supersonic regions are discussed in the section. The weak shock waves or sonic waves, which have been discussed in the last section, are treated in a more general manner by using the Riemann invariants, by which the reactions of acoustic waves in various situations are presented. The non-adiabatic flows are introduced by means of the influence coefficients, which allow not only the influence of heat transfer, but also the influence of friction and changes in area be taken into account. The discussions on isentropic flows and flows through convergent-divergent nozzles are provided for practical application.

9.3.1 Weak Waves, Characteristics, and the Riemann Invariants

As shown previously, weak waves are isentropic, so that the pressure is only a function of one state variable, say $p = p(\rho)$. It follows then

$$\frac{\partial p}{\partial x} = \frac{\mathrm{d}p}{\mathrm{d}\rho}\frac{\partial \rho}{\partial x} = a^2\frac{\partial \rho}{\partial x}. \tag{9.3.1}$$

With this, the local balances of mass and linear momentum for a one-dimensional plane wave traveling in the x-direction read respectively

$$\frac{\partial \rho}{\partial t}+\rho\frac{\partial u}{\partial x}+u\frac{\partial \rho}{\partial x}=0, \qquad \rho\frac{\partial u}{\partial t}+u\frac{\partial u}{\partial x}=-a^2\frac{\partial \rho}{\partial x}. \tag{9.3.2}$$

The fluid is assumed to be at rest initially, through which a wave passes, which induces a change in the fluid density, pressure, and velocity, which are expressed respectively by

$$\rho=\rho_0+\rho', \qquad p=p_0+p', \qquad u=0+u'. \tag{9.3.3}$$

The quantities ρ_0, p_0, and $u_0=0$ are the values of ρ, p, and u in the quiescent state, and the primed quantities are the perturbations with $\rho'/\rho_0 \ll 1$, $p'/p_0 \ll 1$ and $u'/a_0 \ll 1$, where a_0 is the speed of sound in the undisturbed state. Substituting these expressions into Eq. (9.3.2) yields

$$\frac{\partial \rho'}{\partial t}+\rho_0\frac{\partial u'}{\partial x}=0, \qquad \rho_0\frac{\partial u'}{\partial t}+a_0^2\frac{\partial \rho'}{\partial x}=0, \tag{9.3.4}$$

which is a linearized form of Eq. (9.3.2) for a weak wave.

Since ρ_0 is constant, Eq. (9.3.4) may be recast alternatively as

$$\begin{aligned}
&\frac{\partial}{\partial t}\left(\rho_0+\rho'\right)+\rho_0\frac{\partial}{\partial x}\left(0+u'\right)=0, &&\longrightarrow \frac{\partial \rho}{\partial t}+\rho_0\frac{\partial u}{\partial x}=0,\\
&\rho_0\frac{\partial}{\partial t}\left(0+u'\right)+a_0^2\frac{\partial}{\partial x}\left(\rho_0+\rho'\right)=0, &&\longrightarrow \rho_0\frac{\partial u}{\partial t}+a_0^2\frac{\partial \rho}{\partial x}=0.
\end{aligned} \tag{9.3.5}$$

Dividing the first equation by ρ_0 and the second equation by $\rho_0 a_0$ respectively gives

$$\frac{\partial}{\partial t}\left(\frac{\rho}{\rho_0}\right)+a_0\frac{\partial}{\partial x}\left(\frac{u}{a_0}\right)=0, \qquad \frac{\partial}{\partial t}\left(\frac{u}{a_0}\right)+a_0\frac{\partial}{\partial x}\left(\frac{\rho}{\rho_0}\right)=0, \tag{9.3.6}$$

which are combined together to obtain

$$\begin{aligned}
&\frac{\partial}{\partial t}\left(\frac{u}{a_0}+\frac{\rho}{\rho_0}\right)+a_0\frac{\partial}{\partial x}\left(\frac{u}{a_0}+\frac{\rho}{\rho_0}\right)=0,\\
&\frac{\partial}{\partial t}\left(\frac{u}{a_0}-\frac{\rho}{\rho_0}\right)-a_0\frac{\partial}{\partial x}\left(\frac{u}{a_0}-\frac{\rho}{\rho_0}\right)=0,
\end{aligned} \tag{9.3.7}$$

indicating that the material derivatives of the quantities inside the parenthesis should vanish, in which the convective speed of material derivative is the speed of sound taking place along the x-direction. Integrating these equations results in

$$\begin{aligned}
&\frac{u}{a_0}+\frac{\rho}{\rho_0}=C_1, \text{ along } x-a_0t=\text{constant},\\
&\frac{u}{a_0}-\frac{\rho}{\rho_0}=C_2, \text{ along } x+a_0t=\text{constant},
\end{aligned} \tag{9.3.8}$$

where C_1 and C_2 are constants. The lines described by $x-a_0t$ and $x+a_0t$ are called the *characteristics*, along which the terms $u/a_0+p/\rho_0$ and $u/a_0-p/\rho_0$ are called the *Riemann invariants*, which are constant. Typical characteristics passing through a location x and the corresponding Riemann invariants are shown in Fig. 9.9a, in which one of the characteristics is running forwards, while the other is backward-running.

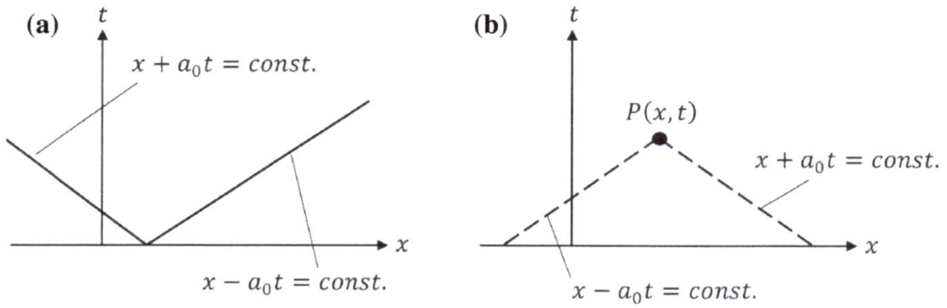

Fig. 9.9 Illustrations of the characteristics and Riemann invariants in the (x, t)-plane. **a** Typical profiles. **b** Evaluation of the field variables at an arbitrary point P

It follows from Eq. (9.2.1)$_3$ that

$$\frac{p}{p_0} = \left(\frac{\rho}{\rho_0}\right)^{\gamma} = \left(1 + \frac{\rho'}{\rho_0}\right)^{\gamma} \sim 1 + \gamma\frac{\rho'}{\rho_0}, \tag{9.3.9}$$

in which the assumption that $\rho'/\rho_0 \ll 1$ has been used. Substituting this expression into Eq. (9.3.8) gives

$$\begin{aligned} \frac{u}{a_0} + \frac{1}{\gamma}\frac{p}{p_0} &= C_1, \text{ along } x - a_0 t = \text{constant}, \\ \frac{u}{a_0} - \frac{1}{\gamma}\frac{p}{p_0} &= C_2, \text{ along } x + a_0 t = \text{constant}, \end{aligned} \tag{9.3.10}$$

which are the alternative forms of the Riemann invariants. There exist two forms of the Riemann invariants. Depending on the problem under consideration, one of the two forms may be chosen to establish a solution procedure. Equations (9.3.8) and (9.3.10) may be used to determine the velocity, density, and pressure at any values of x and t, provided that the values of u, ρ, and p as functions of x are known at some time, e.g. at $t = 0$. For example, let point $P(x, t)$ be any arbitrary point in the (x, t)-plane, through which two characteristics, which originate along the $t = 0$ axis, pass, as shown in Fig. 9.9b. The associated Riemann invariants of two characteristics may be evaluated by the known conditions at $t = 0$. Then, the Riemann invariants at point P deliver two algebraic equations for the unknowns $\{u, \rho\}$, or $\{u, p\}$.

The next subsection is devoted to the applications of characteristics and the associated Riemann invariants to some selected problems.

9.3.2 Illustrations of Characteristics and the Riemann Invariants

Weak shock tubes. Consider a weak wave released in a relatively long tube, as shown in Fig. 9.10a, in which a diaphragm is equipped at $x = 0$. The tube is called a *shock tube*. The gas to the left side of the diaphragm is initially maintained at pressure p_1 which is slightly larger than that to the right side, as shown in Fig. 9.10b.

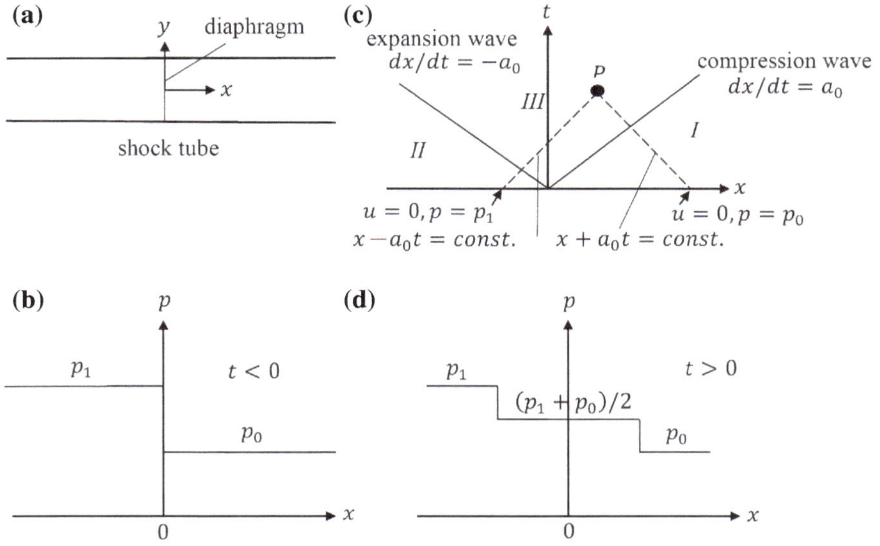

Fig. 9.10 The application of the characteristics and the Riemann invariants for a weak shock wave in a shock tube. **a** The shock tube with the geometric configurations. **b** The initial pressure distribution for $t < 0$. **c** The (x, t)-diagram for the compression and expansion waves. **d** The pressure distribution for $t > 0$

The gases at the two sides are the same, only their states are different.[8] At $t = 0$, the diaphragm breaks, so that a weak pressure wave is released from the vicinity of diaphragm, and two regions of the shock tube tend to equalize their pressures. It is required to determine the pressure and velocity of gas as functions of x and t.

Applying the characteristics and the Riemann invariants to the problem yields the (xt)-diagram shown in Fig. 9.10c. At $t = 0$, a compression wave emanates from the origin and travels into the right-side (low-pressure) region, while an expansion wave travels in the reverse direction to the region of high pressure (the left-side region). Since the shock wave is weak, two waves travel with the speed of sound a_0, with their slopes identified as a_0 and $-a_0$, respectively, for the compression and expansion waves shown in Fig. 9.10c. Three regions inside the shock tube are identified. Region I is the portion of positive x-axis which has yet been affected by the compression wave, with the gas velocity and pressure identified to be null and p_0, respectively. Region II is the portion of negative x-axis, which is the counterpart of region I, unaffected by the expansion wave with vanishing gas velocity and pressure p_1. In-between is region III, which is the portion of x-axis and has been influenced by both the compression and expansion waves. Since the gas velocity and pressure must be continuous across $x = 0$, both the positive and negative portions of x-axis in this region experience the same pressure and velocity, whose values can be determined by

[8] The analysis can be extended to different gases with different properties and states.

using an arbitrary point P in this region, through which two characteristics originating from the x-axis at $t = 0$ pass. By using the known conditions along the x-axis at $t = 0$, the associated Riemann invariants are given by

$$\frac{u}{a_0} + \frac{1}{\gamma}\frac{p}{p_0} = \frac{1}{\gamma}\frac{p_1}{p_0}, \qquad \frac{u}{a_0} - \frac{1}{\gamma}\frac{p}{p_0} = -\frac{1}{\gamma}, \tag{9.3.11}$$

along the characteristics $x - a_0 t = \text{constant}$ and $x + a_0 t = \text{constant}$, respectively, in which the conditions $u = 0$, $p = p_1$ at $t = 0$, $x < 0$, and $u = 0$, $p = p_0$ at $t = 0$, $x > 0$ have been used. The solutions to two algebraic equations are obtained as

$$\frac{u}{a_0} = \frac{1}{2\gamma}\left(\frac{p_1}{p_0} - 1\right), \qquad \frac{p}{p_0} = \frac{1}{2}\left(\frac{p_1}{p_0} + 1\right), \tag{9.3.12}$$

for the velocity and pressure in region *III*. As indicated by the first equation, $u/a_0 > 0$ for $p_1/p_0 > 1$, so that the gas moves along the positive x-direction, coinciding to the facts that gas particles tend to follow compression waves and move away from expansion waves as described in Sect. 9.2.1. The second equation indicates that the pressure in region *III* is simply an arithmetic average of the pressures in the other two regions, with its distribution shown in Fig. 9.10d for $t > 0$. This reveals that a compression wave with amplitude $(p_1 - p_0)/2$ travels along the positive x-direction with speed of sound a_0, while an expansion wave with the same amplitude and speed travels along the negative x-axis.

Wave reflections at wall. When a weak wave strikes a solid boundary, it will be reflected. Compression waves will be reflected as compression waves of same strength, and expansion waves will also likely be reflected as identical expansion waves. To demonstrate these, consider a shock tube which is exactly the same as before, except that its one end is closed, as shown in Fig. 9.11a. The corresponding (x, t)-diagram is shown in Fig. 9.11b. Before the compression wave strikes the end, the physical processes of compression and expansion waves are the same as discussed previously. After the compression wave has stroked the tube end, there exists a reflected wave traveling in the negative x-direction with speed of sound a_0, and the whole tube space is then divided into four regions. Regions $I \sim III$ are the same as before, and region *IV* is the portion of positive x-axis which has been passed through by the reflection wave.

To determine the gas state in region *IV*, an arbitrary point $P(x, t)$ in the (x, t)-plane and the two characteristics $\xi = x - a_0 t = \text{constant}$ and $\eta = x + a_0 t = \text{constant}$ passing through point P are displayed in Fig. 9.11b. The characteristic corresponding to $\xi = \text{constant}$ comes from region *III*, with the velocity and pressure determined by Eq. (9.3.12). Thus, this characteristic may be terminated at any point in region *III* where the values of the Riemann invariants may be obtained. The characteristic corresponding to $\eta = \text{constant}$ runs parallel to the line of reflected wave, and eventually reaches the tube end, from which another characteristic corresponding to $\xi_1 = \text{constant}$ starts due to the presence of reflected wave. It is noted that at the moment there is no information about whether the reflected wave is compressive or expansive.

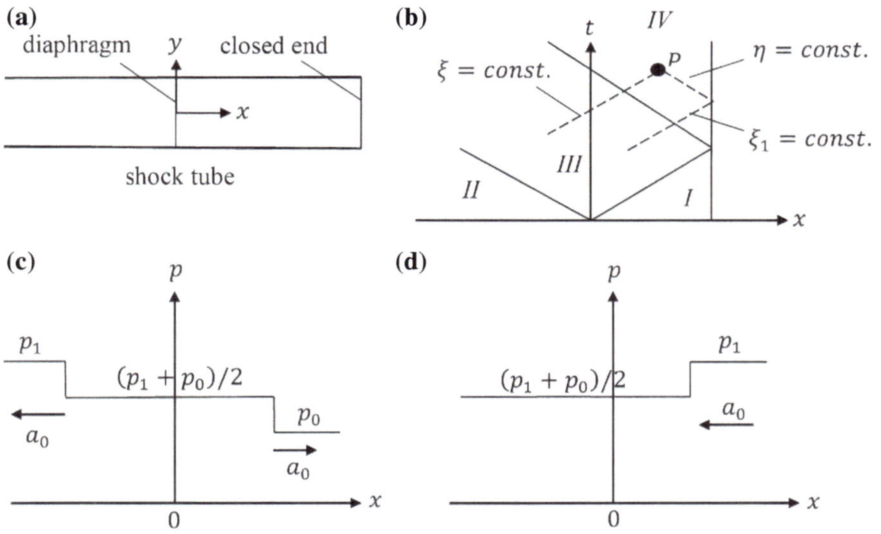

Fig. 9.11 The features of a weak wave in a shock tube with one closed end. **a** The shock tube with the geometric configurations. **b** The (x, t)-diagram and characteristics for the compression and expansion waves. **c** The pressure distribution before the wave reflection. **d** The pressure distribution after the wave reflection

Applying Eq. $(9.3.10)_2$ to η = constant yields

$$\frac{u}{a_0} - \frac{1}{\gamma}\frac{p}{p_0} = -\frac{1}{\gamma}\frac{p_w}{p_0}, \tag{9.3.13}$$

where p_w is the pressure at the tube end, and the conditions $u = 0$ and $p = p_w$ on the tube end have been used. The terms p and u in this equation belong to region IV, which are unknown quantities. In addition, applying Eq. $(9.3.10)_1$ to ξ_1 = constant gives

$$-\frac{1}{\gamma}\frac{p_w}{p_0} = \frac{1}{2\gamma}\left(\frac{p_1}{p_0} - 1\right) + \frac{1}{2\gamma}\left(\frac{p_1}{p_0} + 1\right), \tag{9.3.14}$$

in which Eq. (9.3.12) has been used. This equation can only be satisfied by $p_w = p_1$. With this, the characteristic corresponding to η = constant, i.e., Eq. (9.3.13), becomes

$$\frac{u}{a_0} - \frac{1}{\gamma}\frac{p}{p_0} = -\frac{1}{\gamma}\frac{p_1}{p_0}, \tag{9.3.15}$$

and the Riemann invariant along the characteristic ξ = constant, by using Eq. (9.3.14) with $p_w = p_1$, is given by

$$\frac{u}{a_0} + \frac{1}{\gamma}\frac{p}{p_0} = \frac{1}{\gamma}\frac{p_1}{p_0}. \tag{9.3.16}$$

It should be noted that the terms u and p in the last two equations are respectively the gas velocity and pressure in region IV. The solutions to two algebraic equations are given by

$$u = 0, \qquad p = p_1, \tag{9.3.17}$$

showing that the gas velocity in region *IV* vanishes and the gas pressure equals that in region *II*. The first result is based on the fact that the boundary condition at the closed end requires zero velocity. The second result shows that the reflected wave is a compression one. Initially, the pressure in the right side of tube is p_0. As the first wave passes toward the closed end, the pressure jumps to $(p_1 + p_0)/2$, as shown in Fig. 9.11c. The compression wave is then reflected by the closed end and passes through the right side of tube again, which gives a positive pressure differential $(p_1 + p_0)/2$, so that the pressure of gas in the region which has been passed by the reflected wave becomes p_1 again.

Since the obtained results have no restriction on whether $p_1 > p_0$ or $p_1 < p_0$, they are valid for both compression and expansion waves. Compression waves are reflected as compression waves with same strength, and so behave the expansion waves. It has been established in Sect. 9.2.4 that fluid particles tend to follow a compression wave and move away from an expansion wave. This fact justifies Eq. (9.3.17)$_1$ for a compression wave reflected as a compression one, and an expansion wave reflected as an expansion one.

Wave Reflection and Refraction at Interface. When a wave encounters an interface between two dissimilar gases, some wave part is transmitted through the interface, and the other part is reflected by the interface. To demonstrate these, consider a shock tube in which two different gases are separated by an interface, which exists part way down the tube, as shown in Fig. 9.12a. Initially, the velocity is null everywhere, and the pressure is p_1 for $x < 0$ and p_0 for $x > 0$. Two gases may be different, or simply the same gas with different temperatures, so that the speeds of sound are different, as denoted respectively by a_{01} and a_{02}, with the corresponding specific-heat ratios γ_1 and γ_2. To trigger a weak wave, a diaphragm is equipped in the region before the interface. The (x, t)-diagram describing the sequence of events resulted from the bursting of diaphragm is shown in Fig. 9.12b. For simplicity, it is assumed that when the wave traveling in the positive x-direction hits the gaseous interface, it is partly transmitted and partly reflected there, by which four tube regions are identified. Region *I* represents the initial gas state locating to the right of the diaphragm, although the physical properties are discontinuous at the gaseous interface, with vanishing velocity and pressure p_0. Region *II* marks the initial gas state to the left of the diaphragm, with null velocity and pressure p_1. Region *III* is the portion of shock tube, which is influenced by the passage of waves resulted from the bursting of diaphragm, whose velocity and pressure are determined by using Eq. (9.3.12). Region *IV* is the portion on the two sides of gaseous interface, which is influenced by the passages of reflected and refracted waves initiated at the gaseous interface. It is further divided into two subregions: region *IV-a* is left to the interface, which has been passed through by the reflected wave, while region *IV-b* is right to the interface, which has been passed through by the refracted wave.

To determine the velocity and pressure in region *IV*, an arbitrary point $P(x, t)$ on the interface in the (x, t)-diagram is used. For the characteristics $\xi =$ constant and $\eta =$ constant which pass through point P, each characteristic lies entirely in the domain of one gas only. The characteristic corresponding to $\xi =$ constant may be terminated anywhere in region *III*, while that corresponding to $\eta =$ constant may

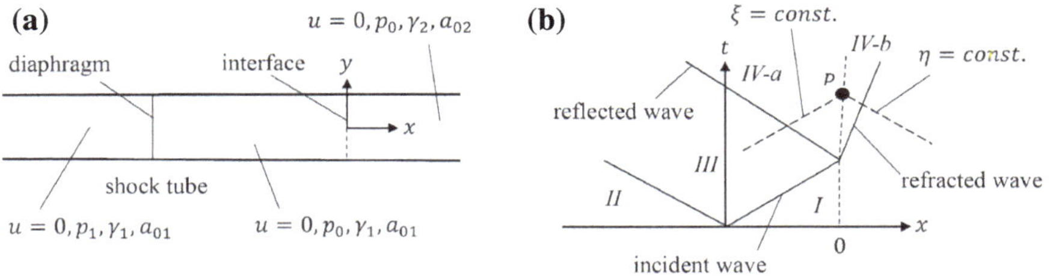

Fig. 9.12 Wave reflection and refraction at a gaseous interface. **a** The shock tube with the geometric configurations. **b** The (x, t)-diagram and characteristics for the reflection and refracted waves

be terminated anywhere in region *IV*. Since the physical observations require that the velocity and pressure must be continuous across the interface at all times, the velocities and pressures in regions *IV-a* and *IV-b* must be the same. However, the interface may move after the impact of incident wave, so that its location may not be necessary at its initial position.

By using Eqs. (9.3.10)$_2$ and (9.3.12), two Riemann invariants along ξ = constant and η = constant are given respectively by

$$\begin{aligned} &\frac{u}{a_{01}} + \frac{1}{\gamma_1}\frac{p}{p_0} = \frac{1}{2\gamma_1}\left(\frac{p_1}{p_0} - 1\right) + \frac{1}{2\gamma_1}\left(\frac{p_1}{p_0} + 1\right) = \frac{1}{\gamma_1}\frac{p_1}{p_0}, \\ &\frac{u}{a_{02}} - \frac{1}{\gamma_2}\frac{p}{p_0} = -\frac{1}{\gamma_2}, \end{aligned} \tag{9.3.18}$$

where u and p are the velocity and pressure in region *IV*. It is noted that the conditions in region *III* are used as the known conditions for the first equation, while the conditions in region *I*, i.e., the undisturbed gas state, are used as the known conditions for the second equation. The solutions to two algebraic equations are obtained as

$$\frac{u}{a_{01}} = \frac{p_1/p_0 - 1}{\gamma_1 + \gamma_2 a_{01}/a_{02}}, \qquad \frac{p}{p_0} = \frac{p_1/p_0 + (\gamma_1/\gamma_2)(a_{02}/a_{01})}{1 + (\gamma_1/\gamma_2)(a_{02}/a_{01})}. \tag{9.3.19}$$

For the pressure ratio $p_1/p_0 > 1$, it follows from the first equation that the gas velocity u in region *IV* is positive, showing that the gaseous interface will move in the positive x-direction, and the incident wave is a compression one, which is followed by the fluid particles. As this incident wave is reflected by the gaseous interface, it is also a compression wave, as already demonstrated previously, although its strength may not be the same as that of incident wave, for the interface is not a solid one. Let the pressure differential across the reflected wave be denoted by Δp_r, it follows from Eq. (9.3.12)$_2$ that

$$\frac{\Delta p_r}{p_0} = \frac{p}{p_0} - \frac{1}{2}\left(\frac{p_1}{p_0} + 1\right) = \frac{[1 - (\gamma_1/\gamma_2)(a_{02}/a_{01})](p_1/p_0 - 1)}{2[1 + (\gamma_1/\gamma_2)(a_{02}/a_{01})]}, \tag{9.3.20}$$

in which Eq. (9.3.19)$_2$ has been used. If $a_{02}/a_{01} \ll 1$, which corresponds to a high-density gas behind the interface, the above equation reduces to $p = p_1$, which has been obtained for a solid end boundary. Thus, as the density difference across the interface increases, the considered circumstance approaches an impermeable boundary for perfect reflection.

Similarly, let the pressure differential across the transmitted or refracted wave be denoted by Δp_t. It follows that

$$\frac{\Delta p_t}{p_0} = \frac{p}{p_0} - 1 = \frac{(p_1/p_0 - 1)}{1 + (\gamma_1/\gamma_2)(a_{02}/a_{01})}, \tag{9.3.21}$$

in which Eq. (9.3.19)$_2$ has been used. Equations (9.3.20) and (9.3.21) represent respectively the strengths of reflected and refracted waves, which depend on the nature of interface. If two gases are the same, i.e., $\gamma_1 = \gamma_2 = \gamma$ and $a_{01} = a_{02} = a_0$, two equations deliver that there is no reflected wave, and the refracted wave is nothing else than the initial incident wave. For the limiting case $a_{02}/a_{01} \to 0$, there exists only a reflected wave. In-between both a reflected and a refracted waves exist.

Piston Problem. Figure 9.13 shows a cylinder or a circular tube, in which a piston slides. Initially, the piston and the gas ahead of it are stationary. At $t = 0$, the piston starts to move at constant velocity U, which triggers the gas ahead of it to move, with possible occurrence of pressure waves. It is required to determine the velocity and pressure ahead of the piston after the motion has started. To achieve these, the (x, t)-diagram for the events is shown in Fig. 9.13b, in which the left straight line represents the instantaneous location of piston, while the right straight line represents the locations of wave front traveling with speed of sound a_0 in the positive x-direction, which results from the impulsive acceleration of piston at $t = 0$. For simplicity, it is assumed that $U/a_0 \ll 1$ in the context of linearization, so that the piston will always be very close to $x = 0$, for which the boundary condition $u = U$ on $x = 0$ is assumed.

The entire tube region is divided into two parts: region I contains the undisturbed gas, which is stationary with pressure p_0. Region II is the space between the wave front and piston, whose velocity and pressure are determined by using an arbitrary point $P(x, t)$ in this region, through which two characteristics ξ = constant and η = constant pass, as shown in Fig. 9.13c. The characteristic corresponding to η = constant enters region I, where the values of u and p are known. The characteristic corresponding to ξ = constant runs parallel to the wave front and eventually encounters the piston, and terminates there, where the fluid velocity is known, but the pressure remains undetermined. From this, another characteristic η_1 = constant is drawn from the point where ξ = constant terminates.

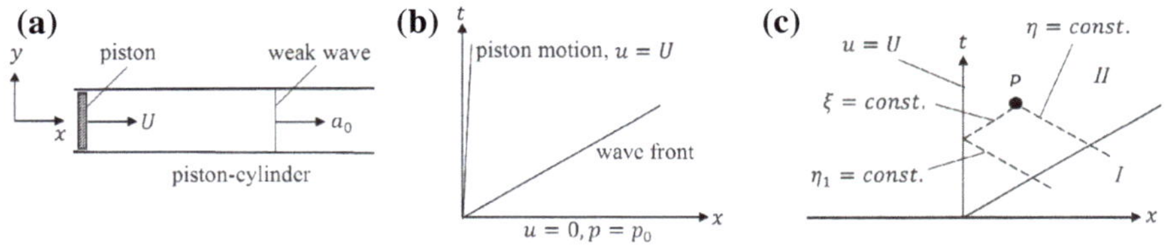

Fig. 9.13 Weak waves induced by a sudden motion of the piston in a cylinder. **a** The geometric configurations. **b** The (x, t)-diagram for the piston motion and wave front. **c** The (x, t)-diagram for the characteristics and the Riemann invariants

Let the pressure on the piston surface be denoted by p_p, with which the Riemann invariant along $\xi =$ constant is given by

$$\frac{u}{a_0} + \frac{1}{\gamma}\frac{p}{p_0} = \frac{U}{a_0} + \frac{1}{\gamma}\frac{p_p}{p_0}. \tag{9.3.22}$$

Similarly, the Riemann invariant along $\eta_1 =$ constant is identified to be

$$\frac{u}{a_0} - \frac{1}{\gamma}\frac{p}{p_0} = \frac{U}{a_0} - \frac{1}{\gamma}\frac{p_p}{p_0} = -\frac{1}{\gamma}, \tag{9.3.23}$$

with which Eq. (9.3.22) becomes

$$\frac{u}{a_0} + \frac{1}{\gamma}\frac{p}{p_0} = 2\frac{U}{a_0} + \frac{1}{\gamma}, \tag{9.3.24}$$

and the Riemann invariant along $\eta =$ constant is then obtained as

$$\frac{u}{a_0} - \frac{1}{\gamma}\frac{p}{p_0} = -\frac{1}{\gamma}. \tag{9.3.25}$$

The solutions to the last two algebraic equations are given by

$$u = U, \qquad \frac{p}{p_0} = \gamma\frac{U}{a_0} + 1, \tag{9.3.26}$$

indicating that the gas velocity in region *II* is everywhere the same as that of piston. It also shows that the pressure there is greater than the initial pressure p_0 by an amount which is proportional to the piston speed U.

Finite-Strength Shock Tubes. Consider a shock tube in which two different gases are separated initially by a diaphragm at $x = 0$, as shown in Fig. 9.14a. The gas velocity is initially null everywhere, with pressures p_4 left to the diaphragm and $p_1 < p_4$ right to the diaphragm, as shown in Fig. 9.14b. The pressure difference $p_4 - p_1$ is assumed to be finite, so that a linear theory is no longer valid. At $t = 0$, the diaphragm breaks, triggering a compression wave with finite strength traveling in the positive x-direction. This wave, as indicated by the results in Sect. 9.2.4, will steepen as it travels, so that eventually a shock wave will develop, as shown in Fig. 9.14c. Equally, the expansion waves traveling in the negative x-direction will also be triggered. They tend to smooth out during the propagation. The corresponding (x, t)-diagram for the events is shown in Fig. 9.14d, in which the shock wave is represented by a single line discontinuity, while the expansion waves extend over a substantial portion of the x-axis, and are represented by an expansion fan, which consists of a series of lines emanating from the origin. The interface between two gases is initially at $x = 0$ and may move due to the influence of waves, as also shown in the figure.

The entire tube region is divided into four parts. While regions *I* and *IV* are those portions of the tube which are yet affected by the waves, region *II* consists of gas 1, and is the tube portion that is affected by the passage of compression wave, while region *III* consists of gas 4 denoting the tube portion which is affected by the passage of expansion waves. The principal interest is the determination of the strength of the shock wave. The solution procedure is that by using the Galilean transformation,

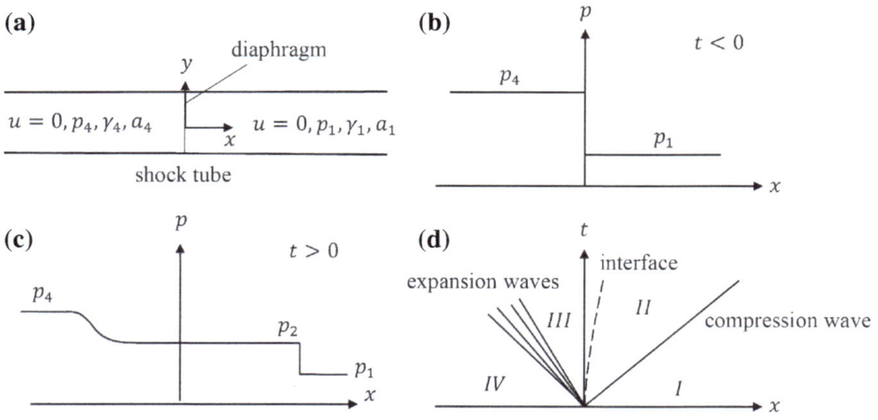

Fig. 9.14 The features of a finite-amplitude wave in a shock tube. **a** The shock tube with the geometric configurations. **b** The initial pressure distribution for $t < 0$. **c** The pressure distribution for $t > 0$. **d** The (x, t)-diagram showing a compression wave and a series of expansion waves

an expression for u_2 in terms of p_2/p_1 may be obtained. An analogue procedure is followed to determine u_3 in terms of the pressure ratio p_4/p_3 crossing the expansion waves. Since the velocities and pressures at the interface between regions II and III must be the same, these conditions are then used to obtain an equation relating the pressure ratios p_2/p_1 across the shock wave to the initial pressure ratio p_4/p_1 across the diaphragm. These steps are discussed separately in the following.

For the compression wave (shock wave), let u_1' and u_2' be the gas velocities respectively in regions I and II under the Galilean transformation, under which the shock wave is stationary, with u_1' the equivalent incoming flow velocity and u_2' the equivalent flow velocity leaving the shock wave, whose relations are established in Sect. 9.2.4. For the considered circumstance, the gas velocity in region I is in fact null, which can be accomplished by using

$$u_1 = u_2' - u_2' = 0, \qquad u_2 = u_1' - u_2' = u_1' \left(1 - \frac{u_2'}{u_1'}\right), \tag{9.3.27}$$

with which the shock wave is now moving with velocity u_2' through a stationary gas, in which the gas velocity behind the shock wave becomes u_2. It follows from the Rankine-Hugoniot equations that the relation between u_1' and u_2' is known, which is used together with Eq. (9.3.27) to obtain

$$u_2 = u_1' \left[1 - \frac{(\gamma_1 + 1)/(\gamma_1 - 1) + p_1/p_2}{1 + (\gamma_1 + 1)/(\gamma_1 - 1)(p_1/p_2)}\right], \tag{9.3.28}$$

with p_1 and p_2 the pressures respectively in regions I and II, as referred to Fig. 9.14c. Since $u_1' = a_1 \mathcal{M}_{a1}'$, where $\mathcal{M}_{a1}'$ is the Mach number of the approaching flow to a stationary shock wave, it follows from Eq. (9.2.54) that

$$(\mathcal{M}_{a1}')^2 = \frac{\gamma_1 + 1}{2\gamma_1}\left(\frac{p_1}{p_2} - 1\right) + 1, \tag{9.3.29}$$

by which Eq. (9.3.28) is recast in the form

$$u_2 = a_1 \sqrt{\frac{\gamma_1 + 1}{2\gamma_1}\left(\frac{p_1}{p_2} - 1\right) + 1}\left[1 - \frac{(\gamma_1+1)/(\gamma_1-1) + p_1/p_2}{1 + (\gamma_1+1)/(\gamma_1-1)(p_1/p_2)}\right], \tag{9.3.30}$$

which is simplified to

$$u_2 = a_1 \sqrt{\frac{2(p_1/p_2 - 1)^2}{\gamma_1[(\gamma_1 - 1) + (\gamma_1 + 1)(p_1/p_2)]}}. \tag{9.3.31}$$

Next, since the expansion waves tend to smooth out and spread themselves over substantial distances, the expansion from p_4 to p_3 takes place continuously, which may be approximated by a very large number of weak expansion waves, each of which is isentropic. It follows from Eq. (9.2.17)$_1$ that

$$\frac{\mathrm{d}u}{a} = -\frac{\mathrm{d}\rho}{\rho}, \tag{9.3.32}$$

since the expansion waves travel in the negative x-direction. For isentropic flows, the local sonic velocity a is given by

$$a^2 = \gamma_4 \frac{p}{\rho} = a_4^2 \left(\frac{\rho}{\rho_4}\right)^{\gamma_4 - 1}, \tag{9.3.33}$$

with which Eq. (9.3.32) is recast alternatively as

$$\mathrm{d}u = -\frac{a_4}{\rho_4^{(\gamma_4-1)/2}} \rho^{(\gamma_4-3)/2} \mathrm{d}\rho. \tag{9.3.34}$$

Integrating this equation with $u = 0$ and $\rho = \rho_4$ yields

$$u = -\frac{2a_4}{\gamma_4 - 1}\left[\left(\frac{\rho}{\rho_4}\right)^{(\gamma_4-1)/2} - 1\right], \tag{9.3.35}$$

which is the value of local velocity u in the expansion waves. Replacing the local density ρ in this equation by the local pressure p through the isentropic law gives

$$u = \frac{2a_4}{\gamma_4 - 1}\left[1 - \left(\frac{p}{p_4}\right)^{(\gamma_4-1)/(2\gamma_4)}\right]. \tag{9.3.36}$$

Applying this equation to the trailing edge of expansion waves with $p = p_3$ and $u = u_3$ results in

$$u_3 = \frac{2a_4}{\gamma_4 - 1}\left[1 - \left(\frac{p_3}{p_4}\right)^{(\gamma_4-1)/(2\gamma_4)}\right]. \tag{9.3.37}$$

Since $u_2 = u_3$ at the gaseous interface, Eqs. (9.3.31) and (9.3.37) imply that

$$\frac{2a_4}{\gamma_4 - 1}\left[1 - \left(\frac{p_2}{p_4}\right)^{(\gamma_4-1)/(2\gamma_4)}\right] = a_1 \sqrt{\frac{2(p_1/p_2 - 1)^2}{\gamma_1[(\gamma_1 - 1) + (\gamma_1 + 1)(p_1/p_2)]}}, \tag{9.3.38}$$

in which p_3 has been replaced by p_2, for two pressures are the same. Solving the above equation for p_4 then yields

$$\frac{p_4}{p_1} = \frac{p_2}{p_1}\left\{1 - \frac{(\gamma_4 - 1)(a_1/a_4)(p_1/p_2 - 1)}{\sqrt{2\gamma_1[(\gamma_1 - 1) + (\gamma_1 + 1)(p_1/p_2)]}}\right\}^{-2\gamma_4/(\gamma_4-1)}. \tag{9.3.39}$$

For small-amplitude compression waves, a linear theory with $p_1/p_2 = 1 - \varepsilon$ can be constructed, with which this equation delivers that $p_4/p_1 = 1 + 2\varepsilon$, coinciding to the results obtained in Sect. 9.3.1.

Let $\mathcal{M}_{as}$ be the Mach number of shock wave propagating through the stationary gas in region I. It follows from the Galilean transformation used previously that $\mathcal{M}_{as} = \mathcal{M}'_{a2}$, where u'_2 is the propagating speed of shock wave, and $(u'_1 - u'_2)$ is the gas velocity behind the shock wave. With these, Eqs. (9.2.51) and (9.2.54) respectively become

$$\mathcal{M}_{as} = \sqrt{\frac{1 + [(\gamma_1 - 1)/2](\mathcal{M}'_{a1})^2}{\gamma_1(\mathcal{M}'_{a1})^2 - (\gamma_1 - 1)/2}}, \qquad \mathcal{M}'_{a1} = \sqrt{1 + \frac{\gamma_1 + 1}{2\gamma_1}\left(\frac{p_1}{p_2} - 1\right)}, \tag{9.3.40}$$

where p_1 and p_2 are referred to Fig. 9.14d. Combining two equations results in

$$\mathcal{M}_{as} = \sqrt{\frac{(\gamma_1 - 1) + (\gamma_1 + 1)(p_2/p_1)}{2\gamma_1}}, \tag{9.3.41}$$

showing that $\mathcal{M}_{as} \to 1$ as $p_2/p_1 \to 1$. In other words, for weak shock waves the wave front travels at the speed of sound, which coincides to the result derived in Sect. 9.2.1. The equation also shows that the Mach number of a strong shock wave can be considerably greater than unity.

9.3.3 Non-adiabatic Flows, the Fanno and Rayleigh Lines

The effects of heat transfer with the surrounding, external body force acting on the fluid, and variation in flow cross-section on the flow characteristics are taken into account in this subsection. However, the flow is still assumed to be one-dimensional, so that the fluid properties in any streamwise location are considered to be the average values at that location. Consider the flow configurations shown in Fig. 9.15a, in which the flow area at location x is denoted by A, and that at $x + \mathrm{d}x$ is $A + \mathrm{d}A$. During the segment $\mathrm{d}x$ an infinitesimal amount of external force $\delta f(x)$ and an infinitesimal amount of heat transfer $\delta q(x)$ take place.

Applying the conservations of mass and linear momentum to the infinitesimal control-volume $A\mathrm{d}x$ yields respectively

$$\frac{\mathrm{d}\rho}{\rho} + \frac{\mathrm{d}u}{u} = -\frac{\mathrm{d}A}{A}, \qquad \rho u\,\mathrm{d}u = -\mathrm{d}p + \delta f, \tag{9.3.42}$$

where ρ, u, and p are respectively the average density, velocity and pressure at location x. Dividing the second equation by p gives

$$\gamma\mathcal{M}_a^2\frac{\mathrm{d}u}{u} + \frac{\mathrm{d}p}{p} = \frac{\delta f}{p}, \tag{9.3.43}$$

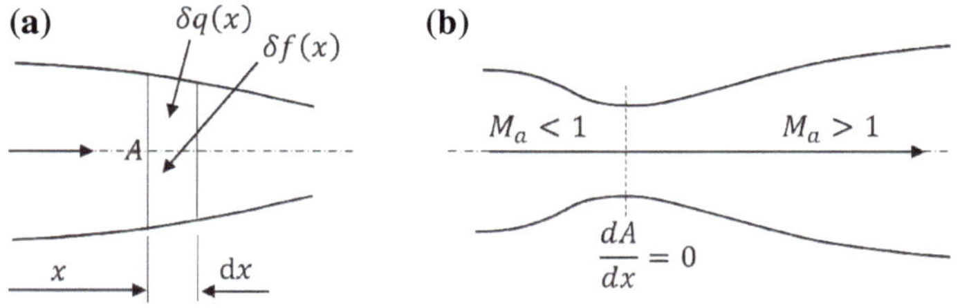

Fig. 9.15 Illustrations of one-dimensional non-adiabatic flows. **a** The geometric configurations. **b** A flow through a typical convergent-divergent nozzle

in which the identity $a^2 = \gamma p/\rho$ has been used, where $\mathcal{M}_a$ represents the local Mach number with $\mathcal{M}_a = u/a$. The thermal-energy equation for the control-volume, with the assumption that the considered fluid is an ideal gas, reads the form

$$c_p \mathrm{d}T + u\,\mathrm{d}u = \delta q, \tag{9.3.44}$$

which is divided by $c_p T$ to obtain

$$\frac{\mathrm{d}T}{T} + \frac{u}{c_p}\frac{\mathrm{d}u}{T} = \frac{\delta q}{c_p T}, \quad \longrightarrow \quad \frac{\mathrm{d}T}{T} + (\gamma-1)\,\mathcal{M}_a^2 \frac{\mathrm{d}u}{u} = \frac{\delta q}{c_p T}, \tag{9.3.45}$$

in which the properties of ideal gas have been used. Equations (9.3.42)$_1$, (9.3.43) and (9.3.45) are the differential balances of mass, linear momentum, and energy for the consider gas flow, which are supplemented by the differential state equation given by

$$\frac{\mathrm{d}p}{p} - \frac{\mathrm{d}\rho}{\rho} - \frac{\mathrm{d}T}{T} = 0. \tag{9.3.46}$$

The four equations represent four algebraic equations for the differentials $\{\mathrm{d}u, \mathrm{d}\rho, \mathrm{d}p, \mathrm{d}T\}$ in terms of the locale values of $\{u, \rho, p, T, \mathcal{M}_a, f, q, A\}$. They may be solved to yield the expressions for each of the differential quantity separately.

Eliminating the terms $\mathrm{d}\rho/\rho$ and $\mathrm{d}T/T$ in Eq. (9.3.46) respectively by using Eqs. (9.3.43) and (9.3.45)$_2$, and adding Eq. (9.3.42)$_1$ to the resulting equation yields

$$\frac{\mathrm{d}u}{u} = \frac{1}{\mathcal{M}_a^2 - 1}\left(\frac{\mathrm{d}A}{A} + \frac{\delta f}{p} - \frac{\delta q}{c_p T}\right), \tag{9.3.47}$$

showing that a change in speed $\mathrm{d}u$ may be accomplished by a change in $\mathrm{d}A$, or via the influence of external force δf or an amount of heat transfer δq. The coefficients $\mathrm{d}A/A$, $\delta f/p$ and $\delta q/(c_p T)$ are called the *influence coefficients*, for they represent the influence of some external excitations on the field variables. Similarly, combining Eq. (9.3.42)$_1$ with Eq. (9.3.47) gives

$$\frac{\mathrm{d}p}{p} = -\frac{\gamma\mathcal{M}_a^2}{\mathcal{M}_a^2 - 1}\frac{\mathrm{d}A}{A} - \frac{1 + (\gamma-1)\mathcal{M}_a^2}{\mathcal{M}_a^2 - 1}\frac{\delta f}{p} + \frac{\gamma\mathcal{M}_a^2}{\mathcal{M}_a^2 - 1}\frac{\delta q}{c_p T}. \tag{9.3.48}$$

Finally, replacing the term $\mathrm{d}u/u$ in Eq. (9.3.45)$_2$ by Eq. (9.3.47) leads to

$$\frac{\mathrm{d}T}{T} = -\frac{(\gamma-1)\mathcal{M}_a^2}{\mathcal{M}_a^2 - 1}\frac{\mathrm{d}A}{A} - \frac{(\gamma-1)\mathcal{M}_a^2}{\mathcal{M}_a^2 - 1}\frac{\delta f}{p} + \frac{\gamma\mathcal{M}_a^2 - 1}{\mathcal{M}_a^2 - 1}\frac{\delta q}{c_p T}. \tag{9.3.49}$$

Equations (9.3.47)–(9.3.49) are the general expressions of the differential changes in the fluid velocity, pressure and temperature in non-adiabatic flows.

As a special case, consider an adiabatic flow without any external body force, for which Eq. (9.3.47) reduces to

$$\frac{\mathrm{d}u}{u}=\frac{1}{\mathcal{M}_a^2-1}\frac{\mathrm{d}A}{A}. \tag{9.3.50}$$

This result indicates that for subsonic flows where $\mathcal{M}_a<1$, the flow can be accelerated, provided that the area of flow passage is reduced. On the contrary, for supersonic flows $\mathcal{M}_a>1$, the area of flow passage must be enlarged in order to increase the flow speed. These conclusions may give a nozzle shape shown in Fig. 9.14b, which is used to accelerate the flow speed from subsonic to supersonic regions. The location where $\mathrm{d}A=0$ is referred to as the *throat* of nozzle, at which $\mathcal{M}_a=1$. Under the same circumstances, Eq. (9.3.48) is simplified to

$$\frac{\mathrm{d}p}{p}=-\frac{\gamma\mathcal{M}_a^2}{\mathcal{M}_a^2-1}\frac{\mathrm{d}A}{A}. \tag{9.3.51}$$

This equation shows that for subsonic flows, a nozzle with reducing cross-sectional area leads to a reducing in the fluid pressure, which is justified by the Bernoulli equation, for a reducing in area increases the fluid velocity, hence giving rise to a pressure decrease. On the other hand, for supersonic flows, a reducing in area gives a positive increase in the fluid pressure. Finally, for an adiabatic flow without any external body force, Eq, (9.3.49) reduces to

$$\frac{\mathrm{d}T}{T}=-\frac{(\gamma-1)\mathcal{M}_a^2}{\mathcal{M}_a^2-1}\frac{\mathrm{d}A}{A}, \tag{9.3.52}$$

showing that for subsonic flows, the fluid temperature can be decreased if the area of flow passage is reduced, provided that $\gamma>1$, which is valid for air. For supersonic flows, the area of flow passage must be enlarged to obtain a drop of the fluid temperature.

Another demonstration is a flow in a constant-area channel without any external body force, for which Eqs. (9.3.47) and (9.3.48) are simplified respectively to

$$\frac{\mathrm{d}u}{u}=-\frac{1}{\mathcal{M}_a^2-1}\frac{\delta q}{c_pT},\qquad \frac{\mathrm{d}p}{p}=\frac{\gamma\mathcal{M}_a^2}{\mathcal{M}_a^2-1}\frac{\delta q}{c_pT}, \tag{9.3.53}$$

showing that for subsonic flows, the fluid velocity can be increased by transferring heat from the surrounding to the fluid, while for supersonic flow, heat must be removed from the fluid in order to have an increase in the fluid velocity.[9] In parallel, heat needs to be provided to the fluid in order to have an increase in pressure for supersonic flows, while heat should be removed from the fluid for subsonic flows. For the same circumstances, Eq. (9.3.49) reduces to

$$\frac{\mathrm{d}T}{T}=\frac{\gamma\mathcal{M}_a^2-1}{\mathcal{M}_a^2-1}\frac{\delta q}{c_pT}, \tag{9.3.54}$$

[9]Here δq is defined to be positive, if it is transferred from the surrounding to the fluid.

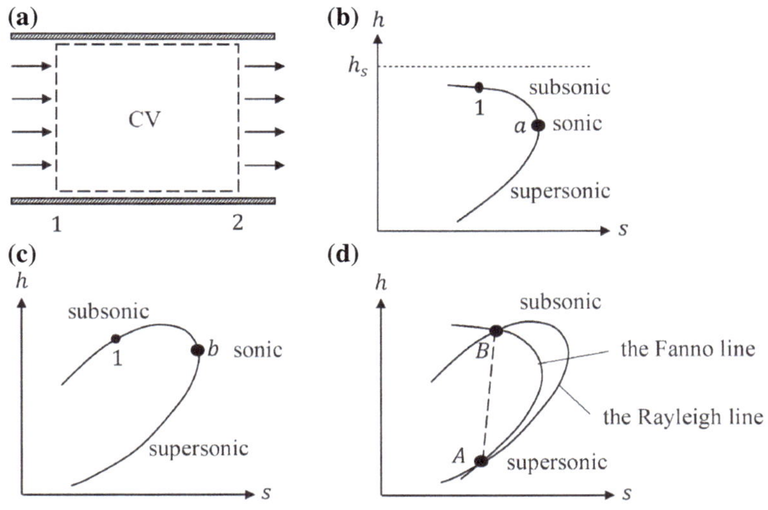

Fig. 9.16 Steady compressible flows in a constant-area conduit. **a** The geometric configurations. **b** The Fanno line for adiabatic flows with external body force. **c** The Rayleigh line for flows with heat transfer. **d** The application of the Fanno and Rayleigh lines for the formation of a shock wave

implying that the effect of transferring heat to the fluid is to increase the fluid temperature for subsonic flows. For supersonic flows in the range $1/\sqrt{\gamma} < \mathcal{M}_a < 1$, the temperature will be increased under the same condition. On the other hand, for an adiabatic flow in a constant-area conduit, Eq. (9.3.49) becomes

$$\frac{\mathrm{d}T}{T} = -\frac{(\gamma - 1)\mathcal{M}_a^2}{\mathcal{M}_a^2 - 1}\frac{\delta f}{p}, \tag{9.3.55}$$

showing that the external forces such as friction tend to increase the fluid temperature for subsonic flows, while the fluid temperature is decreased for supersonic flows. The equations derived in this section are expressed in terms of the differential values of fluid properties. They can be integrated to obtain the expressions for finite changes of fluid properties.

For flows in a constant-area conduit, in addition to the equations derived previously, there exist two graphic methods to obtain the variations in the fluid properties, which are called the *Fanno line* and the *Rayleigh line*,[10] corresponding respectively to the circumstances with frictional and heat transfer effects. To demonstrate the concepts, consider a control-volume in a constant-area conduit shown in Fig. 9.16a, for which the balances of mass, linear momentum in the x-direction, and thermal energy for a steady flow through the conduit read respectively

$$\begin{aligned} &\rho_1 u_1 = \rho_2 u_2 = \frac{\dot{m}}{A}, \qquad p_1 - p_2 - \frac{F_R}{A} = \frac{\dot{m}}{A}(u_2 - u_1), \\ &h_1 + \frac{1}{2}u_1^2 + \frac{\dot{Q}}{\dot{m}} = h_2 + \frac{1}{2}u_2^2, \end{aligned} \tag{9.3.56}$$

[10]Gino Girolamo Fanno, 1882–1962, an Italian mechanical engineer, who developed the Fanno line.

where $\dot{m}$ is the mass flow rate, F_R represents the external force, and $\dot{Q}$ is the amount of heat transfer. The fluid is considered an ideal gas, for which the state equation is given by

$$h = h(s, \rho), \qquad s = s(p, \rho), \tag{9.3.57}$$

where h and s are respectively the specific enthalpy and specific entropy of fluid. For adiabatic flows with external body force, Eqs. $(9.3.56)_{1,3}$ and (9.3.57) define a locus of states. For example, let state 1 of the fluid be known, which is represented by point 1 on the h–s diagram, as shown in Fig. 9.16b. For a chosen value of u_2, the values of ρ_2, h_2, s_2 and p_2 are determined by using respectively Eqs. $(9.3.56)_1$, $(9.3.56)_3$ and (9.3.57), so that state 2 may be determined and represented by a point on the h–s diagram, at which the value of F_R is then determined by Eq. $(9.3.56)_2$. Repeating this procedure for various values of u_2 yields different points on the h–s diagram, among which the curve is the Fanno line, which represents the locus of states at section 2 for a flow starting with the known conditions at state 1 by changing the amount of frictional force F_R.

The Fanno line has three distinct features. Point a marks the maximum specific entropy of fluid, corresponding to $\mathcal{M}_a = 1$. The part above point a is asymptotic to the specific stagnation enthalpy $h_s = h_1 + u_1^2/2 = h_2 + u_2^2/2$, representing the region of subsonic flows. The part below point a represents the region for supersonic flows. For an adiabatic flow, the second law of thermodynamics requires that the specific entropy increases during the flow. So, as starting from either the subsonic or supersonic region, the Mach number reaches the limiting value of unity for the condition of maximum entropy. Consequently, under adiabatic constant-area condition, a subsonic flow can never become supersonic, and in the absence of discontinuity (i.e., no occurrence of shock waves), a supersonic flow cannot become subsonic. Such a restriction is referred to as *choking*.

The Rayleigh line is constructed in a similar manner, except that there exists no external body force, but the influence of heat transfer is taken into account. It is the locus of points representing the states for flows under these conditions. A typical Rayleigh line is shown in Fig. 9.16c, in which point b corresponds to the maximum specific entropy, at which the Mach number is unity. The part of line above point b is devoted to subsonic flows, while that below point b represents the states in supersonic region. The entropy change in the flow is positive for heating and negative for cooling processes in both subsonic and supersonic flows.[11] However, choking still exists, and a subsonic or a supersonic flow can never become supersonic or subsonic by a heating process, respectively.

The applications of the Fanno and Rayleigh lines can be demonstrated by studying the formation of a normal shock wave, as shown in Fig. 9.16d. Point A represents the flow state ahead a shock wave, through which both lines are drawn. Points on the Fanno line represent various possible fluid states in an adiabatic flow, whereas points on the Rayleigh line represent various fluid states in a flow with no frictional effect. Since a normal shock wave is neither adiabatic nor frictionless, the fluid state behind

[11] This is so, for the boundary-layer friction is assumed to be absent.

a shock wave must be the intersection of the Fanno and Rayleigh lines, which is point B in the figure. It is noted that the specific entropy at point B is greater than that in point A, showing that the formation of a normal shock wave is not isentropic. The Fanno and Rayleigh lines provide only qualitative descriptions of the characteristics of compressible flows. Detailed quantitative descriptions need to be obtained by the mathematical equations derived previously.

9.3.4 Isentropic Flows

For steady isentropic flows, Wq. (9.1.3) reduces to

$$\rho(\boldsymbol{u}\cdot\nabla)h = (\boldsymbol{u}\cdot\nabla)p, \qquad (\boldsymbol{u}\cdot\nabla)p = -\rho(\boldsymbol{u}\cdot\nabla)\left(\frac{1}{2}\boldsymbol{u}\cdot\boldsymbol{u}\right), \tag{9.3.58}$$

where the second equation is obtained by taking inner product of the Euler equation with the fluid velocity $\boldsymbol{u}$. Combing two equations yields

$$\rho(\boldsymbol{u}\cdot\nabla)\left(h + \frac{1}{2}\boldsymbol{u}\cdot\boldsymbol{u}\right) = 0, \tag{9.3.59}$$

showing that the term inside the parenthesis is constant along each streamline, which is nothing else than the specific stagnation enthalpy h_s, i.e., $h_s = h + \boldsymbol{u}\cdot\boldsymbol{u}/2$. It corresponds to the specific enthalpy which the fluid would have at vanishing velocity. Since $h = c_p T$ for ideal gases, this expression is extended for the stagnation enthalpy to obtain the *stagnation temperature* T_s, so that

$$c_p T + \frac{1}{2}\boldsymbol{u}\cdot\boldsymbol{u} = c_p T_s, \tag{9.3.60}$$

in which T_s represents to the temperature that the fluid would have if it were brought to rest isentropically. It follows immediately that

$$\frac{T_s}{T} = 1 + \frac{u^2}{2c_p T}, \qquad \longrightarrow \qquad \frac{T_s}{T} = 1 + \frac{\gamma-1}{2}\mathcal{M}_a^2, \tag{9.3.61}$$

in which the properties of ideal gas have been used.

Since for isentropic flows the relation between temperature, density, and pressure ratios is given by

$$\frac{T_s}{T} = \left(\frac{p_s}{p}\right)^{(\gamma-1)/\gamma} = \left(\frac{\rho_s}{\rho}\right)^{\gamma-1}, \tag{9.3.62}$$

the quantities p_s and ρ_s are thus termed respectively the *stagnation pressure* and *stagnation density*. Substituting Eq. (9.3.61) into the above equation results in

$$\frac{p_s}{p} = \left(1 + \frac{\gamma-1}{2}\mathcal{M}_a^2\right)^{\gamma/(\gamma-1)}, \qquad \frac{\rho_s}{\rho} = \left(1 + \frac{\gamma-1}{2}\mathcal{M}_a^2\right)^{1/(\gamma-1)}. \tag{9.3.63}$$

Equations (9.3.62) and (9.3.63) give the temperature, density, and pressure ratios of an ideal gas in an isentropic flow in terms of the stagnation temperature, density, and pressure, and the Mach number. For any two points in an isentropic flow field, if the state at point 1 is known, the values of T_s, ρ_s and p_s are then determined by using these equations, which can be used subsequently to determine the values of T, ρ and p at another point 2. The explicit expressions for the temperature, density, and pressure ratios between any two points in an isentropic flow field can be derived by integrating the obtained results, which is left as an exercise.

9.3.5 Flows Through Nozzle

Consider a compressible flow through a convergent-divergent nozzle shown in Fig. 9.15b again. It follows from the previous results that the flow is subsonic in the convergent section, and is accelerated until the throat, where the local Mach number becomes unity. Since the flow is adiabatic and the frictional losses may be considered to be negligible, the flow in the convergent section may be approximated to be isentropic. With this, the temperature, pressure, and density at the throat, by using Eqs. (9.3.61)$_2$ and (9.3.63), are obtained as

$$\frac{T_s}{T_*} = \frac{\gamma+1}{2}, \qquad \frac{p_s}{p_*} = \left(\frac{\gamma+1}{2}\right)^{\gamma/(\gamma-1)}, \qquad \frac{\rho_s}{\rho_*} = \left(\frac{\gamma+1}{2}\right)^{1/(\gamma-1)}, \tag{9.3.64}$$

where the conditions at the throat are denoted by the subscript "$*$". The values of T_s, p_s, and ρ_s can be determined by using Eqs. (9.3.61)$_2$ and (9.3.63) if the flow conditions at a specific location in the convergent section, e.g. the flow inlet, are known. The data is then substituted into the above equations to obtain the values of T_*, p_*, and ρ_*.

Formulating the conservation of mass between any arbitrary flow section in the convergent section and throat yields

$$\frac{A}{A_*} = \frac{\rho_*}{\rho}\frac{\mathcal{M}_{a*}a_*}{\mathcal{M}_a} = \frac{1}{\mathcal{M}_a}\frac{\rho_*\rho_s}{\rho_s\rho}\sqrt{\frac{T_*T_s}{T_sT}}, \tag{9.3.65}$$

where $\mathcal{M}_{a*}$ is the local Mach number at the throat, which assumes the value of unity, and a_* is the local sonic velocity at the throat with $a_*/a = \sqrt{T_*/T}$, as indicated by the last equation in the fifth footnote in the chapter. Substituting Eqs. (9.3.61)$_2$, (9.3.63)$_2$ and (9.3.64)$_{1,3}$ into the above equation gives

$$\frac{A}{A_*} = \frac{1}{\mathcal{M}_a}\left(\frac{2}{\gamma+1}\right)^{(\gamma+1)/[2(\gamma-1)]}\left(1+\frac{\gamma-1}{2}\mathcal{M}_a^2\right)^{(\gamma+1)/[2(\gamma-1)]}, \tag{9.3.66}$$

which is further simplified to

$$\frac{A}{A_*} = \frac{1}{\mathcal{M}_a}\left[\frac{2}{\gamma+1}\left(1+\frac{\gamma-1}{2}\mathcal{M}_a^2\right)\right]^{(\gamma+1)/[2(\gamma-1)]}. \tag{9.3.67}$$

This equation relates the local flow area to that at the throat region. Let the mass flow rate be denoted by $\dot{m}$, which is identified to be

$$\dot{m} = \rho_* u_* A_* = \frac{\rho_*\rho_s}{\rho_s}\left(\mathcal{M}_{a*}a_*\right)A_* = \rho_s\sqrt{\gamma R T_s}\left(\frac{2}{\gamma+1}\right)^{(\gamma+1)/[2(\gamma-1)]}, \tag{9.3.68}$$

in which $\mathcal{M}_{a*} = 1$, $a_* = \sqrt{\gamma R T_s T_*/T_s}$ and Eq. (9.3.64)$_3$ have been used. Substituting the ideal gas state equation $\rho_s = p_s/(RT_s)$ into this equation results in

$$\dot{m} = \frac{p_s A_*}{\sqrt{T_s}} \sqrt{\frac{\gamma}{R} \left(\frac{2}{\gamma + 1} \right)^{(\gamma+1)/(\gamma-1)}}, \tag{9.3.69}$$

showing that the mass flow rate through a nozzle is proportional to the throat area A_*, as expected, and is also proportional to the stagnation pressure of gas, and is inversely proportional to the square root of stagnation temperature.

In the divergent section, the flow may or may not be further accelerated to become supersonic from the throat region, with typical pressure and Mach number distributions are shown respectively in Figs. 9.17a and b. The entire flow state in the nozzle depends on the pressure ratio p_2/p_1, where p_1 is the fluid pressure at the inlet of convergent section, while p_2 is the pressure at the exit of divergent section. If $p_2/p_1 = 1$, there will be no flow through the nozzle, corresponding to curve A. If $p_2/p_1 > p_*/p_1$, the flow will be accelerated in the convergent section until the maximum fluid velocity is reached at the throat region, which is smaller than the sonic velocity, and is then decelerated in the divergent section, as shown by curve B. The value of p_2/p_1 can further be reduced, until the sonic condition at the throat is reached. In this circumstance, the flow is accelerated in the convergent section, but still decelerated in the divergent section, as indicated by curve C. If the pressure ratio p_2/p_1 is further reduced, the flow conditions in the convergent section, e.g. the sonic condition at the throat and the amount of mass flow rate, remain the same as before, but the flow will be accelerated in the divergent section. Such a characteristic is shown e.g. by curve G, in which the flow is accelerated in the entire convergent-divergent nozzle, and can be approximated to be isentropic, for which the equations derived in Sect. 9.3.4 are applicable. For those values of the pressure ratio p_2/p_1 corresponding to curves D, E, F between curves C and G, no flow pattern can be found to satisfy the isentropic conditions, for irreversible (and hence non-isentropic) shock waves take place somewhere in the divergent section. In these circumstances, normal shock waves are frequently encountered, and the equations derived in Sect. 9.2.4 may be used to relate the fluid conditions ahead and behind the shock waves.

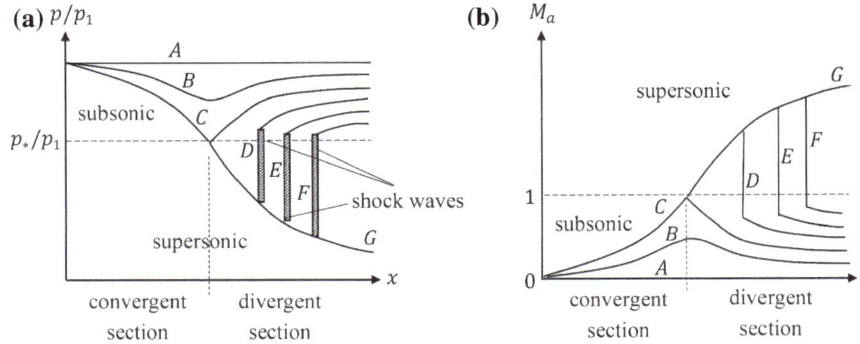

Fig. 9.17 Flow states in the convergent and divergent sections of a nozzle. **a** The distributions of the pressure ratio. **b** The distribution of the local Mach number

9.4 Multi-dimensional Flows

Two-dimensional and three-dimensional steady flows in subsonic and supersonic regions are discussed in this section. The governing equations are derived first, followed by the discussions on the Janzen-Rayleigh expansion. The theory of small perturbation is introduced to study the Prandtl-Glauert rule, which relates subsonic compressible flows to the corresponding incompressible flows. Ackeret's theory for supersonic flows is also explored. A discussion on the exact solution to the Prandtl-Meyer flow problem is provided. The section ends with a discussion on the influence of fluid compressibility on drag and lift.

9.4.1 Irrotational Motions

It follows from Crocco's equation that irrotational flows are also isentropic, for which the pressure gradient can be expressed as

$$\nabla p = \frac{\mathrm{d}p}{\mathrm{d}\rho}\nabla\rho = a^2\nabla\rho, \tag{9.4.1}$$

for $p = p(\rho)$ in isentropic flows. With this, the balance of linear momentum reads

$$\begin{aligned} &\frac{\partial \boldsymbol{u}}{\partial t} + (\boldsymbol{u}\cdot\nabla)\boldsymbol{u} = -\frac{a^2}{\rho}\nabla\rho, \\ \longrightarrow\ &\frac{1}{2}\frac{\partial}{\partial t}(\boldsymbol{u}\cdot\boldsymbol{u}) + \boldsymbol{u}\cdot[(\boldsymbol{u}\cdot\nabla)\boldsymbol{u}] = \frac{a^2}{\rho}\frac{\partial\rho}{\partial t} + a^2\nabla\cdot\boldsymbol{u}, \end{aligned} \tag{9.4.2}$$

where the second equation is obtained by taking inner product of the first equation with $\boldsymbol{u}$, and the conservation of mass has been used. Taking time derivative of the unsteady Bernoulli equation for irrotational flows, i.e., Eq. (7.3.15), yields

$$\frac{\partial^2\phi}{\partial t^2} + \frac{1}{2}\frac{\partial}{\partial t}(\boldsymbol{u}\cdot\boldsymbol{u}) = -\frac{\partial}{\partial t}\left(\int\frac{\mathrm{d}p}{\rho}\right) = -\frac{\mathrm{d}}{\mathrm{d}\rho}\left(\int\frac{a^2}{\rho}\mathrm{d}\rho\right)\frac{\partial\rho}{\partial t} = -\frac{a^2}{\rho}\frac{\partial\rho}{\partial t}, \tag{9.4.3}$$

by which Eq. (9.4.2)$_2$ is simplified to

$$\frac{1}{2}\frac{\partial}{\partial t}(\boldsymbol{u}\cdot\boldsymbol{u}) + \boldsymbol{u}\cdot[(\boldsymbol{u}\cdot\nabla)\boldsymbol{u}] = -\frac{\partial^2\phi}{\partial t^2} - \frac{1}{2}\frac{\partial}{\partial t}(\boldsymbol{u}\cdot\boldsymbol{u}) + a^2\nabla\cdot\boldsymbol{u}. \tag{9.4.4}$$

Expressing the term $\nabla\cdot\boldsymbol{u}$ in this equation gives

$$\nabla\cdot\boldsymbol{u} = \frac{1}{a^2}\left\{\boldsymbol{u}\cdot[(\boldsymbol{u}\cdot\nabla)\boldsymbol{u}] + \frac{\partial}{\partial t}\left(\frac{\partial\phi}{\partial t} + \boldsymbol{u}\cdot\boldsymbol{u}\right)\right\}, \tag{9.4.5}$$

which is valid for irrotational motions of a compressible fluid. The equation which needs to be satisfied by the velocity potential function ϕ is obtained directly from this equation, which is given viz.,

$$\nabla^2\phi = \frac{1}{a^2}\left\{\nabla\phi\cdot[(\nabla\phi\cdot\nabla)\nabla\phi] + \frac{\partial}{\partial t}\left(\frac{\partial\phi}{\partial t} + \nabla\phi\cdot\nabla\phi\right)\right\}, \tag{9.4.6}$$

which is no longer a Laplace equation. However, in a limiting case in which $a^2 \to \infty$, it becomes a Laplace equation. In such a case, the assumption that $a^2 \to \infty$ corresponds to incompressible flows. It is seen that for constant fluid density the equation governing the velocity potential function is linear, while for variable fluid density the equation becomes nonlinear. In addition, Eq. (9.4.6) represents a formidable analytic problem for any specific flow which is to be solved. The encountered difficulty in obtaining exact solutions led to the development of approximate methods, two of which are discussed separately in the next two subsections.

9.4.2 The Janzen-Rayleigh Expansion

For steady flows, Eq. (9.4.6) is simplified to

$$\nabla^2\phi = \frac{1}{a^2}\nabla\phi\cdot[(\nabla\phi\cdot\nabla)\nabla\phi], \qquad \frac{\partial^2\phi}{\partial x_i x_i} = \frac{1}{a^2}\frac{\partial\phi}{\partial x_i}\frac{\partial\phi}{\partial x_j}\frac{\partial^2\phi}{\partial x_i x_j}, \tag{9.4.7}$$

where the right-hand-side represents the compressible effect varying as a^{-2}, and vanishes as $a \to \infty$. In view of these, an approximate solution to a slightly compressible flow could be sought in which the first correction due to the fluid compressibility varies as the square of the Mach number. Hence, the solution to this equation is assumed to be in the form

$$\phi(x, y, z) = U\sum_{n=0}^{\infty}\mathcal{M}_{a\infty}^{2n}\phi_n(x, y, z), \tag{9.4.8}$$

in which a uniform flow with velocity U approaching a body is considered, and $\mathcal{M}_{a\infty}$ is the local Mach number of approaching flow in the region far away from the body, at which the local sonic velocity is a_∞. Substituting this expression into Eq. (9.4.7) yields

$$U\sum_{n=0}^{\infty}\mathcal{M}_{a\infty}^{2n}\frac{\partial^2\phi_n}{\partial x_i x_i} = \frac{U^3}{a^2}\sum_{n=0}^{\infty}\mathcal{M}_{a\infty}^{2n}\frac{\partial\phi_n}{\partial x_i}\sum_{n=0}^{\infty}\mathcal{M}_{a\infty}^{2n}\frac{\partial\phi_n}{\partial x_j}\sum_{n=0}^{\infty}\mathcal{M}_{a\infty}^{2n}\frac{\partial^2\phi_n}{\partial x_i x_j}, \tag{9.4.9}$$

which is expressed alternatively as

$$\sum_{n=0}^{\infty}\mathcal{M}_{a\infty}^{2n}\frac{\partial^2\phi_n}{\partial x_i x_i} = \frac{a_\infty^2}{a^2}\mathcal{M}_{a\infty}^2\sum_{n=0}^{\infty}\mathcal{M}_{a\infty}^{2n}\frac{\partial\phi_n}{\partial x_i}\sum_{n=0}^{\infty}\mathcal{M}_{a\infty}^{2n}\frac{\partial\phi_n}{\partial x_j}\sum_{n=0}^{\infty}\mathcal{M}_{a\infty}^{2n}\frac{\partial^2\phi_n}{\partial x_i x_j}. \tag{9.4.10}$$

For the considered problem, the thermal-energy equation, i.e., Eq. (9.3.59), reduces to

$$\frac{1}{2}\boldsymbol{u}\cdot\boldsymbol{u} + \frac{a^2}{\gamma-1} = \frac{1}{2}U^2 + \frac{a_\infty^2}{\gamma-1}, \tag{9.4.11}$$

in which the properties of ideal gas have been used. It follows form this equation that

$$\frac{a^2}{a_\infty^2} = 1 + \frac{\gamma-1}{2}\left(\mathcal{M}_{a\infty}^2 - \frac{\boldsymbol{u}\cdot\boldsymbol{u}}{a_\infty^2}\right). \tag{9.4.12}$$

Replacing $\boldsymbol{u}$ by $\nabla\phi$ in this equation gives

$$\begin{aligned}\frac{a^2}{a_\infty^2} &= 1+\frac{\gamma-1}{2}\mathcal{M}_{a\infty}^2\left[1-\left(\sum_{n=0}^{\infty}\mathcal{M}_{a\infty}^{2n}\frac{\partial\phi_n}{\partial x_i}\right)^2\right]\\ &= 1+\frac{\gamma-1}{2}\mathcal{M}_{a\infty}^2\left[1-\left(\frac{\partial\phi_0}{\partial x_i}\right)^2\right]+\mathcal{O}\left(\mathcal{M}_{a\infty}^4\right),\end{aligned} \tag{9.4.13}$$

whose inverse is obtained as

$$\frac{a_\infty^2}{a^2} = 1-\frac{\gamma-1}{2}\mathcal{M}_{a\infty}^2\left[1-\left(\frac{\partial\phi_0}{\partial x_i}\right)^2\right]+\mathcal{O}\left(\mathcal{M}_{a\infty}^4\right). \tag{9.4.14}$$

Substituting this equation into Eq. (9.4.10) results in

$$\begin{aligned}\sum_{n=0}^{\infty}\mathcal{M}_{a\infty}^{2n}\frac{\partial^2\phi_n}{\partial x_i x_i} &= \mathcal{M}_{a\infty}^2\left\{1-\frac{\gamma-1}{2}\mathcal{M}_{a\infty}^2\left[1-\left(\frac{\partial\phi_0}{\partial x_i}\right)^2\right]+\mathcal{O}\left(\mathcal{M}_{a\infty}^4\right)\right\}\cdot\\ &\quad \sum_{n=0}^{\infty}\mathcal{M}_{a\infty}^{2n}\frac{\partial\phi_n}{\partial x_i}\sum_{n=0}^{\infty}\mathcal{M}_{a\infty}^{2n}\frac{\partial\phi_n}{\partial x_j}\sum_{n=0}^{\infty}\mathcal{M}_{a\infty}^{2n}\frac{\partial^2\phi_n}{\partial x_i x_j},\end{aligned} \tag{9.4.15}$$

which is assumed to be uniformly valid in $\mathcal{M}_{a\infty}^2$, so that the coefficients of like powers must be balanced on two sides. This leads to a sequence of equations representing the coefficients of $\mathcal{M}_{a\infty}^0$, $\mathcal{M}_{a\infty}^2$, $\mathcal{M}_{a\infty}^4$, etc., which are given respectively as

$$\begin{aligned}\frac{\partial^2\phi_0}{\partial x_i x_i} &= 0,\\ \frac{\partial^2\phi_1}{\partial x_i x_i} &= \frac{\partial\phi_0}{\partial x_i}\frac{\partial\phi_0}{\partial x_j}\frac{\partial^2\phi_0}{\partial x_i x_j},\\ \frac{\partial^2\phi_2}{\partial x_i x_i} &= -\frac{\gamma-1}{2}\left[1-\left(\frac{\partial\phi_0}{\partial x_i}\right)^2\right]\frac{\partial\phi_0}{\partial x_i}\frac{\partial\phi_0}{\partial x_j}\frac{\partial^2\phi_0}{\partial x_i x_j}+\frac{\partial\phi_1}{\partial x_i}\frac{\partial\phi_0}{\partial x_j}\frac{\partial^2\phi_0}{\partial x_i x_j}\\ &\quad +\frac{\partial\phi_0}{\partial x_i}\frac{\partial\phi_1}{\partial x_j}\frac{\partial^2\phi_0}{\partial x_i x_j}+\frac{\partial\phi_0}{\partial x_i}\frac{\partial\phi_0}{\partial x_j}\frac{\partial^2\phi_1}{\partial x_i x_j}, \qquad \text{etc.}\end{aligned} \tag{9.4.16}$$

The equation to be solved for ϕ_0 represents an incompressible-flow problem, corresponding to $\mathcal{M}_{a\infty} \to 0$. The equation for ϕ_1 is linear, although it is non-homogeneous. Having solved ϕ_0, the right-hand-side of Eq. $(9.4.16)_2$ will become an explicit function of the spatial coordinates, so that the solution to ϕ_1 may be obtained. Likewise, having obtained ϕ_0 and ϕ_1, the right-hand-side of Eq. $(9.4.16)_3$ will become an explicit function of the spatial coordinates again, and the solution to ϕ_2 may equally be obtained. By following this procedure, the solutions of $\{\phi_0, \phi_1, \phi_2, \cdots\}$ may be obtained sequently, and each solution contributes a term in Eq. (9.4.8). The accuracy of solution ϕ depends on the number of terms that are included. This expansion in solution is referred to as the *Janzen-Rayleigh expansion*. It is recognized, however, that the emerging differential equations become complicated rapidly, and it is not practically reasonable to conduct the solution by using more than two or three contributions, implying that the obtained approximate solution is only valid for the compressible flows with the Mach numbers smaller than 0.5. The advantage of the Janzen-Rayleigh expansion, however, is that it is valid for any shape of body, not just restricted to slender bodies.

9.4.3 Theory of Small Perturbation

Consider a steady uniform flow approaching a body which is sufficiently slender, so that it induces a small perturbation to the free stream. The velocity potential function may thus be written as

$$\phi(x, y, z) = Ux + \Phi, \qquad \frac{\|\nabla\Phi\|}{U} \ll 1, \tag{9.4.17}$$

where the approaching uniform flow is assumed to be in the positive x-direction, and Φ represents the contribution to the velocity potential function, which results from the perturbation induced by the presence of slender body, with the assumption that the perturbation velocity is much smaller than U. Substituting this expression into Eq. (9.4.6) yields

$$\nabla^2\Phi = \frac{1}{a^2}\left(U\boldsymbol{e}_x + \nabla\Phi\right)\cdot\left[\left(U\boldsymbol{e}_x + \nabla\Phi\right)\cdot\nabla\right]\left(U\boldsymbol{e}_x + \nabla\Phi\right). \tag{9.4.18}$$

Since the perturbation velocity is small, a linearized approximation to Eq. (9.4.18) is possible, i.e.,

$$\nabla^2\Phi = \frac{U^2}{a^2}\frac{\partial^2\Phi}{\partial x^2}, \tag{9.4.19}$$

which, when compared to the Laplace equation of ϕ for ideal fluids, contains only one correction contribution for fluid compressibility on its right-hand-side, and the correction term coincides to the direction of free stream. The derivation of Eq. (9.4.19) is given in the following.

Since

$$\boldsymbol{u}\cdot\boldsymbol{u} = \left(U + u'\right)^2 + \left(v'\right)^2 + \left(w'\right)^2 \sim U^2 + 2Uu', \tag{9.4.20}$$

under a linearized approximation, where u', v', and w' are respectively the perturbation velocity components in the x-, y- and z-directions, substituting this expression into Eq. (9.4.11) gives

$$Uu' + \frac{a^2}{\gamma-1} = \frac{a_\infty^2}{\gamma-1}, \qquad \longrightarrow \qquad a^2 = a_\infty^2\left[1 - (\gamma-1)\frac{Uu'}{a_\infty^2}\right], \tag{9.4.21}$$

which is substituted subsequently into Eq. (9.4.19) to obtain

$$\nabla^2\Phi = \frac{U^2}{a_\infty^2}\left[1 - (\gamma-1)\frac{U}{a_\infty^2}\frac{\partial\Phi}{\partial x}\right]^{-1}\frac{\partial^2\Phi}{\partial x^2}. \tag{9.4.22}$$

Applying a linearization approximation to this equation results in

$$\nabla^2\Phi = \frac{U^2}{a_\infty^2}\frac{\partial^2\Phi}{\partial x^2}. \tag{9.4.23}$$

In three-dimensional rectangular coordinate system, this equation is generalized as

$$\left(1-\mathcal{M}_{a\infty}^{2}\right)\frac{\partial^2\Phi}{\partial x^2}+\frac{\partial^2\Phi}{\partial y^2}+\frac{\partial^2\Phi}{\partial z^2}=0, \tag{9.4.24}$$

showing that for subsonic flows it becomes elliptic with no real characteristic. On the contrary, this equation becomes hyperbolic for supersonic flows, having real characteristics. This observation justifies the fact that shock waves can only take place in supersonic flows. While the Janzen-Rayleigh expansion is valid for any shape of body with the restriction that the Mach number must be less than 0.5, the small-perturbation theory is only valid for slender bodies in both subsonic and supersonic flows, and is invalid for flows with nearly unity Mach number due to the linearization, as will be discussed later.

For compressible frictionless flows, the integration of the pressure around a body surface gives rise to the lift and drag acting on that body. Hence, the prime interest of study is the determination of the pressure field of fluid, which is conventionally expressed in terms of the dimensionless pressure coefficient given by

$$C_p=\frac{2(p-p_\infty)}{\rho_\infty U^2}=\frac{2}{\gamma\mathcal{M}_{a\infty}^{2}}\left(\frac{p}{p_\infty}-1\right), \tag{9.4.25}$$

where the quantities with the subscript "∞" are referred to the flow conditions of free stream. Since the flow is assumed to be irrotational, it is also isentropic. Substituting the isentropic law into the thermal-energy equation between the far-away region and region near the body yields

$$\frac{1}{2}\boldsymbol{u}\cdot\boldsymbol{u}+\frac{a_\infty^2}{\gamma-1}\left(\frac{p}{p_\infty}\right)^{(\gamma-1)/\gamma}=\frac{1}{2}U^2+\frac{a_\infty^2}{\gamma-1}, \tag{9.4.26}$$

by which the pressure ratio is obtained as

$$\frac{p}{p_\infty}=\left[1+\frac{\gamma-1}{2a_\infty^2}\left(U^2-\boldsymbol{u}\cdot\boldsymbol{u}\right)\right]^{\gamma/(\gamma-1)}. \tag{9.4.27}$$

Substituting this equation into Eq. (9.4.25) gives

$$C_p=\frac{2}{\gamma\mathcal{M}_{a\infty}^{2}}\left\{\left[1+\frac{\gamma-1}{2a_\infty^2}\left(U^2-\boldsymbol{u}\cdot\boldsymbol{u}\right)\right]^{\gamma/(\gamma-1)}-1\right\}, \tag{9.4.28}$$

expressing the local value of pressure coefficient in terms of the local velocity for compressible, frictionless, and adiabatic flows.

In the context of small-perturbation theory, the velocity term in Eq. (9.4.28) is expressed as

$$U^2-\boldsymbol{u}\cdot\boldsymbol{u}=-2Uu', \tag{9.4.29}$$

as indicated by Eq. (9.4.20), by which Eq. (9.4.28) is brought to the form

$$C_p=\frac{2}{\gamma\mathcal{M}_{a\infty}^{2}}\left\{\left[1-(\gamma-1)\frac{Uu'}{a_\infty^2}\right]^{\gamma/(\gamma-1)}-1\right\}. \tag{9.4.30}$$

The terms inside the bracket of this equation, by using a first-order approximation, can be simplified to

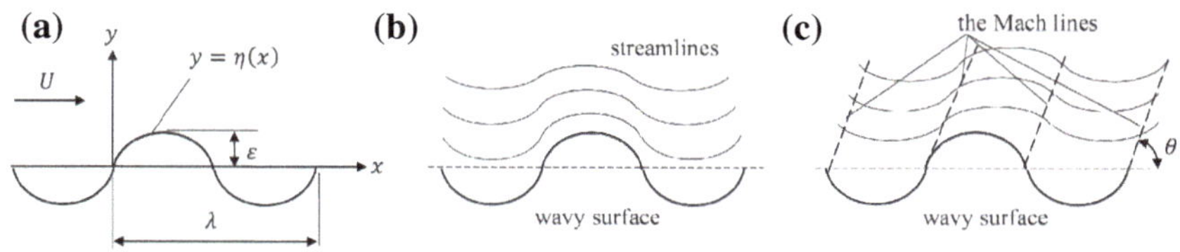

Fig. 9.18 A steady, uniform and compressible flow with velocity U over a wavy boundary. **a** The geometric configurations. **b** The flow pattern in a subsonic case. **c** The flow pattern in a supersonic case. Solid lines: streamlines; dashed lines: the Mach lines

$$\left[1-(\gamma-1)\frac{Uu'}{a_\infty^2}\right]^{\gamma/(\gamma-1)}=1-\gamma\frac{Uu'}{a_\infty^2}, \tag{9.4.31}$$

so that Eq. (9.4.30) becomes

$$C_p=-2\frac{u'}{U}. \tag{9.4.32}$$

This simple result is used in conjunction with the approximate solutions to Eq. (9.4.24). The following three subsections are devoted to the applications of small-perturbation theory.

9.4.4 Flows over Wavy Boundary

Consider a uniform and compressible flow with velocity U over a sinuous surface, as shown in Fig. 9.18a, in which the wavy surface is described by

$$y=\eta(x)=\varepsilon\sin\frac{2\pi x}{\lambda}, \tag{9.4.33}$$

where ε/λ is assumed to be small compared with unity in the context of linearization. For the considered two-dimensional flow, the perturbation velocity potential function Φ should satisfy Eq. (9.4.24), which is subject to the boundary condition on $y-\eta(x)=0$ given by

$$\frac{v'}{U+u'}=\frac{\mathrm{d}y}{\mathrm{d}x}=\frac{2\pi\varepsilon}{\lambda}\cos\frac{2\pi x}{\lambda}, \tag{9.4.34}$$

resulted from the fact that the wavy surface is a streamline. In the context of linearization theory, the term on the left-hand-side may be simplified to v'/U, so that the boundary condition becomes

$$v'=\frac{\partial\Phi}{\partial y}(x,\eta)=\frac{2\pi\varepsilon U}{\lambda}\cos\frac{2\pi x}{\lambda}. \tag{9.4.35}$$

Expanding the left-hand-side in a Taylor series about $\eta=0$, and taking linearization of the resulting equation yields

$$\frac{\partial\Phi}{\partial y}(x,0)=\frac{2\pi\varepsilon U}{\lambda}\cos\frac{2\pi x}{\lambda}. \tag{9.4.36}$$

Another boundary condition is that the perturbation velocity in the region far away from the body should assume a finite value, which is given by

$$\frac{\partial \Phi}{\partial x}(x, y \to \infty) = \text{finite}. \tag{9.4.37}$$

Equations (9.4.24), (9.4.36) and (9.4.37) define a boundary-value problem for the perturbation velocity potential function Φ in the context of small-perturbation theory.

For subsonic flows, Eq. (9.4.24) is of elliptic-type, and it is convenient to replace the spatial coordinate x by a new variable ξ defined by

$$\xi \equiv \frac{x}{\sqrt{1 - \mathcal{M}_{a\infty}^2}}, \tag{9.4.38}$$

so that Eqs. (9.4.24), (9.4.36) and (9.4.37) can be recast as

$$\frac{\partial^2 \Phi}{\partial \xi^2} + \frac{\partial^2 \Phi}{\partial y^2} = 0, \qquad \frac{\partial \Phi}{\partial y}(\xi, 0) = \frac{2\pi\varepsilon U}{\lambda} \cos\left(\frac{2\pi}{\lambda}\sqrt{1 - \mathcal{M}_{a\infty}^2}\xi\right),$$
$$\frac{\partial \Phi}{\partial \xi}(\xi, y \to \infty) = \text{finite}. \tag{9.4.39}$$

The solution to Φ is given by

$$\Phi(\xi, y) = [C_1 \cos(\alpha\xi) + C_2 \sin(\alpha\xi)] \exp(-\alpha y), \tag{9.4.40}$$

where C_1 and C_2 are undetermined constants, and α is an undetermined function. This is done so, because the ξ-domain, in view of Eq. (9.4.39)$_2$, should be trigonometric, while the y-domain is semi-infinite. Imposing Eq. (9.4.39)$_2$ to the proposed solution yields

$$-\alpha C_1 = \frac{2\pi\varepsilon U}{\lambda}, \qquad C_2 = 0, \qquad \alpha = \frac{2\pi}{\lambda}\sqrt{1 - \mathcal{M}_{a\infty}^2}, \tag{9.4.41}$$

with which Eq. (9.4.40) becomes

$$\Phi(\xi, y) = -\frac{U\varepsilon}{\sqrt{1 - \mathcal{M}_{a\infty}^2}} \cos\left(\frac{2\pi}{\lambda}\sqrt{1 - \mathcal{M}_{a\infty}^2}\xi\right) \exp\left(-\frac{2\pi}{\lambda}\sqrt{1 - \mathcal{M}_{a\infty}^2}y\right). \tag{9.4.42}$$

Replacing ξ in this equation by x with Eq. (9.4.38) results in

$$\Phi(x, y) = -\frac{U\varepsilon}{\sqrt{1 - \mathcal{M}_{a\infty}^2}} \cos\frac{2\pi x}{\lambda} \exp\left(-\frac{2\pi}{\lambda}\sqrt{1 - \mathcal{M}_{a\infty}^2}y\right), \tag{9.4.43}$$

showing that the perturbation to the free stream is in phase with the wall, and gives the flow pattern shown in Fig. 9.18b. The perturbation falls out exponentially with distance from the wall surface. The value of u' can be obtained by using Eq. (9.4.43), and it follows that

$$\max\left(\frac{u'}{U}\right) = \frac{(2\pi/\lambda)\varepsilon}{\sqrt{1 - \mathcal{M}_{a\infty}^2}} \ll 1, \tag{9.4.44}$$

for $u'/U \ll 1$ in the context of linearized theory. Substituting Eq. (9.4.43) into Eq. (9.4.32) results in

$$C_p = -\frac{(4\pi/\lambda)\varepsilon}{\sqrt{1 - \mathcal{M}_{a\infty}^2}} \sin\frac{2\pi x}{\lambda} \exp\left(-\frac{2\pi}{\lambda}\sqrt{1 - \mathcal{M}_{a\infty}^2}y\right). \tag{9.4.45}$$

At the wavy surface, $y \sim 0$, so that the pressure coefficient there, denoted by C_{pw}, is obtained as

$$C_{pw} = -\frac{(4\pi/\lambda)\varepsilon}{\sqrt{1-\mathcal{M}_{a\infty}^2}}\sin\frac{2\pi x}{\lambda}. \tag{9.4.46}$$

Comparing this equation with Eq. (9.4.45) shows that on the wavy surface the pressure assumes a minimum at the highest points of crests and a maximum at the lowest points of troughs.

For supersonic flows, Eq. (9.4.24) is of hyperbolic-type, and is expressed alternatively as

$$\frac{\partial^2\Phi}{\partial x^2} - \frac{1}{\mathcal{M}_{a\infty}^2 - 1}\frac{\partial^2\Phi}{\partial y^2} = 0, \tag{9.4.47}$$

which is a one-dimensional wave equation, to which a general solution is of the form

$$\Phi(x, y) = f_1\left(x - \sqrt{\mathcal{M}_{a\infty}^2 - 1}y\right) + f_2\left(x + \sqrt{\mathcal{M}_{a\infty}^2 - 1}y\right), \tag{9.4.48}$$

where f_1 and f_2 are any differential functions. The solution of f_1 represents a wave sloping downstream and away from the wall (right-running waves), so that the perturbations generated by the wavy wall will travel downstream only by this solution. The solution of f_2 represents the signals traveling upstream as they move away from the wall (left-running waves), which must be rejected, for it has no physical meaning in the considered circumstance, to be discussed later. With these, the general solution is given by

$$\Phi(x, y) = f_1\left(x - \sqrt{\mathcal{M}_{a\infty}^2 - 1}y\right). \tag{9.4.49}$$

Imposing Eq. (9.4.36) to the solution yields

$$f_1(x) = -\frac{U\varepsilon}{\sqrt{\mathcal{M}_{a\infty}^2 - 1}}\sin\frac{2\pi x}{\lambda}, \tag{9.4.50}$$

so that Eq. (9.4.49) becomes

$$\Phi(x, y) = -\frac{U\varepsilon}{\sqrt{\mathcal{M}_{a\infty}^2 - 1}}\sin\left[\frac{2\pi}{\lambda}\left(x - \sqrt{\mathcal{M}_{a\infty}^2 - 1}y\right)\right], \tag{9.4.51}$$

which also satisfies Eq. (9.4.37). The pressure coefficient is then determined as

$$C_p = \frac{(4\pi/\lambda)\varepsilon}{\sqrt{\mathcal{M}_{a\infty}^2 - 1}}\cos\left[\frac{2\pi}{\lambda}\left(x - \sqrt{\mathcal{M}_{a\infty}^2 - 1}y\right)\right], \tag{9.4.52}$$

whose value on the wavy surface is obtained as

$$C_{pw} = \frac{(4\pi/\lambda)\varepsilon}{\sqrt{\mathcal{M}_{a\infty}^2 - 1}}\cos\frac{2\pi x}{\lambda}. \tag{9.4.53}$$

Equations (9.4.51) and (9.4.52) indicate that the velocity components and pressure are constant along the lines described by

$$x - \sqrt{\mathcal{M}_{a\infty}^2 - 1}y = \text{constant}, \tag{9.4.54}$$

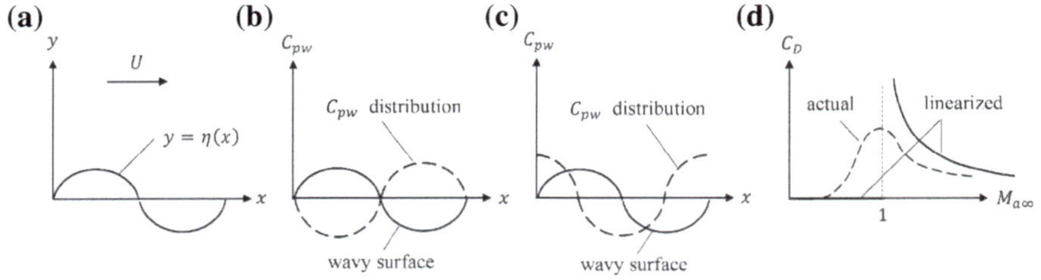

Fig. 9.19 The distribution of pressure coefficient in a section of the wavy surface. **a** The geometric configurations. **b** In subsonic flows. **c** In supersonic flows. **d** The relation between the total drag coefficient and the Mach number of incoming flow

whose slope is obtained as

$$\frac{dy}{dx} = \frac{1}{\sqrt{\mathcal{M}_{a\infty}^2 - 1}} = \tan\theta, \qquad \longrightarrow \qquad \theta = \sin^{-1}\left(\frac{1}{\mathcal{M}_{a\infty}}\right), \tag{9.4.55}$$

where θ represents the inclination angle of the lines with respect to the x-axis. This result shows that the lines along which the flow parameters are constant are in fact the Mach lines.[12] In other words, signals propagating along the Mach lines remain undisturbed. The corresponding flow field is shown in Fig. 9.18c.

For supersonic flows, Eq. (9.4.53) indicates that the pressure on the wavy wall is proportional to $\cos(2\pi x/\lambda)$, so that its peaks are $\pi/2$ out of phase with that of wavy surface, giving rise to a drag force on the wall. On the contrary, Eq. (9.4.46) shows that the phases of pressure and wavy surface are the same in subsonic flows. The pressure distributions on a section of the wavy surface, shown in Fig. 9.19a, are illustrated respectively in Figs. 9.19b and c for subsonic and supersonic flows. Since in subsonic flows the distribution of C_{pw} is symmetric about each geometric peak, there is no drag force. The lack of symmetry of C_{pw} in supersonic flows gives rise to a drag, which is called the *wave drag*. Figure 9.19d illustrates the theoretical estimations on the drags on a body in terms of the Mach number of approaching uniform flow, in which C_D is the actual measured drag coefficient, represented by the dashed line. In the supersonic-flow region, the theoretical drag becomes infinite as the Mach number approaches unity. This is so, because the assumptions used in establishing the linearized theory are no longer valid for sonic flows. This difficulty can be resolved by using a transonic theory, which retains some important terms which are neglected in the linearized theory, resulting in finite values of C_D. The curve of actual drag in the figure shows this result, which also contains the contribution of friction drag. Owing to this, a viscous drag force exists even in subsonic flows, which becomes relatively insignificant for slender bodies compared to the wave drag.

[12]For a thin body interacting with a supersonic flow, each segment of the body boundary acts as a disturbance to the flow adjacent to it. As indicated previously, a disturbance propagates along a Mach line inclined at an angle θ with respect to the flow direction. Thus, the Mach lines above the body boundary are right running and those below the body boundary are left-running. For the considered wavy boundary, only the right running Mach lines take place.

9.4.5 The Prandtl-Glauert Transformation for Subsonic Flows

It may be possible to transform all subsonic-flow problems to the equivalent incompressible-flow problems by means of a simple transformation, which is referred to as the *Prandtl-Glauert transformation*.[13] For subsonic flows over a body whose surface is described by $y = f(x)$, the formulation of small-perturbation theory for the perturbation velocity potential function Φ is given by

$$\frac{\partial^2 \Phi}{\partial x^2} + \frac{1}{1 - \mathcal{M}_{a\infty}^2} \frac{\partial^2 \Phi}{\partial y^2} = 0, \qquad \frac{\partial \Phi}{\partial y}(x, 0) = U \frac{\mathrm{d} f}{\mathrm{d} x}(x), \\ \frac{\partial \Phi}{\partial x}(x, y \to \infty) = \text{finite}. \tag{9.4.56}$$

A new velocity potential function Φ^* and a new spatial coordinate η are introduced as

$$\Phi = \frac{1}{\sqrt{1 - \mathcal{M}_{a\infty}^2}} \Phi^*, \qquad \eta = \sqrt{1 - \mathcal{M}_{a\infty}^2}\, y, \tag{9.4.57}$$

with which Eq. (9.4.56) becomes

$$\frac{\partial^2 \Phi^*}{\partial x^2} + \frac{\partial^2 \Phi^*}{\partial \eta^2} = 0, \qquad \frac{\partial \Phi^*}{\partial \eta}(x, 0) = U \frac{\mathrm{d} f}{\mathrm{d} x}(x), \qquad \frac{\partial \Phi^*}{\partial x}(x, \eta \to \infty) = \text{finite}. \tag{9.4.58}$$

Equation (9.4.58) shows that the problem to be solved in the (x, y)-plane corresponds simply to the problem of an irrotational motion of an incompressible fluid about the body whose surface is also described by $\eta = f(x)$.

It is possible to use the theory of ideal fluids to solve the corresponding ideal-flow problem. Having obtained the solution of Φ^*, the pressure coefficient is determined by using Eq. (9.4.32), viz.,

$$C_p^* = -\frac{2}{U} \frac{\partial \Phi^*}{\partial x}, \tag{9.4.59}$$

and the pressure coefficient corresponding to Φ is then given by

$$C_p = -\frac{2}{U} \frac{\partial \Phi}{\partial x} = -\frac{1}{\sqrt{1 - \mathcal{M}_{a\infty}^2}} \frac{2}{U} \frac{\partial \Phi^*}{\partial x}. \tag{9.4.60}$$

It follows immediately that

$$C_p(x, y) = \frac{C_p^*(x, y)}{\sqrt{1 - \mathcal{M}_{a\infty}^2}}. \tag{9.4.61}$$

As a result, the pressure distribution around a body in a subsonic compressible flow may be obtained by the pressure distribution of the corresponding incompressible and irrotational flow. The rule connecting two pressure distributions, i.e., Eq. (9.4.61), is referred to as the Prandtl-Glauert transformation. It establishes the effect of fluid compressibility in subsonic flows and shows that any subsonic-flow problem, in the context of linearized theory, may be solved by using Eq. (9.4.61), provided that the corresponding ideal-flow problem may be solved.

[13] Hermann Glauert, 1892–1934, a British aerodynamicist, who contributed to the *Prandtl-Glauert singularity* from the Prandtl-Glauert transformation for transonic flows.

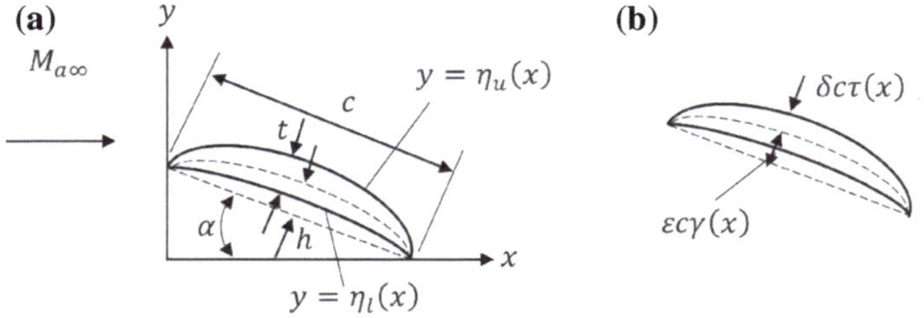

Fig. 9.20 Ackeret's theory for supersonic flows past an airfoil. **a** The geometric configurations and parameters. **b** The half-thickness and half-camber of airfoil

9.4.6 Ackeret's Theory for Supersonic Flows

Consider a supersonic flow through a thin cambered airfoil at an angle of attack α with respect to the free stream, as shown in Fig. 9.20a. The airfoil is characterized by its chord c, maximum thickness t and maximum camber h, with the upper and lower surfaces described respectively by $y = \eta_u(x)$ and $y = \eta_l(x)$. The Mach number of approaching supersonic flow in the region far away from the airfoil is denoted by $\mathcal{M}_{a\infty} > 1$. In the context of linearized theory, the formulation of perturbation velocity potential function with the associated boundary conditions are given by

$$\frac{\partial^2 \Phi}{\partial x^2} - \frac{1}{\mathcal{M}_{a\infty}^2 - 1}\frac{\partial^2 \Phi}{\partial y^2} = 0, \quad \frac{\partial \Phi}{\partial y}(x, 0) = U\frac{\mathrm{d}\eta}{\mathrm{d}x}(x), \quad \frac{\partial \Phi}{\partial x}(x, y \to \infty) = \text{finite}, \tag{9.4.62}$$

where $\eta(x) = \eta_u(x) \cup \eta_l(x)$. Since the boundary conditions on $\eta_u(x)$ and $\eta_l(x)$ are different in general, it is convenient to decompose Φ into the upper and lower parts, Φ_u and Φ_l, corresponding to the solutions subject to the boundary conditions on $\eta_u(x)$ and $\eta_l(x)$, respectively. It follows form the previous discussions that the solutions to Φ_u and Φ_l are of the forms

$$\Phi_u(x, y) = f\left(x - \sqrt{\mathcal{M}_{a\infty}^2 - 1}\,y\right), \qquad \Phi_l(x, y) = g\left(x + \sqrt{\mathcal{M}_{a\infty}^2 - 1}\,y\right), \tag{9.4.63}$$

where f and g are two undetermined functions, in which the left-running solution to Φ_u is omitted in order to satisfy the condition that signals can travel only downstream in supersonic flows, so that the lines along which signals travel must slope downstream as they move away from the airfoil. The same reason is used to omit the left-running solution to Φ_l. Substituting Eqs. $(9.4.62)_{2,3}$ to two solutions separately yields

$$f'(x) = -\frac{U}{\sqrt{\mathcal{M}_{a\infty}^2 - 1}}\frac{\mathrm{d}\eta_u}{\mathrm{d}x}(x), \qquad g'(x) = \frac{U}{\sqrt{\mathcal{M}_{a\infty}^2 - 1}}\frac{\mathrm{d}\eta_l}{\mathrm{d}x}(x). \tag{9.4.64}$$

By using Eq. (9.4.32), the pressure coefficient on the upper surface, C_{pu}, and that on the lower surface, C_{pl}, are obtained as

$$C_{pu} = -\frac{2}{U}f'(x) = \frac{2}{\sqrt{\mathcal{M}_{a\infty}^2 - 1}}\frac{\mathrm{d}\eta_u}{\mathrm{d}x}, \qquad C_{pl} = -\frac{2}{U}g'(x) = -\frac{2}{\sqrt{\mathcal{M}_{a\infty}^2 - 1}}\frac{\mathrm{d}\eta_l}{\mathrm{d}x}, \tag{9.4.65}$$

showing that the local value of pressure coefficient is proportional to the local slope of airfoil surface.

For the consider airfoil, the drag and lift coefficients are of prime interest. Since no viscous force is taken into account, the lift coefficient can be expressed as

$$C_L = \frac{2}{\rho_\infty U^2}\frac{1}{c}\int_0^c (p_{wl} - p_{wu})\,\mathrm{d}x, \tag{9.4.66}$$

where ρ_∞ is the fluid density in the region far away from the airfoil, p_{wl} and p_{wu} are respectively the pressure distributions on the lower and upper airfoil surfaces. By using the pressure coefficients, this equation reduces to

$$C_L = \frac{1}{c}\int_0^c \left(C_{pl} - C_{pu}\right)\mathrm{d}x = \frac{4\alpha}{\sqrt{\mathcal{M}_{a\infty}^2 - 1}}, \tag{9.4.67}$$

in which the geometric conditions $\eta_l(c) = \eta_u(c) = 0$ and $\eta_l(0) = \eta_u(0) = \alpha c$ have been used. This result shows that the lift coefficient acting on a supersonic airfoil depends only on the Mach number of flow and the angle of attack of airfoil, and is independent of the camber and thickness of airfoil. However, for subsonic airfoils, it follows from Eqs. (7.5.116) and (9.4.61) that

$$C_L = \frac{2\pi}{\sqrt{1 - \mathcal{M}_{a\infty}^2}}\left(1 + 0.77\frac{t}{c}\right)\sin\left(\alpha + 2\frac{h}{c}\right), \tag{9.4.68}$$

which depends greatly on the airfoil thickness and camber. Equations (9.4.67) and (9.4.68) reveal the influence of fluid compressibility on the lift coefficient in supersonic and subsonic flows when compared to Eq. (7.5.116), which is only valid for incompressible flows. Similarly, the drag coefficient C_D is determined as

$$C_D = \frac{2}{\rho_\infty U^2}\frac{1}{c}\int_0^{\alpha c} (p_{wl} - p_{wu})\,\mathrm{d}y = \frac{2}{c\sqrt{\mathcal{M}_{a\infty}^2 - 1}}\int_0^c \left[\left(\frac{\mathrm{d}\eta_l}{\mathrm{d}x}\right)^2 + \left(\frac{\mathrm{d}\eta_u}{\mathrm{d}x}\right)^2\right]\mathrm{d}x, \tag{9.4.69}$$

in which it is noted that

$$\mathrm{d}y = \left(\frac{\mathrm{d}y}{\mathrm{d}x}\right)\mathrm{d}x, \qquad \frac{\mathrm{d}y}{\mathrm{d}x} = \frac{\mathrm{d}\eta_l}{\mathrm{d}x}, \qquad \frac{\mathrm{d}y}{\mathrm{d}x} = \frac{\mathrm{d}\eta_u}{\mathrm{d}x}, \tag{9.4.70}$$

where the last two equations are devoted respectively to the lower and upper surfaces of airfoil. Since the integration in Eq. (9.4.69) is positive, it follows that the drag coefficient is non-vanishing for nontrivial airfoil shape, which is a result already obtained in Sect. 9.4.4.

It is interesting to estimate how the airfoil thickness and camber affect the wave drag. Let the thickness and camber of an airfoil be parameterized as

$$\delta = \frac{t}{c}, \qquad \varepsilon = \frac{h}{c}, \tag{9.4.71}$$

with which the half-thickness function $\tau(x)$ and half-camber function $\gamma(x)$ of an airfoil are so defined that the local values of airfoil half-thickness and half-camber are given respectively by $\delta c\tau(x)$ and $\varepsilon c\gamma(x)$, as shown in Fig. 9.20b. Due to the geometric configurations, these two functions must lie in the ranges given by

$$0 \le \tau(x) \le \frac{1}{2}, \qquad 0 \le \gamma(x) \le 1, \tag{9.4.72}$$

with which the upper and lower surfaces of airfoil can be described respectively by

$$\eta_u(x) = \alpha(c-x) + \varepsilon c\gamma(x) + \delta c\tau(x), \qquad \eta_l(x) = \alpha(c-x) + \varepsilon c\gamma(x) - \delta c\tau(x), \tag{9.4.73}$$

which are expressed by the line integral through the mean thickness of airfoil plus/minus the half-thickness, respectively. Substituting these expressions into Eq. (9.4.69) gives rise to

$$C_D = \frac{2}{c\sqrt{\mathcal{M}_{a\infty}^2 - 1}} \int_0^c \left[2\alpha^2 + 2\varepsilon^2 c^2 \left(\gamma'\right)^2 + 2\delta^2 c^2 \left(\tau'\right)^2 - 4\alpha\varepsilon c\gamma'\right] \mathrm{d}x, \tag{9.4.74}$$

where the primes denote differentiations with respect to x. Integrating this equation results in

$$C_D = \frac{4\alpha^2}{\sqrt{\mathcal{M}_{a\infty}^2 - 1}} + \frac{4\varepsilon^2 c}{\sqrt{\mathcal{M}_{a\infty}^2 - 1}} \int_0^c \left(\gamma'\right)^2 \mathrm{d}x + \frac{4\delta^2 c}{\sqrt{\mathcal{M}_{a\infty}^2 - 1}} \int_0^c \left(\tau'\right)^2 \mathrm{d}x, \tag{9.4.75}$$

since $\gamma(0) = \gamma(c) = 0$. Equation (9.4.75) shows that the drag coefficient of an airfoil in a supersonic flow is increased as the airfoil camber and thickness increase. On the contrary, the lift coefficient, as indicated by Eq. (9.4.67), is independent of the camber and thickness. Thus, a supersonic airfoil should be as straight and as thin as possible to obtain a minimum drag coefficient and a maximum lift coefficient. Moreover, airfoils with sharp corners are preferable to rounded corners in supersonic flight. The derived results are known as *Ackeret's theory*.[14]

9.4.7 The Prandtl-Meyer Flow

Consider a steady, two-dimensional supersonic flow of a compressible fluid approaching a sharp bend in a boundary, as shown in Fig. 9.21a, in which the boundary bends in such a direction that an expansion is required to turn the fluid. This flow is referred to as the *Prandtl-Meyer flow*. Let the Mach number of approaching supersonic flow be denoted by $\mathcal{M}_{a\infty}$, which is in parallel to the horizontal boundary. The flow experiences a sudden change in the direction as it just passes the sharp corner, so that the fluid velocity must be deflected gradually toward the inclined direction, in order to satisfy the boundary conditions. Since this deflection is opposite in sense to that shown to be necessary for shock waves, it may be said that an expansion, rather than a compression, will take place. This expansion is a continuous process, which can be approximated by a large number of very weak expansion waves, known as the *Prandtl-Meyer fan*.

Let point P be an arbitrary point in the expansion fan, at which the local Mach number is $\mathcal{M}_a$, and the deflection of fluid velocity relative to its original direction is θ, as shown in Fig. 9.21a. The inclination of the Mach wave passing point P is

---

[14]Jakob Ackeret, 1898–1981, a Swiss aeronautical engineer, who is recognized as one of the foremost aeronautical experts of the twentieth century.

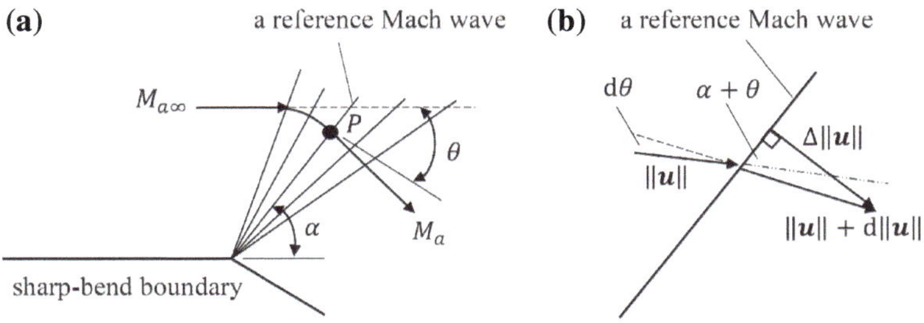

Fig. 9.21 The configuration of the Prandtl-Meyer flow. **a** The Prandtl-Meyer fan. **b** The velocity change across a typical Mach wave in the Prandtl-Meyer fan

denoted by the angle α. It follows that the inclination of the first Mach wave, i.e., the leading Mach wave, is identified to be

$$\alpha_\infty = \sin^{-1}\left(\frac{1}{\mathcal{M}_{a\infty}}\right). \tag{9.4.76}$$

Since the pressure gradient must be normal to each of the Mach lines, the change in fluid velocity must also be normal to the Mach lines. Let the fluid velocity approaching a reference Mach line be denoted by $\boldsymbol{u}$, and $\|\Delta\boldsymbol{u}\|$ be the change in $\|\boldsymbol{u}\|$ which is caused by the Mach wave, as shown in Fig. 9.21b. The fluid velocity emerging from the reference Mach wave will have a magnitude $\|\boldsymbol{u}\| + \mathrm{d}\|\boldsymbol{u}\|$ and is deflected through an angle $\mathrm{d}\theta$ when compared to the deflection angle θ of $\boldsymbol{u}$. It is assumed that $\|\Delta\boldsymbol{u}\|$ is infinitesimally small, so that the limit of an infinite number of the Mach waves is approached. With these, $\mathrm{d}\theta$ is approximated as

$$\mathrm{d}\theta = \frac{\|\Delta\boldsymbol{u}\|}{\|\boldsymbol{u}\| + \mathrm{d}\|\boldsymbol{u}\|}\cos(\alpha+\theta) \sim \frac{\|\Delta\boldsymbol{u}\|}{\|\boldsymbol{u}\|}\cos(\alpha+\theta), \tag{9.4.77}$$

in which all second-order terms are assumed to vanish identically for simplicity. Since Fig. 9.21b implies that

$$\|\boldsymbol{u}\| + \mathrm{d}\|\boldsymbol{u}\| = \|\boldsymbol{u}\| + \|\Delta\boldsymbol{u}\|\sin(\alpha+\theta), \tag{9.4.78}$$

substituting this equation into Eq. (9.4.77) yields

$$\mathrm{d}\theta = \frac{\mathrm{d}\|\boldsymbol{u}\|}{\|\boldsymbol{u}\|}\cot(\alpha+\theta). \tag{9.4.79}$$

The total inclination of the reference Mach wave is $\alpha + \theta$, which is described by

$$\sin(\alpha+\theta) = \frac{1}{\mathcal{M}_a}, \quad \longrightarrow \quad \cot(\alpha+\theta) = \sqrt{\mathcal{M}_a^2 - 1}, \tag{9.4.80}$$

which is substituted into Eq. (9.4.79) to give

$$\mathrm{d}\theta = \sqrt{\mathcal{M}_a^2 - 1}\,\frac{\mathrm{d}\|\boldsymbol{u}\|}{\|\boldsymbol{u}\|}, \tag{9.4.81}$$

which is the local turning angle of fluid velocity at the considered location.

With $\|\boldsymbol{u}\| = a\mathcal{M}_a$, where a is the local sonic velocity, it follows that

$$\frac{\mathrm{d}\|\boldsymbol{u}\|}{\|\boldsymbol{u}\|} = \frac{\mathrm{d}a}{a} + \frac{\mathrm{d}\mathcal{M}_a}{\mathcal{M}_a}. \tag{9.4.82}$$

For the considered problem, the energy equation reads the form

$$\frac{1}{2}\|\boldsymbol{u}\|^2 + \frac{a^2}{\gamma - 1} = \frac{a_0^2}{\gamma - 1}, \tag{9.4.83}$$

where a_0 is the speed of sound in stationary fluid. Multiplying this equation by $(\gamma - 1)/a^2$ yields

$$\frac{\gamma - 1}{2}\mathcal{M}_a^2 + 1 = \frac{a_0^2}{a^2}, \qquad \longrightarrow \qquad 2a\,\mathrm{d}a = \frac{-a_0^2(\gamma - 1)\mathcal{M}_a \mathrm{d}\mathcal{M}_a}{\{1 + [(\gamma - 1)/2]\mathcal{M}_a^2\}^2}, \tag{9.4.84}$$

by which the term $\mathrm{d}\|\boldsymbol{u}\|/\|\boldsymbol{u}\|$ in Eq. (9.4.82) is obtained as

$$\frac{\mathrm{d}\|\boldsymbol{u}\|}{\|\boldsymbol{u}\|} = \frac{1}{1 + [(\gamma - 1)/2]\mathcal{M}_a^2} \frac{\mathrm{d}\mathcal{M}_a}{\mathcal{M}_a}. \tag{9.4.85}$$

Substituting this equation into Eq. (9.4.81) yields

$$\mathrm{d}\theta = \frac{\sqrt{\mathcal{M}_a^2 - 1}}{1 + [(\gamma - 1)/2]\mathcal{M}_a^2} \frac{\mathrm{d}\mathcal{M}_a}{\mathcal{M}_a}, \tag{9.4.86}$$

which is integrated to obtain

$$\begin{aligned} &\theta = f(\mathcal{M}_a) - f(\mathcal{M}_{a\infty}), \\ &f(\mathcal{M}_a) = \sqrt{\frac{\gamma + 1}{\gamma - 1}} \tan^{-1}\left(\sqrt{\frac{\gamma - 1}{\gamma + 1}}(\mathcal{M}_a^2 - 1)\right) - \tan^{-1}\left(\sqrt{\mathcal{M}_a^2 - 1}\right), \end{aligned} \tag{9.4.87}$$

where $f(\mathcal{M}_a)$ is the *Prandtl-Meyer function*. The solution shows that $\theta = 0$ for $\mathcal{M}_a = \mathcal{M}_{a\infty}$, which represents a monotonically increasing function of $\mathcal{M}_a$ for $\mathcal{M}_a > \mathcal{M}_{a\infty}$. Thus, the minimum value of $\theta = 0$ takes place at $\mathcal{M}_{a\infty} = 1$, while the maximum value of θ occurs if $\mathcal{M}_a \to \infty$, which is given viz.,

$$\theta_{max} = \frac{\pi}{2}\left(\sqrt{\frac{\gamma + 1}{\gamma - 1}} - 1\right), \tag{9.4.88}$$

and is only a function of γ. For air, the value of γ is 1.4, so that the maximum flow direction is given by $\theta_{max} = 130°$. The Prandtl-Meyer flow is considered an exact solution to the equations of two-dimensional compressible flows.

9.5 Effect of Fluid Compressibility on Drag and Lift

The variations in the drag and lift coefficients of an object in a supersonic flow have been discussed previously by using the mathematical formulations, e.g. Eqs. (9.4.67) and (9.4.69). Further mathematical discussions are beyond the scope of the book, and a qualitative description will be given.

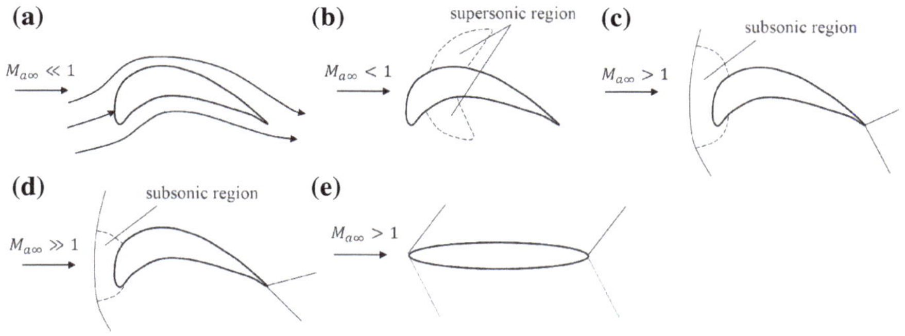

Fig. 9.22 The development of shock waves in the vicinity of an airfoil as the Mach number $\mathcal{M}_{a\infty}$ of free stream increases. **a** Very small values of $\mathcal{M}_{a\infty}$. **b** Intermediate values of $\mathcal{M}_{a\infty} < 1$. **c** Large values of $\mathcal{M}_{a\infty} > 1$. **d** Large values of $\mathcal{M}_{a\infty} \gg 1$. **e** Flow pattern around a pointed-nose body with $\mathcal{M}_{a\infty} > 1$

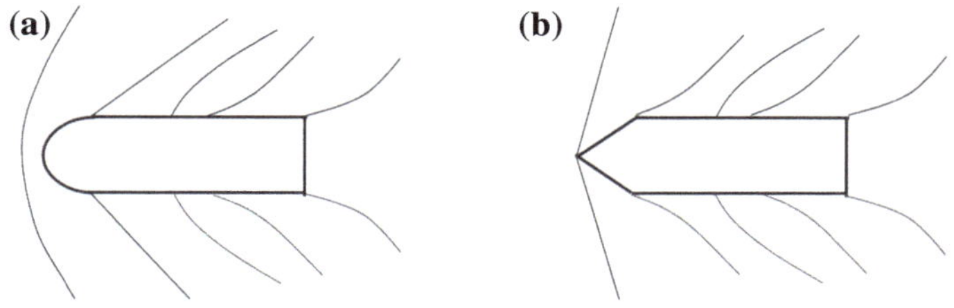

Fig. 9.23 Conical shock waves of a body in a supersonic flow. **a** A rounded-nose body. **b** A pointed-nose body

Figure 9.22 shows the successive flow patterns of a compressible fluid through a body as the free stream velocity is increased from a subsonic region to a supersonic region, or by using the Galilean transformation, a body moving with a velocity from a subsonic to a supersonic region. At low subsonic velocity, the flow pattern does not differ from that of an incompressible flow, and the corresponding streamlines shown in Fig. 9.22a are similar to those of an incompressible flow. The fluid velocity assumes its largest value somewhere near the middle of the upper and lower body surfaces, where the streamlines are closest to one another. When the velocity of free stream is increased, but still nowhere near the local sonic velocity, the local flow over the body may become supersonic in the regions of high local fluid velocity, at which shock waves may form, as shown in Fig. 9.22b. Rapid changes in the velocity and pressure across these shock waves give rise to a sharp rise of the drag coefficient. Further increase in the free stream velocity strengthens the shock waves on the upper and lower body surfaces, and at the same time moves them rearward. This gives that the major portion of flow field immediately surrounding the body becomes supersonic, although the Mach number of free stream is still less than unity. After the free stream has attained a fairly high supersonic velocity, a leading-edge shock wave will develop in addition to two oblique shock waves at the body tail, as shown in Fig. 9.22c. The pattern of shock wave at the leading edge depends on the geometric configurations

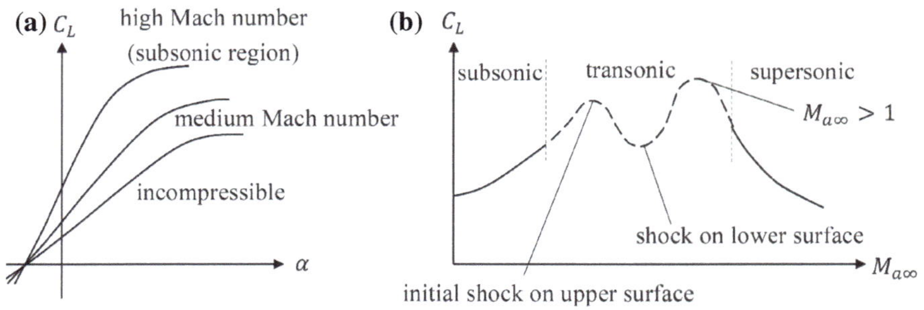

Fig. 9.24 Variations in the lift coefficient of an airfoil. **a** In subsonic-flow region. **b** In the entire range of the Mach number of the free stream

of the nose of body. The leading-edge shock wave is ahead a rounded nose, and the flow behind this shock wave is essentially supersonic, except in an even smaller region between the shock wave and rounded nose of body, as shown in Fig. 9.22d. If the nose is pointed, the leading-edge shock wave will be an attached one, as shown in Fig. 9.22e, and the entire flow field around the body is supersonic.

Essentially, the generated shock waves are conical in three-dimensional circumstances, as shown in Fig. 9.23. Experimental studies on drag tests in supersonic wind tunnels reveal that the drag coefficient for a given body rises rapidly when the Mach number of flow is in the neighborhood of unity, i.e., the flow is in the transonic region. As the Mach number increases further, the drag coefficient falls gradually and tends to approach a constant asymptotically. The value of drag coefficient of a body drops steadily if the nose becomes successively more pointed, while the flow pattern in the rear of body remains unchanged. In a supersonic flow, a sharp-pointed nose creates a narrow shock wave front which tends to minimize the drag. On the contrary, the wake resistance in the rear is relatively insignificant as compared with the shock front, for the low pressure behind the rear of body is physically limited to null.

As shown by the Prandtl-Glauert transformation, the effect of fluid compressibility is to cause an increase in the lift coefficient, which is given by

$$C_L = \frac{C_{L,\text{incomp}}}{\sqrt{1 - \mathcal{M}_{a\infty}^2}}, \tag{9.5.1}$$

where $\mathcal{M}_{a\infty}$ is the Mach number of the subsonic free stream. This equation is valid up to the point of stall, as shown in Fig. 9.24a. As the free stream velocity increases from a subsonic to a supersonic regions, a typical variation in the lift coefficient in relation with $\mathcal{M}_{a\infty}$ from experimental data is shown in Fig. 9.24b. While the lift coefficients in the subsonic and supersonic regions follow simple rules, its value in the transonic region depends on the occurrence of local shock waves as well as the accompanying conditions in the vicinity of airfoil. The unstable lift coefficient in the transonic region reveals the difficulty in stable transonic flight control.

Remarks on lift:

Without loss of generality, the lift of a body in the earth's environment may be thought of as an action in against the influence of gravity. There exist not many

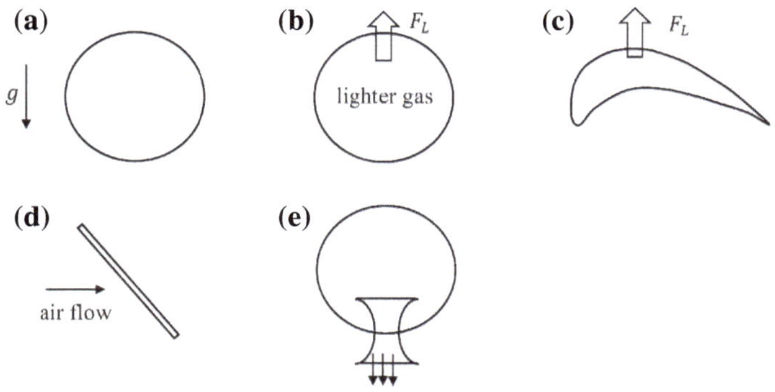

Fig. 9.25 Applications of the fundamental disciplines in fluid mechanics for lift generation. **a** The original shape of a body. **b** Lift generated by buoyancy. **c** Lift generated by airfoil. **d** Life generated by plate with an angle of attack. **e** Thrust generated by a gas flow through a nozzle

available ways that can be used to overcome the influence of gravity. For example, consider a hollow spherical container shown in Fig. 9.25a, which experiences the gravitational acceleration pointing downwards. If the sphere is assumed to be able to move vertically upwards, a lift force is required, which can be accomplished by filling the sphere by a gas whose density is smaller than the density of surrounding air, as shown in Fig. 9.25b. In this case, the lift is generated via buoyancy. Another possibility is that the shape of sphere needs to be transformed into an airfoil, so that a lift can be generated via the pressure difference between the upper and lower airfoil surfaces, as shown in Fig. 9.25c. Naturally, the lift can only be generated when there exists a relative motion between the airfoil and surrounding air, which is accomplished e.g. by associating a jet engine to the airfoil in practice. It is also possible to transform the spherical tank into a plate and place the plate with an angle of attack to the approaching flow, by which a lift can be generated, as shown in Fig. 9.25d. A typical example is kite, and parachute is a variation of this concept in generating lift. Newton's third law of motion can equally be followed to overcome the influence of gravity by using the reaction, e.g. the reaction acting on a rocket or a missile, as shown in Fig. 9.25e, although in such a case, thrust may be a more appropriate terminology. All these methods are the applications of the fundamental disciplines of fluid mechanics.

It follows from classical physics that earth is a giant magnet, and it is theoretically possible to use the electromagnetic theory to generate a magnetic reaction in against the earth's magnetic field to overcome the influence of gravity. Unfortunately, the average magnetic field strength of earth is nearly 6×10^{-5} Tesla, and the maximum magnetic field strength occurs at two poles, which assumes a value smaller than 10^{-4} Tesla. The magnetic strength of earth is so small, that it is practically impossible to use the earth's magnetic field to generate a reaction to overcome the influence of gravity. It may be of interest to look for other possible methods to counterbalance the influence of gravity, or to understand how gravity is transported to look for possible countermeasures to localize or isolate the influence of gravity.

9.6 Exercises

9.1 The equation governing the fluid velocity induced by a finite-amplitude forward-running disturbance in a one-dimensional circumstance is given by

$$\frac{\mathrm{D}u}{\mathrm{D}t} = \frac{\partial u}{\partial t} + (u+a)\frac{\partial u}{\partial x} = 0, \qquad a^2 = \frac{\mathrm{d}p}{\mathrm{d}\rho}.$$

Show that the steepness of wave front, $\partial u/\partial x$, satisfies the relation

$$\frac{\mathrm{D}u}{\mathrm{D}t}\left(\frac{\partial u}{\partial x}\right) \sim \left(\frac{\partial u}{\partial x}\right)^2,$$

and find the value of the proportional constant in the relation. If the steepness of wave front at $t = 0$ is given by

$$\frac{\partial u}{\partial x}(t=0) = S,$$

determine the time duration required for $\partial u/\partial x \to \infty$. Show also that the condition $S < 0$ must hold for the formation of a shock wave.

9.2 The equation describing the propagation of sound waves is the same as that for shallow-liquid waves. Thus, there exists an analogy between sound waves in a gaseous medium and surface waves on a liquid. Find the corresponding physical quantities in this analogy and determine the value of γ, i.e., the specific-heat ratio, which makes the analogy complete.

9.3 An ideal gas flowing in a constant-section conduit is heated from the surrounding, and the flow is approximated to be one-dimensional. If all external forces are neglected, the Mach number of gas is described by

$$\frac{\mathrm{d}\mathcal{M}_a}{\mathcal{M}_a} = \beta\frac{\delta q}{c_p T},$$

where β is an influence coefficient. Determine the expression of β. Use the resulting equation and the equation

$$\frac{\mathrm{d}p}{p} = \frac{\gamma\mathcal{M}_a^2}{\mathcal{M}_a^2 - 1}\frac{\delta q}{c_p T},$$

to obtain a differential relation between p and $\mathcal{M}_a$. Integrating the obtained relation to derive an expression for the pressure ratio p_2/p_1 between any two sections 1 and 2 of the flow, whose Mach numbers are respectively $\mathcal{M}_{a1}$ and $\mathcal{M}_{a2}$.

9.4 Prove that

$$\frac{\mathrm{d}u}{u} = \frac{1}{2d}\frac{f\gamma\mathcal{M}_a^2}{1-\mathcal{M}_a^2}\mathrm{d}x,$$

for the one-dimensional flows satisfying the conditions of the Fanno line, where d is the diameter of a conduit, f represents the friction factor, and $\mathrm{d}x$ denotes an infinitesimal conduit segment. Show that the flow is either accelerated or decelerated by the friction, if the flow is subsonic or supersonic, respectively.

9.5 Consider a compressible flow in a constant-area conduit with friction and heat transfer. To maintain a constant subsonic Mach number, should heat be added to or removed from the flow? Repeat the discussion for a supersonic flow.

9.6 For a compressible fluid in a one-dimensional isentropic flow, derive the expressions of pressure, density, and temperature ratios between any two sections 1 and 2, whose Mach numbers are given respectively by $\mathcal{M}_{a1}$ and $\mathcal{M}_{a2}$.

9.7 Show that the equation that needs to be satisfied by the velocity potential function ϕ for a steady, two-dimensional irrotational motion of an inviscid compressible fluid is given by

$$\left(1-\frac{u^2}{a^2}\right)\frac{\partial^2\phi}{\partial x^2}-2\frac{uv}{a^2}\frac{\partial^2\phi}{\partial x\partial y}+\left(1-\frac{v^2}{a^2}\right)\frac{\partial^2\phi}{\partial y^2}=0,$$

where u and v are the velocity components in the x- and y-directions, respectively.

9.8 Find the differential equation which needs to be satisfied by ϕ_3 in the series of the Janzen-Rayleigh expansion.

9.9 Consider a two-dimensional channel with wavy walls, as shown in the figure, through which a subsonic potential flow takes place. The wavy surface is described by

$$y=d+a\sin\frac{2\pi x}{\lambda},\qquad a\ll d.$$

In the context of linearized approximation, determine the velocity potential function ϕ and the pressure coefficient along the channel centerline.

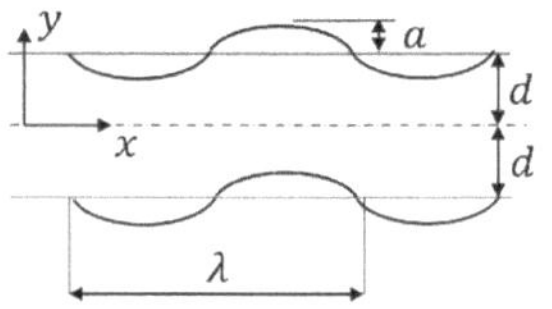

9.10 A double-wedge airfoil is shown in the figure. Use Ackeret's theory to determine the drag coefficient of the airfoil in a horizontally uniform flow with velocity U.

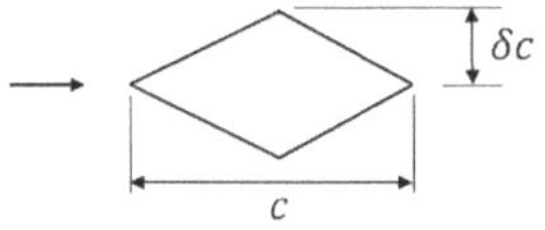

Further Reading

J.D. Anderson, *Hypersonic and High Temperature Gas Dynamics* (McGraw-Hill, Singapore, 1989)
J.D. Anderson, *Modern Compressible Flow* (McGraw-Hill, Singapore, 1990)
R.H. Barnard, D.R. Philpott, *Aircraft Flight*, 3rd edn. (Prentice-Hall, New Jersey, 2004)

R. Courant, K.O. Friedrichs, *Supersonic Flow and Shock Waves* (Springer, Berlin, 1976)
I.G. Currie, *Fundamental Mechanics of Fluids*, 2nd edn. (McGraw-Hill, Singapore, New York, 1993)
M. Eckert, *The Dawn of Fluid Dynamics* (Wiley-VCH, Weinheim, 2006)
R.W. Fox, P.J. Pritchard, A.T. McDonald, *Introduction to Fluid Mechanics*, 7th edn. (Wiley, New York, 2009)
J.E.A. John, *Gas Dynamics*, 2nd edn. (Allyn and Bacon, Massachusetts, 1984)
A.M. Kuethe, C.Y. Chow, *Foundation of Aerodynamics: Bases of Aerodynamic Design*, 3rd edn. (Wiley, New York, 1976)
H.W. Liepmann, A. Roshko, *Elements of Gas Dynamics* (Wiley, New York, 1957)
B.R. Munson, D.F. Young, T.H. Okiishi, *Fundamentals of Fluid Mechanics*, 3rd edn. (Wiley, New York, 1990)
R.H.F. Pao, *Fluid Mechanics* (Wiley, New York, 1961)
R.H.F. Pao, *Fluid Dynamics* (Charles E Merrill Books, Columbus, 1967)
M.A. Saad, *Compressible Fluid Flow* (Prentice-Hall, New Jersey, 1985)
A.H. Shapiro, *The Dynamics and Thermodynamics of Compressible Fluid Flow, Volume 1* (Ronald Press, New York, 1953)
R.S. Shevell, *Fundamentals of Flight*, 2nd edn. (Prentice-Hall, New Jersey, 1989)
P.A. Thompson, *Compressible-Fluid Dynamics* (McGraw-Hill, New York, 1972)
D.J. Tritton, *Physical Fluid Dynamics* (Oxford University Press, Oxford, 1988)
M. Van Dyke, *Perturbation Methods in Fluid Mechanics* (The Parabolic Press, Stanford, 1975)
M. Van Dyke, *An Album of Fluid Motion* (The Parabolic Press, Stanford, 1988)

Open-Channel Flows 10

Open channels are conduits in which a fluid has a free surface or its boundary is exposed essentially to the atmosphere, and open-channel flows are referred to as liquid flows with free surfaces. An open channel is not completely filled by a liquid in general, which introduces the concept of wetted perimeter. The motion of liquid in an open channel is almost always turbulent and unaffected by the surface tension, and is usually driven by the gravitational effect, with a hydrostatic pressure distribution in the vertical direction. Natural drainage of water through numerous creek and river systems, and flows in canals are typical examples. This chapter is devoted to the fundamental concepts in discussing the characteristics of open-channel flows.

The general characteristics and classifications of open-channel flows are first introduced, followed by the velocity distribution in a cross-section, which results essentially from experimental study. The concepts of specific energy and critical flow depth are useful to the understanding of open-channel flows and are used subsequently to study the characteristics of selected steady uniform, rapidly varied and gradually varied open-channel flows. The analogy between open-channel and compressible flows is discussed at the end.

10.1 General Features and Classifications

An *open channel* is a conduit in which a liquid flows with a free surface in contact with the atmosphere. When compared to the flow in a closed conduit such as a pipe, the flow in an open channel is driven essentially by the influence of gravity, and the non-uniform pressure distribution is caused by its own weight. As a result, the slope of an open channel becomes dominant for the flow characteristics. Conventionally, open channels are classified as *natural* or *artificial*. Natural open channels such as

© Springer International Publishing AG 2019

C. Fang, *An Introduction to Fluid Mechanics*, Springer Textbooks in Earth Sciences, Geography and Environment, https://doi.org/10.1007/978-3-319-91821-1_10

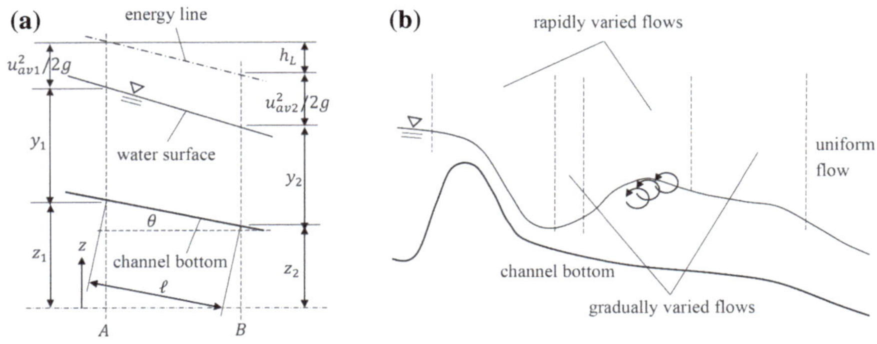

Fig. 10.1 A typical open-channel flow with the physical configurations. **a** Illustrations of the energy line, water surface, and channel bottom in a channel reach. **b** Various stages of an open-channel flow

the passage of a river vary in size and are usually irregular in cross-section and in other hydraulic properties. On the other hand, artificial open channels, e.g. canals are generally built for engineering purpose, whose hydraulic properties are more amenable to calculations. Since many complex physical conditions are encountered in open-channel flows, their mechanics is much more complicated than that of pipe flows. Flow conditions in open channels become further complicated if a constant rising and falling of the free liquid surface in both time and space takes place. The boundary surfaces of open channels vary from the polished ones for test flumes to those of rough and irregular beds of natural streams, and thus the surface roughness varies with the position of free liquid surface, for which reliable experimental data is rather difficult to secure. Hence, there is a greater degree of uncertainty in the determination of friction factors in open-channel flows, and the governing equations are generally empirical or semi-empirical to a large extent.

Figure 10.1a shows a typical reach of an open channel, in which a water flow is slightly convergent downstream, but may be assumed first to be parallel. It is further assumed that the flow has a uniform velocity distribution in the vertical direction and that the slope of channel is small. The water surface coincides then with the *piezo-metric line*, i.e., the line of pressure head, which is also the water depth, termed the *piezo-metric head*. The *energy line* of flow is at a distance of one velocity head above the water surface.[1] The energy loss between sections A and B is defined as the drop in the energy line between the same sections. Since the slopes of energy line, water surface, and channel bottom shown in the figure are different in general, the slope of energy line is called the *hydraulic slope* S defined by

$$S \equiv \frac{h_L}{\ell}, \tag{10.1.1}$$

[1]This holds only for flows with uniform velocity distribution. For non-uniform flows, the kinetic energy coefficient must be introduced.

where ℓ is the distance measured along the channel, and h_L is the total energy difference expressed in terms of head. The *slope of water surface* is denoted by S_w, and the *slope of channel bottom* is given by $S_0 = \sin\theta$, where θ is the inclined angle of channel bottom with respect to a horizontal line. It will be shown later that for uniform open-channel flows, three slopes assume the same value.

The typical flow patterns at different stages in an open channel are shown in Fig. 10.1b. An open-channel flow is conventionally classified based on either the change in the flow conditions in time and space, or the relative effect of viscosity and gravity as compared to the inertia effect. An open-channel flow is *steady* if its volume flow rate at a specific cross-section remains invariant with respect to time. The flows become *unsteady* if this is not the case, e.g. flood waves and surges. Open-channel flows are almost always unsteady to a certain extent, but in most problems they may be approximated to be steady. Open-channel flows may equally be classified as *uniform*, if the flow conditions remain the same at every cross-section along the channel, which can be accomplished by using a prismatic shape. A uniform flow may be either steady or unsteady, depending on whether or not the volume flow rate changes with time, although unsteady uniform flows rarely occur in nature. An open-channel flow is termed *varied*, if the flow conditions change from section to section along the channel length. A varied flow may be steady or unsteady, which is further classified as *rapidly varied* or *gradually varied*. A rapidly varied flow is characterized by an abrupt change in the flow conditions over a relatively short distance, e.g. *hydraulic jump* is a typical example. Flow conditions in gradually varied flows change gradually along the length of channel.

The flow state in an open channel is governed physically by the effect of fluid viscosity and gravity relative to the inertia effect. It may be *laminar*, *transitional*, or *turbulent*, depending on the relative significance between the viscous and inertia effects. Conventionally, the flow state is characterized by the Reynolds number, viz.,

$$\mathcal{R}_e = \frac{\rho u_{av} d_h}{\mu}, \tag{10.1.2}$$

where u_{av} represents the average flow velocity at a given cross-section, d_h denotes the hydraulic diameter of that section, and ρ and μ are respectively the density and viscosity of fluid. Experimental studies for water flow in a wide open channel show that the flow is laminar for $\mathcal{R}_e < 4000$, while it becomes turbulent for $\mathcal{R}_e > 11000$. In-between is the flow in the transitional region. However, most open-channel flows are turbulent. Laminar open-channel flows take place very rarely, although these flows are known to exist, e.g. thin sheets of water flow over a plane surface. The flow velocity may be classified as *subcritical*, *critical*, or *supercritical*, depending on the relative significance between the gravity and inertia effects, which is described by the Froude number, viz.,

$$\mathcal{F}_r = \frac{u_{av}}{\sqrt{g y_h}}, \qquad y_h = \frac{A}{B_w}, \tag{10.1.3}$$

where g is the gravitational acceleration, and y_h represents the *hydraulic depth* at a given cross-section, which is the linear dimension obtained by dividing the cross-sectional area A by the width of water surface B_w at the same cross-section. For example, for rectangular channels, the hydraulic depth and the flow depth are the same. An open-channel flow is said to be subcritical for $\mathcal{F}_r < 1$, at which the flow is described as *tranquil*. In subcritical states, the effect of gravity is more predominant than the inertia effect, so that the flow conditions upstream are affected by the downstream conditions. An open-channel flow is said to be critical if $\mathcal{F}_r = 1$, corresponding to $u_{av} = \sqrt{gy_h}$, which is called the *critical velocity* and is just the same as the propagating velocity of small-amplitude gravity waves, as already obtained in Eq. (7.7.20). In the circumstances in which $\mathcal{F}_r > 1$, flows are referred to as supercritical, frequently described as *rapid* or *torrential*. In these states, the effect of inertia force becomes more predominant and the flow velocity is so rapid that small changes in the downstream conditions cannot affect those in the upstream region.

The subcritical, critical, and supercritical states of an open-channel flow are somewhat analogous to the subsonic, sonic, and supersonic states of a compressible flow. This analogy will be discussed in Sect. 10.6.

10.2 Cross-Sectional Velocity Distributions

The velocities in an open-channel flow are not uniformly distributed in a cross-section. This results from the presence of channel boundary and free liquid surface. A typical pattern of the velocity distribution represented by the *lines of equal velocity* in a section of an open channel is shown in Fig. 10.2a. The maximum velocity occurs at the point or points which are least affected by the channel boundary and liquid surface. Experimental studies show that in ordinary channels, the maximum velocity in a vertical line occurs nearly at point A, which is at a distance of $1/20 \sim 1/4$ of the flow depth below the free liquid surface. However, the pattern of velocity is quite irregular, depending on the shape of cross-section, channel roughness, and channel alignment. The effect of channel roughness is to cause an increase in the curvature of velocity distribution curve, as shown in Fig. 10.2b. Owing to the centrifugal action on the flowing liquid, the velocity increases greatly at the convex side of section located on a bend. It follows from a large number of vertical distribution curves obtained by actual field measurements that the velocity at nearly 0.6 flow depth below the

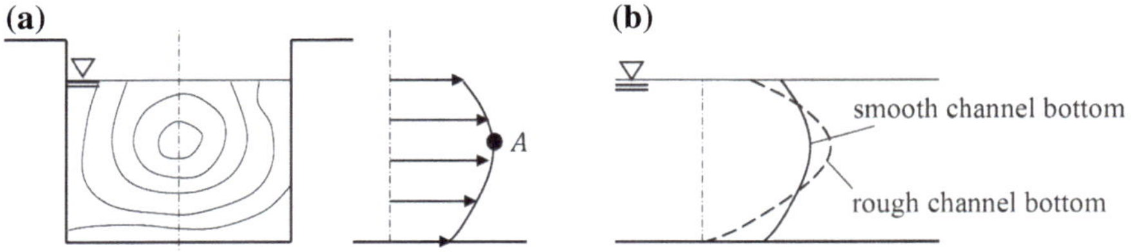

Fig. 10.2 A typical velocity distribution at a given section of an open channel. **a** The lines of equal velocity. **b** Typical vertical velocity profiles for smooth and rough channel surfaces

free liquid surface was found to be very close to the average velocity in the vertical section. The average value of the velocities measured at nearly $0.2 \sim 0.8$ depth from the free liquid surface approximates better to the average velocity.

Theoretical studies show that with slight revisions in the constants, the Prandtl universal velocity distribution law for turbulent pipe flows agrees well with the vertical velocity distribution at a cross-section in a straight wide open channel, where the boundary layer is fully developed. That is, the velocity distribution curves can be represented by the logarithmic equations. Since the velocity distributions are not uniform at a given cross-section, both the kinetic energy coefficient α and momentum coefficient β are greater than unity. Experimental studies show that the values of α vary in the range of $1.03 \sim 1.36$, and those of β vary approximately in the range of $1.01 \sim 1.12$ for fairly straight prismatic channels.

10.3 Specific Energy and Critical Depth

Under the assumption of a steady and uniform velocity profile across any section of the channel, the one-dimensional energy equation between sections A and B shown in Fig. 10.1a is given by

$$\frac{p_1}{\gamma} + \alpha_1 \frac{u_{av1}^2}{2g} + z_1 = \frac{p_2}{\gamma} + \alpha_2 \frac{u_{av2}^2}{2g} + z_2 + h_L, \tag{10.3.1}$$

where α_1 and α_2 are respectively the kinetic energy coefficients at sections A and B. For simplicity, their values are chosen to be unity. Since $p_1/\gamma = y_1$ and $p_2/\gamma = y_2$, it follows that

$$y_1 + \frac{u_{av1}^2}{2g} + S_0 \ell = y_2 + \frac{u_{av2}^2}{2g} + h_L, \tag{10.3.2}$$

in which $S_0 = (z_1 - z_2)/\ell$ has been used. Unfortunately, this equation alone is not sufficient to deliver useful results for the analysis of open-channel flows, for the determination of head loss h_L is difficult. Since $h_L = \ell S$, where S is the hydraulic slope, the above equation can be recast alternatively as

$$y_1 - y_2 = \frac{1}{2g}\left(u_{av2}^2 - u_{av1}^2\right) + (S - S_0)\,\ell. \tag{10.3.3}$$

For the special case in which there is no head loss in a horizontal channel, both the hydraulic slope and the slope of channel bottom become null, so that Eq. (10.3.3) is simplified to

$$y_1 - y_2 = \frac{1}{2g}\left(u_{av2}^2 - u_{av1}^2\right), \tag{10.3.4}$$

showing that the total energy of flow is conservative, which is free to be transformed between the kinetic and potential energies.

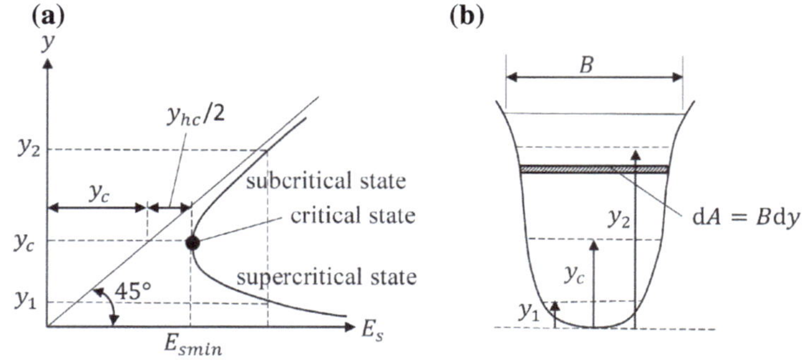

Fig. 10.3 Illustrations of the specific energy curve and critical depth y_c of an open-channel flow. **a** The specific energy curve. **b** The cross-section of channel with the dimensions

The *specific energy* E_s at a channel section is defined as the energy per unit liquid weight at the same section measured from the channel bottom, which is simply the sum of flow depth and velocity head given by[2]

$$E_s = y + \frac{u_{av}^2}{2g} = y + \frac{Q^2}{2gA^2}, \tag{10.3.5}$$

where Q is the volume flow rate with $Q = u_{av}A$. This equation shows that for a given channel and a given volume flow rate, the specific energy is only a function of the flow depth, for the cross-sectional area A can be expressed as a function of y. Hence, Eq. (10.3.5) may be displayed graphically by the so-called *specific energy curve*, in which the specific energy is plotted against the flow depth for a given volume flow rate at a given channel section, as shown in Fig. 10.3a. The specific energy assumes its minimum value E_{smin}, below which the given volume flow rate cannot exist. The flow depth corresponding to E_s is called the *critical depth* y_c, at which the average velocity is referred to as the *critical velocity* u_{avc}. If the flow depth is greater than y_c, the specific energy increases with the flow depth. The portion of curve above the critical point approaches asymptotically the 45-degree line, i.e., $E_s = y$. The velocity at a flow depth greater than y_c is less than u_{avc} for a given value of Q and is therefore referred to as a *subcritical velocity*. On the other hand, if the flow depth is less than y_c, the specific energy increases as the flow depth decreases. The portion of curve below y_c approaches asymptotically the E_s-axis toward the right. Since the velocity in this region is greater than u_{avc}, it is called a *supercritical velocity*. The curve also shows that a given value of Q (and hence a given value of E_s) can occur at two possible flow depths: the *lower state* y_1 and the *upper state* y_2 locating respectively on the lower and upper portions of curve. For $E_s = E_{smin}$, there exists only a single value of the flow depth.

[2]The concept of specific energy was first introduced by Boris A. Bahkmeteff in 1912 and has proved useful in providing satisfactory explanations for open-channel flow phenomena.

To determine the value of y_c, taking derivative of Eq. (10.3.5) with respect to y yields

$$\frac{\mathrm{d}E_s}{\mathrm{d}y} = 1 - \frac{Q^2}{gA_c^3}\frac{\mathrm{d}A_c}{\mathrm{d}y} = 0, \tag{10.3.6}$$

where A_c represents the cross-sectional area at the critical state. Since $\mathrm{d}A = B\mathrm{d}y$ and $u_{avc} = Q/A_c$, as shown in Fig. 10.3b, it is found that

$$\frac{u_{avc}^2}{2g} = \frac{A_c}{2B} = \frac{y_{hc}}{2}, \tag{10.3.7}$$

where B is the channel top width, and y_{hc} represents the *critical hydraulic depth*. This equation shows that at the critical state, the velocity head is one-half of the critical hydraulic depth, giving rise to $u_{avc}/\sqrt{gy_{hc}} = 1$. In other words, as implied by Eq. (10.1.3), the Froude number at the critical state is unity. For rectangular channels, $y_{hc} = y_c$, so that Eq. (10.3.7) reduces to

$$\frac{u_{avc}^2}{2g} = \frac{y_c}{2}, \qquad \longrightarrow \qquad u_{avc} = \sqrt{gy_c}, \tag{10.3.8}$$

and the minimum specific energy is obtained as

$$E_{smin} = \frac{3}{2}y_c. \tag{10.3.9}$$

The last two equations are valid for prismatic channels with rectangular cross-section and show that at the critical state, the velocity head is simply one-half of the critical depth, and the minimum specific energy is 1.5 times larger than the critical depth.

When a flow occurring at or near the critical depth, a relatively small change in the specific energy results in a large change in the flow depth in either direction; hence, the flow at the critical state is quite unstable. Field observations reveal that the free liquid surface undulates excessively when a flow takes place near its critical state.

10.4 Analysis of Steady Flows

10.4.1 Uniform Depth Flows

A uniform open-channel flow can occur only in a prismatic channel laid on a uniform slope, for which the volume flow rate, flow area, and depth, and consequently the flow velocity remains constant at every section along the channel reach. In addition, the energy line, water surface, and channel bottom are all parallel, i.e., $S = S_0 = S_w$. A uniform flow may be accomplished by carefully controlling the components of gravity force acting on the liquid body to be just counterbalanced by the resistance of flow, as shown in Fig. 10.4a. When a liquid enters the channel from the left end, the velocity and hence the flow resistance are smaller than the gravity force, giving rise to an accelerating flow pattern. The velocity and resistance increase gradually until a balance between the resistance and gravity force is reached. Henceforth, the

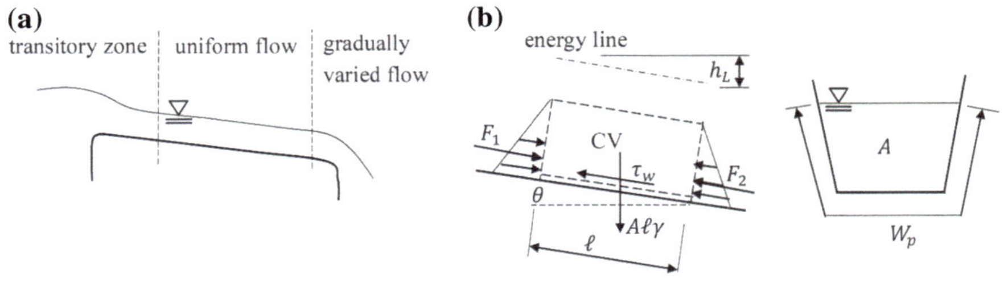

Fig. 10.4 A steady uniform flow in an open channel. **a** Illustrations of the transitory zone, uniform flow, and gradually varied flow regions. **b** A finite control-volume analysis between any two sections of a uniform flow

flow becomes uniform. The region before the uniform flow is called the *transitory zone*, in which the flow is of a gradually varied type. If the channel length is shorter than the transitory zone, a uniform flow cannot be attained.

Applying Newton's second law of motion to the control-volume shown in Fig. 10.4b for a uniform flow in the flow direction yields

$$A\ell\gamma\sin\theta - W_p\ell\tau_w = 0, \tag{10.4.1}$$

in a unit depth perpendicular to the page, where W_p represents the wetted perimeter of cross-section, and τ_w is the shear stress at the channel boundary surface. Since $\sin\theta = h_L/\ell = S = S_0 = S_w$, it follows that

$$\tau_w = \gamma\frac{A}{W_p}S = \gamma r_h S, \qquad r_h = \frac{A}{W_p}, \tag{10.4.2}$$

where r_h is the *hydraulic radius*.[3] On the other hand, in pipe flows the shear stress on the pipe wall is identified to be

$$\tau_w = f\rho\frac{u_{av}^2}{8}, \tag{10.4.3}$$

where f is the friction factor, which can be determined by using the Moody chart. Applying the analogy between Eqs. (10.4.2) and (10.4.3) results in

$$\gamma r_h S = f\rho\frac{u_{av}^2}{8}, \quad\longrightarrow\quad u_{av} = \sqrt{\frac{8g}{f}}\sqrt{r_h S} = C\sqrt{r_h S}, \quad C = \sqrt{\frac{8g}{f}}. \tag{10.4.4}$$

Equation (10.4.4)$_2$ is known as the *Chézy formula* for uniform open-channel flows for the determination of average flow velocity, and the dimensional constant C is called the *Chézy coefficient*.[4]

---

[3]The hydraulic diameter d_h, which is defined in Eq. (8.2.39)$_2$ and used in Eq. (10.1.2), is four time larger than the hydraulic radius.

[4]Antoine de Chézy, 1718–1798, a French hydraulics engineer, who is known for his contributions in open-channel flows.

Previously, an analogy between pipe and open-channel flows has been applied. This was done soon, because the friction factor of highly turbulent flows is a function of the surface roughness alone, although it is essentially dependent on the surface roughness of boundary and the Reynolds number of flow. As discussed in Sect. 8.6.8, extensive tests have firmly established the validity of extending the turbulent flow theory in circular pipes to conduits of non-circular cross-sections with aspect ratios smaller than 4. The hydraulic radius was also shown to be able to account adequately for the difference in the cross-sectional shapes of circular pipes and conduits of non-circular cross-section. It was reasonable to expect that such an extension may be valid for open-channel flows.

Although the Chézy coefficient is related to the friction factor, it is difficult to assign a correct value to the surface roughness of an open channel, especially for natural streams. Thus, the determination of the Chézy coefficient has been a subject of much discussion. In addition, as the aspect ratio of cross-section becomes greater than 4, test data from pipe flows does not apply too well. Despite these, the Chézy formula is widely used for engineering purpose, and many empirical formulations, based on a series of very elaborated measurements in laboratory and in nature, have been proposed to determine the Chézy coefficient. Three principal approaches are discussed in the following.

- *The Kutter-Ganguillet formula.* The value of the Chézy coefficient is expressed in terms of the hydraulic slope S, hydraulic radius r_h, and a roughness factor n given by
$$C = \frac{41.65 + 0.00281/S + 1.811/n}{1 + (41.56 + 0.00281/S)(n/\sqrt{r_h})}, \tag{10.4.5}$$
in the British units as the square root of feet per second. The roughness factor n is known as the *Kutter coefficient*. This empirical formula usually yields satisfactory results, and many tables and nomographs have been made available for the numerical solutions to Eq. (10.4.5), which can be obtained from any handbook of hydraulic engineering.
- *The Bazin formula.* The Chézy coefficient is expressed in terms of the hydraulic radius and the so-called *Bazin's roughness factor* m, viz.,
$$C = \frac{157.6}{1 + m/\sqrt{r_h}}, \tag{10.4.6}$$
in the British units. In general, the Bazin formula is found to be less satisfactory than the Kutter-Ganguillet formula for the same circumstances.
- *The Manning formula.* The average velocity u_{av} is expressed in terms of a roughness factor n, which is called the *Manning coefficient*,[5] the hydraulic radius r_h, and hydraulic slope S given by
$$u_{av} = \frac{1.486}{n} r_h^{2/3} S^{1/2}, \qquad u_{av} = \frac{1}{n} r_h^{2/3} S^{1/2}, \tag{10.4.7}$$

[5]Robert Manning, 1816–1897, an Irish hydraulic engineer, who is best known for the introduction of the Manning formula.

Table 10.1 The Manning coefficients in different open channels.

Wetted perimeter	n	Wetted perimeter	n
Artificial lined channels		*Natural channels*	
Glass	0.010	Clean and straight	0.030
Finished concrete	0.012	Sluggish with deep pools	0.040
Unfinished concrete	0.014	Major rivers	0.035
Clay tie	0.014	*Floodplains*	
Brickwork	0.015	Pasture, farmland	0.035
Rubble masonry	0.025	Light brush	0.050
Excavated earth channels		Heavy brush	0.075
Clean	0.022	Trees	0.15
Gravelly	0.025		
Stony, cobbles	0.035		

in the British and SI units, respectively, where the Manning coefficient is a dimensional constant. The Manning formula has become the most widely used empirical formula in determining the average velocity of a uniform flow in an open channel, owing both to its simplicity in mathematical form and to the satisfactory results it yields in practice. Comparing the Manning formula with the Chézy formula gives

$$C = 1.486\frac{r_h^{1/6}}{n}, \qquad C = \frac{r_h^{1/6}}{n}, \tag{10.4.8}$$

respectively in the British and SI units. The values of the Manning coefficient are found to be approximately equal to those of n in the Kutter-Ganguillet formula under normal ranges of channel slope and hydraulic radius. The values of the Manning coefficient in the SI units for open channels of various types are summarized in Table 10.1.[6] These values are only shown for demonstration. For practical purpose, reference to any handbook of hydraulic engineering should be undertaken. The selection of a correct value to n is a difficult task, since there is no exact rule in guiding a proper selection. The Manning coefficient is frequently interpreted as a measure of the surface roughness in an open channel, which is similar to ε in a circular pipe. Hence, the ratio $r_h^{1/6}/n$ may be regarded as a relative roughness parameter comparable to d/ε in a pipe.

10.4.2 Rapidly Varied Flows with Varied Depths

Changes in flow depth from an upper stage to a lower stage, or vice versa, frequently occur in open channels. If it takes place rapidly over a relatively short distance, the flow is said to be rapidly varied. The flows under a sluice gate and the hydraulic

[6] Data quoted from Chow, V.T., *Open Channel Hydraulics*, McGraw-Hill, New York, 1959.

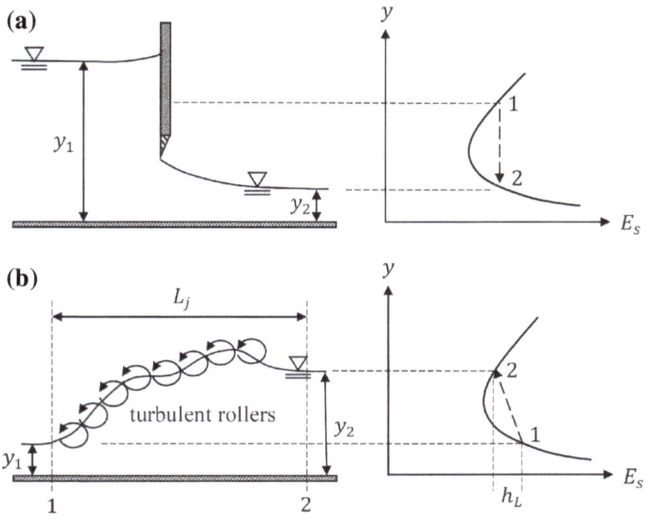

Fig. 10.5 Rapidly varied flows with varied flow depths. **a** A flow under a sluice gate with two alternative depths. **b** A hydraulic jump with two conjugate depths

jump shown respectively in Figs. 10.5a and b are typical examples. For simplicity, a prismatic channel with rectangular cross-section is used for the discussions. In the first case, water flows from an upper stage to a lower stage and forms a convergent flow pattern, in which the dissipated energy is extremely small. The specific energies at sections 1 and 2 remain almost the same, as shown in Fig. 10.5a, where the depths y_1 and y_2 are called the *alternative depths*, which are defined as the two possible flow depths corresponding to the same specific energy. The velocity at section 1 is subcritical, and the corresponding Froude number is less than unity. On the other hand, the velocity at section 2 is in the supercritical region, whose Froude number is greater than unity.

The hydraulic jump in the second case occurs in an open channel if a rapid change in flow initiates from a lower stage to an upper stage. The divergent flow is accompanied by the formation of extremely turbulent rollers on the sloping surface of jump. In such a circumstance, a relatively large amount of energy is dissipated by the turbulent rollers. Hence, the specific energy in the flow immediately downstream from the hydraulic jump is appreciably less than that entering the jump. The energy loss in the jump is denoted by h_L, and the flow depths before and after the jump are no longer the alternative depths. Rather, they are referred to as the *conjugate depths*. Specifically, y_1 and y_2 are respectively called the *lower* and *higher conjugate depths*. Field observations indicate that y_1 is always less than y_c, while y_2 is always greater than y_c.

For a hydraulic jump, it follows from the conservations of mass and linear momentum that

$$u_{av1}y_1 = u_{av2}y_2, \qquad \frac{1}{2}\gamma y_1^2 - \frac{1}{2}\gamma y_2^2 = \rho y_2 u_{av2}^2 - \rho y_1 u_{av1}^2, \tag{10.4.9}$$

by which the ratio of two conjugate depths is obtained as

$$\frac{y_2}{y_1} = \frac{1}{2}\left[\sqrt{1+8\left(\frac{u_{av1}^2}{gy_1}\right)} - 1\right] = \frac{1}{2}\left[\sqrt{1+8\left(\mathcal{F}_{r1}^2\right)} - 1\right], \tag{10.4.10}$$

in which $\mathcal{F}_{r1} > 1$. This result shows that the ratio y_2/y_1 is a function of the Froude number at the lower conjugate depth. The accuracy of this equation has been confirmed well by experimental data. Equation (10.4.10) can be converted to yield an expression for y_1/y_2 given by

$$\frac{y_1}{y_2} = \frac{1}{2}\left[\sqrt{1+8\left(\frac{u_{av2}^2}{gy_2}\right)} - 1\right] = \frac{1}{2}\left[\sqrt{1+8\left(\mathcal{F}_{r2}^2\right)} - 1\right], \tag{10.4.11}$$

where $\mathcal{F}_{r2} < 1$. The last two equations are useful in the analysis and design of a hydraulic jump. A supplementary information may be provided by the energy loss in the jump. It follows from the energy equation that

$$y_1 + \frac{u_{av1}^2}{2g} = y_2 + \frac{u_{av2}^2}{2g} + h_L, \qquad \longrightarrow \qquad \frac{h_L}{y_1} = 1 - \frac{y_2}{y_1} + \frac{\mathcal{F}_{r1}^2}{2}\left[1 - \left(\frac{y_1}{y_2}\right)^2\right], \tag{10.4.12}$$

in which Eqs. (10.4.10) and (10.4.11) have been used. In practice, this energy loss is dissipated over the *length of hydraulic jump* L_j, which is defined as the distance measured from the upstream face of jump to a point on the water surface immediately downstream from the surface roller. The determination of L_j needs to be conducted by experimental studies.

Hydraulic jump is introduced in practice to dissipate the initial kinetic energy of a flow below a spillway, outlet works, chute, or channel structure where the flow velocity is supercritical. Essentially, a hydraulic jump is housed within a *stilling basin*, which is a concrete-paved structure. If the floor of stilling basin is horizontal, the amount of energy dissipated in the hydraulic jump equals the difference in the specific energies at two conjugate depths and depends on the flow conditions at the lower conjugate depth.

10.4.3 Gradually Varied Flows

Gradually varied flows are steady but non-uniform, in which the flow depth, cross-sectional area, hydraulic radius, channel roughness, and channel bottom slope vary gradually along the channel reach, so that it is plausible to assume that the rate of energy loss at a given section is the same as that for a uniform flow having the same velocity and hydraulic radius at the same section. With these, consider the gradually

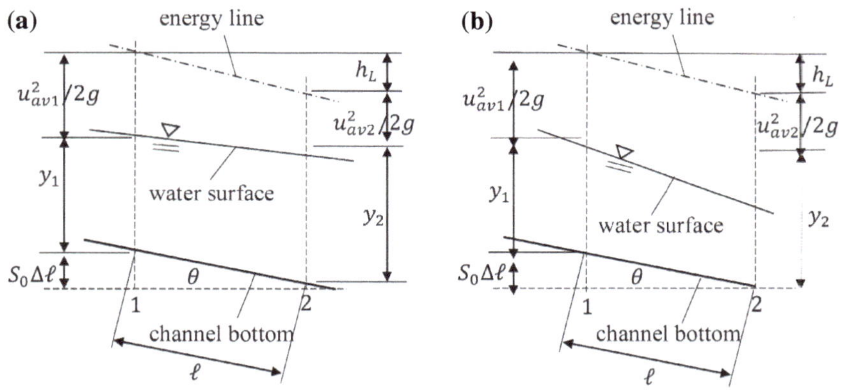

Fig. 10.6 Gradually varied flows in an open channel. **a** An accelerated flow. **b** A retarded flow

varied flows shown in Figs. 10.6a and b for an accelerated and a retarded cases, respectively. It is assumed that the changes in flow depth and velocity are small, so that the surface profiles over the channel reach $\Delta\ell$ may be straight. The energy equation between sections 1 and 2 reads

$$y_1 + \frac{u_{av1}^2}{2g} + S_0\ell = y_2 + \frac{u_{av2}^2}{2g} + S\ell, \tag{10.4.13}$$

in which $h_L = S\ell$, where S is the hydraulic slope, and S_0 represents the slope of channel bottom. Solving ℓ from this equation yields

$$\ell = \frac{1}{S - S_0}\left[\left(y_1 + \frac{u_{av1}^2}{2g}\right) - \left(y_2 + \frac{u_{av2}^2}{2g}\right)\right], \tag{10.4.14}$$

in which S can be determined by using the Manning formula for the average conditions between two sections, in which

$$S = \left(\frac{\bar{n}\bar{u}_{av}}{\bar{r}_h^{2/3}}\right)^2; \qquad \bar{n} = \frac{n_1 + n_2}{2}, \quad \bar{u}_{av} = \frac{u_{av1} + u_{av2}}{2}, \quad \bar{r}_h = \frac{r_{h1} + r_{h2}}{2}, \tag{10.4.15}$$

should be used in the Manning formula.

In the calculation, a channel section is selected with the known flow depth and velocity, say y_1 and u_{av1}. A channel reach is then chosen, with y_2 slightly different from y_1, and the corresponding u_{av2} is determined by using the conservation of mass. With these, the values of $\bar{n}$, $\bar{r}_h$ and $\bar{u}_{av}$ are determined, which are substituted into Eq. (10.4.15) to determine the value of S for the chosen channel reach. The value of ℓ is then determined by using Eq. (10.4.14) for the chosen reach. The same procedure is repeated for another short reach and so on. The water surface profile can then be made up of a series of straight segments. If the differences between the ys are taken to be sufficiently small, a fairly accurate profile of the free liquid surface may be obtained.

10.5 Dynamic Similarity for Free-Surface Flows

Open-channel flows are associated with free surface, in which the gravity force plays an important role in the flow features. As indicated in Sect. 6.5.2, the relative significance of gravity force in an open-channel flow is indexed by the dimensionless Froude number, which is proportional to the ratio between the inertia and gravity forces in the flow. A small value of the Froude number indicates that the gravity force predominates in the flow, whereas a large Froude number indicates that the inertia force is significant. Previously, the Froude number was used in defining whether the flow velocity in an open channel is subcritical, critical, or supercritical. Its role is very similar to the Mach number in compressible flows, as will be shown in the next section. By using the values of the Froude number, the physical significance of an open-channel flow may better be interpreted.

Since for open-channel flows an experimental studies in laboratory scale are intensively conducted, the condition of dynamic similarity in a model study may be satisfied by maintaining a constant Froude number at the corresponding points in the prototype and model to obtain the Froude similitude. At the same time, the geometric similarity between the boundaries of flows must also be maintained to establish a corresponding Froude model.

10.6 Analogy Between Open-Channel and Compressible Flows

Despite the apparent differences between liquids and gases, there exist many similar features associated with a liquid flow in an open channel and a flow of a compressible fluid. Consider first the conservation of mass. For a steady flow of a liquid in an open channel per unit width, the physical law reads

$$u_{av} y = \text{constant}, \tag{10.6.1}$$

which is similar to the same physical law for a steady flow of a compressible fluid given by

$$u_{av} \rho = \text{constant}. \tag{10.6.2}$$

In two equations, y represents the flow depth in the open channel, which is taken to be analogous to the density ρ of a compressible fluid.

In addition to the analogy of conservation of mass, the conservation of linear momentum shares equally an analogy. The balance of linear momentum of a compressible flow, in the context of one-dimensional frictionless circumstance, is given by

$$u_{av} \mathrm{d}u_{av} + \frac{\mathrm{d}p}{\rho} = 0, \quad \longrightarrow \quad \frac{\mathrm{d}\rho}{\rho} = -\frac{1}{\mathrm{d}p/\mathrm{d}\rho} u_{av} \mathrm{d}u_{av} = -\frac{1}{c^2} u_{av} \mathrm{d}u_{av}, \tag{10.6.3}$$

in which $c^2 = \mathrm{d}p/\mathrm{d}\rho$ has been used. On the other hand, the specific energy in an open-channel flow is given by

$$y + \frac{u_{av}^2}{2g} = E_s, \tag{10.6.4}$$

which is differentiated to obtain

$$dy + \frac{1}{g}u_{av}du_{av} = 0. \tag{10.6.5}$$

Since $u_{av} = \sqrt{gy}$, substituting this expression into the above equation yields

$$\frac{dy}{y} = -\frac{1}{u_{avc}^2}u_{av}du_{av}, \tag{10.6.6}$$

which is similar to Eq. (10.6.3)$_2$, where y plays the same role as ρ does.

It follows from Eqs. (10.6.3)$_2$ and (10.6.6) that the critical velocity in an open-channel flow is analogous to the local sonic velocity in a compressible flow. The critical velocity in a liquid flow equals the propagating velocity of a small surface wave, as discussed in Sect. 7.7.3, whereas the sonic velocity is the propagating velocity of a small pressure disturbance in a gas. The phenomena associated with the subcritical, critical, and supercritical velocities in a liquid flow are analogous to those associated respectively with the subsonic, sonic, and supersonic velocities in a gas flow. The conditions of a liquid flow in a hydraulic jump are equally similar to those in a compression shock wave in a compressible flow. Flows entering a hydraulic jump must have a supercritical velocity and a subcritical flow depth. A large amount of initial energy of the flow is dissipated in the jump, so that the flow leaves a jump with a subcritical velocity and a supercritical flow depth. Similarly, a gas flow before a compression shock wave must have a supersonic velocity and a low pressure, whereas a gas flow behind a shock wave moves with subsonic velocity and a high static pressure.

From the perspective of thermodynamics, there is an increase in entropy across a compression shock wave in a compressible flow. A reduce in the specific energy in a hydraulic jump is also indexed by an increase in entropy of a liquid flow.

10.7 Exercises

10.1 Consider a uniform flow in a triangular channel with side angles of α, as shown in the figure. The flow rate is denoted by Q, while the channel slope is S, and the channel material is concrete. Find the required dimension y in order to have a constant Q.

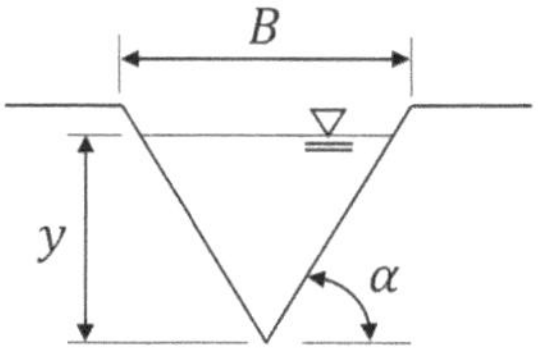

10.2 Water flows in a circular pipe with diameter d at a depth $0 \leq y \leq d$, as shown in the figure. The pipe is laid on a constant slope of S, and the Manning coefficient is n. Determine the depth where the maximum flow rate takes place. Show also that for certain flow rates there exist two possible flow depths corresponding to the same flow rate.

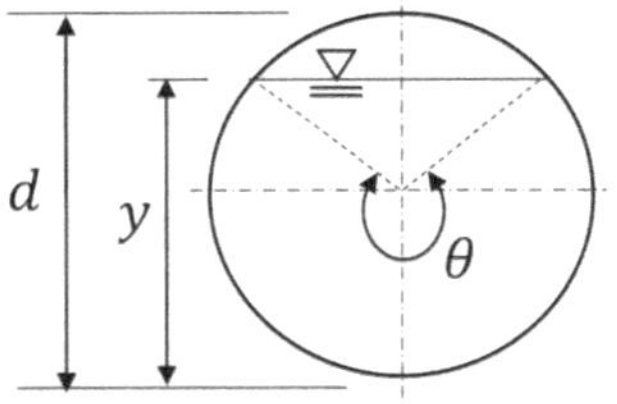

10.3 Water flows uniformly in a rectangular channel with width b and depth y. Determine the aspect ratio b/y for the best hydraulic cross-section, which gives the minimum cross-sectional area for all values of y.

10.4 The hydraulic radius of a vertical sluice gate in a wide rectangular channel shown in the figure is given by

$$r_h = y, \qquad B \gg 2y,$$

where B is the channel width and y is the flow depth. The flow immediately downstream from the gate is essentially a jet that possesses a vena contracta, and the distance from the gate to the vena contracta is approximated as the same as the gate opening h, with the contraction coefficient C_t. Find the distance from the vena contracta to a section b downstream, i.e., Δx, where the flow depth is h_b. The flow depth at the vena contracta is denoted by $h_v = C_t h$ under a given value of the flow rate Q per unit channel width. The channel bottom slope is S_0 and the Manning coefficient is n.

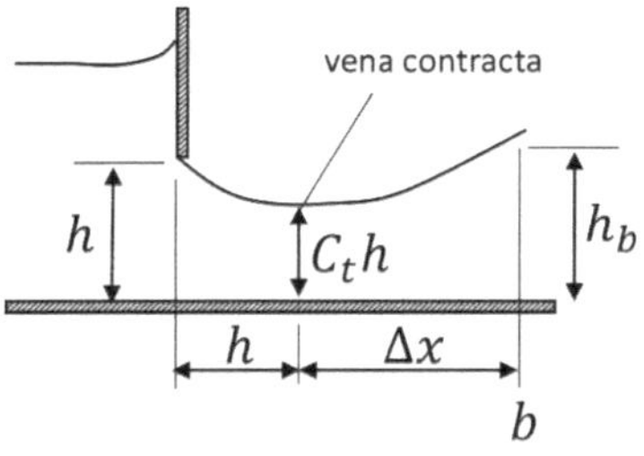

10.5 Consider Fig. 10.5b with Eq. $(10.4.12)_1$, show that the head loss of a hydraulic jump in a horizontal rectangular channel is given by

$$h_L = \frac{(y_2 - y_1)^3}{4y_1 y_2},$$

by combining the energy and momentum equations for a finite control-volume embracing the hydraulic hump.

Further Reading

R.D. Blevins, *Applied Fluid Dynamics Handbook* (Van Nostrand Reinhold, New York, 1984)
V.T. Chow, *Open Channel Hydraulics* (McGraw-Hill, New York, 1959)
R.W. Fox, P.J. Pritchard, A.T. McDonald, *Introduction to Fluid Mechanics*, 7th edn. (Wiley, New York, 2009)
R.H. French, *Open Channel Hydraulics* (McGraw-Hill, New York, 1985)
F.M. Henderson, *Open Channel Flow* (Macmillan, New York, 1966)
B.R. Munson, D.F. Young, T.H. Okiishi, *Fundamentals of Fluid Mechanics*, 3rd edn. (Wiley, New York, 1990)
R.H.F. Pao, *Fluid Mechanics* (Wiley, New York, 1961)
A.H. Shapiro, *The Dynamics and Thermodynamics of Compressible Fluid Flow, Volume 1* (Ronald Press, New York, 1953)
D.J. Tritton, *Physical Fluid Dynamics* (Oxford University Press, Oxford, 1988)
F.M. White, *Fluid Mechanics* (McGraw-Hill, New York, 1986)

Essentials of Thermodynamics 11

Thermodynamics is the science of energy dealing with heat, work, and other forms of energy, their transformations, and their relationships with properties of substances. In this chapter, only the thermodynamics based on the macroscopic description (classical thermodynamics) is addressed, which provides a fundamental knowledge not only to an energy perspective in understanding the motions of fluids, but also to access other branches of thermodynamics, such as irreversible thermodynamics, rational thermodynamics, or continuum thermodynamics. The results from the thermodynamics based on the microscopic description (statistical thermodynamics or statistical mechanics) are provided for selected topics to gain a clearer picture of the underlying physics in atomic and molecular scales.

The fundamental concepts are introduced first, with the focus on the system, system variables, thermodynamic equilibrium, process, cycle, and the state equations of pure substance, ideal and real gases. The derivation of ideal gas state equation based on the kinetic theory of gas is discussed to show the limitations of equation. The transient energies as work and heat are introduced subsequently to illuminate their differences from the energies that could be stored inside a system. A clear understanding of work and heat is a crucial point in clarifying the characteristics of various thermodynamic problems. The four laws of thermodynamics with the corresponding macroscopic properties are introduced in due order: the zeroth law with the empirical temperature and equality of temperature, the first law with internal energy as a macroscopic property, the second law with entropy from both the macroscopic and microscopic interpretations, and the third law with the absolute entropy and absolute zero of the thermodynamic temperature scale.

After the discussions on the second law, two entropy principles, namely the Coleman-Noll and Müller-Liu approaches, and their applications in deriving constitutive or closure models of simple substances are outlined to illustrate the concept of continuum thermodynamics to complete the discussions in Sect. 5.6.1.

© Springer International Publishing AG 2019

C. Fang, *An Introduction to Fluid Mechanics*, Springer Textbooks in Earth Sciences, Geography and Environment, https://doi.org/10.1007/978-3-319-91821-1_11

The thermodynamic relations, including the thermodynamic potential functions, the Maxwell relations, and general expressions of differential changes of thermodynamic properties for simple substances are derived at the end. The obtained results allow the indirectly measurable variables of a system, e.g. the entropy, to be related to those which can be measured directly. The simplified results for ideal gas are derived as an application of the thermodynamic relations.

11.1 Fundamental Concepts

11.1.1 Scope of Thermodynamics

Thermodynamics is a combination of two Greek words: "thermo" and "dynamics." While the first word means "therme (heat)," the second word is interpreted as "strength (power)." Thus, thermodynamics may be understood as the science of "heat power," and is defined as the science of *energy* and *entropy*, or alternatively as the science that deals with *heat*, *work*, and other forms of energy, their transformations, and their relations with the properties of the substances that involve. These substances are called specifically the *working substances*, whose changes in properties are used as a direct or an indirect measure of a considered energy transformation process, for which the concepts of system, surrounding, control-mass, control-volume, microscopic and macroscopic approaches, and the Lagrangian and Eulerian descriptions described in Sect. 2.3 can be applied. Specifically, the *classical thermodynamics* is a macroscopic equilibrium description without referring to the atomic and molecular structures of working substance, while the *statistical thermodynamics* or *statistical mechanics* is a microscopic description in which the

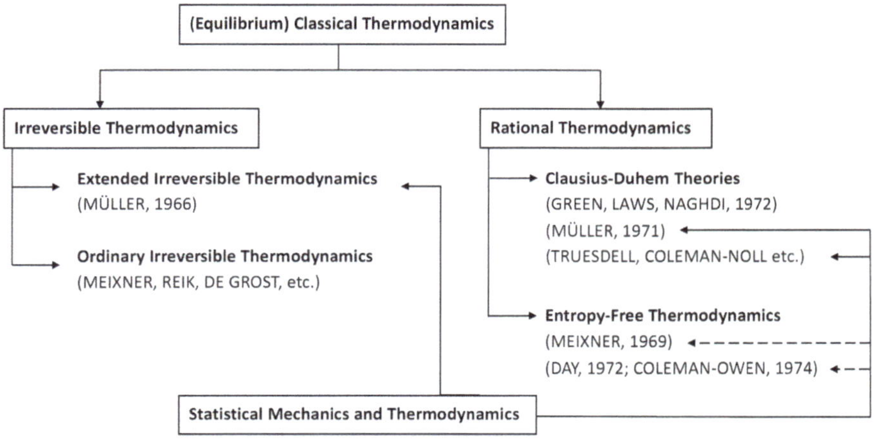

Fig. 11.1 A simplified hierarchy of the macroscopic thermodynamics with supplement from the microscopic thermodynamics

atomic and molecular structures of working substance with relations to the macroscopically measurable quantities are studied. Thermodynamics can also be classified as *irreversible thermodynamics*, *rational thermodynamics*, or *continuum thermodynamics*, depending on the treatments of *entropy production* in a process. Figure 11.1 illustrates a simplified hierarchy of thermodynamics.[1] In the context of this chapter, only the (equilibrium) classical thermodynamics is discussed, for it provides a fundamental knowledge not only to an energy perspective of fluid motion, but also to access other branches of thermodynamics. In parallel, for selected topics the findings from the statistical thermodynamics, statistical mechanics, or kinetic theory of gas are supplemented to deepen the physical understanding. To meet the requirement of a macroscopic description, all working substances considered in the classical thermodynamics are assumed a priori a kind of continuum.

11.1.2 Thermodynamic System and Variable

A *thermodynamic system*, either in a control-mass or a control-volume base, contains a working substance or a combination of several working substances, whose directly or indirectly observable variables are called the *thermodynamic variables*. All variables are classified into two categories: the *intensive* and *extensive variables*. Intensive variables are those whose magnitudes do not depend on the extent of working substance, e.g. the pressure or temperature, while the magnitudes of extensive variables depend on the extent of working substance, e.g. the mass or volume. It is often convenient to refer to extensive variables in terms of their values per unit mass of system, which gives rise to the *specific variables*; i.e., a specific variable is obtained by dividing the corresponding extensive variable by the mass of the system. For example, the specific volume is defined as the volume per unit mass of a system. Obviously, specific variables are intensive variables, and the conventional notation system is that extensive variables are expressed by using capital letters, while the specific and intensive variables are denoted by using the corresponding small letters. This notation system is used throughout the chapter. The thermodynamic variables of a system may also be classified into two categories: the *state* or *process variables*. State variables are those whose values are determined once the system state is prescribed, which are also called *point functions*. The values of process variables depend, on the contrary, on the time and space successions that the system has passed. Process variables are also termed *path functions*. Thermodynamic variables depending only on the state of a system are also termed *thermodynamic properties*.

In the following sections, the findings of classical thermodynamics will be introduced subsequently. It should be pointed out that thermodynamics applies to all types of systems in macroscopic aggregation and sets limits (inequalities) on permissible physical processes. It establishes relationships among apparently unrelated properties

[1] Information quoted from Hutter, K., The Foundations of thermodynamics, its basic postulates and implications: a review of modern thermodynamics, *Acta Mechanica* 27, 1–54, 1977.

for its generality. Thermodynamics is not based on a new and particular law of nature; it instead reflects a commonality or universal feature of all laws. It is the study of the restrictions on the possible properties of matter that follow from the symmetry properties of the fundamental laws of physics.

11.1.3 Thermodynamic Equilibrium, Process, and Cycle

If a thermodynamic system suffers a change with respect to its surrounding, it is usually seen to undergo a change in state. For example, the ascending air bubble in a still water shown in Fig. 2.2b undergoes a change in the air state. If, after a time, the air bubble stops ascending by some means, it will be found to reach a state where no further change in state takes place. In such a circumstance, the air bubble is said to come to *thermodynamic equilibrium*.[2] In general, a system is in thermodynamic equilibrium if it is in equilibrium regarding of all possible changes of state. All properties of the system are prescribed, and no net changes of the properties can be observed. The conditions of thermodynamic equilibrium of a system involve vanishing thermal, mechanical (work-like), chemical, and other possible interactions with its surrounding. For example, for the previously discussed ascending air bubble, a *thermal equilibrium* between the air bubble and surrounding still water is reached if there exists no heat transfer in-between. Likewise, a *mechanical equilibrium* is obtained if all the mechanical interactions between the air bubble and water cease. A *chemical equilibrium* requires vanishing chemical reactions and diffusions between the air bubble and water.

As similar to the theory of mechanics, the theory of classical thermodynamics defines several kinds of equilibrium, whose stabilities can be defined in a similar manner. A thermodynamic system is said to be in *stable*, *unstable*, or *neutral equilibrium* according to the definitions in classical mechanics. However, in the strictest sense, neither classical mechanics nor classical thermodynamics knows unstable equilibrium, for equilibrium is defined in terms of macroscopic variables which are large-scale averages of the quantities experiencing fluctuations in the microscopic level. Although in large-scale average the microscopic fluctuations may be relatively unimportant, these fluctuations, however small, are sufficient to destroy unstable equilibrium. Thus, equilibrium is itself a macroscopic concept in terms of macroscopic quantities in the context of thermodynamics.

Since in thermodynamic equilibrium no further change between a system and its surrounding takes place, all macroscopic quantities in both the system and surrounding assume finite and fixed values, for which the system and surrounding are said to be in a specific *state*. Conversely, a state of a thermodynamic system may be determined by prescribing the values of thermodynamic variables which are functions of state. They will have a particular set of values for a particular state of the

[2]Thermodynamic equilibrium can be defined mathematically by vanishing entropy production or by maximum thermodynamic potential, as will be discussed respectively in Sects. 11.6.1 and 11.8.4.

system. Since many of them can be related to one another in some way, the minimum number of state variables in prescribing a specific state depends on the nature of system and the conditions imposed on it. Hence, the characteristics of working substances contained in a system need to be studied, which will be discussed in the next subsection.

A *thermodynamic process* is said to take place if a system undergoes a series of changes in its state. A process assumes a beginning point and an end point, corresponding to the initial and final states of system. A process is called *reversible* if and only if its direction can be reversed by an infinitesimal change in the conditions without any net change in both the system and surrounding. Occasionally, a process may be reversed by a finite change. However, such a case is not referred to as a reversible process, for thermodynamic reversibility requires two conditions to be fulfilled: The process must be *quasi-static* or *quasi-equilibrium* and without *hysteresis*. A quasi-static process is carried out so slowly that every state through which a system passes may be considered an equilibrium state. This implies that the process should be carried out infinitely slowly, so that a system has a sufficient time to adjust itself to the changing surrounding in order not to depart significantly from equilibrium with it. Every state that the system passes through will be an equilibrium state, and the process can be reversed at any time by reversing the operations on the system. If a process is reversed by a finite change, some irreversibility will be induced in the system and/or surrounding. As an example, consider a rapid compression of the air contained inside a cylinder-piston device, which sets up sound or shock waves, giving rise to regions with different temperatures and pressures in the air. The sound or shock waves cannot be extracted by moving the piston out again. A reversible process should be without any hysteresis. That is, a system, instead of proceeding a different path, can retrace its previous path if the process is reversed. For example, under the linear elastic limit, a steel bar bears no hysteresis in a series of compressions and extensions, while a soil does possess hysteresis under the same operation conditions. Irreversibilities in a process are classified as *internal* or *external*. Internal irreversibility takes place inside a system, e.g. dissipation of fluid, while external irreversibility takes place on the boundary between a system and its surrounding, e.g. heat transfer via finite temperature difference or frictional effect.

In the context of classical thermodynamics, the following processes are of prime importance:

- *isobaric process* $\longleftrightarrow$ pressure $p =$ constant;
- *isothermal process* $\longleftrightarrow$ temperature $T =$ constant;
- *isochoric process* $\longleftrightarrow$ volume $V =$ constant;
- *adiabatic process* $\longleftrightarrow$ amount of heat transfer $Q = 0$;
- *isentropic process* $\longleftrightarrow$ entropy $S =$ constant; and
- *isenthalpic process* $\longleftrightarrow$ enthalpy $H =$ constant.

These processes are essentially the special cases of a more general *polytropic process*, which will be discussed separately in the forthcoming sections.

A *thermodynamic cycle* comprises of more than two thermodynamic processes, whose beginning and end points are the same. That is, a system and its surrounding restore to their initial states after the system has completed a thermodynamic cycle. A reversible thermodynamic cycle can be reversed by an infinitesimal change in the conditions without causing any net change in both the system and surrounding. The concept of thermodynamic cycle can be used to evaluate the ideal performances of many energy conversion devices such as power plants, internal combustion engines, and gas turbines.

11.1.4 Pure Substance and Indicator Diagram

A *pure substance* is one that has homogeneous and invariable chemical composition, which may exist in more than one phase with the same chemical composition in all possible phases. A *phase* is defined as a quantity of substance that is homogeneous throughout. In each phase, the substance may exist at different states. For example, a liquid water, a mixture of liquid water and water vapor (steam), and a mixture of ice and liquid water are all pure substances. A liquid water assumes only one (liquid) phase and may exist at different temperatures under one atmospheric pressure (different states). On the other hand, air is not a pure substance in the strictest sense, for its chemical composition will change if a phase change takes place. However, air and other mixtures of gases can be considered pure substances as a first approximation, if they experience no change in phase during a change in state. A *simple compressible substance* designates a substance whose transient energy as a kind of work accomplished by the changes in volume is dominant when compared to other possible work forms resulted from surface tension, magnetic or electrical effect, etc. In other words, for simple compressible substances only the work induced by the changes in system volume is taken into account. A more detailed discussion on the topic is given in Sect. 11.2.1.

The minimum number of thermodynamic variables required to identify the state of a thermodynamic system corresponds to the degrees of freedom of that system, which is the number of independent variables given by

$$N = C + 2 - P, \tag{11.1.1}$$

where N is the degrees of freedom, C denotes the number of compositions of the system, and P stands for the number of phases appearing in the system. For example, for a mixture of liquid water and steam, $N = 1 + 2 - 2 = 1$, indicating that only the value of one single independent thermodynamic variable needs to be prescribed in order to identify the state of mixture. On the other hand, for a gas or a superheated steam, $N = 1 + 2 - 1 = 2$, showing that two independent thermodynamic variables need to be prescribed to identify the state of the working substance. Hence, the state of a thermodynamic system may be represented in a graphical manner. Consider the air bubble shown in Fig. 2.2b as the thermodynamic system again. The state of air bubble is identified by prescribing the values of two independent thermodynamic variables, say the pressure p and volume V, which is represented by a single point

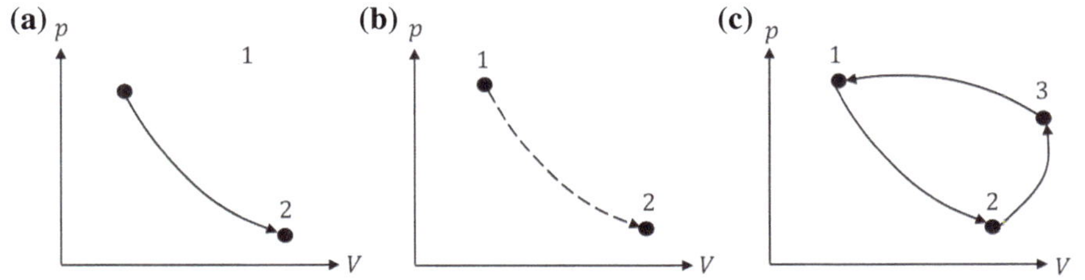

Fig. 11.2 p–V indicator diagrams for processes and cycles. **a** A reversible process from state 1 to state 2 by a solid line with arrow. **b** An irreversible process from state 1 to state 2 by a dashed line with arrow. **c** A reversible cycle consisting of three reversible thermodynamic processes $1 \to 2$, $2 \to 3$, and $3 \to 1$

in a two-dimensional diagram spanned by p and V, as shown by point 1 or point 2 in Fig. 11.2a.[3] Such a diagram is referred to as an *indicator diagram* or a *phase diagram*.

The indicator diagram may be used to represent a thermodynamic process or a thermodynamic cycle. For example, it is assumed that the air bubble is released at state 1, and the ascending motion stops at state 2. If the ascending motion takes place so slowly that at each instant the state of air bubble does not depart significantly from equilibrium, it may be approximated by a reversible thermodynamic process by using a solid line connecting points 1 and 2, with the line arrow denoting the process direction, as shown in Fig. 11.2a. Every point on this solid line represents a state of air bubble at a specific instant. On the contrary, if the ascending motion takes place so rapidly that it is no longer a reversible process, its graphic illustration in the indicator diagram can still be accomplished by using a dashed line connecting points 1 and 2, as shown in Fig. 11.2b. In an irreversible process, every point on the dashed line, however, does not represent a state of air bubble. The dashed line with arrow indicates merely that the air bubble goes from state 1 to state 2 without any information in-between. Figure 11.2c illustrates graphically a thermodynamic cycle consisting of three thermodynamic processes $1 \to 2$, $2 \to 3$ and $3 \to 1$ in the p–V diagram.

11.1.5 Thermodynamic Surface, Ideal and Real Gases

The most encountered working substances in classical thermodynamics are water and air. While the former is used in e.g. a steam power plant via the *Rankine cycle*, the latter is applied in evaluating the ideal performance of internal combustion engines in e.g. the *Otto cycle* or *Diesel cycle*. Since water in steam power plant and other

[3]Figure 11.2a is the p–V indicator diagram, or simply the p–V diagram. Other frequently used indicator diagrams in thermodynamics are the p–T, T–S, and H–S diagrams, as will be discussed later.

applications may exist in solid, liquid, or vapor phase, its two-dimensional indicator diagram needs to be expanded to a three-dimensional generalization, and possibly the points and lines in the two-dimensional indicator diagram become respectively the lines and surfaces in the three-dimensional generalization. Conventionally, the three coordinate variables in the three-dimensional generalization of the indicator diagram are pressure p, volume V, and temperature T. The typical p–V–T phase diagram of water is shown in Fig. 11.3, in which the surface is known as the *thermodynamic surface*. Each point on the thermodynamic surface represents a specific state of water, whose pressure, volume, and temperature assume fixed values.

The line between ice S and liquid water L in the projected p–T diagram is known as the *fusion line*, at which ice transforms to liquid water and vice versa. Specifically, the processes at which ice transforms to liquid water and liquid water transforms to ice are called respectively *fusion* and *freezing*. The line between liquid water and steam V is referred to as the *vaporization line*, at which liquid water transforms to water vapor, called *evaporation*, and water vapor transforms to liquid water, called *condensation*. The line between ice and steam is known as the *sublimation line*, at which a phase change between solid and vapor water takes place. The processes in which ice transforms to steam are termed *sublimation*, while those for steam to ice are termed *solidification*. The intersection of three lines, which is essentially a line on the thermodynamic surface, marks the *triple point* of water, at which its solid, liquid, and vapor phases coexist. Water in the vaporization, fusion, and sublimation lines is referred to as *saturated*, whose pressure and temperature are called respectively the *saturation pressure* and *saturation temperature*. Each saturation pressure has its own saturation temperature, and vice versa. They assume fixed values during a phase change. Ice, liquid water, and steam at saturated states are referred to respectively as *saturated solid*, *saturated liquid*, and *saturated vapor*. Water vapor assuming a temperature larger than the corresponding saturation temperature at a given pressure is called *superheated*. Similarly, liquid water assuming a temperature lower than the saturation temperature of a given pressure is termed *subcooled* or *compressed*.

To every point on the thermodynamic surface, the state of water is fixed, except those in the bell-shaped region, which is more visible in the projected p–V diagram. In this region, saturated liquid water and saturated steam coexist. Hence, water in this region assumes two phases, whose state needs to be determined by supplementing an additional information, namely the *quality* x, which is defined by

$$x \equiv \frac{m_g}{m_f + m_g}, \tag{11.1.2}$$

where m_f and m_g represent respectively the masses of saturated liquid water and saturated water vapor. With this, any thermodynamic variable of a liquid-water and water-vapor mixture at a given quality, α_x, can be determined as

$$\alpha_x = \alpha_f + x\left(\alpha_g - \alpha_f\right) = \alpha_f + x\alpha_{fg}, \qquad \alpha_{fg} \equiv \alpha_g - \alpha_f, \tag{11.1.3}$$

where α_f and α_g represent the values of α at the saturated liquid and saturated vapor states, respectively. The quality is an intensive property, which is only meaningful in the two-phase region. It is used frequently in steam power plant to evaluate the water

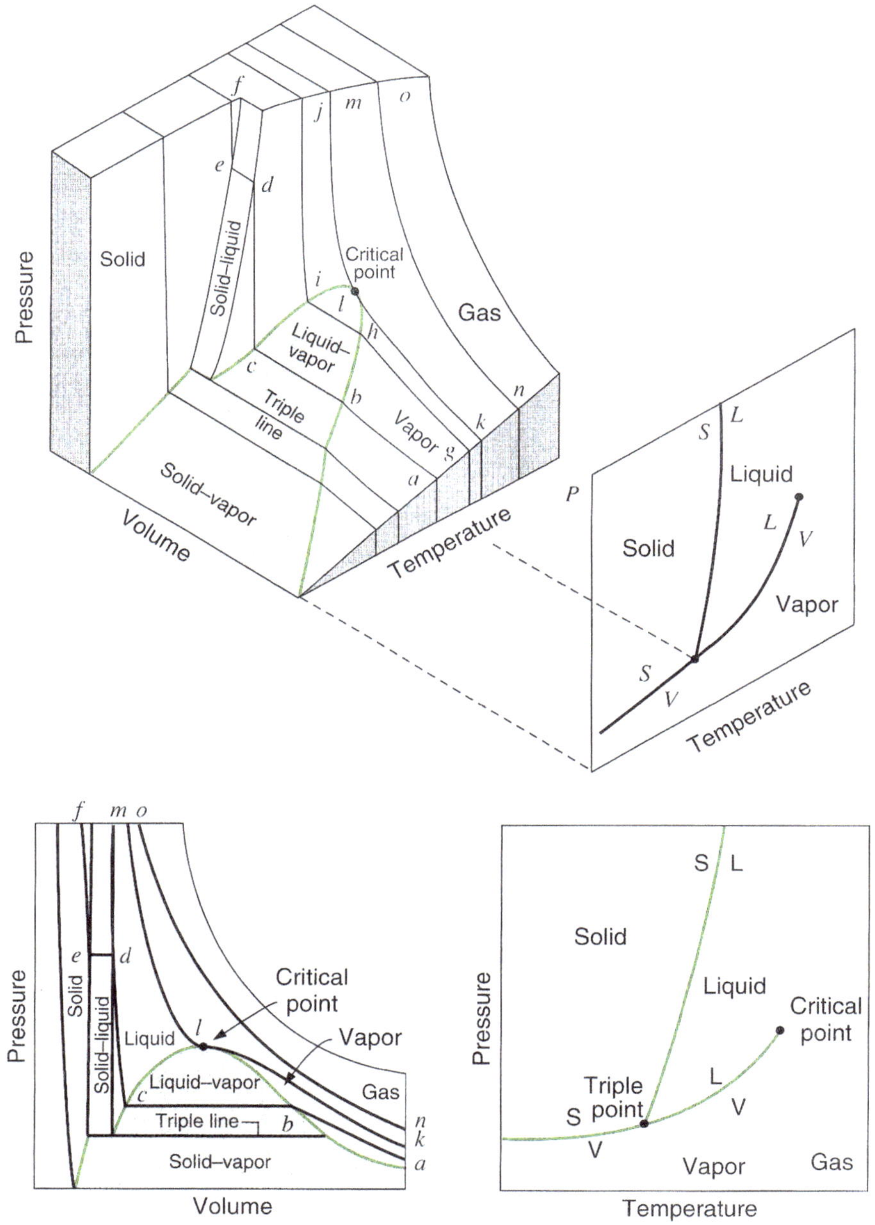

Fig. 11.3 Thermodynamic surface of water in the three-dimensional p–V–T diagram with the projected two-dimensional p–V and p–T diagrams. Quoted from Borgnakke C., Sonntag, R.E., *Fundamentals of Thermodynamics*, 7th ed., John Wiley & Sons, New York, 2009. Used with permission

Table 11.1 Pressures, temperatures and specific volumes at the critical and triple points of water, and the saturation properties corresponding to 1 atmospheric pressure

	Pressure (kPa)	Temperature (°C)	Specific volume (m^3/kg)
Triple point	0.6113	0.01	0.001 (sat.liquid); 206.132 (sat. vapor)
Critical point	$22.09 \cdot 10^3$	374.14	0.003155
(sat.) 1 atm	101.325	(sat.)99.6	0.001044 (sat. liquid); 1.6729 (sat. vapor)

state entering the steam turbine. It should be as high as possible to avoid mechanical erosion of the blades of steam turbine. The left margin of bell-shaped region marks the states of saturated liquid, while the right margin marks the states of saturated vapor. Two margins approach gradually a single point, which is referred to as the *critical point*, above which a phase change of water cannot take place. For example, consider a subcooled liquid which is initially at some point in the liquid part of isothermal line d-c-b-a in Fig. 11.3. As the pressure decreases, the liquid water approaches the state of saturation liquid, i.e., point c, from which a phase change to steam begins. During the phase change process, both the pressure and temperature assume fixed values, for they are respectively the saturation pressure and temperature. The phase change ends at point b, where all liquid water has transformed to saturated water vapor. On the contrary, if the liquid water assumes an initial state whose pressure is larger than that of critical point, e.g. line m-n, decreasing in pressure will not cause any phase change of the liquid water even the pressure becomes so small. The liquid water under such a circumstance is not stable. Table 11.1 summarizes the values of some intensive properties at the triple and critical points of pure water, and under 1 atmospheric pressure.[4]

All pure substances exhibit similar characteristics as those shown in Fig. 11.3, except that the slopes of fusion lines in the projected p–T diagrams are positive. It is due to the fact that water is the only substance on earth which expands at freezing. Theoretically, every substance has its own p–V–T diagram, despite if one can find it. Such a complicated thermodynamic surface cannot be described by using a single state equation. In practice, thermodynamic properties of common working substances have been summarized in the *table of thermodynamic properties* for reference.

For dry air and other gases, their states can be described by using the ideal gas state equation, as discussed in Sect. 2.7, provided that the pressure is low (and hence the density is low) but the temperature is high. The ideal gas state equation can equally be used for unsaturated moist air and saturated air before condensation. The restrictions to the ideal gas state equation result from that the gas molecules are of infinitesimal size and have large molecular mean free path, so that the intermolecular interactions can be neglected. The origins of these restrictions result from the assumptions used in the kinetic theory of gas, as will be discussed in the next subsection.

[4]Data quoted from *ASME Steam Tables: Compact Edition*, American Society of Mechanical Engineers, 1st ed., 2006.

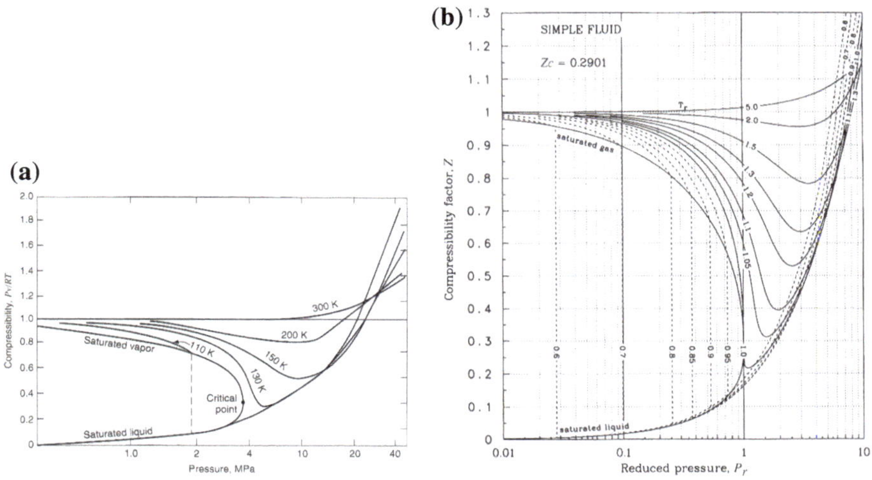

Fig. 11.4 Charts of compressibility factor. **a** For nitrogen. **b** A generalized chart of simple fluids. Quoted from Borgnakke C., Sonntag, R.E., *Fundamentals of Thermodynamics*, 7th ed., John Wiley & Sons, New York, 2009. Used with permission

Gases which depart from the restrictions of ideal gas are called real gases. Many efforts have been made to propose appropriate state equations for their characteristics, and most of them are semiempirical, and each state equation of real gas has its own limitations. A more convenient approach in evaluating the behavior of a real gas is the concept of compressibility factor Z given in Eq. (2.7.3), whose deviation from unity becomes a measure of the deviation from the ideal gas state equation. The typical compressibility factor chart of nitrogen is shown in Fig. 11.4a. It is seen that at all values of temperature, $Z \to 1$ as $p \to 0$, implying that the nitrogen behavior is very close to that predicted by using the ideal gas state equation as the pressure approaches null. In addition, at temperature of 300 K and above, the value of Z is nearly unity up to $p \sim 10\,\text{MPa}$, which indicates that the ideal gas state equation delivers accurate predictions in this operation range. On the contrary, at lower temperatures or at extremely high pressures, the value of Z deviates significantly from unity. At low temperatures, the nitrogen molecules tend to be pulled together due to the enhanced intermolecular attraction caused by moderate densities. This gives rise to that $Z < 1$ in general. At high pressures, very high-density forces of repulsion take place, yielding generally that $Z > 1$.

Although the charts of compressibility factors of other pure substances are quantitatively different, their tendencies are similar. This implies that the charts of compressibility factor of these substances can be put on a common base, which can be accomplished by defining the *reduced pressure* p_r and *reduced temperature* T_r given by

$$p_r \equiv \frac{p}{p_c}, \qquad T_r \equiv \frac{T}{T_c}, \tag{11.1.4}$$

where p_c and T_c are respectively the pressure and temperature at the critical point. Replacing the pressure by p_r and temperature by T_r in Fig. 11.4a yields the generalized chart of compressibility factor, as shown in Fig. 11.4b, which is valid for all substances with simple and essentially spherical molecules. Correlations for substances with more complicated molecular structures are reasonable close, except near saturation or at high density. Hence, Fig. 11.4b is representative for the average behavior of a number of simple substances.

11.1.6 Kinetic Theory of Ideal Gas

The assumptions in deriving the ideal gas state equation by using the kinetic theory of gas are as follows:

1. Any small gas sample consists of an enormous number of particles N, which are all identical and inert for any one chemical species. Let m be the mass of each particle, so that the mass of a gas sample consisting of N particles is mN. If the molar mass (or equivalently the molecular weight) of gas sample is M, the number of moles n of a gas sample is given by $n = mN/M$. The number of particles of a gas sample per mole is *Avogadro's number*,[5] N_A, with $N_A = M/m = 6.0221 \times 10^{23}$.
2. The particles of an ideal gas sample are assumed to resemble small hard spheres which are in perpetual random motion. The mean free path is sufficiently large compared to the particle diameter.
3. There exists no interaction among the particles, except when the particles are in collision with one another or with a solid wall. All collisions are considered smooth and perfectly elastic. Particles maintain themselves in rectilinear motions before and after collisions.
4. Without the influence of external force field, the particles are distributed uniformly throughout a container, and the *number density*, i.e., the number of particles per unit volume, is constant. In any small volume element $\mathrm{d}v$ there are $\mathrm{d}N$ particles, so that $\mathrm{d}N = (N/V)\mathrm{d}v$, where $\mathrm{d}v$ must be so chosen to satisfy the continuum hypothesis for a chosen gas sample.
5. Particles have no preferred directions on their velocities, so that at any instant the probability of finding a particle moving in a specific direction is the same with respect to all directions.
6. Particles at any instant move with different speeds. Some gas particles may move slowly, while a few may move rather rapidly, so that the speed spectrum varies from null to light speed. Let $\mathrm{d}N_w$ represent the number of particles with speeds between w and $w + \mathrm{d}w$, which is assumed to be constant at equilibrium, although these particles may perpetually collide with one another or with a solid wall to change their speeds.

[5]Lorenzo Romano Amedeo Carlo Avogadro, 1776–1856, an Italian scientist, who contributed to the molecular theory known as *Avogadro's law*.

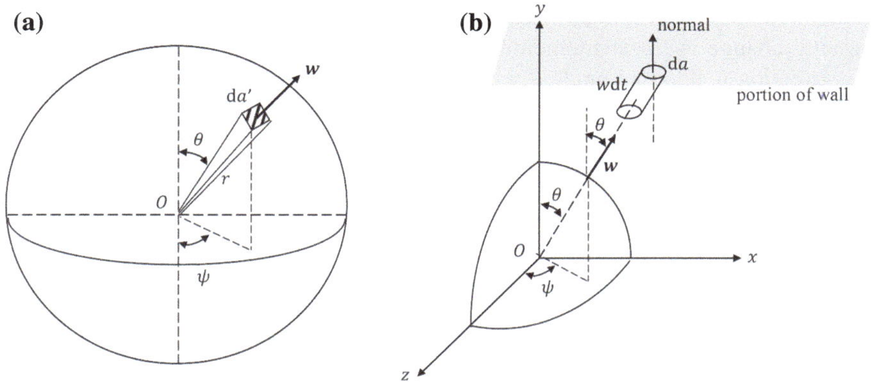

Fig. 11.5 Derivation of the ideal gas state equation by using the kinetic theory of gas. **a** The solid angle $\mathrm{d}\Omega$. **b** The gas particles inside an infinitesimal cylindrical volume

Let $\boldsymbol{w}$ be the velocity of a gas particle originating from point O through an infinitesimal surface element $\mathrm{d}a'$ on a sphere, as shown in Fig. 11.5a. The surface element $\mathrm{d}a'$, in terms of the spherical coordinates $\{r, \theta, \psi\}$, is given by

$$\mathrm{d}a' = (r\,\mathrm{d}\theta)\,(r\sin\theta\,\mathrm{d}\psi)\,, \tag{11.1.5}$$

by which the *solid angle* $\mathrm{d}\Omega$, which is the angle formed by lines radiating from point O and touching the edges of $\mathrm{d}a'$, is defined as

$$\mathrm{d}\Omega \equiv \frac{\mathrm{d}a'}{r^2} = \sin\theta\,\mathrm{d}\theta\,\mathrm{d}\psi, \tag{11.1.6}$$

whose maximum value is 4π, for the surface area of a sphere with radius r is $4\pi r^2$. The fraction of the $\mathrm{d}N_w$ particles with directions lying in $\mathrm{d}\Omega$ is $\mathrm{d}\Omega/4\pi$, so that the number of particles with speed range $\mathrm{d}w$ and angle ranges $\mathrm{d}\theta$ and $\mathrm{d}\psi$, $\mathrm{d}N_{w,\theta,\psi}$, is given by

$$\mathrm{d}N_{w,\theta,\psi} = \mathrm{d}N_w\frac{\mathrm{d}\Omega}{4\pi}, \tag{11.1.7}$$

since the particles have no preferred moving direction. The particles approach a small area element $\mathrm{d}a$ of the container wall, many of which undergo collisions along the moving trajectories. However, only those particles inside a cylindrical volume $\mathrm{d}v$ whose side length is $w\mathrm{d}t$ are taken into account, where $\mathrm{d}t$ denotes a very short time interval in which no inter-particle collision occurs, as shown in Fig. 11.5b. It follows that

$$\mathrm{d}v = (w\,\mathrm{d}t)\cos\theta\,\mathrm{d}a, \tag{11.1.8}$$

so that the number of particles inside this infinitesimal cylindrical volume which strike $\mathrm{d}a$ during $\mathrm{d}t$ is given by

$$\mathrm{d}N_{w,\theta,\psi}\frac{\mathrm{d}v}{V}, \tag{11.1.9}$$

where V represents the total volume of container. Again, Eq. (11.1.9) is so obtained that the gas particles have no preferred directions.

Since the collision between any two particles is assumed to be perfectly elastic, the total change in linear momentum of a particle with speed w striking the wall in a direction making an angle θ with the normal to the wall is $(-2mw\cos\theta)$ per collision, for only the normal component of velocity contributes to the momentum change. Combining Eqs. (11.1.7)–(11.1.9) yields the total linear momentum change per unit time and per unit area given by

$$\left(\frac{\mathrm{d}N_w}{4\pi}\sin\theta\,\mathrm{d}\theta\,\mathrm{d}\psi\right)\left(\frac{1}{V}w\,\mathrm{d}t\cos\theta\,\mathrm{d}a\right)(-2mw\cos\theta), \tag{11.1.10}$$

which needs to be integrated with respect to all directions to give the pressure $\mathrm{d}p_w$ exerted by the wall on those gas particles having the number $\mathrm{d}N_w$. Conversely, the pressure $\mathrm{d}p_w$ exerted by the gas particles on the wall is obtained as

$$\begin{aligned}\mathrm{d}p_w &= mw^2\frac{\mathrm{d}N_w}{V}\left(\frac{1}{2\pi}\int_0^{2\pi}\mathrm{d}\psi\int_0^{\pi/2}\cos^2\theta\,\sin\theta\,\mathrm{d}\theta\right),\\ \longrightarrow\ pV &= \frac{1}{3}m\int_0^{\infty}w^2\mathrm{d}N_w.\end{aligned} \tag{11.1.11}$$

The integration in Eq. $(11.1.11)_2$ can be expressed by using the average of the square of particle speed, $<w^2>$, defined by

$$<w^2>\equiv\frac{1}{N}\int_0^{\infty}w^2\mathrm{d}N_w, \tag{11.1.12}$$

so that

$$pV=\frac{Nm}{3}<w^2>. \tag{11.1.13}$$

Comparing this equation with Eq. $(2.7.1)_1$ gives

$$\frac{Nm}{3}<w^2>=n\bar{R}T. \tag{11.1.14}$$

Since the *average kinetic energy* of gas particles is $(m<w^2>)/2$, as motivated by the theory of rigid body dynamics, it follows that the temperature T is identified to be

$$T=\frac{2N}{3n\bar{R}}\left(\frac{1}{2}m<w^2>\right), \tag{11.1.15}$$

showing that the temperature of an ideal gas is proportional to the average kinetic energy of gas particles.

In the context of kinetic theory of gas, gas particles are assumed to be non-interacting ones, so the potential energy of interparticle interaction may be neglected. The only non-vanishing energy form of particles is the translational kinetic energy. Other possible energy forms such as vibrational and rotational energies are absent. Thus, the *internal energy U* of an ideal monatomic gas is the sum of the kinetic energies of its consisting particles given by

$$U=\sum_j\frac{1}{2}mw_j^2=N\left(\frac{1}{2}m<w^2>\right)=\frac{3}{2}n\bar{R}T, \tag{11.1.16}$$

in which Eq. (11.1.15) has been used. This result shows that the internal energy is only proportional to the thermodynamic temperature T, which is in agreement with experimental outcomes, as will be discussed in a detailed manner in Sect. 11.8.5. On the other hand, since $n = N/N_A$, it is found that

$$k \equiv \frac{\bar{R}}{N_A} = 1.3807 \times 10^{-23}\,\frac{\mathrm{J}}{\mathrm{K}}, \tag{11.1.17}$$

where k represents *Boltzmann's constant*. With this, the average kinetic energy of gas particles, the internal energy, and pressure are expressed alternatively as

$$\frac{1}{2}m <w^2>= \frac{3}{2}kT, \qquad U = \frac{3}{2}\frac{N}{N_A}\bar{R}T = \frac{3}{2}NkT, \qquad p = \frac{n\bar{R}T}{V} = \frac{NkT}{V}. \tag{11.1.18}$$

The obtained results apply only for monatomic gases, in which the gas particles are assumed to be infinitesimal, hard, and spherical, and each gas particle is unaffected by its neighbors.

The assumptions of point mass and non-interacting gas particles in the kinetic theory of ideal gas need to be revised for real gases. For the gas particles with finite volumes and non-vanishing interparticle interactions, the ideal gas state equation may be revised to

$$\left(p + \frac{n^2 a}{V^2}\right)(V - nb) = n\bar{R}T, \tag{11.1.19}$$

where constant a accounts for the cohesive forces between the gas particles, which decreases the measured value of pressure, while constant b accounts for the volumes occupied by the gas particles themselves inside the system volume V. Equation (11.1.19) is the well-known van der Waal state equation for real gases.

11.1.7 Microscopic Perspective of Internal Energy

Temporarily, the energy of a thermodynamic system is defined as the capacity to produce an effect. The rigorous definition of energy will be given in Sect. 11.4.3. Macroscopically, the energy of a system can be classified into two categories: the energies which are stored inside the system and the energies which can be transferred across the system boundary. The stored energies are further classified into three categories: the potential and kinetic energies associated with the system as a whole, which are similar to their counterparts in classical mechanics, and the *internal energy* which depends partly on the temperature of system, as indicated by Eq. (11.1.16). The transferred energy is classified either as *heat* or *work*, which will be discussed in the next section.

Microscopically, the internal energy is closely related to the atomic and molecular structures of matter. By using the molecular point of view, three general energy forms associated with the gas molecules are identified as follows:

- the intermolecular potential energy, caused by the intermolecular interactions;
- the molecular kinetic energy, caused by the translational velocity of each individual gas molecule; and

- the intermolecular energy associated within each individual gas molecule, caused by the atomic and molecular structures and the related intermolecular interactions.

The intermolecular potential energy depends on the magnitudes of intermolecular forces and the positions of molecules relative to one another at any instant of time. Essentially, it is impossible to determine accurately the magnitude of this energy form, for the exact molecular configurations and orientations at any time or the exact intermolecular interactions are yet known. For gases with very low densities, the gas molecules are so widely spaced that the average mean free path is large, yielding negligible intermolecular potential energy. This is exactly what has been assumed in the kinetic theory of ideal gas. For gases at low or moderate densities, the gas molecules are widely spaced, and it is plausible to take into account only two-molecule or three-molecule interactions as the contributions to this energy form, so that it is possible to determine, with reasonable accuracy, the intermolecular potential energy of a system composed of fairly simple molecules. The second energy form is similar to the translational kinetic energy in classical mechanics and depends on the mass and velocity of each individual gas molecule. It can be determined by using the equations in mechanics, either classical or quantum.

The third energy form results from a number of contributions. For example, consider a monoatomic gas such as helium. Each helium atom corresponds to a helium molecule. Associated with each individual helium atom are the electronic energy resulted from both orbital angular momentum of the electrons about the nucleus and spin of the electrons about their axes.[6] However, the electronic energy is very small compared with the translational kinetic energy. For gases with more complex molecules, e.g. those composed of two or three atoms, a gas molecule may rotate about its center of gravity, or the consisting atoms may vibrate with respect to one another. These yield respectively the rotational and vibrational energy modes of a gas molecule. In some circumstances, there may exist an interaction between these two modes. The situation becomes more complicated if the molecular structure is more complex, or the gas molecule possesses a three-dimensional structure, for which additional energy modes need to be accounted for by using the microscopic approach.

11.2 Work and Heat

11.2.1 Definition of Work

In classical physics, work is conventionally defined as a force F acting through a displacement $\mathrm{d}x$ which is in the same direction of F, i.e.,

$$W = \int_1^2 F\,\mathrm{d}x, \tag{11.2.1}$$

[6] Atoms also assume nuclear energy, which is considered a constant if nuclear reaction does not present.

where numbers 1 and 2 mark respectively the start and end points of position x. In classical thermodynamics, work is defined as *that is done by a system if the sole effect on the surrounding could be the raising of a weight*. In the definition, a weight was not actually raised or a force was not actually acted through a given distance. The major concern is that the sole effect external to the system could equivalently be the raising of a weight. In contrast to the energy stored in a system, work is a form of *transient energy*, i.e., the energy which is transferred across a system boundary. So, work is a kind of *boundary phenomenon*. For example, consider an electric motor connected to a battery. If both the motor and battery are chosen as the thermodynamic system, the rotation of motor shaft corresponds equivalently to a raising of a weight, which is the sole effect done by the system to its surrounding. Thus, work is identified to cross the boundary of the system. If only the battery is chosen as the system, the sole effect external to the system could also be the raising of a weight, provided that the electrical motor is an ideal one. The flowing of electrical charges across a system boundary is recognized as work.

Within the present book, work done by the surrounding on a system is considered positive and vice versa.[7] The symbol W is used to denote work, with its unit given by Joule (J) in the SI system. The time rate of change of work is called *power*, with Watt (W) the corresponding SI unit, i.e., $1\,\text{W} = 1\,\text{J/s}$. In the British unit system, power is expressed in terms of *horse power* (hp), and $1\,\text{hp} = 746\,\text{W}$. The corresponding horse power in the SI system is the *Pferdestärke* (ps), with $1\,\text{ps} = 0.9863\,\text{hp} \sim 736\,\text{W}$. The *specific work* is denoted conventionally by using the small letter w.

11.2.2 Work by Moving Boundary of a System

Consider a gas contained inside a piston-cylinder device as the system, as shown in Fig. 11.6a. If the piston is assumed to move so slowly rightward, the gas state at each instant of time departs not significantly form equilibrium, so that a quasi-equilibrium process is accomplished when the piston moves from state 1 to sate 2. Let $\mathrm{d}\ell$ denote an infinitesimal displacement of the piston; then the corresponding infinitesimal work δW done by the gas on its surrounding is given by

$$\delta W = -pA\,\mathrm{d}\ell = -p\,\mathrm{d}v, \tag{11.2.2}$$

where p stands for the gas pressure with the minus sign coinciding to the sign convention of work. Integrating this equation from state 1 into state 2 yields the total amount of work accomplished between two states, which is given by

$${}_1W_2 = W_1^2 = \int_1^2 \delta W = -\int_1^2 p\,\mathrm{d}v, \tag{11.2.3}$$

which is exactly the area under the process curve in the p–V diagram, as shown in the figure. If states 1 and 2 are connected by different process curves, e.g. by those

[7] Conventionally, a positive work is defined as that done by a system on its surrounding. The sign convention defined here is adopted to coincide with that used to derive the general balance equations in Sect. 5.2.3.

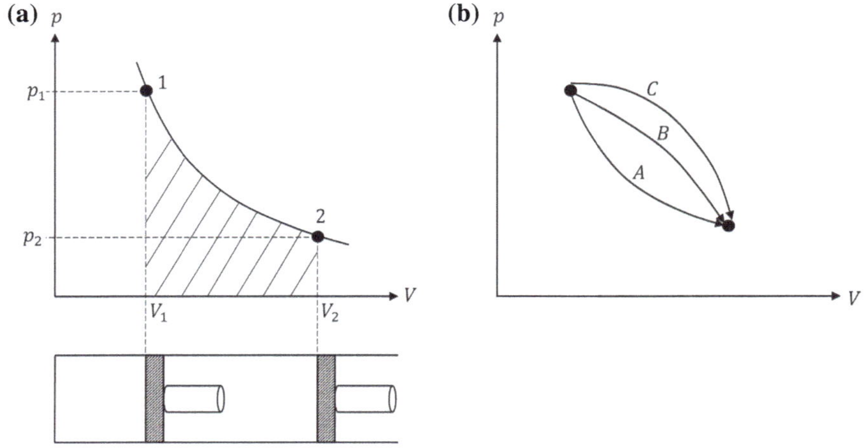

Fig. 11.6 Work conducted by a moving boundary of a simple compressible substance. **a** An expansion process of a piston-cylinder system in the p–V diagram. **b** Three different processes between two states of a thermodynamic system

shown in Fig. 11.6b, the area will be different in different curves A, B, and C. This implies that work is dependent on the process through which the system undergoes, so that work is a path function, not a point function. Mathematically, differentials of point and path functions correspond respectively to exact and inexact differentials, and the symbol "δ" is used to denote an inexact differential of a path function, e.g. δW, in contrast to the symbol "d" used to denote an exact differential of a point function such as dv in Eq. (11.2.3). Since work is a path function, it is meaningless and inappropriate to write $W_2 - W_1$ on the left-hand-side of Eq. (11.2.3). Instead, the expression such as ${}_1W^2$ or W_1^2 is used to denote the difference in a path function between any two states of a system.

Essentially, the relation between p and V needs to be prescribed in order to accomplish the integration in Eq. (11.2.3). This can be obtained either from the experimental outcomes in a graphic form, or from an analytical study. For convenience, consider a *polytropic process*, which is described by

$$pV^n = \text{constant}, \tag{11.2.4}$$

where the value of exponent n may vary from $-\infty$ to ∞. Substituting this expression into Eq. (11.2.3) results in

$$W_1^2 = \left\{ \begin{array}{ll} -p_1 V_1 \ln \dfrac{V_2}{V_1}, & n = 1, \\ -\dfrac{p_2 V_2 - p_1 V_1}{1 - n}, & n \neq 1. \end{array} \right\} \tag{11.2.5}$$

For isobaric process, $W_1^2 = -p\,(V_2 - V_1)$; for isochoric process, $W_1^2 = 0$.

The work done by a moving boundary is in fact the work done on a resisting moving boundary. For a quasi-equilibrium process, the pressures inside the system

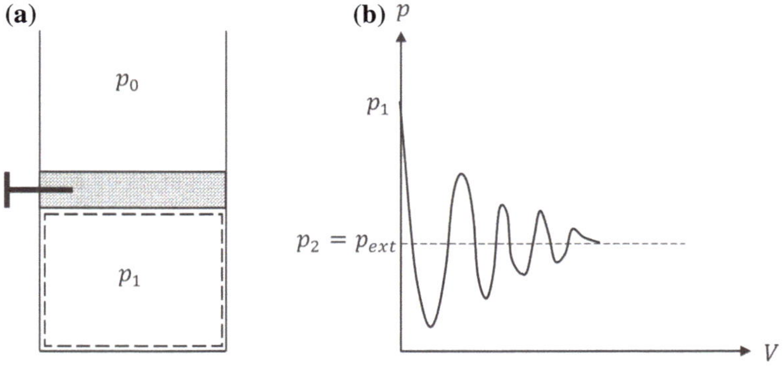

Fig. 11.7 Work in a non-equilibrium process. **a** The configurations of a piston-cylinder device. **b** The gas pressure in a non-quasi-equilibrium expansion

and in the surrounding are exactly the same, and the work can be evaluated by using Eq. (11.2.3). For a non-equilibrium process, the work can no longer be evaluated by using the same equation, for the system state is undetermined at any instant of time. To evaluate the work done on the resisting boundary in such a circumstance, the pressure p may be replaced by the pressure p_{ext} that is experienced by the surrounding on the resisting boundary, with which Eq. (11.2.3) becomes

$$W_1^2 = -\int_1^2 p_{ext} \mathrm{d}v. \tag{11.2.6}$$

For example, consider the piston-cylinder device with a lock pin shown in Fig. 11.7a. The gas contained inside the cylinder is considered a system. After the removal of pin, the system experiences a rapid expansion, with the average gas pressure shown in Fig. 11.7b. The system is at the same time exposed to a boundary pressure p_{ext}, which is given by

$$p_{ext} = p_0 + m_p g/A, \tag{11.2.7}$$

where p_0 is the surrounding atmospheric pressure, and m_p and A represent respectively the mass and cross-sectional area of piston. If the initial gas pressure p_1 is greater than p_{ext}, the piston will move upward, and vice versa. The non-equilibrium expansion ends eventually when the gas pressure approaches an equilibrium state with p_{ext}. Since the work done by the system during this process is done against the force resisting the movement of system boundary, and also since p_{ext} remains constant during the process, the work done by the system is then evaluated as

$$W_1^2 = -\int_1^2 p_{ext} \mathrm{d}v = -p_{ext} (V_2 - V_1). \tag{11.2.8}$$

11.2.3 Other Work Forms

Equation (11.2.3) is valid for simple compressible substances, whose work can only be accomplished by a change in system volume. There exist other types of systems in which work can be defined at a moving boundary. Consider a stretched wire under a given wire tension T as the system. If the length of wire changes by an amount $\mathrm{d}\ell$, the work done on the system is then given by

$$\delta W = T\,\mathrm{d}\ell. \tag{11.2.9}$$

Similar expressions can be found for the works done due to the surface tension σ in a liquid film and the work done due to an electrical current driven by an electrical potential difference ξ, which are given respectively by

$$\delta W = \sigma\,\mathrm{d}a, \qquad \delta W = \xi\,\mathrm{d}z, \tag{11.2.10}$$

where $\mathrm{d}a$ denotes the area change in liquid film, and $\mathrm{d}z$ represents the amount of flowing electrical charges. It is quite possible to have more than one work form in a given process. In such a circumstance, the general expression of work can be generalized as

$$\delta W = -p\,\mathrm{d}V + T\,\mathrm{d}\ell + \sigma\,\mathrm{d}a + \xi\,\mathrm{d}z + \cdots = \sum \alpha \times \mathrm{d}\beta, \tag{11.2.11}$$

where α is any intensive property with β the corresponding extensive property, so that their product contributes a work form. The intensive property α may be recognized as the driving force that causes a change in the related extensive property β. The time rate of change of Eq. (11.2.11) is given by

$$\dot{W} = \frac{\delta W}{\mathrm{d}t} = -p\dot{V} + T\dot{\ell} + \sigma\dot{A} + \xi\dot{z} + \cdots = \sum \alpha \times \dot{\beta}, \tag{11.2.12}$$

provided that all intensive properties do not vary with time. Other work modes can equally be identified in the processes that are not quasi-equilibrium, e.g. the stress work in a viscous fluid.

As an example, consider a metallic wire with initial length ℓ_0 which is stretched by a wire tension T. The normal stress σ and normal strain ϵ are obtained as

$$\sigma = \frac{T}{A} = E\epsilon, \qquad \epsilon = \frac{\mathrm{d}\ell}{\ell_0}, \tag{11.2.13}$$

where A and E are respectively the cross-sectional area and Young's modulus of wire, for the wire is assumed to be perfectly linear elastic, so that Hooke's law can be applied to relate its stress and strain. Substituting Eq. (11.2.13) into Eq. (11.2.9) yields

$$\delta W = T\,\mathrm{d}\ell = AE\ell_0\epsilon\,\mathrm{d}\epsilon, \qquad \longrightarrow \qquad W = \frac{AE\ell_0}{2}\epsilon^2, \tag{11.2.14}$$

which corresponds to the strain energy stored in the wire, resulted from the work done by the surrounding on the wire.

Remarks on Work:

Although work is a kind of energy, it is a transient energy, not an energy stored in a system. Work is a path function and exists only on the boundary of a system in a

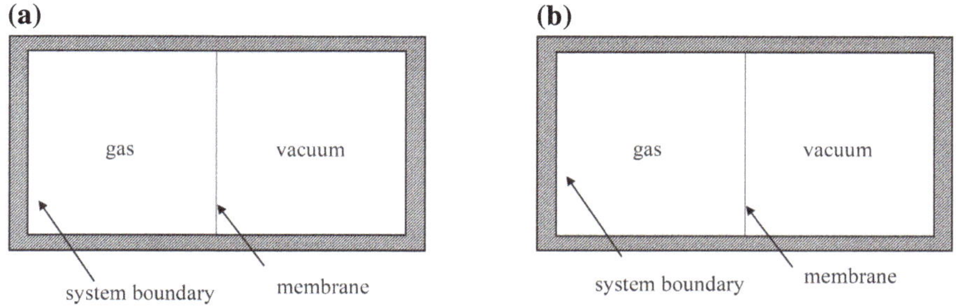

Fig. 11.8 Work as a boundary phenomenon on a system boundary in a free expansion of a gas. **a** Both the gas and vacuum portions are chosen as the system. **b** Only the gas portion is chosen as the system

change in state. It is meaningless to state that a system possesses a certain amount of work. To show this, consider a rigid tank whose left part is filled by a gas, and the right part is completely vacuum. Two parts are separated by a membrane, as shown in Fig. 11.8a, in which the whole space occupied by the vacuum and gas parts is considered a system. If the membrane breaks, a gas flow is triggered inside the system, and eventually the whole tank is filled with the gas. Although in such a circumstance, the gas which was initially in the left part did expand, no work can be identified on the system boundary, provided that any work associated with the rupturing of membrane is neglected. Such an expansion is called a *free expansion* of a gas. Similarly, consider only the gas part as the system. After the rupturing of membrane, the gas does expand. However, the expansion process is a non-equilibrium one, and the work done by the moving boundary needs to be determined by using Eq. (11.2.6). Since there exists no resistance on the system boundary during the free expansion, no work on the boundary can be identified.

Consider a rigid tank which is filled by a gas. A shaft with a paddle on one end and a pulley connected to a weight on the other end is equipped with the tank, as shown in Fig. 11.9a. The chosen system is displayed by the dashed box. Since the system boundary intersects the shaft, work can be associated there with the shearing force in the rotating shaft. On the other hand, if all the tank, shaft, paddle, pulley, and weight are chosen as the system, as shown in Fig. 11.9b, no work crosses the system boundary as the weight moves downward, although a change in potential energy within the system presents.

As a summary, work is a transient energy and a boundary phenomenon. As work crosses the system boundary, it becomes indistinguishable, for it is transformed to an energy stored inside the system. For example, consider a cup of water which is placed on the ground in a rainy day. Without loss of generality, the water surface is the system boundary and the water content inside the cup may be though of as the energy stored inside the system. Rain droplets just crossing the water surface may be thought of as a kind of transient energy. Rain droplets become indistinguishable after they crossed the water surface, for they become part of the water inside the cup.

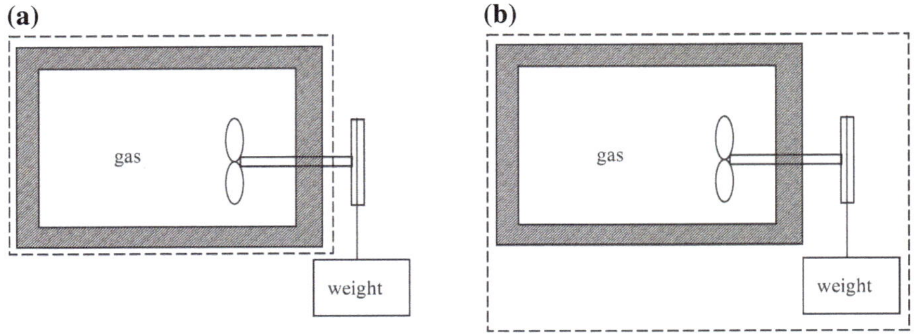

Fig. 11.9 Work on system boundary. **a** Work is identified. **b** No work is identified

11.2.4 Definition of Heat

Consider a Gedankenexperiment as follows: A block of hot copper with temperature T^* is placed in a well-isolated cup of water with temperature $T < T^*$. After a certain period of time, both the copper block and water come to an equilibrium state, and assume a temperature T' with $T < T' < T^*$. Both substances experience a temperature variation and a change in state, which results from an energy transfer from the copper to the water. Thus, in the context of classical thermodynamics, heat is defined as the *form of energy that is transferred across the boundary of a system at a given temperature to another system (or the surrounding) at a lower temperature by virtue of temperature difference.*

As similar to work, heat is a kind of transient energy and a boundary phenomenon, which is denoted conventionally by using the capital letter Q, with Joule as its SI unit, and the amount of heat transfer per unit mass is denoted by using the small letter q. There exist two common units for heat: the *calorie* (cal), which is defined as the amount of heat to raise 1 g water from 14.5 to 15.5 °C, and the *British thermal unit* (Btu), which is defined as the amount of heat to raise 1 pound of water from 59.5°F to 60.5°F. One calorie equals 4.186 J and one Btu equals 1055.06 J. The time rate of change of Q, denoted by $\dot{Q}$, assumes the SI unit in Watt. An infinitesimal amount of heat is expressed by δQ, and the expression ${}_1Q^2$ or Q_1^2 is applied to denote the amount of heat that is transferred during a process from state 1 to state 2 to illuminate its characteristics as a path function. Equally, it is meaningless to state that a body contains a certain amount of heat, for once heat crossed a system boundary, it becomes indistinguishable. Heat transferred from the surrounding to the system is defined to be positive, and vice versa. A process in which Q vanishes is called *adiabatic*.

Although heat and work are both boundary phenomena, a transit energy on a system boundary may be identified as heat or work, depending on the choice of system boundary. For example, consider an electrical heater connected to a battery, as shown in Fig. 11.10. If only the gas contained inside the electrical heater is considered the system, the energy released from the heater to the gas is identified as heat, as shown

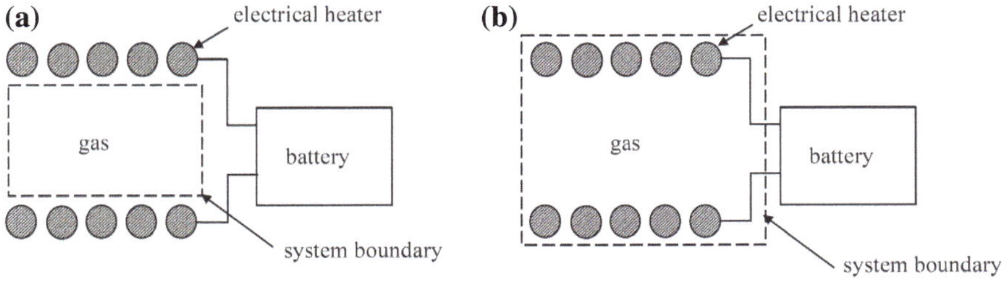

Fig. 11.10 Boundary phenomenon as heat or work. **a** Heat identified on the boundary. **b** Work identified on the boundary

in Fig. 11.10a. On the other hand, if the heater and gas are considered the system, the electrical current crossing the system boundary is identified as work, as shown in Fig. 11.10b. A simple method to identify whether a transient energy on a system boundary is heat or work is to place an isolation layer on the system boundary. If the boundary phenomenon remains effectively unchanged, it is work; otherwise it is heat.

In the context of classical thermodynamics, all processes are assumed to be quasi-equilibrium, so that it takes infinite time to accomplish a heat transfer process. In reality, a heat transfer process takes place at finite time duration, mainly via the mechanisms of *conduction*, *convection*, and *radiation*, or even through the phase change of a working substance, such as the *latent heat* of water vapor.

11.3 Zeroth Law and Temperature

11.3.1 The Zeroth Law

The zeroth law of thermodynamics concerns with the thermal properties of thermodynamic systems in thermal equilibrium, which gives rise to the concept of temperature.

11.1 (The zeroth law) *If two systems are separately in thermal equilibrium with a third system, they must also be in thermal equilibrium with each other.*

Two systems in thermal equilibrium with each other indicate that no change in any observable thermal properties of both systems presents. This suggests that there should exist some property of both systems which assumes exactly the same value, implying the existence of the *empirical temperature* of the working substance contained inside a system.

11.3.2 Empirical Temperature

Construct three frictionless piston-cylinder devices as three thermodynamics systems, and each piston-cylinder device contains a gaseous working substance, which is a simple compressible one. Three systems are marked by the numbers 1, 2, and 3, and their states are identified by prescribing the values of pressure p and volume V. For convenience, system 3 is taken as the reference system, whose pressure p_3 and volume V_3 are kept constant. Now, system 1 is required to be in thermal equilibrium with system 3. This imposes a constraint on the state of system 1, so that p_1 and V_1 are no longer independent of each other. If a particular value of p_1 is chosen, then V_1 will be uniquely determined. Mathematically, there must be a fixed relationship between $\{p_1, V_1, p_3, V_3\}$, which can be expressed as

$$\mathcal{F}_1\left(p_1, V_1, p_3, V_3\right) = 0. \tag{11.3.1}$$

Similarly, requiring system 2 to be in thermal equilibrium with system 3 yields

$$\mathcal{F}_2\left(p_2, V_2, p_3, V_3\right) = 0. \tag{11.3.2}$$

Solving two equations for p_3 gives

$$\begin{aligned} &p_3 = f_1\left(p_1, V_1, V_3\right), \qquad p_3 = f_2\left(p_2, V_2, V_3\right), \\ &\longrightarrow f_1\left(p_1, V_1, V_3\right) = f_2\left(p_2, V_2, V_3\right), \end{aligned} \tag{11.3.3}$$

which implies that

$$p_1 = p_1\left(p_2, V_1, V_2, V_3\right). \tag{11.3.4}$$

If systems 1 and 2 are in thermal equilibrium with each other, it follows from the zeroth law that

$$\mathcal{F}_3\left(p_1, V_1, p_2, V_2\right) = 0, \tag{11.3.5}$$

which is solved to obtain p_1 given by

$$p_1 = p_1\left(p_2, V_1, V_2\right), \tag{11.3.6}$$

which shows that p_1 is determined by three variables p_2, V_1, and V_2. Comparing this equation with Eq. (11.3.4) shows that V_3 is irrelevant, so that Eq. $(11.3.3)_3$ reduces to

$$\Phi_1\left(p_1, V_1\right) = \Phi_2\left(p_2, V_2\right). \tag{11.3.7}$$

Equation (11.3.7) shows the condition that needs to be satisfied if systems 1 and 2 are in thermal equilibrium with each other. It follows that when two (or more) systems are in thermal equilibrium, there exists for each system a function of its state, which assumes a common value for all systems. Thus, for any system in thermal equilibrium with a given reference state, it follows that

$$\Phi\left(p, V\right) = \Theta = \text{constant}, \tag{11.3.8}$$

where Θ assumes the same value for all such systems. This equation is called the *equation of state* of a system, and Θ is termed an *empirical temperature*, or simply *temperature*.[8]

With the empirical temperature, the zeroth law can be restated as: "*when two systems have equality of (empirical) temperature with a third system, they in turn have equality of (empirical) temperature with each other.*" Obviously, the empirical temperature of a substance depends on its properties, and this gives the foundation of *thermometer*.

---

[8]It will be shown in Sect. 11.6.2 that an empirical temperature of a substance can be derived by using the approach of continuum thermodynamics.

11.3.3 Temperature Scales

The empirical temperature is a qualitative description. It is desirable to define the empirical temperature in such a quantitative way that its values form an ordered sequence corresponding to the ideas of hotness of substance, which results in a *scale* of temperature. To establish a particular empirical temperature scale, some system with appropriate thermometric properties must be selected, to which a convenient method of assigning numerical values for the temperatures should be adopted. Let x denote the thermometric property of a substance that is used to express quantitatively the value of temperature. The simplest possible procedure to define a temperature scale is to assume a linear relation between Θ and x given by

$$\Theta(x) = ax, \tag{11.3.9}$$

where a is the proportional constant. Its value is fixed either by choosing the value of temperature at one reference point or by choosing the size of unit so that a given number of units shall be between two fixed points. Either procedure will lead to a unique scale for any one thermometer. However, measurements of a temperature by using different thermometers will be different in general, for the chosen thermometric properties in different thermometers may vary with temperature in quite different manners. The widely used temperature scales are introduced in the following.

The Celsius scale. Water is chosen as the working substance, and its triple point is set to equal 0.01 °C, with °C representing the degree of Celsius.[9] The ice and steam points of water assume respectively the numerical values of 0 and 100, as motivated by experiments.[10] The Celsius scale is in fact a *centigrade scale*, which is defined as one which has one hundred units between the ice and steam points of water. Centigrade scales may be based on any suitable thermometric property of any convenient system. Essentially, centigrade scales will not agree with one another, except at 0 and 100 °C, where they must coincide by definition.

The Fahrenheit scale. It is similar to the Celsius scale, except that the numerical values of 32°F and 212°F are assigned respectively to the ice and steam points of water, with °F representing the degree of Fahrenheit.[11] The Fahrenheit scale is not a centigrade scale, although it still depends on the thermometric properties of substance. The conversion between the Celsius and Fahrenheit scales is given by °F = (9/5) °C+32.

The ideal gas scale. To improve the accuracy of temperature measurement, it is desirable to construct a temperature scale which does not depend on the thermometric property of a substance. In the search of such a scale, it was found that the disagreement was small among measurements based on the behavior of gases. Let

---

[9] Anders Celsius, 1701–1744, a Swedish physicist and mathematician, who proposed the Celsius temperature scale which bears his name.

[10] Ice point is the temperature at which ice melts under one atmospheric pressure. Steam point is the temperature at which water boils under one atmospheric pressure.

[11] Daniel Gabriel Fahrenheit, 1686–1736, a German physicist, who proposed the Fahrenheit scale, which is the first standardized temperature scale widely used.

the state of a gas be described by the pressure p and volume V. The simplest way of constructing a temperature scale is to keep one of two variables fixed and to take temperature as proportional to the other. The constant a in Eq. (11.3.9) is so chosen to give 100 units between the ice and steam points of water. For a *constant pressure gas thermometer* and a *constant volume gas thermometer*, it follows respectively that

$$\Theta = \frac{100}{V_s - V_i} V, \qquad \Theta = \frac{100}{p_s - p_i} p, \tag{11.3.10}$$

where the subscripts "s" and "i" represent respectively the steam and ice points. It was found that all gases give the same value of a temperature in the low-density limit $\rho \to 0$ (hence the low pressure limit $p \to 0$), if the temperature is evaluated by using Eq. (11.3.10). This temperature scale is referred to as the *ideal gas scale.*

The thermodynamic temperature scale. Since all gases give the same temperature in the low pressure limit, it follows from the ideal gas state equation that

$$T = \lim_{p \to 0} \frac{pv}{R}, \tag{11.3.11}$$

where R is the gas constant. Since T represents the thermodynamic temperature scale or *absolute temperature scale*, this equation implies that the determination of thermodynamic temperature scale is almost always ultimately based on gas thermometry.[12] The value of R is so chosen that a numerical value of 273.16 is assigned to the thermodynamic temperature scale of water at the triple point. This is done so, for in practice it is more convenient to use only one reference point for temperature, instead of two reference points, as what have been done in the Celsius and Fahrenheit scales. The unit defined for the thermodynamic temperature scale is called Kelvin which is expressed by the capital letter K. In other words, one Kelvin is the fraction of 273.16 of the thermodynamic temperature scale of the triple point of water. The thermodynamic temperature determined by gas thermometry is thus given by

$$\mathrm{K} = 273.16 \frac{\lim_{p \to 0}(pv)_T}{\lim_{p \to 0}(pv)_{triple}}, \tag{11.3.12}$$

so that the relation between the Celsius and thermodynamic temperature scales is obtained as

$$\mathrm{K} = {}^\circ\mathrm{C} + 273.15. \tag{11.3.13}$$

The thermodynamic temperature assuming 0 K is called *the absolute zero.*

The corresponding thermodynamic temperature in the British unit system is called the *Rankine scale*, with the capital letter R as its denotation. The relation between the Rankine and Fahrenheit scales is given by

$$\mathrm{R} = {}^\circ\mathrm{F} + 459.67. \tag{11.3.14}$$

[12] A more rigorous definition of the thermodynamic temperature can be defined by using the Carnot cycle, as will be discussed in Sect. 11.5.2.

11.4 First Law and Internal Energy

The first law of thermodynamics is simply the balance of energy, which has been discussed by using a control-volume analysis in Sect. 5.3.4 in both differential and integral forms. In this section, the first law will be formulated by using a control-mass analysis to illuminate its physical implications and in particular to show the macroscopic existence of internal energy as a natural consequence. The illustrations of the first law in both the control-mass and control-volume formulations will be given at the end.

11.4.1 Joule's Experiment

Consider a gas inside a rigid tank as the control-mass system. Let the system undergo a cycle that is made of two processes. In the first process, work is done on the system by a paddle that turns as the weight is lowered, as shown in Fig. 11.11a. The system is restored to its initial state by the second process, in which an amount of heat is transferred to the surrounding, until the cycle has been completed, as shown in Fig. 11.11b. The measurements of work and heat are accomplished during the cycle for a wide variety of systems and for various amounts of work and heat. Comparing the measured amounts of work and heat shows that they are always proportional to each other, which can be expressed as

$$J\left|\oint \delta Q\right| = \left|\oint \delta W\right|, \tag{11.4.1}$$

where J is the proportional coefficient depending on the units used for heat and work, $\oint \delta Q$ represents the net amount of heat transfer during the cycle, and $\oint \delta W$ denotes the amount of net work during the same cycle. If both heat and work are expressed in Joule, J assumes a unity value. If heat and work are evaluated respectively in terms of calorie and Joule, then J assumes the value of 4.186. Thus, in the SI system,

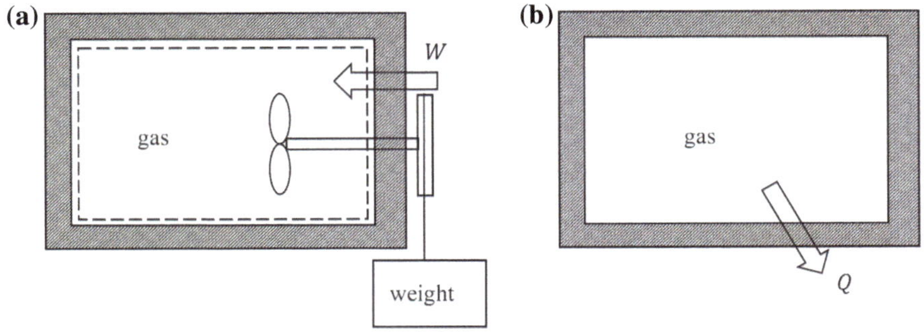

Fig. 11.11 Illustration of Joule's experiment. **a** Work done on the control-mass system. **b** Heat transferred from the control-mass system

Eq. (11.4.1) reduces to

$$\left|\oint \delta Q\right| = \left|\oint \delta W\right|, \quad \longrightarrow \quad \oint \delta Q + \oint \delta W = 0, \tag{11.4.2}$$

showing a direct quantitative equivalence between work and heat, and may be considered the basic statement of first law. The second equation is a generalization of the first equation, if the sign conventions of heat and work defined previously are used.

The experiment described in Fig. 11.11 is known as *Joule's experiment*. In fact, Joule produced heating in various thermally isolated systems by performing work on them, which were accomplished by many means, e.g. viscous dissipation in liquids, friction between solids, and electrical heating. He found that if the only effect of work was to produce heating, then in all cases the amount of work and the corresponding amount of heat were in a fixed proportion to each other, implying a direct equivalence between heat and work as forms of energy.

11.4.2 Control-Mass Formulation for a Process

Consider a control-mass system undergoing a process from state 1 to state 2, which is denoted by process A. The system is brought back to its initial state by other two different processes B and C, so that two cycles can be constructed: the cycles consisting of $A + B$ and $A + C$. Applying Eq. (11.4.2) to two cycles yields

$$\begin{aligned} \int_1^2 \delta Q_A + \int_1^2 \delta W_A + \int_2^1 \delta Q_B + \int_2^1 \delta W_B = 0, \\ \int_1^2 \delta Q_A + \int_1^2 \delta W_A + \int_2^1 \delta Q_C + \int_2^1 \delta W_C = 0, \end{aligned} \tag{11.4.3}$$

where the superscript indicates to which process the indexed quantity is evaluated. Combining two equations gives

$$\int_1^2 (\delta Q + \delta W)_B = \int_1^2 (\delta Q + \delta W)_C , \tag{11.4.4}$$

showing that although work and heat are path functions, the quantity $\delta Q + \delta W$ is a differential of a point function of the system and, therefore, is the differential of a property of the system. This property is defined as the *total energy* or simply *energy* of the control-mass system, denoted conventionally by the symbol E.

The differential in total energy, $\mathrm{d}E$, is then given by

$$\mathrm{d}E = \delta Q + \delta W, \tag{11.4.5}$$

with which Eq. (11.4.4) is expressed alternatively as

$$\Delta E = E_2 - E_1 = Q_1^2 + W_1^2, \tag{11.4.6}$$

which is the first law for a control-mass system undergoing a process from state 1 to state 2. The unit mass formulations of Eqs. (11.4.5) and (11.4.6) are expressed respectively as

$$\mathrm{d}e = \delta q + \delta w, \qquad \Delta e = e_2 - e_1 = q_1^2 + w_1^2. \tag{11.4.7}$$

11.4.3 Internal Energy and Enthalpy

In the context of classical thermodynamics, the energy E of a system consists of three parts: the bulk kinetic and potential energies of system, and the internal energy U representing all other energy contributions which cannot be classified as the first two. The macroscopic existence of internal energy follows directly from the first law, with its microscopic interpretation already discussed in Sect. 11.1.7. Specifically, E is expressed as

$$E = U + KE + PE, \tag{11.4.8}$$

where U is associated with the thermodynamic state of system, and KE and PE are given by

$$KE = \frac{1}{2} m \left(\boldsymbol{u} \cdot \boldsymbol{u} \right), \qquad PE = mgz, \tag{11.4.9}$$

as motivated by the theory of classical mechanics, where m denotes the mass of system, and $\boldsymbol{u}$ and z are respectively the bulk velocity and elevation of the center of mass of the system. With these, Eqs. (11.4.5) and (11.4.6) are recast alternatively in the forms

$$\begin{aligned} &\mathrm{d}U + \mathrm{d}\left(KE\right) + \mathrm{d}\left(PE\right) = \delta Q + \delta W, \\ &\left(U_2 - U_1\right) + \frac{1}{2} m \left(|\boldsymbol{u}_2|^2 - |\boldsymbol{u}_1|^2 \right) + mg\left(z_2 - z_1\right) = Q_1^2 + W_1^2, \end{aligned} \tag{11.4.10}$$

which are the first laws for a control-mass system undergoing a process in differential and integral forms. For a stationary control-mass system on the reference datum of elevation, two equations reduce to

$$\mathrm{d}U = \delta Q + \delta W, \qquad U_2 - U_1 = Q_1^2 + W_1^2, \tag{11.4.11}$$

or alternatively to

$$\mathrm{d}u = \delta q + \delta w, \qquad u_2 - u_2 = q_1^2 + w_1^2, \tag{11.4.12}$$

per unit mass. It is noted that either Eq. (11.4.11) or (11.4.12) only delivers a change in internal energy, not its absolute value. The internal energy of a system at a specific state should be determined with respect to a reference state, at which the internal energy is known or prescribed. For water, the triple point is chosen as the reference state, at which its internal energy vanishes. For ideal gas, the reference state is chosen at the absolute zero, at which the internal energy is null.[13]

For a globally stationary control-mass system undergoing an isobaric process, the first law reads

$$Q_1^2 = U_2 - U_1 - W_1^2 = \left(U_2 - U_1\right) + p\left(V_2 - V_1\right) = \left(U_2 + p_2 V_2\right) - \left(U_1 + p_1 V_1\right), \tag{11.4.13}$$

which shows that the amount of heat transfer is determined in terms of the change in the quantity $(U + pV)$ between the initial and final states of process. Since U, p,

[13]These two reference states are also used for vanishing values of entropy.

and V are all state functions, their combinations are also state functions. Hence, the *enthalpy* H is defined as

$$H \equiv U + pV, \qquad h = u + pv = u + \frac{p}{\rho}, \tag{11.4.14}$$

where h represents the value of H per unit mass, called the *specific enthalpy*. Substituting these expressions into Eq. (11.4.13) yields

$$Q_1^2 = \Delta H = H_2 - H_1, \qquad q_1^2 = \Delta h = h_2 - h_1, \tag{11.4.15}$$

showing that the amount of heat transfer equals the difference in enthalpy between the initial and final states of process. Again, instead of the absolute value, only a difference in enthalpy is defined by the first law. Although enthalpy was derived by using the specific isobaric process, it is in fact a thermodynamic property, so that its applications are not related to, or dependent on, any process that may take place.

Remarks on the Control-Mass Formulation:

Taking time rate of change of Eq. (11.4.5) results in

$$\dot{E} = \dot{Q}_1^2 + \dot{W}_1^2, \tag{11.4.16}$$

which is the first law as a rate equation. It coincides exactly to the general balance statement in integral form given in Eq. (5.2.2). That is, the time rate of change of the energy of a system equals all the powers supplied to the system via the contributions of production, flux, and supply. It follows from classical physics that energy can neither be created nor destroyed, so that the production $\mathcal{P}$ vanishes. The flux $\mathcal{F}$ consists of two terms: the heat transfer rate and power done by the normal stress on the system boundary as work. The supply $\mathcal{S}$ contains only the contribution of gravitational acceleration, which is considered the bulk potential energy of system.

The previous discussions were based on the assumption that the mass of a system remains fixed and identifiable. In fact, it follows from Einstein's relativistic theory that the mass and energy of a substance can be related by the equation given by

$$E = mc^2, \tag{11.4.17}$$

where c is the speed of light. Thus, the mass of a control-mass system does change when its energy changes. However, the change in mass is extremely small, so that it can be neglected in all engineering applications. For example, consider a rigid vessel which is filled by a 1-kg stoichiometric mixture of a hydrocarbon fuel and air. After the combustion has completed, the temperature has been increased, and an amount of heat transfer of 2900 kJ must be removed to restore the system to the initial state. It follows from the first law and Eq. (11.4.17) that a small change in the system mass, i.e., $\Delta m = 3.23 \times 10^{-11}$ kg takes place. This extremely small change in mass cannot be detected by even most accurate chemical balance. A fractional change in mass of this magnitude is far beyond the accuracy required in essentially all engineering applications. Hence, application of the first law in the control-mass formulation will not introduce any significant error into most thermodynamic circumstances, and the concept of control-mass system can be used even though the energy changes.

11.4.4 Specific Heats

For a control-mass system containing a simple compressible substance undergoing a quasi-equilibrium process, the first law reads

$$\delta Q = \mathrm{d}U - \delta W = \mathrm{d}U + p\mathrm{d}V. \tag{11.4.18}$$

The *specific heat* is defined as the amount of heat that is required to raise the temperature of a substance by one degree in unit mass base.[14] Practically, there exist two specific processes to accomplish a temperature rise, as discussed in the following.

1. An isochoric process, in which the volume of system is kept fixed, so that Eq. (11.4.18) reduces to

$$\delta Q = \mathrm{d}U, \longrightarrow c_V \equiv \frac{1}{m}\left(\frac{\delta Q}{\delta T}\right)_V = \frac{1}{m}\left(\frac{\partial U}{\partial T}\right)_V = \left(\frac{\partial u}{\partial T}\right)_V, \tag{11.4.19}$$

   where c_V is the *specific heat at constant volume.*
2. An isobaric process, in which the pressure of system assumes a constant value, and the amount of heat transfer is obtained as

$$\delta Q = \mathrm{d}H, \longrightarrow c_p \equiv \frac{1}{m}\left(\frac{\delta Q}{\delta T}\right)_p = \frac{1}{m}\left(\frac{\partial H}{\partial T}\right)_p = \left(\frac{\partial h}{\partial T}\right)_p, \tag{11.4.20}$$

where c_p is the *specific heat at constant pressure.*

Since all variables in Eqs. $(11.4.19)_2$ and $(11.4.20)_2$ are thermodynamic properties, both specific heats are also thermodynamic properties, although the derivations were conducted by considering two specific processes. For solids and liquids which are nearly incompressible, it follows that

$$\mathrm{d}h = \mathrm{d}u + \mathrm{d}\,(pv) \sim \mathrm{d}u + v\mathrm{d}p, \tag{11.4.21}$$

which can be approximated by

$$\mathrm{d}h \sim \mathrm{d}u \sim c\,\mathrm{d}T, \tag{11.4.22}$$

because for solid and liquids, the specific volume is very small. In the above equation, c is either the constant volume or constant pressure specific heat. In many processes involving solids or liquids, it is not necessary to make a distinction between c_p and c_V. Integrating Eq. (11.4.22) yields

$$(h_2 - h_1) \sim (u_2 - u_1) \sim \int_1^2 c\,\mathrm{d}T. \tag{11.4.23}$$

The specific heat of a substance depends essentially on temperature, and the integration in Eq. (11.4.23) requires a known relation between c and T, which results mainly from experimental calibration. In practice, unless the process occurs at low temperature or over a wide range of temperatures, c can be approximated as a constant, which gives a simple result that $\Delta h \sim \Delta u \sim c(T_2 - T_1)$.

[14]The *heat capacity* is used to denote the same amount of heat without unit mass base.

In general, for simple substances the specific internal energy u depends on two independent thermodynamic properties. For a low-density gas, u depends primarily on T and much less on the second property such as p or v. In the limiting case as an ideal gas, u is only a function of temperature.[15] For an ideal gas, the state equation, specific internal energy, and specific enthalpy read the forms

$$pv = RT, \qquad u = u(T), \qquad h = h(T), \tag{11.4.24}$$

where the last equation is a derived result, for $h = u + pv = u + RT = h(T)$. Substituting these expressions into Eqs. (11.4.19)$_2$ and (11.4.20)$_2$ results respectively in

$$c_{V0} = \frac{\mathrm{d}u}{\mathrm{d}T}, \quad \mathrm{d}u = c_{V0}\mathrm{d}T; \qquad c_{p0} = \frac{\mathrm{d}h}{\mathrm{d}T}, \quad \mathrm{d}h = c_{p0}\mathrm{d}T, \tag{11.4.25}$$

where c_{V0} and c_{p0} represent the specific heats at constant volume and constant pressure for ideal gases. Combining Eq. (11.4.24) with Eq. (11.4.25) shows that

$$c_{p0} - c_{V0} = R. \tag{11.4.26}$$

Although both c_{p0} and c_{V0} are functions of temperature for an ideal gas, their difference is always a constant, i.e., the gas constant. For air approximated as an ideal gas, $c_{p0} = 1.0046$ kJ/kg-K, $c_{V0} = 0.7176$ kJ/kg-K, and $R = 0.287$ kJ/kg-K in the SI system may be used for calculations.

11.4.5 Control-Volume Formulation for a Steady Process

The global first law for a control-volume has been given in either Eq. (5.3.28) or (5.3.30). If it is further assumed that there exist no external energy source and no power done by the shear stresses, Eq. (5.3.30) is then simplified to

$$\dot{E}_{CV} = \dot{Q}_{CV} + \dot{W}_{CV} + \sum \dot{m}h_{tot,i} - \sum \dot{m}h_{tot,o}, \tag{11.4.27}$$

with the correspondences given by

$$\begin{aligned} \dot{E}_{CV} &= \frac{\mathrm{d}}{\mathrm{d}t}\int_V \rho e \,\mathrm{d}v = \frac{\partial}{\partial t}\int_V \rho e \,\mathrm{d}v, \\ \sum \dot{m}h_{tot,o} - \sum \dot{m}h_{tot,i} &= \int_A \left(h + \frac{1}{2}\boldsymbol{u}\cdot\boldsymbol{u} + gz\right)(\rho\boldsymbol{u}\cdot\boldsymbol{n})\,\mathrm{d}a, \end{aligned} \tag{11.4.28}$$

where $h_{tot} = h + (\boldsymbol{u}\cdot\boldsymbol{u})/2 + gz$, known as the *total specific enthalpy*, and the subscripts "i" and "o" represent respectively the intake and discharge flows. The term $\dot{W}_{CV}$ contains mostly the work via shaft into the control-volume. Equation (11.4.27) is the conventional formulation of first law for a control-volume in classical thermodynamics. Similarly, the global balance of mass given in Eq. (5.3.2) is expressed alternatively as

$$\dot{m}_{CV} = \sum \dot{m}_i - \sum \dot{m}_o, \tag{11.4.29}$$

[15]This has been confirmed by Joule's experiments. In Sect. 11.8.5, this result will be derived by using the thermodynamic relations for simple compressible substance.

with

$$\dot{m}_{CV} = \frac{\mathrm{d}}{\mathrm{d}t}\int_V \rho\,\mathrm{d}v = \frac{\partial}{\partial t}\int_V \rho\,\mathrm{d}v, \qquad \sum \dot{m}_o - \sum \dot{m}_i = \int_A (\rho\boldsymbol{u}\cdot\boldsymbol{n})\,\mathrm{d}a. \tag{11.4.30}$$

For a steady-state process with uniform-flow assumption on the control-surfaces, Eqs. (11.4.29) and (11.4.27) reduce respectively to

$$\sum \dot{m}_i = \sum \dot{m}_o, \qquad \dot{Q}_{CV} + \dot{W}_{CV} = \sum \dot{m}h_{tot,o} - \sum \dot{m}h_{tot,i}, \tag{11.4.31}$$

which are further simplified to

$$\dot{m}_i = \dot{m}_o = \dot{m}, \quad q_{CV} + w_{CV} = h_{tot,o} - h_{tot,i}, \quad q_{CV} = \frac{\dot{Q}_{CV}}{\dot{m}}, \quad w_{CV} = \frac{\dot{W}_{CV}}{\dot{m}}, \tag{11.4.32}$$

provided that there exist only a single intake and a single discharge flows, where q_{CV} and w_{CV} represent respectively the amounts of heat and work transfer per unit mass flow rate.

11.4.6 Control-Volume Formulation for a Transient Process

In classical thermodynamics, a transient process cannot be treated directly, for it is not a quasi-equilibrium process, and at every instant of time, the system deviates significantly from an equilibrium state. However, an approximation to a transient process is possible, which is described in the following.

Consider a control-volume undergoing a transient process which begins at state 1 and ends at state 2 in a time duration t. The balance of mass given in Eq. (11.4.29) is then integrated over time t to obtain

$$(m_2 - m_1)_{CV} = \sum m_i - \sum m_o, \tag{11.4.33}$$

with

$$\int_0^t \dot{m}_{CV}\mathrm{d}t = (m_2 - m_1)_{CV}, \quad \int_0^t \sum \dot{m}_i\,\mathrm{d}t = \sum m_i, \quad \int_0^t \sum \dot{m}_o\,\mathrm{d}t = \sum m_o. \tag{11.4.34}$$

The term $(m_2 - m_1)_{CV}$ represents the difference in mass of the control-volume between the final and initial states, while the terms $\sum m_i$ and $\sum m_o$ denote respectively the total masses that have flowed into and leaved the control-volume during the time duration t. This equation is considered the balance of mass for a control-volume undergoing a transient process.

Similarly, integrating Eq. (11.4.27) gives rise to

$$\begin{aligned}\left[m_2\left(u_2 + \frac{1}{2}\boldsymbol{u}_2\cdot\boldsymbol{u}_2 + gz_2\right) - m_1\left(u_1 + \frac{1}{2}\boldsymbol{u}_1\cdot\boldsymbol{u}_1 + gz_1\right)\right]_{CV} &= Q_{CV} + W_{CV} \\ + \sum m_i\left(h_i + \frac{1}{2}\boldsymbol{u}_i\cdot\boldsymbol{u}_i + gz_i\right) - \sum m_o\left(h_o + \frac{1}{2}\boldsymbol{u}_o\cdot\boldsymbol{u}_o + gz_o\right),&\end{aligned} \tag{11.4.35}$$

with

$$\begin{aligned}
&\int_0^t \dot{E}_{CV}\mathrm{d}t = \left[m_2\left(u_2 + \frac{1}{2}\boldsymbol{u}_2\cdot\boldsymbol{u}_2 + gz_2\right) - m_1\left(u_1 + \frac{1}{2}\boldsymbol{u}_1\cdot\boldsymbol{u}_1 + gz_1\right)\right]_{CV},\\
&\int_0^t \dot{Q}_{CV}\,\mathrm{d}t = Q_{CV}, \qquad \int_0^t \dot{W}_{CV}\,\mathrm{d}t = W_{CV},\\
&\int_0^t \sum \dot{m}_i\left(h_i + \frac{1}{2}\boldsymbol{u}_i\cdot\boldsymbol{u}_i + gz_i\right)\mathrm{d}t = \sum m_i\left(h_i + \frac{1}{2}\boldsymbol{u}_i\cdot\boldsymbol{u}_i + gz_i\right),\\
&\int_0^t \sum \dot{m}_o\left(h_o + \frac{1}{2}\boldsymbol{u}_o\cdot\boldsymbol{u}_o + gz_o\right)\mathrm{d}t = \sum m_o\left(h_o + \frac{1}{2}\boldsymbol{u}_o\cdot\boldsymbol{u}_o + gz_o\right).
\end{aligned} \tag{11.4.36}$$

Equation (11.4.36)$_1$ represents the difference in total energy of control-volume between the final and initial states; Eq. (11.4.36)$_2$ denotes the total amounts of heat and work transfer, while Eqs. (11.4.36)$_3$ and (11.4.36)$_4$ express the total energies that are transferred by the total intake mass $\sum m_i$ and total discharged mass $\sum m_o$.

It is seen that in the established approximation to a transient process, instead of taking into account the detailed time variation of every property of the control-volume directly, only the difference in every property between the beginning and end states of process is accounted for. Such an approximation is not an exact transient analysis and should be referred to as a *lump analysis*. The concept can equally be formulated for a control-mass undergoing a transient process.

Based on the previous discussions, the first law can be summarized as follows:

11.2 (The first law) *The time rate of change of the total energy of a thermodynamic system, either in control-mass or control-volume formulation, equals the sum of all time rates of energies delivered from the surrounding to the system in heat and work, and other possible energy forms.*

11.4.7 Illustrations of First Law

Consider a cylinder-piston device filled with nitrogen as the control-mass system, which assumes pressure p_1, temperature T_1, and volume V_1 at an initial state. The piston is then driven to compress the nitrogen until a state with p_2 and T_2 is reached. The compression process is assumed to be a quasi-equilibrium one, and the amount of work done on the system is denoted by W_1^2. It is required to determine the amount of heat transfer Q_1^2 during the process. It follows from Eq. (11.4.6) that

$$\Delta E \sim \Delta U = Q_1^2 + W_1^2, \qquad \longrightarrow \qquad Q_1^2 = m\,(u_2 - u_1) - W_1^2, \tag{11.4.37}$$

with the assumption that the system is globally stationary. The nitrogen is further assumed to be an ideal gas, so that its mass m can be obtained by using the ideal gas state equation, which is given by

$$m = \frac{pV}{RT} = \frac{p_1V_1}{RT_1}, \qquad R = \frac{\bar{R}}{M}, \tag{11.4.38}$$

where $\bar{R}$ is the universal gas constant, and M denotes the molecular weight of nitrogen having the value of 28. Substituting this equation into Eq. (11.4.37)$_2$ gives rise to

$$Q_1^2 = \frac{p_1 V_1 M}{\bar{R} T_1} c_{V0} \left(T_2 - T_1\right) - W_1^2. \tag{11.4.39}$$

The value of Q_1^2 should be negative, when the numerical values of all quantities are substituted, for the compression increases the temperature of nitrogen which is higher than that of surrounding, so that heat will be transferred from the system to the surrounding. Such a heat transfer, by using the sign convention defined previously, is negative.

Consider air with pressure p_1 and temperature T_1 entering a well-insulated nozzle with velocity V_1, as shown in Fig. 11.12a. The air leaves the nozzle at pressure p_2 and velocity V_2. It is required to determine the temperature of air when it leaves the nozzle, i.e., temperature T_2. Since a flow process is considered, construct the control-volume as shown by the dashed line in the figure, through which the air flows. Further, it is assumed that the air flow is steady, so that Eq. (11.4.32)$_2$ reduces to

$$h_{tot,1} = h_{tot,2}, \tag{11.4.40}$$

because $\dot{E}_{CV} = 0$ for a steady process, $\dot{Q}_{CV} = 0$ for a well-insulated nozzle, $\dot{W}_{CV} = 0$ for there exists no shaft, and $\dot{m}_1 = \dot{m}_2$ as implied by the balance of mass. Expanding this equation yields

$$h_1 + \frac{1}{2} V_1^2 + g z_1 = h_2 + \frac{1}{2} V_2^2 + g z_2. \tag{11.4.41}$$

With the assumptions that $z_1 \sim z_2$ and air is an ideal gas, this equation reduces to

$$c_{p0} \left(T_2 - T_1\right) = \frac{1}{2} \left(V_1^2 - V_2^2\right), \longrightarrow T_2 = T_1 + \frac{1}{2 c_{p0}} \left(V_1^2 - V_2^2\right). \tag{11.4.42}$$

Substituting the numerical values of T_1, V_1, and V_2 into this equation gives the value of T_2.

Consider a filling process shown in Fig. 11.12b, in which a rigid vessel is connected to a pipe via a valve, through which air flows into the tank at constant pressure p_i and constant temperature T_i from the pipe. Initially the tank is completely evacuated, and the valve is then opened to allow air flowing into the tank, until the pressure in

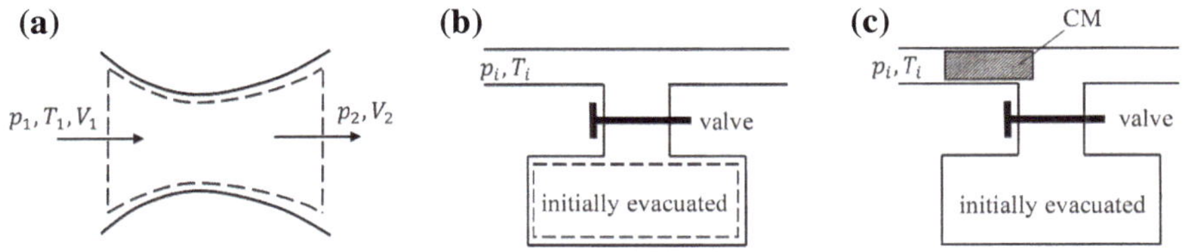

Fig. 11.12 Illustrations of the first law. **a** An air flow through a well-insulated nozzle. **b** The filling process of air from a pipe to an initially evacuated tank by using a control-volume formulation. **c** The filling process in **b** is solved by using a control-mass formulation

the tank is p_i, followed by which the valve is closed. The filling process is assumed to be adiabatic, and the kinetic and potential energies are negligible. It is required to determine the air temperature in the tank after the filling process is completed. The filling process is a transient one, to which the space inside the tank is considered a control-volume, as illustrated by the dashed line in the figure. Let the beginning and end states of filling process be denoted by 1 and 2 respectively with state 1 corresponding to state i. It follows from Eq. (11.4.35) that

$$(m_2 u_2)_{CV} = m_i h_i, \tag{11.4.43}$$

because $Q_{CV} = 0$ in an adiabatic process, $W_{CV} = 0$ for the control-volume associated with no shaft, $m_{1CV} = 0$ for the tank is initially empty, and there exists only a single intake flow without any discharged flow. It follows from Eq. (11.4.33) that $m_{2CV} = m_i$, so that Eq. (11.4.43) becomes

$$(u_2)_{CV} = h_i, \qquad \longrightarrow \qquad (T_2)_{CV} = \frac{c_{p0}}{c_{v0}} T_i, \tag{11.4.44}$$

provided that the air is assumed to be an ideal gas. Since $c_{p0} > c_{v0}$ in general, the air temperature inside the tank after the filling process is greater than that of air in the supply pipe.

The problem can be solved equally by using a control-mass formulation. Consider the total air amount that enters the evacuated space as the control-mass, as shown in Fig. 11.12c. The air flow inside the pipe exerts work on the boundary of the control-mass, which is identified to be

$$W_1^2 = m p_i v_i, \tag{11.4.45}$$

which is known as the *flow work*. Substituting this expression into Eq. (11.4.11)$_2$ gives

$$U_2 - U_1 = Q_1^2 + W_1^2 = m p_i v_i, \tag{11.4.46}$$

so that

$$U_2 = m u_2 = m\left(u_i + p_i v_i\right) = m h_i, \qquad \longrightarrow \qquad u_2 = h_i, \tag{11.4.47}$$

which coincides to Eq. (11.4.44)$_1$.

11.5 Second Law and Entropy

Although the second law has been formulated as a global balance of entropy for a control-volume given in either Eq. (5.3.36) or (5.3.37) in Sect. 5.3.5, it can be formulated for a control-mass system, and there exist various macroscopic statements about the second law in different circumstances, some of which will be introduced in this section to illuminate the physical implications. The macroscopic and microscopic interpretations of entropy from statistical mechanics will be discussed to show its underlying physics. The illustrations of the second law, together with the applications of the conservation of mass and the first law, will be demonstrated at the end.

11.5.1 Heat Engine, Refrigerator, and Classical Statements

Heat engine is an abstract device that operates in a cycle and performs a net work and a net heat transfer. In other words, heat engine transforms heat into mechanical work. Practical accomplishments of heat engine are e.g. the Otto or Diesel engine. Equally, refrigerator is an abstract device that operates in a cycle and has heat transferred to it from a low-temperature body and heat transferred from it to a high-temperature body, though which work is required. In other words, refrigerator transfers heat from a low-temperature region to a high-temperature region. Practical accomplishments of refrigerator are e.g. refrigerator used in air conditioning. Specifically, a refrigerator denotes a system in which the removal of heat from a cooler region is the objective, while a *heat pump* attributes to the desired delivery of heat to a hotter region. The schematic illustrations of heat engine and refrigerator are shown respectively in Figs. 11.13a and b. The term *thermal reservoir* is used to denote a body to which and from which heat can be transferred indefinitely without any change in its (empirical) temperature Θ. Thermal reservoir is only an abstract but convenient concept in analyzing thermodynamic problems. It does not exist in the physical world, although ocean and atmosphere are good approximations to the concept.

As shown in Fig. 11.13a, the *thermal efficiency* of a heat engine is denoted by η_{th}, which is defined by

$$\eta_{th} \equiv \frac{\text{desire}}{\text{cost}} = \frac{|W_{net}|}{|Q_H|} = \frac{|Q_H| - |Q_L|}{|Q_H|} = 1 - \frac{|Q_L|}{|Q_H|} < 1, \tag{11.5.1}$$

in which $|Q_H|$ and $|Q_L|$ represent respectively the absolute values of the amounts of heat transfer with the Θ_H- and Θ_L-reservoirs, and the first law implies that $|W_{net}| = |Q_H| - |Q_L|$. The thermal efficiency of a heat engine is always smaller than unity, and conventionally $\eta_{th} = 0.35 \sim 0.5$ for a power plant; $\eta_{th} = 0.30 \sim 0.35$ and $\eta_{th} = 0.35 \sim 0.45$ for an Otto engine and a Diesel engine, respectively. For refrigerators and heat pumps, as referred to Fig. 11.13b, their efficiencies are evaluated by using the *coefficient of performance* (COP), β, which is defined in the same way as η_{th}, in which the desires are respectively $|Q_L|$ and $|Q_H|$, with the expense of $|W_{in}|$. Thus,

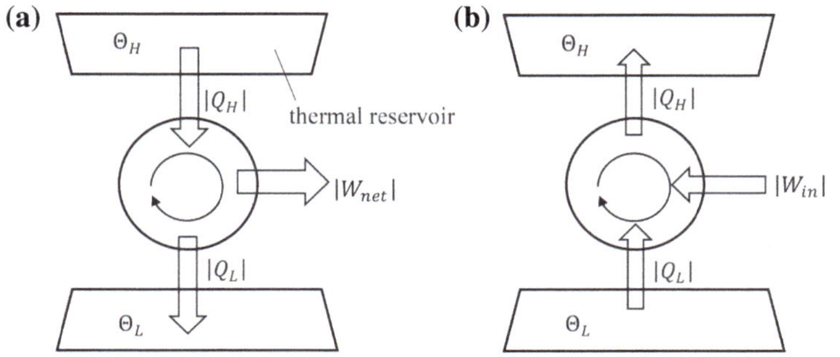

Fig. 11.13 Schematic illustrations of heat engine, refrigerator, and thermal reservoir. **a** Heat engine. **b** Refrigerator

the COPs of a refrigerator and a heat pump are given respectively by

$$\beta_{ref} = \frac{1}{|Q_H| / |Q_L| - 1} > 1, \quad \beta_{hp} = \frac{1}{1 - |Q_L| / |Q_H|} > 1, \quad \beta_{hp} - \beta_{ref} = 1. \tag{11.5.2}$$

A household refrigerator may have a COP of about 2.5, whereas that of a deep-freezing unit will be closer to unity. For a heat pump operating over a moderate temperature range, its COP will be around 4, with the value decreasing sharply as the operating temperature range is broadened.

The classical statements of second law are made to heat engine and refrigerator. Specifically, the *Kelvin-Planck statement* is made for heat engine, while the *Clausius statement* is made for refrigerator.

11.3 (The Kelvin-Planck statement) *It is impossible to construct a device which operates in a cycle and produces no effect other than the raising of a weight when exchanging heat with a single reservoir.*[16]

11.4 (The Clausius statement) *It is impossible to construct a device which operates in a cycle and produces no effect other than the transfer of heat from a cooler body to a hotter body.*

The Kelvin-Planck statement indicates that it is impossible to construct a heat engine that operates in a cycle, receiving a given amount of heat from a high-temperature body, and doing an equal amount of work. In other words, heat cannot be converted into work completely, and the thermal efficiency of any heat engine should always be less than unity. On the contrary, work can be converted completely into heat; e.g. frictional work can be converted completely into heat. The Clausius statement indicates that it is impossible to construct a refrigerator or a heat pump that operates in a cycle without an input of work. In other words, heat transfer from a low-temperature region to a high-temperature region cannot accomplish spontaneously. Hence, the COP of any refrigerator or heat pump must be less than infinity.

It is noted that both statements are based on experimental evidences and are negative descriptions. Both statements are equivalent. Violation of one statement will lead to violation of the other statement. For example, if the Clausius statement is untrue, a refrigerator by which heat can flow spontaneously from a cooler body to a hotter body can be constructed. This refrigerator is then combined with a normal heat engine, as shown in Fig. 11.14, in which the amount of heat which is exchanged with the Θ_L-reservoir in the refrigerator is maintained to be the same as that in the heat engine. Choosing both the refrigerator and heat engine as a new system shows that this system, operating in a cycle, exchanges heat with a single reservoir, i.e., the Θ_H-reservoir, and produces a net amount of work $|W_{net}| = |Q_H| - |Q_L|$. This conclusion violates the Kelvin-Planck statement. The fact that violation of the Kelvin-Planck statement leads to violation of the Clausius statement is left as an

[16]Max Karl Ernst Ludwig Planck, 1858–1947, a German theoretical physicist, who was the Nobel Prize Winner in Physics in 1918 for his discovery of energy quanta.

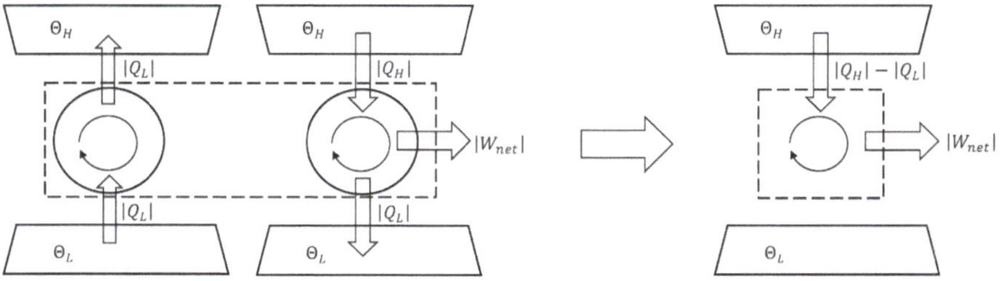

Fig. 11.14 Equivalence between the Kelvin-Planck and Clausius statements by showing that the untruth of latter statement leads to the untruth of former statement

exercise. Both proofs together show that the truth of either form of the second law is both a necessary and sufficient condition for the truth of the other.

The first and second laws imply that it is impossible to construct a device with perpetual motion. The first law does not allow *perpetual motion of the first kind*: A device cannot operate continuously by creating its own energy, for energy is a conserved quantity. The second law forbids *perpetual motion of the second kind*: A device cannot be made which runs continuously by using the internal energy of a single reservoir, although this would not violate the first law. A further possible way of constructing a device in perpetual motion would be to remove all dissipative effects such as friction, viscosity, or electrical resistance, so that the motion, once started in the device, could persist. This is called *perpetual motion of the third kind*, which is equally impossible in any system governed by classical physical laws, although it violates neither the first nor the second laws.

11.5.2 Carnot's Cycle, Carnot's Theorem, and Thermodynamic Temperature

Consider a heat engine operating between a high-temperature reservoir and a low-temperature reservoir in a cycle consisting of reversible processes. The heat engine becomes a refrigerator if the cycle is reversed. Specifically, consider a reversible cycle consisting of the following four reversible processes[17]:

1. A reversible isothermal process, in which an amount of heat is transferred from a high-temperature reservoir with temperature Θ_H (process $1 \to 2$: an isothermal expansion);

[17] A reversible process should be one in which all factors rendering reversibility must be removed, e.g. the effect of friction, unstrained expansion, heat transfer via finite temperature difference, mixing of different substances. Irreversible factors can be classified as *internal* and *external irreversibilities* with respect to the system under consideration. A quasi-equilibrium process is less constrained than a reversible process between the same states.

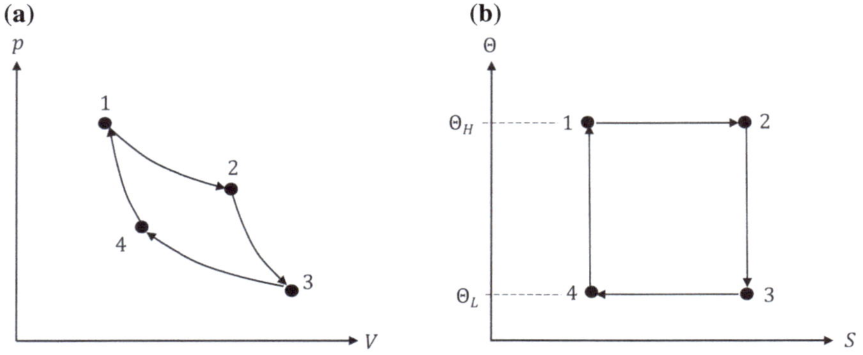

Fig. 11.15 Schematic illustrations of the Carnot cycle. **a** The p–V diagram. **b** The Θ–S diagram

2. A *reversible adiabatic process*,[18] in which the temperature of working substance decreases from the high temperature Θ_H to the low temperature Θ_L (process $2 \to 3$: an adiabatic expansion);
3. A reversible isothermal process, in which an amount of heat is transferred to the low-temperature reservoir with temperature Θ_L (process $3 \to 4$: an isothermal compression); and
4. A reversible adiabatic process, in which the temperature of working substance increases from Θ_L to Θ_H (process $4 \to 1$: an adiabatic compression).

The cycle so constructed is called the *Carnot cycle*,[19] with its graphic representations in the two-dimensional p–V and Θ–S diagrams shown respectively in Figs. 11.15a and b. An engine operating in the Carnot cycle is called a *Carnot engine*. Refrigerators and heat pumps can also operate in the Carnot cycle reversely.

The Carnot theorem states that *no engine operating between two reservoirs can be more efficient than a Carnot engine operating under the same conditions.* In other words, the thermal efficiency of a Carnot engine marks the ideal maximum limit. To show this, consider a heat engine operating between the Θ_H- and Θ_L-reservoirs, whose thermal efficiency η_{th} is larger than the thermal efficiency η_C of a Carnot engine, as shown in Fig. 11.16a. It follows that

$$\eta_{th} = \frac{|W_{net}|}{|Q_H|} > \eta_C = \frac{|W_{net,C}|}{|Q_{H,C}|}, \qquad \longrightarrow \qquad |W_{net}| > |W_{net,C}|, \tag{11.5.3}$$

for both engines receive the same amount of heat from the T_H-reservoir, i.e., $|Q_H| = |Q_{H,C}|$, and hence $|Q_L| < |Q_{L,C}|$. Now, the Carnot engine is reversed by part of the work $|W_{net}|$ produced by the heat engine, so that $|Q_H| = |Q_{H,C}|$ (in reverse

[18]Later, it will be shown that a reversible adiabatic process corresponds to an isentropic process, in which the thermodynamic property, the entropy S, remains fixed.

[19]Nicolas Léonard Sadi Carnot, 1796–1832, a French military engineer, who is often described as "Father of Thermodynamics," whose work led eventually to the formulation of second law.

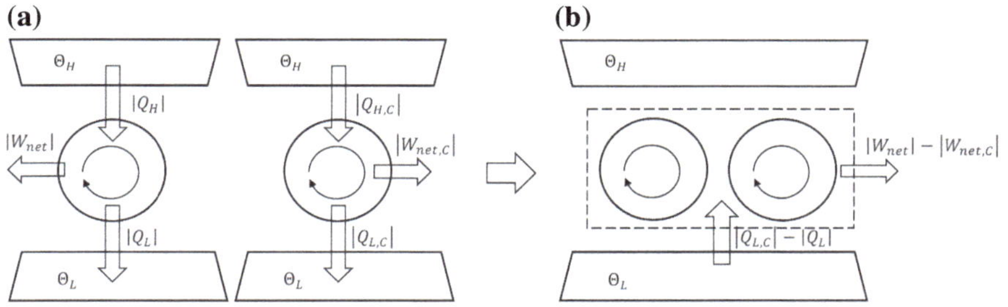

Fig. 11.16 Illustration of the Carnot theorem. **a** A heat engine which is more efficient than a Carnot engine between the same reservoirs. **b** Violation of the Kelvin-Planck statement

direction) is maintained. Combining both engines as the new system shows that it exchanges heat only with the Θ_L-reservoir and produces a net work, as shown in Fig. 11.16b, which violates the Kelvin-Planck statement. Thus, it is concluded that

$$\eta_{th} \leq \eta_C, \tag{11.5.4}$$

for any heat engine. If the heat engine were replaced by any reversible engine R, it follows that $\eta_C \geq \eta_{th,R}$. Since both engines in this case are reversible, the Carnot engine could have been used to drive the engine R reversely, giving that $\eta_C \leq \eta_{th,R}$. It follows immediately that $\eta_{th,R} = \eta_C$. Thus, a corollary of the Carnot theorem reads: "*All reversible engines operating between the same reservoirs have the same thermal efficiency*." This motivates that the thermal efficiency of any reversible engine operating between two reservoirs must only be a function of the (empirical) temperatures of two reservoirs, so that

$$\frac{|Q_H|}{|Q_L|} = f\left(\Theta_H, \Theta_L\right), \tag{11.5.5}$$

where f is a universal function of Θ_H and Θ_L.

Consider two Carnot engines denoted by C_1 and C_2, which operate respectively between the reservoirs at Θ_1 and Θ_2, and Θ_2 and Θ_3, as shown in Fig. 11.17a. Let C_1 absorb $|Q_1|$ at Θ_1 and reject $|Q_2|$ at Θ_2, and C_2 is so adjusted that it absorbs $|Q_2|$ at Θ_2 and rejects $|Q_3|$ at Θ_3. With these, Eq. (11.5.5) implies that

$$\frac{|Q_1|}{|Q_2|} = f_1\left(\Theta_1, \Theta_2\right), \qquad \frac{|Q_2|}{|Q_3|} = f_2\left(\Theta_2, \Theta_3\right). \tag{11.5.6}$$

Since there exists no net heat transfer at Θ_2, the Θ_2-reservoir is superfluous, which may be bypassed while Eq. (11.5.6) is still valid. Furthermore, the Carnot engines, C_1 and C_2, and Θ_2-reservoir may be considered a single system, as shown in Fig. 11.17b, which absorbs heat $|Q_1|$ at Θ_1 and rejects heat $|Q_3|$ at Θ_3, for which

$$\frac{|Q_1|}{|Q_3|} = f_3\left(\Theta_1, \Theta_3\right), \tag{11.5.7}$$

is obtained. Combining Eqs. (11.5.6) and (11.5.7) gives

$$f_3\left(\Theta_1, \Theta_3\right) = f_1\left(\Theta_1, \Theta_2\right) f_2\left(\Theta_2, \Theta_3\right). \tag{11.5.8}$$

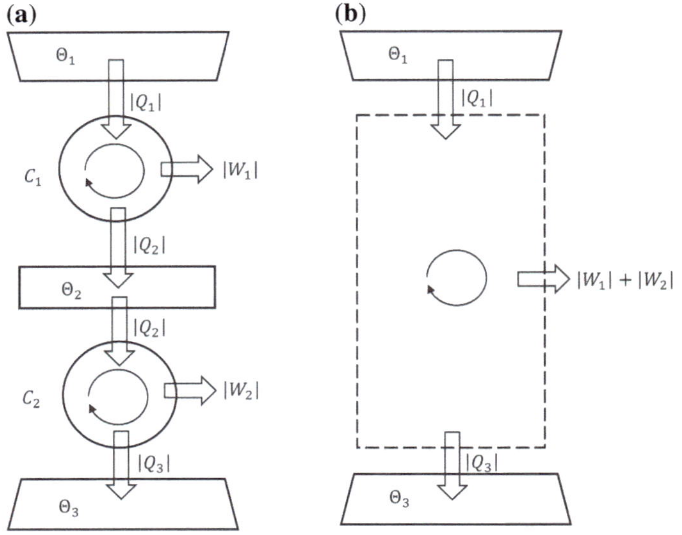

Fig. 11.17 Thermodynamic temperature in terms of the Carnot engine. **a** Two Carnot engines in a serial connection. **b** An equivalent Carnot engine between two reservoirs at the highest and lowest temperatures

This equation can only be fulfilled if the fs factorize in the form of $T\ (\Theta_1/\Theta_2)$, where Ts represent the universal quantities depending only on the empirical temperatures. Thus, it is required that

$$\frac{|Q_1|}{|Q_2|} = \frac{T_1}{T_2}, \tag{11.5.9}$$

which defines the thermodynamic temperature apart from the proportional constant which fixes the size of unit. In other words, the thermodynamic temperature is defined as that *the ratio of the thermodynamic temperatures of two reservoirs equals the ratio of the amounts of heat exchanged at those reservoirs by a reversible engine operating between them.* This temperature is that appearing in the ideal gas state equation, and the measurements of thermodynamic temperature are usually ultimately based on gas thermometry, as discussed in Sect. 11.3.3.

In terms of the thermodynamic temperature, the thermal efficiency of a heat engine and COPs of a refrigerator and a heat pump operating in the Carnot cycle are given by

$$\eta_C = 1 - \frac{T_L}{T_H}, \qquad \beta_{ref,C} = \frac{T_L}{T_H - T_L}, \qquad \beta_{hp,C} = \frac{T_H}{T_H - T_L}, \tag{11.5.10}$$

where T_H and T_L are the thermodynamic temperatures of high-temperature and low-temperature reservoirs, respectively. Equation (11.5.10) gives the maximum efficiencies of these devices.

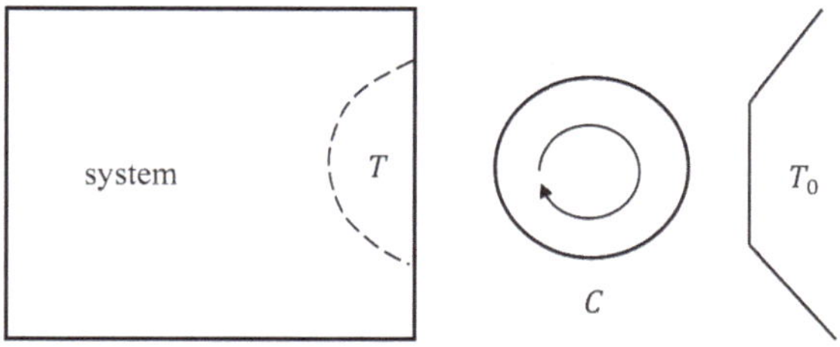

Fig. 11.18 The configurations in deducing the Clausius theorem

11.5.3 Clausius' Theorem and Entropy

Consider a Carnot engine C which is in contact with a system at temperature T and a reservoir at constant temperature T_0, as shown in Fig. 11.18. Let the Carnot engine undergo a cycle which consists of the following reversible processes:

1. C is at T_0.
2. C is compressed (or expanded) adiabatically until its temperature is T, which is the temperature of the part of system in contact with C.
3. C is in contact with the system to absorb or supply heat by an isothermal change at T.
4. C is expanded (or compressed) adiabatically until its temperature is T_0.
5. C is placed in contact with the reservoir and compressed (or expanded) isothermally at T_0 unit it is brought again to its initial state.

This is done so, for the Carnot engine can be generalized to a complex one which executes its cycle in infinitesimal steps without any assumptions about the uniqueness of its adiabatics or about whether the working substance can depart from the specified cycle.

Let the heat supplied to the Carnot engine at T in one journey be denoted by δQ, so that the corresponding heat absorbed by the reservoir is obtained as

$$\frac{T_0}{T}\delta Q, \quad \longrightarrow \quad T_0 \oint \frac{\delta Q}{T} \leq 0, \tag{11.5.11}$$

where the second equation represents the heat absorbed by the reservoir in one complete cycle, as constrained by the Kelvin-Planck statement. Since T_0 must be necessarily positive, it follows that

$$\oint \frac{\delta Q}{T} \leq 0, \tag{11.5.12}$$

for any cycle, including a reversible one. On the contrary, if the considered cycle was reversible, it would be possible to reverse the cycle to derive

$$\oint \frac{\delta Q}{T} \geq 0. \tag{11.5.13}$$

Comparing this equation with Eq. (11.5.12) shows that for a reversible cycle,

$$\oint \frac{\delta Q}{T} = 0, \tag{11.5.14}$$

must hold. Equations (11.5.12) and (11.5.14) together form the **Clausius theorem**, which states that *for any cycle*, $\oint(\delta Q/T) \leq 0$, *where the equality necessarily holds for a reversible one*.

To apply the Clausius theorem, consider a system undergoing a process A from state 1 to state 2. The system is brought again to state 1 by two different processes B and C. It is assumed that all three processes are reversible, so that two reversible cycles are constructed. For the cycle $1 \to A \to 2 \to B \to 1$ and the cycle $1 \to A \to 2 \to C \to 1$, the Clausius theorem implies that

$$\int_1^2 \left(\frac{\delta Q}{T}\right)_A + \int_2^1 \left(\frac{\delta Q}{T}\right)_B = 0,$$
$$\int_1^2 \left(\frac{\delta Q}{T}\right)_A + \int_2^1 \left(\frac{\delta Q}{T}\right)_C = 0. \tag{11.5.15}$$

Combining two equations yields

$$\int_1^2 \left(\frac{\delta Q}{T}\right)_B = \int_1^2 \left(\frac{\delta Q}{T}\right)_C, \tag{11.5.16}$$

showing that the quantity $\delta Q/T$ is not path-dependent, and must be a differential of a thermodynamic property. The *infinitesimal change in entropy* $\mathrm{d}S$ is then defined by

$$\mathrm{d}S \equiv \left(\frac{\delta Q}{T}\right)_{rev}. \tag{11.5.17}$$

Since S is a function of state, $\mathrm{d}S$ must be an exact differential. However, δQ is not an exact differential, so that the fraction $1/T$ serves as an integrating factor to make $\delta Q/T$ an exact differential from the mathematical perspective. Equation (11.5.17) defines the thermodynamic property, the entropy S, from the macroscopic perspective. However, only its change between any two states is defined, not its absolute value. The reference state in which $S = 0$ for air and water is the same as those for internal energy. For a reversible adiabatic process, it is seen from Eq. (11.5.17) that $\mathrm{d}S$ vanishes. Such a process is called an *isentropic process*, in which the entropy remains unchanged.

Although $\mathrm{d}S$ in Eq. (11.5.17) is defined in terms of a reversible process, the entropy is a state function, whose change in accompanying a given change in state must always be the same, whatsoever the change of state may occur. Only if the state change takes place reversibly, can the entropy change related to the heat transfer be given by

$$\Delta S = \int \left(\frac{\delta Q}{T}\right)_{rev}. \tag{11.5.18}$$

To show the integration of $\delta Q/T$ for an irreversible process, consider two cycles used previously. It is assumed now that process C is an irreversible one, so that applying the Clausius theorem to the cycle $1 \to A \to 2 \to C \to 1$ yields

$$\int_1^2 \left(\frac{\delta Q}{T}\right)_A + \int_2^1 \left(\frac{\delta Q}{T}\right)_C < 0,$$
$$\longrightarrow \int_1^2 \left(\frac{\delta Q}{T}\right)_{irr} < \int_1^2 \left(\frac{\delta Q}{T}\right)_{rev} = S_2 - S_1 = \Delta S, \tag{11.5.19}$$

or alternatively,

$$\mathrm{d}S > \left(\frac{\delta Q}{T}\right)_{irr}, \tag{11.5.20}$$

which indicates that the integration of $\delta Q/T$ in an irreversible process between any two states is less than the entropy change between the same states. Combining Eq. (11.5.17) with Eq. (11.5.20) results in

$$\mathrm{d}S \geq \frac{\delta Q}{T}, \tag{11.5.21}$$

which is a general expression for an infinitesimal change in entropy in a process, where the equality necessarily applies if the change in state is reversible. This equation may be thought of as the focal point of second law.

For example, for an *isolated system*, i.e., a system which is completely isolated from its surrounding, $\delta Q = 0$, so that Eq. (11.5.21) becomes

$$\mathrm{d}S \geq 0, \tag{11.5.22}$$

indicating that the entropy cannot decrease, which is known as the *law of increase of entropy*. A particular application of this law is that it can be used to determine the equilibrium configuration of an isolated system. Since in approaching equilibrium the entropy of system can only increase, the final equilibrium configuration is the one in which the entropy is as large as possible. In addition, this law provides a natural direction to the time sequence of natural events. In the context of Newtonian mechanics, all processes are reversible, since the equations remain indifferent by replacing t by $-t$. On the other hand, the inevitable sequence to events, or alternatively the "arrow of time," is indicated by the increase of entropy. All changes in nature are part of the irreversible progresses toward universal equilibrium. Thus, the natural direction of events results from there not being thermodynamic equilibrium throughout the universe. As long as temperature and density differences exist, natural evolution will continue and events will be directed forward toward equilibrium.

11.5.4 Implications of Entropy as a Macroscopic Property

The first law for a control-mass undergoing a process between any two states is given in Eq. $(11.4.10)_1$. This equation applies for any process, including an irreversible one. For a reversible process, the work due to the moving boundary and the amount of heat transfer can be determined by

$$\delta W = -p\,\mathrm{d}V, \qquad \delta Q = T\mathrm{d}S, \tag{11.5.23}$$

so that the first law becomes

$$\mathrm{d}U = T\mathrm{d}S - p\,\mathrm{d}V, \tag{11.5.24}$$

if the control-mass is globally stationary. Although Eq. (11.5.24) has been derived by using a reversible process, it may be applied for any process, however accomplished, for all quantities in the equation are only functions of state, and the integration of equation must be independent of the process. Obviously, for irreversible processes,

Eq. (11.5.23) is no longer valid, but Eq. (11.5.24) is still applicable. In an irreversible process, it follows from the Clausius theorem that $\delta Q_{irr} < T\mathrm{d}S$, so that the first law yields $\delta W_{irr} > -p\,\mathrm{d}V$, as expected, since in the presence of irreversibility the total work done should be greater than that which would be required to effect the same change in volume of the system without irreversibility. Similarly, it is also possible to derive

$$\mathrm{d}H = T\mathrm{d}S + V\mathrm{d}p, \tag{11.5.25}$$

which is valid for any process between any two states. Equations (11.5.24) and (11.5.25) are called the $T\mathrm{d}S$ equations, which are valid for any simple compressible substance. As motivated by Eq. (11.5.24), the general form of first law for a control-mass can be given by

$$\mathrm{d}U = T\mathrm{d}S + \sum x_i \mathrm{d}X_i, \tag{11.5.26}$$

if other mechanisms that produce work exist, where x_i represents an intensive variable, with X_i its conjugate extensive variable. Since entropy is an extensive variable, as implied by its definition, the thermodynamic temperature T must be its corresponding intensive variable. The product $T\mathrm{d}S$ is entirely similar to the work terms and may be grouped with them. Hence, a condensed form of Eq. (11.5.26) may be given by

$$\mathrm{d}U = \sum x_i \mathrm{d}X_i, \tag{11.5.27}$$

where the summation necessarily includes the product $T\mathrm{d}S$ which is relevant to all systems.

The specific heats defined in Eqs. (11.4.19)$_2$ and (11.4.20)$_2$ can be expressed in terms of entropy change. Since $\delta Q = T\mathrm{d}S$ for a reversible process, it follows that

$$c_V = \frac{T}{m}\left(\frac{\partial S}{\partial T}\right)_V = T\left(\frac{\partial s}{\partial T}\right)_V, \qquad c_p = \frac{T}{m}\left(\frac{\partial S}{\partial T}\right)_p = T\left(\frac{\partial s}{\partial T}\right)_p, \tag{11.5.28}$$

for the specific heats at constant volume and constant pressure, respectively. For ideal gases, the $T\mathrm{d}S$ equations reduce to

$$\mathrm{d}S = m\,\mathrm{d}s = m\left(c_{V0}\frac{\mathrm{d}T}{T} + R\frac{\mathrm{d}V}{V}\right) = m\left(c_{p0}\frac{\mathrm{d}T}{T} - R\frac{\mathrm{d}p}{p}\right), \tag{11.5.29}$$

so that a finite change in the specific entropy between any two states is obtained as

$$\Delta s = \frac{\Delta S}{m} = c_{V0}\ln\frac{T_2}{T_1} + R\ln\frac{V_2}{V_1} = c_{p0}\ln\frac{T_2}{T_1} - R\ln\frac{p_2}{p_1}. \tag{11.5.30}$$

As a macroscopic description, entropy can be used as a measure of the degeneration of energy of a system. As indicated by the first law, the work that can be extracted from a system in an infinitesimal change in state is $\delta W = \delta Q - \mathrm{d}U$, where δQ is related to the entropy change given by $\delta Q \le T_0\mathrm{d}S$ with T_0 representing the temperature at which heat is supplied, so that δW must satisfy $\delta W \le T_0\mathrm{d}S - \mathrm{d}U$. For a given change in state, both $\mathrm{d}U$ and $\mathrm{d}S$ are prescribed, so that a maximum work could be extracted from the system if the process is reversible. In such a case, the total entropy change of both the system and its surrounding must vanish, for in any process involving reversible exchange of heat with the surrounding, $\mathrm{d}S_{sys} = -\mathrm{d}S_{surr}$, as indicated by the law of increase of entropy. On the other hand, for an irreversible process, the entropy change of surrounding is given by $\mathrm{d}S_{surr} = -\delta Q/T_0$, provided that there exists no irreversibility there, while the entropy change of system must satisfy $\mathrm{d}S_{sys} \geq \delta Q/T_0$, so that the law of increase of entropy is fulfilled. The work that can be extracted from the system becomes less than that which could be extracted if the same change in state had been made reversibly. It becomes clear that associated with the increase in entropy is the "loss" of some energy which could have been used for work. It is said that the energy is degraded in that it is less useful for work.

For example, consider two bodies which are marked by numbers 1 and 2, having respectively temperatures T_1 and T_2 with $T_1 > T_2$. Two bodies are connected via a thermal resistance, and a small amount of heat Q is allowed to be transferred from body 1 to body 2. The total entropy change of two bodies is given by

$$\Delta S = \Delta S_1 + \Delta S_2 = Q\left(\frac{1}{T_2} - \frac{1}{T_1}\right) > 0, \tag{11.5.31}$$

since $T_2 < T_1$. This expression indicates that the entropy will continue to increase as long as the heat transfer brings the bodies toward equilibrium. Now the small amount of heat is guided into a Carnot engine which is placed between two bodies, with T_0 representing the temperature of its coldest thermal reservoir. The work which can be extracted from the Carnot engine is obtained as

$$W = Q\left(1 - \frac{T_0}{T_1}\right). \tag{11.5.32}$$

If the same amount of heat is allowed to flow first from body 1 to body 2 and then guided to the Carnot engine, the obtainable amount of work from the Carnot engine becomes

$$W' = Q\left(1 - \frac{T_0}{T_2}\right) < W. \tag{11.5.33}$$

It follows that the energy has become degraded in the irreversible heat conduction process to the extent that the obtainable useful work has been decreased by

$$W - W' = T_0 \Delta S, \tag{11.5.34}$$

showing that the increase in entropy in an irreversible process is a measure of the extent to which the energy becomes degraded in that change in state. Thus, in order to extract a maximum amount of useful work from a system or a set of systems, the change in state must be performed reversibly, so that the total entropy of the system and its surrounding is conserved.

An extension of this example is the determination of the final equilibrium temperature of two bodies. They can be allowed to reach thermal equilibrium by either heat conduction or by operating a Carnot engine in-between and extracting work. In the first case, the total internal energies of two bodies are conserved, i.e., $U_1 + U_1 =$ constant, for which the final equilibrium temperature $T_f^{\Delta U=0}$ is obtained as

$$T_f^{\Delta U=0} = \frac{C_1 T_1 + C_2 T_2}{C_1 + C_2}, \tag{11.5.35}$$

where C_1 and C_2 are respectively the heat capacities of bodies 1 and 2, which are taken as constants for simplicity. For the second case, the total entropy $S_1 + S_2 =$ constant, with the extracted work given by $W = -\Delta(U_1 + U_2)$. In the considered isentropic process, the final equilibrium temperature $T_f^{\Delta S=0}$ is determined to be

$$T_f^{\Delta S=0} = T_1^{C_1/(C_1+C_2)} T_2^{C_2/(C_1+C_2)} < T_f^{\Delta U=0}. \tag{11.5.36}$$

The difference in the final equilibrium temperature between two cases corresponds to the lower value of the total internal energy which results from work having been done.

11.5.5 Entropy from Statistical Mechanics

Up to this point, only the macroscopic descriptions of entropy have been discussed. It has been shown that the equilibrium state of an isolated system is that whose entropy takes its maximum value, so that the maximization of the macroscopic entropy is the condition for determining the equilibrium configuration. An alternative approach is to apply the probability theory at the microscopic level to the various possible configurations of a system to seek the specific configuration whose probability is the greatest. Such an approach is referred to as the *statistical mechanics* or *statistical thermodynamics*, by which the microscopic interpretation of entropy will be given.

Consider a monatomic gas inside an adiabatic rigid cubic vessel with side length L. The gas is assumed to be an ideal one, and there exist totally N gas particles, which are weakly interacting or quasi-independent of one another; i.e., the interaction between particles is only considered at collision, and all particles are indistinguishable; i.e., they have neither a preferred location in space nor a preferred velocity. The translational kinetic energy in the x-direction of each particle is given by

$$\epsilon_x = \frac{1}{2} m \dot{x}^2 = \frac{p_x^2}{2m}, \qquad p_x = m\dot{x}, \tag{11.5.37}$$

where m is the mass of particle, whose linear momentum in the x-direction is denoted by p_x. Each particle is further assumed to be free to move between two planes of the vessel, so that the *Bohr-Sommerfeld form* of quantum mechanics shows that in a complete cycle of the particle motion with a total distance $2L$,[20] the product of p_x with $2L$ equals an integer n_x times *Planck's constant* $\hbar$, namely,

$$2 p_x L = n_x \hbar, \tag{11.5.38}$$

where n_x is interpreted as the *quantum number* of a particle having linear momentum p_x. Substituting this equation into Eq. (11.5.37) yields

$$\epsilon_x = n_x^2 \frac{\hbar^2}{8mL^2}, \qquad \longrightarrow \qquad n_x = \frac{L}{\hbar}\sqrt{8m\epsilon_x}, \tag{11.5.39}$$

[20] Niels Henrik David Bohr, 1885–1962, a Danish physicist, who contributed to the foundational understanding of atomic structure and quantum theory and was the Nobel Prize Winner in Physics in 1922.

showing that the value of ϵ_x is discrete, corresponding to the integer value of n_x. However, if n_x changes by unity, the corresponding change in ϵ_x is extremely small, for n_x is itself exceedingly large.[21] Taking into account all three components of the linear momentum of a particle yields the total translational kinetic energy in a cubic of side L, viz.,

$$\begin{aligned}\epsilon = \epsilon_x + \epsilon_y + \epsilon_z = \frac{p_x^2 + p_y^2 + p_z^2}{2m} = \frac{\hbar^2}{8mL^2}\left(n_x^2 + n_y^2 + n_z^2\right) \\ = \frac{\hbar^2}{8mV^{2/3}}\left(n_x^2 + n_y^2 + n_z^2\right),\end{aligned} \quad (11.5.40)$$

where V denotes the volume of cubic vessel.

The specification of three integers n_x, n_y, and n_z corresponds to a specification of the quantum state of a particle inside the vessel, in which all states characterized by $n_x^2 + n_z^2 + n_z^2 =$ constant will give the same energy. It is also possible to have different combinations of n_x, n_y, and n_z to reach the same energy level. The number of possible combinations of ns corresponding to an energy level ϵ_i is referred to as the *degeneracy* G_i, which, in any actual case, is an enormous number. For the particles of an ideal gas, there exists only a finite number of discrete energy levels. It is of prime interest in the context of statistical mechanics to determine the population of these energy levels at equilibrium, i.e., the number of particles N_i corresponding to the energy level ϵ_i. It can be shown that the degeneracy G_i corresponding to an energy level ϵ_i is very much larger than the number of particles N_i occupying that energy level at room temperature. Since $G_i \gg N_i$, it is unlikely that at room temperature more than one particle will occupy the same quantum state at any one time. The relation between an energy level ϵ_i and its corresponding degeneracy G_i and the number of particles (population) at that energy level, N_i, is shown schematically in Fig. 11.19a.

However, at one instant of time, some particles move rapidly and some slowly, so that the particles are distributed in a large number of different quantum states. The particles collide with one another and with the vessel walls as time goes on or

[21] This can be shown by considering a cubic box filled with a gaseous helium at 300 K, whose side length is assumed to be 10 cm. It follows from the kinetic theory of gas that the average translational kinetic energy of a monatomic ideal gas is $3kT/2$. Since each gas particle has three degrees of freedom with no preferred direction in velocity, the average energy associated with each translational degree of freedom is $kT/2$, so that

$$\epsilon_x = \frac{1}{2}kT \sim 2.1 \times 10^{-21}\ \text{J}.$$

The corresponding change in n_x, by using Eq. (11.5.39)$_2$, is given by

$$n_x = \frac{L}{\hbar}\sqrt{8m\epsilon_x} \sim 10^9.$$

Hence, the change in energy caused by a change in n_x by unity is so small that for most practical purposes, the energy may be assumed to vary continuously.

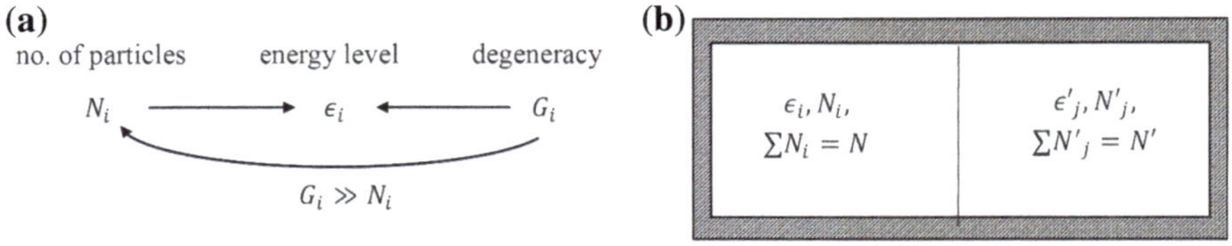

Fig. 11.19 Entropy from statistical mechanics. **a** The relation between the population N_i and the corresponding energy level ϵ_i and degeneracy G_i. **b** A canonical ensemble in the statistical mechanics

emit or absorb photons, so that each particle undergoes many changes in its quantum state. As a fundamental assumption in the statistical mechanics, it is assumed that *all quantum states have equal likelihood of being occupied*. Thus, the probability that a particle may find itself in a given quantum state is the same for all quantum states. With these, consider the N_i particles in any of the G_i quantum states associated with the energy level ϵ_i, for which any single particle would have G_i choices in occupying G_i quantum states. A second particle would have the same possibility, and so on. The total number of ways that N_i distinguishable particles could be distributed among G_i quantum states would be $G_i^{N_i}$. For indistinguishable particles, the number of permutations of N_i particles is $N_i!$. Therefore, the number of ways that N_i indistinguishable quasi-independent particles can be distributed among G_i quantum states is given by

$$\frac{G_i^{N_i}}{N_i!}. \tag{11.5.41}$$

The specification of all N_is of an ideal gas inside a vessel with volume V, total particles $N = \sum N_i$, and internal energy U at any instant of time is understood as a description of a macro-state of that gas.

The number of ways, Ω, by which a macro-state may be achieved is then given by a product of the term given in Eq. (11.5.41), i.e.,

$$\Omega = \prod \frac{G_i^{N_i}}{N_i!}, \tag{11.5.42}$$

over all possible values of i, where Ω is referred to as the *thermodynamic probability*, or alternatively the *accessible microstates* of a particular macro-state. The larger the value of Ω is, the greater will be the probability of finding the system in that macro-state. If the values of V, N, and U are held constant, the equilibrium state of a system will correspond to that macro-state in which Ω assumes the maximum value. To find this value, taking logarithm to Eq. (11.5.42) gives

$$\ln \Omega = \sum N_i \ln \frac{G_i}{N_i} + N, \tag{11.5.43}$$

which is subject to

$$N = \sum N_i = \text{constant}, \qquad U = \sum N_i \epsilon_i = \text{constant}. \tag{11.5.44}$$

In deriving Eq. (11.5.43), *Stirling's approximation* has been used.[22] In Eq. (11.5.43), both ϵ_is and G_is are constants, while all N_is are variables.

Since $\mathrm{d}N = 0$, it follows from Eq. (11.5.43) that

$$\mathrm{d}\,(\ln \Omega) = \sum \ln \frac{G_i}{N_i} \mathrm{d}N_i. \tag{11.5.45}$$

Taking total differential to Eq. (11.5.44) yields respectively

$$\sum \mathrm{d}N_i = 0, \qquad \sum \epsilon_i \mathrm{d}N_i = 0, \tag{11.5.46}$$

which are the constraints to Eq. (11.5.45). By using the method of the Lagrangian multipliers, Eq. (11.5.46) is recast as

$$\sum (\ln \lambda)\, \mathrm{d}N_i = 0, \qquad -\sum \beta \epsilon_i \mathrm{d}N_i = 0, \tag{11.5.47}$$

where $(\ln \lambda)$ and $(-\beta)$ are the Lagrangian multipliers of Eqs. $(11.5.46)_1$ and $(11.5.46)_2$, respectively. Substituting these expressions into Eq. (11.5.45) gives

$$\ln \frac{G_i}{N_i} + \ln \lambda - \beta \epsilon_i = 0, \qquad \longrightarrow \qquad N_i = \lambda\, G_i \exp\left(-\beta \epsilon_i\right), \tag{11.5.48}$$

for $\mathrm{d}N_i$ can take any arbitrary value independently of any other $\mathrm{d}N_j$. It is seen that the population N_i of any energy level at equilibrium is proportional to the degeneracy of that energy level and varies exponentially with the energy level. Taking summation of Eq. (11.5.48) over all energy levels yields

$$\sum N_i = \lambda \sum G_i \exp\left(-\beta \epsilon_i\right), \qquad \longrightarrow \qquad \lambda = \frac{N}{\sum G_i e^{-\beta \epsilon_i}}, \tag{11.5.49}$$

by which the *Zustandsumme* (sum over states) or alternatively the *partition function* Z is defined by

$$Z \equiv \sum G_i \exp\left(-\beta \epsilon_i\right), \tag{11.5.50}$$

so that

$$\lambda = \frac{N}{Z}, \qquad \longrightarrow \qquad N_i = \frac{N}{Z} G_i \exp\left(-\beta \epsilon_i\right). \tag{11.5.51}$$

Consider a thermally isolated composite system consisting of two ideal monatomic gases separated by a diathermic wall, as shown in Fig. 11.19b, which is known as a *canonical ensemble* in the statistical mechanics. The thermodynamic probability Ω of the composite system is simply the product of the thermodynamic probabilities of two ideal gases, so that $(\ln \Omega)$ takes the form

$$\ln \Omega = \sum N_i \ln \frac{G_i}{N_i} + N + \sum N'_j \ln \frac{G'_j}{N'_j} + N', \tag{11.5.52}$$

[22]James Stirling, 1692–1770, a Scottish mathematician, whose works are known as the *Stirling numbers*, *Stirling permutations*, and *Stirling approximation*.

which is subject to the constraints given by

$$\begin{aligned} &\sum N_i = N = \text{constant}, \qquad \sum N'_j = N' = \text{constant}, \\ &\sum N_i\epsilon_i + \sum N'_j\epsilon_j = U = \text{constant}, \end{aligned} \tag{11.5.53}$$

in which the total energy of composite system is maintained to be constant. By using the method of the Lagrangian multipliers, Eqs. (11.5.52) and (11.5.53) are recast alternatively as

$$\begin{aligned} &\sum \ln \frac{G_i}{N_i} \mathrm{d}N_i + \sum \ln \frac{G'_j}{N'_j} \mathrm{d}N'_j = 0, \\ &(\ln \lambda) \sum \mathrm{d}N_i = 0, \quad \left(\ln \lambda'\right) \sum \mathrm{d}N'_j = 0, \quad -\beta \sum \epsilon_i \mathrm{d}N_i - \beta \sum \epsilon'_j \mathrm{d}N'_j = 0, \end{aligned} \tag{11.5.54}$$

in which no distinction between β and β' is made, for this Lagrangian multiplier is related to the gas temperature only, and two ideal gases are in thermal equilibrium in the presence of a diathermic wall. With a similar procedure described previously, it follows from these equations that

$$N_i = \lambda G_i \exp\left(-\beta\epsilon_i\right), \qquad N'_j = \lambda G'_j \exp\left(-\beta\epsilon'_j\right). \tag{11.5.55}$$

The law of increase of entropy of an isolated system indicates that the entropy of an isolated system undergoing a spontaneous, irreversible process always increases and has the maximum value which is in consistent with its internal energy and volume when equilibrium is reached. This fact is equally reflected microscopically by an increase in the thermodynamic probability to the maximum value. As a result, there must exist a relation between the macroscopic entropy S and microscopic thermodynamic probability Ω. In view of Fig. 11.19b, the total entropy S of composite system is given by

$$S = S_A + S_B, \tag{11.5.56}$$

with its microscopic description in terms of the thermodynamic probability given viz.,

$$\Omega = \Omega_A \Omega_B, \tag{11.5.57}$$

where the subscripts A and B denote respectively the gases in the left and right parts. If the relation between S and Ω is expressed by $S = f(\Omega)$, where f is any differential function, Eqs. (11.5.56) and (11.5.57) together imply that

$$f\left(\Omega_A \Omega_B\right) = f\left(\Omega_A\right) + f\left(\Omega_B\right). \tag{11.5.58}$$

Differentiating this equation first with respect to Ω_A and then to Ω_B yields

$$f'\left(\Omega_A \Omega_B\right) + \Omega_A \Omega_B f''\left(\Omega_A \Omega_B\right) = 0, \qquad \longrightarrow \qquad \frac{f''(\Omega)}{f'(\Omega)} = -\frac{1}{\Omega}, \tag{11.5.59}$$

which is integrated to obtain

$$f\left(\Omega\right) = k' \ln \Omega + \Omega_0, \tag{11.5.60}$$

where k' is an arbitrary constant and Ω_0 is the integration constant. Substituting this expression into $S = f(\Omega)$ gives

$$S = k' \ln \Omega + S_0, \tag{11.5.61}$$

where S_0 is also an integration constant, which is conventionally chosen to be zero to correspond to a statistical probability of unity for a completely ordered state.

To determine the value of k', consider a closed system. It follows from Eqs. (11.5.43) and $(11.5.48)_2$ that

$$\mathrm{d}(\ln \Omega) = \beta \mathrm{d}\left(\sum \epsilon_i N_i\right) - (\ln \lambda)\, \mathrm{d}\left(\sum N_i\right) = \beta \mathrm{d}U, \tag{11.5.62}$$

for $\mathrm{d}(\sum N_i) = 0$ and $\sum \epsilon_i N_i = U$. If the volume of system is kept constant, the value of β is then determined to be

$$\beta = \left(\frac{\mathrm{d}\ln \Omega}{\mathrm{d}U}\right)_V = \frac{1}{k'}\frac{\mathrm{d}}{\mathrm{d}U}\left(k' \ln \Omega\right)_V = \frac{1}{k'}\left(\frac{\partial S}{\partial U}\right)_V, \tag{11.5.63}$$

in which Eq. (11.5.61) has been used. Applying the first law to the system yields

$$\mathrm{d}U = -p\,\mathrm{d}V + T\mathrm{d}S, \tag{11.5.64}$$

provided that the process under consideration is reversible. If the reversible process is assumed to take place at constant volume, it follows from the above equation that

$$\frac{1}{T} = \left(\frac{\partial S}{\partial U}\right)_V, \tag{11.5.65}$$

which provides an important link between the classical thermodynamics and statistical mechanics. Since both S and U in this equation can be computed by using the statistical mechanics, the derivative on the right-hand-side gives the reciprocal of the Kelvin temperature, and hence the temperature as a macroscopic concept is introduced into the statistical mechanics. Substituting Eq. (11.5.65) into Eq. (11.5.63) gives

$$\beta = \frac{1}{k'T}. \tag{11.5.66}$$

Further investigation on Eq. (11.5.66) could be accomplished by prescribing the energy levels in terms of the partition function. Substituting Eq. (11.5.66) into Eq. $(11.5.51)_2$ leads to

$$N_i = \frac{N}{Z} G_i \exp\left(-\frac{\epsilon_i}{k'T}\right), \qquad Z = \sum G_i \exp\left(-\frac{\epsilon_i}{k'T}\right) = Z(V, T). \tag{11.5.67}$$

The determination of partition function Z is the key part in the statistical mechanics, by which different macroscopic thermodynamic properties can be expressed in terms of the quantities in the statistical mechanics. To show this, taking partial derivative of Z in Eq. $(11.5.67)_2$ with respect to T at constant V gives

$$\left(\frac{\partial Z}{\partial T}\right)_V = \sum G_i \frac{\epsilon_i}{k'T^2} \exp\left(-\frac{\epsilon_i}{k'T}\right) = \frac{UZ}{Nk'T^2}, \tag{11.5.68}$$

giving rise to

$$U = Nk'T^2 \left(\frac{\partial \ln Z}{\partial T} \right)_V , \tag{11.5.69}$$

which is the expression of internal energy in terms of the partition function. Similarly, combining Eqs. (11.5.43), (11.5.61) and $(11.5.67)_1$ yields

$$S = Nk' \ln \frac{Z}{N} + \frac{U}{T} + Nk'. \tag{11.5.70}$$

The *Helmholtz function* F is defined as $F \equiv U - TS$, which, by using Eqs. (11.5.69) and (11.5.70), can be expressed as

$$F = -k'T \left(N \ln Z - \ln N! \right) . \tag{11.5.71}$$

Taking total differential of this equation yields

$$\mathrm{d}F = -p \,\mathrm{d}V - S\mathrm{d}T, \qquad \longrightarrow \qquad p = -\left(\frac{\partial F}{\partial V} \right)_T = Nk'T \left(\frac{\partial \ln Z}{\partial V} \right)_T , \tag{11.5.72}$$

in which Eq. (11.5.71) has been used. Equations (11.5.69)–(11.5.72) show that the values of U, S, F, and p, which are all macroscopic properties, can be obtained by using the microscopic approach, provided that the partition function Z is prescribed as a known function of T and V.

The partition function Z has already been given in Eq. $(11.5.67)_2$, where the summation should be taken over all energy levels. This equation corresponds to

$$Z = \sum_{\text{levels}} G_i \exp\left(-\frac{\epsilon_i}{k'T} \right) = \sum_{\text{states}} \exp\left(-\frac{\epsilon_i}{k'T} \right), \tag{11.5.73}$$

over all quantum states. For an ideal monatomic gas inside an adiabatic rectangular box whose side lengths are a, b and c in the x-, y- and z-directions, respectively, consider only the translational kinetic energy of a gas particle, so that the energy of any quantum state j, by using Eq. (11.5.40), is identified to be

$$\epsilon_i = \frac{\hbar^2}{8m} \left(\frac{n_x^2}{a^2} + \frac{n_y^2}{b^2} + \frac{n_z^2}{c^2} \right), \tag{11.5.74}$$

where n_x, n_y, and n_z are the quantum numbers specifying the various possible quantum states. Substituting this expression into Eq. (11.5.73) yields

$$Z = \sum_{n_x=1}^{\infty} \exp\left(-\frac{\hbar^2}{8mk'T} \frac{n_x^2}{a^2} \right) \sum_{n_y=1}^{\infty} \exp\left(-\frac{\hbar^2}{8mk'T} \frac{n_y^2}{b^2} \right) \sum_{n_z=1}^{\infty} \exp\left(-\frac{\hbar^2}{8mk'T} \frac{n_z^2}{c^2} \right), \tag{11.5.75}$$

which can be simplified to

$$Z = \int_0^{\infty} \exp\left(-\frac{\hbar^2}{8mk'T} \frac{n_x^2}{a^2} \right) \mathrm{d}n_x \int_0^{\infty} \exp\left(-\frac{\hbar^2}{8mk'T} \frac{n_y^2}{b^2} \right) \mathrm{d}n_y \cdot \int_0^{\infty} \exp\left(-\frac{\hbar^2}{8mk'T} \frac{n_z^2}{c^2} \right) \mathrm{d}n_z, \tag{11.5.76}$$

because the values of n_x, n_y, and n_z that give rise to appreciable values of the energy are extremely large, so that a change in these values by unity produces a change in energy which is exceeding small. Conducting the integrals gives

$$\begin{aligned} Z &= \left(\frac{a}{2}\sqrt{\frac{8\pi mk'T}{\hbar^2}}\right)\left(\frac{b}{2}\sqrt{\frac{8\pi mk'T}{\hbar^2}}\right)\left(\frac{c}{2}\sqrt{\frac{8\pi mk'T}{\hbar^2}}\right) \\ &= V\left(\frac{2\pi mk'T}{\hbar^2}\right)^{3/2} = Z(V, T), \end{aligned} \tag{11.5.77}$$

where $V = abc$. This equation delivers the partition function of an ideal monatomic gas, in which only the translational kinetic energies of gas particles are taken into account.

Substituting Eq. (11.5.77) into Eq. $(11.5.72)_2$ results in

$$p = \frac{N}{V}k'T. \tag{11.5.78}$$

Comparing this equation with Eq. $(11.1.18)_3$ from the kinetic theory of gas shows that the proportional constant k' is nothing else than Boltzmann's constant. Similarly, substituting Eq. (11.5.77) into Eqs. (11.5.69) and (11.5.70) yields respectively

$$U = \frac{3}{2}NkT, \qquad S = Nk\left[\frac{3}{2}\ln T + \ln\frac{V}{N} + \ln\left(\frac{2\pi mk}{\hbar^2}\right)^{3/2} + \frac{5}{2}\right]. \tag{11.5.79}$$

The first equation coincides exactly to Eq. (11.1.16) derived by using the kinetic theory of gas and shows that when gas particles having three translational degrees of freedom come to statistical equilibrium, the energy per gas particle is $3kT/2$. The second equation can further be expressed for a unity mole of gas, in which $N = N_A$ and $N_A k = \bar{R}$, so that

$$\bar{s} = \bar{c}_{V0}\ln T + \bar{R}\ln\bar{v} + \bar{R}\ln\left[\frac{(2\pi mk/\hbar^2)^{3/2}}{N_A}\right] + \frac{5}{2}\bar{R}. \tag{11.5.80}$$

Expressing Eq. (11.5.30) in terms of mole base gives

$$\bar{s} = \bar{c}_{V0}\ln T + \bar{R}\ln\bar{v} + \bar{s}_0. \tag{11.5.81}$$

Comparing the last two equations indicates that the reference value $\bar{s}_0$ can be computed by using the statistical mechanics.

It can be further shown that for an ideal monatomic gas consisting of a large number of indistinguishable, quasi-independent particles in equilibrium, the average internal energy per gas particle equals the product of energy modes and $kT/2$. This result is known as the *principle of equipartition of energy*. For an ideal monatomic gas, which has three energy modes as three translational kinetic energies, it can be shown by using the statistical mechanics that

$$\bar{u} = N_A\left(\frac{3kT}{2}\right) = \frac{3}{2}\bar{R}T, \qquad \bar{c}_{V0} = \left(\frac{\partial\bar{u}}{\partial T}\right)_V = \frac{3}{2}\bar{R}, \qquad \gamma \equiv \frac{\bar{c}_{p0}}{\bar{c}_{V0}} = \frac{5}{3}, \tag{11.5.82}$$

for $\bar{c}_{p0} = \bar{c}_{V0} + \bar{R}$ with $\bar{c}_{p0} = 5\bar{R}/2$, where γ is the specific heat ratio. For diatomic gases, it is found that

$$\bar{u} = \frac{5}{2}\bar{R}T, \qquad \bar{c}_{V0} = \frac{5}{2}\bar{R}, \qquad \bar{c}_{p0} = \frac{7}{2}\bar{R}, \qquad \gamma \equiv \frac{c_{p0}}{c_{v0}} = \frac{7}{3}, \tag{11.5.83}$$

if the gas particles are assumed to be in dumbbell shape with two additional rotational degrees of freedom.

Work and Heat from the Statistical Mechanics:

The statistical mechanics can be applied to derive the boundary phenomena as work and heat. It is assumed here that the vessel used in the previous discussions is a piston-cylinder device, so that a change in system volume is allowed. The energy level ϵ_i of individual particles undergoing only three translational energy modes is given in Eq. (11.5.40). Let B_i represent the sum of the squares of the quantum numbers corresponding to the ith energy level, so that

$$\epsilon_i = \frac{\hbar}{8m} B_i V^{-2/3}. \tag{11.5.84}$$

If the set of quantum numbers corresponding to ϵ_i is given, then B_i is known, so that the energy level depends only on the system volume. Taking logarithm to this equation yields

$$\ln \epsilon_i = \ln \frac{\hbar^2}{2m} + \ln B_i - \frac{2}{3} \ln V, \tag{11.5.85}$$

which is differentiated to obtain

$$\frac{\mathrm{d}\epsilon_i}{\epsilon_i} = -\frac{2}{3}\frac{\mathrm{d}V}{V}, \qquad \longrightarrow \qquad \mathrm{d}\epsilon_i = -\frac{2\epsilon_i}{3}\frac{\mathrm{d}V}{V}. \tag{11.5.86}$$

Multiplying this equation by N_i and taking summation over all possible energy levels of the resulting equation gives

$$\sum N_i \mathrm{d}\epsilon_i = -\frac{2}{3}\frac{\sum N_i \epsilon_i}{V}\mathrm{d}V = -\frac{2}{3}\frac{U}{V}\mathrm{d}V. \tag{11.5.87}$$

Since the microscopic expressions of pressure and internal energy from both the kinetic theory of gas and statistical mechanics are given by $p = NkT/V$ and $U = 3NkT/2$ under the circumstance that each particle assumes only three translational energy modes, substituting these expressions into Eq. (11.5.87) leads to

$$p = \frac{2}{3}\frac{U}{V}, \tag{11.5.88}$$

which is substituted into Eq. (11.5.87) to obtain

$$\sum N_i \mathrm{d}\epsilon_i = -p\,\mathrm{d}V, \tag{11.5.89}$$

showing that a change in system volume causes changes in the values of energy levels without alternating their corresponding populations. That is, work on a system only changes the values of energy levels.

Similarly, it follows from Eq. (11.5.62) that

$$\mathrm{d}\,(\ln \Omega) = \beta \sum \epsilon_i \mathrm{d}N_i, \tag{11.5.90}$$

provided that the populations of particles N_i change but the energy levels ϵ_i remain unchanged. Since $k\,\mathrm{d}(\ln\Omega) = \mathrm{d}S$, $k\beta\sum\epsilon_i\mathrm{d}N_i = \mathrm{d}S$ and $k\beta = 1/T$, this equation can be expressed alternatively as

$$\sum\epsilon_i\mathrm{d}N_i = T\mathrm{d}S, \tag{11.5.91}$$

showing that a reversible heat transfer only changes the populations of energy levels without changing the values of energy levels. Consequently, the expression $\mathrm{d}U = \sum N_i\mathrm{d}\epsilon_i + \sum\epsilon_i\mathrm{d}N_i$ is nothing else than the first law in the macroscopic description, with $\sum N_i\mathrm{d}\epsilon_i = -p\,\mathrm{d}V$ and $\sum\epsilon_i\mathrm{d}N_i = T\mathrm{d}S$.

11.5.6 Entropy as System Disorder and System Information

Based on the law of increase of entropy of an isolated system and the relation between entropy and microscopic thermodynamic probability, it follows that the entropy of a system is a measure of system disorder. If energy is to be extracted from a system as efficiently as possible, the energy should be stored in an ordered form. A mechanical energy stored in a spring is an ideal example. Thermal energy is also useful in extracting work, provided that the associated temperature is high, for T is the intensive variable associated with S. When energy is degraded in an irreversible process, it takes a less ordered form; e.g. in a mechanical friction process, the ordered mechanical energy is dissipated as the disordered molecular motions of heat. This concept applies equally for a "heat flow" down a temperature gradient, where the non-equilibrium ordering of thermal energy, corresponding to a finite temperature difference, is reduced. In other words, an increase in entropy of an isolated system indicates an increase in microscopic thermodynamic probability, which gives rise to more disordered molecular motions, so that the system disorder becomes increased.

The microscopic disordering of particles during an irreversible process arises from the fact that the motions of individual particles are free from the control of any human activity. Such a concept has been demonstrated by using the well-known Gedankenexperiment, *Maxwell's demon*. Since the postulation of this Gedankenexperiment, there have been attempts to disprove the second law by proposing a perpetual motion device of the second kind. Of particular interest is the concept of *information* proposed by Szilard in 1929,[23] which, without loss of generality, is a fundamental amount of entropy, and gives another interpretation for the number of ways by which a particular macro-state of a system may be achieved. If more information of a system could be known, the thermodynamic probability corresponding to a specific macro-state would be reduced.

Let the information of a system be denoted by I. A convenient measure of I conveyed when the number of ways reduced from Ω_0 to Ω_1 is given by

$$I = k\ln\frac{\Omega_0}{\Omega_1}, \tag{11.5.92}$$

[23] Leo Szilard, 1898–1964, a Hungarian-born American physicist, who conceived the nuclear chain reaction in 1933, which led eventually to the Manhattan Project that built an atomic bomb.

which shows that the more reduction of entropy is, the more information will be. Thus, information can be defined as *negentropy*. It follows then from Boltzmann's equation that

$$I = S_0 - S_1, \qquad S_1 = S_0 - I, \tag{11.5.93}$$

which can be interpreted that the entropy of a system is reduced by the amount of information about the state of a system. Or conversely, *entropy measures the lack of information about the exact state of a system*, as proposed by Brillouin. For example, consider an isothermal compression of an ideal gas containing N particles from a volume V_0 to a volume V_1, for which the reduction in entropy, from the macroscopic description, is obtained as

$$S_0 - S_1 = Nk \ln \frac{V_0}{V_1}. \tag{11.5.94}$$

However, a decrease in volume corresponds to a decrease in the number of ways in achieving this state, for there are fewer microstates with position coordinates in the smaller volume. Before the compression, each particle may be within the volume V_0, so that the number of locations that each particle could occupy is $V_0/\Delta V$, where ΔV is some arbitrary small volume. After the compression, each particle is to be found within V_1, with a smaller possible number of locations given by $V_1/\Delta V$. With these, it follows that for each particle,

$$I_p = k \ln \frac{\Omega_0}{\Omega_1} = k \ln \frac{V_0}{V_1}, \tag{11.5.95}$$

which is summed up over the entire gas of N particles to obtain

$$I = \sum I_p = Nk \ln \frac{V_0}{V_1}, \tag{11.5.96}$$

which coincides to Eq. (11.5.94). The increase in information as a result of compression is seen to be identical with the corresponding entropy reduction.

Remarkable progresses and advances of information theory have been made in deepening the understanding and applicability of information storage and information ensure. The discussions on these topics are beyond the scope of the book. Interesting readers can find some related literature in the further reading at the end of chapter.

11.5.7 Control-Mass and Control-Volume Formulations for a Process

The entropy change of a control-mass system undergoing a process is given in Eq. (11.5.21). This equation can be expressed alternatively as

$$\mathrm{d}S = \frac{\delta Q}{T} + \delta S_{gen}, \quad \delta S_{gen} \geq 0, \qquad \longrightarrow \qquad \dot{S} = \frac{\dot{Q}}{T} + \dot{S}_{gen}, \tag{11.5.97}$$

in which δS_{gen} is termed the *entropy generation* or *entropy production* induced by all internal and external irreversibilities in a process. The equality in the expression of $\delta S_{gen} \geq 0$ applies for a reversible process, while the greater than sign is valid for an irreversible process. This expression can also be recognized as another form of the second law. It is noted that Eq. (11.5.97)$_2$ corresponds to the general balance of entropy given in Eq. (5.2.2), in which the production $\mathcal{P} = \dot{S}_{gen}$, the flux $\mathcal{F}$ is accomplished via the heat transfer rate on the system boundary, and the supply $\mathcal{S} = 0$ for simplicity.

It follows from Eq. (11.5.97)$_1$ that

$$\delta Q = T\mathrm{d}S - T\delta S_{gen}, \tag{11.5.98}$$

which shows that the amount of heat transfer during a reversible process is simply the product $T\mathrm{d}S$, while in an irreversible process it is less than that for the reversible case under the same entropy change $\mathrm{d}S$. Similarly, substituting Eq. (11.5.98) into the $T\mathrm{d}S$ equations and subsequently the resulting equation into the first law yields

$$\delta W = -p\,\mathrm{d}V - T\delta S_{gen}, \tag{11.5.99}$$

showing that the amount of work in an irreversible process is reduced by an amount proportional to the entropy generation. This point has been discussed in Sect. 11.5.6 in terms of the arguments based on the Carnot engine, and the product $T\delta S_{gen}$ is often called the *lost work*, although it is not a real work or a lost energy, but rather a lost opportunity in extracting work.

For a control-volume system undergoing a process, the second law has been formulated as a global balance of entropy given in either Eq. (5.3.36) or (5.3.37). If it is further assumed that there exists no external entropy supply, then Eq. (5.3.36) reduces to

$$\dot{S}_{CV} = \sum \dot{m}_i s_i - \sum \dot{m}_o s_o + \sum \frac{\dot{Q}}{T} + \dot{S}_{gen}, \qquad \dot{S}_{gen \geq 0}, \tag{11.5.100}$$

with the correspondences given by

$$\dot{S}_{CV} = \frac{\mathrm{d}}{\mathrm{d}t}\int_V \eta\rho\,\mathrm{d}v = \frac{\partial}{\partial t}\int_V \eta\rho\,\mathrm{d}v, \quad \sum \dot{m}_o s_o - \sum \dot{m}_i s_i = \int_A \eta\,(\rho\boldsymbol{u}\cdot\boldsymbol{n})\,\mathrm{d}a,$$
$$\sum \frac{\dot{Q}}{T} = -\int_A \phi_\eta \cdot \boldsymbol{n}\mathrm{d}a, \qquad \dot{S}_{gen} = \int_V \rho\pi_\eta\,\mathrm{d}v. \tag{11.5.101}$$

For a steady process, Eq. (11.5.100) may be simplified to

$$\sum \dot{m}_o s_o - \sum \dot{m}_i s_i = \sum \frac{\dot{Q}}{T} + \dot{S}_{gen}, \tag{11.5.102}$$

which becomes

$$s_o - s_i = \sum \frac{q}{T} + s_{gen}, \tag{11.5.103}$$

provided that there exist only a single intake flow and a single discharge flow. For an adiabatic process, it follows from the above equation that

$$s_o = s_i + s_{gen} \geq s_i. \tag{11.5.104}$$

A similarity between fluid mechanics and classical thermodynamics is that the Bernoulli equation described in Sect. 7.3 corresponds to an isentropic flow. To show this, Eq. (11.5.103) is recast in the differential form

$$\delta q = T\,\mathrm{d}s - T\delta s_{gen}, \qquad \longrightarrow \qquad q = \mathrm{d}h - v\,\mathrm{d}p - T\delta s_{gen}, \tag{11.5.105}$$

in which the $T\mathrm{d}S$ equation has been used. Integrating this equation from the intake state "i" to the discharged state "o" yields

$$q_i^o = \int_i^o \delta q = (h_o - h_i) - \int_i^o v\,\mathrm{d}p - \int_i^o T\delta s_{gen}, \tag{11.5.106}$$

which is substituted into the first law, i.e., Eq. $(11.4.32)_2$ to obtain

$$w_i^o = w_{CV} = \int_i^o v\,\mathrm{d}p + \int_i^o T\delta s_{gen} + \frac{1}{2}\left(\boldsymbol{u}_o \cdot \boldsymbol{u}_o - \boldsymbol{u}_i \cdot \boldsymbol{u}_i\right) + g\left(z_o - z_i\right). \tag{11.5.107}$$

Under the assumptions that the flow is incompressible and isentropic, and there exists no shaft work in the control-volume, this equation reduces to

$$\frac{p_i}{\rho} + \frac{\boldsymbol{u}_i \cdot \boldsymbol{u}_i}{2} + gz_i = \frac{p_o}{\rho} + \frac{\boldsymbol{u}_o \cdot \boldsymbol{u}_o}{2} + gz_o, \tag{11.5.108}$$

which coincides to Eq. (7.3.9). This shows that the assumption of isentropic flow in classical thermodynamics is used to accomplish an incompressible frictionless flow in fluid mechanics.

For a control-volume system undergoing a transient process, the lump analysis described in Sect. 11.4.6 can be applied. Integrating Eq. (11.5.100) with respect to a given time duration gives

$$[m_2 s_2 - m_1 s_1]_{CV} = \sum m_i s_i - \sum m_e s_e + \sum \frac{Q}{T} + \Delta S_{gen}, \tag{11.5.109}$$

in which

$$\begin{aligned}
&\int_0^t \dot{S}_{CV}\,\mathrm{d}t = [m_2 s_2 - m_1 s_1]_{CV}\,, \\
&\int_0^t \sum \dot{m}_i s_i\,\mathrm{d}t = \sum m_i s_i, \qquad \int_0^t \sum \dot{m}_o s_o\,\mathrm{d}t = \sum m_o s_o, \\
&\int_0^t \sum \frac{\dot{Q}}{T}\,\mathrm{d}t = \sum \frac{Q}{T}, \qquad \int_0^t \dot{S}_{gen}\mathrm{d}t = \Delta S_{gen},
\end{aligned} \tag{11.5.110}$$

and the notations defined in Eqs. (11.4.33) and (11.4.34) have been used.

Based on the established results, the second law can be summarized as follows:

11.5 (The second law) *The time rate of change of entropy generation of a thermodynamic system, either in control-mass or control-volume formulation, must be greater*

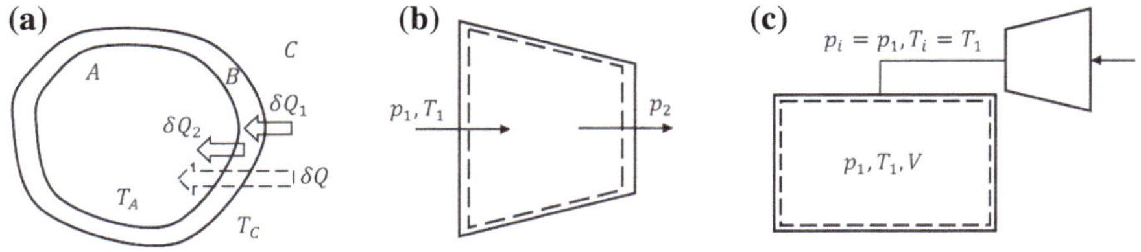

Fig. 11.20 Illustrations of the second law. **a** A heat transfer via a finite temperature difference in a control-mass system. **b** The work required for an air compressor in a steady-flow process. **c** The work required in a transient filling process of air into a rigid tank via an air compressor

than or equal to zero, with the equality and greater than sign devoting respectively to reversible and irreversible processes.[24]

11.5.8 Illustrations of Second Law

Consider one kilogram of air contained inside a cylinder, which is fitted with a frictionless piston with pressure p_1 and temperature T_1. The air is allowed to expand to $p_2 < p_1$ in a reversible adiabatic process. It is required to determine the amount of work during the process. The air is assumed to be an ideal gas, so that the first law per unit mass reads

$$u_2 - u_1 = q_1^2 + w_1^2, \qquad \longrightarrow \qquad w_1^2 = u_2 - u_1 = c_{V0}\,(T_2 - T_1)\,. \tag{11.5.111}$$

For a reversible adiabatic process, the second law per unit mass reduces to

$$\Delta s = s_2 - s_1 = c_{p0}\ln\frac{T_2}{T_1} - R\ln\frac{p_2}{p_1} = 0, \qquad \longrightarrow \qquad T_2 = T_1\frac{p_2}{p_1}\exp\left(\frac{R}{c_{p0}}\right), \tag{11.5.112}$$

which is substituted into Eq. (11.5.111)$_2$ to obtain

$$w_1^2 = T_1\left[c_{V0}\frac{p_2}{p_1}\exp\left(\frac{R}{c_{p0}}\right) - 1\right]. \tag{11.5.113}$$

Since $p_2 < p_1$, $c_{p0} - c_{V0} = R$ and $c_{p0}/c_{V0} = \gamma = 1.4$ for air, Eq. (11.5.113) shall yield a negative value of w_1^2, indicating that the work has been done by the system to the surrounding.

Consider a body A with temperature T_A and its surrounding C with temperature T_C, as shown in Fig. 11.20a. There exists a wall B with finite thickness between A

[24]It can also be formulated as "*the time rate of change of total entropy of an isolated system must be greater than or equal to zero.*" So far, the macroscopic definition of entropy is provided based on the traditional treatments. A more mathematical abstract in defining the *empirical entropy* can be accomplished by using the *Carathéodory formulation of second Law*. Constantin Carathéodory, 1873–1950, a Greek mathematician. He pioneered the axiomatic formulation of thermodynamics along a purely geometrical approach in 1909.

and C, and it is assumed that $T_C > T_A$, so that an amount of heat transfer δQ from C through B toward A presents. The state of wall is assumed to be unchanged during the heat transfer process, but is not uniform; i.e., its temperatures at the contact surfaces with A and C are respectively T_A and T_C. It is required to determine the entropy generation induced by the heat transfer process in the wall. The wall B is chosen as the control-mass system, for which the first law reads

$$\mathrm{d}E = 0 = \delta Q_1 - \delta Q_2, \qquad \longrightarrow \qquad \delta Q_1 = \delta Q_2 = \delta Q. \tag{11.5.114}$$

For the control-mass system, the second law reads

$$\mathrm{d}S = 0 = \frac{\delta Q_1}{T_C} - \frac{\delta Q_2}{T_A} + \delta S_{gen}, \tag{11.5.115}$$

which is recast alternatively as

$$\delta S_{gen} = \delta Q \left(\frac{1}{T_A} - \frac{1}{T_C} \right), \tag{11.5.116}$$

in which Eq. (11.5.114) has been used. Since $T_C > T_A$, it is seen that $\delta S_{gen} > 0$. For the circumstance in which $T_C < T_A$, the same conclusion is obtained, for the heat transfer is now accomplished from A through B toward C. This example demonstrates the direction of heat transfer from a higher-temperature domain to a lower-temperature domain as a natural consequence of the second law.

Consider air entering a compressor, as shown in Fig. 11.20b. The air enters the compressor with p_1 and T_1, while it leaves the compressor with pressure p_2. If the compression process is assumed to be steady, reversible, and adiabatic, it is required to determine the work to accomplish the compression process. To obtain this, construct the control-volume system, as shown by the dashed line in the figure, for which the first law reads

$$\begin{aligned} & q_1^2 + w_1^2 + \left(h_1 + \frac{1}{2}\boldsymbol{u}_1 \cdot \boldsymbol{u}_1 + g z_1 \right) - \left(h_2 + \frac{1}{2}\boldsymbol{u}_2 \cdot \boldsymbol{u}_2 + g z_2 \right) = 0, \\ \longrightarrow \quad & w_1^2 = h_2 - h_1 = c_{p0}\,(T_2 - T_1)\,, \end{aligned} \tag{11.5.117}$$

in which air as an ideal gas, $z_1 = z_2 = 0$ and $\|\boldsymbol{u}_1\| \sim \|\boldsymbol{u}_2\| \sim 0$ have been assumed. Similarly, it follows from the second law that

$$\Delta s = s_2 - s_1 = 0 = c_{p0} \ln \frac{T_2}{T_1} - R \ln \frac{p_2}{p_1}, \qquad \longrightarrow \qquad T_2 = T_1 \frac{p_2}{p_1} \exp\left(\frac{R}{c_{p0}} \right). \tag{11.5.118}$$

Substituting this result into Eq. (11.5.117) results in

$$w_1^2 = T_1 \left[c_{p0} \frac{p_2}{p_1} \exp\left(\frac{R}{c_{p0}} \right) - 1 \right], \tag{11.5.119}$$

which must assume positive values, for work is required to compress air from a low pressure to a high pressure.

Consider a tank connected to an air compressor, as shown in Fig. 11.20c. Initially, the air pressure and temperature inside the tank are p_1 and T_1 respectively and the tank volume is denoted by V. The compressor is started to charge the tank up to a pressure $p_2 \gg p_1$, and then it shuts off. It is required to determine the air temperature T_2 inside the tank after the filling process, and the amount of work required to fill the tank. For simplicity, the filling process is assumed to be a reversible and adiabatic one, and construct the control-volume system, as shown by the dashed line in the figure. Since the filling process is unsteady, it follows from the mass balance in a transient process that

$$(m_2 - m_1)_{CV} = m_i, \tag{11.5.120}$$

where m_i represents the total amount of air mass which is delivered to the tank during the filling process. Similarly, the first and second laws read respectively

$$\begin{aligned}(m_2u_2 - m_1u_1)_{CV} &= Q + W + m_ih_i,\\ (m_2s_2 - m_1s_1)_{CV} &= \sum \frac{Q}{T} + m_is_i + \Delta S_{gen},\end{aligned} \tag{11.5.121}$$

which are simplified to

$$W = (m_2u_2 - m_1u_1)_{CV} - m_ih_i, \qquad m_2s_2 = m_1s_1 + m_is_i = m_2s_1, \tag{11.5.122}$$

in which Eq. (11.5.120) has been used, and it is noted that $s_i = s_1$ and $h_i = h_1$. It follows from the second equation that $s_1 = s_2$ in the CV, so that

$$\begin{aligned}&\Delta s_{CV} = (s_2 - s_1)_{CV} = 0 = c_{p0}\ln\frac{T_2}{T_1} - R\ln\frac{p_2}{p_1},\\ &\longrightarrow T_2 = T_1\frac{p_2}{p_1}\exp\left(\frac{R}{c_{p0}}\right).\end{aligned} \tag{11.5.123}$$

With this, the air masses inside the CV before and after the filling process, by using the ideal gas state equation, are determined to be

$$m_1 = \frac{p_1V}{RT_1}, \qquad m_2 = \frac{p_2V}{RT_2}, \qquad m_i = m_2 - m_1 = \frac{V}{R}\left(\frac{p_2}{T_2} - \frac{p_1}{T_1}\right). \tag{11.5.124}$$

Substituting Eqs. (11.5.123) and (11.5.124) into Eq. $(11.5.122)_1$ results in

$$W = \frac{p_1V}{R}\left\{c_{V0}\left(\frac{p_2}{p_1} - 1\right) + c_{p0}\left[1 - \exp\left(-\frac{R}{c_{p0}}\right)\right]\right\}, \tag{11.5.125}$$

which should assume positive values, for $p_2 \gg p_1$. This result coincides to the physical observation, because work needs to be delivered to the control-volume to complete the filling process.

11.6 Entropy Principles and Continuum Thermodynamics

11.6.1 Entropy Principles

In Sect. 5.6, the material equations of substances have been derived by following some universal principles, and Eq. (5.6.22) was obtained for viscous thermoelastic fluids. Further investigations on this equation may be possible via the second law of thermodynamics. It follows that Eq. (5.3.43) must be satisfied for all admissible physical processes. In other words, every permissible choice of the material equations specifies a system of field equations, including the balances of mass, linear and angular momentums, and energy, whose solutions must conform the second law; i.e., the entropy production must be nonnegative. The accomplishment of this objective is called the *continuum thermodynamics*,[25] for which two approaches will be introduced in the context of *entropy principle*. The main differences between two approaches are the postulates of entropy supply and entropy flux, and the treatments of balance equations in the second law.

The Coleman-Noll approach.[26] The relations between the entropy flux and entropy supply $\{\phi_\eta, s_\eta\}$ and heat flux and external energy supply $\{\boldsymbol{q}, \zeta\}$ are specified by following the Duhem-Truesdell relations associated with the absolute temperature T given by

$$\phi_\eta = \frac{\boldsymbol{q}}{T}, \qquad s_\eta = \frac{\zeta}{T}. \tag{11.6.1}$$

When incorporating these relations into the second law, it is assumed that the balance of linear momentum accommodates non-vanishing external supply terms which can be prescribed arbitrarily, so that they can take any value to render their effect when necessary. This point will be explored in the next subsection.

The Müller-Liu approach. To softening the assumptions made in the Coleman-Noll approach, Müller formulated a weaker form of the entropy principle, which are summarized in the following.

1. In every material, there exists a specific entropy η, which is an additive quantity and should satisfy the local balance statement given in Eq. (5.3.43).
2. The specific entropy η and its flux ϕ_η are considered material quantities, for which the same material laws hold as for the remaining material quantities in accordance with the *rule of equipresence*.
3. The entropy production π_η must be nonnegative for all thermodynamic processes.
4. The external supply terms appearing in the balance equations cannot influence the material responses.
5. There exist special material singular surfaces between two continua, across which the (empirical) temperature and the tangential velocity are continuous. The singular surfaces are called the *ideal walls*.

[25]Continuum thermodynamics can be classified into several subdisciplines. For a detailed discussion, see e.g. Muschik, W., Papenfuss, C., Ehrentraut, H., A sketch of continuum thermodynamics, *J. Non-Newtonian Fluid Mech.*, 96, 255–290, 2001.

[26]A more precise description of this approach is the Duhem-Truesdell relations with the Coleman-Noll exploitation.

11.6.2 Continuum Thermodynamics

To explore the difference between two entropy principles, consider a viscous heat-conducting compressible fluid as the working substance, whose material responses, as indicated by Eq. (5.6.22), are prescribed by

$$\mathcal{C} = \mathcal{C}(\rho, \boldsymbol{D}, T, \text{grad}\, T), \qquad \mathcal{C} \in \{\epsilon, \eta, \boldsymbol{q}, \boldsymbol{t}\}. \tag{11.6.2}$$

Further investigations on the above relations are first investigated by using the Coleman-Noll approach, followed by the Müller-Liu approach.

The Coleman-Noll approach. The material responses, as implied by Eq. (11.6.2), are prescribed by the balance of energy and entropy given respectively by

$$\rho\dot{\epsilon} = -\text{div}\,\boldsymbol{q} + \text{tr}\,(\boldsymbol{D}\boldsymbol{t}) + \rho\zeta, \qquad \rho\dot{\eta} + \text{div}\left(\frac{\boldsymbol{q}}{T}\right) - \rho\frac{\zeta}{T} = \rho\pi_\eta \geq 0, \tag{11.6.3}$$

in which Eq. (11.6.1) has been substituted into the balance of entropy. Combining two equations yields

$$\rho\,(T\dot{\eta} - \dot{\epsilon}) + \text{tr}\,(\boldsymbol{D}\boldsymbol{t}) - \frac{\boldsymbol{q}\cdot\text{grad}\,T}{T} \geq 0, \tag{11.6.4}$$

which is expanded in the form

$$\begin{aligned}
&\rho\left(T\frac{\partial\eta}{\partial\boldsymbol{D}} - \frac{\partial\epsilon}{\partial\boldsymbol{D}}\right)\cdot\dot{\boldsymbol{D}} + \rho\left(T\frac{\partial\eta}{\partial T} - \frac{\partial\epsilon}{\partial T}\right)\dot{T} \\
&+\rho\left(T\frac{\partial\eta}{\partial(\text{grad}\,T)} - \frac{\partial\epsilon}{\partial(\text{grad}\,T)}\right)\cdot(\text{grad}\,T)^{\cdot} \\
&+\left[-\rho^2\left(T\frac{\partial\eta}{\partial\rho} - \frac{\partial\epsilon}{\partial\rho}\right)\boldsymbol{I} + \boldsymbol{t}\right]\cdot\boldsymbol{D} - \frac{\boldsymbol{q}\cdot(\text{grad}\,T)}{T} \geq 0,
\end{aligned} \tag{11.6.5}$$

in which the time rate of changes of Eq. (11.6.2) and the balance of mass have been substituted by using the chain rule of differentiation. This inequality must be satisfied by all admissible thermodynamic processes. In obtaining Eq. (11.6.5), the balance of linear momentum has not been accounted for, for it is assumed that the external supplies in the balance statement can be so selected that it holds identically.

The inequality (11.6.5) can be written in the form as

$$\boldsymbol{a}\cdot\boldsymbol{x} + \Gamma \geq 0, \tag{11.6.6}$$

with the vectors $\boldsymbol{a}$ and $\boldsymbol{b}$ being in the n-dimensional space and the scalar Γ given by

$$\boldsymbol{a} = \left[\rho\left(T\frac{\partial\eta}{\partial\boldsymbol{D}} - \frac{\partial\epsilon}{\partial\boldsymbol{D}}\right), \rho\left(T\frac{\partial\eta}{\partial T} - \frac{\partial\epsilon}{\partial T}\right), \rho\left(T\frac{\partial\eta}{\partial(\text{grad}\,T)} - \frac{\partial\epsilon}{\partial(\text{grad}\,T)}\right)\right], \tag{11.6.7}$$

$$\boldsymbol{x} = \left[\dot{\boldsymbol{D}}, \dot{T}, (\text{grad}\,T)^{\cdot}\right], \qquad \Gamma = \left[-\rho^2\left(T\frac{\partial\eta}{\partial\rho} - \frac{\partial\epsilon}{\partial\rho}\right)\boldsymbol{I} + \boldsymbol{t}\right]\cdot\boldsymbol{D} - \frac{\boldsymbol{q}\cdot(\text{grad}\,T)}{T}.$$

It is readily to verify that $\boldsymbol{a}$ is independent of $\boldsymbol{x}$, and since $\boldsymbol{x}$ can take any values, inequality (11.6.6) cannot be fulfilled unless

$$\boldsymbol{a} = \boldsymbol{0}, \qquad \Gamma \geq 0. \tag{11.6.8}$$

The proof of the independency between $\boldsymbol{a}$ and $\boldsymbol{x}$ is left as an exercise.

The condition $\boldsymbol{a} = \boldsymbol{0}$ yields

$$T\frac{\partial \eta}{\partial \alpha} = \frac{\partial \epsilon}{\partial \alpha}, \qquad \alpha \in \{\boldsymbol{D}, T, \text{grad}\, T\}, \tag{11.6.9}$$

which must hold as identities. Conducting mixed derivatives to the above equation with respect to each member of α gives

$$\frac{\partial \epsilon}{\partial \boldsymbol{D}} = \boldsymbol{0}, \qquad \frac{\partial \eta}{\partial \boldsymbol{D}} = \boldsymbol{0}, \qquad \frac{\partial \epsilon}{\partial (\text{grad}\, T)} = \boldsymbol{0}, \qquad \frac{\partial \eta}{\partial (\text{grad}\, T)} = \boldsymbol{0}, \tag{11.6.10}$$

which, by referring to Eq. (11.6.2), shows that

$$\epsilon = \epsilon\,(\rho, T)\,, \qquad \eta = \eta\,(\rho, T)\,, \qquad \frac{\partial \epsilon}{\partial T} = T\frac{\partial \eta}{\partial T}. \tag{11.6.11}$$

Further investigations may become possible by applying the condition $\Gamma \geq 0$ in thermodynamic equilibrium, which, in view of Eq. (11.6.2), is defined to be a thermodynamic process with vanishing entropy production with uniformly distributed temperature and velocity fields. That is,

$$\pi_\eta|_E = 0, \qquad \longleftrightarrow \qquad \text{grad}\, T|_E = \boldsymbol{0}, \;\; \boldsymbol{D}|_E = \boldsymbol{0}, \tag{11.6.12}$$

where the symbol "$|_E$" denotes that the indexed quantity is evaluated at thermodynamic equilibrium. Imposing thermodynamic equilibrium on Γ shows that

$$\Gamma|_E = 0, \qquad \Gamma|_E = \text{minimum}, \tag{11.6.13}$$

which leads to

$$\left(\frac{\partial \Gamma}{\partial \boldsymbol{D}}\right)\Big|_E = \boldsymbol{0}, \qquad \left(\frac{\partial \Gamma}{\partial (\text{grad}\, T)}\right)\Big|_E = \boldsymbol{0}, \tag{11.6.14}$$

and that the matrix

$$\left.\begin{pmatrix} \dfrac{\partial^2 \Gamma}{\partial \boldsymbol{D} \partial \boldsymbol{D}} & \dfrac{\partial^2 \Gamma}{\partial \boldsymbol{D} \partial (\text{grad}\, T)} \\ \dfrac{\partial^2 \Gamma}{\partial (\text{grad}\, T) \partial \boldsymbol{D}} & \dfrac{\partial^2 \Gamma}{\partial (\text{grad}\, T) \partial (\text{grad}\, T)} \end{pmatrix}\right|_E, \tag{11.6.15}$$

must be positive semi-definite. Equation (11.6.14) gives

$$\boldsymbol{t}|_E = \rho^2\left(T\frac{\partial \eta}{\partial \rho} - \frac{\partial \epsilon}{\partial \rho}\right)\boldsymbol{I} \equiv -p\boldsymbol{I}, \qquad \boldsymbol{q}|_E = \boldsymbol{0}, \tag{11.6.16}$$

where p is a scalar used as an abbreviation for the scalar expressions in front of $\boldsymbol{I}$. Equation (11.6.16) shows that a viscous, heat-conducting fluid is isotropic, whose stress tensor in equilibrium is essentially determined, provided that ϵ and η are known. In parallel, the equilibrium heat flux vanishes identically.

Substituting Eq. (11.6.16) into Eq. (11.6.11) yields

$$\epsilon = \epsilon\,(\rho, T)\,, \qquad \eta = \eta\,(\rho, T)\,, \qquad \frac{\partial \eta}{\partial \rho} = \frac{1}{T}\left(\frac{\partial \epsilon}{\partial \rho} - \frac{p}{\rho^2}\right), \qquad \frac{\partial \eta}{\partial T} = \frac{1}{T}\frac{\partial \epsilon}{\partial T}, \tag{11.6.17}$$

by which the total differential of ϵ is obtained as

$$\mathrm{d}\epsilon = T\,\mathrm{d}\eta + \frac{p}{\rho^2}\,\mathrm{d}\rho, \qquad \longrightarrow \qquad \mathrm{d}\eta = \frac{1}{T}\left[\mathrm{d}\epsilon + p\,\mathrm{d}\left(\frac{1}{\rho}\right)\right], \tag{11.6.18}$$

where the second equation is simply one of the $T\mathrm{d}s$ equations. Hence, Eq. (11.6.17)$_3$ may be thought of as the definition of pressure, provided that the material equations of ϵ and η are known. Specifically, the material equation for the pressure, e.g. $p = p(\rho, T)$, is referred to as the *thermal state equation*, while the expressions $\epsilon = \epsilon(\rho, T)$ and $\eta = \eta(\rho, T)$ are termed the *caloric state equations*. It follows that the material equations are not prescribed independently of each other; all the more, the caloric state equations determine then the thermal state equation.

The extra stress tensor $\boldsymbol{T}$ is introduced as

$$\boldsymbol{T} = \boldsymbol{t} + p\boldsymbol{I}, \qquad \boldsymbol{T}|_E = \boldsymbol{0}, \tag{11.6.19}$$

so that the residual entropy inequality, i.e., Eq. (11.6.7)$_3$, becomes

$$\Gamma = \boldsymbol{T}\cdot\boldsymbol{D} - \frac{\boldsymbol{q}\cdot\operatorname{grad}T}{T} \geq 0. \tag{11.6.20}$$

For simplicity, based on the rule of equipresence, $\boldsymbol{T}$ and $\boldsymbol{q}$ are proposed as

$$\boldsymbol{T} = \boldsymbol{T}\left(\rho, T, \boldsymbol{D}\right), \qquad \boldsymbol{q} = \boldsymbol{q}\left(\rho, T, \operatorname{grad}T\right), \tag{11.6.21}$$

whose most general isotropic expressions are given by

$$\boldsymbol{T} = a_1\boldsymbol{I} + a_2\boldsymbol{D} + a_3\boldsymbol{D}^2, \qquad \boldsymbol{q} = -k\operatorname{grad}T, \tag{11.6.22}$$

with

$$a_i = a_i\left(\rho, T, I^1_{\boldsymbol{D}}, I^2_{\boldsymbol{D}}, I^3_{\boldsymbol{D}}\right), \qquad k = k\left(\rho, T, \|\operatorname{grad}T\|\right), \tag{11.6.23}$$

where $a_1(\rho, T, 0, 0, 0) = 0$. Equations (11.6.22) and (11.6.23) represent the most general material equations for the extra stress tensor and heat flux vector of a viscou heat conducting compressible fluid. The coefficients in Eq. (11.6.23) can be restricted by using Eq. (11.6.15).

To explore the idea, it is assumed for simplicity that $\boldsymbol{T}$ and $\boldsymbol{q}$ depend linearly on $\boldsymbol{D}$ and grad T respectively so that Eq. (11.6.22) reduces to

$$\boldsymbol{T} = \kappa I^1_{\boldsymbol{D}}\boldsymbol{I} + 2\mu\boldsymbol{D}, \qquad \boldsymbol{q} = -k\operatorname{grad}T, \tag{11.6.24}$$

where the coefficients κ, μ, and λ are respectively the bulk and shear viscosities, and the thermal conductivity, which are all functions of ρ and T. These expressions are suitable for the Fourier heat-conducting Newtonian fluids. Substituting Eq. (11.6.24) into Eq. (11.6.20) results in

$$\Gamma = \kappa\left(I^1_{\boldsymbol{D}}\right)^2 + 2\mu\boldsymbol{D}\cdot\boldsymbol{D} + k\frac{\|\operatorname{grad}T\|^2}{T} \geq 0, \tag{11.6.25}$$

which is recast alternatively as

$$\Gamma = \kappa\frac{x^2}{2} + \mu\frac{y^2}{2} + k\frac{z^2}{2} \geq 0, \tag{11.6.26}$$

with x, y, and z defined by

$$x \equiv \sqrt{2} I_D^1, \qquad y \equiv \sqrt{4 \boldsymbol{D} \cdot \boldsymbol{D}}, \qquad z \equiv \sqrt{\frac{2}{T}} \| \operatorname{grad} T \|, \tag{11.6.27}$$

in which it is noted that $\Gamma|_E = 0$ at $x = y = z = 0$. Applying the condition that $\Gamma|_E$ assumes a minimum value with respect to Eq. (11.6.26) gives

$$\left.\frac{\partial \Gamma}{\partial x}\right|_E = 0, \qquad \left.\frac{\partial \Gamma}{\partial y}\right|_E = 0, \qquad \left.\frac{\partial \Gamma}{\partial z}\right|_E = 0, \tag{11.6.28}$$

and the matrix

$$\begin{pmatrix} \dfrac{\partial^2 \Gamma}{\partial x^2} & 0 & 0 \\ 0 & \dfrac{\partial^2 \Gamma}{\partial y^2} & 0 \\ 0 & 0 & \dfrac{\partial^2 \Gamma}{\partial z^2} \end{pmatrix} = \begin{pmatrix} \kappa & 0 & 0 \\ 0 & \mu & 0 \\ 0 & 0 & k \end{pmatrix}, \tag{11.6.29}$$

must be positive semi-definite. These two equations can only be fulfilled with

$$\kappa = \kappa(\rho, T) \geq 0, \qquad \mu = \mu(\rho, T) \geq 0, \qquad k = k(\rho, T) \geq 0, \tag{11.6.30}$$

showing that the bulk and shear viscosities and the thermal conductivity in a linear, heat-conducting fluid are compatible with the second law, provided that they are functions of density and temperature with positive values.

It is recognized that the Coleman-Noll approach in exploring the second law restricts considerations to the analysis of open systems, for the external supplies in the balance of linear momentum can be so chosen that they do not affect the exploitation of entropy inequality. This assumption, although mathematically convenient, may be physically questionable, for the physical world may not be so general as to allow arbitrarily large or small external supplies. Additionally, a knowledge of the thermodynamic temperature is required, which was taken over from the classical thermodynamics for simple systems. For complex systems, a priori existence of the absolute temperature may not be appropriate. Last, the validity of the relations between the entropy flux and entropy supply, and heat flux and energy supply given in Eq. (11.6.1) is not automatically ascertained. These relations also demand a priori existence of the absolute temperature.

The Müller-Liu approach. For demonstration, a heat-conducting compressible fluid is considered the working substance, whose material responses are prescribed by

$$\mathcal{C} = \mathcal{C}(\rho, \Theta, \operatorname{grad} \Theta), \qquad \mathcal{C} \in \left\{\epsilon, \eta, \boldsymbol{q}, \boldsymbol{t}, \boldsymbol{\phi}_\eta\right\}, \tag{11.6.31}$$

where Θ is an empirical temperature, the entropy flux ϕ is a material quantity, and $\boldsymbol{D}$ is omitted in $\mathcal{C}$ for simplicity. The balances of mass, linear momentum, and energy read respectively

$$\dot{\rho} + \rho(\operatorname{div} \boldsymbol{u}) = 0, \qquad \rho \dot{\boldsymbol{u}} - \operatorname{div} \boldsymbol{t} - \rho \boldsymbol{b} = \boldsymbol{0}, \qquad \rho \dot{\epsilon} + \operatorname{div} \boldsymbol{q} - \operatorname{tr}(\boldsymbol{t} \boldsymbol{D}) - \rho \zeta = 0, \tag{11.6.32}$$

in which ρ, Θ, and $\boldsymbol{u}$ are considered independent fields. A thermodynamically permissible process should be one which satisfies the second law and Eq. (11.6.32) simultaneously. This can be achieved by considering Eq. (11.6.32) as the constraints of the second law via the method of the Lagrangian multipliers, viz.,

$$\begin{aligned}&\rho\dot{\eta}+\operatorname{div}\boldsymbol{\phi}_{\eta}-\rho s_{\eta}-\lambda^{\rho}\left[\dot{\rho}+\rho(\operatorname{div}\boldsymbol{u})\right]-\boldsymbol{\lambda}^{u}\cdot\left[\rho\dot{\boldsymbol{u}}-\operatorname{div}\boldsymbol{t}-\rho\boldsymbol{b}\right]\\&-\lambda^{\epsilon}\left[\rho\dot{\epsilon}+\operatorname{div}\boldsymbol{q}-\operatorname{tr}(\boldsymbol{t}\boldsymbol{D})-\rho\zeta\right]\geq 0,\end{aligned}\tag{11.6.33}$$

where λ^{ρ}, $\boldsymbol{\lambda}^{u}$, and λ^{ϵ} are respectively the Lagrangian multipliers of the balances of mass, linear momentum, and energy. It readily verified that the second law and Eq. (11.6.32) imply Eq. (11.6.33), and vice versa, i.e., satisfying Eq. (11.6.33) for unrestricted fields and satisfying simultaneously the second law and Eq. (11.6.32) are equivalent.[27]

Substituting Eq. (11.6.31) into Eq. (11.6.33) by using the chain rule of differentiation yields

$$\begin{aligned}&\rho\left(\frac{\partial\eta}{\partial\Theta}-\lambda^{\epsilon}\frac{\partial\epsilon}{\partial\Theta}\right)\dot{\theta}+\rho\left(\frac{\partial\eta}{\partial\rho}-\lambda^{\epsilon}\frac{\partial\epsilon}{\partial\rho}-\frac{\lambda^{\rho}}{\rho}\right)\dot{\rho}\\&+\rho\left(\frac{\partial\eta}{\partial(\operatorname{grad}\Theta)}-\lambda^{\epsilon}\frac{\partial\epsilon}{\partial(\operatorname{grad}\Theta)}\right)(\operatorname{grad}\Theta)^{\cdot}+\left(\frac{\partial\boldsymbol{\phi}_{\eta}}{\partial\rho}-\lambda^{\epsilon}\frac{\partial\boldsymbol{q}}{\partial\rho}+\boldsymbol{\lambda}^{u}\frac{\partial\boldsymbol{t}}{\partial\rho}\right)\cdot\operatorname{grad}\rho\\&+\left(\frac{\partial\boldsymbol{\phi}_{\eta}}{\partial(\operatorname{grad}\Theta)}-\lambda^{\epsilon}\frac{\partial\boldsymbol{q}}{\partial(\operatorname{grad}\Theta)}+\boldsymbol{\lambda}^{u}\frac{\partial\boldsymbol{t}}{\partial(\operatorname{grad}\Theta)}\right)\cdot\operatorname{grad}(\operatorname{grad}\Theta)-\rho\boldsymbol{\lambda}^{u}\cdot\boldsymbol{u}\\&+\left(\frac{\partial\boldsymbol{\phi}_{\eta}}{\partial\Theta}-\lambda^{\epsilon}\frac{\partial\boldsymbol{q}}{\partial\Theta}+\boldsymbol{\lambda}^{u}\frac{\partial\boldsymbol{t}}{\partial\Theta}\right)\cdot\operatorname{grad}\Theta+\lambda^{\epsilon}\operatorname{tr}\left[\left(\boldsymbol{t}-\rho\frac{\lambda^{\rho}}{\lambda^{\epsilon}}\boldsymbol{I}\right)\boldsymbol{D}\right]\\&-\rho s_{\eta}+\rho\boldsymbol{\lambda}^{u}\cdot\boldsymbol{b}+\rho\lambda^{\epsilon}\zeta\geq 0.\end{aligned}\tag{11.6.34}$$

Since it is assumed that the external supplies cannot influence the material responses, it follows that

$$s_{\eta}=\lambda^{\epsilon}\zeta+\boldsymbol{\lambda}^{u}\cdot\boldsymbol{b},\tag{11.6.35}$$

which serves as an identity to determine the entropy supply. When compared with Eq. $(11.6.1)_2$, Eq. (11.6.35) is more general than that postulated in the Duhem-Truesdell relations.

In a similar procedure as the Coleman-Noll approach, inequality (11.6.34) can be recast alternatively as

$$\boldsymbol{a}\cdot\boldsymbol{x}+\Gamma\geq 0,\tag{11.6.36}$$

with

$$\boldsymbol{x}=\left[\dot{\Theta},\,\dot{\rho},\,(\operatorname{grad}\Theta)^{\cdot},\,\operatorname{grad}\rho,\,\operatorname{grad}(\operatorname{grad}\Theta),\,\boldsymbol{D}\right].\tag{11.6.37}$$

Since $\boldsymbol{x}$ can take any value at a fixed material point, i.e., it is possible to reconstruct an admissible thermodynamic process with arbitrary $\boldsymbol{x}$; the necessary and sufficient condition to fulfill inequality (11.6.36) is that $\boldsymbol{a}=\boldsymbol{0}$ and $\Gamma\geq 0$. Or alternatively,

[27] The proof can be found in Liu, I.S., Method of Lagrange multipliers for exploitation of the entropy principle, *Archive of Rational Mechanics and Analysis*, 46, 131–148, 1972.

$$\frac{\partial \eta}{\partial \Theta} - \lambda^\epsilon \frac{\partial \epsilon}{\partial \Theta} = 0, \qquad \frac{\partial \eta}{\partial \rho} - \lambda^\epsilon \frac{\partial \epsilon}{\partial \rho} - \frac{\lambda^\rho}{\rho} = 0,$$

$$\frac{\partial \eta}{\partial (\text{grad}\,\Theta)} - \lambda^\epsilon \frac{\partial \epsilon}{\partial (\text{grad}\,\Theta)} = \mathbf{0}, \qquad \frac{\partial \boldsymbol{\phi}_\eta}{\partial \rho} - \lambda^\epsilon \frac{\partial \boldsymbol{q}}{\partial \rho} + \boldsymbol{\lambda}^u \frac{\partial \boldsymbol{t}}{\partial \rho} = \mathbf{0},$$

$$\text{sym}\left\{\frac{\partial \boldsymbol{\phi}_\eta}{\partial (\text{grad}\,\Theta)} - \lambda^\epsilon \frac{\partial \boldsymbol{q}}{\partial (\text{grad}\,\Theta)} + \boldsymbol{\lambda}^u \frac{\partial \boldsymbol{t}}{\partial (\text{grad}\,\Theta)}\right\} = \mathbf{0}, \; \boldsymbol{t} = \rho \frac{\lambda^\rho}{\lambda^\epsilon} \boldsymbol{I} = -p\boldsymbol{I}, \tag{11.6.38}$$

must hold. It is found that the Lagrangian multipliers, as they are determined alone by the material quantities and themselves, can only depend on the independent material quantities. Hence, inequality (11.6.36) is equally linear in $\dot{\boldsymbol{u}}$, which gives rise to

$$\boldsymbol{\lambda}^u = \mathbf{0}, \tag{11.6.39}$$

showing that the linear momentum equation does not modify the exploitation of entropy principle, at least not in the considered restricted theory of a heat-conducting compressible fluid. This result also confirms the assumption used in the Coleman-Noll approach in exploring the entropy inequality. Last, Eq. $(11.6.38)_6$ indicates that the stress tensor is isotropic and becomes determined once λ^ϵ and λ^ρ are known.

The entropy flux and heat flux are further assumed to be objective, so that their representations, as motivated by Eq. $(11.6.31)_1$, are proposed as

$$\boldsymbol{\phi}_\eta = -\phi'_\eta\left(\rho, \Theta, \|\text{grad}\,\Theta\|^2\right) \text{grad}\,\Theta, \qquad \boldsymbol{q} = -q'\left(\rho, \Theta, \|\text{grad}\,\Theta\|^2\right) \text{grad}\,\Theta, \tag{11.6.40}$$

where the minus signs are introduced for convenience. Substituting these expressions into Eqs. $(11.6.38)_{2-6}$ yields

$$\begin{aligned} &\phi'_\eta \boldsymbol{I} + 2\frac{\partial \phi'_\eta}{\partial \|\text{grad}\,\Theta\|^2} (\text{grad}\,\Theta \otimes \text{grad}\,\Theta) \\ &= \lambda^\epsilon q' \boldsymbol{I} + 2\lambda^\epsilon \frac{\partial q'}{\partial \|\text{grad}\,\Theta\|^2} (\text{grad}\,\Theta \otimes \text{grad}\,\Theta), \end{aligned} \tag{11.6.41}$$

in which Eq. (11.6.39) has been used. It follows that

$$\phi'_\eta = \lambda^\epsilon q', \qquad \frac{\partial \lambda^\epsilon}{\partial \|\text{grad}\,\Theta\|^2} = 0, \tag{11.6.42}$$

for (grad Θ) can take any value. The Eq. (11.6.42) shows that the entropy flux is collinear with the heat flux, whereby the factor is simply the Lagrangian multiplier λ^ϵ, which is not a function of (grad Θ). Substituting Eqs. (11.6.39), (11.6.40) and (11.6.42) into Eq. $(11.6.38)_4$ gives

$$\lambda^\epsilon \frac{\partial q'}{\partial \rho} + \frac{\partial \lambda^\epsilon}{\partial \rho} q' = \lambda^\epsilon \frac{\partial q'}{\partial \rho}, \qquad \longrightarrow \qquad \frac{\partial \lambda^\epsilon}{\partial \rho} q' = 0, \tag{11.6.43}$$

showing that λ^ϵ is no longer permitted to be a function of ρ, because $q' \neq 0$ in general. With these, it is concluded that

$$\boldsymbol{\phi}_\eta = \lambda^\epsilon(\Theta)\, \boldsymbol{q}, \tag{11.6.44}$$

which approaches the Duhem-Truesdell relation that $\lambda^\epsilon(\Theta) = 1/\Theta$, if Θ represents the absolute temperature. However, at the present stage $\lambda^\epsilon(\Theta)$ is still a materially dependent function of the empirical temperature Θ. Further investigations are accomplished by considering two heat-conducting compressible fluids which are separated by a material singular surface, through which the empirical temperature is continuous, as indicated by the Müller-Liu entropy principle. For convenience, let all the quantities belonging to one fluid be denoted by the superscript "+", while those of the other fluid be denoted by the superscript "−". The differences in entropy and heat fluxes across the material singular surface are given by

$$(\boldsymbol{\phi}_\eta \cdot \boldsymbol{n})^+ - (\boldsymbol{\phi}_\eta \cdot \boldsymbol{n})^- = (\lambda^\epsilon \boldsymbol{q} \cdot \boldsymbol{n})^+ - (\lambda^\epsilon \boldsymbol{q} \cdot \boldsymbol{n})^- = 0, \qquad (\boldsymbol{q} \cdot \boldsymbol{n})^+ - (\boldsymbol{q} \cdot \boldsymbol{n})^- = 0, \tag{11.6.45}$$

where $\boldsymbol{n}$ is the unit normal of material singular surface at the evaluation point. These equations imply that

$$\left(\lambda^{\epsilon+} - \lambda^{\epsilon-}\right) \boldsymbol{q} \cdot \boldsymbol{n} = 0, \qquad \longleftrightarrow \qquad \lambda^{\epsilon+}(\Theta) = \lambda^{\epsilon-}(\Theta), \tag{11.6.46}$$

provided that $\boldsymbol{q} \cdot \boldsymbol{n} \neq 0$. Since two fluids can arbitrarily be chosen with different material responses, Eq. (11.6.46)$_2$ shows that Θ must be a temperature scale which is independent of the material properties. This motivates the existence of absolute temperature, as will be shown later. The expression $\lambda^\epsilon(\Theta)$ is referred to as the *coldness function* or simply *coldness* of a material, and its reciprocal is denoted by

$$\tilde{\Theta} = \frac{1}{\lambda^\epsilon(\Theta)}. \tag{11.6.47}$$

Differentiating Eq. (11.6.38)$_1$ with respect to (grad Θ) and Eq. (11.6.38)$_3$ with respect to Θ yields

$$\frac{\partial^2 \eta}{\partial(\text{grad}\,\Theta)\partial\Theta} = \lambda^\epsilon \frac{\partial^2 \epsilon}{\partial\Theta\partial(\text{grad}\,\Theta)} + \frac{\partial\lambda^\epsilon}{\partial\Theta}\frac{\partial\epsilon}{\partial(\text{grad}\,\Theta)}, \tag{11.6.48}$$

showing that ϵ cannot be a function of (grad Θ), for $\partial\lambda^\epsilon/\partial\Theta \neq 0$ in general. This leads to $\partial\epsilon/\partial(\text{grad}\,\Theta) = \boldsymbol{0}$, which is substituted into Eq. (11.6.38)$_3$ to obtain that η is independent of (grad Θ). The same result can also be found for λ^ρ. With these, the functional dependencies of η, ϵ, and λ^ρ are obtained as

$$\eta = \eta\,(\rho, \Theta)\,, \qquad \epsilon = \epsilon\,(\rho, \Theta)\,, \qquad \lambda^\rho = \lambda^\rho\,(\rho, \Theta)\,, \tag{11.6.49}$$

by which the combination of Eq. (11.6.38)$_1$ with Eq. (11.6.38)$_2$ gives

$$\mathrm{d}\eta = \lambda^\epsilon \left[\frac{\partial\epsilon}{\partial\Theta}\mathrm{d}\Theta + \left(\frac{\partial\epsilon}{\partial\rho} + \frac{\lambda^\rho}{\rho\lambda^\epsilon}\right)\mathrm{d}\rho\right] = \frac{1}{\tilde{\Theta}}\left[\mathrm{d}\epsilon + p\,\mathrm{d}\left(\frac{1}{\rho}\right)\right], \tag{11.6.50}$$

in which Eq. (11.6.38)$_6$ has been used. This equation is one of the $T\mathrm{d}s$ equations. Taking cross differentiation of the coefficients in the equation yields

$$\frac{\mathrm{d}(\ln\lambda^\epsilon)}{\mathrm{d}\Theta} = \frac{1}{\lambda^\epsilon}\frac{\mathrm{d}\lambda^\epsilon}{\mathrm{d}\Theta} = \frac{\partial p/\partial\Theta}{(\partial\epsilon/\partial\rho)\rho^2 - p}, \tag{11.6.51}$$

which is integrated to obtain

$$\tilde{\Theta}(\Theta) = \tilde{\Theta}(\Theta_0)\exp\left\{-\int_{\Theta_0}^{\Theta}\frac{\partial p/\partial\xi}{(\partial\epsilon/\partial\rho)\rho^2 - p}\mathrm{d}\xi\right\}, \tag{11.6.52}$$

in which it is noted that $\tilde{\Theta}$ is independent of the material properties. Substituting the ideal gas state equation into this equation shows that $\tilde{\Theta}(\Theta) = \tilde{\Theta}(\Theta_0)$, which motivates that

$$\tilde{\Theta}(\Theta) = T, \tag{11.6.53}$$

where T represents the Kelvin temperature. Hence, the absolute temperature is a derived result, not an assumed quantity. From now on, the empirical temperature Θ is replaced by the absolute temperature T for clarity.

The residual entropy inequality Γ is expressed alternatively as

$$\Gamma = -\frac{\boldsymbol{q} \cdot \text{grad}\, T}{T^2} \geq 0, \tag{11.6.54}$$

which should assume a minimum value at thermodynamic equilibrium with vanishing entropy production. The necessary conditions to achieve these are given by

$$\left[\frac{\partial \Gamma}{\partial(\text{grad}\, T)}\right]\Big|_E = \mathbf{0}, \qquad \left[\frac{\partial^2 \Gamma}{(\partial(\text{grad}\, T))^2}\right]\Big|_E : \text{ positive semi-definite.} \tag{11.6.55}$$

Applying Eq. $(11.6.55)_1$ to Eq. (11.6.54) gives rise to

$$\boldsymbol{q}|_E = \mathbf{0}, \tag{11.6.56}$$

showing that the heat flux vanishes in thermodynamic equilibrium. Within the isotropic representation, the stress tensor $\boldsymbol{t}$ and heat flux vector $\boldsymbol{q}$ are in the forms

$$\boldsymbol{t} = -p(\rho, T)\boldsymbol{I}, \qquad \boldsymbol{q} = -q'(\rho, T, \|\text{grad}\, T\|^2)\text{grad}\, T, \tag{11.6.57}$$

which are substituted into Eq. $(11.6.55)_2$ to obtain

$$q'(\rho, T, 0) \geq 0, \tag{11.6.58}$$

showing that q' is nonnegative at grad $T = \mathbf{0}$, which is compatible with the entropy principle.

Although the same results for a heat-conducting compressible fluid have been obtained by either the Duhem-Truesdell relations with the Coleman-Noll approach, or the Müller-Liu entropy principle, the latter approach was formulated with weaker assumptions. It was also proved that the linear momentum balance does influence the exploitation of entropy principle in the context of the Müller-Liu approach, and the absolute temperature was not assumed a priori to exist; rather, it has been proved to be the inverse of the Lagrangian multiplier of the energy balance, which is independent of the material properties. Furthermore, the Duhem-Truesdell relations of entropy flux and entropy supply become proved statements. These facts mediate to the model equations derived by using the Müller-Liu approach strengthened credibility. It is likely that for more complex materials, both approaches do not necessarily furnish the same results. This point will be exploited in a more detailed manner by studying the characteristics of granular flows in the next chapter.

11.7 Third Law and Absolute Zero

The third law concerns with the limiting behavior of thermodynamic systems as the temperature approaches the absolute zero, which is given in the following[28]:

11.6 (The third law) *The contribution to the entropy of a system by each aspect which is in internal thermodynamic equilibrium tends to vanish as the temperature tends to zero.*

The phrase "by each aspect of a system" means a part of the system or a process in it which interacts only weakly with the rest of system, making an essentially independent contribution to the properties of the whole. The formulation delivers an absolute base to measure the entropy of each substance, and the entropy relative to this base is known as the *absolute entropy*. Since experiments can only determine the differences in entropy, and in the given formulation the third law states that the entropy due to each aspect of all systems takes the same value at the absolute zero, the choice of zero for the universal constant in the Boltzmann equation brings the third law into agreement with it, so that the ground state in quantum mechanics is completely in order, with $\Omega_0 = 1$ in Eq. (11.5.60) and $S_0 = 0$ in Eq. (11.5.61). In other words, the ground state of any system is *non-degenerated*. However, it should be noted that in deriving the Boltzmann equation, the integration constant S_0 in Eq. (11.5.61) is chosen to be zero. The essential point of third law is that the integration constant is the same for all systems, although it is strictly a matter of convenience to set it equal to zero.

Another formulation of the third law is that "*it is impossible to reach the absolute zero for any systems by any means*". Such a conclusion can readily be obtained by referring to Fig. 11.17a. Essentially, an infinite number of the Carnot engines can be allocated, and the temperature in the low-temperature reservoir of each Carnot engine can be decreased continuously. Unfortunately, the Kelvin-Planck statement asserts that the thermal efficiency of a Carnot engine cannot be unity, so the temperature in the low-temperature reservoir can only be approached to the absolute zero, but cannot become the absolute zero.

Remarks on the Four Laws of Thermodynamics:

The zeroth law delivers the concepts of empirical temperature and equality of temperature between any two systems. The first law implies the macroscopic existence of internal energy and restates the balance of energy of a system, as implied by the physical fact that energy is a conserved quantity. The second law implies the macroscopic existence of entropy, and may be formulated as a balance of entropy of a system. With the law of increase of entropy of an isolated system, the second law indicates the "directions" of the time evolutions of all natural events. The third law defines the absolute entropy and the impossibility of reaching the absolute zero by

[28]The formulation is quoted from Sir Francis Simon, 1893–1956, a German and later British physical chemist and physicist, who made a significant contribution to the creation of the atomic bomb by separating the isotope Uranium-235.

any means. In practice, the first and second laws are of great importance, which may be combined into a single statement which reads: "*the energy of whole universe is conserved, while the entropy always increases*," as proposed by Clausius.

11.8 Thermodynamic Relations

11.8.1 Thermodynamic Potentials

By using the first law, two functions of state with dimensions of energy have been defined: the internal energy and enthalpy. Other important functions of state may be defined and used to determine the equilibrium states of thermodynamic systems under various constraints. They are called the *thermodynamic potential functions*. For a simple compressible substance, there are four important thermodynamic potential functions, which are given by

$$U, \qquad H \equiv U + pV, \qquad F \equiv U - TS, \qquad G \equiv H - TS = U - TS + pV, \tag{11.8.1}$$

where F is called the Helmholtz function, and G is termed the *Gibbs function*.[29] These definitions are appropriate for a thermodynamic system subject to work by hydrostatic pressure only, i.e., work is induced by volume change. For system involving other work forms, the analogous functions may be obtained by replacing $-p$ and V by other appropriate pair of variables, e.g. those discussed in Sect. 11.2.3. As indicated by the definitions, four potential functions are extensive quantities.

Taking total differential of Eq. (11.8.1) yields respectively

$$\begin{aligned} \mathrm{d}U &= T\mathrm{d}S - p\mathrm{d}V, \quad \mathrm{d}H = T\mathrm{d}S + V\mathrm{d}p, \\ \mathrm{d}F &= -S\mathrm{d}T - p\mathrm{d}V, \, \mathrm{d}G = -S\mathrm{d}T + V\mathrm{d}p, \end{aligned} \tag{11.8.2}$$

which are the differential forms of four potential functions. Each expression has two terms on the right corresponding to two degrees of freedom for simple compressible substances. Each term derives from two pairs of fundamental variables, namely (T, S) and (p, V). Since four potential functions are point functions, their total differentials are exact ones, and Eq. (11.8.2) also shows that each potential function has a different pair of fundamental variables as its natural or proper independent arguments. That is, $U = U(S, V)$, $H = H(S, p)$, $F = F(T, V)$, and $G = G(T, p)$. If any one of the potential functions is explicitly known in terms of its proper arguments, the complete information of a system is known, for any of the parameters of state may be determined from the given function.

[29]Josiah Willard Gibbs, 1839–1903, an American scientist, who made important contributions to the statistical mechanics and invented the modern vector calculus.

For example, consider a thermodynamic system, whose Helmholtz function $F = F(T, V)$ is known. It follows from Eq. (11.8.2)$_3$ that

$$S = -\left(\frac{\partial F}{\partial T}\right)_V, \qquad p = -\left(\frac{\partial F}{\partial V}\right)_T, \tag{11.8.3}$$

with which the expressions of U, H, and G are obtained from their definitions, viz.,

$$\begin{aligned} U &= -T^2\left[\frac{\partial}{\partial T}\left(\frac{F}{T}\right)\right]_V, \qquad H = F - T\left(\frac{\partial F}{\partial T}\right)_V - V\left(\frac{\partial F}{\partial V}\right)_T, \\ G &= -V^2\left[\frac{\partial}{\partial V}\left(\frac{F}{V}\right)\right]_T, \end{aligned} \tag{11.8.4}$$

where the first equation is known as the *Gibbs-Helmholtz equation*, which expresses U in terms of F. The other important characteristics of Eq. (11.8.1) are that it is always possible to calculate how a potential function changes if the system goes from one state to another, provided that a suitable information is given. For example, the change in G from state (T_0, p_0) to state (T_0, p_1) is obtained as

$$G(T_0, p_1) - G(T_0, p_0) = \int_{p_0}^{p_1}\left(\frac{\partial G}{\partial p}\right)_T \mathrm{d}p = \int_{p_0}^{p_1} V \mathrm{d}p, \tag{11.8.5}$$

for which the information of $V = V(p)$ is required.

For a thermally isolated system, it follows from the first law that $\mathrm{d}U = \delta W$, showing that the decrease in internal energy equals the work done by the system. If the process is further assumed to be a reversible one, the work done by the system becomes $-\mathrm{d}U = p\,\mathrm{d}V$. On the other hand, if the process is an isochoric one, the first law reads $\mathrm{d}U = \delta Q$, indicating that an increase in internal energy equals the amount of heat which is absorbed. With this, the specific heat at constant volume becomes

$$c_V = T\left(\frac{\partial s}{\partial T}\right)_V = \left(\frac{\partial u}{\partial T}\right)_V. \tag{11.8.6}$$

Similarly, for a system undergoing an isentropic process, the change in enthalpy is related to the change in pressure. Specifically, if the process is further assumed to be isobaric, an increase in enthalpy equals the amount of heat supplied to the system, namely $\mathrm{d}H = \delta Q$, so that

$$c_p = T\left(\frac{\partial s}{\partial T}\right)_p = \left(\frac{\partial h}{\partial T}\right)_p. \tag{11.8.7}$$

Equations (11.8.6) and (11.8.7) are the general definitions of specific heats at constant volume and constant pressure.

It follows from Eq. (11.8.1)$_3$ that the decrease in the Helmholtz function is the maximum amount of mechanical work that may be extracted from a system undergoing an isothermal process. Thus, the Helmholtz function is also called the *Helmholtz free energy*. If the process was irreversible, the amount of work would be less than $p\,\mathrm{d}V$. If the process is isothermal, the extracted work may be greater than, equal to, or less than ΔU, depending on the direction of heat transfer process. Hence, F is a useful energy function for isothermal processes. On the other hand, for an isochoric process, $\mathrm{d}F$ is related to the change in temperature of the system. Similarly, the Gibbs function of a system undergoing a reversible process with constant temperature and pressure remains constant. This condition applies to many physical and chemical changes, and a constant Gibbs function may then be used to represent the constraints on the system.

11.8.2 The Legendre Differential Transformation

For a thermodynamic system with n degrees of freedom, as motivated by Eq. $(11.8.2)_1$, the expression of $\mathrm{d}U$ contains the contributions of $T\mathrm{d}S$ and other $(n-1)$ work-like terms, each of which is in the form of $x_i \mathrm{d}X_i$, where x_i is any intensive variable, with X_i its corresponding extensive variable. The system has thus $2n$ primary variables to form n conjugate pairs whose products have the dimension of energy, e.g. (T, S) and (p, V). Since each potential function is constructed by a pair of conjugate variables, the system with n degrees of freedom has then 2^n potential functions corresponding to the twofold choices offered by each pair.

To obtain these potentials in a systematic way, the expression of $\mathrm{d}U$ is first written down, which consists of the term $T\mathrm{d}S$ plus all the work terms and has as its independent variables the extensive members of conjugate pairs. One then picks out the terms in which the wrong member of conjugate pair is the independent variable and adds to or subtracts from $\mathrm{d}U$ the differential of the product of the conjugate pair to remove the unwanted term and replace it by the required one. This procedure yields a new differential expression, which is still associated with n terms but with a different set of the independent variables. The obtained expression is exact and has the dimension of energy, which becomes a new potential function. Such a procedure is referred to as the *Legendre differential transformation.*

To explore the idea, consider a wire which is subject to tension and hydrostatic pressure, so that the first law reads

$$\mathrm{d}U = T\mathrm{d}S + \mathcal{T}\mathrm{d}\ell - p\mathrm{d}V, \qquad \longrightarrow \qquad U = U(S, \ell, V), \tag{11.8.8}$$

with the independent variables S, ℓ, and V, and the system has three degrees of freedom. A new potential function e.g. in terms of (T, ℓ, p) is required. Adding the terms $-\mathrm{d}(TS)$ and $\mathrm{d}(pV)$ to the left-hand-side of Eq. $(11.8.8)_1$ gives

$$\mathrm{d}U - \mathrm{d}(TS) + \mathrm{d}(pV) = \mathrm{d}(U - TS + pV) = -S\mathrm{d}T + \mathcal{T}\mathrm{d}\ell + V\mathrm{d}p, \tag{11.8.9}$$

by which a new potential function G' is obtained as

$$\mathrm{d}G' = -S\mathrm{d}T + \mathcal{T}\mathrm{d}\ell + V\mathrm{d}p, \qquad \longrightarrow \qquad G' = U - TS + pV = G'(T, \ell, p). \tag{11.8.10}$$

11.8.3 The Maxwell Relations

For a simple compressible substance, it follows from Eq. $(11.8.2)_1$ that

$$\left(\frac{\partial U}{\partial S}\right)_V = T, \qquad \left(\frac{\partial U}{\partial V}\right)_S = -p, \tag{11.8.11}$$

which are obtained by taking partial derivatives of U with respect to its proper variables S and V, respectively. Differentiating two equations again with respect to the opposite variables yields

$$\frac{\partial^2 U}{\partial V \partial S} = \left(\frac{\partial T}{\partial V}\right)_S \quad \frac{\partial^2 U}{\partial S \partial V} = -\left(\frac{\partial p}{\partial S}\right)_V . \tag{11.8.12}$$

Since the double partial derivatives in these equations are communicative, for U is assumed to be a continuous function with respect to V and S, it is found that

$$\left(\frac{\partial T}{\partial V}\right)_S = -\left(\frac{\partial p}{\partial S}\right)_V . \tag{11.8.13}$$

This result can equally be identified directly from Eq. (11.8.2)$_1$, for $\mathrm{d}U$ is an exact differential, so that the coefficients of proper variables must satisfy Eq. (11.8.13).

Similar results can be obtained by conducting the same procedures to H, F, and G, which are given by

$$\left(\frac{\partial T}{\partial V}\right)_p = -\left(\frac{\partial p}{\partial S}\right)_T, \quad \left(\frac{\partial T}{\partial p}\right)_S = \left(\frac{\partial V}{\partial S}\right)_p, \quad \left(\frac{\partial T}{\partial p}\right)_V = \left(\frac{\partial V}{\partial S}\right)_T . \tag{11.8.14}$$

Equations (11.8.13) and (11.8.14) are known as the *Maxwell relations*. Their usefulness lies in the transformation of variables they make possible, in particular the transformations of those variables which cannot be measured directly. Although the Maxwell relations have been deduced in the form which is appropriate to a system subject to work by hydrostatic pressure, similar expressions hold for any system with two degrees of freedom. In view of this, the conjugate pair (T, S) applies to any system, but (p, V) needs to be replaced by their corresponding variables such as those described in Sect. 11.2.3.

For systems with more than two independent proper variables, the number of the Maxwell relations becomes much greater. A system with n degrees of freedom has 2^n potential functions; each of the conjugate pairs yields $n(n-1)/2$ Maxwell relations. In practice, it is easier to consider each problem separately and construct when necessary the potential functions which give the required differential coefficients.

11.8.4 General Conditions of Thermodynamic Equilibrium

Consider a system which interacts with its surrounding, in which an amount of heat is transferred from the surrounding to the system. The entropy change of system is related to the transferred heat given by

$$\delta Q \leq T_0 \mathrm{d}S, \tag{11.8.15}$$

where T_0 represents the temperature of surrounding. If the surrounding exerts a pressure p_0 on the system boundary, which is the only source of work, it follows that $\delta W = -p_0 \mathrm{d}V$. Substituting this expression and Eq. (11.8.15) into the first law yields

$$\mathrm{d}U \leq T_0 \mathrm{d}S - p_0 \mathrm{d}V, \quad \longrightarrow \quad \mathrm{d}A = \mathrm{d}U + p_0 \mathrm{d}V - T_0 \mathrm{d}S \leq 0, \tag{11.8.16}$$

where A is defined as

$$A \equiv U + p_0 V - T_0 S, \tag{11.8.17}$$

which is referred to as the *availability* of system, and the temperature T_0 and pressure p_0 are referred to the surrounding, not to the system. The implication of Eq. $(11.8.16)_2$ is that in any natural process the availability of a system cannot increase, for the process is essentially irreversible. It follows that the general condition for equilibrium of a system in a given surrounding is that its availability becomes a minimum, or

$$\mathrm{d}A = \mathrm{d}U + p_0 \mathrm{d}V - T_0 \mathrm{d}S = 0, \tag{11.8.18}$$

must hold for all possible infinitesimal displacements from equilibrium.

The availability of a system represents the maximum work that may be extracted from the system in a given surrounding. To demonstrate this, consider a gas contained inside a cylinder-piston device as the system with pressure p which differs in general from the surrounding pressure p_0. The device is also thermally isolated from the surrounding, so that its temperature T differs from the surrounding temperature T_0. The greatest possible amount of work that can be extracted from the system in a process may be obtained if the process is reversible. Thus, for a small reversible change in state, the first law and availability imply that

$$\mathrm{d}U = T\mathrm{d}S - p\mathrm{d}V, \qquad \mathrm{d}A = (T - T_0)\,\mathrm{d}S - (p - p_0)\,\mathrm{d}V. \tag{11.8.19}$$

Now, a Carnot engine is placed between the system and surrounding. Since the process is reversible, the total entropy of system and surrounding remains unchanged, and the work done by the engine, δW_C, is given by

$$\delta W_C = \delta Q - \delta Q_0 = (T - T_0)\,\mathrm{d}S, \tag{11.8.20}$$

where δQ and δQ_0 represent the amounts of heat transfer at T and T_0, respectively. Comparing this equation with Eq. $(11.8.19)_2$ shows that the first term on the right-hand side of Eq. $(11.8.19)_2$ is the maximum work that can be obtained from the system due to heat transfer, and the term $(p - p_0)\mathrm{d}V$ denotes the net mechanical work done on the piston. Work can thus be extracted from the system, as long as $T \neq T_0$ and $p \neq p_0$, so that the value of A is decreased. Consequently, the term $(A - A_{min})$ is nothing else than the maximum amount of work which may be extracted from the system in a given surrounding.

The general condition for equilibrium is that the availability must assume a minimum value. Since $\mathrm{d}A$ is given by Eq. $(11.8.19)_2$, the terms $\mathrm{d}S$ and $\mathrm{d}V$ must be zero in an infinitesimal displacement from equilibrium to have $\mathrm{d}A = 0$, for S and V are independent degrees of freedom. Four special cases are discussed in the following.

- Thermally isolated and isochoric systems. The system temperature T differs T_0, so that $\mathrm{d}S = 0$ must hold in order to obtain a vanishing value of the first term on the right-hand-side of Eq. $(11.8.19)_2$. In such a circumstance, S of the system assumes a maximum value. Since $\mathrm{d}V = 0$, the second term vanishes identically, but the pressure p is not defined. With these, Eq. (11.8.19)$_2$ reduces to

$$\mathrm{d}A = 0, \tag{11.8.21}$$

under the conditions that

$$\mathrm{d}S = 0, \qquad \mathrm{d}V = 0, \qquad \mathrm{d}U = 0, \tag{11.8.22}$$

where the last equation is motivated by Eq. (11.8.2)$_1$.

- Thermally isolated and isobaric systems. The first term on the right-hand-side of Eq. (11.8.19)$_2$ vanishes if $\mathrm{d}S = 0$. The second term becomes null if $p = p_0$ is required. With these, the conditions for $\mathrm{d}A = 0$ are given by

$$\mathrm{d}S = 0, \qquad \mathrm{d}p = 0, \qquad \mathrm{d}H = 0, \tag{11.8.23}$$

where the last equation is obtained by using Eq. (11.8.2)$_2$.

- Isochoric systems. In order to obtain a vanishing value of the first term, T must be equal to T_0; i.e., the system is not thermally isolated, but is in thermal equilibrium with its surrounding in any infinitesimal reversible process. However, since $\mathrm{d}V = 0$, the second term vanishes identically, while p remains not directly defined. With these, the associated conditions for $\mathrm{d}A = 0$ are given by

$$\mathrm{d}T = 0, \qquad \mathrm{d}V = 0, \qquad \mathrm{d}F = 0, \tag{11.8.24}$$

where the last equation is motivated by Eq. (11.8.2)$_3$.

- Isobaric systems. As similar to the second and third cases, $T = T_0$ and $p = p_0$ are required in order to obtain vanishing values of $\mathrm{d}A$. With these, the associated conditions for $\mathrm{d}A = 0$ are obtained as

$$\mathrm{d}T = 0, \qquad \mathrm{d}p = 0, \qquad \mathrm{d}G = 0, \tag{11.8.25}$$

where the last equation is obtained from Eq. (11.8.2)$_4$.

The obtained four sets of conditions for equilibrium are summarized in the following:

$$\left\{\begin{array}{l} U = U(S, V),\ \mathrm{d}S = 0,\ \mathrm{d}V = 0,\ \mathrm{d}U = 0, \\ H = H(S, p),\ \mathrm{d}S = 0,\ \mathrm{d}p = 0,\ \mathrm{d}H = 0, \\ F = F(T, V),\ \mathrm{d}T = 0,\ \mathrm{d}V = 0,\ \mathrm{d}F = 0, \\ G = G(T, p),\ \mathrm{d}T = 0,\ \mathrm{d}p = 0,\ \mathrm{d}G = 0. \end{array}\right\} \tag{11.8.26}$$

Each potential function in Eq. (11.8.26) is expressed in terms of its proper variables. The four sets of equilibrium conditions are entirely equivalent in the sense that they lead to identical physical results. In arriving at the general conditions for equilibrium, no restriction has been made to the internal complexity of system. For example, $\mathrm{d}U$ may contain other variables related to the degrees of freedom which are internal to the system in addition to (T, S) and (p, V). However, the corresponding terms of these internal degrees of freedom do not appear in the equation because the system as a whole changes its internal energy only by means of heat and work with the surrounding. Thus, the conditions of equilibrium place restrictions implicitly on those variables which are implicit in the potential functions. This gives rise to that each equilibrium condition contains three contributions: the first two for the proper

Table 11.2 Conditions for a stable equilibrium of a thermodynamic system

Specified variables	Equilibrium condition
(T, p)	G minimum
(T, V)	F minimum
(U, V)	S minimum
(S, V)	U minimum
$(S.p)$	H minimum
(G, T)	p minimum
(G, p)	T minimum
(F, T)	V minimum
(F, V)	T minimum
(U, S)	V minimum
(H, S)	p minimum
(H, p)	S minimum

variables and the last one for the internal degrees of freedom. In other words, the first two terms are imposed externally as external conditions, leaving the last term for the determination of internal degrees of freedom. For systems without internal degrees of freedom, the first two terms are sufficient to determine the conditions of equilibrium.

Comparing Eq. (11.8.26) with Eq. $(11.8.16)_2$ and the definitions of four potential functions shows the nature of the extreme implied by itself. For example, for the Helmholtz function, it follows from Eqs. $(11.8.2)_{1,3}$ that

$$\mathrm{d}F = \mathrm{d}U - S\mathrm{d}T - T\mathrm{d}S, \tag{11.8.27}$$

which is substituted into Eq. $(11.8.16)_2$ to obtain

$$\mathrm{d}A = \mathrm{d}F + S\mathrm{d}T + p_0\mathrm{d}V \leq 0, \tag{11.8.28}$$

for $T = T_0 =$ constant and $p = p_0 =$ constant. In order to make A a minimum, three sets of conditions need to be fulfilled: (a) F must be a minimum under the given values of T and V; (b) V must be a minimum under the given values of T and F; or (c) T must be a minimum under the given values of V and F. Similar results can be obtained by using the same procedure to other potential functions, with the results summarized in Table 11.2, in which the conditions in the first three rows are often used.

The results summarized in the table are all equivalent and based on the law of increase of entropy. Each result represents the simplest way of applying the law under given conditions. On the contrary, if the chosen potential function, instead of assuming a minimum vale, assumes a maximum vale, the results are still applicable, although the reached equilibrium is not a stable one. This circumstance is used intensively in the treatment of phase change and the underlying physics of chemical thermodynamics.

11.8.5 Applications to Simple Compressible Substances

The main advantages of the Maxwell relations are that the variables which cannot be directly observed or measured, such as entropy, can be related to those variables which can be measured directly. To demonstrate this, some examples are discussed.

Partial derivatives of specific heats. Taking partial derivative of c_p with respect to pressure under isothermal condition yields

$$\left(\frac{\partial c_p}{\partial p}\right)_T = T\frac{\partial}{\partial p}\Big)_T\left(\frac{\partial s}{\partial T}\right)_p = T\frac{\partial}{\partial T}\Big)_p\left(\frac{\partial s}{\partial p}\right)_T = -T\left(\frac{\partial^2 v}{\partial T^2}\right)_p, \quad (11.8.29)$$

in which Eqs. (11.8.7) and $(11.8.14)_1$ have been used. Similarly, it is possible to obtain

$$\left(\frac{\partial c_V}{\partial v}\right)_T = T\left(\frac{\partial^2 p}{\partial T^2}\right)_v. \quad (11.8.30)$$

The variations of c_p and c_v with respect to the variations in pressure and specific volume under isothermal condition may be determined, provided that the state equation of substance is known.

Difference in specific heats. Expressing the specific entropy as $s = s(T, v)$ and taking total derivative of this expression gives

$$\mathrm{d}s = \left(\frac{\partial s}{\partial T}\right)_v \mathrm{d}T + \left(\frac{\partial s}{\partial v}\right)_T \mathrm{d}v, \quad (11.8.31)$$

which motivates that

$$\left(\frac{\partial s}{\partial T}\right)_p = \left(\frac{\partial s}{\partial T}\right)_v + \left(\frac{\partial s}{\partial v}\right)_T\left(\frac{\partial v}{\partial T}\right)_p. \quad (11.8.32)$$

By using this equation and Eqs. (11.8.6) and (11.8.7), it is found that

$$c_p - c_V = T\left(\frac{\partial s}{\partial v}\right)_T\left(\frac{\partial v}{\partial T}\right)_p = T\left(\frac{\partial p}{\partial T}\right)_v\left(\frac{\partial v}{\partial T}\right)_p, \quad (11.8.33)$$

in which the Maxwell relations have been used. This result can further be simplified by using the *isobaric expansivity* β_p and *isothermal compressibility* κ_T of a substance defined by

$$\beta_p \equiv \frac{1}{V}\left(\frac{\partial V}{\partial T}\right)_p, \qquad \kappa_T \equiv -\frac{1}{V}\left(\frac{\partial V}{\partial p}\right)_T, \quad (11.8.34)$$

so that Eq. (11.8.33) becomes

$$c_p - c_V = \frac{vT\beta_p^2}{\kappa_T}. \quad (11.8.35)$$

Specific heat ratio. The specific heat ratio γ is defined by

$$\gamma \equiv \frac{c_p}{c_V}, \quad (11.8.36)$$

which is expressed alternatively as

$$\gamma = \frac{\kappa_T}{\kappa_S} = \frac{(\partial V/\partial p)_T}{(\partial V/\partial p)_S}, \quad (11.8.37)$$

where κ_S is the *adiabatic compressibility* defined by

$$\kappa_S \equiv -\frac{1}{V}\left(\frac{\partial V}{\partial p}\right)_S. \tag{11.8.38}$$

Similar results can be obtained for *permittivities*, *Young's moduli*, or *magnetic susceptibilities*. In these expressions, the differentials of the intensive variables with respect to the associated extensive variables involve, which are called the *stiffnesses* of a substance. The reciprocal differential is known as a *compliance coefficient*.

Differential changes in specific internal energy and specific enthalpy. Expressing the specific internal energy and specific enthalpy as

$$u = u(T, v), \qquad h = h(T, p), \tag{11.8.39}$$

and taking total differential to both expressions yields

$$\mathrm{d}u = \left(\frac{\partial u}{\partial T}\right)_v \mathrm{d}T + \left(\frac{\partial u}{\partial v}\right)_T \mathrm{d}v, \qquad \mathrm{d}h = \left(\frac{\partial h}{\partial T}\right)_p \mathrm{d}T + \left(\frac{\partial h}{\partial p}\right)_T \mathrm{d}p, \tag{11.8.40}$$

which, by using the $T\mathrm{d}S$ equations, are brought to the forms

$$\mathrm{d}u = c_V \mathrm{d}T + \left[T\left(\frac{\partial s}{\partial v}\right)_T - p\right]\mathrm{d}v, \qquad \mathrm{d}h = c_p \mathrm{d}T + \left[v + T\left(\frac{\partial s}{\partial p}\right)_T\right]\mathrm{d}p, \tag{11.8.41}$$

in which Eqs. (11.8.6) and (11.8.7) have been used. Substituting the Maxwell relations into these equations gives

$$\mathrm{d}u = c_V \mathrm{d}T + \left[T\left(\frac{\partial p}{\partial T}\right)_v - p\right]\mathrm{d}v, \qquad \mathrm{d}h = c_p \mathrm{d}T + \left[v - T\left(\frac{\partial v}{\partial T}\right)_p\right]\mathrm{d}p, \tag{11.8.42}$$

which are used in experiments to evaluate the changes in specific internal energy and specific enthalpy of a working substance, provided that a reference state is known.

Differential change in specific entropy . Expressing the specific entropy as functions of (T, v) and (T, p) and taking total derivative of the expressions yield

$$\mathrm{d}s = \left(\frac{\partial s}{\partial T}\right)_v \mathrm{d}T + \left(\frac{\partial s}{\partial v}\right)_T \mathrm{d}v, \qquad \mathrm{d}s = \left(\frac{\partial s}{\partial T}\right)_p \mathrm{d}T + \left(\frac{\partial s}{\partial p}\right)_T \mathrm{d}p. \tag{11.8.43}$$

Substituting Eqs. (11.8.6), (11.8.7) and the Maxwell relations into these expressions results in

$$\mathrm{d}s = c_V\frac{\mathrm{d}T}{T} + \left(\frac{\partial p}{\partial T}\right)_v \mathrm{d}v, \qquad \mathrm{d}s = c_p\frac{\mathrm{d}T}{T} - \left(\frac{\partial v}{\partial T}\right)_p \mathrm{d}p, \tag{11.8.44}$$

which are the general expressions of the differential change in specific entropy for simple compressible substances.

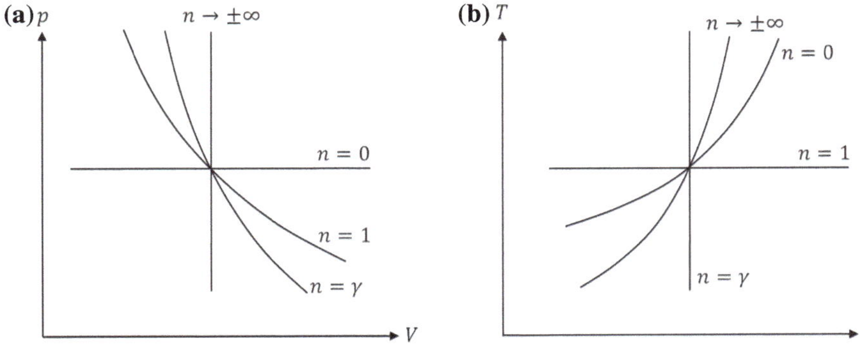

Fig. 11.21 Illustrations of the polytropic processes with variations in the exponent n. **a** In the p–V diagram. **b** In the T–S diagram

For ideal gases, the obtained results can further be simplified by using the ideal gas state equation. With $pv = RT$, it follows that

$$\left(\frac{\partial c_p}{\partial p}\right)_T = 0, \quad \left(\frac{\partial c_V}{\partial v}\right)_T = 0, \quad c_p - c_V = R, \qquad \gamma = 1 + \frac{R}{c_V},$$
$$\mathrm{d}u = c_V \mathrm{d}T, \qquad \mathrm{d}h = c_p \mathrm{d}T, \qquad \mathrm{d}s = c_V \frac{\mathrm{d}T}{T} + R\frac{\mathrm{d}v}{v}, \quad \mathrm{d}s = c_p \frac{\mathrm{d}T}{T} - R\frac{\mathrm{d}p}{p}, \tag{11.8.45}$$

showing that while $\{c_p, c_V, u, h\}$ are only functions of temperature, s is a function of either $\{T, v\}$ or $\{T, p\}$.

For an ideal gas at constant temperature in an adiabatic process, it follows from the ideal gas state equation that

$$pV = \text{constant}, \qquad \longrightarrow \qquad \frac{\mathrm{d}p}{p} + \frac{\mathrm{d}V}{V} = 0, \tag{11.8.46}$$

which is equivalent to

$$\left(\frac{\partial p}{\partial V}\right)_T = -\frac{p}{V}, \qquad \left(\frac{\partial p}{\partial V}\right)_S = -\gamma\frac{p}{V}. \tag{11.8.47}$$

If it is further assumed that the specific heats are constant, so that Eq. (11.8.47) may be integrated to obtain

$$pV^\gamma = \text{constant}, \qquad TV^{\gamma-1} = \text{constant}, \qquad T^\gamma p^{1-\gamma} = \text{constant}. \tag{11.8.48}$$

Comparing Eq. $(11.8.48)_1$ with Eq. (11.2.4) shows that $n = \gamma$ corresponds to a reversible adiabatic (isentropic) process. Different values of the exponent n in Eq. (11.2.4) thus correspond to different polytropic processes. Specifically,

$$\begin{aligned} n &= 0, && \longleftrightarrow\ p = \text{constant}, && \longleftrightarrow\ \text{isobaric process}, \\ n &= 1, && \longleftrightarrow\ T = \text{constant}, && \longleftrightarrow\ \text{isothermal process}, \\ n &= \gamma, && \longleftrightarrow\ S = \text{constant}, && \longleftrightarrow\ \text{isentropic process}, \\ n &\to \pm\infty && \longleftrightarrow\ V = \text{constant}, && \longleftrightarrow\ \text{isochoric process}, \end{aligned} \tag{11.8.49}$$

are found. The schematic illustrations of these polytropic processes in the two-dimensional p–V and T–S indicator diagrams are shown in Fig. 11.21. It is noted that the slope of isothermal curve is less than that of isentropic curve in the p–V diagram, as implied by Eq. (11.8.47). A similar result is obtained in the T–S diagram.

11.9 Exercises

11.1 A piston-cylinder device shown in the figure contains air at pressure p_1 and temperature T_1. The device configuration allows the air to be cooled to the surrounding temperature $T_0 < T_1$. Derive the conditions that should be fulfilled if the final state of air permits the piston to just rest on the stop. In this circumstance, what is the specific work done by the air during the process?

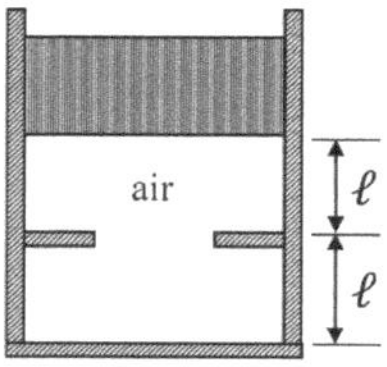

11.2 A rigid tank is divided by a membrane into two rooms, and both rooms contain an ideal gas, as shown in the figure. Initially, the pressure, volume, and temperature in room A are respectively p_A, V_A, and T_A, and the gas mass in room B is m_B with pressure p_B and temperature T_B. The membrane then ruptures and heat transfer takes place so that two ideal gases come to a uniform temperature T_u. Find the amount of heat transfer during the process.

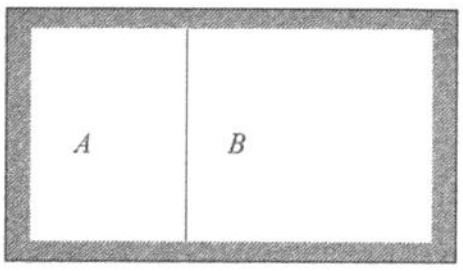

11.3 An air pistol contains compressed air in a small cylindrical volume, as shown in the figure. Initially, the volume, pressure, and temperature of air are given respectively by V_1, T_1, and p_1. A bullet with mass m acts as a piston initially held by a pin. When released, the air expands in an isothermal process and assumes pressure p_2 when the bullet just leaves the cylinder.

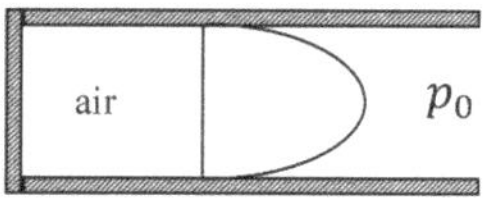

(a) Find the final volume and mass of air.
(b) Determine the work done by the air and the work done on the atmosphere.
(c) Determine the work done to the bullet and the bullet exit velocity.

11.4 Two air streams are combined into a single one, as shown in the figure. Two streams are at the same pressure p_i, with volume flow rates and temperatures Q_1 and T_1 for one flow, and Q_2 and T_2 for the other. They mix without any heat transfer to produce an exit flow at p_i. Neglect the kinetic energies of flows, and find the temperature and volume flow rate of the exit flow.

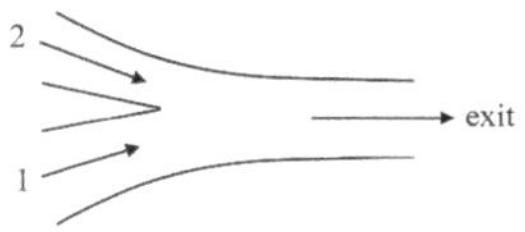

11.5 A mass-loaded piston-cylinder device shown in the figure contains air at pressure p_1, temperature T_1, and volume V_1 in the initial state. The cylinder volume up to the stop is V_t. An air line with pressure p_i and temperature T_i is connected to the device by a valve that is then opened until a final pressure p_2 inside the device is reached, at which the temperature is T_2. Find the air mass that enters the device, and the amounts of work and heat transfer during the filling process.

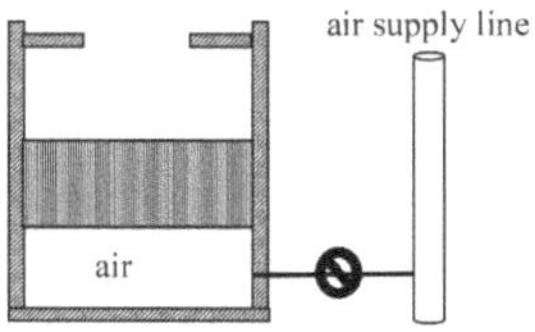

11.6 A piston-cylinder device made of steel with mass m_s contains ammonia with mass m_a and pressure p_1. Both ammonia and steel are at temperature T_i initially. Some stops are placed so that the minimum volume inside the device is V_s. The whole system is now cooled down to T_f by heat transfer to the ambient environment at temperature $T_0 < T_f$. It is assumed that during the cooling process both steel and ammonia assume the same temperature simultaneously. Find the work, heat transfer, and the total entropy generation in the process.

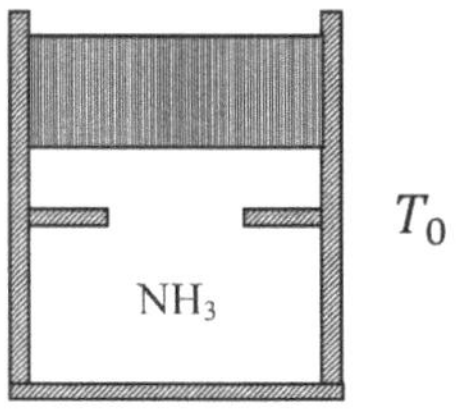

11.7 Show that violation of the Kelvin-Planck statement leads to violation of the Clausius statement of the second law.

11.8 Consider the fourth problem again. Find the total rate of entropy generation.

11.9 Consider the fifth problem again. Determine the total entropy generation in the filling process.

11.10 Prove that inequality (11.6.6) can only be fulfilled by $\boldsymbol{a} = \mathbf{0}$ and $\Gamma \geq 0$.

11.11 Use the Duhem-Truesdell relations with the Coleman-Noll approach, and the Müller-Liu approach to derive the equilibrium expressions of the stress tensor and heat flux vector for a viscous, heat-conducting incompressible fluid. The material classes given in Eqs. (11.6.2) and (11.6.31) can be used as the starting point.

11.12 Over a certain small range of pressures and temperatures, the state equation of a substance is given by

$$\frac{pv}{RT} = 1 - C\frac{p}{T^4},$$

where C is a constant. Derive the expressions for the changes in specific enthalpy and specific entropy in an isothermal process.

11.13 The *Joule-Thompson coefficient* μ_J is a measure of the direction and magnitude of temperature change with respect to pressure of a gaseous substance in a *throttling process*, in which the enthalpy remains fixed. For any three properties x, y, and z, use the relation

$$\left(\frac{\partial x}{\partial y}\right)_z \left(\frac{\partial y}{\partial z}\right)_x \left(\frac{\partial z}{\partial x}\right)_y = -1,$$

to show that

$$\mu_J = \left(\frac{\partial T}{\partial p}\right)_h = \frac{RT^2}{pc_p}\left(\frac{\partial Z}{\partial T}\right)_p,$$

where Z is the compressibility factor of the gaseous substance.

11.14 Derive the expressions for

$$\left(\frac{\partial T}{\partial v}\right)_u, \quad \left(\frac{\partial h}{\partial s}\right)_v,$$

which do not contain the properties h, u, or s.

11.15 A solid is assumed to have uniform properties in all directions with its volume V given by $V = \ell_x \ell_y \ell_z$, where $\{\ell_x, \ell_z, \ell_z\}$ are the side lengths in the $\{x, y, z\}$-directions, respectively. Show that the isobaric volume expansivity $\beta_p = 3\alpha_T$, where α_T represents the coefficient of thermal expansion given by

$$\alpha_T = \frac{1}{\ell}\left(\frac{\delta \ell}{\delta T}\right)_p.$$

Further Reading

C.J. Adkins, *Equilibrium Thermodynamics*, 3rd edn. (Cambridge University Press, Cambridge, 1983)

P. Atkins, *Four Laws That Drive the Universe* (Oxford University Press, Oxford, 2007)

A. Bejan, *Advanced Engineering Thermodynamics*, 3rd edn. (Wiley, New York, 2006)

A. Ben-Naim, *A Farewell to Entropy: Statistical Thermodynamics Based on Information* (World Scientific Publishing, Singapore, 2008)

C. Borgnakke, R.E. Sonntag, *Fundamentals of Thermodynamics*, 7th edn. (Wiley, New York, 2009)

H.B. Callen, *Thermodynamics and an Introduction to Thermostatics*, 2nd edn. (Wiley, New York, 1985)

J.S. Dugdale, *Entropy and Its Physical Meaning* (Taylor & Francis, London, 1996)

J.B. Fenn, *Engines, Energy and Entropy* (W.H. Freeman and Company, New York, 1982)

W.M. Haddad, V.S. Chellaboina, S.G. Nersesov, *Thermodynamics: A Dynamical Systems Approach* (Princeton University Press, Princeton, 2005)

K. Hutter, k. Jönk, *Continuum Methods of Physical Modeling* (Springer, Berlin, 2004)

H. Leff, A.F. Rex, *Maxwell's Demon: Entropy, Information, Computing* (Adam Hilger, Bristol, 1990)

I. Müller, W.H. Müller, *Fundamentals of Thermodynamics and Applications* (Springer, Berlin, 2009)

I. Müller, T. Ruggeri, *Rational Extended Thermodynamics*, 2nd edn. (Springer, Berlin, 1998)

I. Müller, W. Weiss, *Entropy and Energy* (Springer, Berlin, 2005)

D.R. Olander, *General Thermodynamics* (Taylor & Francis, London, 2008)

H.C. Öttinger, *Beyond Equilibrium Thermodynamics* (Wiley, New York, 2005)

J.P. Sethna, *Statistical Mechanics: Entropy, Order Parameters, and Complexity* (Oxford University Press, Oxford, 2006)

C.L. Tien, J.H. Lienhard, *Statistical Thermodynamics (Holt* (Rinehart & Winston, New York, 1971)

C. Truesdell, *Rational Thermodynamics* (McGraw-Hill, New York, 1969)

C. Truesdell, *A First Course in Rational Continuum Mechanics, Volume 1* (Academic, New York, 1977)

C. Truesdell, R.G. Muncaster, *Fundamentals of Maxwell's Kinetic Theory of a Simple Monatomic Gas* (Academic, New York, 1980)

C. Truesdell, W. Noll, *The Non-linear Field Theories of Mechanics* (Springer, Berlin, 1992)

M.W. Zemansky, R.H. Dittman, *Heat and Thermodynamics*, 7th edn. (McGraw-Hill, Singapore, 1997)

Granular Flows 12

Granular matters are collections of a large amount of discrete solid particles with interstices filled with a fluid, and granular flows are granular matters in flowing state. In contrast to simple fluids such as water or air which can be dealt with by classical fluid mechanics, granular flows exhibit significant non-Newtonian features. The evolutions of grain configurations as well as the interstitial fluids influence to a large extent the macroscopic features, which are referred to as the microstructural effects. Granular flows may be macroscopically considered complex rheological fluids, whose features are significantly affected by the microscopic time- and space-dependent internal structures. In other words, granular flows assume multiple time and length scales.

This chapter is devoted to a study of granular flows with interstices filled essentially with a gas. The intention is to demonstrate the applications of the mature disciplines of fluid mechanics and continuum thermodynamics to the study of the characteristics of complex flows. The approach can equally be used for other complex matters. First, a general description of granular matters and granular flows is provided, followed by a discussion on the distinct features of granular matters. The phase transition in a laminar flow and the characteristics of a turbulent flow with weak turbulent intensity are studied in due course to show their macroscopic features compared with simple fluids.

12.1 Granular Matters and Granular Flows

12.1.1 Definition of Granular Matter

In year 1644, René Descartes wrote in his book entitled *Principles of Philosophy*, a statement to characterize the difference between solids and liquids,[1] which reads: "a

[1]Or Renatus Cartesius in Latin, hence the name Cartesian Coordinates, 1596–1650, a French philosopher, mathematician, and scientist, who is dubbed "Father of Modern Western Philosophy".

© Springer International Publishing AG 2019

C. Fang, *An Introduction to Fluid Mechanics*, Springer Textbooks in Earth Sciences, Geography and Environment, https://doi.org/10.1007/978-3-319-91821-1_12

body is liquid when it is divided into several smaller parts that move separately, and it is solid when all its parts are in contact." Granular matters were probably not on Descartes' mind, but in some sense, they might as well have been since under some situations, granular matters exhibit fluid-like behavior, while in other circumstances, they do have solid-like behavior. The conventional definition of *granular matter* or *granular material* is that *granular matters are collections of a large number of discrete solid particles with interstices filled with a fluid or a gas. The solid particles need not to have the same size or have the same properties like density and shape.*

From this perspective, granular matters are in the most general sense multiphase, multi-constituent mixtures constituting of solid particles having different properties and of interstitial fluid or gas. The interstitial fluid/gas affects significantly the macroscopic behavior. However, if the interstitial fluid, e.g. air, plays an insignificant role in the transportation processes such as momentum transport, the materials are refereed to as *dry granular matters*, which may be treated as dispersed single-phase ones. Furthermore, if only one constituent exists, single-phase and single-constituent granular matters present, and dry granular mixtures are referred to those which are single-phase, but with multi-constituents. On the other hand, if the mass of interstitial fluid or its momentum is comparable to those of solid particles, the interaction between the fluid and solid phases may be significant, for the interaction may provide the driving force for the motion of solid phase. The dynamic responses of granular matters might thus be very complex, and so, aspects of conventional fluid mechanics, plasticity, solid mechanics, and non-Newtonian rheology might be involved to describe the characteristics of such matters.

For dispersed single-phase granular matters the volume that the solid particles do not occupy is called the *pore* or *empty space*. Complementary to this pore space is the quantity, called the *volume fraction*, which is multifaceted and can be defined for each solid constituent if more than one exists. A single-constituent continuum theory with/without some additional variable accounting for the role of microstructures, e.g. the evolution of grain configuration or pore space at the macro-level can be applied to describe the macroscopic behavior. For multiphase granular matters, the whole pore space or only part of it might be filled with a fluid or a gas, and are called *saturated* and *unsaturated*, respectively. Continuum theories for mixtures or multiphase continua should be revised so that the degrees of multiphase compositions can be appropriately accounted for.

Granular matters are encountered in many different forms and various contexts in everyday life. Rice, muesli, washing powder, and sand are typical examples. The first application of granular matters was the hourglass or sand clock. These devices were in common use by the end of thirteenth century e.g. to measure the speed of ships. They were further used during the Middle Age by scholars to regulate their studies and by the clergy to time their sermons. Modern applications of granular matters can be found e.g. in the area of soil mechanics, problems concerned with foundation strength, retaining walls, slope stability, which are of major importance to both civil and agricultural engineering. In the fields of production technology and process engineering, feeding and discharging particulate materials into and from any kind of storage systems are typical operations of bulk solids handling giving rise to flows of granular matters. In new and emerging technical fields such as ultrastructural processing of ceramics, new methods of Xerography and powder metal forming, many problems involving flows of granular matters remain. Closely related to these industrial applications are problems arising in geophysical and environmental contexts. Snow, rock, or powder avalanches, debris or pyroclastic flows, and the formation of dunes are typical examples.

12.1.2 Distinct Features of Granular Matters

When compared with the behavior of conventional fluids and solids, granular matters exhibit a number of distinctive features, which are not common to "ordinary" fluids or solids. Granular matters may behave somewhat like solids or fluids or even like gases under different external excitations. Their behavior can also in a process suddenly change from a fluid-like state to a solid-like state, and very often repeatedly. A short discussion on a number of distinctive features of granular matters is provided in the following.

Force chains. In static circumstances under the influence of gravity, the granulate weights of a granular matter is balanced by the strong and weak force chains distributed inside itself, which assume "tree-branch"-like forms. The force chains result essentially from the long-term enduring frictional contacts among the granulates. In dynamic circumstances, e.g. a granular matter under a simple shear, the force chains near the central region remain nearly unchanged, while those near the boundaries experience time and spatial variations, either in strong or weak form, which result mainly from the short-term instantaneous inelastic collisions among the granulates. There exist thus twofold grain-grain interactions, and the formations of strong and weak force chains, as well as the long- and short-term grain-grain interactions, have been confirmed in experiments, e.g. by using the photo-elastic technique. Twofold grain-grain interactions give rise to various macroscopic phenomena of a granular matter, although they belong to the microscopic interactions. There must exist some mesoscale mechanisms to transport the microscopic interactions to the macroscopic phenomena. In other words, a granular matter is a substance with multiple time and spatial scales in contrast to simple substances.

Dilatancy. Deformations in a granular matter are always accompanied by volume changes. This was first reported by Reynolds in 1885 with the statement: "A strongly compacted granular material placed in a flexible envelope invariably sees its volume increase when the envelope is deformed. If the envelope is unstretchable but deformable, no deformation is possible until the applied force breaks the envelope or fractures the granular material." The increase in volume during deformation can be interpreted by a purely geometric argument. If an array of identical spherical grains at densest packing is subject to a load such as a shear, it follows from the geometric considerations that particles must ride one over another. An increase in volume of the bulk material will take place. This observation becomes one of the great principles of the physics of granular matters, which is known as *Reynolds' dilatancy principle*.

Reynolds' statement is essentially static; however, it holds equally dynamically: Under rapid shearing, a *dispersive pressure* develops, as called by Bagnold in 1954.[2] He observed that in a dynamic situation, a volume expansion will be accomplished with a corresponding dispersive pressure and recognized that the presence of these internal pressure forces are responsible for the tendency of volume expansion. The dilatant behavior leads to the normal stress differences in granular matters, which are similar to those in nonlinear solids and fluids. Hence, although the density of solid particles can in some case be constant, the bulk material cannot be considered density preserving because of the dilatancy, for which the Navier-Stokes or power law constitutive functions may not be suitable. Rheological models must be introduced to account for these characteristics.

Internal angles of friction. Granulates can be piled up in a heap (triangular piles in 2D and circular cones in 3D) at rest. The surface angle θ of heap is called the *angle of response*, which is the limiting angle below which the heap stays unchanged and above which the surface granulates move down as avalanches until θ is reconstituted again. The interior yielding behavior is conventionally described by the general *Mohr-Coulomb yield criterion*,[3] i.e., yielding will occur on a plane element at an interior point, if the shear and normal stresses acting on the plane element are related by $|\tau| = c + \sigma \tan\phi$, where c is the cohesion, and ϕ represents the *static internal angle of friction*. It is generally assumed that $\theta = \phi$, and $\sigma > -c/\tan\phi$. For dynamic situations, experiments show that when dry granular matters are sheared quite rapidly, the ratio of shear-to-normal stresses remains nearly constant and is close to the values observed during quasi-static deformations but somewhat smaller. As a rule of thumb, the *dynamic internal angle of friction* is approximately 3°-4° smaller than the corresponding static internal angle of friction. This feature is basically rate-independent and plastic. For dry granular matters, the adhesion c is simply neglected. If the interstitial fluid exists or the granulates are charged, due to the viscosity of interstitial fluid or electrostatic charging, the interparticle adhesive forces appear, and non-vanishing cohesion presents.

Fluidization. Avalanches can travel longer distances than one would expect on the base that the loss in potential energy from initiation to run out is balanced by the work done due to basal sliding. It is widely accepted that flowing granular matters possess some fluid-like features. The most accepted interpretation is that in a very thin layer immediately above the sliding surface, strong shearing gives rise to enhanced collisions among the granulates, resulting in an increase in the mean particle distance and subsequently reducing the effective friction angle, so that the shear stress at the base is reduced. The supports for the interpretation have been provided by various experiments. This feature is basically viscous, and theories for it require higher-order closure relations not only for eddy-eddy interactions but also for particle-particle and possibly eddy-particle interactions.

---

[2]Ralph Alger Bagnold, 1896–1990, a British general, who is generally considered to have been a pioneer of desert exploration and laid the foundations for the research on sand transport by wind in his influential book entitled "*The Physics of Blown Sand and Desert Dunes*", which is still a main reference in the field of granular matter.

[3]Christian Otto Mohr, 1835–1918, a German civil engineer. Charles-Augustin de Coulomb, 1736–1806, a French military engineer and physicist.

Arch formation. In a granular flow, the granulates may have possibility to form sufficiently strong arch structures to sustain the impact of incoming granulates, so that the flow will be prohibited, and the granular matter experiences a change from fluid to solid state. The reverse circumstance may also take place if the arch structures break, so that discontinuous flows may occur, which are closely related to the properties of granulates such as the surface roughness, shape or size, and the geometry of flow passage. This feature may be described by using higher-order phase-change models, for the chemical compositions of granulates and bulk material remains unchanged. The phase change of a granular flow may sometimes cause failures of silos/hoppers, for unexpected high stresses on hopper walls may occur.

Turbulent fluctuation. Due to the long- and short-term grain-grain interactions at the micro-scale, the macroscopic quantities of a granular flow experience time and spatial fluctuations in their values. Such a characteristic is similar to a turbulent motion of the Newtonian fluids. This implies that the macroscopic quantities need to be decomposed into the mean and fluctuating parts by some means, and the fluctuating parts may combine with one another to form ergodic terms, which have significant influence on the macroscopic momentum and energy transportations. In addition, the turbulent fluctuations also imply that the conventional no-slip boundary condition is no longer valid for granular flows, and solid boundaries may act as the energy sources/sinks of the turbulent kinetic energies of granular flows. A better understanding of turbulent motions of granular flows may provide an improvement to prevent the bed erosion/deposition in avalanches and debris flows.

Particle segregation. When a granular matter consisting of several sorts of solid particles is deformed, there remains a tendency that the granulates having the same or similar properties tend to collect themselves in some part. In a gravity-driven shear flow with a free surface, it is observed that larger particles move toward the free surface, while smaller particles gather at the lower part of flow layer. This phenomenon is called the *particle segregation*, or *reverse/inverse grading* in geological context. Experimental studies show that four factors may give rise to particle segregation: (a) difference in particle size, (b) difference in particle shape, (c) difference in particle density, and (d) difference in particle resilience. Not all factors are of equal importance. The most important one is the size difference. The first interpretation of particle segregation was proposed by Bagnold in 1954. Although many efforts have been made to clarify the underlying physical mechanisms, it remains not completely clear.[4]

12.1.3 Granular Flows

A granular flow is a granular matter in a flowing state. During a flowing process, the characteristics of phase transition, and turbulent fluctuation may be the most two

[4]In 1988, Savage and Lun used the concept of information entropy originating from the statistical thermodynamics to derive a relatively successful theory for particle segregation, known as the *random fluctuating sieve mechanism*.

important features which need to be studied. However, such complicated macroscopic phenomena imply that it is inappropriate to consider a granular flow to be a simple fluid, for the microstructural effect in the micro- and meso-scales results subsequently in the distinct rheological features in the macro-scale. The phenomena of phase transition and turbulent fluctuation in dry granular flows will be considered in Sects. 12.2 and 12.3, respectively.

12.1.4 Modelings of Granular Flows

Granular matters are physically discrete. The grain-grain interactions, which manifest in collisional and frictional contact, are significant during flow processes, and may vary with respect to space and time. There exist three mechanisms for the generation of stress for dry granular flows: (a) the dry Coulomb-type rubbing friction, (b) the transport of momentum by particle translation between contacts, and (c) the dispersive momentum transport by collisional interactions. Since the stress tensor is considered a constitutive variable, the major difference between various theories lies in how these three contributions are taken into account. All three mechanisms are essentially effective; but in some circumstance, only single one or two are relatively significant. In different flow circumstances, different theories are proposed to take the dominant mechanisms into account. For instance, at high solid concentrations and low shear rates, the granulates are probably in close contact, so that a quasi-static, rate-independent Coulomb-type stress tensor is appropriate. On the contrary, at very low concentrations and high shear rates, the granulates are likely to be in contact in a very short period of time, and the mean free paths are relatively large in comparison with the granulate diameter. The momentum transport gained by grain-grain collisions is significant, and the bulk material will in some way behave like a dilute gas. In such a case, the stress generation is mainly due to the grain-grain collisions. If both the concentrations and shear rates are large, the momentum transfer occurs as a result of collisional interactions, for the empty space is too small to permit grain transport among collisions. Such a flow is said to be in the *grain inertia regime*. For multi-constituent granular matters, the picture becomes more complicated, for the interactions between different constituents may also provide additional contributions to the stress tensor. Modelings of granular flows are classified into three main catalogs as follows.

Molecular and event-driven dynamics (discrete element method). Since dry granular matters are composed of discrete solid particles, and the bulk properties are recognized as the mean values of particle motions and interparticle interactions, it is possible to study the macroscopic behavior by simulating the dynamic behaviors of consisting particles. This approach is referred to as the direct numerical simulation, which is akin to the direct numerical simulation (DNS) of the Navier-Stokes equation in turbulence theory. The calculations are typically conducted for a fixed number of spherical particles that are usually bounded by a box with stationary or periodic boundaries. The initial velocities and positions of particles are assigned randomly, and the equation of motion of an individual particle is simply Newton's second law of motion. Hence, the system is fundamentally a many-particle one, whose exact state in each time step can be obtained by integrating the equations of motion for all particles simultaneously. The key concern of approach is how the interparticle interactions, e.g. collision, friction and adhesion, and the particle rotations are accounted for.

If the *hard-sphere* approximation is used,[5] the resulting approach is known as the *event-driven dynamics* (ED), in which the momentum loss during collisions is characterized solely by means of the coefficient of elastic restitution, at least when particle rotations are neglected. The mechanisms of restitution of elastic energy and friction are treated as if they were completely decoupled, and the dry friction is essentially modeled by using Coulomb's law. The ED model is also the principle behind various pile-synthesizing methods, including the *dynamics of contacts*, *Monte Carlo*, and *steepest descent*. If the *soft-sphere* approximation is used, the resulting prototypical algorithm is referred to as the *molecular dynamics* (MD), in which the term "soft-sphere" means that the friction and elastic restitution come into play only when the spheres penetrate into one another, and the magnitude of interaction depends on the penetration depth. The essence of this approach revolves around the deformation of spheres. Thus, how long the spheres remain in contact is of paramount importance. Although most ED/MD simulations are performed for dry granular matters, the stickiness due to the humidity of interstitial fluid may make it necessary to account for the cohesion of the viscous nature of surrounding fluid. In such a case, the Navier-Stokes equation can be applied to model the fluid phase. What remains is to appropriately incorporate the interactions between the granulates and fluid. In both approximations, each individual particle requests at least 6 degrees of freedom (3 for position and 3 for velocity) to identify its exact state. Unfortunately, all direct numerical simulations originating from this approach can hardly handle real practical problems which involve hundreds of thousands in fact millions of particles, in which the interactions between a particle and its neighbors are far from simple. Despite this, the MD/ED approach provides some insights into the formulations of theories, much the same as experimental outcomes.

Statistical mechanics and kinetic theory. It is accepted that all matters are composed of atoms and molecules, and the bulk properties are the statistically averaged descriptions of the motions consisting of atoms and molecules. The statistical mechanics can thus be applied to study the behavior of a granular matter. The main difference between the statistical mechanics and ED/MD approach is that only a small amount of particles can be handled in the latter approach, while in the former approach, the number of particles becomes theoretically infinite. The key concern to the statistical mechanics lies in what kind of energies an individual particle or a group of particles can assume, how these energies can be modeled, and which probability distribution function should be used. In addition, as discussed in Sect. 11.5.5, the particles in the context of statistical mechanics should be in equilibrium at absolute zero, which is a very stringent assumption and might make the theories inapplicable

[5]The term "hard-sphere" does not necessarily imply that the collisions are perfectly elastic. Rather, it means that there is no interpenetration or deformation during impact.

for problems of technical relevance. This approach has been used to establish various theories, in which the appropriate Hamiltonian for the system including an interaction potential was proposed. The equations of motion of system were obtained by using a projection operator technique.

On the other hand, the kinetic theory of gas is used to describe the granular flows in the grain inertial regime. The governing equation is the Boltzmann equation, and the key concern is how the particle-particle interactions can be proposed. Conventionally, the interactions are expressed by the collision operators accounting for the energy loss during collisions. In the Boltzmann equation, different orders of moments appear, and the number of taken moments defines the complexity of theory. The important results are the evolution equations for the field variables such as density, velocity, and *granular temperature*,[6] and the formulations of constitutive variables are required. The kinetic-theory-like approaches that have been used for granular matters are extensions of the ideas of the *Brownian motion*, *Grad's thirteen moment method*, *relaxation models*, and *Enskog's dense gas theory*.

Continuum mechanics. Under specific assumptions, the behavior of a granular matter can be described by using the methods of continuum mechanics and thermodynamics. In this approach, only the behavior and properties on the macroscopic length scales, i.e., the lengths of several particle diameters, can be described. All theories based on this approach are essentially phenomenological. The closure conditions for the theories are accomplished by using the entropy principles of continuum thermodynamics, e.g. those described in Sect. 11.6.2.

Theories based on the continuum thermodynamical approach can be divided into two categories. The bulk density ρ can be expressed as the product of the true mass density of grains, γ, and the volume fraction ν viz., $\rho = \gamma\nu$. The balance of mass reduces to the balance of volume fraction, if the grains are incompressible, i.e., $\gamma =$ constant. For problems including heat transfer, the number of conventional balance equations is five, i.e., the balances of volume fraction, linear momentum, and internal energy, which equals the number of unknown primitive fields, i.e., the volume fraction, velocity, and temperature. The problem may be mathematically well-posed, and one has the chance to obtain the solutions to the primitive fields by integrating the equations simultaneously with the appropriately formulated boundary conditions, for which the *continuum mechanical models without additional balance laws* are established. Adequate closure conditions for the stress tensor and other phenomenological quantities need to be formulated appropriately.

For compressible grains, γ varies as ν does. Two variables are genuinely independent and so the empty space affects the kinematics of granular matters at the macroscopic scale by encompassing several microstructural elements. If the "material point" of a homogenized continuum is applied, all the events happening inside

[6]The granular temperature is defined as one-third of the mean fluctuation kinetic energy of a granular matter, which was introduced by Blinowski and Ogawa in 1978. See e.g. Ogawa, S., Multitemperature theory of granular materials. In: Cowin, S.C., Satake, M. (eds) Proc. Continuum Mechanical and Statistical Approaches in the Mechanics of Granular Materials. US-Japan Seminar, Sendai, Japan, 1978.

the material point are considered microstructural effects, which may be accounted for by introducing some *internal variables* at the macroscopic level. For example, the decomposition of $\rho = \gamma\nu$ provides an internal variable, namely the volume fraction ν, whose evolution may describe the microstructural effects induced by the configuration of pore space. It follows from mathematical completeness that a new equation should be supplemented. Theories based on this perspective are referred to as the *continuum mechanical models with additional balance laws*. Theories with additional balance laws can further be classified into two subclasses: Additional constitutive functions are introduced for the internal variables, or additional field equations in the form of either evolution or balance type are proposed for the internal variables. The effect of pore space is then represented by the internal variable ν with different supplementary equations for its evolution.

Although the volume fraction plays a significant role in the macroscopic behavior and is conventionally considered an internal variable in the context of continuum mechanics, it cannot take all the microstructural effects into account. As shown in Fig. 12.1a, the geometry of pore space, not just its volume fraction, may be significant. Let ω be a representative volume element (RVE) at material point $\boldsymbol{x}$ and $\chi_s(\boldsymbol{\xi})$ represent a characteristic function of the solid phase. One may introduce the variables

$$\nu \equiv \int_\omega \chi_s(\boldsymbol{\xi})\,\mathrm{d}\omega, \qquad \boldsymbol{\nu}_n = \int_\omega \underbrace{\boldsymbol{\xi}\otimes\boldsymbol{\xi}\otimes\cdots\otimes\boldsymbol{\xi}}_{n}\,\chi_s(\boldsymbol{\xi})\,\mathrm{d}\omega, \quad \chi_s(\boldsymbol{\xi}) = \left\{\begin{matrix} 0, & \boldsymbol{\xi}\in\mathcal{NS}, \\ 1, & \boldsymbol{\chi}\in\mathcal{S}, \end{matrix}\right\} \tag{12.1.1}$$

where $\mathcal{S}$ represents the solid phase, while $\mathcal{NS}$ denotes the non-solid phase. It is seen from this equation that one can introduce infinitely many variables that may be applied to account for all microstructural effects, and the volume fraction is only a special case, i.e., the zeroth-order term of $\boldsymbol{\nu}_n$. The number of introduced internal variables depends on the nature of practical problems and the preciseness that one wishes. The volume fraction assumes merely that $\boldsymbol{\nu}_n$ is not relevant for all $n \geq 1$, e.g. the internal variables accounting for the rotational motion of granulates.

It should be pointed out that in the context of continuum mechanics, the use of internal variables to account for the microstructural effects means a

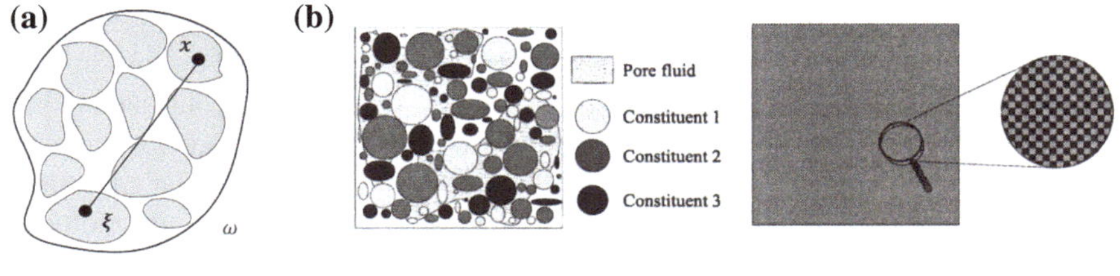

Fig. 12.1 Continuum approach to granular matter. **a** A representative volume element (RVE) ω at material point $\boldsymbol{x}$. **b** The internal structure of a multiphase system constituting of discrete identifiable solid granulates which are dispersed in a gaseous and a fluid phases, with a uniform and idealized mixture model after the homogenization process

"homogenization" or "smearing" process, which is shown in Fig. 12.1b. The volume fraction is used to describe the distribution of pore space, and the pore space is assumed to be a continuous function of some parameters with its actual state disregarded for each "material point". This means that, even though the pore space is discrete in nature, instead of the actual pore space state, only the "averaged pore space" of a material point is considered. From this perspective, it is plausible to assume that ν is a continuous function of material points.

Remarks on granular- and fluid-systems:
Three approaches for granular matters are extensions of their counterparts for simple fluids. Since the constituents of a granular system and a simple fluid are not the same, an intercomparison is discussed in the following.

- **Grain size**. The size of a grain in a granular system is not of the same order as that of a molecule in a fluid. For a granular system composed of sands, the sand grain is of an order of 10^{18} times more massive and voluminous than a water molecule. Although both systems can be treated in accordance with the laws of classical mechanics, the grain size has an important bearing on the applicability of continuum hypothesis in comparison with the characteristic length of a flow field.
- **Energy loss**. Although both the trajectories of grains and molecules can be described by using classical mechanics, a molecule is essentially a quantum-mechanical object, which is able to undergo completely elastic collisions. On the contrary, the grains are totally classical and in each collision the kinetic energy is lost, which is then transformed into heat of the colliding grains. The fact that in granular systems the grain-grain collisions do not conserve kinetic energy leads to a strikingly different macroscopic behavior from what would be expected for a molecular fluid. The circumstance holds equally even though the inelastic energy loss per collision is extremely small.
- **Grain shape**. Unlike molecules, grains are not identical particles. No two grains look precisely alike. It is expected that a continuous spectrum of the grain size exists in a granular system which introduces significant complications into the finding of a succinct theoretical description. To overcome the difficulty, the concept of *continuous diversity* to be developed.
- **Grain-grain interaction**. Since the grains are not exactly spherical and due to the surface roughness, frictional forces exist, it is expected that the grain-grain force can not be diametric. This indicates that grains experience mutual torques in most collisions, and their rotations and/or spins must be taken into account in any real granular system. Quite contrary, although the molecules in a fluid are not necessarily spherical, many of them are reasonably round, and in the molecular case, there is no analog to the macroscopic friction forces. Typically, the molecule-molecule force in a fluid has a repulsive core due to the exclusion principle and a weak but relatively long-range attractive piece which is responsible for the macroscopic phenomena like surface tension. The van der Waals force governs the evaluation of fluid parameters such as viscosity. On the contrary, for granular systems, if the grains are dry and/or the viscosity of interstitial fluids is very small, it is usually

assumed that the attractive forces (cohesion) are not significant, and only the repulsive grain-grain interactions are dominant.

- **Validity of continuum hypothesis**. Because of the size of grains, the number density of a granular matter is much smaller than that in a fluid. For example, a cubic mm of water contains about 10^{19} molecules, but the same volume of sand might contain nearly about 10 grains or less. For a hydrodynamic system, macroscopic quantities such as velocity could change significantly over a distance of 1 mm. Since the number density is so huge, smaller volumes which are far smaller than 1 mm^3 still contain a large number of molecules across whose linear dimensions the change in velocity is expected to be very small. For granular systems, due to very small number densities, the field quantities such as velocity change rapidly across small dimensions (e.g. 1 mm), and it is no longer clear if the continuum picture applies. This raises doubts about the validity of continuum hypothesis in granular systems. Fortunately, if the only length scale in a granular flow were the grain size itself, it would be expected that the continuum hypothesis can be applied to both hydrodynamic and granular systems, for there exists no way by which two systems could be distinguished. However, every real granular flow involves at least two other independent length scales, i.e., the size of container confining the system and the length scale over which the kinetic energy of grains is degraded. For the first length scale, its ratio to the molecular diameter in a hydrodynamic system might be of an order of 10^8, with the corresponding ratio of a typical sand grain system only of an order of 10^3. Hence, granular systems are always *lumpy* in a sense which can never be removed by scaling. For the second length scale, it has been shown that if the inelastic behavior of grains is not small, the average number of collisions of grains will not be large, so that the second length scale will be only a few multiples of the grain diameter. At least as far as energy transport is concerned, it is not difficult to conceive situations where substantial changes in the macroscopic quantities can occur over distances of small numbers of grain diameters.

In the forthcoming two sections, the characteristics of a dry granular laminar flow experiencing a phase transition and the characteristics of a dry granular turbulent flow with weak turbulent intensity will be investigated to demonstrate the applications of the mature disciplines of fluid mechanics and continuum thermodynamics to the study of complex phenomena.

12.2 Phase Transition in a Laminar Dense Flow

12.2.1 Introduction

The states of dry granular flows are conventionally characterized by the value of the inertia number I defined by

$$I \equiv \sqrt{2}\|\boldsymbol{D}\|d/\sqrt{p/\gamma}, \tag{12.2.1}$$

with $\|\boldsymbol{D}\| = \sqrt{\mathrm{tr}\boldsymbol{D}^2}$, where $\boldsymbol{D}$ is the symmetric part of velocity gradient, d is the particle diameter, p represents the pressure of flow, and γ denotes the true mass density of grains. The inertia number I is often used to denote the microstructural effects. As I increases, the microstructures experience sequences of transition, giving rise to three different macroscopic states. Small values of I correspond to the *quasi-static state*, in which the grains are locked into a kind of elastic networks and interact with one another through the long-term enduring contacts. For large values of I, elastic networks nearly vanish and the grains are colliding intensively with one another during a short period of time; such a situation is referred to as the *collisional state*. Between two extremes is the *dense state*, corresponding to intermediate values of I, in which both the enduring contacts and instantaneous collisions are significant.

In a quasi-static state, the long-term enduring contacts among the grains dominate, and the interparticle frictional forces become significant, giving rise to a rate-independent stress relation, which may be accounted for by e.g. the Coulomb plastic model given by

$$\tau_s = \sigma_s \tan\phi_s, \tag{12.2.2}$$

in a two-dimensional situation, where τ_s and σ_s are respectively the shear and normal stresses on the evaluation plane, and ϕ_s represents the static internal friction angle. The subscript "s" denotes that the indexed quantity is evaluated at quasi-static state. On the contrary, in a collisional state, stresses are generated essentially from the momentum transfers of grains during the short-term collisions, which may be accounted for by using e.g. the kinetic/dense gas theory. The simplest model is Bagnold's rheological model given by

$$\sigma_d = g(\nu)\mu_1\left(\frac{\mathrm{d}u}{\mathrm{d}y}\right)^2\cos\phi_d, \qquad \tau_d = g(\nu)\mu_1\left(\frac{\mathrm{d}u}{\mathrm{d}y}\right)^2\sin\phi_d = \sigma_d\tan\phi_d, \tag{12.2.3}$$

in a two-dimensional flow, where τ_d and σ_d are the shear and normal stresses on the evaluation plane, ν is the volume fraction, μ_1 represents a material constant, which is analogous to the fluid viscosity, u denotes the velocity component orthogonal to the coordinate y, and ϕ_d is the dynamic internal friction angle. The subscript "d" denotes that the indexed quantity is evaluated at collisional state, and Eq. (12.2.3) implies that dry granular flows in collisional state exhibit non-Newtonian stress characteristics. Since the ratios of the shear-to-normal stresses in both cases can be expressed in terms of the effective friction angles, i.e., ϕ_s and ϕ_d in the quasi-static and collisional states, respectively, and the difference between the normal stress and pressure under normal conditions is less than 5%, τ is essentially determined once the effective friction angle and pressure are known. Such an idea is equally extended for flows in dense state: The effective friction angle is measured, and is used to determine the shear stress in various viscoplastic constitutive models. Unfortunately, these models only apply for dense flows and are not appropriate for flows in quasi-static or collisional state.

An alternative for the determination of shear and normal stresses in dense state is a linear combination of Eqs. (12.2.2) and (12.2.3), i.e.,

$$\sigma = \sigma_s + \sigma_d, \qquad \tau = \tau_s + \tau_d = \sigma_s\tan\phi_s + \sigma_d\tan\phi_d = \sigma\tan\phi, \tag{12.2.4}$$

for a plane flow, where ϕ is the effective friction angle in dense state. Equation (12.2.4) is valid for the whole range of inertia number, provided that $\{\phi_s, \phi_d, \sigma_s, \sigma_d\}$ are known. Paradoxically, this advantage is exactly the drawback of approach: For a dry granular flow in dense state, its microstructures, e.g. the distributions of grains, vary continuously between the long-term elastic networks and short-term free collisions. The contributions of σ_s and σ_d to σ (and hence the pressure p) cannot be known a priori, but should depend on the exact flow states. Although the pressure can be measured in experiments, the relative contributions of σ_s and σ_d to p remain unknown.

Since the grains having long-term enduring contacts exhibit solid-like characteristics, and those having short-term instantaneous collisions exhibit fluid-like features, and the transition between two extremes is a continuous function of space and time, the individual contribution of σ_s or σ_d to p may be indicated by using an internal variable, called the order parameter φ, by which a thermodynamically consistent constitutive model may be established. The model can be extended to flows in quasi-static and collisional states as φ approaches to its extreme values. This approach will be discussed in the following.

12.2.2 Pressure-Ratio Order Parameter

The pressure in a dry granular dense flow consists of the static part, p_s, which results from the long-term contacts between the grains, and the dynamic part p_d resulted from the short-term collisions, i.e.,

$$p = p_s + p_d, \tag{12.2.5}$$

with $p_s \sim \sigma_s$, $p_d \sim \sigma_d$, for under normal conditions the difference between σ_s and p_s (and hence σ_d and p_d) is less than 5%. The relative contributions of p_s and p_d to p depend on the flow state and are indexed by the order parameter defined by

$$\varphi \equiv (p_s/p)^{1/2}. \tag{12.2.6}$$

The extreme values of $\varphi = 1$ and $\varphi = 0$ correspond then to the flows in quasi-static and collisional states, respectively. For flows in dense state, φ varies between two extremes. With these, Eq. (12.2.4) is expressed alternatively as

$$\begin{aligned} & p = p_s + p_d, \quad \tau = \varphi^2 p \tan\phi_s + (1-\varphi^2) p \tan\phi_d, \\ \longrightarrow \quad & \tan\phi = \varphi^2 \tan\phi_s + (1-\varphi^2)\tan\phi_d. \end{aligned} \tag{12.2.7}$$

Since p can be measured in experiments, ϕ and τ on the evaluation plane can essentially be obtained, provided that φ is determined. It is possible to use Eq. (12.2.7) to identify the stress state for flows in quasi-static, dense, or collisional states by changing the value of φ.

As the strain rate increases, φ decreases from its maximum value of unity toward its minimum value of zero. This is due to a sequence of change in the microstructure: The distribution of grains changes from an inelastic network toward a dispersive free collision, which leads to a macroscopic phase transition. Specifically, it gives rise to a higher-order solid–fluid phase transition in dense flows.[7] The higher-order phase transition gives different contributions of p_s and p_d to p, and hence, φ is coupled with other field quantities.

12.2.3 Balance Equations and Constitutive Class

The balance equations of an isotropic dry granular dense flow are given by

$$\text{mass balance} \quad 0 = \dot{\rho} + \rho \nabla \cdot \boldsymbol{v}, \tag{12.2.8}$$

$$\text{linear momentum balance} \quad \mathbf{0} = \rho \dot{\boldsymbol{v}} - \nabla \cdot \boldsymbol{t}^{\mathrm{T}} - \rho \boldsymbol{b}, \tag{12.2.9}$$

$$\text{angular momentum balance} \quad \mathbf{0} = \boldsymbol{t} - \boldsymbol{t}^{\mathrm{T}}, \tag{12.2.10}$$

$$\text{order-parameter balance} \quad 0 = \dot{\varphi} + \nabla \cdot \boldsymbol{f} - \xi, \tag{12.2.11}$$

$$\text{internal energy balance} \quad 0 = \rho \dot{e} - \boldsymbol{t} \cdot \boldsymbol{D} + \nabla \cdot \boldsymbol{q} - \rho r, \tag{12.2.12}$$

$$\text{internal friction balance} \quad \mathbf{0} = \mathring{\boldsymbol{Z}} - \boldsymbol{\Phi}, \quad (\mathring{\boldsymbol{Z}} \equiv \dot{\boldsymbol{Z}} - [\boldsymbol{\Omega}, \boldsymbol{Z}]), \tag{12.2.13}$$

$$\text{entropy balance} \quad 0 = \rho \dot{\eta} + \nabla \cdot \boldsymbol{\phi} - \rho s - \pi, \qquad \pi \geq 0, \tag{12.2.14}$$

where ρ is the bulk density of granular matter, $\boldsymbol{v}$ represents the velocity, $\boldsymbol{t}$ is the Cauchy stress tensor, $\boldsymbol{b}$ denotes the specific body force, $\boldsymbol{f}$ and ξ are the flux and production in φ, respectively, e is the specific internal energy, $\boldsymbol{q}$ is the heat flux, r denotes the specific energy supply, $\boldsymbol{Z}$ represents an Euclidean frame-indifferent, second-rank symmetric tensor (a spatial internal variable) describing the frictional and non-conservative forces inside a RVE, $\boldsymbol{\Omega}$ denotes any orthogonal rotation of RVE, $\boldsymbol{\Phi}$ is a tensor-valued constitutive relation for the production of $\dot{\boldsymbol{Z}}$, η is the specific entropy, $\boldsymbol{\phi}$ represents the entropy flux, s is the specific entropy supply, and π is the entropy production. The identity $[\boldsymbol{A}, \boldsymbol{B}] = \boldsymbol{AB} - \boldsymbol{BA}$ holds for two arbitrary second-rank tensors $\boldsymbol{A}$ and $\boldsymbol{B}$.

Equations (12.2.8), (12.2.9) and (12.2.14) are the conventional balances of mass, linear momentum, and entropy, respectively. Since the material is not considered micro-polar or Cosserat-type, and the effects of particle rotation and surface couple are excluded, the balance of angular momentum reduces to the symmetry of the Cauchy stress tensor. In Eq. (12.2.11), the order parameter φ is considered an independent field quantity with its time evolution described by a general balance equation without the external supply term, for φ is regarded as an internal variable. Reduced forms of Eq. (12.2.11) can be found in other viscoplastic theories with order parameter. To account for the effect of plasticity, the internal friction and other non-conservative forces inside a granular micro-continuum are represented by $\boldsymbol{Z}$, whose time evolution is described by Eq. (12.2.13), which is a phenomenological generalization of the Mohr-Coulomb friction criterion. In Eq. (12.2.13), $\mathring{\boldsymbol{Z}}$ is the so-called corotational objective derivative of $\boldsymbol{Z}$. It reduces to the Jaumann derivative if $\boldsymbol{\Omega}$ is chosen to be $\boldsymbol{W}$, the skew-symmetric part of velocity gradient. Equations (12.2.8)–(12.2.9) and (12.2.11)–(12.2.13) form a mathematically likely well-posed system for the unknown field ρ, $\boldsymbol{v}$, φ, $\boldsymbol{Z}$, and θ (the empirical temperature),

[7] Unlike the first-order phase transition, the chemical composition of a granular matter experiences no change during the solid-fluid transition. Only some parts of the granular body behave like a fluid, while the other parts exhibit a solid-like feature.

provided that the constitutive equations can be expressed as functions of the field quantities.[8]

The constitutive class is proposed as

$$\mathcal{Q} = \{\rho, \boldsymbol{g}_1, \theta, \boldsymbol{g}_2, \varphi, \boldsymbol{g}_3, \boldsymbol{L}, \boldsymbol{Z}\}, \tag{12.2.15}$$

for the constitutive variables

$$\mathcal{C} = \mathcal{C}(\mathcal{Q}), \qquad \mathcal{C} \in \{\boldsymbol{t}, \boldsymbol{f}, \xi, e, \boldsymbol{q}, \boldsymbol{\Phi}, \eta, \boldsymbol{\phi}\}, \tag{12.2.16}$$

where $\boldsymbol{g}_1 = \nabla\rho$, $\boldsymbol{g}_2 = \nabla\theta$, $\boldsymbol{g}_3 = \nabla\varphi$, and $\boldsymbol{L} = \nabla\boldsymbol{v}$. Equation (12.2.15) is proposed based on Truesdell's equipresence principle; the omission of velocity goes back to the requirement of material frame-indifference. Although the exclusion of $\dot{\theta}$ in Eq. (12.2.15) will cause an indifference between the empirical and absolute temperatures and leads to an infinite propagation speed of a small thermal disturbance in the material, it is rather rare and unimportant for most granular matters like soils and is omitted for simplicity. Within the material objectivity, Eq. (12.2.15) can be expressed as

$$\mathcal{Q} = \{\rho, \boldsymbol{g}_1, \theta, \boldsymbol{g}_2, \varphi, \boldsymbol{g}_3, \boldsymbol{D}, \boldsymbol{Z}\}, \tag{12.2.17}$$

where ρ and $\boldsymbol{g}_1$ are used to describe the elastic effect, and $\boldsymbol{L}$ and $\boldsymbol{Z}$ are used to capture the viscous and plastic effects, respectively. This relation defines the constitutive class used in the analysis, with which Eq. (12.2.10) holds identically.

12.2.4 Thermodynamic Analysis

Exploitation of the entropy inequality. The second law of thermodynamics requires that the entropy production π in Eq. (12.2.14) should be non-negative during a physical process. A physically admissible process should be one in which entropy inequality (12.2.14), balance equations (12.2.8), (12.2.9) and (12.2.11)–(12.2.13) as well as constitutive relations (12.2.16) and (12.2.17) hold simultaneously. This can be achieved by regarding the balance equations as the constraints of Eq. (12.2.14) via the method of the Lagrange multipliers, viz.,

$$\begin{aligned}\pi = {}& \rho\dot{\eta} + \nabla\cdot\boldsymbol{\phi} - \rho s - \lambda^\rho(\dot{\rho} + \rho\nabla\cdot\boldsymbol{v}) - \boldsymbol{\lambda}^v\cdot(\rho\dot{\boldsymbol{v}} - \nabla\cdot\boldsymbol{t} - \rho\boldsymbol{b}) - \lambda^\varphi(\dot{\varphi} + \nabla\cdot\boldsymbol{f} - \xi) \\ & - \boldsymbol{\lambda}^Z\cdot(\dot{\boldsymbol{Z}} - [\boldsymbol{\Omega}, \boldsymbol{Z}] - \boldsymbol{\Phi}) - \lambda^e(\rho\dot{e} - \boldsymbol{t}\cdot\boldsymbol{D} + \nabla\cdot\boldsymbol{q} - \rho r) \geq 0,\end{aligned} \tag{12.2.18}$$

where λ^ρ, $\boldsymbol{\lambda}^v$, λ^φ, $\boldsymbol{\lambda}^Z$, and λ^e are the Lagrange multipliers of balances of mass, linear momentum, order parameter, internal friction, and internal energy, respectively. The material behavior is required to be independent of the external supplies, so that

$$-\rho s + \rho\boldsymbol{\lambda}^v\cdot\boldsymbol{b} + \rho\lambda^e r = 0, \tag{12.2.19}$$

which is an identity for the entropy supply and is more general than the classical Duhem-Truesdell relations. Since $\dot{\theta}$ is not included in Eq. (12.2.17), it is plausible to

[8] While the balance of angular momentum can be fulfilled by the prescription of constitutive class, the balance of entropy is an inequality which will be used in the thermodynamic analysis.

introduce the Helmholtz free energy ψ in the form of $\psi \equiv e - \theta\eta$ and to conjecture that $\lambda^e = 1/\theta$.[9] Substituting these and Eq. (12.2.19) into Eq. (12.2.18) yields

$$\begin{aligned}\pi = &\rho\theta^{-1}\left(\dot{e}-\dot{\theta}\eta-\dot{\psi}\right)+\nabla\cdot\boldsymbol{\phi}-\lambda^{\rho}(\dot{\rho}+\rho\nabla\cdot\boldsymbol{v})-\boldsymbol{\lambda}^{v}\cdot(\rho\dot{\boldsymbol{v}}-\nabla\cdot\boldsymbol{t})-\lambda^{\varphi}\left(\dot{\varphi}+\nabla\cdot\boldsymbol{f}-\xi\right)\\ &-\boldsymbol{\lambda}^{Z}\cdot(\dot{\boldsymbol{Z}}-[\boldsymbol{\Omega},\boldsymbol{Z}]-\boldsymbol{\Phi})-\theta^{-1}\left(\rho\dot{e}-\boldsymbol{t}\cdot\boldsymbol{D}+\nabla\cdot\boldsymbol{q}\right)\geq 0,\end{aligned} \tag{12.2.20}$$

which, by incorporating with Eqs. (12.2.16) and (12.2.17) by using the chain rule of differentiation, is recast in the form

$$\begin{aligned}\pi = &-\left(\rho\theta^{-1}\psi_{,\rho}+\lambda^{\rho}\right)\dot{\rho}-\rho\theta^{-1}\left(\eta+\psi_{,\theta}\right)\dot{\theta}-\left(\rho\theta^{-1}\psi_{,\varphi}+\lambda^{\varphi}\right)\dot{\varphi}\\ &-\rho\theta^{-1}\psi_{,\boldsymbol{D}}\cdot\dot{\boldsymbol{D}}-\left(\rho\theta^{-1}\psi_{,\boldsymbol{Z}}+\boldsymbol{\lambda}^{Z}\right)\cdot\dot{\boldsymbol{Z}}-\rho\theta^{-1}\left(\psi_{,g_1}\dot{\boldsymbol{g}}_1+\psi_{,g_2}\dot{\boldsymbol{g}}_2+\psi_{,g_3}\dot{\boldsymbol{g}}_3\right)\\ &+\left(\boldsymbol{\phi}_{,\rho}+\boldsymbol{\lambda}^{v}\boldsymbol{t}_{,\rho}-\lambda^{\varphi}\boldsymbol{f}_{,\rho}-\theta^{-1}\boldsymbol{q}_{,\rho}\right)\cdot\boldsymbol{g}_1\\ &+\left(\boldsymbol{\phi}_{,\theta}+\boldsymbol{\lambda}^{v}\boldsymbol{t}_{,\theta}-\lambda^{\varphi}\boldsymbol{f}_{,\theta}-\theta^{-1}\boldsymbol{q}_{,\theta}\right)\cdot\boldsymbol{g}_2\\ &+\left(\boldsymbol{\phi}_{,\varphi}+\boldsymbol{\lambda}^{v}\boldsymbol{t}_{,\varphi}-\lambda^{\varphi}\boldsymbol{f}_{,\varphi}-\theta^{-1}\boldsymbol{q}_{,\varphi}\right)\cdot\boldsymbol{g}_3\\ &+\left(\boldsymbol{\phi}_{,g_1}+\boldsymbol{\lambda}^{v}\boldsymbol{t}_{,g_1}-\lambda^{\varphi}\boldsymbol{f}_{,g_1}-\theta^{-1}\boldsymbol{q}_{,g_1}\right)\cdot\nabla\boldsymbol{g}_1\\ &+\left(\boldsymbol{\phi}_{,g_2}+\boldsymbol{\lambda}^{v}\boldsymbol{t}_{,g_2}-\lambda^{\varphi}\boldsymbol{f}_{,g_2}-\theta^{-1}\boldsymbol{q}_{,g_2}\right)\cdot\nabla\boldsymbol{g}_2\\ &+\left(\boldsymbol{\phi}_{,g_3}+\boldsymbol{\lambda}^{v}\boldsymbol{t}_{,g_3}-\lambda^{\varphi}\boldsymbol{f}_{,g_3}-\theta^{-1}\boldsymbol{q}_{,g_3}\right)\cdot\nabla\boldsymbol{g}_3\\ &+\left(\boldsymbol{\phi}_{,\boldsymbol{D}}+\boldsymbol{\lambda}^{v}\boldsymbol{t}_{,\boldsymbol{D}}-\lambda^{\varphi}\boldsymbol{f}_{,\boldsymbol{D}}-\theta^{-1}\boldsymbol{q}_{,\boldsymbol{D}}\right)\cdot\nabla\boldsymbol{D}\\ &+\left(\boldsymbol{\phi}_{,\boldsymbol{Z}}+\boldsymbol{\lambda}^{v}\boldsymbol{t}_{,\boldsymbol{Z}}-\lambda^{\varphi}\boldsymbol{f}_{,\boldsymbol{Z}}-\theta^{-1}\boldsymbol{q}_{,\boldsymbol{Z}}\right)\cdot\nabla\boldsymbol{Z}\\ &-\lambda^{\rho}\rho\nabla\cdot\boldsymbol{v}-\rho\boldsymbol{\lambda}^{v}\cdot\dot{\boldsymbol{v}}+\lambda^{\varphi}\xi+\theta^{-1}\boldsymbol{t}\cdot\boldsymbol{D}+[\boldsymbol{Z},\boldsymbol{\lambda}^{Z}]\cdot\boldsymbol{\Omega}+\boldsymbol{\lambda}^{Z}\cdot\boldsymbol{\Phi}\geq 0,\end{aligned} \tag{12.2.21}$$

where the identity $\boldsymbol{\lambda}^Z\cdot[\boldsymbol{\Omega},\boldsymbol{Z}]=[\boldsymbol{Z},\boldsymbol{\lambda}^Z]\cdot\boldsymbol{\Omega}$ has been used.

Let $\boldsymbol{\mathcal{X}}=\{\dot{\boldsymbol{v}},\dot{\rho},\dot{\theta},\dot{\varphi},\dot{\boldsymbol{D}},\dot{\boldsymbol{Z}},\dot{\boldsymbol{g}}_1,\dot{\boldsymbol{g}}_2,\dot{\boldsymbol{g}}_3,\nabla\boldsymbol{g}_1,\nabla\boldsymbol{g}_2,\nabla\boldsymbol{g}_3,\nabla\boldsymbol{D},\nabla\boldsymbol{Z}\}$. It follows that inequality (12.2.21) is in the form

$$\boldsymbol{a}\cdot\boldsymbol{\mathcal{X}}+b\geq 0, \tag{12.2.22}$$

where the vector $\boldsymbol{a}$ and scalar b are functions of Eq. (12.2.17), but not of $\boldsymbol{\mathcal{X}}$. Hence, Eq. (12.2.22) is linear in $\boldsymbol{\mathcal{X}}$, and since $\boldsymbol{\mathcal{X}}$ can take any values, it would be possible to violate the inequality unless

$$\boldsymbol{a}=\boldsymbol{0},\qquad\text{and}\qquad b\geq 0, \tag{12.2.23}$$

where the first condition yields the *Liu identities* and the second condition gives the residual entropy inequality. Specifically, condition $\boldsymbol{a}=\boldsymbol{0}$ yields the Lagrangian multipliers $\{\boldsymbol{\lambda}^v,\lambda^\rho,\lambda^\varphi,\boldsymbol{\lambda}^Z\}$ given by

$$\boldsymbol{\lambda}^v=\boldsymbol{0},\ \lambda^\rho=-\rho\theta^{-1}\psi_{,\rho},\ \lambda^\varphi=-\rho\theta^{-1}\psi_{,\varphi},\ \boldsymbol{\lambda}^Z=-\rho\theta^{-1}\psi_{,\boldsymbol{Z}}, \tag{12.2.24}$$

and the restrictions

$$\boldsymbol{0}=\psi_{,g},\qquad g\in\left\{\boldsymbol{g}_1,\boldsymbol{g}_2,\boldsymbol{g}_3,\boldsymbol{D}\right\}, \tag{12.2.25}$$

[9] The exclusion of $\dot{\theta}$ in Eq. (12.2.17) only leads to $\lambda^e=\hat{\lambda}^e(\theta)$. The specific form of $\lambda^e=1/\theta$ can only be derived for simple substances. This conjecture is motivated by previous works.

on the free energy ψ. Moreover, the identities

$$\begin{aligned} \mathbf{0} &= \operatorname{sym}\left(\boldsymbol{\phi}_{,\boldsymbol{g}} - \lambda^{\varphi}\boldsymbol{f}_{,\boldsymbol{g}} - \theta^{-1}\boldsymbol{q}_{,\boldsymbol{g}}\right), \quad \boldsymbol{g} \in \{\boldsymbol{g}_1, \boldsymbol{g}_2, \boldsymbol{g}_3\}, \\ \mathbf{0} &= \boldsymbol{\phi}_{,A} - \lambda^{\varphi}\boldsymbol{f}_{,A} - \theta^{-1}\boldsymbol{q}_{,A}, \qquad A \in \{\boldsymbol{D}, \boldsymbol{Z}\}, \end{aligned} \tag{12.2.26}$$

among $\boldsymbol{\phi}$, $\boldsymbol{t}$, $\boldsymbol{f}$, and $\boldsymbol{q}$ are obtained, in which Eq. (12.2.24)$_1$ has been used, where $\operatorname{sym}(\boldsymbol{A})$ denotes the symmetric part of tensor $\boldsymbol{A}$. Lastly, it is found that

$$\eta = -\rho\theta^{-1}\psi_{\rho}, \qquad \boldsymbol{Z}\boldsymbol{\lambda}^{Z} = \boldsymbol{\lambda}^{z}\boldsymbol{Z}. \tag{12.2.27}$$

By using Eq. (12.2.24)$_1$, the residual entropy inequality reads

$$\begin{aligned} \pi = &\left(\boldsymbol{\phi}_{,\rho} - \lambda^{\varphi}\boldsymbol{f}_{,\rho} - \theta^{-1}\boldsymbol{q}_{,\rho}\right)\cdot\boldsymbol{g}_1 + \left(\boldsymbol{\phi}_{,\theta} - \lambda^{\varphi}\boldsymbol{f}_{,\theta} - \theta^{-1}\boldsymbol{q}_{,\theta}\right)\cdot\boldsymbol{g}_2 \\ &+ \left(\boldsymbol{\phi}_{,\varphi} - \lambda^{\varphi}\boldsymbol{f}_{,\varphi} - \theta^{-1}\boldsymbol{q}_{,\varphi}\right)\cdot\boldsymbol{g}_3 - \lambda^{\rho}\rho\nabla\cdot\boldsymbol{v} + \lambda^{\varphi}\xi + \theta^{-1}\boldsymbol{t}\cdot\boldsymbol{D} + \boldsymbol{\lambda}^{Z}\cdot\boldsymbol{\Phi} \geq 0. \end{aligned} \tag{12.2.28}$$

Extra entropy flux. In the context of rational extended thermodynamics, all fluxes contribute to the entropy flux, so that the extra entropy flux vector $\boldsymbol{k}$ is defined as

$$\boldsymbol{k} \equiv \boldsymbol{\phi} - \theta^{-1}\boldsymbol{q} - \boldsymbol{\lambda}^{v}\boldsymbol{t} - \lambda^{\varphi}\boldsymbol{f} = \boldsymbol{\phi} - \theta^{-1}\boldsymbol{q} - \lambda^{\varphi}\boldsymbol{f}, \tag{12.2.29}$$

which is a measure of the deviation between the entropy flux and the effects of other fluxes. It reduces to a measure of the degree of collinearity between the entropy and heat fluxes when no other fluxes are present. It follows from Eqs. (12.2.16), (12.2.17), (12.2.24)$_4$ and (12.2.25) that the functional dependencies of ψ and λ^{φ} are identified to be

$$\psi = \hat{\psi}(\rho, \theta, \varphi, \boldsymbol{Z}), \qquad \lambda^{\varphi} = -\rho\theta^{-1}\psi_{,\varphi} = \hat{\lambda}^{\varphi}(\rho, \theta, \varphi, \boldsymbol{Z}). \tag{12.2.30}$$

Since substituting Eqs. (12.2.29) and (12.2.30)$_2$ into Eq. (12.2.26)$_2$ gives $\boldsymbol{k} \neq \hat{\boldsymbol{k}}(\cdot, \boldsymbol{D})$, a general linear isotropic expression of $\boldsymbol{k}$, in view of Eq. (12.2.17), is given viz.,

$$\begin{aligned} &\boldsymbol{k} = \vartheta_1\boldsymbol{g}_1 + \vartheta_2\boldsymbol{g}_2 + \vartheta_3\boldsymbol{g}_3 + \vartheta_4\boldsymbol{Z}\boldsymbol{g}_1 + \vartheta_5\boldsymbol{Z}^2\boldsymbol{g}_1 + \vartheta_6\boldsymbol{Z}\boldsymbol{g}_2 + \vartheta_7\boldsymbol{Z}^2\boldsymbol{g}_2 + \vartheta_8\boldsymbol{Z}\boldsymbol{g}_3 + \vartheta_9\boldsymbol{Z}^2\boldsymbol{g}_3, \\ &\vartheta_i = \hat{\vartheta}_i(\rho, \theta, \varphi),\ i = 1\text{–}9. \end{aligned} \tag{12.2.31}$$

Incorporating Eqs. (12.2.29), (12.2.30)$_2$ and (12.2.31) into Eq. (12.2.26)$_1$ gives

$$\begin{aligned} &\mathbf{0} = \operatorname{sym}\left(\boldsymbol{k}_{,\boldsymbol{g}_1}\right) = \vartheta_1\boldsymbol{I} + \vartheta_4\boldsymbol{Z} + \vartheta_5\boldsymbol{Z}^2, \quad \mathbf{0} = \operatorname{sym}\left(\boldsymbol{k}_{,\boldsymbol{g}_2}\right) = \vartheta_2\boldsymbol{I} + \vartheta_6\boldsymbol{Z} + \vartheta_7\boldsymbol{Z}^2, \\ &\mathbf{0} = \operatorname{sym}\left(\boldsymbol{k}_{,\boldsymbol{g}_3}\right) = \vartheta_3\boldsymbol{I} + \vartheta_8\boldsymbol{Z} + \vartheta_9\boldsymbol{Z}^2, \end{aligned} \tag{12.2.32}$$

where $\boldsymbol{I}$ is the second-rank identity tensor. Since $\boldsymbol{Z}$ is symmetric and can vary independently, the only possibility to fulfill Eq. (12.2.32) is that $\vartheta_1 = \vartheta_2 = \cdots = \vartheta_9 = 0$, so that

$$\boldsymbol{k} = \mathbf{0}, \tag{12.2.33}$$

showing that the entropy flux $\boldsymbol{\phi}$ is not collinear with the heat flux $\boldsymbol{q}$ by an amount of $-\lambda^{\varphi}\boldsymbol{f}$, which results from the variations in the order parameter: an index of the effect of phase transition.

Substituting Eqs. (12.2.30)$_2$ and (12.2.33) into Eq. (12.2.26)$_2$ yields

$$\varphi_{,\boldsymbol{Z}}\boldsymbol{f} = \mathbf{0}, \qquad \longrightarrow \qquad \lambda^{\varphi} = \hat{\lambda}^{\varphi}(\rho, \theta, \varphi), \tag{12.2.34}$$

for $\boldsymbol{f}$ does not vanish in general. In view of Eqs. (12.2.24), (12.2.29) and (12.2.33), residual entropy inequality (12.2.28) can be recast in the form

$$\begin{aligned}\theta\pi = \; & \theta\lambda^{\varphi}_{,\rho}\boldsymbol{f}\cdot\boldsymbol{g}_1 + \theta\left(\lambda^{\varphi}_{,\theta}\boldsymbol{f} - \theta^{-2}\boldsymbol{q}\right)\cdot\boldsymbol{g}_2 + \theta\lambda^{\varphi}_{,\varphi}\boldsymbol{f}\cdot\boldsymbol{g}_3 \\ & + (\boldsymbol{t} + p\boldsymbol{I})\cdot\boldsymbol{D} - \rho\psi_{,\varphi}\xi - \rho\psi_{,\boldsymbol{Z}}\cdot\boldsymbol{\Phi} \geq 0,\end{aligned} \tag{12.2.35}$$

in which $p = \rho^2\psi_{,\rho}$, known as the thermodynamic pressure, has been used. It is determined once ψ is prescribed.

Thermodynamic equilibrium. Thermodynamic equilibrium is defined to be a time-independent process with uniform thermodynamic field quantities and vanishing entropy production, viz.,

$$\pi|_{\mathrm{E}} = 0, \tag{12.2.36}$$

where the subscript "E" indicates that the indexed quantity is evaluated in thermodynamic equilibrium. The equilibrium and dynamic sets of Eq. (12.2.17) are defined as

$$\mathcal{Q}|_{\mathrm{E}} = \left(\rho, \boldsymbol{g}_1, \theta, \boldsymbol{0}, 1, \boldsymbol{0}, \boldsymbol{0}, \boldsymbol{Z}\right); \qquad \mathcal{Q}^{\mathrm{D}} = (\boldsymbol{g}_2, \boldsymbol{g}_3, \boldsymbol{D}), \quad \mathcal{Q}^{\mathrm{D}}|_{\mathrm{E}} = \boldsymbol{0}, \tag{12.2.37}$$

so that in thermodynamic equilibrium, the gradients of temperature, order parameter, and velocity should vanish, and φ assumes its equilibrium value, namely $\varphi|_{\mathrm{E}} = 1$, while density may experience spatial variation due to the compressible effect of grains. With these, it is plausible to decompose each constitutive variable $\mathcal{C}$ into its equilibrium part $\mathcal{C}|_{\mathrm{E}}$ and dynamic part $\mathcal{C}^{\mathrm{D}}$,viz.,

$$\mathcal{C} = \mathcal{C}|_{\mathrm{E}} + \mathcal{C}^{\mathrm{D}}; \qquad \mathcal{C}|_{\mathrm{E}} = \mathcal{C}|_{\mathrm{E}}(\mathcal{Q}|_{\mathrm{E}}), \quad \mathcal{C}^{\mathrm{D}} = \mathcal{C}^{\mathrm{D}}(\mathcal{Q}^{\mathrm{D}}), \quad \mathcal{C}^{\mathrm{D}}|_{\mathrm{E}} = 0. \tag{12.2.38}$$

Moreover, the entropy production π assumes its global minimum value at an equilibrium state. Under sufficiently smoothness, π has to satisfy the conditions

$$\pi_{,\boldsymbol{g}_2}|_{\mathrm{E}} = \boldsymbol{0}, \qquad \pi_{,\boldsymbol{g}_3}|_{\mathrm{E}} = \boldsymbol{0}, \qquad \pi_{,\boldsymbol{D}}|_{\mathrm{E}} = \boldsymbol{0}, \tag{12.2.39}$$

and that the Hessian matrix of π with respect to $\{\boldsymbol{g}_2, \boldsymbol{g}_3, \boldsymbol{D}\}$ at thermodynamic equilibrium is positive semi-definite. While the first condition restricts the equilibrium forms of constitutive variables, the second condition constrains the signs of material parameters in them. For simplicity, only Eqs. (12.2.39) and (12.2.36) will be dealt with.

First, applying Eq. (12.2.36) to Eq. (12.2.35) gives

$$0 = \theta\lambda^{\varphi}_{,\rho}\boldsymbol{f}|_{\mathrm{E}}\cdot\boldsymbol{g}_1 - \rho\psi_{,\varphi}\xi|_{\mathrm{E}} - \rho\psi_{,\boldsymbol{Z}}\cdot\boldsymbol{\Phi}|_{\mathrm{E}}, \tag{12.2.40}$$

which constrains $\boldsymbol{f}|_{\mathrm{E}}$, $\xi|_{\mathrm{E}}$, and $\boldsymbol{\Phi}|_{\mathrm{E}}$ when ψ is prescribed, for λ^{φ} is determined via Eq. (12.2.30)$_2$. Second, combining Eq. (12.2.35) with Eq. (12.2.39) results in

$$\begin{aligned}&\boldsymbol{0} = \theta\lambda^{\varphi}_{,\rho}\boldsymbol{f}_{,\boldsymbol{g}_2}|_{\mathrm{E}}\cdot\boldsymbol{g}_1 + \theta\left(\lambda^{\varphi}_{,\theta}\boldsymbol{f}|_{\mathrm{E}} - \theta^{-2}\boldsymbol{q}|_{\mathrm{E}}\right) - \rho\psi_{,\varphi}\xi_{,\boldsymbol{g}_2}|_{\mathrm{E}} - \rho\psi_{,\boldsymbol{Z}}\cdot\boldsymbol{\Phi}_{,\boldsymbol{g}_2}|_{\mathrm{E}},\\ &\boldsymbol{0} = \theta\lambda^{\varphi}_{,\rho}\boldsymbol{f}_{,\boldsymbol{g}_3}|_{\mathrm{E}}\cdot\boldsymbol{g}_1 + \theta\lambda^{\varphi}_{,\varphi}\boldsymbol{f}|_{\mathrm{E}} - \rho\psi_{,\varphi}\xi_{,\boldsymbol{g}_3}|_{\mathrm{E}} - \rho\psi_{,\boldsymbol{Z}}\cdot\boldsymbol{\Phi}_{,\boldsymbol{g}_3}|_{\mathrm{E}},\\ &\boldsymbol{0} = \theta\lambda^{\varphi}_{,\rho}\boldsymbol{f}_{,\boldsymbol{D}}|_{\mathrm{E}}\boldsymbol{g}_1 + \boldsymbol{t}|_{\mathrm{E}} + p\boldsymbol{I} - \rho\psi_{,\varphi}\xi_{,\boldsymbol{D}}|_{\mathrm{E}} - \rho\psi_{,\boldsymbol{Z}}\boldsymbol{\Phi}_{,\boldsymbol{D}}|_{\mathrm{E}}.\end{aligned} \tag{12.2.41}$$

Equations (12.2.40) and (12.2.41) are the constraints that should be satisfied by $\boldsymbol{q}|_{\mathrm{E}}$, $\boldsymbol{\Phi}|_{\mathrm{E}}$, $\boldsymbol{f}|_{\mathrm{E}}$, and $\xi|_{\mathrm{E}}$. It is noted that Eqs. (12.2.40) and (12.2.41) hold essentially for compressible flows. For incompressible flows, ρ and $\nabla\rho$ are no longer independent arguments in the constitutive class and should be deleted from Eq. (12.2.17). One can perform the analysis again and the same results, namely Eqs. (12.2.40) and (12.2.41), can equally be obtained, provided that the pressure p is no longer determined by $p = \rho^2\psi_{,\rho}$, but should be considered an independent field to be determined by the momentum equation.

Simplifications. Since φ and $\boldsymbol{Z}$ are production-like internal quantities with vanishing net fluxes in equilibrium, it is plausible to assume that

$$\boldsymbol{f}|_{\mathrm{E}} = \boldsymbol{0}, \qquad \xi|_{\mathrm{E}} = 0, \qquad \boldsymbol{\Phi}|_{\mathrm{E}} = \boldsymbol{0}, \tag{12.2.42}$$

so that

$$\begin{aligned} &\boldsymbol{f} = \hat{\boldsymbol{f}}(\rho, \theta, \varphi, \boldsymbol{g}_3, \boldsymbol{D}), \qquad \xi = \hat{\xi}(\rho, \theta, \varphi, \boldsymbol{g}_3, \boldsymbol{D}), \qquad \boldsymbol{\Phi} = \hat{\boldsymbol{\Phi}}(\rho, \theta, \varphi, \boldsymbol{Z}, \boldsymbol{D}), \\ &\boldsymbol{q} = \hat{\boldsymbol{q}}(\rho, \theta, \varphi, \boldsymbol{g}_2), \end{aligned} \tag{12.2.43}$$

which are motivated by the principle of phase separation. In Eq. (12.2.43), it is assumed that $\boldsymbol{f}$ and ξ are generated mainly due to the spatial variation of order parameter φ and strain rate $\boldsymbol{D}$, $\boldsymbol{\Phi}$ is caused by the strain rate $\boldsymbol{D}$, while $\boldsymbol{q}$, by using Fourier's law, results from non-vanishing temperature gradient $\boldsymbol{g}_2$. With these and the assumption of incompressible granular flow,[10] Eq. (12.2.41) reduces to

$$\boldsymbol{q}|_{\mathrm{E}} = \boldsymbol{0}, \qquad \xi_{,\boldsymbol{g}_3}|_{\mathrm{E}} = \boldsymbol{0}, \qquad \boldsymbol{t}|_{\mathrm{E}} = -p\boldsymbol{I} + \rho\psi_{,\varphi}\xi_{,\boldsymbol{D}}|_{\mathrm{E}} + \rho\psi_{,\boldsymbol{Z}}\boldsymbol{\Phi}_{,\boldsymbol{D}}, \tag{12.2.44}$$

while Eq. (12.2.40) holds identically. The first equation indicates a vanishing equilibrium heat flux, the second one illustrates that the spatial gradient of φ has no influence on ξ, which, as will be shown later, becomes a justification and a connection to the phase transition theories like the *Ginzburg-Landau model*. The last equation shows that the equilibrium stress tensor is not spherical due to the second and third terms on its RHS, which result from the continuous variations of grain distributions (phase transition contribution) and frictional effect among the grains (internal friction contribution). Two contributions are the foundations to the existence of a granular heap in static equilibrium.

12.2.5 Rheological Constitutive Model

It is assumed that $\boldsymbol{t}^{\mathrm{D}}$, $\boldsymbol{f}^{\mathrm{D}}$, and $\boldsymbol{q}^{\mathrm{D}}$ depend explicitly and quasi-linearly on $\boldsymbol{D}$, $\boldsymbol{g}_3$, and $\boldsymbol{g}_2$, respectively, while ξ^{D} depends explicitly and quasi-linearly on φ in the forms

$$\begin{aligned} &\boldsymbol{t}^{\mathrm{D}} = \lambda \mathrm{tr}(\boldsymbol{D})\boldsymbol{I} + 2\mu\boldsymbol{D}, \qquad \boldsymbol{f}^{\mathrm{D}} = -\ell\boldsymbol{g}_3, \qquad \boldsymbol{q}^{\mathrm{D}} = -\kappa\boldsymbol{g}_2, \\ &\xi^{\mathrm{D}} = -\varsigma\left(c_0 + c_1\varphi + c_2\varphi^2 + \cdots\right), \end{aligned} \tag{12.2.45}$$

where the coefficient λ is similar to the bulk viscosity, μ is the dynamic viscosity, ℓ denotes the diffusion coefficient, κ represents the thermal conductivity, ς stands for

[10] From the perspective of practical application, this assumption is justified in most cases, although in most dry granular flows γ can be considered a constant, but ν experiences a variation, so that a non-uniform bulk density field presents.

the relaxation coefficient, and $c_0, c_1, c_2 \dots$ are material coefficients, whose functional dependencies are given by

$$\begin{aligned}\lambda, \mu &= \text{funct.}(\rho, \theta, \varphi, I_D^1, I_D^2, I_D^3, I_Z^1, I_Z^2, I_Z^3), \ell = \text{funct.}(\rho, \theta, \varphi, I_D^1, I_D^2, , I_D^3),\\ \kappa &= \text{funct.}(\rho, \theta, \varphi), \varsigma = \text{funct.}(\rho, \theta, I_D^1, I_D^2, I_D^3).\end{aligned} \tag{12.2.46}$$

These coefficients can be determined essentially by comparing numerical results with experimental outcomes. This may be a cumbersome procedure and possibly involves inverse techniques. It is noted that the explicit expression of ξ^{D} on φ, i.e., Eq. (12.2.45)$_4$, is proposed by using a power series, where the relation of $c_0 + c_1 + c_2 + \cdots = 0$ must hold to fulfill restriction (12.2.42)$_2$. For implementation of the constitutive equations for incompressible granular dense flows, the specific forms of $\boldsymbol{\Phi}$, ξ, and μ need to be prescribed.

Plastic model. Different prescriptions of $\boldsymbol{\Phi}$ allow different plastic characteristics entering the constitutive formulations. Specifically, $\boldsymbol{\Phi}$ is assumed to be a Coulomb-type plastic model given by

$$\boldsymbol{\Phi} = \boldsymbol{R}(\rho, \theta, \varphi, \boldsymbol{Z})\|\boldsymbol{D}\|, \tag{12.2.47}$$

which is used to denote the rate-independent stress contribution. With this, Eq. (12.2.44)$_3$ takes the form

$$\begin{aligned}\boldsymbol{t}|_{\mathrm{E}} &= -p\boldsymbol{I} + \rho\psi_{,\boldsymbol{Z}}\boldsymbol{R}(\rho, \theta, \varphi, \boldsymbol{Z})\mathrm{dir}(\boldsymbol{D}) = -p\boldsymbol{I} + \sqrt{2}\tau_0\mathrm{dir}(\boldsymbol{D}),\\ &= -p\boldsymbol{I} + \sqrt{2}\varphi^2 p\tan\phi_s\mathrm{dir}(\boldsymbol{D}),\end{aligned} \tag{12.2.48}$$

with $\mathrm{dir}(\boldsymbol{D}) = \boldsymbol{D}/\|\boldsymbol{D}\|$. In deriving Eq. (12.2.48), Eqs. (12.2.2), (12.2.6), (12.2.42)$_2$ and (12.2.45)$_4$ have been used, and the contribution of $\rho\psi_{,\boldsymbol{Z}}\boldsymbol{R}(\rho, \theta, \varphi, \boldsymbol{Z})$ is chosen to be $\sqrt{2}\tau_0$, where τ_0 is the yield stress as the critical static shear stress for impending flows to merge the conventional Coulomb plasticity.[11] It is noted that Eq. (12.2.48) is not (Fréchet) differentiable at $\boldsymbol{D} = \boldsymbol{0}$ and should be regularized when applied to flows in quasi-static state.

Production of order parameter. Since the variations in the grain distribution are modeled by φ, and such a microstructural state transition gives rise to a macroscopic phase transition without an actual change in the chemical composition, it is a higher-order phase transition process. Hence, the second-order Ginzburg-Landau phase transition model is used to describe the production of order parameter, viz.,

$$\left.\begin{aligned}\xi &= -\varsigma\mathcal{F}_{,\varphi},\\ \mathcal{F} &= (I - I_c)\,\varphi^2 + I_c\varphi^4/2,\end{aligned}\right\} \quad\longrightarrow\quad \xi = -2\varsigma\left[(I - I_c)\varphi + I_c\varphi^3\right], \tag{12.2.49}$$

where $\mathcal{F}$ is a free energy-like potential assuming a minimum value between 0 and 1, I_c is the critical value of I, over which the flow is in collisional state. Typical value of I_c is of an order of 10^{-1} and is regarded as a constant in the study. Since $\varphi = 1$ and $I = 0$ in equilibrium states, Eq. (12.2.49)$_2$ reduces to a vanishing $\xi|_{\mathrm{E}}$, which satisfies

[11] In doing so, the internal friction can only enter the Cauchy stress tensor via the prescription of yield stress, hence, the evolution of internal friction is decoupled from other field equations.

restriction (12.2.42)$_2$. In addition, Eq. (12.2.49)$_2$ also fulfills condition (12.2.45)$_4$, provided that $c_1 = 2(I - I_c)$, $c_3 = 2I_c$, and $c_0 = c_2 = c_4 = \cdots = 0$ are chosen. These results are significant, because the derived constraints given in Eqs. (12.2.42)$_2$ and (12.2.45)$_4$ deliver the thermodynamic justifications not only to the proposed model, but also to other order-parameter-based models.

Viscosity. The shear resistance between different material layers, which is influenced by the microstructure (grain distribution), is measured by the viscosity assumed in the form

$$\mu = \mu_0\, g(\varphi)\, \Xi(I_D^1, I_D^2, I_D^3), \qquad \longrightarrow \qquad \mu = \mu_0 \frac{\varphi_1}{1-\varphi^2}\sqrt{I_D^2}, \tag{12.2.50}$$

where φ_1 and μ_0 are constants, Ξ is an unprescribed function, which reduces to $\Xi = \sqrt{I_D^2}$ for isochoric flows, $g(\varphi)$ represents an undetermined function of φ describing the influence of grain distribution on the viscosity. The specification of $g(\varphi) = \varphi_1/(1-\varphi^2)$ is a derived result, with the derivations summarized in the following.

The viscosity for dry granular matter near the most compacted state is assumed to take the form

$$\mu = \mu_0\, g(\rho)\, \Xi(I_D^1, I_D^2, I_D^3), \qquad g(\rho) = \frac{1}{1-\rho/\rho_c}, \tag{12.2.51}$$

where ρ_c is the value of ρ at the most compacted state. Equation (12.2.51) is conventionally used in the kinetic-theory-based hydrodynamic models in high-density limit and is suitable to account for the dynamic stress responses in the study. However, a direct application of Eq. (12.2.51) into Eq. (12.2.45)$_1$ yields two drawbacks: (a) It is not appropriate for incompressible flows, and (b) its value experiences strong variations when ρ is very close to ρ_c, giving rise to strong fluctuations in the values of stress. To overcome the difficulty, the functional dependency of Eq. (12.2.51)$_2$ on ρ needs to be transformed to φ.

It follows from Eqs. (12.2.1) and (12.2.58)$_1$ that

$$\varphi^2 = 1 - \frac{I}{I_c} = 1 - \frac{\sqrt{\gamma}d}{\sqrt{p}I_c}\left|\frac{\mathrm{d}u}{\mathrm{d}y}\right|, \qquad \longrightarrow \qquad p = \frac{\gamma d^2}{(1-\varphi^2)^2 I_c^2}\left(\frac{\mathrm{d}u}{\mathrm{d}y}\right)^2, \tag{12.2.52}$$

for dense flows. Granular flows near the highest density limit have the pressure essentially generated by the short-term instantaneous collisions, which can be approximated by Bagnold's rheological model, viz.,

$$p_d = \rho_0 \frac{1}{1-\rho/\rho_c}\left(\frac{\mathrm{d}u}{\mathrm{d}y}\right)^2, \qquad \longrightarrow \qquad (1-\varphi^2)p = \rho_0\frac{1}{1-\rho/\rho_c}\left(\frac{\mathrm{d}u}{\mathrm{d}y}\right)^2, \tag{12.2.53}$$

in which Eqs. (12.2.5) and (12.2.6) have been used, where ρ_0 denotes the initial bulk density. Comparing Eq. (12.2.52) with Eq. (12.2.53) yields

$$g(\rho) = \frac{1}{1-\rho/\rho_c} = \frac{\varphi_1}{1-\varphi^2} = g(\varphi), \tag{12.2.54}$$

with $\varphi_1 = \gamma d^2/(\rho_0 I_c^2)$. Equation (12.2.54) is then used in the formulation of viscosity in Eq. (12.2.50)$_2$ to account for the dynamic stress responses. For convenience, it is expressed as

$$\varphi^2 = 1 - \varphi_1 \left(1 - \frac{\rho}{\rho_c}\right), \tag{12.2.55}$$

which, together with Eq. (12.2.58)$_1$, shows that φ as well as ρ decrease as strain rate increases, a phenomenon known as dilatancy. This can equally be recognized from

$$\rho = \rho_c - eI, \qquad \text{with } e = \rho_c/(\varphi_1 I_c), \tag{12.2.56}$$

as derived by combining Eqs. (12.2.54), (12.2.55) and (12.2.58)$_1$. This linear relation has been confirmed in experiments. Since it has been reported that $e/\gamma \approx 0.3$, $I_c \approx 0.2$, and $\rho_c \approx 0.8$, it follows that φ_1 assumes an approximated value of 8 in the study. With this and Eq. (12.2.55), it is verified that the variation in φ can account for about 12% density variation in the present model, although the granular flows are assumed a priori incompressible.

With Eqs. (12.2.38), (12.2.42), (12.2.44)–(12.2.45) and (12.2.48)–(12.2.50), the complete thermodynamically consistent rheological constitutive model with a pressure-ratio order parameter for an incompressible, isochoric dry granular dense flow is obtained as

$$\begin{aligned} \boldsymbol{t} &= -(p - \lambda\,\mathrm{tr}\boldsymbol{D})\,\boldsymbol{I} + \sqrt{2}\varphi^2\, p \tan\phi_s\, \mathrm{dir}(\boldsymbol{D}) + 2\mu_0 \frac{\varphi_1}{1-\varphi^2}\sqrt{I_{\boldsymbol{D}}^2}, \\ \boldsymbol{f} &= -\ell\,\nabla\varphi, \qquad\qquad \xi = -\varsigma\left[2(I - I_c)\varphi + 2I_c\varphi^3\right], \\ \boldsymbol{\Phi} &= \boldsymbol{R}(\rho, \theta, \varphi, \boldsymbol{Z})\|\boldsymbol{D}\|, \qquad \boldsymbol{q} = -\kappa\nabla\theta. \end{aligned} \tag{12.2.57}$$

The complete field equations for the unknown fields $\{p, \boldsymbol{v}, \varphi, \boldsymbol{Z}, \theta\}$ can be obtained in principle by substituting this equation into Eqs. (12.2.8)–(12.2.9) and (12.2.11)–(12.2.13).

Remarks:

1. In the work of da Cruz, et al., the order parameter and effective friction angle are respectively given by

$$\varphi = \begin{cases} \sqrt{(I_c - I)/I_c}, & I \le I_c, \\ 0, & I > I_c, \end{cases} \qquad \tan\phi = \tan\phi_s + \frac{I}{I_c}(\tan\phi_d - \tan\phi_s), \tag{12.2.58}$$

which are the alternative expressions of Eqs. (12.2.6) and (12.2.7). However, Eq. (12.2.58) has been proposed without thermodynamic justifications. More interesting is that the effective friction angle is found to increase linearly with the inertia number I in dense flows (and hence Eq. (12.2.58)$_2$ is called the friction law). This indicates that the state transition process is indeed second-order and justifies the application of the second-order Ginzburg-Landau phase transition model to the determination of ξ in Eq. (12.2.49).

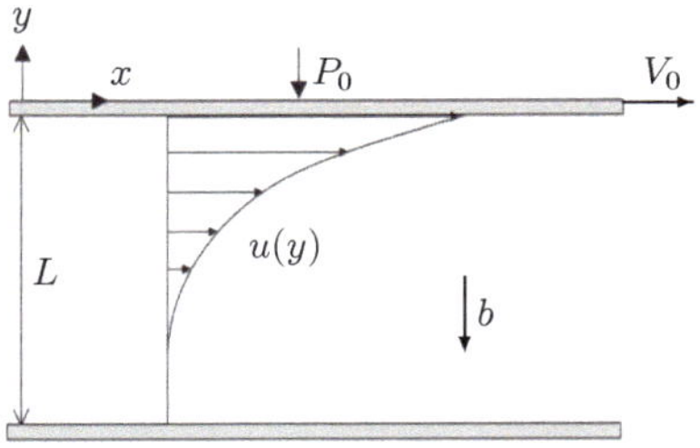

Fig. 12.2 Sketch of a dry granular dense matter in a horizontal shearing and the coordinates

2. Substituting Eqs. $(12.2.57)_{2,3}$ into Eq. (12.2.11) gives the balance of order parameter as

$$\dot{\varphi} = \nabla \cdot (\ell \nabla \varphi) - 2\varsigma \left[(I - I_c)\varphi + I_c \varphi^3 \right], \tag{12.2.59}$$

which is a kind of time-dependent *Ginzburg-Landau equation*. The equation can also be derived by using the variational approach, viz.,

$$\dot{\varphi} = -\varsigma \frac{\delta F}{\delta \varphi}, \quad \text{with } F = \int_B \left(\mathcal{F} + \frac{\ell}{2\varsigma} |\nabla \varphi|^2 \right) dv, \tag{12.2.60}$$

in which B denotes the material body. While the first volume integral in Eq. $(12.2.60)_2$ represents the local short-term interaction contributions to the potential function F with $\mathcal{F}$ given by Eq. $(12.2.49)_1$, the second volume integral shows the contributions of non-local long-term interaction. These illustrate that Eq. (12.2.59) is justified by both the thermodynamical and variational approaches.

12.2.6 Numerical Simulations

Consider a two-dimensional, steady, isothermal, incompressible dry granular dense flow between two infinite parallel plates separated by a distance L with the coordinates shown in Fig. 12.2, where the lower plate is stationary, while the upper plate moves at a constant speed V_0. The flow is triggered by the motion of upper plate, to which a constant pressure P_0 is applied to fit the experimental setup.

Field equations and boundary conditions. It follows from the geometric configuration that a parallel flow prevails, i.e.,

$$\boldsymbol{v} = [u(y), 0, 0], \quad \boldsymbol{b} = [0, -b, 0], \quad \varphi = \varphi(y), \quad p = p(y), \quad \boldsymbol{Z}_{ij} = Z_{ij}(y), \tag{12.2.61}$$

where $u(y)$ is the velocity component in the x-direction, and Z_{ij} is the component of $\boldsymbol{Z}$ with $\{i, j\}$=1-2. Substituting Eq. (12.2.61) into Eqs. (12.2.8), (12.2.9) and (12.2.59) yields the reduced balance equations in the forms

$$0 = \frac{\mathrm{d}t_{xy}}{\mathrm{d}y}, \quad 0 = \frac{\mathrm{d}t_{yy}}{\mathrm{d}y} - \rho b, \quad 0 = \frac{\ell}{\varsigma} \frac{\mathrm{d}^2 \varphi}{\mathrm{d}y^2} - 2(I - I_c)\varphi - 2I_c \varphi^3, \tag{12.2.62}$$

in which the diffusivity ℓ is considered a constant. The first two equations are the balances of linear momentum in the x- and y-directions, respectively, while the last equation is the reduced time evolution of order parameter. The associated constitutive equations are given by

$$t_{xy} = \varphi^2\, p\, \tan\phi_s + \frac{\mu_0\,\varphi_1}{1-\varphi^2}\left(\frac{\mathrm{d}u}{\mathrm{d}y}\right)^2, \qquad t_{yy} = -p, \tag{12.2.63}$$

which are obtained by substituting Eq. (12.2.61) into Eq. (12.2.57)$_1$. Incorporating Eq. (12.2.63)$_2$ into Eq. (12.2.62)$_2$ gives the linear profile of $p(y)$ in the form of $p(y) = P_0 - \rho b y$, which may be used in Eq. (12.2.63)$_1$ to determine t_{xy}, as will be shown later. Of particular importance is that Eq. (12.2.63)$_1$ takes an alternative form, viz.,

$$t_{xy} = p\,\tan\phi_s + \tilde{\mu}\sqrt{\gamma p}\,\frac{\mathrm{d}u}{\mathrm{d}y}, \qquad \tilde{\mu} = \frac{d}{I_c}\left(\frac{\mu_0}{\rho_0} - \tan\phi_s\right), \tag{12.2.64}$$

in which Eqs. (12.2.1), (12.2.53) and (12.2.58)$_1$ have been used. The first term on the RHS of the first equation denotes the rate-independent stress contribution, while the second term illustrates the dynamic contribution. Since $\tilde{\mu} > 0$, it follows from the second equation that $\mu_0/\rho_0 > \tan\phi_s$, which justifies the frictional law for dense flows, and also the derived stress model given in Eq. (12.2.57)$_1$.

Substituting Eq. (12.2.63)$_1$ into Eq. (12.2.62)$_1$ gives the field equation for $u(y)$ as

$$0 = \frac{\mathrm{d}}{\mathrm{d}y}\left[\varphi^2\, p\, \tan\phi_s + \frac{\mu_0\,\varphi_1}{1-\varphi^2}\left(\frac{\mathrm{d}u}{\mathrm{d}y}\right)^2\right]. \tag{12.2.65}$$

This Eq. (12.2.62)$_3$ are respectively the governing differential equations for $u(y)$ and $\varphi(y)$, with the boundary conditions given by

$$\left.\frac{\mathrm{d}\varphi}{\mathrm{d}y}\right|_{y=0} = -cI, \quad \left.\frac{\mathrm{d}\varphi}{\mathrm{d}y}\right|_{y=-L} = cI; \qquad u|_{y=0} = V_0, \;\; u|_{y=-L} = 0. \tag{12.2.66}$$

In doing so, it is assumed that the grains and two plate surfaces are sufficiently rough, so that the grains adhere to the solid plates due to the frictional effect, and the conventional no-slip condition for $u(y)$ holds. The Neumann boundary conditions for $\varphi(y)$ given in Eqs. (12.2.66)$_{1,2}$ are motivated by the facts that the solid boundaries behave as energy sources/sinks for the fluctuating kinetic energies of grains, which are the dominant causes of a solid-fluid transition, so that the order-parameter gradients normal to the boundaries are assumed to be proportional to the inertia number with the proportionality c pointing downward/upward on the upper/lower planes, respectively.

Non-dimensionalization. Define the dimensionless quantities

$$\begin{aligned}
&\bar{y} = \frac{y}{d}, \quad \bar{L} = \frac{L}{d}, \quad \bar{u} = \frac{u}{\sqrt{bd}}, \quad \bar{V}_0 = \frac{V_0}{\sqrt{bd}},\\
&\bar{p} = \frac{p}{\rho b d}, \quad \alpha = \frac{\ell}{\varsigma d^2}, \quad \beta = \frac{\mu_0\varphi_1}{\rho_0 d^2}, \quad \bar{c} = dc,
\end{aligned} \tag{12.2.67}$$

with which Eqs. (12.2.62)$_3$ and (12.2.65) become

$$\begin{aligned}
&0 = \alpha\frac{\mathrm{d}^2\varphi}{\mathrm{d}\bar{y}^2} - 2\,(I - I_c)\,\varphi - 2I_c\varphi^3, \qquad I = \sqrt{\frac{\gamma}{\rho p}}\frac{\mathrm{d}\bar{u}}{\mathrm{d}\bar{y}},\\
&0 = \frac{\mathrm{d}}{\mathrm{d}\bar{y}}\left[\varphi^2\bar{p}\tan\phi_s + \frac{\beta}{1-\varphi^2}\left(\frac{\mathrm{d}\bar{u}}{\mathrm{d}\bar{y}}\right)^2\right],
\end{aligned} \tag{12.2.68}$$

which are associated with the dimensionless boundary conditions given by

$$\left.\frac{d\varphi}{d\bar{y}}\right|_{\bar{y}=0} = -\bar{c}I, \quad \left.\frac{d\varphi}{d\bar{y}}\right|_{\bar{y}=-\bar{L}} = \bar{c}I; \qquad \bar{u}|_{\bar{y}=0} = \bar{V}_0, \quad \bar{u}|_{\bar{y}=-\bar{L}} = 0. \qquad (12.2.69)$$

Numerical method. The two-point nonlinear BVP given in Eqs. (12.2.68) and (12.2.69) are solved numerically to obtain $\bar{u}(\bar{y})$ and $\varphi(\bar{y})$, for which the iterative method is used. To this end, Eq. (12.2.68) is recast in the form

$$\begin{aligned}
\frac{d^2\varphi}{d\bar{y}^2} &= \frac{2\,(I-I_c)}{\alpha}\varphi + \frac{2I_c}{\alpha}\varphi^3, \\
\frac{d^2\bar{u}}{d\bar{y}^2} &= \frac{\varphi^2(1-\varphi^2)\tan\phi_s}{2\beta d\bar{u}/d\bar{y}} - \frac{\varphi(1-\varphi^2)(\bar{P}_0-\bar{y})\tan\phi_s}{\beta}\frac{d\varphi/d\bar{y}}{d\bar{u}/d\bar{y}} - \frac{\varphi}{1-\varphi^2}\frac{d\varphi}{d\bar{y}}\frac{d\bar{u}}{d\bar{y}},
\end{aligned} \qquad (12.2.70)$$

in which $\bar{p} = \bar{P}_0 - \bar{y}$ has been used. In Eq. (12.2.70), the highest derivatives of φ and $\bar{u}$ are expressed in terms of their lower derivatives. An iterative procedure is constructed as

$$\begin{aligned}
\frac{d^2\varphi^{k+1}}{d\bar{y}^2} &= \frac{2\,(I-I_c)}{\alpha}\varphi^k + \frac{2I_c}{\alpha}(\varphi^k)^3, \\
\frac{d^2\bar{u}^{k+1}}{d\bar{y}^2} &= \frac{(\varphi^k)^2(1-(\varphi^k)^2)\tan\phi_s}{2\beta d\bar{u}^k/d\bar{y}} - \frac{\varphi^k(1-(\varphi^k)^2)(\bar{P}_0-\bar{y}^k)\tan\phi_s}{\beta}\frac{d\varphi^k/d\bar{y}}{d\bar{u}^k/d\bar{y}} \\
&\quad - \frac{\varphi^k}{1-(\varphi^k)^2}\frac{d\varphi^k}{d\bar{y}}\frac{d\bar{u}^k}{d\bar{y}},
\end{aligned} \qquad (12.2.71)$$

where the superscript "k" denotes the iteration step. Equation (12.2.71) and boundary conditions

$$\left.\frac{d\varphi^{k+1}}{d\bar{y}}\right|_{\bar{y}=0} = -\bar{c}I, \quad \left.\frac{d\varphi^{k+1}}{d\bar{y}}\right|_{\bar{y}=-\bar{L}} = \bar{c}I; \qquad \bar{u}^{k+1}|_{\bar{y}=0} = \bar{V}_0, \quad \bar{u}^{k+1}|_{\bar{y}=-\bar{L}} = 0, \qquad (12.2.72)$$

define two linear differential boundary value problems for φ^{k+1} and $\bar{u}^{k+1}$. By using the finite-difference method, two linear algebraic equation systems can be deduced and solved for each iterative step $(k+1)$. Hence, sequences of solutions $\varphi^{(0)}(\bar{y})$, $\varphi^{(1)}(\bar{y}), \varphi^{(2)}(\bar{y}), \ldots$ and $\bar{u}^{(0)}(\bar{y}), \bar{u}^{(1)}(\bar{y}), \bar{u}^{(2)}(\bar{y}), \ldots$ are determined in the following manner: If the initial $\varphi^{(0)}(\bar{y})$ and $\bar{u}^{(0)}(\bar{y})$ are given, then $\varphi^{(1)}(\bar{y})$, $\varphi^{(2)}(\bar{y})$, … and $\bar{u}^{(1)}(\bar{y}), \bar{u}^{(2)}(\bar{y}), \ldots$ are calculated successively as the solutions to the boundary value problems.

To achieve a better convergence, the method of successive under-relaxation is used. That is, for the $(k+1)$ iterative step, the estimated values of φ^{k+1} and $\bar{u}^{k+1}$: $\tilde{\varphi}^{k+1}$ and $\tilde{\bar{u}}^{k+1}$ are determined, and then φ^{k+1} and $\bar{u}^{k+1}$ are revised by

$$\varphi^{k+1} = \varphi^k + \omega(\tilde{\varphi}^{k+1} - \varphi^k), \quad \bar{u}^{k+1} = \bar{u}^k + \omega(\tilde{\bar{u}}^{k+1} - \bar{u}^k), \qquad 0 < \omega < 1, \qquad (12.2.73)$$

where ω is a under-relaxation parameter, whose value should be so chosen that convergent iterations are reached. For the iteration procedure, the dimensionless profiles

$$\varphi^{(0)} = 1.0, \qquad \bar{u}^{(0)} = 1 + \bar{y}/\bar{L}, \qquad (12.2.74)$$

are used as the initial solutions to Eq. (12.2.71).

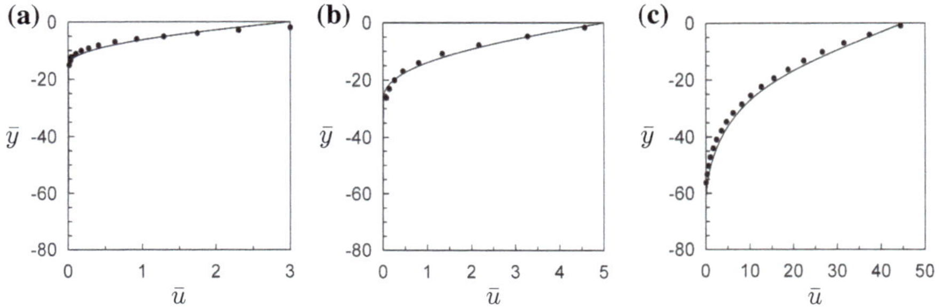

Fig. 12.3 Calculated profiles of $\bar{u}$ under $\alpha = 4$, $\beta = 12$, $\phi_s = 18°$, $I_c = 0.2$ and $\bar{c} = 0$. **a** $\bar{P}_0 = 10$, $\bar{V}_0 = 3$. **b** $\bar{P}_0 = 50$, $\bar{V}_0 = 5$. **c** $\bar{P}_0 = 50$, $\bar{V}_0 = 45$. Solid lines: present model; dots: DEM simulations

Numerical results. The calculated profiles of $\bar{u}(\bar{y})$ and $\varphi(\bar{y})$ for the variations in the parameters $\bar{V}_0$, $\bar{P}_0$, ϕ_s, α, β and I_c are shown in Figs. 12.3-12.9, in which the horizontal axes denote the calculated values of $\bar{u}(\bar{y})$ or $\varphi(\bar{y})$, while the vertical axes are the distance $\bar{y}$ apart from the upper plate. Figure 12.3 illustrates the calculated profiles of $\bar{u}(\bar{y})$ and $\varphi(\bar{y})$ compared with the results from DEM simulations to estimate the model validity,[12] where $\alpha = 4$, $\beta = 12$, $\phi_s = 18°$, $I_c = 0.2$, and $\bar{c} = 0$ are assigned with different prescriptions of $\bar{P}_0$ and $\bar{V}_0$, whose values are chosen to be the same as those used in the DEM simulations. As $\bar{P}_0$ and $\bar{V}_0$ increase, the gains interlock with one another more efficiently, and the shear on the upper plane can be transmitted more efficiently toward the lower plane, as shown by the $\bar{u}$-profiles with increasing amplitude. The entire granular shear layer can be divided into two regions. In the region near the upper plane, the grains are colliding strongly with one another, exhibiting fluid-like characteristics. Outside this region, the grains remain almost stationary and behave like a bulk solid due to the long-term enduring contacts. The transition from solid to fluid behavior emerges gradually as approaching the upper plane. The correspondence between the calculated results and results from DEM simulations shows the validity of the proposed model.

The calculated of $\bar{u}(\bar{y})$ and $\varphi(\bar{y})$ for variations in $\bar{V}_0$ are displayed in Fig. 12.4, in which $\bar{P}_0 = 50$, $\alpha = 4$, $\beta = 12$, $\phi_s = 18°$, $I_c = 0.2$ and $\bar{c} = 0$. As $\bar{V}_0$ increases, the shear on the upper plane becomes larger, and its transmission toward the lower plate enhances gradually, which results in the velocity profiles with increasing amplitude, as shown in Fig. 12.4a. The profiles of $\varphi(\bar{y})$ are shown in Fig. 12.4b, in which φ decreases monotonically from its maximum value on the lower plane toward the upper plane. As $\bar{V}_0$ increases, the decreasing tendency of φ becomes more obvious. Again, the whole layer is divided into two regions: In the lower region, φ remains almost unchanged, corresponding to the dense packing of grains with solid-like

[12]The DEM simulation results are quoted from Volfson, D., Tsimging, L.D., Aranson, I.S., Partially fluidized shear granular flow: Continuum theory and molecular dynamics simulations, *Physics Review E*, 68, 021301, 2003.

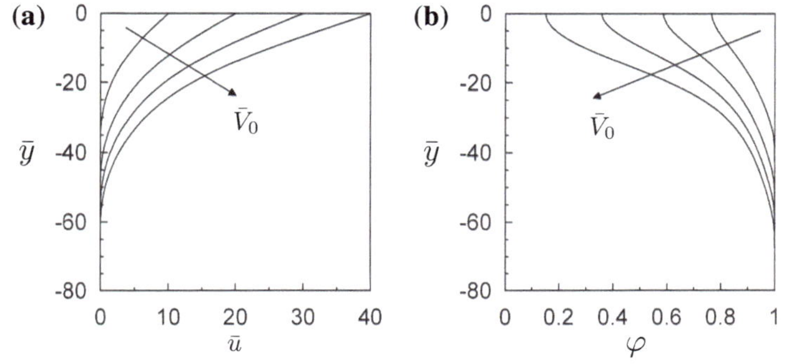

Fig. 12.4 Calculated profiles of $\bar{u}$ and φ for variations in $\bar{V}_0$ (=10, 20, 30, 40) under $\bar{P}_0 = 50$, $\alpha = 4$, $\beta = 12$, $\phi_s = 18°$, $I_c = 0.2$ and $\bar{c} = 0$. **a** Profiles of $\bar{u}(\bar{y})$. **b** Profiles of $\varphi(\bar{y})$

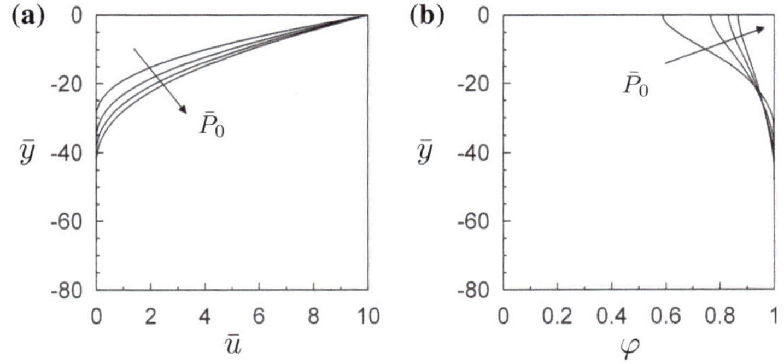

Fig. 12.5 Calculated profiles of $\bar{u}$ and φ for variations in $\bar{P}_0$ (=25, 50, 75, 100) under $\bar{V}_0 = 10$, $\alpha = 4$, $\beta = 12$, $\phi_s = 18°$, $I_c = 0.2$ and $\bar{c} = 0$. **a** Profiles of $\bar{u}(\bar{y})$. **b** Profiles of $\varphi(\bar{y})$

behavior. In the upper region, φ experiences strong variations, which indicate that the dense packing among the grains is broken, and the grains are colliding strongly with one another. Wherever φ is smaller, the collisions are stronger, resulting in fluid-like behavior. The transition from solid to fluid states can be recognized through the continuous profiles of φ.

Calculations for variations in $\bar{P}_0$ have been carried out and the obtained profiles of $\bar{u}(\bar{y})$ and $\varphi(\bar{y})$ are shown in Fig. 12.5, in which $\bar{V}_0 = 10$, $\alpha = 4$, $\beta = 12$, $\phi_s = 18°$, $I_c = 0.2$ and $\bar{c} = 0$. Increasing $\bar{P}_0$ corresponds to increase the normal pressure on the upper plane. Larger values of $\bar{P}_0$ indicate that larger normal pressures are applied on the upper plate and push the grains to interlock stronger with one another. This gives more solid-like behavior of the granular body. There exists a thin layer near the upper plane, in which the grains are colliding with one another due the significant shear there. Below this layer, the grains are interlocked and behave as a solid. In contrast to Fig. 12.4a, the increasing tendency of velocity amplitude as $\bar{P}_0$ increases is due to the motions of bulk grain solid, not the motions of individual grains, as those shown in Fig. 12.5a. The influences of static friction angle ϕ_s on the calculated

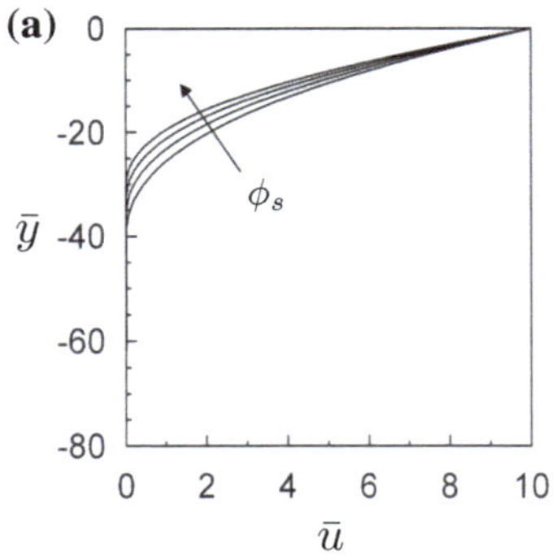

Fig. 12.6 Calculated profiles of $\bar{u}$ and φ for the variations in ϕ_s (=16°, 18°, 20°, 22°) under $\bar{V}_0 = 10$, $\bar{P}_0 = 50$, $\alpha = 4$, $\beta = 12$, $I_c = 0.2$ and $\bar{c} = 0$. **a** Profiles of $\bar{u}(\bar{y})$. **b** Profiles of $\varphi(\bar{y})$

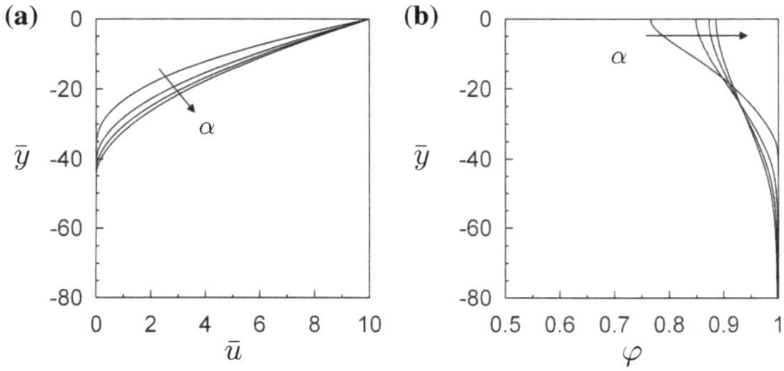

Fig. 12.7 Calculated profiles of $\bar{u}$ and φ for variations in α (=4, 40, 80, 120) under $\bar{V}_0 = 10$, $\bar{P}_0 = 50$, $\beta = 12$, $\phi_s = 18°$, $I_c = 0.2$ and $\bar{c} = 0$. **a** Profiles of $\bar{u}(\bar{y})$. **b** Profiles of $\varphi(\bar{y})$

$\bar{u}$- and φ-profiles are illustrated in Fig. 12.6, where $\bar{V}_0 = 10$, $\bar{P}_0 = 50$, $\alpha = 4$, $\beta = 12$, $I_c = 0.2$ and $\bar{c} = 0$. Increasing ϕ_s corresponds to increase the yield shear stress, and the transmission of shear becomes less efficient, so that the upper fluid-like layer becomes thinner, as shown in Fig. 12.6a. This tendency is also manifest in the profiles of φ, as shown in Fig. 12.6b. As ϕ_s increases, there exists only a thin layer underneath the upper plane, in which the grains behave like a fluid. Outside this layer, the grains remain as a stationary bulk solid.

Calculations have been conducted for variations in α, and the results are shown in Fig. 12.7, where $\bar{V}_0 = 10$, $\bar{P}_0 = 50$, $\beta = 12$, $\phi_s = 18°$, $I_c = 0.2$, and $\bar{c} = 0$. Increasing α tends to enhance the "diffusion of φ" under fixed values of d and ς, and to cause the fluid-like characteristic be more significant. Such a tendency is recognized in Fig. 12.7, in which as α increases, the production of φ on the upper plane can penetrate more efficiently toward the lower plane via a better diffusion, resulting in a thicker fluidized layer with larger velocity amplitude. Figure 12.8 illustrates the obtained results of $\bar{u}(\bar{y})$ and $\varphi(\bar{y})$ under variations in β, in which $\bar{V}_0 = 10$, $\bar{P}_0 = 50$,

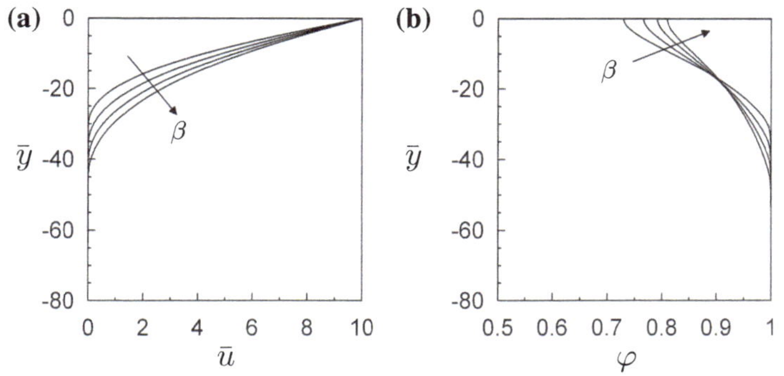

Fig. 12.8 Calculated profiles of $\bar{u}$ and φ for variations in β (=10, 12, 14, 16) under $\bar{V}_0 = 10$, $\bar{P}_0 = 50$, $\alpha = 4$, $\phi_s\, 18°$, $I_c = 0.2$ and $\bar{c} = 0$. **a** Profiles of $\bar{u}(\bar{y})$. **b** Profiles of $\varphi(\bar{y})$

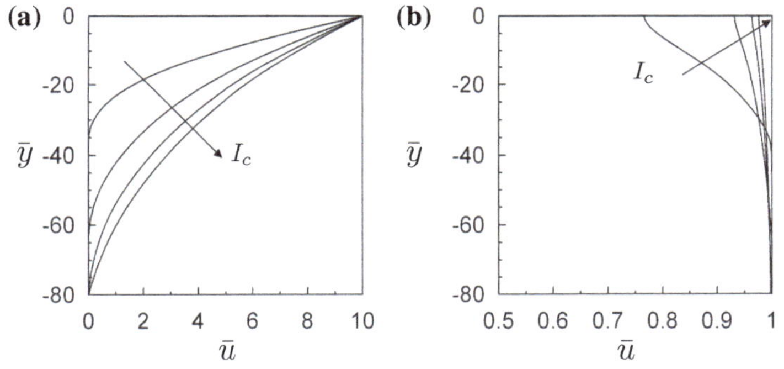

Fig. 12.9 Calculated profiles of $\bar{u}$ and φ for variations in I_c (=0.2, 0.4, 0.6, 0.8) under $\bar{V}_0 = 10$, $\bar{P}_0 = 50$, $\alpha = 4$, $\beta = 12$, $\phi_s = 18°$ and $\bar{c} = 0$. **a** Profiles of $\bar{u}(\bar{y})$. **b** Profiles of $\varphi(\bar{y})$

$\alpha = 4$, $\phi_s = 18°$, $I_c = 0.2$, and $\bar{c} = 0$ are chosen. The parameter β is a measure of viscosity, and as it increases, the adhesion between granular layers enhances gradually, so that the shear on the upper plate can better be transmitted toward the lower plate. This gives rise to thicker fluidization layers near the upper plane, in which the grains move with larger speeds. The thickness of fluidized layer increases as β increases due to the enhanced viscous adhesion.

Since the parameter I_c denotes the critical inertia number between the solid and fluid state of a dry granular dense flow, the calculations for its variations have been conducted and the results are displayed in Fig. 12.9, in which $\bar{V}_0 = 10$, $\bar{P}_0 = 50$, $\alpha = 4$, $\beta = 12$, $\phi_s = 18°$, and $\bar{c} = 0$. Larger values of I_c correspond to smaller shear resistances, under which a larger portion of the granular body can be set in flow. Such a tendency is revealed in Fig. 12.9b, in which as I_c increases, the fluidized layer becomes thicker, and the grains are moving with larger speed shown in Fig. 12.9a. Only a small portion near the lower plate remains stationary and behaves as a solid body. For large values of I_c, e.g. $I_c = 0.8$, the velocity profile across the channel is likely linear, and the granular body is similar to a Newtonian fluid.

Conclusions. During the motion, a dry granular dense flow experiences a sequence of microstructure transition exhibiting an increase in the shear stress, it is considered a material with second-order phase transition from solid to fluid behavior. It is found that the solid–fluid phase transition contributes to the entropy flux by the amount of $-\lambda^{\varphi} \boldsymbol{f}$. The classical selection of entropy flux is no longer valid due the changes in the microstructures. The non-collinearity between the heat and entropy fluxes adds the complexity of thermodynamic analysis and enters the equilibrium expressions of constitutive quantities, as shown in Eqs. (12.2.40) and (12.2.41). Of particular interest is the kinematic evolution equation of φ, which can be derived from either the thermodynamic or variational approach. Although it is used frequently in other order-parameter-based constitutive models, the thermodynamic consistencies of Eqs. (12.2.11) and (12.2.59) have been established successfully, so that the phase transition model found its root on the thermodynamic consistency. These results deliver a criterion to verify the thermodynamic consistencies of other order-parameter-based constitutive models.

Numerical simulations show that the granular flow in a simple plane shear tends to have two distinct regions. In the region near the upper plane, the grains are colliding strongly with one another and behave like a fluid. In the region near the lower plane, the long-term enduring contacts among the grains dominate, and the grains behave like a bulk solid. The transition from the solid-like to fluid-like regions becomes obvious when approaching the upper plane. Although the relative thicknesses of fluidized and solid regions vary as the model parameters vary, the above tendency holds and corresponds well to the results from the DEM simulations. These findings suggest that the established model is able to distinguish different states of granular flows by varying the value of order parameter.

12.3 A Turbulent Flow with Weak Intensity

12.3.1 Introduction

The microstructural grain-grain interaction of a dry granular matter results from the long-term enduring frictional contact and sliding, and short-term inelastic collision. While a flow in quasi-static state and a flow in collisional state are defined when the dominant grain-grain interactions are the long-term and short-term ones, respectively, a dense flow is characterized by twofold grain-grain interactions with equal significance. Twofold grain-grain interactions induce time and spatial fluctuations of the macroscopic behavior, a phenomenon similar to turbulent flows of the Newtonian fluids. However, the turbulent fluctuation in a dry granular dense matter is distinct from that in the Newtonian fluids in three aspects: (a) It emerges from twofold grain-grain interactions, in contrast to that from incoming flow instability, instability in transition region or flow geometry in the Newtonian fluids; (b) It emerges even at slow speed, in contrast to that in the Newtonian fluids, which is characterized by the critical Reynolds number; and (c) while turbulent fluctuation induces most energy production with anisotropic eddies and energy dissipation with fairly isotropic eddies at the scales similar to the integral and Kolmogorov scales in the Newtonian fluids, respectively, granular eddies at the inertia subrange, or the Taylor microeddies, are barely recognized. A dry granular dense flow can thus be considered a rheological fluid with significant kinetic energy dissipation, since the turbulent fluctuation induces an energy cascade from the stress power at the mean scale toward the thermal dissipation at the subsequent length and timescales. The conventional Reynolds-filter process is applied to decompose the variables into the mean and fluctuating parts, yielding the mean balance equations with ergodic fluctuating terms, which need to be expressed as functions of the mean fields, and are referred to as the closure relations.

Studies on the turbulent characteristics of dry granular systems, embracing equilibrium and non-equilibrium regimes, are so far yet complete. Various models for slow creeping and dense laminar flows and for collisional flows have been developed. The turbulence models based on Prandtl's mixing length have been proposed to account for the turbulent viscosity. Although the influence of velocity fluctuation on the linear momentum balance was taken into account via Reynolds' stress, the influence of fluctuating kinetic energy was not considered, with the formulations constructed without energy-entropy consideration. The energy-entropy balance, however, is an important part to the mean flow characteristics and turbulence realizability conditions. Attempts in taking the fluctuating kinetic energy into account were accomplished by using the granular temperature. However, only the equilibrium closure relations were obtained; numerical simulations of benchmark problems compared with experimental outcomes were missing or insufficient. The granular coldness, a similar concept to the granular temperature, was extended to account for the influence of turbulent fluctuation induced by twofold grain-grain interactions. A kinematic equation was used to describe the time evolution of turbulent kinetic energy with the turbulent dissipation considered a closure relation, yielding a zeroth-order closure model. It has been shown that while the mean porosity and velocity profiles coincided to the experimental measurements, the turbulent dissipation demonstrated a similarity to that of the Newtonian fluids in turbulent shear flows. Insufficiencies were, however, identified. First, the zeroth-order model was more appropriate for turbulent flows with weak intensity, in which the turbulent eddy evolutions at various length and timescales were hardly taken into account. Second, the phenomenological granular coldness was used for both the turbulent kinetic energy and dissipation implicitly, resulting in an unclear role played by solid boundaries. The goal of analysis is to establish a first-order closure model to account for the influence of turbulent eddy evolution and to illustrate the roles played by solid boundaries.

12.3.2 Mean Balance Equations and Turbulent State Space

Following the balance equations for laminar motion and the Reynolds-filter process, the mean balance equations for turbulent motion are given by

$$0 = \dot{\bar{\gamma}}\bar{\nu} + \bar{\gamma}\dot{\bar{\nu}} + \bar{\gamma}\bar{\nu}\operatorname{div}\bar{\boldsymbol{v}}, \tag{12.3.1}$$

$$\mathbf{0} = \bar{\gamma}\bar{\nu}\dot{\bar{\boldsymbol{v}}} - \operatorname{div}(\bar{\boldsymbol{t}} + \boldsymbol{R}) - \bar{\gamma}\bar{\nu}\bar{\boldsymbol{b}}, \tag{12.3.2}$$

Table 12.1 Variables and parameters in the mean balance equations

$\bar{\boldsymbol{b}}$	Mean specific body force;	$\bar{\boldsymbol{D}}$	Symmetric part of mean velocity gradient;
$\bar{e}$	Mean specific internal energy;	$\bar{f}$	Mean production associated with $\bar{\nu}$;
f^{ε}	Production associated with $\bar{\gamma}\bar{\nu}\varepsilon$;	$\bar{\boldsymbol{h}}$	Mean flux associated with $\bar{\nu}$;
k	Specific turbulent kinetic energy;	$\boldsymbol{K}^T$	Flux associated with $\bar{\gamma}\bar{\nu}k$;
$\boldsymbol{K}^{\varepsilon}$	Flux associated with $\bar{\gamma}\bar{\nu}\varepsilon$;	$\bar{\boldsymbol{q}}$	Mean heat flux;
$\boldsymbol{Q}$	Turbulent heat flux;	$\bar{r}$	Mean specific energy supply;
$\boldsymbol{R}$	Reynolds' stress;	$\bar{s}$	Mean specific entropy supply;
t	Transpose;	$\bar{\boldsymbol{t}}$	Mean Cauchy stress;
$\bar{\boldsymbol{v}}$	Mean velocity;	$\bar{\boldsymbol{Z}}$	Mean internal friction;
α	Arbitrary quantity;	$\bar{\alpha}$	Time-averaged value of α;
α'	Fluctuating value of α;	$\dot{\alpha}$	Material derivative of α with respect to $\bar{\boldsymbol{v}}$;
$\bar{\gamma}$	Mean true mass density of solid grains;	ε	Specific turbulent dissipation;
$\bar{\eta}$	Mean specific entropy;	$\bar{\nu}$	Mean (solid) volume fraction;
$\bar{\pi}$	Mean entropy production;	$\bar{\phi}$	Mean entropy flux;
ϕ'	Turbulent entropy flux;	$\bar{\boldsymbol{\Omega}}$	Any mean orthogonal rotation of a RVE;
∇	Nabla operator		

$$\boldsymbol{0} = \bar{\boldsymbol{t}} - \bar{\boldsymbol{t}}^{\mathrm{T}}, \tag{12.3.3}$$
$$0 = \bar{\gamma}\bar{\nu}\dot{\bar{e}} - \bar{\boldsymbol{t}} \cdot \bar{\boldsymbol{D}} + \operatorname{div}(\bar{\boldsymbol{q}} + \boldsymbol{Q}) - \bar{\gamma}\bar{\nu}\varepsilon - \bar{\gamma}\bar{\nu}\bar{r}, \tag{12.3.4}$$
$$0 = \bar{\gamma}\bar{\nu}\dot{\bar{\eta}} + \operatorname{div}(\bar{\boldsymbol{\phi}} + \boldsymbol{\phi}') - \bar{\gamma}\bar{\nu}\bar{s} - \bar{\pi}, \tag{12.3.5}$$
$$0 = \dot{\bar{\nu}} + \bar{\nu}\nabla \cdot \bar{\boldsymbol{v}} - \nabla \cdot \bar{\boldsymbol{h}} - \bar{f}, \tag{12.3.6}$$
$$\boldsymbol{0} = \mathring{\bar{\boldsymbol{Z}}} - \bar{\boldsymbol{\Phi}}, \quad (\mathring{\bar{\boldsymbol{Z}}} \equiv \dot{\bar{\boldsymbol{Z}}} - [\bar{\boldsymbol{\Omega}}, \bar{\boldsymbol{Z}}]), \tag{12.3.7}$$
$$0 = \bar{\gamma}\bar{\nu}\dot{k} - \boldsymbol{R} \cdot \bar{\boldsymbol{D}} - \operatorname{div} \boldsymbol{K}^T + \bar{\gamma}\bar{\nu}\varepsilon, \tag{12.3.8}$$
$$0 = \bar{\gamma}\bar{\nu}\dot{\varepsilon} - \nabla \cdot \boldsymbol{K}^{\varepsilon} - f^{\varepsilon}, \tag{12.3.9}$$

with the abbreviations,

$$\begin{aligned}
&0 = R_{ij} + \overline{\bar{\gamma}\bar{\nu}v_i'v_j'}, 0 = Q_j - \overline{\bar{\gamma}\bar{\nu}e'v_j'}, 0 = R_{ijk} + \overline{\bar{\gamma}\bar{\nu}v_i'v_j'v_k'},\\
&0 = \bar{\gamma}\bar{\nu}\varepsilon - \overline{t_{ij}'\frac{\partial v_i'}{\partial x_j}}, 0 = \phi_j' - \overline{\bar{\gamma}\bar{\nu}\eta'v_j'}, 0 = \bar{\gamma}\bar{\nu}k + \frac{1}{2}R_{ii},\\
&0 = K_j^T - \overline{t_{ij}'v_i'} - \frac{1}{2}R_{iij},
\end{aligned} \tag{12.3.10}$$

in which $\bar{\nu}$ is the mean volume fraction. The variables and parameters arising in Eqs. (12.3.1)–(12.3.10) are defined in Table 12.1. The terms in Eq. (12.3.10) are the ergodic fluctuations; they and other quantities, to be shown later, are the closure quantities which need to be prescribed as functions of the mean fields.

Equations (12.3.1)–(12.3.5) are respectively the conventional mean balances of mass, linear and angular momentums, internal energy and entropy for a fluid continuum in turbulent motion, with the symmetry of the mean Cauchy stress and mean density $\bar{\rho}$ decomposed into $\bar{\rho} = \bar{\gamma}\bar{\nu}$. This introduces the mean volume fraction $\bar{\nu}$, considered an internal variable with its time evolution described by Wilmánski's model given in Eq. (12.3.6) for dry granular dense flows. To account for the rate-independent characteristics, an Euclidean frame-indifferent, stress like, symmetric second-rank tensor $\bar{\boldsymbol{Z}}$ is introduced as an internal variable for the mean frictional and other non-conservative forces inside a granular micro-continuum. It is motivated by statistical mechanics and is a phenomenological generalization of the Mohr-Coulomb model for granular materials at low energy and high-grain volume fraction, with its time evolution described kinematically by Eq. (12.3.7).

Equation (12.3.8), the evolution of turbulent kinetic energy, derived by taking the inner product of velocity with the balance of linear momentum followed by the Reynolds-filter process, is considered. This is done so, for there exists an energy cascade from the mean flow scale toward a smallest (dissipation) scale in turbulent flows. In Eq. (12.3.8), the turbulent kinetic energy is generated via the stress power done by Reynolds' stress and mean shear rate at the mean flow scale, transferred subsequently via the flux $\boldsymbol{K}^T$ at various length and time scales, and eventually dissipated at the smallest scale by the turbulent dissipation.[13] In contrast to the zeroth-order model, the turbulent dissipation is considered an internal variable and an independent field, with its time evolution kinematically described by Eq. (12.3.9). Hence, the proposed closure model is classified as a first-order k-ε model.

The quantities

$$\mathcal{P} = \{\bar{\gamma}, \bar{\nu}, \bar{\boldsymbol{v}}, \bar{\boldsymbol{Z}}, \vartheta^M, \vartheta^T, \varepsilon\}, \quad \mathcal{C} = \{\bar{\boldsymbol{t}}, \boldsymbol{R}, \bar{e}, \bar{\boldsymbol{q}}, \boldsymbol{Q}, \bar{\eta}, \phi^T, \bar{\boldsymbol{h}}, \bar{f}, \bar{\boldsymbol{\Phi}}, k, \boldsymbol{K}^T, \boldsymbol{K}^\varepsilon, f^\varepsilon\}, \tag{12.3.11}$$

are introduced respectively as the primitive mean fields and closure quantities, by which $\mathcal{C}$ should be constructed based on the turbulent state space given by

$$\mathcal{Q} = \{\nu_0, \bar{\nu}, \dot{\bar{\nu}}, \boldsymbol{g}_1, \bar{\gamma}, \boldsymbol{g}_2, \vartheta^M, \boldsymbol{g}_3, \vartheta^T, \boldsymbol{g}_4, \varepsilon, \boldsymbol{g}_5, \bar{\boldsymbol{D}}, \bar{\boldsymbol{Z}}\}, \qquad \mathcal{C} = \hat{\mathcal{C}}(\mathcal{Q}), \tag{12.3.12}$$

in which $\boldsymbol{g}_1 \equiv \operatorname{grad}\bar{\nu}$, $\boldsymbol{g}_2 \equiv \operatorname{grad}\bar{\gamma}$, $\boldsymbol{g}_3 \equiv \operatorname{grad}\vartheta^M$, $\boldsymbol{g}_4 \equiv \operatorname{grad}\vartheta^T$, $\boldsymbol{g}_5 \equiv \operatorname{grad}\varepsilon$ and $\phi^T \equiv \bar{\phi} + \phi'$. In Eq. (12.3.11), ϑ^M is the *material coldness*, which can be shown to be the inverse of an empirical material temperature θ^M for simple materials. The turbulent kinetic energy is expressed conventionally by the *granular temperature* θ^T, or alternatively by the *granular coldness* ϑ^T.[14] The state space given in Eq. (12.3.12) is proposed based on Truesdell's equipresence principle and the principle of frame-indifference, in which ν_0 is the value of $\bar{\nu}$ in the reference configuration. The terms $\{\nu_0, \bar{\nu}, \dot{\bar{\nu}}, \boldsymbol{g}_1, \bar{\gamma}, \boldsymbol{g}_2\}$ are used for elastic effect, corresponding to $\{\bar{\rho}, \dot{\bar{\rho}}, \nabla\bar{\rho}\}$ for complex rheological fluids; $\{\vartheta^M, \boldsymbol{g}_3\}$ represent temperature-dependencies of physical

[13]In the Newtonian fluids, it is called the *Kolmogorov scale*.

[14]The simple relation between θ^M and ϑ^M does not hold in general between θ^T and ϑ^T for dry granular systems. It is only understood that $\vartheta^T = \hat{\vartheta}^T(\theta^T, \dot{\theta}^T)$. Thus, either θ^i or ϑ^i, $i = \{M, T\}$, can be introduced as primitive fields.

properties; $\{\vartheta^T, \boldsymbol{g}_4\}$ stand for influence of the turbulent kinetic energy; $\{\varepsilon, \boldsymbol{g}_5\}$ denote effect of the turbulent dissipation; and $\bar{\boldsymbol{D}}$ and $\bar{\boldsymbol{Z}}$ are for viscous and rate-independent effects, respectively. Due to the principle of frame-indifference, $\{\bar{\boldsymbol{t}}, \bar{\boldsymbol{q}}, \bar{\boldsymbol{h}}, \bar{\boldsymbol{\Phi}}\}$ are objective; since $\boldsymbol{v}'$ (the fluctuating velocity) is also objective, $\{\boldsymbol{R}, \boldsymbol{Q}, \boldsymbol{K}^T, \boldsymbol{K}^\varepsilon\}$ are equally objective. With these, Eq. (12.3.3) holds identically.

12.3.3 Thermodynamic Analysis

Exploitation of the entropy inequality. The turbulence realizability conditions require that, during a physically admissible process, the second law of thermodynamics, with its local form of a non-negative entropy production, and all mean balance equations should be satisfied simultaneously. This can be achieved by regarding the mean balance equations as the constraints of inequality (12.3.5) via the method of the Lagrange multiplier in the context of the Müller-Liu approach, viz.,

$$\begin{aligned}
\bar{\pi} = {}& \bar{\gamma}\bar{\nu}\dot{\bar{\eta}} + \operatorname{div}\boldsymbol{\phi}^T - \bar{\gamma}\bar{\nu}\bar{s} - \lambda^{\bar{\gamma}}\left(\dot{\bar{\gamma}}\bar{\nu} + \bar{\gamma}\dot{\bar{\nu}} + \bar{\gamma}\bar{\nu}\operatorname{div}\bar{\boldsymbol{v}}\right) - \boldsymbol{\lambda}^{\bar{\boldsymbol{v}}}\cdot\left(\bar{\gamma}\bar{\nu}\dot{\bar{\boldsymbol{v}}} - \operatorname{div}(\bar{\boldsymbol{t}} + \boldsymbol{R}) - \bar{\gamma}\bar{\nu}\bar{\boldsymbol{b}}\right)\\
& - \lambda^{\bar{e}}\left(\bar{\gamma}\bar{\nu}\dot{\bar{e}} - \bar{\boldsymbol{t}}\cdot\bar{\boldsymbol{D}} + \operatorname{div}(\bar{\boldsymbol{q}} + \boldsymbol{Q}) - \bar{\gamma}\bar{\nu}\varepsilon - \bar{\gamma}\bar{\nu}\bar{r}\right)\\
& - \lambda^{\bar{\nu}}\left(\dot{\bar{\nu}} + \bar{\nu}\nabla\cdot\bar{\boldsymbol{v}} - \nabla\cdot\bar{\boldsymbol{h}} - \bar{f}\right) - \boldsymbol{\lambda}^{\bar{\boldsymbol{Z}}}\cdot\left(\mathring{\bar{\boldsymbol{Z}}} - \bar{\boldsymbol{\Phi}}\right)\\
& - \lambda^{k}\left(\bar{\gamma}\bar{\nu}\dot{k} - \boldsymbol{R}\cdot\bar{\boldsymbol{D}} - \operatorname{div}\boldsymbol{K}^T + \bar{\gamma}\bar{\nu}\varepsilon\right) - \lambda^{\varepsilon}\left(\bar{\gamma}\bar{\nu}\dot{\varepsilon} - \nabla\cdot\boldsymbol{K}^{\varepsilon} - f^{\varepsilon}\right) \geq 0,
\end{aligned} \tag{12.3.13}$$

with $\{\lambda^{\bar{\gamma}}, \boldsymbol{\lambda}^{\bar{\boldsymbol{v}}}, \lambda^{\bar{e}}, \lambda^{\bar{\nu}}, \boldsymbol{\lambda}^{\bar{\boldsymbol{Z}}}, \lambda^k, \lambda^\varepsilon\}$ the Lagrange multipliers corresponding to Eqs. (12.3.1)–(12.3.2), (12.3.4) and (12.3.6)–(12.3.9), respectively. Since $\dot{\vartheta}^M$ and $\dot{\vartheta}^T$ are not considered in Eq. (12.3.12), it is assumed that

$$\lambda^{\bar{e}} = \vartheta^M, \qquad \lambda^k = \vartheta^T, \qquad \vartheta^M\psi^T \equiv \vartheta^M\bar{e} + \vartheta^T k - \bar{\eta}, \tag{12.3.14}$$

where ψ^T is the specific *turbulent Helmholtz free energy*. Since material behavior is assumed to be independent of external supplies, it follows that $(-\bar{\gamma}\bar{\nu}\bar{r} + \bar{\gamma}\bar{\nu}\boldsymbol{\lambda}^{\bar{\boldsymbol{v}}}\cdot\bar{\boldsymbol{b}} + \vartheta^M\bar{\gamma}\bar{\nu}\bar{r}) = 0$, an equation determining the mean entropy supply. Substituting these Eqs. (12.3.11)–(12.3.12) and (12.3.14) into Eq. (12.3.13) yields

$$\begin{aligned}
\bar{\pi} = {}& -\left(\bar{\gamma}\bar{\nu}\vartheta^M\psi^T_{,\bar{\nu}} + \lambda^{\bar{\gamma}}\bar{\gamma} + \lambda^{\bar{\nu}}\right)\dot{\bar{\nu}} - \bar{\gamma}\bar{\nu}\vartheta^M\psi^T_{,\dot{\bar{\nu}}}\ddot{\bar{\nu}} - \bar{\gamma}\bar{\nu}\vartheta^M\psi^T_{,\boldsymbol{g}_1}\cdot\left(\nabla\dot{\bar{\nu}} - \boldsymbol{g}_1\nabla\bar{\boldsymbol{v}}\right)\\
& - \left(\bar{\gamma}\bar{\nu}\vartheta^M\psi^T_{,\bar{\gamma}} + \lambda^{\bar{\gamma}}\bar{\nu}\right)\dot{\bar{\gamma}} - \bar{\gamma}\bar{\nu}\left(\vartheta^M\psi^T_{,\vartheta^T} - k\right)\dot{\vartheta}^T\\
& - \bar{\gamma}\bar{\nu}\vartheta^M\psi^T_{,\boldsymbol{g}_2}\cdot\dot{\boldsymbol{g}}_2 - \bar{\gamma}\bar{\nu}\left(\vartheta^M\psi^T_{,\vartheta^M} - \bar{e} + \psi^T\right)\dot{\vartheta}^M - \bar{\gamma}\bar{\nu}\vartheta^M\psi^T_{,\boldsymbol{g}_3}\cdot\dot{\boldsymbol{g}}_3\\
& - \bar{\gamma}\bar{\nu}\vartheta^M\psi^T_{,\boldsymbol{g}_4}\cdot\dot{\boldsymbol{g}}_4 - \bar{\gamma}\bar{\nu}\left(\vartheta^M\psi^T_{,\varepsilon} + \lambda^\varepsilon\right)\dot{\varepsilon}\\
& - \bar{\gamma}\bar{\nu}\vartheta^M\psi^T_{,\boldsymbol{g}_5}\cdot\dot{\boldsymbol{g}}_5 - \bar{\gamma}\bar{\nu}\vartheta^T\psi^T_{,\bar{\boldsymbol{D}}}\cdot\dot{\bar{\boldsymbol{D}}} - \left(\bar{\gamma}\bar{\nu}\vartheta^M\psi^T_{,\bar{\boldsymbol{Z}}} + \boldsymbol{\lambda}^{\bar{\boldsymbol{Z}}}\right)\cdot\dot{\bar{\boldsymbol{Z}}}\\
& + \sum_g\left(\bar{\phi}^T_{,g} + \boldsymbol{\lambda}^{\bar{\boldsymbol{v}}}(\bar{\boldsymbol{t}} + \boldsymbol{R})_{,g} + \lambda^{\bar{\nu}}\bar{\boldsymbol{h}}_{,g} - \vartheta^M(\bar{\boldsymbol{q}} + \boldsymbol{Q})_{,g} + \vartheta^T\boldsymbol{K}^T_{,g} + \lambda^\varepsilon\boldsymbol{K}^\varepsilon_{,g}\right)\cdot\nabla g\\
& + \sum_{\boldsymbol{g}}\left(\bar{\boldsymbol{\phi}}^T_{,\boldsymbol{g}} + \boldsymbol{\lambda}^{\bar{\boldsymbol{v}}}(\bar{\boldsymbol{t}} + \boldsymbol{R})_{,\boldsymbol{g}} + \lambda^{\bar{\nu}}\bar{\boldsymbol{h}}_{,\boldsymbol{g}} - \vartheta^M(\bar{\boldsymbol{q}} + \boldsymbol{Q})_{,\boldsymbol{g}} + \vartheta^T\boldsymbol{K}^T_{,\boldsymbol{g}} + \lambda^\varepsilon\boldsymbol{K}^\varepsilon_{,\boldsymbol{g}}\right)\cdot\nabla\boldsymbol{g}
\end{aligned}$$

$$+\sum_A \left(\bar{\phi}^T_{,A} + \boldsymbol{\lambda}^{\bar{\boldsymbol{v}}}(\bar{\boldsymbol{t}} + \boldsymbol{R})_{,A} + \lambda^{\bar{\nu}}\bar{\boldsymbol{h}}_{,A} - \vartheta^M(\bar{\boldsymbol{q}} + \boldsymbol{Q})_{,A} + \vartheta^T \boldsymbol{K}^T_{,A} + \lambda^{\varepsilon}\boldsymbol{K}^{\varepsilon}_{,A}\right) \cdot \nabla \boldsymbol{A}$$

$$+\left(\vartheta^M \bar{\boldsymbol{t}} + \vartheta^T \boldsymbol{R} - \left(\lambda^{\bar{\gamma}}\bar{\gamma}\bar{\nu} + \lambda^{\bar{\nu}}\bar{\nu}\right)\boldsymbol{I}\right) \cdot \bar{\boldsymbol{D}} - \bar{\gamma}\bar{\nu}\boldsymbol{\lambda}^{\bar{\boldsymbol{v}}} \cdot \dot{\boldsymbol{v}} + \lambda^{\bar{\nu}}\bar{f} + [\boldsymbol{\lambda}^{\bar{\boldsymbol{Z}}}, \bar{\boldsymbol{Z}}] \cdot \bar{\boldsymbol{\Omega}} + \boldsymbol{\lambda}^{\bar{\boldsymbol{Z}}} \cdot \bar{\boldsymbol{\Phi}}$$

$$+(\vartheta^M - \vartheta^T)\bar{\gamma}\bar{\nu}\varepsilon + \lambda^{\varepsilon} f^{\varepsilon} \geq 0, \tag{12.3.15}$$

with $g \in \{\nu_0, \bar{\nu}, \dot{\bar{\nu}}, \bar{\gamma}, \vartheta^M, \vartheta^T, \varepsilon\}$; $\boldsymbol{g} \in \{\boldsymbol{g}_1, \boldsymbol{g}_2, \boldsymbol{g}_3, \boldsymbol{g}_4, \boldsymbol{g}_5\}$; $\boldsymbol{A} \in \{\bar{\boldsymbol{D}}, \bar{\boldsymbol{Z}}\}$; $\boldsymbol{I}$ the second-rank identity tensor; and $\dot{\boldsymbol{g}}_1 = \text{grad}\,\dot{\bar{\nu}} - \boldsymbol{g}_1(\text{grad}\,\bar{\boldsymbol{v}})$ has been used. The inequality (12.3.15) is expressed alternatively as

$$\bar{\pi} = \boldsymbol{a} \cdot \boldsymbol{\mathcal{X}} + b \geq 0, \tag{12.3.16}$$

$$\boldsymbol{\mathcal{X}} = \left\{ \begin{array}{l} \dot{\bar{\boldsymbol{v}}}, \ddot{\bar{\nu}}, \dot{\bar{\gamma}}, \dot{\vartheta}^M, \dot{\vartheta}^T, \dot{\varepsilon}, \dot{\boldsymbol{g}}_2, \dot{\boldsymbol{g}}_3, \dot{\boldsymbol{g}}_4, \dot{\boldsymbol{g}}_5, \dot{\bar{\boldsymbol{D}}}, \dot{\bar{\boldsymbol{Z}}}, \bar{\boldsymbol{\Omega}}, \\ \text{grad}\,\nu_0, \text{grad}\,\dot{\bar{\nu}}, \text{grad}\,\boldsymbol{g}_1, \text{grad}\,\boldsymbol{g}_2, \text{grad}\,\boldsymbol{g}_3, \text{grad}\,\boldsymbol{g}_4, \text{grad}\,\boldsymbol{g}_5, \text{grad}\,\bar{\boldsymbol{D}}, \\ \text{grad}\,\bar{\boldsymbol{Z}} \end{array} \right\}.$$

Since $\boldsymbol{\mathcal{X}}$ is the set of independent variations of $\mathcal{Q}$, and $\boldsymbol{a}$ and b are functions of Eq. (12.3.12), but not of $\boldsymbol{\mathcal{X}}$, inequality (12.3.16) is linear in $\boldsymbol{\mathcal{X}}$. Since $\boldsymbol{\mathcal{X}}$ can take any value, it would be possible to violate Eq. (12.3.16) unless

$$\boldsymbol{a} = \boldsymbol{0}, \qquad \text{and} \qquad b \geq 0. \tag{12.3.17}$$

The condition $(12.3.17)_1$ leads to

$$\psi^T_{,y} = \boldsymbol{0}, \qquad y \in \{\dot{\bar{\nu}}, \boldsymbol{g}_2, \boldsymbol{g}_3, \boldsymbol{g}_4, \boldsymbol{g}_5, \bar{\boldsymbol{D}}\}, \tag{12.3.18}$$

$$\begin{array}{lll} \boldsymbol{\lambda}^{\bar{\boldsymbol{v}}} = \boldsymbol{0}, & \lambda^{\bar{\gamma}} = -\bar{\gamma}\vartheta^M \psi^T_{,\bar{\gamma}}, & \boldsymbol{\lambda}^{\bar{\boldsymbol{Z}}} = -\bar{\gamma}\bar{\nu}\vartheta^M \psi^T_{,\bar{\boldsymbol{Z}}}, \\ \lambda^{\varepsilon} = -\vartheta^M \psi^T_{,\varepsilon}, & \boldsymbol{\lambda}^{\bar{\boldsymbol{Z}}}\bar{\boldsymbol{Z}} = \bar{\boldsymbol{Z}}\boldsymbol{\lambda}^{\bar{\boldsymbol{Z}}}, & \end{array} \tag{12.3.19}$$

$$\bar{e} = \psi^T + \vartheta^M \psi^T_{,\vartheta^M}, \quad k = \vartheta^M \psi^T_{,\vartheta^T}; \tag{12.3.20}$$

and the equations

$$\boldsymbol{0} = \boldsymbol{\phi}^T_{,\nu_0} + \lambda^{\bar{\nu}}\bar{\boldsymbol{h}}_{,\nu_0} - \vartheta^M(\bar{\boldsymbol{q}} + \boldsymbol{Q})_{,\nu_0} + \vartheta^T \boldsymbol{K}^T_{,\nu_0} + \lambda^{\varepsilon}\boldsymbol{K}^{\varepsilon}_{,\bar{\nu}_0}, \tag{12.3.21}$$

$$\boldsymbol{0} = \boldsymbol{\phi}^T_{,A} + \lambda^{\bar{\nu}}\bar{\boldsymbol{h}}_{,A} - \vartheta^M(\bar{\boldsymbol{q}} + \boldsymbol{Q})_{,A} + \vartheta^T \boldsymbol{K}^T_{,A} + \lambda^{\varepsilon}\boldsymbol{K}^{\varepsilon}_{,A}, \tag{12.3.22}$$

$$\boldsymbol{0} = \text{sym}\left\{\boldsymbol{\phi}^T_{,\boldsymbol{x}} + \lambda^{\bar{\nu}}\bar{\boldsymbol{h}}_{,\boldsymbol{x}} - \vartheta^M(\bar{\boldsymbol{q}} + \boldsymbol{Q})_{,\boldsymbol{x}} + \vartheta^T \boldsymbol{K}^T_{,\boldsymbol{x}} + \lambda^{\varepsilon}\boldsymbol{K}^{\varepsilon}_{,\boldsymbol{x}}\right\}, \tag{12.3.23}$$

$$\boldsymbol{0} = \boldsymbol{\phi}^T_{,\dot{\bar{\nu}}} + \lambda^{\bar{\nu}}\bar{\boldsymbol{h}}_{,\dot{\bar{\nu}}} - \vartheta^M(\bar{\boldsymbol{q}} + \boldsymbol{Q})_{,\dot{\bar{\nu}}} + \vartheta^T \boldsymbol{K}^T_{,\dot{\bar{\nu}}} + \lambda^{\varepsilon}\boldsymbol{K}^{\varepsilon}_{,\dot{\bar{\nu}}} - \bar{\gamma}\bar{\nu}\vartheta^M \psi^T_{,\boldsymbol{g}_1}, \tag{12.3.24}$$

where $\boldsymbol{x} \in \{\boldsymbol{g}_2\text{-}\boldsymbol{g}_5\}$. The condition $(12.3.17)_2$ gives the residual entropy inequality in the form

$$\begin{aligned}\bar{\pi} = &\{-\bar{\gamma}\bar{\nu}\vartheta^M\psi^T_{,\bar{\nu}} + \bar{\gamma}^2\vartheta^M\psi^T_{,\bar{\gamma}} - \lambda^{\bar{\nu}}\}\dot{\bar{\nu}} + \{\bar{\nu}(\bar{\gamma}^2\vartheta^M\psi^T_{,\bar{\gamma}} - \lambda^{\bar{\nu}})\boldsymbol{I} \\ &+\vartheta^M\bar{\boldsymbol{t}} + \vartheta^T\boldsymbol{R} + \bar{\gamma}\bar{\nu}\vartheta^M\psi^T_{,\boldsymbol{g}_1} \otimes \boldsymbol{g}_1\} \cdot \bar{\boldsymbol{D}} \\ &+\sum_g\{\boldsymbol{\phi}^T_{,g} + \lambda^{\bar{\nu}}\bar{\boldsymbol{h}}_{,g} - \vartheta^M(\bar{\boldsymbol{q}} + \boldsymbol{Q})_{,g} + \vartheta^T\boldsymbol{K}^T_{,g} + \lambda^{\varepsilon}\boldsymbol{K}^{\varepsilon}_{,g}\} \cdot \operatorname{grad} g \\ &+\{\bar{\gamma}\bar{\nu}\varepsilon(\vartheta^M - \vartheta^T) + \lambda^{\bar{\nu}}\bar{f} + \lambda^{\varepsilon}f^{\varepsilon} - \bar{\gamma}\bar{\nu}\vartheta^M\psi^T_{,\bar{\boldsymbol{Z}}} \cdot \bar{\boldsymbol{\Phi}}\} \geq 0, \end{aligned} \qquad (12.3.25)$$

with g newly defined as $g \in \{\bar{\nu}, \bar{\gamma}, \vartheta^M, \vartheta^T, \varepsilon\}$.

Extra entropy flux. Define the extra entropy flux $\boldsymbol{\xi}$, viz.,

$$\boldsymbol{\xi} \equiv \boldsymbol{\phi}^T - \vartheta^M(\bar{\boldsymbol{q}} + \boldsymbol{Q}) + \vartheta^T\boldsymbol{K}^T + \lambda^{\varepsilon}\boldsymbol{K}^{\varepsilon} + \lambda^{\bar{\nu}}\bar{\boldsymbol{h}} + \boldsymbol{\lambda}^{\bar{v}}(\bar{\boldsymbol{t}} + \boldsymbol{R}), \qquad (12.3.26)$$

which is a measure of the deviation between the entropy flux and other fluxes. In view of Eq. (12.3.12), any vector-valued isotropic vector $\boldsymbol{f}$ should satisfy

$$\begin{aligned}\boldsymbol{Q}(\tau)\hat{\boldsymbol{f}}(\mathcal{Q}) = \hat{\boldsymbol{f}}\Big(&\nu_0(t), \bar{\nu}(t), \dot{\bar{\nu}}(t), \boldsymbol{Q}(\tau)\boldsymbol{g}_1(t), \bar{\gamma}(t), \boldsymbol{Q}(\tau)\boldsymbol{g}_2(t), \\ &\vartheta^M(t), \boldsymbol{Q}(\tau)\boldsymbol{g}_3(t), \vartheta^T(t), \boldsymbol{Q}(\tau)\boldsymbol{g}_4(t), \\ &\varepsilon(t), \boldsymbol{Q}(\tau)\boldsymbol{g}_5(t), \boldsymbol{Q}(\tau)\bar{\boldsymbol{D}}(T)\boldsymbol{Q}(\tau)^{\mathrm{T}}, \boldsymbol{Q}(\tau)\bar{\boldsymbol{Z}}(T)\boldsymbol{Q}(\tau)^{\mathrm{T}}\Big),\end{aligned} \qquad (12.3.27)$$

for all time-dependent rotations $\boldsymbol{Q}(\tau)$ at an *observer time* τ for a specific reference time state t with $\boldsymbol{Q}(\tau)\,\boldsymbol{Q}^{\mathrm{T}}(\tau) = \boldsymbol{Q}^{\mathrm{T}}(\tau)\,\boldsymbol{Q}(\tau) = \boldsymbol{I}$. Differentiating Eq. (12.3.27) with respect to τ yields

$$\begin{aligned}\dot{\boldsymbol{Q}}\boldsymbol{f} = &\boldsymbol{f}_{,(\boldsymbol{Q}\boldsymbol{g}_1)}(\boldsymbol{Q}\boldsymbol{g}_1)^{\cdot} + \boldsymbol{f}_{,(\boldsymbol{Q}\boldsymbol{g}_2)}(\boldsymbol{Q}\boldsymbol{g}_2)^{\cdot} + \boldsymbol{f}_{,(\boldsymbol{Q}\boldsymbol{g}_3)}(\boldsymbol{Q}\boldsymbol{g}_3)^{\cdot} + \boldsymbol{f}_{,(\boldsymbol{Q}\boldsymbol{g}_4)}(\boldsymbol{Q}\boldsymbol{g}_4)^{\cdot} \\ &+\boldsymbol{f}_{,(\boldsymbol{Q}\boldsymbol{g}_5)}(\boldsymbol{Q}\boldsymbol{g}_5)^{\cdot} + \boldsymbol{f}_{,(\boldsymbol{Q}\bar{\boldsymbol{D}}\boldsymbol{Q}^{\mathrm{T}})}(\boldsymbol{Q}\bar{\boldsymbol{D}}\boldsymbol{Q}^{\mathrm{T}})^{\cdot} + \boldsymbol{f}_{,(\boldsymbol{Q}\bar{\boldsymbol{Z}}\boldsymbol{Q}^{\mathrm{T}})}(\boldsymbol{Q}\bar{\boldsymbol{Z}}\boldsymbol{Q}^{\mathrm{T}})^{\cdot}.\end{aligned} \qquad (12.3.28)$$

Since the orthogonality of $\boldsymbol{Q}$ implies that $\dot{\boldsymbol{Q}} = \boldsymbol{S}\boldsymbol{Q}$, with $\boldsymbol{S}$ the skew-symmetric part of $\dot{\boldsymbol{Q}}\boldsymbol{Q}^{\mathrm{T}}$, replacing $\dot{\boldsymbol{Q}}$ by $\boldsymbol{S}\boldsymbol{Q}$ and letting $\boldsymbol{Q} = \boldsymbol{I}$ in Eq. (12.3.28) gives

$$\begin{aligned}\boldsymbol{S}\boldsymbol{f} = &\boldsymbol{f}_{,\boldsymbol{g}_1}\boldsymbol{S}\boldsymbol{g}_1 + \boldsymbol{f}_{,\boldsymbol{g}_2}\boldsymbol{S}\boldsymbol{g}_2 + \boldsymbol{f}_{,\boldsymbol{g}_3}\boldsymbol{S}\boldsymbol{g}_3 + \boldsymbol{f}_{,\boldsymbol{g}_4}\boldsymbol{S}\boldsymbol{g}_4 + \boldsymbol{f}_{,\boldsymbol{g}_5}\boldsymbol{S}\boldsymbol{g}_5 \\ &+\boldsymbol{f}_{,\bar{\boldsymbol{D}}}[\boldsymbol{S}, \bar{\boldsymbol{D}}] + \boldsymbol{f}_{,\bar{\boldsymbol{Z}}}[\boldsymbol{S}, \bar{\boldsymbol{Z}}].\end{aligned} \qquad (12.3.29)$$

Since $\boldsymbol{\xi}$, $\boldsymbol{\phi}^T$, $\bar{\boldsymbol{q}}$, $\boldsymbol{Q}$, $\boldsymbol{K}^T$, $\boldsymbol{K}^{\varepsilon}$, and $\bar{\boldsymbol{h}}$ should satisfy Eq. (12.3.29), it follows from Eqs. (12.3.26) and (12.3.29) that

$$\begin{aligned}&\boldsymbol{S}\boldsymbol{\phi}^T - \vartheta^M\boldsymbol{S}(\bar{\boldsymbol{q}} + \boldsymbol{Q}) + \vartheta^T\boldsymbol{S}\boldsymbol{K}^T + \lambda^{\varepsilon}\boldsymbol{S}\boldsymbol{K}^{\varepsilon} + \lambda^{\bar{\nu}}\boldsymbol{S}\bar{\boldsymbol{h}} = \\ &\quad\sum_g\{\boldsymbol{\phi}^T_{,g} - \vartheta^M(\bar{\boldsymbol{q}} + \boldsymbol{Q})_{,g} + \vartheta^T\boldsymbol{K}^T_{,g} + \lambda^{\varepsilon}\boldsymbol{K}^{\varepsilon}_{,g} + \lambda^{\bar{\nu}}\bar{\boldsymbol{h}}_{,g}\}\boldsymbol{S}\boldsymbol{g} \\ &\quad+\sum_A\{\boldsymbol{\phi}^T_{,A} - \vartheta^M(\bar{\boldsymbol{q}} + \boldsymbol{Q})_{,A} + \vartheta^T\boldsymbol{K}^T_{,A} + \lambda^{\varepsilon}\boldsymbol{K}^{\varepsilon}_{,A} + \lambda^{\bar{\nu}}\bar{\boldsymbol{h}}_{,A}\}[\boldsymbol{S}, \boldsymbol{A}],\end{aligned} \qquad (12.3.30)$$

which reduces to

$$\boldsymbol{S}\boldsymbol{\xi} = \boldsymbol{A}(\boldsymbol{S}\boldsymbol{g}_1) + \boldsymbol{B}(\boldsymbol{S}\boldsymbol{g}_2) + \boldsymbol{C}(\boldsymbol{S}\boldsymbol{g}_3) + \boldsymbol{D}(\boldsymbol{S}\boldsymbol{g}_4) + \boldsymbol{E}(\boldsymbol{S}\boldsymbol{g}_5), \qquad (12.3.31)$$

with the abbreviations,

$$
\begin{aligned}
\boldsymbol{A} &= (\phi^T_{,\boldsymbol{g}_1} - \vartheta^M(\bar{\boldsymbol{q}} + \boldsymbol{Q})_{,\boldsymbol{g}_1} + \vartheta^T \boldsymbol{K}^T_{,\boldsymbol{g}_1} + \lambda^{\varepsilon}\boldsymbol{K}^{\varepsilon}_{,\boldsymbol{g}_1} + \lambda^{\bar{\nu}}\bar{\boldsymbol{h}}_{,\boldsymbol{g}_1}),\\
\boldsymbol{B} &= (\phi^T_{,\boldsymbol{g}_2} - \vartheta^M(\bar{\boldsymbol{q}} + \boldsymbol{Q})_{,\boldsymbol{g}_2} + \vartheta^T \boldsymbol{K}^T_{,\boldsymbol{g}_2} + \lambda^{\varepsilon}\boldsymbol{K}^{\varepsilon}_{,\boldsymbol{g}_2} + \lambda^{\bar{\nu}}\bar{\boldsymbol{h}}_{,\boldsymbol{g}_2}),\\
\boldsymbol{C} &= (\phi^T_{,\boldsymbol{g}_3} - \vartheta^M(\bar{\boldsymbol{q}} + \boldsymbol{Q})_{,\boldsymbol{g}_3} + \vartheta^T \boldsymbol{K}^T_{,\boldsymbol{g}_3} + \lambda^{\varepsilon}\boldsymbol{K}^{\varepsilon}_{,\boldsymbol{g}_3} + \lambda^{\bar{\nu}}\bar{\boldsymbol{h}}_{,\boldsymbol{g}_3}),\\
\boldsymbol{D} &= (\phi^T_{,\boldsymbol{g}_4} - \vartheta^M(\bar{\boldsymbol{q}} + \boldsymbol{Q})_{,\boldsymbol{g}_4} + \vartheta^T \boldsymbol{K}^T_{,\boldsymbol{g}_4} + \lambda^{\varepsilon}\boldsymbol{K}^{\varepsilon}_{,\boldsymbol{g}_4} + \lambda^{\bar{\nu}}\bar{\boldsymbol{h}}_{,\boldsymbol{g}_4}),\\
\boldsymbol{E} &= (\phi^T_{,\boldsymbol{g}_5} - \vartheta^M(\bar{\boldsymbol{q}} + \boldsymbol{Q})_{,\boldsymbol{g}_5} + \vartheta^T \boldsymbol{K}^T_{,\boldsymbol{g}_5} + \lambda^{\varepsilon}\boldsymbol{K}^{\varepsilon}_{,\boldsymbol{g}_5} + \lambda^{\bar{\nu}}\bar{\boldsymbol{h}}_{,\boldsymbol{g}_5}).
\end{aligned}
\tag{12.3.32}
$$

Equation (12.3.31) is further simplified by using the dual vector $\boldsymbol{\omega}$ of $\boldsymbol{S}$, viz.,

$$
\boldsymbol{\omega} \times \boldsymbol{\xi} = \boldsymbol{A}(\boldsymbol{\omega} \times \boldsymbol{g}_1) + \boldsymbol{B}(\boldsymbol{\omega} \times \boldsymbol{g}_2) + \boldsymbol{C}(\boldsymbol{\omega} \times \boldsymbol{g}_3) + \boldsymbol{D}(\boldsymbol{\omega} \times \boldsymbol{g}_4) + \boldsymbol{E}(\boldsymbol{\omega} \times \boldsymbol{g}_5),
\tag{12.3.33}
$$

for all $\boldsymbol{\omega}$. Since $\{\boldsymbol{A}, \boldsymbol{B}, \boldsymbol{C}, \boldsymbol{D}, \boldsymbol{E}\}$ are skew-symmetric, letting $\boldsymbol{\omega}$ in Eq. (12.3.33) be $\boldsymbol{\omega} = 1\boldsymbol{e}_1 + 0\boldsymbol{e}_2 + 0\boldsymbol{e}_3$ leads to

$$
\begin{aligned}
0\boldsymbol{e}_1 - \xi_3\boldsymbol{e}_2 + \xi_2\boldsymbol{e}_3 = \{&-\boldsymbol{A}_{12}(\boldsymbol{g}_1)_3 + \boldsymbol{A}_{13}(\boldsymbol{g}_1)_2\\
&- \boldsymbol{B}_{12}(\boldsymbol{g}_2)_3 + \boldsymbol{B}_{13}(\boldsymbol{g}_2)_2 - \boldsymbol{C}_{12}(\boldsymbol{g}_3)_3 + \boldsymbol{C}_{13}(\boldsymbol{g}_3)_2\\
&- \boldsymbol{D}_{12}(\boldsymbol{g}_4)_3 + \boldsymbol{D}_{13}(\boldsymbol{g}_4)_2 - \boldsymbol{E}_{12}(\boldsymbol{g}_5)_3 + \boldsymbol{E}_{13}(\boldsymbol{g}_5)_2\}\boldsymbol{e}_1\\
&+ \{\boldsymbol{A}_{23}(\boldsymbol{g}_1)_2 + \boldsymbol{B}_{23}(\boldsymbol{g}_2)_2 + \boldsymbol{C}_{23}(\boldsymbol{g}_3)_2 + \boldsymbol{D}_{23}(\boldsymbol{g}_4)_2 + \boldsymbol{E}_{23}(\boldsymbol{g}_5)_2\}\boldsymbol{e}_2\\
&+ \{\boldsymbol{A}_{23}(\boldsymbol{g}_1)_3 + \boldsymbol{B}_{23}(\boldsymbol{g}_2)_3 + \boldsymbol{C}_{23}(\boldsymbol{g}_3)_3 + \boldsymbol{D}_{23}(\boldsymbol{g}_4)_3 + \boldsymbol{E}_{23}(\boldsymbol{g}_5)_3\}\boldsymbol{e}_3,
\end{aligned}
\tag{12.3.34}
$$

with $\boldsymbol{\xi} = (\xi_1, \xi_2, \xi_3)$ with respect to the coordinates spanned by the orthonormal base $\{\boldsymbol{e}_1, \boldsymbol{e}_2, \boldsymbol{e}_3\}$. Comparing the coefficients of $\boldsymbol{e}_1$ gives rise to

$$
\begin{aligned}
0 = &-\boldsymbol{A}_{12}(\boldsymbol{g}_1)_3 + \boldsymbol{A}_{13}(\boldsymbol{g}_1)_2 - \boldsymbol{B}_{12}(\boldsymbol{g}_2)_3 + \boldsymbol{B}_{13}(\boldsymbol{g}_2)_2 - \boldsymbol{C}_{12}(\boldsymbol{g}_3)_3 + \boldsymbol{C}_{13}(\boldsymbol{g}_3)_2\\
&-\boldsymbol{D}_{12}(\boldsymbol{g}_4)_3 + \boldsymbol{D}_{13}(\boldsymbol{g}_4)_2 - \boldsymbol{E}_{12}(\boldsymbol{g}_5)_3 + \boldsymbol{E}_{13}(\boldsymbol{g}_5)_2.
\end{aligned}
\tag{12.3.35}
$$

Similarly, let $\boldsymbol{\omega}$ be $\boldsymbol{\omega} = 0\boldsymbol{e}_1 + 1\boldsymbol{e}_2 + 0\boldsymbol{e}_3$ and $\boldsymbol{\omega} = 0\boldsymbol{e}_1 + 0\boldsymbol{e}_2 + 1\boldsymbol{e}_3$ and perform the calculations again. The obtained results are summarized in the following:

$$
\begin{aligned}
&\begin{bmatrix} 0 & \boldsymbol{A}_{13} & \boldsymbol{A}_{21} \\ \boldsymbol{A}_{32} & 0 & \boldsymbol{A}_{21} \\ \boldsymbol{A}_{32} & \boldsymbol{A}_{13} & 0 \end{bmatrix} \boldsymbol{g}_1 + \begin{bmatrix} 0 & \boldsymbol{B}_{13} & \boldsymbol{B}_{21} \\ \boldsymbol{B}_{32} & 0 & \boldsymbol{B}_{21} \\ \boldsymbol{B}_{32} & \boldsymbol{B}_{13} & 0 \end{bmatrix} \boldsymbol{g}_2 + \begin{bmatrix} 0 & \boldsymbol{C}_{13} & \boldsymbol{C}_{21} \\ \boldsymbol{C}_{32} & 0 & \boldsymbol{C}_{21} \\ \boldsymbol{C}_{32} & \boldsymbol{C}_{13} & 0 \end{bmatrix} \boldsymbol{g}_3\\
&+ \begin{bmatrix} 0 & \boldsymbol{D}_{13} & \boldsymbol{D}_{21} \\ \boldsymbol{D}_{32} & 0 & \boldsymbol{D}_{21} \\ \boldsymbol{D}_{32} & \boldsymbol{D}_{13} & 0 \end{bmatrix} \boldsymbol{g}_4 + \begin{bmatrix} 0 & \boldsymbol{E}_{13} & \boldsymbol{E}_{21} \\ \boldsymbol{E}_{32} & 0 & \boldsymbol{E}_{21} \\ \boldsymbol{E}_{32} & \boldsymbol{E}_{13} & 0 \end{bmatrix} \boldsymbol{g}_5 = \boldsymbol{0}.
\end{aligned}
\tag{12.3.36}
$$

Since $\{\boldsymbol{g}_1, \boldsymbol{g}_2, \boldsymbol{g}_3, \boldsymbol{g}_4, \boldsymbol{g}_5\}$ are independent of one another and do not vanish simultaneously, it follows that

$$
\begin{bmatrix} 0 & \boldsymbol{X}_{13} & \boldsymbol{X}_{21} \\ \boldsymbol{X}_{32} & 0 & \boldsymbol{X}_{21} \\ \boldsymbol{X}_{32} & \boldsymbol{X}_{13} & 0 \end{bmatrix} = \boldsymbol{0}, \quad \longrightarrow \quad \boldsymbol{X} = \boldsymbol{0}; \quad \boldsymbol{X} \in \{\boldsymbol{A}, \boldsymbol{B}, \boldsymbol{C}, \boldsymbol{D}, \boldsymbol{E}\},
\tag{12.3.37}
$$

with which a vanishing $\boldsymbol{\xi}$ is obtained for all $\boldsymbol{S}$.

With $\boldsymbol{\xi} = \mathbf{0}$, Eqs. (12.3.21)–(12.3.24) are simplified to

$$\mathbf{0} = \lambda^{\bar{\nu}}_{,\nu_0}\bar{\boldsymbol{h}} + \lambda^{\varepsilon}_{,\nu_0}\boldsymbol{K}^{\varepsilon}, \qquad \mathbf{0} = \lambda^{\bar{\nu}}_{,\boldsymbol{A}}\bar{\boldsymbol{h}} + \lambda^{\varepsilon}_{,\boldsymbol{A}}\boldsymbol{K}^{\varepsilon},$$
$$\mathbf{0} = \operatorname{sym}\left\{\lambda^{\bar{\nu}}_{,\boldsymbol{x}} \otimes \bar{\boldsymbol{h}} + \lambda^{\varepsilon}_{,\boldsymbol{x}} \otimes \boldsymbol{K}^{\varepsilon}\right\}, \qquad 0 = \lambda^{\bar{\nu}}_{,\dot{\bar{\nu}}}\bar{\boldsymbol{h}} + \bar{\gamma}\bar{\nu}\vartheta^M\psi^T_{,\boldsymbol{g}_1}. \tag{12.3.38}$$

It is seen from Eq. (12.3.38)$_4$ that

$$\lambda^{\bar{\nu}} = \alpha\dot{\bar{\nu}}, \qquad \alpha = \hat{\alpha}(\nu_0, \bar{\nu}, \bar{\gamma}, \boldsymbol{g}_1, \vartheta^M, \vartheta^T, \varepsilon, \bar{\boldsymbol{Z}}); \qquad -\alpha\bar{\boldsymbol{h}} = \bar{\gamma}\bar{\nu}\vartheta^M\psi^T_{,\boldsymbol{g}_1}, \tag{12.3.39}$$

with α a scalar function. Equation (12.3.39)$_3$ determines $\bar{\boldsymbol{h}}$, once ψ^T is prescribed. Since both $\lambda^{\bar{\nu}}$ and λ^{ε} are isotropic scalar functions, Eq. (12.3.38)$_1$ suggests a collinearity between $\bar{\boldsymbol{h}}$ and $\boldsymbol{K}^{\varepsilon}$ via $\boldsymbol{K}^{\varepsilon} = \alpha'\bar{\boldsymbol{h}}$, with α' a scalar function of Eq. (12.3.12). Substituting these into Eqs. (12.3.38)$_{1-3}$ yields

$$-\lambda^{\bar{\nu}} = \alpha'\lambda^{\varepsilon}, \qquad \alpha' = \alpha''\dot{\bar{\nu}}, \qquad \alpha'' = \hat{\alpha}''(\bar{\nu}, \bar{\gamma}, \vartheta^M, \vartheta^T, \varepsilon); \qquad \boldsymbol{K}^{\varepsilon} = \alpha''\dot{\bar{\nu}}\bar{\boldsymbol{h}}. \tag{12.3.40}$$

In deriving Eq. (12.3.40), the relation $\lambda^{\varepsilon} \neq \hat{\lambda}^{\varepsilon}(\cdot, \dot{\bar{\nu}})$ has been used. Equation (12.3.40) is important, for it indicates that not only the flux of turbulent dissipation is collinear to the flux of mean volume fraction with a linear dependency on $\dot{\bar{\nu}}$, justified in view of the dominant long-term grain-grain interaction in weak turbulent motion, but also delivers an identity for $\lambda^{\bar{\nu}}$, once λ^{ε} is obtained from Eq. (12.3.19)$_4$.

The residual entropy inequality (12.3.25) is recast alternatively as

$$\begin{aligned}\bar{\pi} = {} & \left\{\vartheta^M(\bar{p} - \bar{\beta}) - \alpha\dot{\bar{\nu}}\right\}\dot{\bar{\nu}} + \left\{\bar{\boldsymbol{q}} + \boldsymbol{Q} - (\alpha_{,\vartheta^M} + \lambda^{\varepsilon}_{,\vartheta^M}\alpha'')\dot{\bar{\nu}}\bar{\boldsymbol{h}}\right\} \cdot \boldsymbol{g}_3 \\ & - \left\{\boldsymbol{K}^T + (\alpha_{,\vartheta^T} + \lambda^{\varepsilon}_{,\vartheta^T}\alpha'')\dot{\bar{\nu}}\bar{\boldsymbol{h}}\right\} \cdot \boldsymbol{g}_4 - \left\{\alpha_{,\varepsilon} + \lambda^{\varepsilon}_{,\varepsilon}\alpha''\dot{\bar{\nu}}\bar{\boldsymbol{h}}\right\} \cdot \boldsymbol{g}_5 \\ & + \left\{\bar{\nu}(\vartheta^M\bar{p} - \alpha\dot{\bar{\nu}})\boldsymbol{I} + \vartheta^M\bar{\boldsymbol{t}} + \vartheta^T\boldsymbol{R} - \alpha\bar{\boldsymbol{h}} \otimes \boldsymbol{g}_1\right\} \cdot \bar{\boldsymbol{D}} + \bar{\pi}_{int} \geq 0,\end{aligned} \tag{12.3.41}$$

with $\bar{\pi}_{int}$ the internal dissipation given by

$$\begin{aligned}\bar{\pi}_{int} = {} & -(\alpha_{,\bar{\nu}} + \lambda^{\varepsilon}_{,\bar{\nu}}\alpha'')\dot{\bar{\nu}}\bar{\boldsymbol{h}} \cdot \boldsymbol{g}_1 - (\alpha_{,\bar{\gamma}} + \lambda^{\varepsilon}_{,\bar{\gamma}}\alpha'')\dot{\bar{\nu}}\bar{\boldsymbol{h}} \cdot \boldsymbol{g}_2 + \alpha\dot{\bar{\nu}}\bar{f} + \bar{\gamma}\bar{\nu}\varepsilon(\vartheta^M - \vartheta^T) \\ & + \lambda^{\varepsilon} f^{\varepsilon} - \bar{\gamma}\bar{\nu}\vartheta^M\psi^T_{,\bar{\boldsymbol{Z}}} \cdot \bar{\boldsymbol{\Phi}},\end{aligned} \tag{12.3.42}$$

and the abbreviations,

$$\bar{p} \equiv \bar{\gamma}^2\psi^T_{,\bar{\gamma}}, \qquad \bar{\beta} \equiv \bar{\gamma}\bar{\nu}\psi^T_{,\bar{\nu}}, \tag{12.3.43}$$

known respectively as the *turbulent thermodynamic pressure* and *turbulent configurational pressure*, extended from their counterparts in laminar flows.

Thermodynamic equilibrium. Thermodynamic equilibrium is defined to be a time-independent process with uniform vanishing mean entropy production, viz.,

$$\bar{\pi}|_{\mathrm{E}} = 0, \tag{12.3.44}$$

where the subscript "E" indicates that the indexed quantity is evaluated at thermodynamic equilibrium. In view of Eqs. (12.3.12), (12.3.41) and (12.3.42), the definition motivates respectively the equilibrium and dynamic state spaces given by

$$\mathcal{Q}|_{\mathrm{E}} \equiv \left(\nu_0, \bar{\nu}, 0, \boldsymbol{g}_1, \bar{\gamma}, \boldsymbol{g}_2, \vartheta^M, \mathbf{0}, \vartheta^T, \mathbf{0}, \varepsilon, \mathbf{0}, \mathbf{0}, \bar{\boldsymbol{Z}}\right), \quad \mathcal{Q}^D \equiv \left(\dot{\bar{\nu}}, \boldsymbol{g}_3, \boldsymbol{g}_4, \boldsymbol{g}_5, \bar{\boldsymbol{D}}\right). \tag{12.3.45}$$

The dynamic quantities $\{\dot{\bar{\nu}}, \boldsymbol{g}_3, \boldsymbol{g}_4, \boldsymbol{g}_5, \bar{\boldsymbol{D}}\}$ should vanish at an equilibrium state, with $\mathcal{Q}^D|_{\mathrm{E}} = 0$. In addition, under sufficient smoothness, $\bar{\pi}$ has to satisfy that

$$\bar{\pi}_{,a}|_{\mathrm{E}} = 0, \qquad a \in \mathcal{Q}^D, \tag{12.3.46}$$

and the Hessian matrix of $\bar{\pi}$ with respect to $\mathcal{Q}^D$ at thermodynamic equilibrium should be positive semi-definite. Since the latter condition constrains the sign of material parameters in the closure relations, only Eqs. (12.3.44) and (12.3.46) are investigated.

First, applying Eqs. (12.3.44) and (12.3.45)$_1$ to Eqs. (12.3.41)–(12.3.42) yields

$$\lambda^\varepsilon f^\varepsilon|_{\mathrm{E}} = \bar{\gamma}\bar{\nu}\vartheta^M \psi^T_{,\bar{\boldsymbol{Z}}} \cdot \bar{\boldsymbol{\Phi}}|_{\mathrm{E}} - (\vartheta^M - \vartheta^T)\bar{\gamma}\bar{\nu}\varepsilon, \tag{12.3.47}$$

showing that the turbulent dissipation at an equilibrium state results from its production and internal friction. Second, incorporating Eqs. (12.3.45)$_2$ and (12.3.46) into Eqs. (12.3.41) and (12.3.42) gives

$$\begin{aligned}
0 &= \vartheta^M(\bar{p} - \bar{\beta}) - (\alpha_{,\bar{\nu}} + \lambda^\varepsilon_{,\bar{\nu}}\alpha'')\bar{\boldsymbol{h}} \cdot \boldsymbol{g}_1 - (\alpha_{,\bar{\gamma}} + \lambda^\varepsilon_{,\bar{\gamma}}\alpha'')\bar{\boldsymbol{h}} \cdot \boldsymbol{g}_2 + \alpha\bar{f}|_{\mathrm{E}} \\
&\quad - \bar{\gamma}\bar{\nu}\vartheta^T \psi^T_{,\bar{\boldsymbol{Z}}} \cdot \bar{\boldsymbol{\Phi}}_{,\dot{\bar{\nu}}}|_{\mathrm{E}} + \lambda^\varepsilon f^\varepsilon_{,\dot{\bar{\nu}}}|_{\mathrm{E}},
\end{aligned} \tag{12.3.48}$$

$$\boldsymbol{0} = (\bar{\boldsymbol{q}} + \boldsymbol{Q})|_{\mathrm{E}} - \bar{\gamma}\bar{\nu}\vartheta^M \psi^T_{,\bar{\boldsymbol{Z}}} \cdot \bar{\boldsymbol{\Phi}}_{,\boldsymbol{g}_3}|_{\mathrm{E}} + \lambda^\varepsilon f^\varepsilon_{,\boldsymbol{g}_3}|_{\mathrm{E}}, \tag{12.3.49}$$

$$\boldsymbol{0} = -\boldsymbol{K}^T|_{\mathrm{E}} - \bar{\gamma}\bar{\nu}\vartheta^M \psi^T_{,\bar{\boldsymbol{Z}}} \cdot \bar{\boldsymbol{\Phi}}_{,\boldsymbol{g}_4}|_{\mathrm{E}} + \lambda^\varepsilon f^\varepsilon_{,\boldsymbol{g}_4}|_{\mathrm{E}}, \tag{12.3.50}$$

$$\boldsymbol{0} = -\bar{\gamma}\bar{\nu}\vartheta^M \psi^T_{,\bar{\boldsymbol{Z}}} \cdot \bar{\boldsymbol{\Phi}}_{,\boldsymbol{g}_5}|_{\mathrm{E}} + \lambda^\varepsilon f^\varepsilon_{,\boldsymbol{g}_5}|_{\mathrm{E}}, \tag{12.3.51}$$

$$\boldsymbol{0} = \bar{\nu}\vartheta^M \bar{p}\boldsymbol{I} + \vartheta^M \bar{\boldsymbol{t}}|_{\mathrm{E}} + \vartheta^T \boldsymbol{R}|_{\mathrm{E}} - \alpha\bar{\boldsymbol{h}} \otimes \boldsymbol{g}_1 - \bar{\gamma}\bar{\nu}\vartheta^M \psi^T_{,\bar{\boldsymbol{Z}}} \cdot \bar{\boldsymbol{\Phi}}_{,\bar{\boldsymbol{D}}}|_{\mathrm{E}} + \lambda^\varepsilon f^\varepsilon_{,\bar{\boldsymbol{D}}}|_{\mathrm{E}}. \tag{12.3.52}$$

While Eq. (12.3.48) indicates that both the internal friction and production of turbulent dissipation affect the evolutions of mean volume fraction via the variations in $\bar{\boldsymbol{h}}$, $\bar{p}$ and $\bar{\beta}$, justified in view of the grain-grain interactions, it also delivers an equilibrium expression of $\bar{f}$. Equation (12.3.49) denotes the equilibrium mean and turbulent heat fluxes in terms of the internal friction and turbulent dissipation production. It reduces to vanishing mean and turbulent heat fluxes for isothermal flows. The Eq. (12.3.50) provides an equilibrium expression of the flux $\boldsymbol{K}^T$ of turbulent kinetic energy in terms of the internal friction and turbulent dissipation production. Since the turbulent dissipation is a (negative) production of turbulent kinetic energy, Eqs. (12.3.47) and (12.3.50) imply that at an equilibrium state, there exists an energy cascade from the turbulent kinetic energy toward the turbulent dissipation through internal friction, similar to that found in turbulent shear flows of the Newtonian fluids. Equation (12.3.51) delivers a restriction on $f^\varepsilon|_{\mathrm{E}}$, in addition to the result in Eq. (12.3.47). Last, Eq. (12.3.52) gives an expression that should be satisfied by the equilibrium mean Cauchy stress $\bar{\boldsymbol{t}}|_{\mathrm{E}}$ and Reynolds' stress $\boldsymbol{R}|_{\mathrm{E}}$ in terms of the internal friction, production of turbulent dissipation, and mean volume fraction gradient.

Equations (12.3.18), (12.3.20)$_2$, (12.3.39)–(12.3.40) and (12.3.47)–(12.3.52) are the derived equilibrium closure relations of $\{\psi^T, \bar{\boldsymbol{h}}, \bar{f}, \bar{\boldsymbol{t}}, \boldsymbol{R}, \boldsymbol{K}^T, \boldsymbol{K}^\varepsilon, f^\varepsilon, k, \bar{\boldsymbol{\Phi}}\}$, some of which are implicit ones. For incompressible grains, Eqs. (12.3.39)$_3$, (12.3.48) and (12.3.52) reduce to

$$\mathbf{0} = \psi^T_{,\boldsymbol{g}_1}, \tag{12.3.53}$$

$$0 = \vartheta^M(\bar{p} - \bar{\beta}) - \bar{\gamma}\bar{\nu}\vartheta^M \psi^T_{,\bar{\boldsymbol{Z}}} \cdot \bar{\boldsymbol{\Phi}}_{,\dot{\bar{\nu}}}|_{\mathrm{E}} + \lambda^\varepsilon f^\varepsilon_{,\dot{\bar{\nu}}}|_{\mathrm{E}}, \tag{12.3.54}$$

$$\mathbf{0} = \bar{\nu}\vartheta^M \bar{p}\boldsymbol{I} + \vartheta^M \bar{\boldsymbol{t}}|_{\mathrm{E}} + \vartheta^T \boldsymbol{R}|_{\mathrm{E}} - \bar{\gamma}\bar{\nu}\vartheta^M \psi^T_{,\bar{\boldsymbol{Z}}} \cdot \bar{\boldsymbol{\Phi}}_{,\bar{\boldsymbol{D}}}|_{\mathrm{E}} + \lambda^\varepsilon f^\varepsilon_{\bar{\boldsymbol{D}}}|_{\mathrm{E}}, \tag{12.3.55}$$

while Eqs. (12.3.47) and (12.3.49)–(12.3.51) remain unchanged, with $\boldsymbol{K}^\varepsilon = \mathbf{0}$ from Eq. $(12.3.40)_4$. Two sets of the equilibrium closure relations are hence obtained. While Eqs. (12.3.18) and $(12.3.20)_2$ are valid for both circumstances, Eqs. $(12.3.39)_3$, $(12.3.40)_4$ and (12.3.47)–(12.3.52) apply for compressible grains; Eqs. (12.3.49)–(12.3.51) and (12.3.53)–(12.3.55) with vanishing $\bar{\boldsymbol{h}}$, $\bar{f}$ and $\boldsymbol{K}^\varepsilon$ apply for incompressible grains. The obtained results are summarized in Table 12.2.

Remarks:

1. At a thermodynamic equilibrium state, all productions cease. It is plausible to assume that
$$0 = \bar{f}|_{\mathrm{E}}, \qquad \mathbf{0} = \bar{\boldsymbol{\Phi}}|_{\mathrm{E}}, \qquad 0 = f^\varepsilon|_{\mathrm{E}}. \tag{12.3.56}$$
Substituting these into Eq. (12.3.47) gives a vanishing turbulent dissipation at an equilibrium state, with which Eq. (12.3.9) is fulfilled.
2. In the next subsection, a hypoplastic model will be used for $\bar{\boldsymbol{\Phi}}$ with $\bar{\boldsymbol{\Phi}} = \hat{\bar{\boldsymbol{\Phi}}}(\bar{\nu}, \bar{\boldsymbol{D}}, \bar{\boldsymbol{Z}})$, with which Eq. $(12.3.56)_2$ is automatically satisfied, and Eq. (12.3.8) reduces to
$$\nabla\lambda^\varepsilon \cdot f^\varepsilon_{,\boldsymbol{g}_4}|_{\mathrm{E}} + \lambda^\varepsilon \nabla \cdot (f^\varepsilon_{,\boldsymbol{g}_4}|_{\mathrm{E}}) = 0. \tag{12.3.57}$$
It implies a balance between the turbulent dissipation production and granular coldness gradient and yields a restriction that should be fulfilled by ψ^T and $f^\varepsilon|_{\mathrm{E}}$ (i.e., $\lambda^\varepsilon f^\varepsilon_{,\boldsymbol{g}_4}|_{\mathrm{E}}$ should be a solenoidal field of $\boldsymbol{g}_4$).
3. Equations (12.3.52) and (12.3.55), by using Truesdell's equipresence principle, are decomposed respectively as
$$\vartheta^M \bar{\boldsymbol{t}}|_{\mathrm{E}} = -\bar{\nu}\vartheta^M \bar{p}\boldsymbol{I} + \alpha\bar{\boldsymbol{h}} \otimes \boldsymbol{g}_1 + \bar{\gamma}\bar{\nu}\vartheta^M \psi^T_{,\bar{\boldsymbol{Z}}} \cdot \bar{\boldsymbol{\Phi}}_{,\bar{\boldsymbol{D}}}|_{\mathrm{E}}, \qquad \vartheta^T \boldsymbol{R}|_{\mathrm{E}} = -\lambda^\varepsilon f^\varepsilon_{,\bar{\boldsymbol{D}}}|_{\mathrm{E}};$$
$$\vartheta^M \bar{\boldsymbol{t}}|_{\mathrm{E}} = -\bar{\nu}\vartheta^M \bar{p}\boldsymbol{I} + \bar{\gamma}\bar{\nu}\vartheta^M \psi^T_{,\bar{\boldsymbol{Z}}} \cdot \bar{\boldsymbol{\Phi}}_{,\bar{\boldsymbol{D}}}|_{\mathrm{E}}, \qquad \vartheta^T \boldsymbol{R}|_{\mathrm{E}} = -\lambda^\varepsilon f^\varepsilon_{,\bar{\boldsymbol{D}}}|_{\mathrm{E}}, \tag{12.3.58}$$
for compressible and incompressible grains. We choose the mean Cauchy stress to be generated through the mean quantities, with Reynolds' stress mainly induced via the quantities related to turbulent fluctuation (e.g. the turbulent dissipation production), a procedure widely used for the Newtonian fluids. A vanishing equilibrium Reynolds' stress is obtained if the flow is laminar. While Eq. $(12.3.58)_1$ indicates that a dry granular heap with compressible grains at an equilibrium state may be accomplished via either the internal friction or mean volume fraction gradient, Eq. $(12.3.58)_2$ delivers that the internal friction is the only mechanism for a non-spherical equilibrium Cauchy stress, when Wilmánski's model is used for the time evolution of mean volume fraction.
4. In Table 12.2, ψ^T, $\bar{\boldsymbol{h}}$, $\bar{p}$ and $\bar{\beta}$ assume the same functionals at both equilibrium and non-equilibrium states. The results for incompressible grains can be obtained from those of compressible grains with vanishing $\bar{\boldsymbol{h}}$ and $\bar{f}$.

Table 12.2 Obtained thermodynamically consistent equilibrium closure relations

	Compressible grains	Incompressible grains
ψ^T	$\psi^T = \hat{\psi}^T(\nu_0, \bar{\nu}, \nabla\bar{\nu}, \bar{\gamma}, \vartheta^M, \vartheta^T, \varepsilon, \bar{\mathbf{Z}})$;	$\psi^T = \hat{\psi}^T(\nu_0, \bar{\nu}, \bar{\gamma}, \vartheta^M, \vartheta^T, \varepsilon, \bar{\mathbf{Z}})$;
$\bar{p}$	$\bar{p} = \bar{\gamma}^2 \psi^T_{,\bar{\gamma}}$;	$\bar{p} = \bar{\gamma}^2 \psi^T_{,\bar{\gamma}}$;
$\bar{\beta}$	$\bar{\beta} = \bar{\gamma}\bar{\nu}\psi^T_{,\bar{\nu}}$;	$\bar{\beta} = \bar{\gamma}\bar{\nu}\psi^T_{,\bar{\nu}}$;
$\bar{\boldsymbol{h}}$	$-\alpha\bar{\boldsymbol{h}} = \bar{\gamma}\bar{\nu}\vartheta^M \psi^T_{,\boldsymbol{g}_1}$;	$\mathbf{0}$;
	$\alpha = \hat{\alpha}(\nu_0, \bar{\nu}, \boldsymbol{g}_1, \bar{\gamma}, \vartheta^M, \vartheta^T, \varepsilon, \bar{\mathbf{Z}})$;	–
$\bar{f}$	$\bar{f}\|_{\mathrm{E}} = 0$;	$\mathbf{0}$;
$\bar{\gamma}\bar{\nu}k$	$\bar{\gamma}\bar{\nu}k = \bar{\gamma}\bar{\nu}\vartheta^M \psi^T_{,\vartheta^T}$;	$\bar{\gamma}\bar{\nu}k = \bar{\gamma}\bar{\nu}\vartheta^M \psi^T_{,\vartheta^T}$;
$\boldsymbol{K}^T$	$\boldsymbol{K}^T\|_{\mathrm{E}} = -\bar{\gamma}\bar{\nu}\vartheta^M \psi^T_{,\bar{\mathbf{Z}}} \cdot \bar{\mathbf{\Phi}}_{,\boldsymbol{g}_4}\|_{\mathrm{E}} + \lambda^\varepsilon f^\varepsilon_{,\boldsymbol{g}_4}\|_{\mathrm{E}}$;	$\boldsymbol{K}^T\|_{\mathrm{E}} = -\bar{\gamma}\bar{\nu}\vartheta^M \psi^T_{,\bar{\mathbf{Z}}} \cdot \bar{\mathbf{\Phi}}_{,\boldsymbol{g}_4}\|_{\mathrm{E}} + \lambda^\varepsilon f^\varepsilon_{,\boldsymbol{g}_4}\|_{\mathrm{E}}$;
$\boldsymbol{K}^\varepsilon$	$\boldsymbol{K}^\varepsilon = \alpha''\dot{\bar{\nu}}\bar{\boldsymbol{h}}$;	$\mathbf{0}$;
	$\alpha'' = \hat{\alpha}''(\bar{\nu}, \bar{\gamma}, \vartheta^M, \vartheta^T, \varepsilon)$;	–
f^ε	$f^\varepsilon\|_{\mathrm{E}} = 0$;	$f^\varepsilon\|_{\mathrm{E}} = 0$;
$\bar{\boldsymbol{q}}$ and $\boldsymbol{Q}$	$(\bar{\boldsymbol{q}} + \boldsymbol{Q})\|_{\mathrm{E}} = \bar{\gamma}\bar{\nu}\vartheta^M \psi^T_{,\bar{\mathbf{Z}}} \cdot \bar{\mathbf{\Phi}}_{,\boldsymbol{g}_3}\|_{\mathrm{E}} - \lambda^\varepsilon f^\varepsilon_{,\boldsymbol{g}_3}\|_{\mathrm{E}}$;	$(\bar{\boldsymbol{q}} + \boldsymbol{Q})\|_{\mathrm{E}} = \bar{\gamma}\bar{\nu}\vartheta^M \psi^T_{,\bar{\mathbf{Z}}} \cdot \bar{\mathbf{\Phi}}_{,\boldsymbol{g}_3}\|_{\mathrm{E}} - \lambda^\varepsilon f^\varepsilon_{,\boldsymbol{g}_3}\|_{\mathrm{E}}$;
$\bar{\mathbf{\Phi}}$	$\bar{\mathbf{\Phi}}\|_{\mathrm{E}} = \mathbf{0}$;	$\bar{\mathbf{\Phi}}\|_{\mathrm{E}} = \mathbf{0}$;
$\bar{\boldsymbol{t}}$	$\vartheta^M \bar{\boldsymbol{t}}\|_{\mathrm{E}} = -\bar{\nu}\vartheta^M \bar{p}\boldsymbol{I} + \alpha\bar{\boldsymbol{h}} \otimes \boldsymbol{g}_1 + \bar{\gamma}\bar{\nu}\vartheta^M \psi^T_{,\bar{\mathbf{Z}}} \cdot \bar{\mathbf{\Phi}}_{,\bar{\boldsymbol{D}}}\|_{\mathrm{E}}$;	$\vartheta^M \bar{\boldsymbol{t}}\|_{\mathrm{E}} = -\bar{\nu}\vartheta^M \bar{p}\boldsymbol{I} + \bar{\gamma}\bar{\nu}\vartheta^M \psi^T_{,\bar{\mathbf{Z}}} \cdot \bar{\mathbf{\Phi}}_{,\bar{\boldsymbol{D}}}\|_{\mathrm{E}}$;
$\boldsymbol{R}$	$\boldsymbol{R}\|_{\mathrm{E}} = -\lambda^\varepsilon f^\varepsilon_{,\bar{\boldsymbol{D}}}\|_{\mathrm{E}}$;	$\boldsymbol{R}\|_{\mathrm{E}} = -\lambda^\varepsilon f^\varepsilon_{,\bar{\boldsymbol{D}}}\|_{\mathrm{E}}$.

12.3.4 First-Order Closure Model

The closure quantities are assumed to consist of the equilibrium and dynamic responses, viz.,

$$\mathcal{C} = \mathcal{C}|_{\mathrm{E}} + \mathcal{C}^D. \quad \mathcal{C} \in \{\bar{\boldsymbol{t}}, \boldsymbol{R}, \bar{\mathbf{\Phi}}, \boldsymbol{K}^T, f^\varepsilon\}. \tag{12.3.59}$$

Specifically, $\{\bar{\boldsymbol{t}}^D, \boldsymbol{R}^D, (\boldsymbol{K}^T)^D, (f^\varepsilon)^D\}$ are assumed to be quasi-statically dependent on $\mathcal{Q}^D$, viz.,

$$\begin{aligned} \bar{\boldsymbol{t}}^D &= \epsilon^M \dot{\bar{\nu}}\boldsymbol{I} + \lambda^M (tr\,\bar{\boldsymbol{D}})\boldsymbol{I} + 2\mu^M \bar{\boldsymbol{D}}, & \boldsymbol{R}^D &= \epsilon^T \dot{\bar{\nu}}\boldsymbol{I} + \lambda^T (tr\,\bar{\boldsymbol{D}})\boldsymbol{I} + 2\mu^T \bar{\boldsymbol{D}}, \\ (f^\varepsilon)^D &= f_1\dot{\bar{\nu}} - f_2(\boldsymbol{g}_4 \cdot \boldsymbol{g}_4) + f_3(\boldsymbol{g}_5 \cdot \boldsymbol{g}_5), & (\boldsymbol{K}^T)^D &= -f_4\boldsymbol{g}_4, \end{aligned} \tag{12.3.60}$$

with the isotropic scalar functions $\{\epsilon^M, \epsilon^T, f_1, f_2, f_3, f_4\}$ and $\{\lambda^M, \lambda^T, \mu^M, \mu^T\}$ depending respectively on $\{\nu_0, \bar{\nu}, \bar{\gamma}, \vartheta^M, \vartheta^T, \varepsilon\}$ and $\{\nu_0, \bar{\nu}, \bar{\gamma}, \vartheta^M, \vartheta^T, \varepsilon, I^1_{\bar{D}}, I^2_{\bar{D}}, I^3_{\bar{D}}\}$.

In Eq. (12.3.60), by using Truesdell's equipresence principle, the dynamic response of $(\vartheta^M \bar{\boldsymbol{t}} + \vartheta^T \boldsymbol{R})$ is assumed to depend explicitly and linearly on $\{\dot{\bar{\nu}}, \bar{\boldsymbol{D}}\}$, while that of turbulent dissipation production is an explicit and linear function of $\{\dot{\bar{\nu}}, \boldsymbol{g}_4, \boldsymbol{g}_5\}$, and that of flux $\boldsymbol{K}^T$ depends explicitly and linearly on $\boldsymbol{g}_4$, as motivated by Fourier's law. In doing so, the flowing dry granular dense matter is considered a viscous-inelastic isotropic fluid, a so-called Stokes or Reiner-Rivlin fluid. The scalar functions μ^M and μ^T are respectively the *material viscosity* (viscosity in laminar flows) and *phenomenological viscosity* due to turbulent fluctuation, known as the *turbulent viscosity*.

Turbulent Helmholtz free energy. The specific turbulent Helmholtz free energy ψ^T, in view of Eqs. (12.3.12), (12.3.18) and (12.3.48), is assumed to consist of an elastic part, ψ^T_e, and a plastic part, ψ^T_f, viz.,

$$\psi^T = \psi_e^T(\nu_0, \bar{\nu}, \bar{\gamma}, \vartheta^M, \vartheta^T) + \psi_f^T(\varepsilon, I_{\bar{Z}}^1, I_{\bar{Z}}^2, I_{\bar{Z}}^3), \quad (12.3.61)$$

in which the rate-independent and dissipative characteristics are confined within ψ_f^T. For weak turbulent intensity, ψ_e^T is further assumed to be in the form of $\psi_e^T = (\psi_e^T)'(1 + \vartheta^T/\vartheta^M)$, with $(\psi_e^T)'$ expanded in a Taylor series about $\bar{\nu} = \bar{\nu}_m$, where $\bar{\nu}_m$ is the critical mean volume fraction at which shearing is decoupled from dilatation. With these, the specific form of ψ^T is obtained as

$$\psi^T = \alpha_0(\bar{\nu} - \bar{\nu}_m)^2\left(1 + \frac{\vartheta^T}{\vartheta^M}\right) + \psi_f^T(\varepsilon, I_{\bar{Z}}^1, I_{\bar{Z}}^2, I_{\bar{Z}}^3), \quad (12.3.62)$$

with $\alpha_0 = \hat{\alpha}_0(\nu_0, \bar{\gamma}, \vartheta^M)$, a positive constant. This equation delivers that smaller granular coldness results in smaller free energy. It reduces to its counterpart in laminar flows if both ϑ^T and ε vanish. For flows with significant turbulent intensity, higher-order terms of ϑ^T in the Taylor series expansion of ψ_e^T and the coupled terms between ε and $\bar{\boldsymbol{Z}}$ in ψ_f^T need to be elaborated.

Material and turbulent viscosities. The material viscosity μ^M and turbulent viscosity μ^T are proposed as

$$\mu^M = \mu_0\bar{\gamma}^2\left(\frac{\bar{\nu}_m}{\bar{\nu}_\infty - \bar{\nu}}\right)^8 \Xi, \quad \mu^T = \mu_0\bar{\gamma}^2\frac{\vartheta^T}{\vartheta^M}\left(\frac{\bar{\nu}_m}{\bar{\nu}_\infty - \bar{\nu}}\right)^8 \Xi, \quad \Xi = \hat{\Xi}(I_{\bar{D}}^1, I_{\bar{D}}^2, I_{\bar{D}}^3), \quad (12.3.63)$$

with $\mu_0 = \hat{\mu}_0(\nu_0, \vartheta^M)$, a positive constant; and $\bar{\nu}_\infty$ represents the mean volume fraction corresponding to the possible densest packing of grains. Equation $(12.3.63)_1$ is a curve fitting in which the exponent 8 is chosen to fit the experimental results. Equation $(12.3.63)_2$ is motivated by Eq. $(12.3.63)_1$ with a separate linear dependency on ϑ^T, for a flow with weak turbulent intensity deviates slightly from its laminar state. Equation (12.3.63) asserts that the total stress is larger in turbulent flows than in laminar flows, corresponding to the experimental outcomes of the Newtonian fluids in turbulent shear flows. For laminar flows, ϑ^T vanishes, and hence, μ^T vanishes correspondingly.

Hypoplastic model. Different prescriptions of $\bar{\boldsymbol{\Phi}}$ allow different rate-independent characteristics be taken into account. Specifically, a hypoplastic form of $\bar{\boldsymbol{\Phi}}$ is given by

$$\bar{\boldsymbol{\Phi}} = \hat{\bar{\boldsymbol{\Phi}}}(\bar{\nu}, \bar{\boldsymbol{D}}, \bar{\boldsymbol{Z}}) = f_s(\bar{\nu}, I_{\bar{Z}}^1)\Big\{a^2\bar{\boldsymbol{D}} + \check{\bar{\boldsymbol{Z}}}tr(\check{\bar{\boldsymbol{Z}}}\bar{\boldsymbol{D}}) + f_d(\bar{\nu})a(\check{\bar{\boldsymbol{Z}}} + \check{\bar{\boldsymbol{Z}}}^*)\|\bar{\boldsymbol{D}}\|\Big\}, \quad (12.3.64)$$

with $\check{\bar{\boldsymbol{Z}}} = \bar{\boldsymbol{Z}}/\mathrm{tr}(\bar{\boldsymbol{Z}})$, $\check{\bar{\boldsymbol{Z}}}^* = \check{\bar{\boldsymbol{Z}}} - \frac{1}{3}\boldsymbol{I}$ and $\|\bar{\boldsymbol{D}}\| = \sqrt{\mathrm{tr}\bar{\boldsymbol{D}}^2}$, where $\check{\bar{\boldsymbol{Z}}}$ is the versor of $\bar{\boldsymbol{Z}}$, $\check{\bar{\boldsymbol{Z}}}^*$ denotes the deviator of $\check{\bar{\boldsymbol{Z}}}$ and a is a positive constant. The scalar functions f_s and f_d are the stiffness and density factors, denoting strain harding/softing and mean pressure-dependent bulk density, respectively. The constant a is related to the stress state $\bar{\boldsymbol{Z}}_c$ and the frictional angle φ_c at the critical state and can be determined empirically. Equation (12.3.64) is homogeneous of degree zero with respect to $\bar{\boldsymbol{Z}}$ and fulfills Eq. $(12.3.56)_2$. In contrast to conventional elasto-plastic theories, the hypoplastic model given in Eq. (12.3.64) bears two features: (a) Distinction between loading and unloading is automatically accomplished; and (b) elastic/inelastic deformations need not *a priori* be distinguished; information about yield surface and plastic potential is not necessary.

With these, the closure relations of $\{\boldsymbol{K}^T, f^\varepsilon, \boldsymbol{t}, \boldsymbol{R}\}$, in the context of first-order k-ε closure model, for an isochoric, isothermal dry granular dense matter with incompressible grains are obtained as

$$\boldsymbol{K}^T = -f_4\boldsymbol{g}_4, \qquad f^\varepsilon = f_1\dot{\bar{\nu}} - f_2(\boldsymbol{g}_4\cdot\boldsymbol{g}_4) + f_3(\boldsymbol{g}_5\cdot\boldsymbol{g}_5), \quad (12.3.65)$$

$$\begin{aligned}\bar{\boldsymbol{t}} &= -(\bar{\nu}\bar{p} - \epsilon^M\dot{\bar{\nu}} - \lambda^M tr\,\bar{\boldsymbol{D}})\boldsymbol{I} + f_s(\zeta_1\boldsymbol{I} + \zeta_2\bar{\boldsymbol{Z}} + \zeta_3\bar{\boldsymbol{Z}}^2)\\ &\quad + 2\mu_0\bar{\gamma}^2\left(\frac{\bar{\nu}_m}{\bar{\nu}_\infty - \bar{\nu}}\right)^8\sqrt{|I^2_{\bar{\boldsymbol{D}}}|}\,\bar{\boldsymbol{D}},\end{aligned} \quad (12.3.66)$$

$$\boldsymbol{R} = (\epsilon^T\dot{\bar{\nu}} + \lambda^T tr\,\bar{\boldsymbol{D}})\boldsymbol{I} + 2\mu_0\bar{\gamma}^2\frac{\vartheta^T}{\vartheta^M}\left(\frac{\bar{\nu}_m}{\bar{\nu}_\infty - \bar{\nu}}\right)^8\sqrt{|I^2_{\bar{\boldsymbol{D}}}|}\,\bar{\boldsymbol{D}}, \quad (12.3.67)$$

in which the Cayley-Hamilton theorem and the notations

$$\begin{aligned}&c_1 = \psi^T_{f,I^1_{\bar{\boldsymbol{Z}}}},\quad c_2 = \psi^T_{f,I^2_{\bar{\boldsymbol{Z}}}},\quad c_3 = \psi^T_{f,I^3_{\bar{\boldsymbol{Z}}}},\quad (I^1_{\bar{\boldsymbol{Z}}})_{,\bar{\boldsymbol{Z}}} = \boldsymbol{I},\quad (I^2_{\bar{\boldsymbol{Z}}})_{,\bar{\boldsymbol{Z}}} = I^1_{\bar{\boldsymbol{Z}}}\boldsymbol{I} - \bar{\boldsymbol{Z}},\\ &(I^3_{\bar{\boldsymbol{Z}}})_{,\bar{\boldsymbol{Z}}} = I^3_{\bar{\boldsymbol{Z}}}\bar{\boldsymbol{Z}}^{-1},\quad \zeta_1 = a^2(c_1 + c_2I^1_{\bar{\boldsymbol{Z}}}) - c_3a^2I^2_{\bar{\boldsymbol{Z}}}I^3_{\bar{\boldsymbol{Z}}},\qquad \zeta_3 = c_3a^2(I^3_{\bar{\boldsymbol{Z}}})^2,\\ &\zeta_2 = (c_1 + c_2I^1_{\bar{\boldsymbol{Z}}})(I^1_{\bar{\boldsymbol{Z}}})^{-1} - c_2(a^2 + (I^1_{\bar{\boldsymbol{Z}}})^{-2}tr\,\bar{\boldsymbol{Z}}^2) + c_3I^3_{\bar{\boldsymbol{Z}}}(3(I^1_{\bar{\boldsymbol{Z}}})^{-2} + a^2I^1_{\bar{\boldsymbol{Z}}}),\end{aligned} \quad (12.3.68)$$

have been used, where $\bar{\boldsymbol{h}} = \boldsymbol{0}$, $f = 0$, $\boldsymbol{K}^\varepsilon = \boldsymbol{0}$, and $\{\bar{\gamma}\bar{\nu}k, \psi^T, \bar{\boldsymbol{\Phi}}\}$ are given respectively in Eqs. (12.3.20)$_2$, (12.3.62) and (12.3.64), in which the turbulent kinetic energy is determined once the profile of $\bar{\nu}$ is known.

The field equations of primitive mean fields $\{\bar{\boldsymbol{v}}, \bar{\nu}, \vartheta^T, \varepsilon, \bar{\boldsymbol{Z}}\}$ can be obtained by substituting the obtained closure relations into Eqs. (12.3.1)–(12.3.2) and (12.3.4)–(12.3.7), in which $\bar{p}$ should be computed from the mean linear momentum equation. Since the system is mathematically likely well-posed, one has the chance to obtain the solutions to the primitive mean fields. In applying the closure relations, the phenomenological parameters $\{\alpha_0, f_{1-4}, f_s, f_d, a, \epsilon^M, \lambda^M, \epsilon^T, \lambda^T, \mu_0, \zeta_{1-3}, \bar{\nu}_m, \bar{\nu}_\infty\}$ need to be prescribed. These are accomplished by comparing numerical results with experimental outcomes, possibly involved inverse technique.

12.3.5 Numerical Simulations

Field equations and boundary conditions. Consider a fully developed, two-dimensional *stationary turbulent shear flow* down an inclined moving plane, as shown in Fig. 12.10. It is assumed that

$$\begin{aligned}&\bar{\boldsymbol{v}} = [\bar{u}(y), \bar{v}(y), 0],\qquad \bar{\nu} = \bar{\nu}(y),\qquad \bar{p} = \bar{p}(y),\qquad \vartheta^T = \vartheta^T(y),\qquad \varepsilon = \varepsilon(y),\\ &\bar{Z}_{ij} = \bar{Z}_{ij}(y),\end{aligned} \quad (12.3.69)$$

with $\bar{v}/\bar{u} \sim 0$, $u' \neq 0$, $v' \neq 0$, and $\{i, j\} = (x, y)$, where $\{\bar{u}(y), \bar{\nu}(y), \vartheta^T(y), \varepsilon(y), \bar{Z}_{ij}(y)\}$ are respectively the mean velocity component in the x-direction, the mean volume fraction, the granular coldness, the turbulent dissipation, and the mean internal friction components. They are motivated by the assumption that $\alpha_{,x} \ll \alpha_{,y}$ for any quantity α in simple turbulent shear flows.

Fig. 12.10 A gravity-driven stationary flows down an inclined moving plane and the coordinates

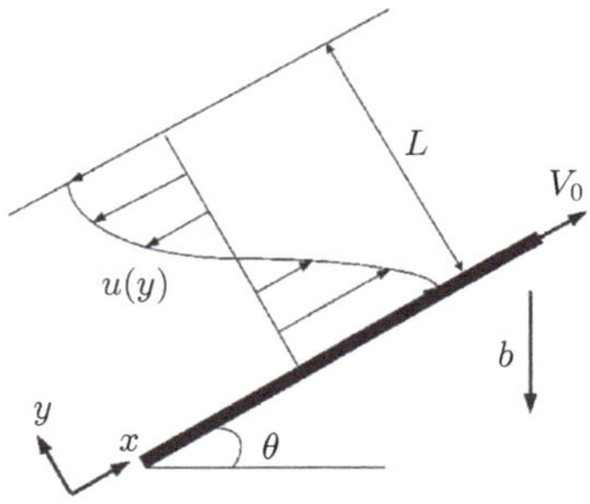

The considered flow corresponds to the *critical state* in which $\dot{\bar{\rho}} = 0$ and $\mathring{\bar{\boldsymbol{Z}}} = \boldsymbol{0}$. Since at the critical state f_d is set to be unity, incorporating this into Eq. (12.3.64) yields

$$\boldsymbol{0} = f_s \left\{ a_c^2 \bar{\boldsymbol{D}} + \check{\bar{\boldsymbol{Z}}} tr(\check{\bar{\boldsymbol{Z}}} \bar{\boldsymbol{D}}) + a_c (\check{\bar{\boldsymbol{Z}}} + \check{\bar{\boldsymbol{Z}}}^*) \| \bar{\boldsymbol{D}} \| \right\}, \tag{12.3.70}$$

with $a_c = \sqrt{8/27} \sin\varphi_c$, which is the value of a at the critical state, and φ_c represents the critical friction angle. Since f_s does not vanish in general, substituting Eq. (12.3.69) into the above equation gives

$$\begin{aligned} &0 = \check{\bar{Z}}_{xx} \check{\bar{Z}}_{xy} s + a_c (2\check{\bar{Z}}_{xx} - \tfrac{1}{3}), \qquad 0 = \check{\bar{Z}}_{yy} \check{\bar{Z}}_{xy} s + a_c (2\check{\bar{Z}}_{yy} - \tfrac{1}{3}), \\ &0 = a_c^2 s + \check{\bar{Z}}_{xy}^2 s + 2 a_c \check{\bar{Z}}_{xy}, \end{aligned} \tag{12.3.71}$$

with $s \equiv \bar{D}_{xy} / \| \bar{\boldsymbol{D}} \| = \bar{D}_{xy} / |\bar{D}_{xy}|$. The only non-trivial solution to Eq. (12.3.71) is $\bar{Z}_{xx} = \bar{Z}_{yy}$ and $\bar{Z}_{xy} = -s\sqrt{8/3} \sin\varphi_c \bar{Z}_{yy}$. Thus, Eq. (12.3.5) is decoupled from other mean balance equations. For further identifications, a specific form of f_s is proposed as

$$f_s = [(1 - \bar{\nu}_s)/(1 - \bar{\nu})]^m, \qquad m = 1, \tag{12.3.72}$$

where $\bar{\nu}_s$ is the minimum mean volume fraction. The exponent m with unity value is justified for most cases. With these, the mean field equations are obtained as

$$\begin{aligned} 0 = \frac{\mathrm{d}}{\mathrm{d}y} \left[\frac{1 - \bar{\nu}_s}{1 - \bar{\nu}} \left(\zeta_2 \bar{Z}_{xy} + \zeta_3 \bar{Z}_{xy}^2 \right) + \mu_0 \bar{\gamma}^2 \left(1 + \frac{\vartheta^T}{\vartheta^M} \right) \left(\frac{\bar{\nu}_m}{\bar{\nu}_\infty - \bar{\nu}} \right)^8 \left(\frac{\mathrm{d}\bar{u}}{\mathrm{d}y} \right)^2 \right] \\ -\bar{\gamma} \bar{\nu} b \sin\theta, \end{aligned} \tag{12.3.73}$$

$$\begin{aligned} 0 = \frac{\mathrm{d}}{\mathrm{d}y} \left[2\alpha_0 \bar{\gamma} \bar{\nu}^2 (\bar{\nu} - \bar{\nu}_m) \left(1 + \frac{\vartheta^T}{\vartheta^M} \right) - \frac{1 - \bar{\nu}_s}{1 - \bar{\nu}} \left(\zeta_1 + \zeta_2 \bar{Z}_{yy} + \zeta_3 \bar{Z}_{yy}^2 \right) \right] \\ +\bar{\gamma} \bar{\nu} b \cos\theta, \end{aligned} \tag{12.3.74}$$

$$0 = \mu_0 \bar{\gamma}^2 \frac{\vartheta^T}{\vartheta^M} \left(\frac{\bar{\nu}_m}{\bar{\nu}_\infty - \bar{\nu}} \right)^8 \left(\frac{\mathrm{d}\bar{u}}{\mathrm{d}y} \right)^3 - f_4 \frac{\mathrm{d}^2 \vartheta^T}{\mathrm{d}y^2} - \bar{\gamma} \bar{\nu} \varepsilon, \tag{12.3.75}$$

$$0 = -f_2 \left(\frac{\mathrm{d}\vartheta^T}{\mathrm{d}y} \right)^2 + f_3 \left(\frac{\mathrm{d}\varepsilon}{\mathrm{d}y} \right)^2, \tag{12.3.76}$$

for the unknown mean fields $\{\bar{u}(y), \bar{\nu}(y), \vartheta^T(y), \varepsilon(y)\}$, where Eqs. (12.3.73) and (12.3.74) are respectively the mean balances of linear momentum in the x- and y-directions, Eq. (12.3.75) is the balance of turbulent kinetic energy, and Eq. (12.3.76) is the balance of turbulent dissipation.

Solid boundaries as energy sources and sinks of the turbulent kinetic energies of grains have been demonstrated in literature, and the conventional no-slip condition for velocity is not valid due to grain-slipping on boundaries. However, in the experiments the flow is accomplished by using a circulating conveyor belt with a transversal groove of trapezoidal shape in the size slightly larger than the grain size, it is assumed that on the plane the no-slip condition is valid, and the mean volume fraction approaches a fixed value. In addition, as motivated by turbulent flows of the Newtonian fluids, the turbulent kinetic energy, represented by an implicit function of the granular coldness in the study, assumes equally a fixed value, with equally a fixed value assigned to the turbulent dissipation. Since the stationary flow is accomplished by discharging a constant mass flux to the plane, the thickness of flow is fixed, and the shear force on the free surface is negligible due to the significant density difference between the grains and air. With these, the normal mean volume fraction and velocity gradients on the free surface should vanish, so that the boundary conditions are prescribed as

$$y=0: \quad \bar{u}=V_0, \ \bar{\nu}=\bar{\nu}_b, \ \vartheta^T=\vartheta_b^T, \ \varepsilon=\varepsilon_b; \qquad y=L: \quad \frac{\mathrm{d}\bar{u}}{\mathrm{d}y}=0, \quad \frac{\mathrm{d}\vartheta^T}{\mathrm{d}y}=0, \tag{12.3.77}$$

with $\{\bar{\nu}_b, \vartheta_b^T, \varepsilon_b\}$ the boundary values of $\{\bar{\nu}, \vartheta^T, \varepsilon\}$ on the solid plane, respectively; L represents the flow thickness, and V_0 denotes the velocity of inclined plane.

Non-dimensionalization. Define the dimensionless parameters, viz.,

$$\begin{aligned}
&\tilde{y}=\frac{y}{L}, \qquad \tilde{u}=\frac{\bar{u}}{V_0}, \qquad \tilde{\nu}=\frac{\bar{\nu}}{\bar{\nu}_m}, \qquad \tilde{\nu}_\infty=\frac{\bar{\nu}_\infty}{\bar{\nu}_m},\\
&\tilde{\nu}_s=\frac{\bar{\nu}_s}{\bar{\nu}_m}, \qquad \tilde{\nu}_b=\frac{\bar{\nu}_b}{\bar{\nu}_m}, \qquad \tilde{\vartheta}^T=\frac{\vartheta^T}{\vartheta^M}, \qquad \tilde{\vartheta}_b^T=\frac{\vartheta_b^T}{\vartheta^M},\\
&\tilde{\varepsilon}=\frac{\bar{\gamma}\bar{\nu}_m L^2\varepsilon}{f_4\vartheta^M}, \quad \tilde{\varepsilon}_b=\frac{\bar{\gamma}\bar{\nu}_m L^2\varepsilon_b}{f_4\vartheta^M}, \quad S_1=\frac{\bar{\nu}_m b L^3}{\mu_0\bar{\gamma}V_0^2}, \quad S_2=\frac{bL}{\alpha_0\bar{\nu}_m^2}, \ \chi=\frac{\bar{\gamma}^2V_0^3\mu_0}{f_4L\vartheta^M},\\
&\xi=\frac{f_2\bar{\gamma}^2\bar{\nu}_m^2L^4}{f_3(f_4)^2}, \quad \Xi_1=\frac{(\zeta_2\bar{Z}_{xy}+\zeta_3\bar{Z}_{xy}^2)L^2}{\mu_0\bar{\gamma}^2V_0^2}, \quad \Xi_2=\frac{(\zeta_1+\zeta_2\bar{Z}_{yy}+\zeta_3\bar{Z}_{yy}^2)}{\alpha_0\bar{\gamma}\bar{\nu}_m^3}.
\end{aligned} \tag{12.3.78}$$

Substituting Eq. (12.3.78) into Eqs. (12.3.73)–(12.3.77) results in the dimensionless mean field equations given by

$$0=\frac{\mathrm{d}}{\mathrm{d}\tilde{y}}\left[\Xi_1\frac{1-\bar{\nu}_m\tilde{\nu}_s}{1-\bar{\nu}_m\tilde{\nu}}+\frac{1+\tilde{\vartheta}^T}{(\tilde{\nu}_\infty-\tilde{\nu})^8}\left(\frac{\mathrm{d}\tilde{u}}{\mathrm{d}\tilde{y}}\right)^2\right]-S_1\tilde{\nu}\sin\theta, \tag{12.3.79}$$

$$0=\frac{\mathrm{d}}{\mathrm{d}\tilde{y}}\left[2\tilde{\nu}^2(\tilde{\nu}-1)(1+\tilde{\vartheta}^T)-\Xi_2\frac{1-\bar{\nu}_m\tilde{\nu}_s}{1-\bar{\nu}_m\tilde{\nu}}\right]+S_2\tilde{\nu}\cos\theta, \tag{12.3.80}$$

$$0=\frac{\chi\tilde{\vartheta}^T}{(\tilde{\nu}_\infty-\tilde{\nu})^8}\left(\frac{\mathrm{d}\tilde{u}}{\mathrm{d}\tilde{y}}\right)^3-\frac{\mathrm{d}^2\tilde{\vartheta}^T}{\mathrm{d}\tilde{y}^2}-\tilde{\nu}\tilde{\varepsilon}, \tag{12.3.81}$$

$$0=-\xi\left(\frac{\mathrm{d}\tilde{\vartheta}^T}{\mathrm{d}\tilde{y}}\right)^2+\left(\frac{\mathrm{d}\tilde{\varepsilon}}{\mathrm{d}\tilde{y}}\right)^2, \tag{12.3.82}$$

for $\{\tilde{u}(\tilde{y}), \tilde{\nu}(\tilde{y}), \tilde{\vartheta}^T(\tilde{y}), \tilde{\varepsilon}(\tilde{y})\}$, associated with the dimensionless boundary conditions in the forms

$$\tilde{y}=0: \quad \tilde{u}=1,\ \tilde{\nu}=\tilde{\nu}_b,\ \tilde{\vartheta}^T=\tilde{\vartheta}^T_b,\ \tilde{\varepsilon}=\tilde{\varepsilon}_b; \qquad \tilde{y}=1: \quad \frac{\mathrm{d}\tilde{u}}{\mathrm{d}\tilde{y}}=0,\ \frac{\mathrm{d}\tilde{\vartheta}^T}{\mathrm{d}\tilde{y}}=0. \tag{12.3.83}$$

Equations (12.3.79)–(12.3.83) define a BVP, in which S_2 denotes the combined effect of gravity and flow thickness; S_1 represents the influence of viscosity under a fixed value of S_2; both Ξ_1 and Ξ_2 denote the effect of hypoplastic-related forces; χ accounts for the characteristics of viscosity with respect to turbulent kinetic energy flux; and ξ denotes the relative significance between turbulent dissipation production and turbulent kinetic energy flux. For implementations of numerical simulation, the values of $\{\bar{\nu}_b, \bar{\nu}_m, \bar{\nu}_\infty, \bar{\nu}_s\}$ are given by

$$\bar{\nu}_b=0.51, \qquad \bar{\nu}_m=0.555, \qquad \bar{\nu}_\infty=0.644, \qquad \bar{\nu}_s=0.25, \tag{12.3.84}$$

followed which

$$\tilde{\nu}_b=0.919, \qquad \tilde{\nu}_\infty=1.16, \qquad \tilde{\nu}_s=0.451. \tag{12.3.85}$$

The values of ϑ^T_b and ε_b on the plane remain undetermined. However, non-vanishing but finite values of ϑ^T_b and ε_b are used in the analyses for simplicity, as motivated by the findings of the Newtonian fluids in turbulent shear flows, and the experimental and field observations of granular systems. The defined BVP and associated boundary conditions are solved by using the iterative method described in Sect. 12.2.6.

Numerical results. The parameter S_1 can be expressed as $S_1=\mathcal{A}S_2$, with $\mathcal{A}=\alpha_0\bar{\nu}_m^3 L^2/(\mu_0\bar{\gamma}V_0^2)$. Numerical tests have shown that only the relative magnitudes of $\tilde{\nu}$-, $\tilde{u}$-, turbulent kinetic energy and dissipation profiles are influenced by the values of $\mathcal{A}$, but the tendencies remain unchanged. In addition, the calculated results are influenced to a small extent by changing the values of χ and ξ. Thus, $\{\mathcal{A}, \chi, \xi\}$ are chosen to be constant in the calculations for simplicity, with $\theta=15.6°$ to match the experimental setup. Since the parameters Ξ_1 and Ξ_2 are of equal importance, their values are set equal. In Figs. 12.11-12.14, the horizontal axes denote the values of $\tilde{\nu}$, $\tilde{u}$, dimensionless turbulent kinetic energy and dissipation, while the vertical axes represent the distance $\tilde{y}$ from the solid plane.

Figure 12.11 illustrates the profiles of $(1-\tilde{\nu})$, $\tilde{u}$, and dimensionless turbulent kinetic energy and dissipation, in which $S_2=0.01$, $\Xi_1=\Xi_2=0.01$, $\tilde{\varepsilon}_b=0.015$, $\chi=0.01$, $\xi=0.5$ and $\tilde{\vartheta}^T_b=[0.01, 0.02, 0.03]$, as indicated by the arrows. The dashed lines represent the laminar flow solutions.[15] For comparison, the profiles of mean porosity $\tilde{\epsilon}$, defined by $\tilde{\epsilon}\equiv 1-\tilde{\nu}$, are presented. Figure 12.11c shows that the mean porosity increases from the solid boundary toward free surface with an "exponential-like" tendency, with larger increasing rate when approaching the free

[15]The experimental results are quoted from Perng, A.T.H., Capart, H., Chou, H.T., Granular configurations, motions, and correlations in slow uniform flows driven by an inclined conveyor belt, *Granular Matter*, 8, 5–17, 2006. The results of zeroth-order model are quoted from Fang, C., Wu, W., On the weak turbulent motions of an isothermal dry granular dense flow with incompressible grains: part II. Complete closure models and numerical simulations, *Acta Geotechica*, 9(5), 739–752, 2014.

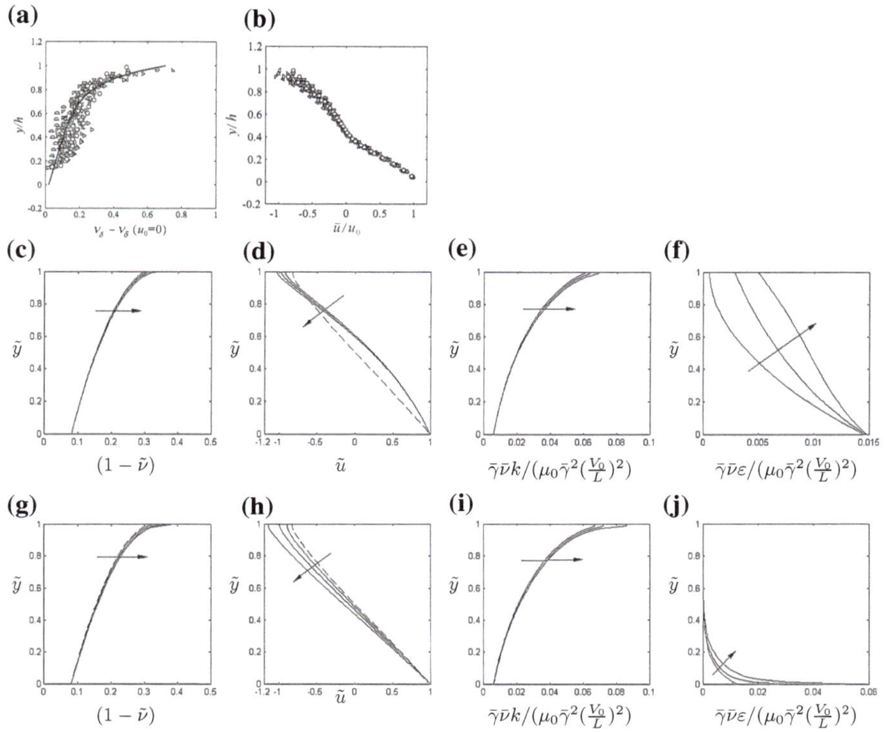

Fig. 12.11 Profiles of $(1-\tilde{\nu})$, $\tilde{u}$, $\bar{\gamma}\bar{\nu}k$ and $\bar{\gamma}\bar{\nu}\varepsilon$, in which $\Xi_1 = \Xi_2 = 0.01$, $S_2 = 0.01$, $\chi = 0.01$, $\tilde{\varepsilon}_b = 0.015$, $\xi = 0.5$, and $\tilde{\vartheta}_b^T = [0.01, 0.02, 0.03]$ indicated by the arrows. **a**, **b**: Experimental results. **c-f**: The results from the first-order model. **g-j**: The results from the zeroth-order model. Dashed lines: laminar flow solutions

surface. This reflects that the grains near the solid boundary are dominated by the long-term grain-grain interaction, in which they form a kind of inelastic network. On the other hand, the grains near the free surface collide with one another intensively, resulting in dominant short-term grain-grain interaction with significant turbulent fluctuation and larger mean porosity. The profiles shown in Fig. 12.11d show that $\tilde{u}$ decreases monotonically form the solid boundary toward free surface in a rather nonlinear way due to the combined influence of gravity and plane shearing, a distinct non-Newtonian feature. The profiles of $(1-\tilde{\nu})$ and $\tilde{u}$ correspond qualitatively to the experimental measurements shown in Figs. 12.11a and b. The bends in the porosity and velocity profiles near the free surface and central region of the test data result from the facts that the grains experience free collisions, causing larger grain mean free path there, while near the central region the grains experience significant sliding. In these two regions, the grains exhibit distinct non-continuum characteristics, which are hardly described by the established model. However, the established model is able to describe the global characteristics of porosity and velocity variations.

The profiles of turbulent kinetic energy and dissipation are shown respectively in Figs. 12.11e and f, in which their values are divided by the mean shear rate

$\mu_0 \bar{\gamma}^2 (V_0/L)^2$. The dimensionless turbulent dissipation decreases from the maximum value on the solid plane toward minimum value on the free surface, while the dimensionless turbulent kinetic energy evolves in a reverse manner. These findings correspond to different dominant grain-grain interactions near the solid boundary and free surface, and the turbulent dissipation profile demonstrates a similarity to that of the Newtonian fluids in turbulent shear flows. Increasing $\tilde{\vartheta}_b^T$ is to provide more turbulent kinetic energy from the solid plane to the granular body, inducing enhanced turbulent dissipation across the flow layer with slightly enhanced turbulent kinetic energy near the free surface, as respectively shown in Figs. 12.11e and f. These are equally illustrated in Figs. 12.11c and d by more convex mean porosity and velocity profiles when $\tilde{\vartheta}_b^T$ increases.

While the profiles of $(1-\tilde{\nu})$ and dimensionless turbulent kinetic energy from the first- and zeroth-order models are similar, those from the first-order model assume smaller amplitudes than the zeroth-order model near the free surface. This results from that a "local" balance is imposed between the turbulent kinetic energy and dissipation in the zeroth-order model, while in the first-order model, the turbulent eddy evolution is taken into account, resulting in more efficient turbulent kinetic energy transfer across the flow layer. The mean shear and stress power of solid plane are more efficiently transferred toward the free surface, giving rise to a discrepancy in the $\tilde{u}$-profiles in the central regions by comparing Fig. 12.11d with Fig. 12.11h. The influence of turbulent eddy evolution is equally manifest in the profiles of dimensionless turbulent dissipation, by comparing Fig. 12.11f with Fig. 12.11j. When compared with the experimental outcomes, the zeroth-order model delivers more accurate estimations on the mean porosity and velocity distributions than the first-order model. This is so, because the turbulent shear flows accomplished in the experiments deviate only slightly from the laminar state, for which the zeroth-order model is more appropriate. The deviations between the first- and zeroth-order models, in particular in the estimated profiles of mean porosity and dimensionless turbulent dissipation, deliver the significant influence of turbulent eddy evolution on the mean flow characteristics.

Figure 12.12 illustrates the influence of variations in S_2 on the profiles of $(1-\tilde{\nu})$, $\tilde{u}$ and dimensionless turbulent kinetic energy and dissipation, in which $\Xi_1 = \Xi_2 = 0.01$, $\tilde{\vartheta}_b^T = 0.02$, $\tilde{\varepsilon}_b = 0.015$, $\chi = 0.01$, $\xi = 0.5$ and $S_2 = [0.001, 0.004, 0.008]$, as indicated by the arrows. Increasing S_2 tends to enhance the gravitational effect, resulting in larger turbulent fluctuation (and hence larger turbulent kinetic energy) near the free surface, while near the solid plane, the turbulent dissipation is more dominant due to the plane shearing, corresponding to different dominant grain-grain interactions. When S_2 is small, the dominant interaction between the grains is the long-term one, yielding that the $(1-\tilde{\nu})$- and $\tilde{u}$-profiles deviate slightly from their laminar counterparts. When S_2 is large, the dominant grain-grain interaction is the short-term one, in particular near the free surface, causing larger discrepancies in the $(1-\tilde{\nu})$- and $\tilde{u}$-profiles from the laminar flow solutions. The difference between two closure models is seen in the profiles of $\tilde{u}$ and dimensionless turbulent dissipation, as shown in Fig. 12.12b with Fig. 12.12f, and Fig. 12.12d with Fig. 12.12h, resulted from the influence of enhanced turbulent eddy evolution for larger values of S_2.

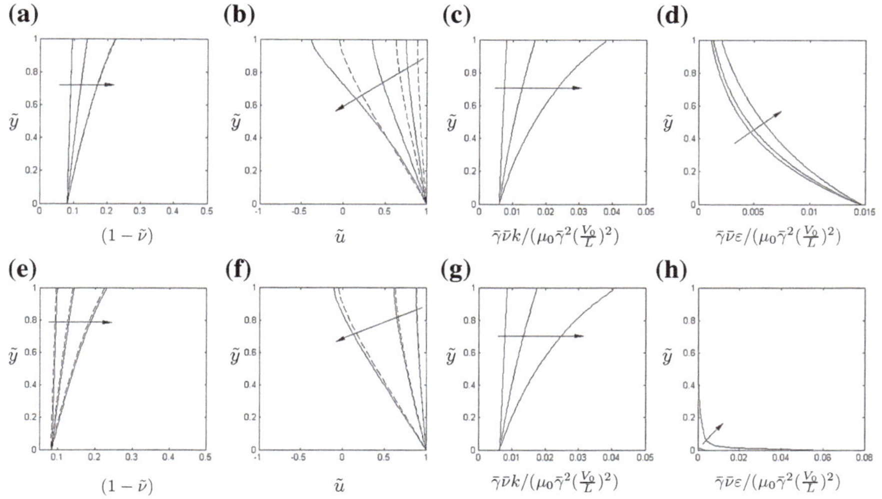

Fig. 12.12 Profiles of $(1-\tilde{\nu})$, $\tilde{u}$, $\bar{\gamma}\bar{\nu}k$ and $\bar{\gamma}\bar{\nu}\varepsilon$ for variations in S_2, in which $\Xi_1 = \Xi_2 = 0.01$, $\chi = 0.01, \tilde{\varepsilon}_b = 0.015, \xi = 0.5, \tilde{\vartheta}_b^T = 0.02$ and $S_2 = [0.001, 0.004, 0.008]$ indicated by the arrows. **a-d**: The results from the first-order model. **e-h**: The results from the zeroth-order model. Dashed lines: laminar flow solutions

Numerical simulations for variations in Ξ_1 and Ξ_2 are summarized in Fig. 12.13, in which $\tilde{\vartheta}_b^T = 0.01$, $\tilde{\varepsilon}_b = 0.015$, $S_2 = 0.01$, $\chi = 0.01$, $\xi = 0.5$ and $\Xi_1 = \Xi_2 = [0.001, 0.005, 0.01]$, as indicated by the arrows. When Ξ_1 and Ξ_2 increase, the hypoplastic effect inside a granular RVE is enhanced, which tends to generate more energy dissipation near the solid boundary where the shearing is maximum, resulting in more intensive turbulent dissipation, as illustrated in Figs. 12.13d and h for the first- and zeroth-order models, respectively. Since the turbulent fluctuation induced by the mean shearing and gravitational influence is fixed through fixed values of S_2 and $\tilde{\vartheta}_b^T$, the enhanced hypoplastic effect is confined in the regions near the solid boundary in the zeroth-order model shown in Fig. 12.13h. The estimated turbulent dissipation from the first-order model, however, is more efficiently distributed across the flow layer due to the influence of turbulent eddy evolution, as shown in Fig. 12.13d. This feature is equally manifest in the $(1-\tilde{\nu})$- and $\tilde{u}$-profiles, as displayed in Fig. 12.13a with Fig. 12.13e, and Fig. 12.13b with Fig. 12.13f, respectively. On the contrary, the profiles of turbulent kinetic energy are only affected slightly by increasing Ξ_1 and Ξ_2 in both models, as shown in Figs. 12.13c and g.

The influence of variations in $\tilde{\varepsilon}_b$ on the profiles of $(1-\tilde{\nu})$, $\tilde{u}$ and dimensionless turbulent kinetic energy and dissipation is summarized in Fig. 12.14, in which $\Xi_1 = \Xi_2 = 0.01, \xi = 0.5, S_2 = 0.01, \chi = 0.01, \tilde{\vartheta}_b^T = 0.01$, and $\tilde{\varepsilon}_b = [0.015, 0.03, 0.045]$, as indicated by the arrows. Since in the zeroth-order model such a parameter variation is not possible, only the results of first-order model are shown. Increasing $\tilde{\varepsilon}_b$ tends to induce more convex turbulent dissipation profiles across the flow layer with larger amplitudes near the solid plane, as shown in Fig. 12.14d. This is due to that

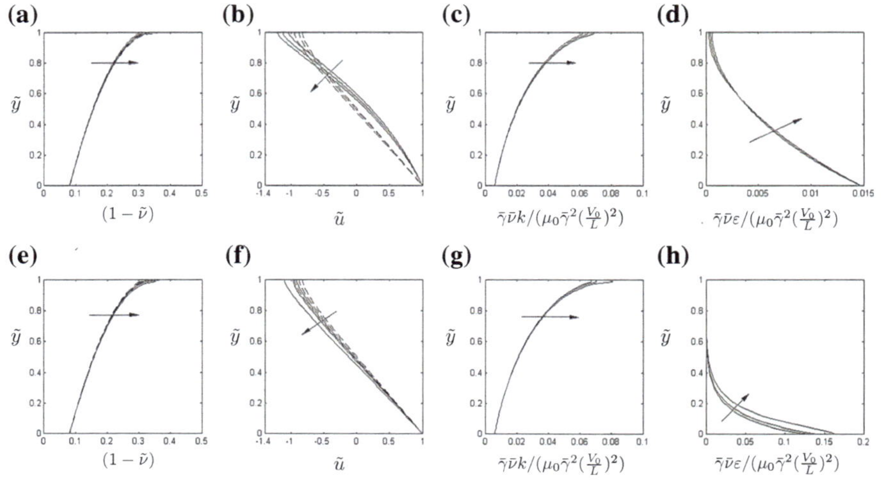

Fig. 12.13 Profiles of $(1-\tilde{\nu})$, $\tilde{u}$, $\bar{\gamma}\bar{\nu}k$ and $\bar{\gamma}\bar{\nu}\varepsilon$ for variations in Ξ_1 and Ξ_2, in which $S_2 = 0.01$, $\chi = 0.01$, $\tilde{\varepsilon}_b = 0.015$, $\xi = 0.5$, $\tilde{\vartheta}_b^T = 0.01$ and $\Xi_1 = \Xi_2 = [0.001, 0.005, 0.01]$ indicated by the arrows. **a**-**d**: The results from the first-order model. **e**-**h**: The results from the zeroth-order model. Dashed lines: laminar flow solutions

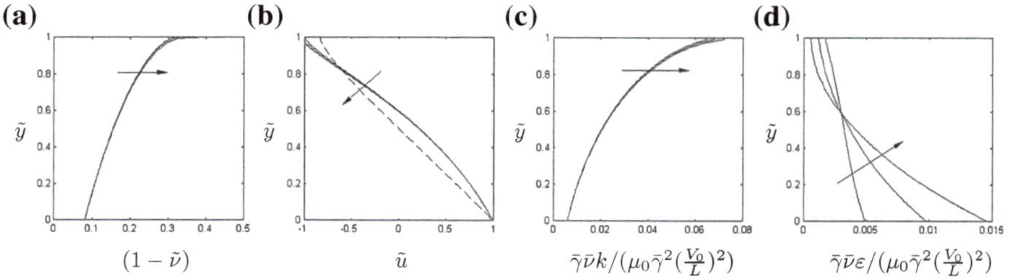

Fig. 12.14 Profiles of $(1-\tilde{\nu})$, $\tilde{u}$, $\bar{\gamma}\bar{\nu}k$ and $\bar{\gamma}\bar{\nu}\varepsilon$ for variations in $\tilde{\varepsilon}_b$, in which $S_2 = 0.01$, $\chi = 0.01$, $\xi = 0.5$, $\tilde{\vartheta}_b^T = 0.01$, $\Xi_1 = \Xi_2 = 0.01$ and $\tilde{\varepsilon}_b = [0.015, 0.03, 0.045]$ indicated by the arrows. **a**-**d**: The results from the first-order model. **e**-**h**: The results from the zeroth-order closure model. Dashed lines: laminar flow solutions

the turbulent dissipation results from the combined interactions between the mean shearing and turbulent eddy evolution at different length and timescales. Increasing $\tilde{\varepsilon}_b$ is to induce more intensive turbulent eddy evolution near the solid boundary, while near the free surface, the turbulent eddy is almost absent. This feature is also reflected by the $(1-\tilde{\nu})$-, $\tilde{u}$- and dimensionless turbulent kinetic energy profiles illustrated in Figs. 12.14a-c, respectively.

For various variations in the parameters, the dimensionless turbulent dissipation estimated by the first-order model evolves from the maximum value on the solid plane toward minimum and *finite* value on the free surface. This finding not only corresponds to the experimental outcomes of the Newtonian fluids in turbulent shear flows, but also is justified, for the observations of slow flows imply fixed and finite values of the turbulent kinetic energy (and hence fixed and finite values of the turbulent dissipation) on the free surface due to the dispersive grain-grain collisions. This phenomenon, however, is barely recognized in the zeroth-order model, for the estimated turbulent dissipation approaches null on the free surface, a physically unsatisfactory result. Moreover, the small differences between the laminar flow solutions and solutions of the zeroth- and first-order models, in particular in the porosity profiles, result from the considered stationary flow. In fact, the considered flow is a simple plane shear flow slightly deviated from a purely laminar one, in which the turbulent characteristics are nearly absent. For other time-dependent flows or flows with complicated geometry, more distinguished differences may be expected.

Conclusions. The mean porosity and velocity profiles correspond qualitatively to the experimental outcomes. Due to the dominant long-term grain-grain interaction near the solid boundary, the grains interlock with one another to form a kind of loosely inelastic network, with more intensive turbulent dissipation. On the contrary, near the free surface the grains are dominated by the short-term grain-grain interaction, resulting in more intensive turbulent fluctuation with more intensive turbulent kinetic energy. These are demonstrated by the more convex mean porosity and velocity profiles between the free and solid boundaries. Although the grains form a kind of inelastic networks and interact with one another via the dominant long-term grain-grain interaction, the turbulent fluctuation, via the turbulent eddy evolutions, may drive the grains to distribute in a manner that the flux of mean volume fraction deviates from that of turbulent fluctuating kinetic energy. This deviation allows more turbulent kinetic energy to be transferred across the flow layer, resulting in significant influence on the mean flow features.

While the turbulent dissipation evolves from its maximum value on the solid plane toward minimum value on the free surface, the turbulent kinetic energy distributes in a reverse manner. These findings correspond not only to different dominant grain-grain interactions, but also to the findings of the Newtonian fluids in stationary turbulent shear flows. Comparison between the numerical and experimental results suggests that the influence of turbulent fluctuation needs be accounted for even the flow speed is small in two aspects: (a) the turbulent nature of dry granular system; and (b) the more accurate estimations on the mean porosity and velocity.

When compared with the zeroth-order model, the first-order model allows the influence of turbulent eddy evolution to be taken into account to some extent. The turbulent kinetic energy and plane shearing are more efficiently transferred across the flow layer, resulting in the velocity profiles with larger amplitudes in the central and upper regions. In contrast to the zeroth-order model, the solid boundary is shown to act as an energy source of the turbulent kinetic energy through the prescription of $\tilde{\vartheta}^T_b$, and as an energy source for the turbulent dissipation through the prescription of $\tilde{\varepsilon}_b$. In addition, the turbulent dissipation assumes a finite value on the free surface, which, when compared with the zeroth-order model, corresponds better to the observations. It is suggested that the zeroth-order model is sufficient to account for the influence of turbulent fluctuation in turbulent creeping flows. For dense flows, the first-order model is more appropriate to account for the influence induced by turbulent eddy evolution, with solid boundaries apparently as energy source and sink of the turbulent kinetic energies of solid grains.

12.4 Exercises

12.1 A cup is filled with dry sand, where the diameter and density of a typical sand particle are nearly 100 μm and 2500 kg/m^3, respectively. Use a simple energy balance from the statistical mechanics to show that a sand sample needs to be heated up to nearly 10^{11} K to allow a single sand particle to be elevated by a particle diameter. This example shows that for a dry granular matter, the influence of temperature variation needs not to be taken into account in normal operation conditions.

12.2 Consider an array of circular disks with diameter r in densest packing circumstance. Show that the maximum increase in area of the array is given by $\Delta A = 0.268rL$ if the array is under a simple plane shear, where L represents the characteristic length of the array.

12.3 Consider a cylindrical container with radius r filled with a dry sand up to the height h. Let the origin of the coordinate z be placed on the sand surface, and z point vertically downward. Use a simple force balance to show that the pressure of sand at a specific value of $z \ll h$ is given by

$$p(z) = p_\infty \left[1 - \exp\left(-\frac{z}{\lambda}\right)\right],$$

where p_∞ is the pressure at the container bottom, λ represents a characteristic length given by $\lambda = r/(2\mu K)$, μ denotes the static friction coefficient between sand and container wall, and K stands for the coefficient of redirection transporting a vertical load into a horizontal direction. This equation is referred to as the *Janssen model*, which describes the pressure distribution in a silo filled with spherical grains.

12.4 Complete the derivation of Eq. (12.2.21) by substituting Eqs. (12.2.16) and (12.2.17) into Eq. (12.2.20) with the chain rule of differentiation.

12.5 Complete the derivation of Eq. (12.3.15) by substituting Eqs. (12.3.11)–(12.3.12) and (12.3.14) into Eq. (12.3.13) with the chain rule of differentiation.

Further Reading

S.J. Antony, W. Hoyle, Y. Ding (eds.), *Granular Materials: Fundamentals and Applications* (The Royal Society of Chemistry, Cambridge, 2004)

I.S. Aranson, L.S. Tsimring, *Granular Pattern* (Oxford University Press, Oxford, 2009)

T. Aste, T.D. Matteo, A. Tordesillas (eds.), *Granular and Complex Materials* (World Scientific, New Jersey, 2007)

D. Bideau, A. Hansen (eds.), *Disorder and Granular Media* (North-Holland, Amsterdam, 1993)

G. Capriz, P. Giovine, P.M. Mariano (eds.), *Mathematical Models of Granular Matter* (Springer, Berlin, 2008)

P. Coussot, *Mudflows Rheology and Dynamics* (A.A. Balkema, Rotterdam, 1997)

D.A. Drew, D.D. Joseph, S.L. Passman (eds.), *Particulate Flows: Processing and Rheology* (Springer, Berlin, 1998)

J. Duran, *Sands, Powders, and Grains: An Introduction to the Physics of Granular Materials* (Springer, Berlin, 2000)

C. Fang, Gravity-driven dry granular slow flows down an inclined moving plane: a comparative study between two concepts of the evolution of porosity. *Rheological Acta* **48**, 971–992 (2009)

C. Fang, Rheological characteristics of solid-fluid transition in dry granular dense flows: a thermodynamically consistent constitutive model with a pressure ratio order parameter. *Int. J. Numer. Anal. Methods Geomech.* **34**(9), 881–905 (2010)

C. Fang, A k-ε turbulent closure model of an isothermal dry granular dense matter. *Contin. Mech. Thermodyn.* **28**(4), 1049–1069 (2016)

K. Hutter, N. Kirchner (eds.), *Dynamic Response of Granular and Porous Materials under Large and Catastrophic Deformations* (Springer, Berlin, 2003)

K. Hutter, K. Wilmánski (eds.), *Kinetic and Continuum Theories of Granular and Porous Media* (Springer, Berlin, 1999)

K. Iwashita, M. Oda (eds.), *Mechanics of Granular Materials: An Introduction* (A.A. Balkema, Rotterdam, 1999)

M. Jakob, O. Hungr (eds.), *Debris-Flow Hazards and Related Phenomena* (Springer, Berlin, 2005)

D. Kolymbas (ed.), *Constitutive Modeling of Granular Materials* (Springer, Berlin, 2000)

A. Mehta, *Granular Physics* (Cambridge University Press, Cambridge, 2007)

T. Pöschel, N. Brilliantov (eds.), *Granular Gas Dynamics* (Springer, Berlin, 2003)

S. Pudasaini, K. Hutter, *Avalanche Dynamics* (Springer, Berlin, 2007)

K.K. Rao, P.R. Nott, *An Introduction to Granular Flows* (Cambridge University Press, Cambridge, 2008)

A.F. Revuzhenko, *Mechanics of Granular Media* (Springer, Berlin, 2006)

G.H. Ristow, *Pattern Formation in Granular Materials* (Springer, Berlin, 2000)

L. Schneider, K. Hutter, *Solid-Fluid Mixtures of Frictional Materials in Geophysical and Geotechnical Context* (Springer, Berlin, 2009)

T. Takahashi, *Debris Flows: Mechanics, Prediction and Countermeasures* (Taylor & Francis, London, 2007)

Orthogonal Curvilinear Coordinates

A

In three-dimensional circumstance, the orthogonal curvilinear coordinates q_1, q_2 and q_3 are defined by

$$q_i = q_i(x_1, x_2, x_3), \qquad i = 1\text{-}3, \tag{A.1}$$

where x_i are the Cartesian coordinates. This equation is assumed to have a unique inverse, so that

$$x_i = x_i(q_1, q_2, q_3), \quad i = 1\text{-}3, \qquad \longrightarrow \qquad \boldsymbol{x} = \boldsymbol{x}(q_j). \tag{A.2}$$

For constant values of q_2 and q_3, the equation $\boldsymbol{x} = \boldsymbol{x}(q_1)$ describes a curve in space which is the coordinate curve q_1, and $\partial\boldsymbol{x}/\partial q_1$ denotes a tangent vector to this curve. It follows that the corresponding unit vector in the direction of increasing q_1 is given by

$$\boldsymbol{e}_1 = \frac{\partial\boldsymbol{x}/\partial q_1}{\|\partial\boldsymbol{x}/\partial q_1\|}, \qquad \longrightarrow \qquad \frac{\partial\boldsymbol{x}}{\partial q_1} = \left\|\frac{\partial\boldsymbol{x}}{\partial q_1}\right\|\boldsymbol{e}_1 = h_1\boldsymbol{e}_1, \qquad h_1 = \left\|\frac{\partial\boldsymbol{x}}{\partial q_1}\right\|. \tag{A.3}$$

Such a procedure can be repeated to obtain $\partial\boldsymbol{x}/\partial q_2 = h_2\boldsymbol{e}_2$ and $\partial\boldsymbol{x}/\partial q_3 = h_3\boldsymbol{e}_3$, where $h_2 = \|\partial\boldsymbol{x}/\partial q_2\|$ and $h_3 = \|\partial\boldsymbol{x}/\partial q_3\|$. The coefficients $\{h_1, h_2, h_3\}$ are called the *metric-scale factors*.

It follows from Eq. (A.2) that

$$\mathrm{d}\boldsymbol{x} = \frac{\partial\boldsymbol{x}}{\partial q_j}\mathrm{d}q_j = \left(h_j\mathrm{d}q_j\right)\boldsymbol{e}_i, \qquad \mathrm{d}\boldsymbol{x}\cdot\mathrm{d}\boldsymbol{x} = \left(h_j\mathrm{d}q_j\right)^2, \tag{A.4}$$

which represents the square of a line element. Consider a volume element in space spanned by the coordinates $\{q_1, q_2, q_3\}$, which is given by

$$\mathrm{d}v = \mathrm{d}\boldsymbol{x}_1\cdot(\mathrm{d}\boldsymbol{x}_2\times\mathrm{d}\boldsymbol{x}_3) = h_1h_2h_3\mathrm{d}q_1\mathrm{d}q_2\mathrm{d}q_3, \tag{A.5}$$

if the line elements $\mathrm{d}\boldsymbol{x}_1$, $\mathrm{d}\boldsymbol{x}_2$, and $\mathrm{d}\boldsymbol{x}_3$ are assumed to be linearly independent to one another, with $\mathrm{d}\boldsymbol{x}_1 = (h_1\mathrm{d}q_1)\boldsymbol{e}_1$, $\mathrm{d}\boldsymbol{x}_2 = (h_2\mathrm{d}q_2)\boldsymbol{e}_2$, and $\mathrm{d}\boldsymbol{x}_3 = (h_3\mathrm{d}q_3)\boldsymbol{e}_3$. For the considered volume element, the corresponding surface elements are obtained as

$$\begin{aligned}&\mathrm{d}\boldsymbol{a}_1 = \mathrm{d}\boldsymbol{x}_2\times\mathrm{d}\boldsymbol{x}_3 = \left(h_2h_3\mathrm{d}q_2\mathrm{d}q_3\right)\boldsymbol{e}_1, \mathrm{d}\boldsymbol{a}_2 = \mathrm{d}\boldsymbol{x}_1\times\mathrm{d}\boldsymbol{x}_3 = \left(h_1h_3\mathrm{d}q_1\mathrm{d}q_3\right)\boldsymbol{e}_2,\\ &\mathrm{d}\boldsymbol{a}_3 = \mathrm{d}\boldsymbol{x}_1\times\mathrm{d}\boldsymbol{x}_2 = \left(h_1h_2\mathrm{d}q_1\mathrm{d}q_2\right)\boldsymbol{e}_3.\end{aligned} \tag{A.6}$$

© Springer International Publishing AG 2019

C. Fang, *An Introduction to Fluid Mechanics*, Springer Textbooks in Earth Sciences, Geography and Environment, https://doi.org/10.1007/978-3-319-91821-1

Let ϕ be any scalar function, $\boldsymbol{v}$ be any vector function, and $\boldsymbol{T}$ be any second-order tensor, which are functions of the orthogonal coordinates $\{q_1, q_2, q_3\}$. The general formulations of gradient, divergence, curl, and the Laplacian and Lagrangian derivative for $\{\phi, \boldsymbol{v}, \boldsymbol{T}\}$ are summarized in the following.

- Gradient of ϕ:

$$\text{grad}\,\phi = \frac{1}{h_1}\frac{\partial \phi}{\partial q_1}\boldsymbol{e}_1 + \frac{1}{h_2}\frac{\partial \phi}{\partial q_2}\boldsymbol{e}_2 + \frac{1}{h_1}\frac{\partial \phi}{\partial q_3}\boldsymbol{e}_3. \tag{A.7}$$

- Gradient of $\boldsymbol{v}$:

$$(\text{grad}\,\boldsymbol{v})_{11} = \frac{1}{h_1}\frac{\partial v_1}{\partial q_1} + \frac{v_2}{h_1 h_2}\frac{\partial h_1}{\partial q_2} + \frac{v_3}{h_1 h_3}\frac{\partial h_1}{\partial q_3}, \tag{A.8}$$

$$(\text{grad}\,\boldsymbol{v})_{22} = \frac{1}{h_2}\frac{\partial v_2}{\partial q_2} + \frac{v_3}{h_2 h_3}\frac{\partial h_2}{\partial q_3} + \frac{v_1}{h_1 h_2}\frac{\partial h_2}{\partial q_1}, \tag{A.9}$$

$$(\text{grad}\,\boldsymbol{v})_{33} = \frac{1}{h_3}\frac{\partial v_3}{\partial q_3} + \frac{v_1}{h_1 h_3}\frac{\partial h_3}{\partial q_1} + \frac{v_2}{h_2 h_3}\frac{\partial h_3}{\partial q_2}, \tag{A.10}$$

$$2(\text{grad}\,\boldsymbol{v})_{32} = \frac{h_3}{h_2}\frac{\partial (v_3/h_3)}{\partial q_2} + \frac{h_2}{h_3}\frac{\partial (v_2/h_2)}{\partial q_3} = 2(\text{grad}\,\boldsymbol{v})_{23}, \tag{A.11}$$

$$2(\text{grad}\,\boldsymbol{v})_{13} = \frac{h_1}{h_3}\frac{\partial (v_1/h_1)}{\partial q_3} + \frac{h_3}{h_1}\frac{\partial (v_3/h_3)}{\partial q_1} = 2(\text{grad}\,\boldsymbol{v})_{31}, \tag{A.12}$$

$$2(\text{grad}\,\boldsymbol{v})_{21} = \frac{h_2}{h_1}\frac{\partial (v_2/h_2)}{\partial q_1} + \frac{h_1}{h_2}\frac{\partial (v_1/h_1)}{\partial q_2} = 2(\text{grad}\,\boldsymbol{v})_{12}, \tag{A.13}$$

in which (grad $\boldsymbol{v}$) is assumed to be symmetric.

- Divergence of $\boldsymbol{v}$:

$$\text{div}\,\boldsymbol{v} = \frac{1}{h_1 h_2 h_3}\left\{\frac{\partial}{\partial q_1}(h_2 h_3 v_1) + \frac{\partial}{\partial q_2}(h_1 h_3 v_2) + \frac{\partial}{\partial q_3}(h_1 h_2 v_3)\right\}. \tag{A.14}$$

- Divergence of $\boldsymbol{T}$:

$$(\text{div}\,\boldsymbol{T})|_1 = \left\{\frac{1}{h_1 h_2 h_3}\left[\frac{\partial}{\partial q_1}(h_2 h_3 T_{11}) + \frac{\partial}{\partial q_2}(h_1 h_3 T_{21}) + \frac{\partial}{\partial q_3}(h_1 h_2 T_{31})\right]\right.$$
$$\left. + \frac{T_{21}}{h_1 h_2}\frac{\partial h_1}{\partial q_2} + \frac{T_{31}}{h_1 h_3}\frac{\partial h_1}{\partial q_3} - \frac{T_{22}}{h_1 h_2}\frac{\partial h_2}{\partial q_1} - \frac{T_{33}}{h_1 h_3}\frac{\partial h_3}{\partial q_1}\right\}\boldsymbol{e}_1, \tag{A.15}$$

$$(\text{div}\,\boldsymbol{T})|_2 = \left\{\frac{1}{h_1 h_2 h_3}\left[\frac{\partial}{\partial q_1}(h_2 h_3 T_{12}) + \frac{\partial}{\partial q_2}(h_1 h_3 T_{22}) + \frac{\partial}{\partial q_3}(h_1 h_2 T_{32})\right]\right.$$
$$\left. + \frac{T_{32}}{h_2 h_3}\frac{\partial h_2}{\partial q_3} + \frac{T_{12}}{h_1 h_2}\frac{\partial h_2}{\partial q_1} - \frac{T_{33}}{h_2 h_3}\frac{\partial h_3}{\partial q_2} - \frac{T_{11}}{h_1 h_2}\frac{\partial h_1}{\partial q_2}\right\}\boldsymbol{e}_2, \tag{A.16}$$

$$(\text{div}\,\boldsymbol{T})|_3 = \left\{\frac{1}{h_1 h_2 h_3}\left[\frac{\partial}{\partial q_1}(h_2 h_3 T_{13}) + \frac{\partial}{\partial q_2}(h_1 h_3 T_{23}) + \frac{\partial}{\partial q_3}(h_1 h_2 T_{33})\right]\right.$$
$$\left. + \frac{T_{13}}{h_1 h_3}\frac{\partial h_3}{\partial q_1} + \frac{T_{23}}{h_2 h_3}\frac{\partial h_3}{\partial q_2} - \frac{T_{11}}{h_1 h_3}\frac{\partial h_1}{\partial q_3} - \frac{T_{22}}{h_2 h_3}\frac{\partial h_2}{\partial q_3}\right\}\boldsymbol{e}_3. \tag{A.17}$$

- Curl of $\boldsymbol{v}$:

$$\text{curl}\,\boldsymbol{v} = \frac{1}{h_1h_2h_3}\begin{vmatrix} h_1\boldsymbol{e}_1 & h_2\boldsymbol{e}_2 & h_3\boldsymbol{e}_3 \\ \dfrac{\partial}{\partial q_1} & \dfrac{\partial}{\partial q_2} & \dfrac{\partial}{\partial q_3} \\ h_1v_1 & h_2v_2 & h_3v_3 \end{vmatrix}. \tag{A.18}$$

- Laplacian of ϕ:

$$\text{lap}\,\phi = \frac{1}{h_1h_2h_3}\left\{\frac{\partial}{\partial q_1}\left(\frac{h_2h_3}{h_1}\frac{\partial\phi}{\partial q_1}\right) + \frac{\partial}{\partial q_2}\left(\frac{h_1h_3}{h_2}\frac{\partial\phi}{\partial q_2}\right) + \frac{\partial}{\partial q_3}\left(\frac{h_1h_2}{h_3}\frac{\partial\phi}{\partial q_3}\right)\right\}. \tag{A.19}$$

- Laplacian of $\boldsymbol{v}$:

$$\begin{aligned}(\text{lap}\,\boldsymbol{v})|_1 = &\left\{\frac{1}{h_1}\frac{\partial\Omega}{\partial q_1} + \frac{1}{h_2h_3}\left[\frac{\partial}{\partial q_3}\left(\frac{h_2}{h_1h_3}\left(\frac{\partial(h_1v_1)}{\partial q_3} - \frac{\partial(h_3v_3)}{\partial q_1}\right)\right)\right.\right.\\ &\left.\left.-\frac{\partial}{\partial q_2}\left(\frac{h_3}{h_1h_2}\left(\frac{\partial(h_2v_2)}{\partial q_1} - \frac{\partial(h_1v_1)}{\partial q_2}\right)\right)\right]\right\}\boldsymbol{e}_1,\end{aligned} \tag{A.20}$$

$$\begin{aligned}(\text{lap}\,\boldsymbol{v})|_2 = &\left\{\frac{1}{h_2}\frac{\partial\Omega}{\partial q_2} + \frac{1}{h_1h_3}\left[\frac{\partial}{\partial q_1}\left(\frac{h_3}{h_1h_2}\left(\frac{\partial(h_2v_2)}{\partial q_1} - \frac{\partial(h_1v_1)}{\partial q_2}\right)\right)\right.\right.\\ &\left.\left.-\frac{\partial}{\partial q_3}\left(\frac{h_1}{h_2h_3}\left(\frac{\partial(h_3v_3)}{\partial q_2} - \frac{\partial(h_2v_2)}{\partial q_3}\right)\right)\right]\right\}\boldsymbol{e}_2,\end{aligned} \tag{A.21}$$

$$\begin{aligned}(\text{lap}\,\boldsymbol{v})|_3 = &\left\{\frac{1}{h_3}\frac{\partial\Omega}{\partial q_3} + \frac{1}{h_1h_2}\left[\frac{\partial}{\partial q_2}\left(\frac{h_1}{h_2h_3}\left(\frac{\partial(h_3v_3)}{\partial q_2} - \frac{\partial(h_2v_2)}{\partial q_3}\right)\right)\right.\right.\\ &\left.\left.-\frac{\partial}{\partial q_1}\left(\frac{h_2}{h_1h_3}\left(\frac{\partial(h_1v_1)}{\partial q_3} - \frac{\partial(h_3v_3)}{\partial q_1}\right)\right)\right]\right\}\boldsymbol{e}_3,\end{aligned} \tag{A.22}$$

where $\Omega = \text{div}\,\boldsymbol{v}$.

- Lagrangian derivative of $\boldsymbol{v}$:

$$\begin{aligned}(\boldsymbol{v}\cdot\nabla)\boldsymbol{v}|_1 = &\frac{1}{h_1}\left\{\left(v_1\frac{\partial v_1}{\partial q_1} + v_2\frac{\partial v_2}{\partial q_1} + v_3\frac{\partial v_3}{\partial q_1}\right) - \frac{v_2}{h_2}\left[\frac{\partial(h_2v_2)}{\partial q_1} - \frac{\partial(h_1v_1)}{\partial q_2}\right]\right.\\ &\left.+\frac{v_3}{h_3}\left[\frac{\partial(h_1v_1)}{\partial q_3} - \frac{\partial(h_3v_3)}{\partial q_1}\right]\right\}\boldsymbol{e}_1,\end{aligned} \tag{A.23}$$

$$\begin{aligned}(\boldsymbol{v}\cdot\nabla)\boldsymbol{v}|_2 = &\frac{1}{h_2}\left\{\left(v_1\frac{\partial v_1}{\partial q_2} + v_2\frac{\partial v_2}{\partial q_2} + v_3\frac{\partial v_3}{\partial q_2}\right) - \frac{v_3}{h_3}\left[\frac{\partial(h_3v_3)}{\partial q_2} - \frac{\partial(h_2v_2)}{\partial q_3}\right]\right.\\ &\left.+\frac{v_1}{h_1}\left[\frac{\partial(h_2v_2)}{\partial q_1} - \frac{\partial(h_1v_1)}{\partial q_2}\right]\right\}\boldsymbol{e}_2,\end{aligned} \tag{A.24}$$

$$(\boldsymbol{v}\cdot\nabla)\boldsymbol{v}|_3 = \frac{1}{h_3}\left\{\left(v_1\frac{\partial v_1}{\partial q_3} + v_2\frac{\partial v_2}{\partial q_3} + v_3\frac{\partial v_3}{\partial q_3}\right) - \frac{v_1}{h_1}\left[\frac{\partial(h_1 v_1)}{\partial q_3} - \frac{\partial(h_3 v_3)}{\partial q_1}\right]\right. \tag{A.25}$$
$$\left. + \frac{v_2}{h_2}\left[\frac{\partial(h_3 v_3)}{\partial q_2} - \frac{\partial(h_2 v_2)}{\partial q_3}\right]\right\}\boldsymbol{e}_3.$$

It follows from Eqs. (1.4.3), (1.4.13), (1.4.23) and (A.3) that

$$\{h_1, h_2, h_3\} = \{1, 1, 1\}, \tag{A.26}$$

for the rectangular coordinate system shown in Fig. 1.1a, and

$$\{h_1, h_2, h_3\} = \{1, r, 1\}, \tag{A.27}$$

for the cylindrical coordinate system shown in Fig. 1.1b, and

$$\{h_1, h_2, h_3\} = \{1, \rho, \rho\sin\theta\}, \tag{A.28}$$

for the spherical coordinate system shown in Fig. 1.1c. The expressions given in Sect. 1.4 can be reproduced by substituting Eqs. (A.26)–(A.28) into Eqs. (A.7)–(A.25).

Solutions to Selected Exercises B

1.1 (b) Let $\boldsymbol{a} = a_i\boldsymbol{e}_i$, $\boldsymbol{b} = b_j\boldsymbol{e}_j$ and $\boldsymbol{c} = c_k\boldsymbol{e}_k$, so that

$$\begin{aligned}\boldsymbol{a}\times(\boldsymbol{b}\times\boldsymbol{c}) &= (a_i\boldsymbol{e}_i)\times\left[\left(b_j\boldsymbol{e}_j\right)\times(c_k\boldsymbol{e}_k)\right] = (a_i\boldsymbol{e}_i)\times\left(b_jc_k\varepsilon_{jkm}\boldsymbol{e}_m\right)\\ &= a_ib_jc_k\varepsilon_{jkm}\varepsilon_{imn}\boldsymbol{e}_n = \left(a_ib_jc_k\right)\left(\delta_{jn}\delta_{ki}-\delta_{ji}\delta_{kn}\right)\boldsymbol{e}_n\\ &= a_ib_jc_k\delta_{jn}\delta_{ki}\boldsymbol{e}_n - a_ib_jc_k\delta_{ji}\delta_{kn}\boldsymbol{e}_n\\ &= (a_ic_i)\left(b_j\boldsymbol{e}_j\right)-(a_ib_i)(c_k\boldsymbol{e}_k) = (\boldsymbol{a}\cdot\boldsymbol{c})\,\boldsymbol{b}-(\boldsymbol{a}\cdot\boldsymbol{b})\,\boldsymbol{c}.\end{aligned} \quad \text{(B.1)}$$

1.3 Let $\boldsymbol{T} = T_{ij}\left(\boldsymbol{e}_i\otimes\boldsymbol{e}_j\right)$ and $\boldsymbol{U} = U_{st}\left(\boldsymbol{e}_s\otimes\boldsymbol{e}_t\right)$, and it is noted that $T_{ij} = T_{ji}$ and $U_{st} = -U_{ts}$. The trace of product $\boldsymbol{TU}$ is obtained as

$$\begin{aligned}\text{tr}\,(\boldsymbol{TU}) = \boldsymbol{T}\cdot\boldsymbol{U} = T_{ij}U_{ji} &= T_{11}U_{11}+T_{12}U_{21}+T_{13}U_{31}+T_{21}U_{12}+T_{22}U_{22}\\ &+T_{23}U_{32}+T_{31}U_{13}+T_{32}U_{23}+T_{33}U_{33} = 0,\end{aligned} \quad \text{(B.2)}$$

because $U_{11} = U_{22} = U_{33} = 0, T_{12} = T_{21}, T_{13} = T_{31}, T_{23} = T_{32}, U_{12} = -U_{21}$, $U_{13} = -U_{31}$ and $U_{23} = -U_{21}$. The same result can also be obtained by using a simple matrix operation, viz.,

$$\begin{aligned}&\text{tr}\,(\boldsymbol{TU}) = \text{tr}\left[(\boldsymbol{TU})^{\mathrm{T}}\right] = \text{tr}\left(\boldsymbol{U}^{\mathrm{T}}\boldsymbol{T}^{\mathrm{T}}\right) = \text{tr}\,(-\boldsymbol{UT}) = -\text{tr}\,(\boldsymbol{TU}),\\ \longrightarrow\quad &\text{tr}\,(\boldsymbol{TU}) = 0.\end{aligned} \quad \text{(B.3)}$$

1.8 (a) It follows from the definition of the antisymmetric part of a second-order tensor that

$$\begin{aligned}&\boldsymbol{T}^{\mathrm{a}} = \tfrac{1}{2}\left(\boldsymbol{T}-\boldsymbol{T}^{\mathrm{T}}\right),\\ \longrightarrow\quad &2\boldsymbol{T}^{\mathrm{a}} = (1-\cos\theta)\left[(\boldsymbol{a}^{\mathrm{w}}\otimes\boldsymbol{a}^{\mathrm{w}})-(\boldsymbol{a}^{\mathrm{w}}\otimes\boldsymbol{a}^{\mathrm{w}})^{\mathrm{T}}\right]+\sin\theta\left(\boldsymbol{U}-\boldsymbol{U}^{\mathrm{T}}\right).\end{aligned} \quad \text{(B.4)}$$

Since $\boldsymbol{U} = -\boldsymbol{U}^{\mathrm{T}}$ and $(\boldsymbol{a}^{\mathrm{w}}\otimes\boldsymbol{a}^{\mathrm{w}}) = (\boldsymbol{a}^{\mathrm{w}}\otimes\boldsymbol{a}^{\mathrm{w}})^{\mathrm{T}}$, it is concluded that

$$2\boldsymbol{T}^{\mathrm{a}} = 2\,(\sin\theta)\,\boldsymbol{U}, \qquad \longrightarrow \qquad \boldsymbol{T}^{\mathrm{a}} = (\sin\theta)\,\boldsymbol{U}. \quad \text{(B.5)}$$

(b) As implied by the above equation, the dual vector of $\boldsymbol{T}^{\mathrm{a}}$ is $(\sin\theta)$ times the dual vector of $\boldsymbol{U}$, which is given by $(\sin\theta)\,\boldsymbol{a}^{\mathrm{w}}$.

© Springer International Publishing AG 2019

C. Fang, *An Introduction to Fluid Mechanics*, Springer Textbooks in Earth Sciences, Geography and Environment, https://doi.org/10.1007/978-3-319-91821-1

The above analysis is based on the given rotation tensor $\boldsymbol{T}$, which can be derived by the following arguments. It is assumed that a rigid body undergoes a right-handed rotation with angle θ about an axis which is in parallel to the unit vector $\boldsymbol{m}$. Let the origin of coordinate system be on the rotational axis, and $\boldsymbol{r}$ be the position vector for a typical point in the body, which can be decomposed into $\boldsymbol{r} = \boldsymbol{r}_m + \boldsymbol{r}_p$, where $\boldsymbol{r}_m$ is in parallel to $\boldsymbol{m}$, while $\boldsymbol{r}_p$ is perpendicular to $\boldsymbol{m}$, whose unit vector is given by $\boldsymbol{p} = \boldsymbol{r}_p/\|\boldsymbol{r}_p\|$. The set of $\{\boldsymbol{m}, \boldsymbol{p}, \boldsymbol{q}\}$ forms an orthonormal base for any vector which rotates an angle about $\boldsymbol{m}$, where $\boldsymbol{q} = \boldsymbol{m} \times \boldsymbol{p}$, so that

$$\boldsymbol{T}\boldsymbol{r}_m = \boldsymbol{r}_m, \qquad \boldsymbol{T}\boldsymbol{r}_p = \|\boldsymbol{r}_p\| \left(\cos\theta\, \boldsymbol{p} + \sin\theta\, \boldsymbol{q}\right), \tag{B.6}$$

so that

$$\boldsymbol{T}\boldsymbol{r} = \boldsymbol{T}\left(\boldsymbol{r}_m + \boldsymbol{r}_p\right) = \boldsymbol{r}\cos\theta + \boldsymbol{r}_m\left(1 - \cos\theta\right) + \sin\theta\left(\boldsymbol{m} \times \boldsymbol{r}\right). \tag{B.7}$$

Since $\boldsymbol{r}_m = (\boldsymbol{r} \cdot \boldsymbol{m})\,\boldsymbol{m}$, substituting this expression into the above equation yields

$$\boldsymbol{T}\boldsymbol{r} = \boldsymbol{r}\cos\theta + (\boldsymbol{m} \cdot \boldsymbol{r})\left(1 - \cos\theta\right)\boldsymbol{m} + \sin\theta\left(\boldsymbol{m} \times \boldsymbol{r}\right). \tag{B.8}$$

By the definitions of dyadic product and dual vector, one obtains respectively $(\boldsymbol{m} \cdot \boldsymbol{r})\,\boldsymbol{m} = (\boldsymbol{m} \otimes \boldsymbol{m})\,\boldsymbol{r}$, and $\boldsymbol{m} \times \boldsymbol{r} = \boldsymbol{U}\boldsymbol{r}$, where $\boldsymbol{m}$ becomes the dual vector of $\boldsymbol{U}$. Substituting these expressions into Eq. (B.8) results in

$$\boldsymbol{T}\boldsymbol{r} = \left[\left(1 - \cos\theta\right)\left(\boldsymbol{m} \otimes \boldsymbol{m}\right) + \cos\theta\boldsymbol{I} + \sin\theta\boldsymbol{U}\right]\boldsymbol{r}, \tag{B.9}$$

which gives the prescribed expression of $\boldsymbol{T}$, where $\boldsymbol{m} = \boldsymbol{a}^{\mathrm{w}}$.

1.11 (d) Let $\boldsymbol{a} = a_j\boldsymbol{e}_j$, so that

$$\begin{aligned}
\nabla \times (\nabla \times \boldsymbol{a}) &= \left(\frac{\partial}{\partial x_m}\boldsymbol{e}_m\right) \times \left[\left(\frac{\partial}{\partial x_i}\boldsymbol{e}_i\right) \times \left(a_j\boldsymbol{e}_j\right)\right] \\
&= \left(\frac{\partial}{\partial x_m}\boldsymbol{e}_m\right) \times \left(\frac{\partial a_j}{\partial x_i}\varepsilon_{ijk}\boldsymbol{e}_k\right) = \frac{\partial^2 a_j}{\partial x_m \partial x_i}\varepsilon_{ijk}\varepsilon_{mks}\boldsymbol{e}_s \\
&= \frac{\partial^2 a_j}{\partial x_m \partial x_i}\left(\delta_{is}\delta_{jm} - \delta_{im}\delta_{js}\right)\boldsymbol{e}_s \\
&= \left(\frac{\partial}{\partial x_i}\boldsymbol{e}_i\right)\left(\frac{\partial a_j}{\partial x_i}\right) - \frac{\partial^2}{\partial x_i^2}\left(a_j\boldsymbol{e}_j\right) \\
&= \nabla\left(\nabla \cdot \boldsymbol{a}\right) - \nabla^2\boldsymbol{a}.
\end{aligned} \tag{B.10}$$

1.12 Let $\boldsymbol{T}$ be expressed as

$$\boldsymbol{T} = \left[\boldsymbol{T}_1, \boldsymbol{T}_2, \boldsymbol{T}_3\right], \tag{B.11}$$

where $\{\boldsymbol{T}_1, \boldsymbol{T}_2, \boldsymbol{T}_3\}$ are column vectors. It follows from the definition of determinant that

$$\det\boldsymbol{T} = \det\left[\boldsymbol{T}_1, \boldsymbol{T}_2, \boldsymbol{T}_3\right] = \boldsymbol{T}_1 \cdot \left(\boldsymbol{T_2} \times \boldsymbol{T}_3\right) = T_{i1}T_{j2}T_{k3}\varepsilon_{ijk}, \tag{B.12}$$

which, by renaming the indices with cyclic interchange, can be recast alternatively as

$$\det \boldsymbol{T} = T_{j1} T_{k2} T_{i3} \varepsilon_{jki} = T_{i3} T_{j1} T_{k2} \varepsilon_{ijk}. \tag{B.13}$$

Implementing all six permutations yields

$$\begin{aligned} \det \boldsymbol{T} &= T_{i1} T_{j2} T_{k3} \varepsilon_{ijk} = T_{i2} T_{j3} T_{k1} \varepsilon_{ijk} = T_{i3} T_{j1} T_{k2} \varepsilon_{ijk} \\ &= -T_{i3} T_{j2} T_{k1} \varepsilon_{ijk} = -T_{i2} T_{j1} T_{k3} \varepsilon_{ijk} = -T_{i1} T_{j3} T_{k2} \varepsilon_{ijk}, \end{aligned} \tag{B.14}$$

showing that

$$6 \det \boldsymbol{T} = T_{ir} T_{js} T_{kt} \varepsilon_{ijk} \varepsilon_{rst}, \qquad \longrightarrow \qquad \det \boldsymbol{T} = \frac{1}{6} T_{ir} T_{js} T_{kt} \varepsilon_{ijk} \varepsilon_{rst}. \tag{B.15}$$

For the second part, it follows from Eq. (B.15)$_2$ that

$$T_{ir} T_{js} T_{kt} \varepsilon_{ijk} \varepsilon_{rst} \delta_{pq} = 3\, \varepsilon_{pjk} \varepsilon_{rst} T_{js} T_{kt} T_{rq} = (6 \det \boldsymbol{T})\, \delta_{pq}. \tag{B.16}$$

Since the definition of tensor inverse implies that

$$\boldsymbol{T}^{-1} \boldsymbol{T} = \boldsymbol{I}, \qquad \left[\boldsymbol{T}^{-1}\right]_{pr} T_{rq} = \delta_{pq}, \tag{B.17}$$

comparing the second equation with Eq. (B.16)$_2$ gives

$$\left[\boldsymbol{T}^{-1}\right]_{pr} = \frac{1}{2 \det \boldsymbol{T}} \varepsilon_{pjk} \varepsilon_{rst} T_{js} T_{kt}. \tag{B.18}$$

1.13 Taking trace of the Cayley-Hamilton theorem for a second-order tensor $\boldsymbol{T}$ yields

$$0 = \operatorname{tr} \left(\boldsymbol{T}^3 - I_{\boldsymbol{T}}^1 \boldsymbol{T}^2 + I_{\boldsymbol{T}}^2 \boldsymbol{T} - I_{\boldsymbol{T}}^3 \boldsymbol{I} \right) = I_{\boldsymbol{T}^3}^1 - I_{\boldsymbol{T}}^1 I_{\boldsymbol{T}^2}^1 + I_{\boldsymbol{T}}^2 I_{\boldsymbol{T}}^1 - 3 I_{\boldsymbol{T}}^3. \tag{B.19}$$

Solving this equation for $I_{\boldsymbol{T}}^3$ gives

$$I_{\boldsymbol{T}}^3 = \frac{1}{3} \left(I_{\boldsymbol{T}^3}^1 - I_{\boldsymbol{T}}^1 I_{\boldsymbol{T}^2}^1 + I_{\boldsymbol{T}}^2 I_{\boldsymbol{T}}^1 \right) = \frac{1}{3} \left[I_{\boldsymbol{T}^3}^1 - \frac{3}{2} I_{\boldsymbol{T}}^1 I_{\boldsymbol{T}^2}^1 + \frac{1}{2} \left(I_{\boldsymbol{T}}^1 \right)^3 \right], \tag{B.20}$$

in which the expression of second scalar invariant has been used. Since

$$I_{\boldsymbol{T}^2}^1 = \left(I_{\boldsymbol{T}}^1 \right)^2 - 2 I_{\boldsymbol{T}}^2, \tag{B.21}$$

substituting this equation into Eq. (B.20) leads to

$$I_{\boldsymbol{T}}^3 = \frac{1}{3} \left[I_{\boldsymbol{T}^3}^1 - \left(I_{\boldsymbol{T}}^1 \right)^3 \right] + I_{\boldsymbol{T}}^2 I_{\boldsymbol{T}}^1. \tag{B.22}$$

For the remaining parts, it follows from the definition of first scalar invariant that

$$\frac{\partial I_{\boldsymbol{T}}^1}{\partial \boldsymbol{T}} = \frac{\partial T_{kk}}{\partial T_{ij}} = \delta_{ik} \delta_{jk} = \delta_{ij} = \boldsymbol{I}. \tag{B.23}$$

Taking derivative of the second invariant with respect to $\boldsymbol{T}$ gives rise to

$$\frac{\partial I_{\boldsymbol{T}}^2}{\partial \boldsymbol{T}} = \frac{1}{2} \frac{\partial}{\partial \boldsymbol{T}} \left[\left(I_{\boldsymbol{T}}^1 \right)^2 - I_{\boldsymbol{T}^2}^1 \right] = I_{\boldsymbol{T}}^1 \frac{I_{\boldsymbol{T}}^1}{\partial \boldsymbol{T}} - \frac{1}{2} \frac{\partial}{\partial \boldsymbol{T}} \left(\boldsymbol{T} \cdot \boldsymbol{T} \right), \tag{B.24}$$

where the last term on the right-hand-side is identified to be

$$\frac{1}{2}\frac{\partial}{\partial \boldsymbol{T}}\left(\boldsymbol{T}\cdot\boldsymbol{T}\right)=\frac{1}{2}\frac{\partial}{\partial T_{ij}}\left(T_{lk}T_{kl}\right)=\frac{1}{2}\left(\delta_{il}\delta_{jk}T_{kl}+\delta_{ik}\delta_{jl}T_{lk}\right)=T_{ji}, \quad (B.25)$$

or alternatively as

$$\frac{\partial I^1_{T^2}}{\partial \boldsymbol{T}}=2\boldsymbol{T}^{\mathrm{T}}. \quad (B.26)$$

Substituting this result and Eq. (B.23) into Eq. (B.24) gives

$$\frac{\partial I^2_T}{\partial \boldsymbol{T}}=I^1_T\boldsymbol{I}-\boldsymbol{T}^{\mathrm{T}}. \quad (B.27)$$

Similarly, taking derivative of Eq. (B.22) with respect to $\boldsymbol{T}$ results in

$$\frac{\partial I^3_T}{\partial \boldsymbol{T}}=\frac{\partial}{\partial \boldsymbol{T}}\left\{\frac{1}{3}\left[I^1_{T^3}-\left(I^1_T\right)^3\right]+I^2_T I^1_T\right\}=\frac{1}{3}\frac{\partial I^1_{T^3}}{\partial \boldsymbol{T}}-\left(I^1_T\right)^2\frac{\partial I^1_T}{\partial \boldsymbol{T}}+I^1_T\frac{\partial I^2_T}{\partial \boldsymbol{T}}+I^2_T\frac{\partial I^1_T}{\partial \boldsymbol{T}}, \quad (B.28)$$

which is recast alternatively as

$$\frac{\partial I^3_T}{\partial \boldsymbol{T}}=\left(\boldsymbol{T}^2-I^1_T\boldsymbol{T}+I^2_T\boldsymbol{I}\right)^{\mathrm{T}}, \quad (B.29)$$

in which the identity

$$\begin{aligned}\frac{\partial I^1_{T^3}}{\partial \boldsymbol{T}}&=\frac{\partial}{\partial T_{ij}}\left(T_{kl}T_{lm}T_{mk}\right)=\delta_{ik}\delta_{jl}T_{lm}T_{mk}+\delta_{il}\delta_{jm}T_{kl}T_{mk}\\&+\delta_{im}\delta_{jk}T_{kl}T_{lm}=3T_{jm}T_{mi}=3\left(\boldsymbol{T}^2\right)^{\mathrm{T}},\end{aligned} \quad (B.30)$$

and Eqs. (B.23) and (B.27) have been used. Substituting the Cayley-Hamilton theorem into Eq. (B.29) leads to

$$\frac{\partial I^3_T}{\partial \boldsymbol{T}}=\boldsymbol{T}^{-\mathrm{T}}I^3_T. \quad (B.31)$$

1.16 The singularities of $f(z)$ are simple poles at $z=0$, $z=i$ and $z=-i$, at which the residues, denoted by $Res(f,z)$, are given respectively by

$$Res\,(f,0)=-1,\quad Res\,(f,i)=1+\frac{1}{2}\cos i,\quad Res\,(f,-i)=-1+\frac{1}{2}\cos i. \quad (B.32)$$

Let C represent any closed contour on the complex plane z. Depending on the relative position between C and the singularities, the following cases are discussed.

- If C does not enclose any of the singularities, then $\oint f(z)\,\mathrm{d}z=0$, as motivated by the Cauchy theorem.
- If C encloses $z=0$ but not $z=\pm i$, then

$$\oint f(z)\,\mathrm{d}z=(2\pi i)Res\,(f,0)=-2\pi i. \quad (B.33)$$

- If C encloses $z = i$ but not $z = 0$ or $z = -i$, then

$$\oint f(z)\,\mathrm{d}z = (2\pi i) Res\,(f, i) = 2\pi i \left(1 + \frac{1}{2}\cos i\right). \tag{B.34}$$

- If C encloses $z = -i$ but not $z = 0$ or $z = i$, then

$$\oint f(z)\,\mathrm{d}z = (2\pi i) Res\,(f, -i) = 2\pi i \left(-1 + \frac{1}{2}\cos i\right). \tag{B.35}$$

- If C encloses $z = 0$ and $z = i$ but not $z = -i$, then

$$\oint f(z)\,\mathrm{d}z = (2\pi i)\left[Res\,(f, 0) + Res\,(f, i)\right] = \pi i \cos i. \tag{B.36}$$

- If C encloses $z = 0$ and $z = -i$ but not $z = i$, then

$$\oint f(z)\,\mathrm{d}z = (2\pi i)\left[Res\,(f, 0) + Res\,(f, -i)\right] = 2\pi i \left(-2 + \frac{1}{2}\cos i\right). \tag{B.37}$$

- If C encloses $z = \pm i$ but not $z = 0$, then

$$\oint f(z)\,\mathrm{d}z = (2\pi i)\left[Res\,(f, i) + Res\,(f, -i)\right] = 2\pi i \cos i. \tag{B.38}$$

- If C encloses all three singularities, then

$$\oint f(z)\,\mathrm{d}z = (2\pi i)\left[Res\,(f, 0) + Res\,(f, i) + Res\,(f, -i)\right] = 2\pi i(-1 + \cos i). \tag{B.39}$$

2.7 In view of the figure, applying Newton's second law of motion to the block along the inclined plane yields

$$mg \sin\theta - F_v = m\frac{\mathrm{d}V}{\mathrm{d}t}, \qquad F_v = \mu \frac{V}{h} A, \tag{B.40}$$

where F_v represents the viscous force acting on the block by the liquid film, and V is the block velocity which is a function of time. Combining two equations gives

$$\frac{\mathrm{d}V}{\mathrm{d}t} + \left(\frac{\mu A}{mh}\right) V - g \sin\theta = 0, \tag{B.41}$$

to which the solution is obtained as

$$V(t) = \frac{mgh \sin\theta}{\mu A}\left[1 - \exp\left(-\frac{\mu A}{mg} t\right)\right]. \tag{B.42}$$

As $t \to \infty$, $V \to mgh \sin\theta/(\mu A)$, which is the terminal velocity of block.

2.8 In view of Fig. 2.5b, the rotational motion of outer cylinder will cease gradually with time as the rope snapped. Applying Newton's second law of motion to the outer cylinder yields

$$m_2 R^2 \frac{\mathrm{d}\omega}{\mathrm{d}t} = -\frac{2\pi R^3 \mu h}{a}\omega, \quad \longrightarrow \quad \frac{\mathrm{d}\omega}{\omega} = -\frac{2\pi R\mu h}{m_2 a}, \tag{B.43}$$

which is integrated to obtain

$$\omega(t) = \omega_0 \exp\left[-\frac{2\pi R\mu h}{m_2 a}t\right], \tag{B.44}$$

in which $\omega(0) = \omega_0$ has been used. The time duration Δt that $\omega(t)$ becomes 1% of ω_0 is obtained as

$$0.01 = \exp\left[-\frac{2\pi R\mu h}{m_2 a}\Delta t\right], \quad \longrightarrow \quad \Delta t = -\frac{m_2 a}{2\pi R\mu h}\ln(0.01). \tag{B.45}$$

2.11 Let θ be a counterclockwise angle with respect to the rotational axis with $\theta < \alpha$. It follows from the geometric configurations that

$$\tau = \mu\frac{\mathrm{d}u}{\mathrm{d}y} \sim \mu\frac{u}{h} = \frac{\mu\,\omega R\sin\theta}{h}, \tag{B.46}$$

where u represents the tangential velocity perpendicular to the page. Integrating this equation over the entire spherical surface in contact with the oil results in the torque T given by

$$T = \int_0^\alpha R\sin\theta\left(\frac{\mu\,\omega R\sin\theta}{h}\right)2\pi R^2\sin\theta\,\mathrm{d}\theta = \frac{2\pi\mu\,\omega R^4}{h}\left(\frac{\cos^3\alpha}{3} - \cos\alpha + \frac{2}{3}\right), \tag{B.47}$$

by which the dimensionless torque $\overline{T}$ is obtained as

$$\overline{T} = \frac{T}{2\pi\mu\,\omega R^4/h} = \frac{\cos^3\alpha}{3} - \cos\alpha + \frac{2}{3}. \tag{B.48}$$

3.1 Let point 1 be on the liquid-free-surface in the reservoir, while point 2 be on the liquid-free-surface in the inclined tube. The pressure difference between two points is given by

$$\Delta p = p_1 - p_2 = \rho_w s g\,(h + \ell\sin\theta)\,, \tag{B.49}$$

where ρ_w is the density of water, and h represents the difference in liquid-free-surfaces in the reservoir before and after the application of Δp. Since the decrease in liquid volume in the reservoir should be the same as the increase in liquid volume in the inclined tube, it follows that

$$\frac{\pi D^2}{4}h = \frac{\pi d^2}{4}\ell, \quad \longrightarrow \quad h = \ell\left(\frac{d}{D}\right)^2, \tag{B.50}$$

which is substituted into Eq. (B.49) to obtain

$$\Delta p = \rho_w s g \ell \left[\sin\theta + \left(\frac{d}{D}\right)^2 \right]. \tag{B.51}$$

Solving this equation for ℓ gives

$$\ell = \frac{\Delta p}{\rho_w s g} \frac{1}{\sin\theta + (d/D)^2}. \tag{B.52}$$

For the remaining part, consider the reservoir to be connected to a normal U-tube manometer, whose deflection is denoted by h'; i.e., h' is the distance between the free liquid surfaces in the reservoir and tube, which is given by

$$h' = \frac{\Delta p}{\rho_w s g}. \tag{B.53}$$

The sensitivity of inclined-tube manometer is then obtained as

$$sens = \frac{\ell + h}{h'} \sim \frac{\ell}{h'} = \frac{1}{\sin\theta + (d/D)^2}, \tag{B.54}$$

showing that the values of s, $\sin\theta$, and d/D should be chosen as small as possible to increase the sensitivity.

3.5 The free-body-diagram of gate is shown in the figure, where W is the gate's weight, B represents the buoyant force acting on the gate, F_B denotes the reaction at point B, F is the hydrostatic force component acting on the curve surface of gate along the horizontal direction, while A_x and A_y are the reaction components at hinge A in the horizontal and vertical directions, respectively. The magnitudes of W, B, and F are are obtained as

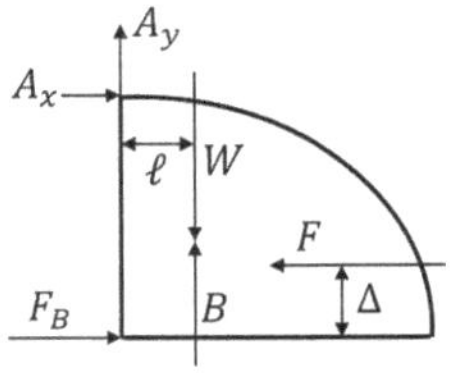

$$W = \frac{\pi R^2 b}{4} s\gamma_w, \qquad B = \frac{\pi R^2 b}{4}\gamma_w, \qquad F = \gamma_w \left(h - \frac{R}{2}\right) Rb, \tag{B.55}$$

where γ_w represents the specific weight of water. Both W and B act at the centroid of gate volume, for the gate and water are assumed to be homogeneous, while F acts at the centroid of the pressure distribution diagram between points A and B. Since the gate is in static equilibrium, taking all the moments with respect to point A to be null yields

$$\begin{gathered} 0 = \sum M_A = F_B R + (1-s)\,\gamma_w \frac{\pi R^2 b}{4} \ell - F\Delta, \\ \ell = \frac{4R}{3\pi}, \quad \Delta = \frac{1}{3}\frac{3h - 2R}{2h - R} R, \end{gathered} \tag{B.56}$$

giving rise to

$$F_B = \frac{\gamma_w Rb}{3}\left[\left(h - \frac{R}{2}\right)\frac{3h - 2R}{2h - R} + (s - 1)R\right], \tag{B.57}$$

in which $s > 1.0$ must be satisfied in order to have a compression at point B. The exercise has been solved by using the concept of buoyancy to determine the vertical component of hydrostatic force acting on the curved surface. It can also be solved by estimating the hydrostatic forces acting on the bottom and curved surfaces of gate, although this approach is a little bit cumbersome.

3.6 It follows from a simple force balance of the elevated water column along the direction of gravitational acceleration that

$$h = \frac{2\sigma\cos\theta}{\gamma R}, \quad \longrightarrow \quad R = \frac{2\sigma\cos\theta}{\gamma h}, \tag{B.58}$$

where h is the column height, R represents the radius of tube, θ is the angle of contact, and σ and γ are respectively the surface tension coefficient and specific weight of water. Substituting the values of $\sigma = 0.073$ N/m, $\gamma = 9.8$ kN/m^3, $\theta \sim 0°$, and $h = 1$ mm quoted from Sect. 3.3.1 into this equation gives $R \sim 14.9$ mm, so that the diameter of tube is nearly 29.8 mm.

3.11 It follows from the geometric configurations and Eq. (3.5.1) that

$$\frac{\partial p}{\partial r} = \rho r\omega^2, \quad \frac{\partial p}{\partial z} = -\rho g, \tag{B.59}$$

so that the pressure differences between points A, B, C, and D are obtained as

$$p_B = p_A + \rho g h, \quad p_C = p_B + \frac{\rho \ell^2\omega^2}{2}, \quad p_C = p_D + \rho g h. \tag{B.60}$$

Combining three equations with $p_D = p_{atm}$ yields

$$p_A = p_{atm} - \frac{\rho \ell^2\omega^2}{2}, \tag{B.61}$$

showing that the minimum pressure of liquid occurs at point A. Larger the angular speed ω is, smaller the pressure at point A will be. With $p_A = p_v$, the maximum angular velocity ω_{max} is obtained as

$$\omega_{max} = \sqrt{\frac{2\,(p_{atm} - p_v)}{\rho}\frac{1}{\ell}}. \tag{B.62}$$

For $\omega > \omega_{max}$, the pressure at point A will be smaller than the vapor pressure of liquid. If, furthermore, p_A is smaller than the saturated value of p_v, the liquid molecules may transform to vapor state to induce *cavitation* in the liquid.

4.1 (a) It follows from the definition of streamlines that

$$\frac{\mathrm{d}y}{\mathrm{d}x} = \frac{v}{u} = \frac{v_0}{u_0\sin[\omega(t - y/v_0)]}, \tag{B.63}$$

which is integrated with a fixed value of t to obtain

$$\frac{u_o v_0}{\omega}\cos\left[\omega\left(t-\frac{y}{v_0}\right)\right]=v_0 x+C, \tag{B.64}$$

where C is the integration constant. Applying the conditions $(x, y) = (0, 0)$ at $t = 0$ and $(x, y) = (0, 0)$ at $t = 2\pi/\omega$ to this solution yields respectively $C = u_0 v_0/\omega$ and $C = 0$, with which two streamlines are identified to be

$$x=\frac{u_0}{\omega}\left[\cos\left(\frac{\omega y}{v_0}\right)-1\right];\qquad x=\frac{u_0}{\omega}\sin\left(\frac{\omega y}{v_0}\right). \tag{B.65}$$

(b) It follows from the definition of pathlines that

$$\frac{\mathrm{d}x}{\mathrm{d}t}=u_0\sin\left[\omega\left(t-\frac{y}{v_0}\right)\right],\qquad \frac{\mathrm{d}y}{\mathrm{d}t}=v_0. \tag{B.66}$$

Integrating the second equation yields $y = v_0 t + C_1$, which is substituted into the first equation to obtain

$$\frac{\mathrm{d}x}{\mathrm{d}t}=-u_0\sin\left(\frac{C_1\omega}{v_0}\right). \tag{B.67}$$

Integrating this equation gives

$$x=-\left[u_0\sin\left(\frac{C_1\omega}{v_0}\right)\right]t+C_2, \tag{B.68}$$

where C_1 and C_2 are integration constants. Applying the conditions $(x, y) = (0, 0)$ at $t = 0$ and $(x, y) = (0, 0)$ at $t = 2\pi/\omega$ to the obtained solutions yield-s respectively $\{C_1, C_2\} = \{0, 0\}$ and $\{C_1, C_2\} = \{-\pi v_0/(2\omega), -\pi u_0/(2\omega)\}$, with which two pathlines are identified to be

$$x=0,\ y=v_0 t;\qquad x=u_0\left(t-\frac{\pi}{2\omega}\right),\ y=v_0\left(t-\frac{\pi}{2\omega}\right),\ \longrightarrow\ y=\frac{v_0}{u_0}x. \tag{B.69}$$

The obtained two streamlines and pathlines are shown graphically in the below figure.

(c) The streakline through the origin at $t = 0$ is the locus of particles at $t = 0$ that previously passed through the origin. Thus, each particle that flows through the origin moves along a straight line, i.e., along its pathline, as indicated by Eq. (B.69), with the slopes lying between $\pm v_0/u_0$. Particles passing through the origin at different times are located on different rays emitting from the origin and at different distances apart from the origin, and the resulting streakline is then shown in the figure. The exercise may be repeated again if the x-component of fluid velocity is prescribed by $u = u_0 \cos[\omega(t - y/v_0)]$.

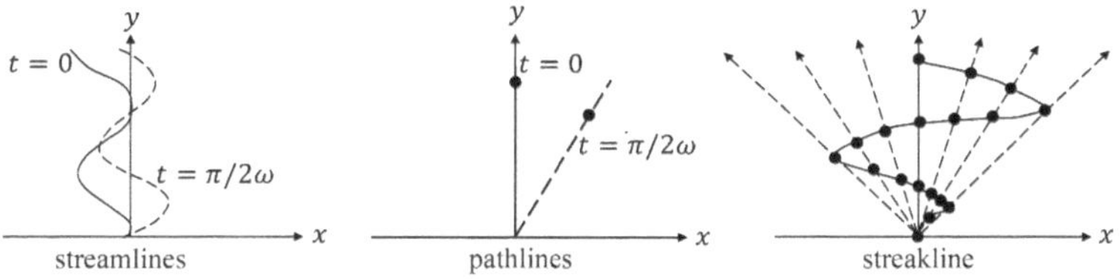

5.1 Let $\mathrm{d}\boldsymbol{x}$ and $\mathrm{d}\boldsymbol{X}$ be the line elements in the present and reference configurations, respectively. Specifically, construct the sets $\{\mathrm{d}\boldsymbol{x}^1, \mathrm{d}\boldsymbol{x}^2, \mathrm{d}\boldsymbol{x}^3\}$ and $\{\mathrm{d}\boldsymbol{X}^1, \mathrm{d}\boldsymbol{X}^2, \mathrm{d}\boldsymbol{X}^3\}$ for three different line elements in $\mathcal{B}^P$ and $\mathcal{B}^R$, which satisfy

$$\mathrm{d}\boldsymbol{x}^i = \boldsymbol{F}\mathrm{d}\boldsymbol{X}^i, \qquad i = 1, 2, 3. \tag{B.70}$$

For an infinitesimal volume element, it follows from the above equation that

$$\mathrm{d}v^P = \left(\mathrm{d}\boldsymbol{x}^1 \times \mathrm{d}\boldsymbol{x}^2\right) \cdot \mathrm{d}\boldsymbol{x}^3 = \left(\boldsymbol{F}\mathrm{d}\boldsymbol{X}^1 \times \boldsymbol{F}\mathrm{d}\boldsymbol{X}^2\right) \cdot \boldsymbol{F}\mathrm{d}\boldsymbol{X}^3. \tag{B.71}$$

Since the identity

$$(\det \boldsymbol{F})\,(\boldsymbol{a} \times \boldsymbol{b}) \cdot \boldsymbol{c} = \det\,(\boldsymbol{F}\boldsymbol{a}, \boldsymbol{F}\boldsymbol{b}, \boldsymbol{F}\boldsymbol{c}) = (\boldsymbol{F}\boldsymbol{a} \times \boldsymbol{F}\boldsymbol{b}) \cdot \boldsymbol{F}\boldsymbol{c}, \tag{B.72}$$

holds for any $\{\boldsymbol{a}, \boldsymbol{b}, \boldsymbol{c}\} \in \mathcal{R}^3$ and any $\boldsymbol{F} \in \mathcal{R}^{3\times3}$, substituting this identity into Eq. (B.71) gives

$$\mathrm{d}v^P = (\det \boldsymbol{F})\left(\mathrm{d}\boldsymbol{X}^1 \times \mathrm{d}\boldsymbol{X}^2\right) \cdot \mathrm{d}\boldsymbol{X}^3 = J\,\mathrm{d}v^R. \tag{B.73}$$

A surface element in the present configuration is given by

$$\mathrm{d}\boldsymbol{a}^P = \mathrm{d}\boldsymbol{x}^1 \times \mathrm{d}\boldsymbol{x}^2 = \boldsymbol{F}\mathrm{d}\boldsymbol{X}^1 \times \boldsymbol{F}\mathrm{d}\boldsymbol{X}^2. \tag{B.74}$$

It follows from Eq. (B.72) that

$$\begin{aligned}(\det \boldsymbol{F})\left(\mathrm{d}\boldsymbol{X}^1 \times \mathrm{d}\boldsymbol{X}^2\right) \cdot \boldsymbol{c} &= \left(\boldsymbol{F}\mathrm{d}\boldsymbol{X}^1 \times \boldsymbol{F}\mathrm{d}\boldsymbol{X}^2\right) \cdot \boldsymbol{F}\boldsymbol{c} \\ &= \left(\mathrm{d}\boldsymbol{x}^1 \times \mathrm{d}\boldsymbol{x}^2\right) \cdot \boldsymbol{F}\boldsymbol{c} = \boldsymbol{F}^{\mathrm{T}}\mathrm{d}\boldsymbol{a} \cdot \boldsymbol{c},\end{aligned} \tag{B.75}$$

which holds for any vector $\boldsymbol{c}$, in which the identity $\boldsymbol{a} \cdot (\boldsymbol{A}\boldsymbol{b}) = (\boldsymbol{A}^{\mathrm{T}}\boldsymbol{a}) \cdot \boldsymbol{b}$, $\forall\{\boldsymbol{a}, \boldsymbol{b}\} \in \mathcal{R}^3$ and $\boldsymbol{A} \in \mathcal{R}^{3\times3}$ has been used. Substituting this equation into Eq. (B.74) results in

$$\mathrm{d}\boldsymbol{a} = J\,\boldsymbol{F}^{-\mathrm{T}}\mathrm{d}\boldsymbol{a}^R, \qquad \mathrm{d}\boldsymbol{a}^R = \mathrm{d}\boldsymbol{X}^1 \times \mathrm{d}\boldsymbol{X}^2. \tag{B.76}$$

The derivation of Eq. (B.72) is left as an additional exercise for interesting readers.

5.7 Multiplying the ith component of differential linear momentum balance with u_j and the jth component with u_i, and combining two resulting equations gives

$$\begin{aligned}&u_i\left[\frac{\partial(\rho u_j)}{\partial t} + \frac{\partial}{\partial x_k}(\rho u_j u_k) - \frac{\partial t_{jk}}{\partial x_k} - \rho b_j\right] \\ &+u_j\left[\frac{\partial(\rho u_i)}{\partial t} + \frac{\partial}{\partial x_k}(\rho u_i u_k) - \frac{\partial t_{ik}}{\partial x_k} - \rho b_i\right] = 0,\end{aligned} \tag{B.77}$$

which can be simplified to

$$\begin{aligned}&\frac{\partial(\rho u_i u_j)}{\partial t} + \frac{\partial}{\partial x_k}(\rho u_i u_j u_k) - 2\,\mathrm{sym}\left[\frac{\partial(u_i t_{jk})}{\partial x_k}\right] \\ &-2\,\mathrm{sym}\left[\frac{\partial u_i}{\partial x_k} t_{jk}\right] - 2\,\mathrm{sym}\left(\rho u_i b_j\right) = 0,\end{aligned} \tag{B.78}$$

in which the local mass balance has been used. Arranging all the terms in this equation yields

$$\frac{\partial}{\partial t}\left(\rho\frac{u_i u_j}{2}\right)+\frac{\partial}{\partial x_k}\left(\rho\frac{u_i u_j}{2}u_k\right) = \text{sym}\left[\frac{\partial(u_i t_{jk})}{\partial x_k}\right]+\text{sym}\left[\frac{\partial u_i}{\partial x_k}t_{jk}\right] \\ +\text{sym}\left(\rho u_i b_j\right). \tag{B.79}$$

Taking of this equation gives

$$\begin{aligned}\frac{\partial}{\partial t}\left(\rho\frac{u_i u_i}{2}\right)+\frac{\partial}{\partial x_k}\left(\rho\frac{u_i u_i}{2}u_k\right) &= \frac{\partial(u_i t_{ik})}{\partial x_k}+\frac{\partial u_i}{\partial x_k}t_{ik}+\rho u_i b_i,\\ \frac{\partial}{\partial t}\left(\rho\frac{\boldsymbol{u}\cdot\boldsymbol{u}}{2}\right)+\text{div}\left(\frac{\rho\boldsymbol{u}\cdot\boldsymbol{u}}{2}\boldsymbol{u}\right) &= \text{div}\,(\boldsymbol{u}\,\boldsymbol{t})-\text{tr}\left[\text{grad}\,(\boldsymbol{u}\,\boldsymbol{t})\right]+\rho\boldsymbol{u}\cdot\boldsymbol{b},\end{aligned} \tag{B.80}$$

which becomes the local balance of kinetic energy, as already given in Eq. (5.3.34). Consequently, the time variation of the physical quantity $(\rho\boldsymbol{u}\cdot\boldsymbol{u}/2)$ is balanced by the production, $-\text{tr}\left[\text{grad}\,(\boldsymbol{u}\,\boldsymbol{t})\right]$, the supply, $\rho\boldsymbol{u}\cdot\boldsymbol{b}$, and the flux, $-(\boldsymbol{u}\,\boldsymbol{t})$.

5.10 Let $F_s = h(x,t) - y = 0$ represent the fluid free surface at all times, so that

$$0=\frac{\mathrm{d}F_s}{\mathrm{d}t}=\frac{\mathrm{d}h(x,t)}{\mathrm{d}t}-\frac{\mathrm{d}y}{\mathrm{d}t}=\frac{\partial h}{\partial t}+\frac{\partial h}{\partial x}u-v, \tag{B.81}$$

for $\mathrm{d}x/\mathrm{d}t = u$ and $\mathrm{d}y/\mathrm{d}t = v$. Since the fluid is assumed to be incompressible, the local balance of mass reduces to $\text{div}\,\boldsymbol{u} = 0$, which is integrated from $y = 0$ to $y = h$ to obtain

$$\begin{aligned}&\int_0^{h(x,t)}\left(\frac{\partial u}{\partial x}+\frac{\partial y}{\partial y}\right)\mathrm{d}y = 0,\\ \longrightarrow\quad &\frac{\partial}{\partial x}\int_0^{h(x,t)} u\,\mathrm{d}y - u(h)\frac{\partial h}{\partial x}+v(h)-v(0)=0,\end{aligned} \tag{B.82}$$

in which Leibnitz's integration rule has been used. Since at the surface of inclined plane, $v(0) = 0$, for the plane is solid, and on the free liquid surface Eq. (B.81) holds, substituting these into Eq. (B.82)$_2$ gives

$$\frac{\partial h}{\partial t}+\frac{\partial}{\partial x}\int_0^{h(x,t)} u\,\mathrm{d}y = 0, \tag{B.83}$$

which can be expressed alternatively as

$$\frac{\partial h}{\partial t}+\frac{\partial Q}{\partial x}=0,\qquad Q=\int_0^{h(x,t)} u\,\mathrm{d}y = 0. \tag{B.84}$$

If Q is expressed by $Q = Q(h)$, then $\partial Q/\partial x = (\mathrm{d}Q/\mathrm{d}h)(\partial h/\partial x)$, with which Eq. (B.84) becomes

$$\frac{\partial h}{\partial t}+C(h)\frac{\partial h}{\partial x}=0,\qquad C(h)=\frac{\mathrm{d}Q}{\mathrm{d}h}, \tag{B.85}$$

which is a one-dimensional wave equation of h, to which a general solution is expressed as $f = F(x - C(h)t)$. This can be verified easily by substituting the solution into Eq. (B.85) to show that the equation holds identically.

5.13 The region occupied by the whole water jet is chosen as the finite control-volume. Applying the integral balance of linear momentum to the CV along the y-direction yields

$$\sum F_y = \frac{\partial}{\partial t}\int_{CV} v\,(\rho\,\mathrm{d}v) + \int_{CS} v\,(\rho\,\boldsymbol{u}\cdot\boldsymbol{n}\,\mathrm{d}a)\,, \tag{B.86}$$

where $\boldsymbol{u}$ is the velocity of water, and v is its component along the y-direction. In this equation, the friction between the water jet and surface of inclined plate is neglected for simplicity. Under the assumptions of steady and uniform flows, Eq. (B.86) reduces to

$$\sum F_y = \rho V^2 A \sin\theta, \;\; V = \|\boldsymbol{u}\|. \tag{B.87}$$

Since the gravitational acceleration is assumed to be perpendicular to the page, $\sum F_y$ only consists of the force resulted from the compression of spring, it follows that

$$k\,(y_0 - L\sin\theta) = \rho V^2 A \sin\theta, \;\; \longrightarrow \;\; \theta = \sin^{-1}\left(\frac{k y_0}{\rho V^2 A + kL}\right), \tag{B.88}$$

if the hinge is subject only the reactions along the x-direction.

5.17 The region inside the tank is chosen as the finite control-volume, and the origin of rectangular coordinates $\{x, y\}$ is located on the moving tank, with x and y the horizontal and vertical coordinates, respectively. Applying the global balance of mass to the CV yields

$$\frac{\partial}{\partial t}\int_{CV} \rho\,\mathrm{d}v + \int_{CS} \rho\,\boldsymbol{u}\cdot\boldsymbol{n}\,\mathrm{d}a = 0, \tag{B.89}$$

which is simplified to

$$\begin{aligned} \frac{\partial M_{CV}}{\partial t} &= -\int_{CS} \rho\,\boldsymbol{u}\cdot\boldsymbol{n}\,\mathrm{d}a = \rho\,(V-U)\,A, \\ \longrightarrow \quad \frac{\mathrm{d}M_{CV}}{\mathrm{d}t} &= \rho\,(V-U)\,A, \end{aligned} \tag{B.90}$$

where M_{CV} is the mass of the water contained inside the CV, and U represents the velocity of tank along the x-direction. They are both functions of time. Applying the global balance of linear momentum in non-inertial reference frame to the CV along the x-direction gives

$$\sum F_x - M_{CV} a_x = \frac{\partial}{\partial t}\int_{CV} \rho\,u\,\mathrm{d}v + \int_{CS} \boldsymbol{u}\,(\rho\,\boldsymbol{u}\cdot\boldsymbol{n})\,\mathrm{d}a, \tag{B.91}$$

which, under the assumptions that water is incompressible and all frictional forces acting on the tank are negligible, is simplified to

$$M_{CV}\frac{\mathrm{d}U}{\mathrm{d}t} = \rho\,(V-U)^2\,A, \tag{B.92}$$

where $a_x = \mathrm{d}U/\mathrm{d}t$, and it is noted that the water velocity inside the CV, within the moving reference frame, is extremely small, so that the volume integration in Eq. (B.91) can be neglected as a first engineering approximation. Substituting Eq. $(B.90)_2$ into Eq. (B.92) yields

$$\begin{aligned} M_{CV}\frac{\mathrm{d}U}{\mathrm{d}t} &= -M_{CV}\frac{\mathrm{d}(V-U)}{\mathrm{d}t} = (V-U)\frac{\mathrm{d}M_{CV}}{\mathrm{d}t}, \\ \longrightarrow \quad \frac{\mathrm{d}(V-U)}{V-U} &= -\frac{\mathrm{d}M_{CV}}{M_{CV}}, \end{aligned} \tag{B.93}$$

which is integrated to obtain

$$M_{CV} = \frac{M_0 V}{V-U}, \tag{B.94}$$

for $M_{CV}(t=0) = M_0$. Substituting this result into Eq. (B.92) gives rise to

$$\frac{\mathrm{d}(V-U)}{(V-U)^3} = -\frac{\rho A}{M_0 V}\mathrm{d}t, \tag{B.95}$$

which is integrated to obtain

$$\frac{U}{V} = 1 - \left(1 + 2\frac{\rho V A}{M_0}t\right)^{-1/2}. \tag{B.96}$$

As $t \to \infty$, $U \to V$, so that the tank will eventually move with constant velocity $U = V$.

5.19 The problem is first solved by constructing the finite control-volume with the coordinates $\{x, y, z\}$, which rotate coherently with the system, as shown in the below figure. The global balance of angular momentum for the CV in non-inertia reference frame reads

$$\begin{aligned} &\sum \boldsymbol{r} \times \boldsymbol{F}_{CS} + \int_{CV} (\boldsymbol{r} \times g)\rho\,\mathrm{d}v + \boldsymbol{T} \\ &- \int_{CV} \boldsymbol{r} \times [2\boldsymbol{\omega} \times \boldsymbol{u} + \boldsymbol{\omega} \times (\boldsymbol{\omega} \times \boldsymbol{r}) + \dot{\boldsymbol{\omega}} \times \boldsymbol{r}]\,\rho\,\mathrm{d}v \\ &= \frac{\partial}{\partial t}\int_{CV} (\boldsymbol{r} \times \boldsymbol{u})\,\rho\,\mathrm{d}v + \int_{CS} (\boldsymbol{r} \times \boldsymbol{u})(\rho\,\boldsymbol{u} \cdot \boldsymbol{n})\mathrm{d}a, \end{aligned} \tag{B.97}$$

where $\boldsymbol{T}$ represents the external torque, g is the gravitational acceleration, and $\boldsymbol{r}$ and $\boldsymbol{u}$ denote respectively the position vector and velocity of a specific point that are measured in in the rotating reference frame. For simplicity, it is assumed that there exists no surface force, the gravitational acceleration points in the z-direction, the flow is steady with respect to the rotating reference frame, and $\boldsymbol{r}$ and $\boldsymbol{u}$ are collinear, so that Eq. (B.97) reduces to

$$\boldsymbol{T} = \int_{CV} \boldsymbol{r} \times [2\boldsymbol{\omega} \times \boldsymbol{u} + \boldsymbol{\omega} \times (\boldsymbol{\omega} \times \boldsymbol{r}) + \dot{\boldsymbol{\omega}} \times \boldsymbol{r}]\,\rho\,\mathrm{d}v. \tag{B.98}$$

For the upper branch, it follows from the geometric configurations that

$$\begin{aligned} \boldsymbol{\omega} &= \omega\boldsymbol{k}, & \boldsymbol{r} &= L\left(\cos\alpha\boldsymbol{k} + \sin\alpha\boldsymbol{i}\right), \\ \boldsymbol{u} &= \frac{Q}{2A}\left(\cos\alpha\boldsymbol{k} + \sin\alpha\boldsymbol{i}\right), & A &= \frac{\pi D^2}{4}, \end{aligned} \tag{B.99}$$

with which Eq. (B.98) becomes

$$\begin{aligned}\boldsymbol{T}_u = \rho A \bigg[&-\left(\frac{L^2\omega Q}{2A} + \frac{\dot{\omega}L^3}{3}\right)\sin\alpha\cos\alpha\boldsymbol{i} \\ &+\left(\frac{L^2\omega Q}{2A} + \frac{\dot{\omega}L^3}{3}\right)\sin^2\alpha\boldsymbol{k} - \frac{\omega^2 L^3}{3}\sin\alpha\cos\alpha\boldsymbol{j}\bigg],\end{aligned} \tag{B.100}$$

in which it is noted that $\mathrm{d}v$ is given by $\mathrm{d}v = A\mathrm{d}r$, and the integration is conducted from $r = 0$ to $r = L$. Similarly, for the lower branch, it is found that

$$\boldsymbol{\omega} = \omega\boldsymbol{k}, \qquad \boldsymbol{r} = L\left(\cos\alpha\boldsymbol{k} - \sin\alpha\boldsymbol{i}\right), \qquad \boldsymbol{u} = \frac{Q}{2A}\left(\cos\alpha\boldsymbol{k} - \sin\alpha\boldsymbol{i}\right), \tag{B.101}$$

so that Eq. (B.98) becomes

$$\begin{aligned}\boldsymbol{T}_l = \rho A \bigg[&\left(\frac{L^2\omega Q}{2A} + \frac{\dot{\omega}L^3}{3}\right)\sin\alpha\cos\alpha\boldsymbol{i} \\ &+\left(\frac{L^2\omega Q}{2A} + \frac{\dot{\omega}L^3}{3}\right)\sin^2\alpha\boldsymbol{k} + \frac{\omega^2 L^3}{3}\sin\alpha\cos\alpha\boldsymbol{j}\bigg].\end{aligned} \tag{B.102}$$

Combining Eqs. (B.100) and (B.102) results in the total torque given by

$$\boldsymbol{T} = \boldsymbol{T}_u + \boldsymbol{T}_l = \rho A\left(\frac{L^2\omega Q}{A} + \frac{2\dot{\omega}L^3}{3}\right)\sin^2\alpha\boldsymbol{k}. \tag{B.103}$$

The steady-state portion of $\boldsymbol{T}$ is identified to be

$$\boldsymbol{T}_{steady} = \rho L^2\omega Q\sin^2\alpha\boldsymbol{k}, \tag{B.104}$$

and the torque needed to provide a non-vanishing angular acceleration is obtained as

$$\boldsymbol{T}_{acc} = \frac{2\rho A\dot{\omega}L^3}{3}\sin^2\alpha\boldsymbol{k}. \tag{B.105}$$

Now, the problem is solved by using the fixed control-volume with stationary coordinates $\{x, y, z\}$ shown in the figure, for which the global balance of angular momentum reads

$$\begin{aligned}\sum \boldsymbol{r}\times\boldsymbol{F}_{CS} + \int_{CV}(\boldsymbol{r}\times g)\rho\,\mathrm{d}v + \boldsymbol{T} = &\frac{\partial}{\partial t}\int_{CV}(\boldsymbol{r}\times\boldsymbol{u})\,\rho\,\mathrm{d}v \\ &+\int_{CS}(\boldsymbol{r}\times\boldsymbol{u})(\rho\boldsymbol{u}\cdot\boldsymbol{n})\mathrm{d}a,\end{aligned} \tag{B.106}$$

where $\boldsymbol{r}$ and $\boldsymbol{u}$ should be expressed in terms of the fixed coordinates. The first two terms on the left-hand-side and the first term on the right-hand-side are neglected based on the assumptions used previously, with which this equation is simplified to

$$\boldsymbol{T} = \int_{CS}(\boldsymbol{r}\times\boldsymbol{u})(\rho\boldsymbol{u}\cdot\boldsymbol{n})\mathrm{d}a. \tag{B.107}$$

It follows form the geometric configurations that for the upper branch,

$$\boldsymbol{r} = L\sin\alpha\boldsymbol{j}, \qquad \boldsymbol{u} = \omega L\boldsymbol{k}, \tag{B.108}$$

so that Eq. (B.107) becomes

$$\boldsymbol{T} = \boldsymbol{T}_{steady} = \rho\omega Q L^2 \sin^2 \alpha \boldsymbol{i}, \tag{B.109}$$

in which the contribution from the lower branch is exactly the same as that of the upper branch, for two branches are symmetric with respect to the rotating axis. The determined torque given in Eq. (B.109) is valid for a steady case. For the accelerating torque on the upper branch, it follows that

$$\|\boldsymbol{T}_{acc}\|_u = I\dot{\omega}, \qquad I = \int \|\boldsymbol{r}\|^2 \, \mathrm{d}m = \int_0^L (\ell \sin\alpha)^2 \rho A \, \mathrm{d}\ell = \frac{\rho A L^3}{3} \sin^2 \alpha. \tag{B.110}$$

Again, due to the symmetry of two branches with respect to the rotating axis, the accelerating torque for the lower branch assumes the same value, so that the total accelerating torque is obtained as

$$\|\boldsymbol{T}_{acc}\| = \frac{2\rho A \dot{\omega} L^3}{3} \sin^2 \alpha. \tag{B.111}$$

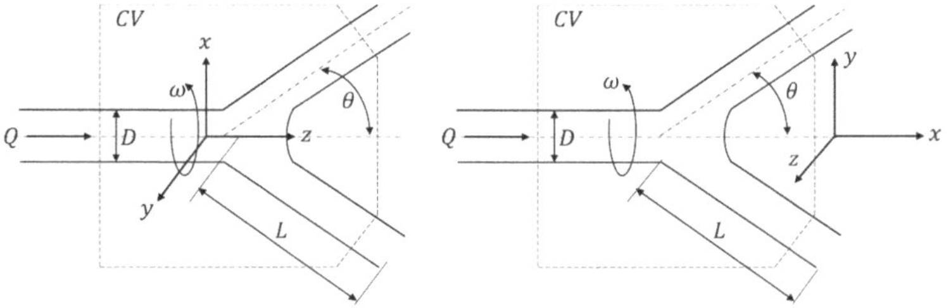

5.21 By using the assumption of incompressible fluid, the local balance of mass reads

$$\nabla \cdot \boldsymbol{u} = 0, \quad \longrightarrow \quad \frac{\partial u}{\partial x} + \frac{\partial v}{\partial y} = 0, \quad \longrightarrow \quad \frac{\partial u}{\partial x} = 0, \tag{B.112}$$

in which the assumption of fully developed flow, i.e., $v = v(x)$, has been used. Integrating the last equation yields $u = u(y)$. Since the no-slip boundary condition implies that $u(x = 0, y) = 0$, it follows that $u = 0$ over the entire flow field. Next, applying the Navier-Stokes equation along the x- and y-directions gives rise respectively to

$$\begin{aligned} u\frac{\partial u}{\partial x} + v\frac{\partial u}{\partial y} &= -\frac{1}{\rho}\frac{\partial p}{\partial x} + \nu\left(\frac{\partial^2 u}{\partial x^2} + \frac{\partial^2 u}{\partial y^2}\right), \\ u\frac{\partial v}{\partial x} + v\frac{\partial v}{\partial y} &= -\frac{1}{\rho}\frac{\partial p}{\partial y} - g + \nu\left(\frac{\partial^2 v}{\partial x^2} + \frac{\partial^2 v}{\partial y^2}\right), \end{aligned} \tag{B.113}$$

in which the assumption of steady flow has been used. With $u = 0$, the first equation reduces to

$$\frac{\partial p}{\partial x} = 0, \tag{B.114}$$

showing that $p = f(y)$ with f any differentiable function. Since $p(x = h, y) = p_0$ on the free surface of liquid film, where p_0 denotes the atmospheric pressure, it follows that $p = p_0$ in the entire flow field. Substituting this and $v = v(x)$ into Eq. (B.113)$_2$ yields

$$\frac{\mathrm{d}^2 v}{\mathrm{d}x^2} = \frac{g}{\nu}, \qquad \longrightarrow \qquad v(x) = \frac{g}{2\nu}x^2 + C_1 x + C_2, \tag{B.115}$$

where C_1 and C_2 are integration constants. Applying the boundary conditions $v(x = 0) = V_0$ and $\mathrm{d}v/\mathrm{d}x(x = h) = 0$ gives respectively $C_1 = -gh/\nu$ and $C_2 = V_0$, with which Eq. (B.115)$_2$ becomes

$$v(x) = \frac{g}{\nu}\left(\frac{x}{2} - h\right)x + V_0. \tag{B.116}$$

The volume flow rate per unit depth perpendicular to the page, q, is given by

$$q = \int_0^h v\,\mathrm{d}x = V_0 h - \frac{gh^3}{3\nu}, \tag{B.117}$$

so that the average flow velocity, v_{av}, is obtained as

$$v_{av} = \frac{q}{h} = V_0 - \frac{gh^2}{3\nu}, \tag{B.118}$$

showing that there will be a net upward liquid flow if $V_0 > gh^2/(3\nu) = V_c$. For example, for water at 1 atmospheric pressure and 20 °C, $\nu \sim 1.004 \times 10^{-6}$ m^2/s, so that for a water film with $h = 1$ mm, $V_c \sim 3.25$ m/s. Thus, the belt speed must be greater than V_c, which is a relatively large speed.

6.5 For the Thompson-type overfall weir, the flow rate Q is influenced by the overfall height h, the gravitational acceleration g, and the density ρ, and dynamic viscosity μ of fluid, so that

$$f\,(Q, h, g, \rho, \mu) = 0, \tag{B.119}$$

which is assumed to be dimensionally homogeneous. By using the MKS system, the dimensional matrix is obtained as

$$\begin{bmatrix} 0 & 0 & 0 & 1 & 1 \\ 3 & 1 & 1 & -3 & -1 \\ -1 & 0 & -2 & 0 & -1 \end{bmatrix}, \tag{B.120}$$

whose rank is 3. Thus, there exist two independent dimensionless products identified to be

$$\Pi_1 = \frac{Q^2}{2gh^5}, \qquad \Pi_2 = \frac{gh^3}{\nu^2}, \tag{B.121}$$

where $\nu = \mu/\rho$, so that Eq. (B.119) is brought to

$$\frac{Q^2}{2gh^5} = f'\left(\frac{gh^3}{\nu^2}\right). \tag{B.122}$$

If the dynamic viscosity does not play a role in the determination of Q, Eq. (B.122) can be simplified to

$$Q \sim \sqrt{2g}\,h^{5/2}. \tag{B.123}$$

For the Poincelet-type overfall weir, the weir width b also influences the volume flow rate Q, so that Eq. (B.119) is extended to

$$f\,(Q, h, b, g, \rho, \mu) = 0. \tag{B.124}$$

By using the MKS system, the rank of the emerging dimensional matrix is 3, so that there exist three dimensionless products, which are obtained as

$$\Pi_1 = \frac{Q^2}{2gh^5}, \qquad \Pi_2 = \frac{gh^3}{\nu^2}, \qquad \Pi_3 = \frac{b}{h}, \tag{B.125}$$

with which

$$Q = \sqrt{2g}\,h^{5/2} f'\left(\frac{gh^3}{\nu^2}, \frac{b}{h}\right). \tag{B.126}$$

It is also possible to request a linear dependency of Q upon b with insignificant role played by μ, so that this equation is simplified to

$$Q \sim \sqrt{2g}\,b\,h^{3/2}. \tag{B.127}$$

Equations (B.123) and (B.127) have been throughly tested, and the two overfall weirs are used intensively in hydraulic engineering to measure the discharges of *open-channel flows*.

6.6 First, it is assumed that $y = f(x_1, x_2, \cdots, x_n)$ is a dimensionally homogeneous equation, but there exists no product of powers of x_i with the same dimension as y. It is further assumed that the rank of argumented matrix $[a_{i0} : a_{ij}]$ is r. It follow from Eq. (6.3.40) that the rank of matrix $[a_{ij}]$ must be smaller than r. For further exploitation, it is assumed that a vanishing determinant of $[a_{i0} : a_{ij}]$ lies in the upper left corner without loss of generality. If $r = m$, where m represents the number of independent fundamental dimensions, then the determinant Δ of $[a_{i0} : a_{ij}]$ is given by

$$\Delta = \begin{vmatrix} a_{10} & a_{11} & a_{12} & \cdots & a_{1(m-1)} \\ a_{20} & a_{21} & a_{22} & \cdots & a_{2(m-1)} \\ \vdots & & & & \vdots \\ a_{m0} & a_{m1} & a_{m2} & \cdots & a_{m(m-1)} \end{vmatrix} \neq 0, \tag{B.128}$$

which may be recast alternatively as

$$\Delta = \sum_{i=1}^{m} A_{i0}a_{i0} = A_{10}a_{10} + A_{20}a_{20} + \cdots + A_{m0}a_{m0}, \tag{B.129}$$

where A_{i0} is the *co-factor* of a_{i0}. The theory of determinants implies that

$$\sum_{i=1}^{m} A_{i0}a_{ik} = 0, \qquad \forall k = 1, 2, \cdots, n. \tag{B.130}$$

Since y is assumed to be dimensionally homogeneous, Eqs. (6.3.5) and (6.3.8) hold for the variables $\alpha_j,\ j = 1, 2, \cdots, m$. Now, a new set of the fundamental units is introduced as

$$\alpha_i = G^{A_{i0}}, \qquad i = 1, 2, \cdots, m, \tag{B.131}$$

where G is any positive real number, with which Eq. $(6.3.8)_2$ becomes

$$K_j = \prod_{i=1}^{m} \alpha_i^{a_{ij}} = G^{\sum_{i=1}^{m} A_{i0}a_{ij}} = 1, \qquad j = 1, 2, \cdots, n, \tag{B.132}$$

showing that all K_js can be made to equal to unity by specific values of α_i. Similarly, the value of K_0 is obtained as

$$K_0 = \prod_{i=1}^{m} \alpha_i^{a_{i0}} = G^{\sum_{i=1}^{m} A_{i0}a_{i0}} \neq 1. \tag{B.133}$$

Substituting Eqs. (B.132) and (B.133) into Eq. (6.3.7) gives

$$K_0 y = f\left(x_1, x_2, \cdots, x_n\right), \qquad K_0 = G^{\sum_{i=1}^{m} A_{i0}a_{i0}}, \tag{B.134}$$

in which K_0 can arbitrarily be assigned since $G > 0$, as implied by that G is any positive real number. It is concluded that Eq. $(B.134)_1$ cannot be a function, which contradicts to the assumption at the beginning. Thus, the initial assumption is incorrect. By these, the proof has been accomplished for the case $r = m$.

Next, consider the case in which $r < m$, for which Δ is of size r, so that there exists an $r \times r$ matrix with its non-vanishing determinant given by

$$\Delta = \begin{vmatrix} a_{10} & a_{11} & a_{12} & \cdots & a_{1(r-1)} \\ a_{20} & a_{21} & a_{22} & \cdots & a_{2(r-1)} \\ \vdots & & & & \vdots \\ a_{m0} & a_{m1} & a_{m2} & \cdots & a_{r(r-1)} \end{vmatrix} \neq 0, \qquad r < m, \tag{B.135}$$

for which the corresponding results given in Eqs. (B.129) and (B.130), by using a similar procedure discussed previously, are obtained respectively as

$$\begin{aligned} \Delta &= \sum_{i=1}^{r} A_{i0}a_{i0} = A_{10}a_{10} + A_{20}a_{20} + \cdots + A_{r0}a_{r0}, \\ &\sum_{i=1}^{r} A_{i0}a_{ik} = 0, \ \forall k = 1, 2, \cdots, n. \end{aligned} \tag{B.136}$$

With $\alpha_i = G^{A_{i0}}$ for $i = 1, 2, \cdots, r$ and $\alpha_j = 1$ for $r < j < m$, one obtains

$$K_j = G^{\sum_{i=1}^{r} A_{i0}a_{ij}} = G^0 = 1, \qquad K_0 = G^{\sum_{i=1}^{r} A_{i0}a_{i0}} \neq 1, \tag{B.137}$$

with which the same conclusion is reached. Letting $a_j = 1$ for all $j > r$ corresponds to a permissible change of the fundamental units. It follows from Eqs. (6.3.24) and (6.3.28) that a dimensionally homogeneous equation in the form $y = f(x_1, x_2, \cdots, x_n)$ can always be brought to a dimensionless form

$$\Pi = F\left(x_1, x_2, \cdots, x_n\right), \tag{B.138}$$

in which Π is dimensionless and F represents a new function. The proof is now complete, QED.

6.10 Expanding the material derivative of $\boldsymbol{u}$ in the given equation gives rise to

$$\frac{\partial \boldsymbol{u}}{\partial t} + (\boldsymbol{u} \cdot \nabla)\, \boldsymbol{u} + 2\boldsymbol{\omega} \times \boldsymbol{u} = -\frac{1}{\rho} \nabla p. \tag{B.139}$$

Defining the scaling variables

$$\boldsymbol{u} = V\bar{\boldsymbol{u}}, \qquad p = (\Delta p)\bar{p}, \qquad \boldsymbol{\omega} = \Omega\bar{\boldsymbol{\omega}} \qquad \boldsymbol{x} = L\bar{\boldsymbol{x}}, \qquad t = \left(\frac{L}{V}\right)\bar{t}, \tag{B.140}$$

where $\{V, \Delta p, \Omega, L, L/V\}$ are respectively the characteristic velocity, pressure differential, angular speed, length and timescales, and substituting these expressions into Eq. (B.139) results in

$$\frac{\partial \bar{\boldsymbol{u}}}{\partial \bar{t}} + \left(\bar{\boldsymbol{u}} \cdot \bar{\nabla}\right) \bar{\boldsymbol{u}} + 2\left(\frac{\Omega L}{V}\right) \bar{\boldsymbol{\omega}} \times \bar{\boldsymbol{u}} = -\left(\frac{\Delta p}{\rho V^2}\right) \bar{\nabla} \bar{p}, \tag{B.141}$$

so that there exist two dimensionless numbers given by

$$\Pi_1 = \frac{\Omega L}{V}, \qquad \Pi_2 = \frac{\Delta p}{\rho V^2}, \tag{B.142}$$

which are the Strouhal number and pressure coefficient, respectively. The last term on the left-hand-side of Eq. (B.141) is the Coriolis force in a rotating reference frame, which leads to the phenomena such as geostrophic flow in atmospheric science.

7.2 The flow inside the inclined pipe reducer is assumed to be incompressible, frictionless, and steady, so the the Bernoulli equation along the pipe centerline (a streamline) between points 1 and 2 reads

$$\frac{p_1}{\rho_w} + \frac{u_1^2}{2} = \frac{p_2}{\rho_w} + \frac{u_2^2}{2} + g\ell, \tag{B.143}$$

where the datum of elevation is located at point 1, ℓ represents the difference in elevation between two points, and ρ_w denotes the density of water. The pressure difference $p_1 - p_2$ is obtained as

$$p_1 - p_2 = \rho_w g\ell + \frac{\rho_w}{2}\left(u_2^2 - u_1^2\right) = \rho_w g\ell + \frac{\rho_w V^2}{2}\left[1 - \left(\frac{d_2}{d_1}\right)^4\right], \tag{B.144}$$

because $u_1 A_1 = u_2 A_2$, as indicated by the continuity equation, where $A_1 = \pi d_1^2/4$, $A_2 = \pi d_2^2/4$ and $u_2 = V$. On the other hand, the same pressure difference can be evaluated by using the manometer, which is given by

$$\begin{aligned} & p_1 - \gamma_w \ell' + \gamma_0 h + \gamma_w \left(\ell' - h - \ell\right) = p_2, \\ \longrightarrow \quad & p_1 - p_2 = (\gamma_w - \gamma_0)\, h + \gamma_w \ell, \end{aligned} \tag{B.145}$$

where γ_w is the specific weight of water, ℓ' represents the difference in elevation between point 1 and the interface between water and manometer liquid with specific gravity γ_0 in left tube. Substituting this equation into Eq. (B.144) results in

$$h = \frac{\rho_w V^2}{2(\gamma_w - \gamma_0)} \left[1 - \left(\frac{d_2}{d_1}\right)^4\right], \tag{B.146}$$

which assumes a positive value, for $\gamma_w > \gamma_0$ and $d_2 < d_1$. If the specific gravity of liquid in the manometer is denoted by s, Eq. (B.146) can be further simplified to

$$h = \frac{V^2}{2g(1-s)} \left[1 - \left(\frac{d_2}{d_1}\right)^4\right]. \tag{B.147}$$

7.8 Let points 1 and 2 be located on the free liquid surfaces in the left and right tubes, respectively, which are connected by a streamline. The flow is assumed to be incompressible and frictionless, but unsteady, so that the Bernoulli equation along a streamline between points 1 and 2 reads

$$\frac{p_1}{\rho} + \frac{u_1^2}{2} + gz_1 = \frac{p_2}{\rho} + \frac{u_2^2}{2} + gz_2 + \int_1^2 \frac{\partial u}{\partial t}\,\mathrm{d}s, \tag{B.148}$$

where $\mathrm{d}s$ represents an infinitesimal line element along the streamline. Let the datum of elevation be located at the equilibrium free liquid surface, i.e., the dashed line, and $z_1 = z$, so that the continuity equation between points 1 and 2 implies that

$$\frac{\pi d_1^2}{4} z = \frac{\pi d_2^2}{4} |z_2|, \qquad \longrightarrow \qquad |z_2| = z \left(\frac{d_1}{d_2}\right)^2. \tag{B.149}$$

The line integral in Eq. (B.148) is evaluated to be

$$\int_1^2 \frac{\partial u}{\partial t}\,\mathrm{d}s = -\ddot{z}(z + \ell_1) - \int_0^L \left(\frac{d_1}{d}\right)^2 \ddot{z}\,\mathrm{d}s - \left(\frac{d_1}{d_2}\right)^2 \ddot{z} \left[\ell_2 - \left(\frac{d_1}{d_2}\right)^2 z\right], \tag{B.150}$$

which is recast alternatively as

$$\begin{aligned}\int_1^2 \frac{\partial u}{\partial t}\,\mathrm{d}s &= -\ddot{z}\left[z + \ell_1 + \int_0^L \left(\frac{d_1}{d}\right)^2 \mathrm{d}s + \left(\frac{d_1}{d_2}\right)^2 \ell_2 - \left(\frac{d_1}{d_2}\right)^4 z\right] \\ &= -\ddot{z}\mathcal{L}(z),\end{aligned} \tag{B.151}$$

where d represents the diameter of shrinking tube with length L, which decreases linearly from d_1 to d_2, so that $d = d_1 - (d_1 - d_2)s/L$, with s the distance measured from the interface between the left and shrinking tubes. It is noted that the terms inside the bracket of this equation are only a function of z, and the fluid velocity inside the U-tube is everywhere in the reverse direction of $\mathrm{d}s$.

Since $p_1 = p_0$ and $p_2 = p_0$, where p_0 represents the atmospheric pressure, substituting these and Eqs. (B.149)$_2$ and (B.151) into Eq. (B.148) yields

$$\frac{\dot{z}^2}{2} + gz = -\ddot{z}\mathcal{L}(z) + \frac{\dot{z}^2}{2}\left(\frac{d_1}{d_2}\right)^4 - \left(\frac{d_1}{d_2}\right)^2 gz, \tag{B.152}$$

which is simplified to

$$\ddot{z}\mathcal{L}(z) + \frac{1}{2}\left[1 - \left(\frac{d_1}{d_2}\right)^4\right]\dot{z}^2 + g\left[1 + \left(\frac{d_1}{d_2}\right)^2\right]z = 0, \tag{B.153}$$

for the oscillating motion of free liquid surfaces. For small values of z, this nonlinear ODE, as an approximation, can be linearized as

$$\ddot{z} + \frac{g[1 + (d_1/d_2)^2]}{\mathcal{L}(0)}z = 0, \qquad \lim_{z\to 0}\mathcal{L}(z) = \mathcal{L}(0), \tag{B.154}$$

for the coefficient in front of $\dot{z}$ is nearly null, so that the natural frequency ω of oscillating motion is obtained as

$$\omega = \sqrt{\frac{g[1 + (d_1/d_2)^2]}{\mathcal{L}(0)}}. \tag{B.155}$$

7.12 (a) The flow is assumed to be two-dimensional, incompressible, and irrotational, which can be obtained by superimposing a source with strength m locating at $(x = \ell, y)$ and a source with same strength locating at $(x = -\ell, y)$. It follows from Table 7.2 that the velocity potential function ϕ is given by

$$\phi = \frac{m}{4\pi}\left\{\ln\left[(x-\ell)^2 + y^2\right] + \ln\left[(x+\ell)^2 + y^2\right]\right\}. \tag{B.156}$$

(b) The velocity component u along the x-direction is obtained by using the definition of ϕ, viz.,

$$u = \frac{\partial\phi}{\partial x} = \frac{m}{4\pi}\left[\frac{2(x-\ell)}{(x-\ell)^2 + y^2} + \frac{2(x+\ell)}{(x+\ell)^2 + y^2}\right], \tag{B.157}$$

whose value at $x = 0$ is obtained as

$$u(x=0) = \frac{m}{4\pi}\left(\frac{-2\ell}{\ell^2 + y^2} + \frac{2\ell}{\ell^2 + y^2}\right) = 0, \tag{B.158}$$

showing that there is no flow through the wall, and the flow on the wall is in parallel to the wall.

(c) The velocity component v in the y-direction is given by

$$v = \frac{\partial\phi}{\partial y} = \frac{m}{\pi}\left[\frac{y}{(x-\ell)^2 + y^2}\right], \tag{B.159}$$

whose value on the wall is obtained as

$$v(x=0) = \frac{m}{\pi}\left(\frac{y}{\ell^2 + y^2}\right). \tag{B.160}$$

The fluid velocity at the wall, V_w, is given by $V_w = \sqrt{u^2(x=0) + v^2(x=0)} = v(x=0)$, and by using the Bernoulli equation, the pressure on the wall, p_w, is obtained as

$$p_0 = p_w + \frac{\rho V_w^2}{2}, \qquad \longrightarrow \qquad p_w = p_0 - \frac{\rho m^2}{2\pi^2}\left(\frac{y}{\ell^2 + y^2}\right)^2. \tag{B.161}$$

7.17 To accomplish the required flow field, a doublet with strength μ is located at a distance ℓ from the origin along the x-axis, and a doublet with strength μ' is located at a distance a^2/ℓ also along the x-axis from the origin, as shown in Fig. 7.25d without the line-distributed sink. For an arbitrary point P with distances ξ and η respectively from μ and μ', the stream function, by using the principle of superposition, is given by

$$\psi(r,\theta) = -\frac{\mu}{4\pi\xi}\sin^2\beta - \frac{\mu'}{4\pi\eta}\sin^2\alpha. \tag{B.162}$$

Let point P locate on the sphere surface with radius $r = a$. The geometric considerations of the triangles $\Delta OP\mu$ and $\Delta OP\mu'$ show that

$$\begin{aligned} &\frac{a}{\sin(\pi-\beta)} = \frac{\xi}{\sin\theta}, \qquad \frac{a}{\sin(\pi-\alpha)} = \frac{\xi}{\sin\theta}, \\ &\xi^2 = a^2 + \ell^2 - 2a\ell\cos\theta,\ \eta^2 = a^2 + \frac{a^4}{\ell^2} - 2\frac{a^3}{\ell}\cos\theta. \end{aligned} \tag{B.163}$$

Substituting Eq. (B.163)$_1$ into Eq. (B.162) yields

$$\psi(a,\theta) = -\frac{a^2\sin\theta}{4\pi\eta^3}\left(\frac{\eta^3\mu}{\xi^3} + \mu'\right), \tag{B.164}$$

showing that $\psi \neq$ constant at $r = a$, unless $\mu' = -(\eta/\xi)^3\mu$ is satisfied. Substituting this condition into Eq. (B.163)$_2$ and eliminating the term $\cos\theta$ gives

$$\mu' = -\left(\frac{a}{\ell}\right)^3\mu, \tag{B.165}$$

with which Eq. (B.162) is simplified to

$$\psi(r,\theta) = -\frac{\mu}{4\pi\xi}\sin^2\beta + \frac{\mu a^3}{4\pi\ell^3\eta}\sin^2\eta, \tag{B.166}$$

which corresponds to a sphere of radius a with a doublet of strength μ locating at $r = \ell$ along the x-axis. The corresponding velocity potential function is then obtained as

$$\phi(r,\theta) = \frac{\mu}{4\pi\xi^2}\cos\beta - \frac{\mu a^3}{4\pi\ell^3\eta^2}\cos\alpha, \tag{B.167}$$

where the first term on the right-hand-side is the contribution due to the doublet at $r = \ell$, while the second term denotes that due to the doublet at $r = a^2/\ell$. Since in Eq. (7.6.78), $\boldsymbol{u}_i$ represents the velocities induced by all singularities of the flow field except that of the doublet at $r = \ell$, Eq. (B.167) may reduce to

$$\phi(r,\theta) = -\frac{\mu a^3}{4\pi\ell^3\eta^2}\cos\alpha, \longrightarrow \phi(x,0) = -\frac{\mu a^3}{4\pi\ell^3(x - a^2/\ell)^2}, \tag{B.168}$$

where the second equation is evaluated along the x-axis, in which $r = x < a^2/\ell$, $\alpha = 0$, and $\eta = x - a^2/\ell$. Substituting Eq. (B.168)$_2$ into Eq. (7.6.78) results in

$$\frac{\partial\boldsymbol{u}_i}{\partial x}(x=\ell,0) = -\frac{3\mu a^3\ell}{2\pi(\ell^2-a^2)^4}\boldsymbol{e}_x, \longrightarrow \boldsymbol{f} = \frac{3\rho\mu^2a^3\ell}{2\pi(\ell^2-a^2)^4}\boldsymbol{e}_x. \tag{B.169}$$

7.20 Substituting the given expression of wave shape into the equations of given potential and kinetic energies per wavelength yields

$$PE = \frac{1}{2}\rho g \varepsilon^2 \int_0^\lambda \sin^2\left[\frac{2\pi}{\lambda}(x - ct)\right] \mathrm{d}x = \frac{1}{4}\rho g \varepsilon^2 \lambda,$$
$$KE = \frac{\rho \pi c^2 \varepsilon^2}{\lambda} \int_0^\lambda \coth\left(\frac{2\pi}{\lambda}\right) \cos^2\left[\frac{2\pi}{\lambda}(x - ct)\right] \mathrm{d}x = \frac{1}{4}\rho g \varepsilon^2 \lambda, \tag{B.170}$$

in which Eq. $(7.7.18)_2$ has been used. Thus, $PE = KE$ is identified.

8.7 For the considered problem, Eq. (8.2.40) is still valid for the velocity component u along the x-direction, except that the boundary conditions should be revised to

$$u(y = 0, t) = U \cos(\omega t), \qquad u(y = h, t) = 0. \tag{B.171}$$

Let u be expressed by $u = Re\left[f(y)\exp(i\omega t)\right]$. Substituting this expression into Eq. (8.2.40) yields

$$\frac{\mathrm{d}^2 f}{\mathrm{d}y^2} - \alpha^2 f = 0, \qquad \alpha = \sqrt{\frac{i\omega}{\nu}}, \tag{B.172}$$

to which the solution is obtained as

$$f(y) = C_1 \exp(\alpha y) + C_2 \exp(-\alpha y), \tag{B.173}$$

which is subject to the boundary conditions given by $f(y = 0) = U$ and $f(y = h) = 0$, as implied by Eq. (B.171). It follows that

$$C_1 = -U\frac{\exp(-\alpha h)}{\exp(\alpha h) - \exp(-\alpha h)}, \qquad C_2 = U\frac{\exp(\alpha h)}{\exp(\alpha h) - \exp(-\alpha h)}, \tag{B.174}$$

with which Eq. (B.173) becomes

$$f(y) = U\frac{\sinh[\alpha(h - y)]}{\sinh(\alpha h)}, \tag{B.175}$$

so that u is given by

$$u(y, t) = Re\left[U\frac{\sinh[\alpha(h - y)]}{\sinh(\alpha h)}\exp(i\omega t)\right]. \tag{B.176}$$

8.9 Substituting Eq. $(8.3.28)_2$ into Eq. (8.3.14) gives

$$M_i = -\frac{4c\mu}{r^4}\int_A \varepsilon_{ijk} x_j x_k \,\mathrm{d}a, \tag{B.177}$$

in which Eq. (8.3.31) has been used. In view of the properties of permutation symbol, it is found that $M_i = 0$, for $\varepsilon_{123}x_2x_3 = -\varepsilon_{132}x_3x_2$, $\varepsilon_{312}x_1x_2 = -\varepsilon_{321}x_2x_1$ and $\varepsilon_{231}x_3x_1 = -\varepsilon_{213}x_1x_3$.

8.12 For a steady, two-dimensional isothermal flow of a Newtonian fluid with constant density and dynamic viscosity, the continuity and the Navier-Stokes equations along the x- and y-directions reduce respectively to

$$\begin{aligned} \frac{\partial u}{\partial x} + \frac{\partial v}{\partial y} &= 0, \\ u\frac{\partial u}{\partial x} + v\frac{\partial u}{\partial y} &= -\frac{1}{\rho}\frac{\partial p}{\partial x} + \frac{\mu}{\rho}\left(\frac{\partial^2 u}{\partial x^2} + \frac{\partial^2 u}{\partial y^2}\right), \\ u\frac{\partial v}{\partial x} + v\frac{\partial v}{\partial y} &= -\frac{1}{\rho}\frac{\partial p}{\partial y} + \frac{\mu}{\rho}\left(\frac{\partial^2 v}{\partial x^2} + \frac{\partial^2 v}{\partial y^2}\right), \end{aligned} \tag{B.178}$$

in which the body forces are omitted for simplicity. Define the scaling variables with the stretched coordinates as follows:

$$\bar{x} = \frac{x}{L}, \qquad \bar{y} = \frac{y}{L}\sqrt{\mathcal{R}_e}, \qquad \bar{u} = \frac{u}{U}, \qquad \bar{v} = \frac{v}{U}\sqrt{\mathcal{R}_e}, \qquad \bar{p} = \frac{p - p_0}{\rho U^2}, \tag{B.179}$$

where $\{L, U\}$ are respectively the characteristic length and velocity scales, p_0 denotes the reference pressure, and $\mathcal{R}_e$ represents the Reynolds number. Substituting these expressions into Eq. (B.178) with a limiting procedure $\mathcal{R}_e \to \infty$ results in

$$\frac{\partial \bar{u}}{\partial \bar{x}} + \frac{\partial \bar{v}}{\partial \bar{y}} = 0, \qquad \bar{u}\frac{\partial \bar{u}}{\partial \bar{x}} + \bar{v}\frac{\partial \bar{u}}{\partial \bar{y}} = -\frac{\partial \bar{p}}{\partial \bar{x}} + \frac{\partial^2 \bar{u}}{\partial \bar{y}^2}, \qquad 0 = -\frac{\partial \bar{p}}{\partial \bar{y}}. \tag{B.180}$$

The dimensional forms of these equations with $\mathcal{R}_e \to \infty$ are then given by

$$\frac{\partial u}{\partial x} + \frac{\partial v}{\partial y} = 0, \qquad u\frac{\partial u}{\partial x} + v\frac{\partial u}{\partial y} = -\frac{1}{\rho}\frac{\mathrm{d}p}{\mathrm{d}x} + \frac{\mu}{\rho}\frac{\partial^2 u}{\partial y^2}, \tag{B.181}$$

which are the boundary-layer equations. For a uniform flow with $u = U(x)$, Eq. $(\text{B.181})_2$ reduces to

$$u\frac{\partial u}{\partial x} + v\frac{\partial u}{\partial y} = U\frac{\mathrm{d}U}{\mathrm{d}x} + \frac{\mu}{\rho}\frac{\partial^2 u}{\partial y^2}, \tag{B.182}$$

in which the Bernoulli equation has been used.

8.16 Substituting Eq. (8.5.55) into Eq. $(8.5.53)_2$ with $\mathcal{P}_r = 1$ yields

$$-\frac{6B\eta}{(a+\eta^2)^4} + \frac{AB\eta}{(a+\eta^2)^4} = 0 \qquad \longrightarrow \qquad A = 6, \tag{B.183}$$

by which it is found that

$$\begin{aligned} f' &= \frac{12\eta}{a+\eta^2} - \frac{12\eta^3}{(a+\eta^2)^2}, \\ f'' &= \frac{12}{a+\eta^2} - \frac{60\eta^2}{(a+\eta^2)^2} + \frac{48\eta^4}{(a+\eta^2)^3}, \\ f''' &= -\frac{144\eta}{(a+\eta^2)^2} + \frac{432\eta^3}{(a+\eta^2)^3} - \frac{288\eta^5}{(a+\eta^2)^4}, \\ \frac{\mathrm{d}}{\mathrm{d}\eta}\left(\frac{f'}{\eta}\right) &= -\frac{48\eta}{(a+\eta^2)^2} + \frac{48\eta^3}{(a+\eta^2)^3}. \end{aligned} \tag{B.184}$$

Substituting these expressions into Eq. (8.5.54)$_2$ gives

$$12aB\int_0^\infty \frac{\eta\,\mathrm{d}\eta}{(a+\eta^2)^5} = \frac{3B}{2a^3} = \frac{1}{2\pi}, \qquad \longrightarrow \qquad B = \frac{a^3}{3\pi}. \tag{B.185}$$

On the other hand, substituting Eq. (B.184) into Eq. (8.5.53)$_1$ gives rise to $B = 96a$, which, when compared with Eq. (B.185)$_2$, results in

$$a = 12\sqrt{2\pi}, \qquad B = \frac{(12\sqrt{2\pi})^3}{3\pi}. \tag{B.186}$$

With these, $f(\eta)$ and $F(\eta)$ are obtained as

$$f(\eta) = \frac{6\eta^2}{12\sqrt{2\pi}+\eta^2}, \qquad F(\eta) = \frac{(12\sqrt{2\pi})^3}{3\pi}\frac{1}{(12\sqrt{2\pi}+\eta^2)^3}. \tag{B.187}$$

The expressions of $\psi(x, r)$, $\theta(x, r)$, and $\eta(x, r)$ can then be obtained by substituting Eqs. (B.187) and (8.5.52) into Eq. (8.5.47).

8.19 Let the radius of a smooth circular pipe be denoted by a. For a fully developed laminar flow, the axial velocity component is given in Eq. (8.2.24)$_2$. Substituting this expression into Eq. (8.6.43) yields

$$\begin{aligned}\frac{1}{2}\alpha(\rho u_{av}\pi a^2)u_{av}^2 &= \rho\pi u_{max}^3\int_0^a\left[1-\left(\frac{r}{a}\right)^2\right]^3 r\,\mathrm{d}r = \frac{\rho\pi a^2}{8}u_{max}^3,\\ \beta(\rho u_{av}\pi a^2)u_{av} &= 2\pi\rho u_{max}^2\int_0^a\left[1-\left(\frac{r}{a}\right)^2\right]^2 r\,\mathrm{d}r = \frac{\rho\pi a^2}{3}u_{max}^2.\end{aligned} \tag{B.188}$$

Since $u_{av} = u_{max}/2$, it follows that $\alpha = 2$ and $\beta = 4/3$. For a fully developed turbulent flow, substituting Eq. (8.6.39) into Eq. (8.6.43) gives

$$\begin{aligned}\frac{1}{2}\alpha(\rho\bar{u}_{av}\pi a^2)\bar{u}_{av}^2 &= \rho\pi\bar{u}_c^3\int_0^a\left(1-\frac{r}{a}\right)^{3/n} r\,\mathrm{d}r = \rho\pi a^2\bar{u}_c^3\frac{n^2}{(n+3)(2n+3)},\\ \beta(\rho\bar{u}_{av}\pi a^2)\bar{u}_{av} &= 2\pi\rho\bar{u}_c^2\int_0^a\left(1-\frac{r}{a}\right)^{2/n} r\,\mathrm{d}r = 2\rho\pi a^2\bar{u}_c^2\frac{n^2}{(n+2)(2n-2)}.\end{aligned} \tag{B.189}$$

Substituting Eq. (8.6.41) into the above expressions results in the expressions of α and β as those given in Eq. (8.6.46). The integrations in Eq. (B.189) may be accomplished by changing the integrated variables.

9.2 The balances of mass and linear momentum for a sound wave in a one-dimensional circumstance are given respectively by

$$\frac{\partial\rho}{\partial t} + \frac{\partial(\rho u_1)}{\partial x_1} = 0, \qquad \frac{\partial u_1}{\partial t} + u_1\frac{\partial u_1}{\partial x_1} = -\frac{1}{\rho}\frac{\partial p}{\partial x_1} = -\frac{1}{\rho}\frac{\mathrm{d}p}{\mathrm{d}\rho}\frac{\partial\rho}{\partial x_1}, \tag{B.190}$$

whose counterparts for a shallow liquid wave are given by

$$\frac{\partial h}{\partial t} + \frac{\partial(hu_1)}{\partial x_1} = 0, \qquad \frac{\partial u_1}{\partial t} + u_1\frac{\partial u_1}{\partial x_1} = -g\frac{\partial h}{\partial x_1}, \tag{B.191}$$

where g is the gravitational acceleration, and h is interpreted as the local liquid depth. It is seen that u_1 in a sound wave corresponds to u_1 in a shallow liquid wave, while ρ in a sound wave plays the same role of h in a shallow liquid wave. Furthermore, equalizing the right-hand-sides of Eqs. (B.190)$_2$ and (B.191)$_2$ yields

$$\frac{1}{\rho}\frac{\mathrm{d}p}{\mathrm{d}\rho} = g. \tag{B.192}$$

The isentropic law, i.e., Eq. (9.1.8), implies that

$$\frac{\mathrm{d}p}{\mathrm{d}\rho} = \text{constant} \times \gamma\rho^{\gamma-1}, \tag{B.193}$$

which is substituted into Eq. (B.192) to obtain

$$\text{constant} \times \gamma\rho^{\gamma-2} = g, \qquad \longrightarrow \qquad \gamma = 2, \tag{B.194}$$

in order to accomplish the analogy. Using $\gamma = 2$ in Eqs. (9.1.8) and (B.194)$_1$ gives

$$2\left(\frac{p_0}{\rho_0^2}\right) = g, \tag{B.195}$$

where p_0 and ρ_0 are respectively the pressure and density of a gas medium in a reference state.

9.9 For the considered circumstance, it follows from Eqs. (9.4.17), (9.4.19) and (9.4.24) that

$$\left(1 - \mathcal{M}_\infty^2\right)\frac{\partial^2\Phi}{\partial x^2} + \frac{\partial^2\Phi}{\partial y^2} = 0, \tag{B.196}$$

should be satisfied by the velocity potential function Φ resulted from the perturbation induced by the wavy channel under a linearized approximation, where $\mathcal{M}_\infty$ represents the Mach number of incoming flow. By using the method of separation variables, Φ is decomposed into

$$\begin{aligned}\Phi = {} & [A\cos(kx) + B\sin(kx)]\cdot \\ & \left[C\exp\left(\sqrt{1-\mathcal{M}_\infty^2}ky\right) + D\exp\left(-\sqrt{1-\mathcal{M}_\infty^2}ky\right)\right],\end{aligned} \tag{B.197}$$

where $\{A, B, C, D\}$ are constants. The associated boundary conditions are given by

$$y = d: \ \frac{\mathrm{d}y}{\mathrm{d}x} = \frac{v_b(x,d)}{u_\infty}, \qquad y = -d: \ \frac{\mathrm{d}y}{\mathrm{d}x} = \frac{v_b(x,-d)}{u_\infty}, \tag{B.198}$$

where v_b is the velocity component in the y-direction on the wavy boundary with $v_b = \partial\Phi/\partial y$, and u_∞ represents the velocity of incoming flow. Substituting these conditions into Eq. (B.197) yields $k = 2\pi/\lambda$, $B = 0$, $D = -A$ and $AC = u_\infty a/\sqrt{1-\mathcal{M}_\infty^2}$, so that Eq. (B.197) becomes

$$\begin{aligned}\Phi = {} & \frac{u_\infty a}{\sqrt{1-\mathcal{M}_\infty^2}}\left[\frac{\exp\left(\frac{2\pi}{\lambda}\sqrt{1-\mathcal{M}_\infty^2}y\right) - \exp\left(\frac{-2\pi}{\lambda}\sqrt{1-\mathcal{M}_\infty^2}y\right)}{\exp\left(\frac{2\pi}{\lambda}\sqrt{1-\mathcal{M}_\infty^2}d\right) + \exp\left(\frac{-2\pi}{\lambda}\sqrt{1-\mathcal{M}_\infty^2}d\right)}\right]\cdot \\ & \cos\left(\frac{2\pi}{\lambda}x\right),\end{aligned} \tag{B.199}$$

by which the velocity potential function ϕ is obtained as

$$\phi = \frac{u_\infty a}{\sqrt{1-\mathcal{M}_\infty^2}} \left[\frac{\exp\left(\frac{2\pi}{\lambda}\sqrt{1-\mathcal{M}_\infty^2}y\right) - \exp\left(\frac{-2\pi}{\lambda}\sqrt{1-\mathcal{M}_\infty^2}y\right)}{\exp\left(\frac{2\pi}{\lambda}\sqrt{1-\mathcal{M}_\infty^2}d\right) + \exp\left(\frac{-2\pi}{\lambda}\sqrt{1-\mathcal{M}_\infty^2}d\right)} \right] \cdot \cos\left(\frac{2\pi}{\lambda}x\right) + u_\infty x. \tag{B.200}$$

Since along the channel centerline, i.e., along the x-axis, $u = \partial\phi/\partial x$ is evaluated at $y = 0$, it follows that $u = u_\infty$, so $p = p_\infty$, where p_∞ is the pressure of incoming flow, as implied by the Bernoulli equation. With this, $C_p = 0$. The same results can be obtained by using Eq. (9.4.32) directly, in which u' should be determined from Φ given in Eq. (B.199).

10.4 The energy equation between the contracta and section b reads

$$S_0\Delta x + \frac{u_1^2}{2g} = \Delta y + \frac{u_b^2}{2g} + S\Delta x, \tag{B.201}$$

where S_0 is the channel bottom slope, u_1 denotes the average velocity at the contracta, u_b represents the average velocity at section b, S is the hydraulic slope, and Δx and Δy are respectively the relative distances between the contracta and section b in the x- and y-directions. Solving Δx from this equation gives

$$\Delta x = \frac{1}{S_0 - S}\left[\Delta y + \left(\frac{u_b^2}{2g} - \frac{u_1^2}{2g}\right)\right], \tag{B.202}$$

which can be used to determine the values of Δx if the values of u_1, u_b and S are known, for $\Delta y = h_b - C_t h$ and the value of S_0 is given. Let Q denote the volume flow rate per unit channel width, so that u_1 and u_b are obtained as

$$u_1 = \frac{Q}{y} = \frac{Q}{C_t h}, \qquad u_b = \frac{Q}{y} = \frac{Q}{h_b}. \tag{B.203}$$

By using the Manning formula, the hydraulic slope S can be expressed as

$$S = \left(\frac{n u_{av}}{r_h^{2/3}}\right)^2. \tag{B.204}$$

Since the hydraulic slope varies gradually from the vena contracta toward section b, it is plausible to consider the average velocity and hydraulic radius on the right-hand-side of Eq. (B.204) as the arithmetic means of those at the contracta and section b, so that

$$S \sim S_{av} = \left[\frac{n(u_1 + u_b)}{8(C_t h + h_b)^{2/3}}\right]^2. \tag{B.205}$$

Substituting Eqs. (B.203) and (B.205) into Eq. (B.202) yields the required expression of Δx.

11.3 (a) Consider the air inside the pistol as the control-mass system, and let the initial state of air be denoted by number 1, while the state at which the bullet just leaves the pistol be denoted by number 2. The air is assumed to be an ideal gas, so that its mass m_a, by using the ideal gas state equation, is obtained as

$$m_a = \frac{p_1 V_1}{R T_1}, \tag{B.206}$$

where R represents the gas constant of air. Since the expansion process is an isothermal one, which can be expressed as $pV = \text{constant}$, the volume V_2 is then given by

$$V_2 = \frac{p_1}{p_2} V_1. \tag{B.207}$$

(b) For the isothermal expansion process, the work done by the air to the surrounding, W_1^2, is determined as

$$W_1^2 = \int_1^2 -p\,\mathrm{d}V = -p_1 V_1 \ln \frac{V_2}{V_1}, \tag{B.208}$$

which should assume a negative value, for $V_2 > V_1$. The work done by the air to the atmosphere, $(W_1^2)_0$, is obtained as

$$(W_1^2)_0 = \int_1^2 -p_0\,\mathrm{d}V = -p_0 (V_2 - V_1), \tag{B.209}$$

where p_0 denotes the atmospheric pressure, which is a constant, and $(W_1^2)_0$ should be smaller than zero.

(c) The difference between W_1^2 and $(W_1^2)_0$ is the work done by the air on the bullet. It is assumed that the bullet is initially at rest and leaves the pistol with velocity V_e. The balance of energy requires that

$$\frac{1}{2} m V_e^2 = \left|W_1^2\right| - \left|(W_1^2)_0\right|, \quad \longrightarrow \quad V_e = \sqrt{\frac{V_1}{m}\left[p_1 \ln \frac{V_2}{V_1} - p_0\left(\frac{V_2}{V_1} - 1\right)\right]}, \tag{B.210}$$

where m is the bullet mass.

11.6 The initial state of ammonia is described by $\{p_1, T_i\}$. During the cooling process, the volume of ammonia is decreased, until a minimum value V_s is reached due to the stops, at which the specific volume is given by $v_s = V_s/m_a$, which is associated with pressure p_1; i.e., it is an isobaric cooling process. Since during the cooling process, ammonia may experience a phase change, it cannot be approximated by an ideal gas, and its properties should be referred to its tables of thermodynamic properties. By using the information of $\{p_1, v_s\}$, the corresponding saturation temperature T_s can be determined. If $T_f < T_s$, ammonia is in the two-phase region, so that its quality x is identified to be

$$v_s = v_f + x\left(v_g - v_f\right), \quad \longrightarrow \quad x = \frac{v_s - v_f}{v_g - v_f}, \tag{B.211}$$

where v_f and v_g represent respectively the specific volumes of the saturation liquid and saturation vapor, corresponding to the state described by $\{p_1, T_s\}$. With the value of x, the values of u and s of ammonia at $T = T_f$ can be determined, which are denoted respectively by u_2 and s_2.
Specifically, let the initial states of the device and ammonia be denoted by number 1, while those after the cooling process be denoted by number 2. The work done by ammonia during the cooling process is obtained as

$$W_1^2 = \int_1^2 -p\,\mathrm{d}V = -p_1 m_a (v_2 - v_1) = -p_1 m_a (v_s - v_1), \qquad \text{(B.212)}$$

where $v_2 = v_s$. The above equation is substituted into the first law to obtain

$$Q_1^2 = m_a (u_2 - u_1) + m_s c \left(T_f - T_i\right) - W_1^2, \qquad \text{(B.213)}$$

where not only the amount of heat transfer with ammonia, but also that with the whole device is taken into account, and c represents the specific heat of steel. The entropy generation of the system and surrounding is then given by

$$S_{gen} = m_a (s_2 - s_1) + m_s c \left(\ln \frac{T_f}{T_i}\right) - \frac{Q_1^2}{T_0} > 0, \qquad \text{(B.214)}$$

for the irreversibilities of cooling process result from the phase change of ammonia and the heat transfer at finite temperature differences.

11.13 Let $\{x, y, z\}$ be $\{T, p, h\}$, respectively, so that

$$\left(\frac{\partial T}{\partial p}\right)_h \left(\frac{\partial p}{\partial h}\right)_T \left(\frac{\partial h}{\partial T}\right)_p = -1. \qquad \text{(B.215)}$$

Substituting this expression into the Joule-Thomson coefficient yields

$$\mu_J = \left(\frac{\partial T}{\partial p}\right)_h = -\frac{1}{(\partial p/\partial h)_T (\partial h/\partial T)_p} = -\frac{1}{c_p}\left(\frac{\partial h}{\partial p}\right)_T, \qquad \text{(B.216)}$$

which, by using Eqs. $(11.8.40)_2$ and $(11.8.41)_2$, is recast in the form

$$\mu_J = \frac{1}{c_p}\left[T\left(\frac{\partial v}{\partial T}_p\right) - v\right]. \qquad \text{(B.217)}$$

It follows from the compressibility factor that

$$pv = ZRT, \quad \longrightarrow \quad \left(\frac{\partial v}{\partial T}\right)_p = \frac{v}{T} + \frac{RT}{p}\left(\frac{\partial Z}{\partial T}\right)_p, \qquad \text{(B.218)}$$

which is substituted into Eq. (B.217) to obtain

$$\mu_J = \left(\frac{\partial T}{\partial p}\right)_h = \frac{RT^2}{pc_p}\left(\frac{\partial Z}{\partial T}\right)_p. \qquad \text{(B.219)}$$

12.1 The energy provided to a single sand particle results in an increase in its translational kinetic energy, which, by using the statistical mechanics, can be approximated as $kT/2$, where k represents the Boltzmann constant. This energy is then transformed in the potential energy as mgd, where m is the mass of sand particle. Both energies are required to be the same, so that

$$\frac{1}{2}kT = mgd, \qquad \longrightarrow \qquad T \sim 10^{11}\text{K},$$

in which the values $k = 1.38 \cdot 10^{23}$ kg-m^2/(Ks^2), $m = 1.3 \cdot 10^{-9}$ kg/m^3, $g = 9.81$ m/s^2, and $d = 10^{-4}$ m have been used.

Index

© Springer International Publishing AG 2019

C. Fang, *An Introduction to Fluid Mechanics*, Springer Textbooks in Earth Sciences, Geography and Environment, https://doi.org/10.1007/978-3-319-91821-1

MIX
Papier aus verantwortungsvollen Quellen
Paper from responsible sources
FSC® C105338

If you have any concerns about our products,
you can contact us on
ProductSafety@springernature.com

In case Publisher is established outside the EU,
the EU authorized representative is:
Springer Nature Customer Service Center GmbH
Europaplatz 3, 69115 Heidelberg, Germany

Printed by Libri Plureos GmbH
in Hamburg, Germany